# Learning OpenCV 3

## *Computer Vision in C++ with the OpenCV Library*

***Adrian Kaehler and Gary Bradski***

Beijing • Boston • Farnham • Sebastopol • Tokyo

**Learning OpenCV 3**
by Adrian Kaehler and Gary Bradski



Published by O'Reilly Media, Inc., 1005 Gravenstein Highway North, Sebastopol, CA 95472.

O'Reilly books may be purchased for educational, business, or sales promotional use. Online editions are also available for most titles (*http://www.oreilly.com/safari*). For more information, contact our corporate/institutional sales department: 800-998-9938 or *corporate@oreilly.com*.

**Editor:** Dawn Schanafelt
**Production Editor:** Kristen Brown
**Copyeditor:** Rachel Monaghan
**Proofreader:** James Fraleigh
**Indexer:** Ellen Troutman
**Interior Designer:** David Futato
**Cover Designer:** Karen Montgomery
**Illustrator:** Rebecca Demarest

December 2016: First Edition

**Revision History for the First Edition**
2016-12-09: First Release
2017-01-20: Second Release
2017-12-15: Third Release
2021-08-20: Fourth Release

See *http://oreilly.com/catalog/errata.csp?isbn=9781491937990* for release details.



978-1-491-93799-0

[LSI]

# Table of Contents

# Preface

This book provides a working guide to the C++ Open Source Computer Vision Library (OpenCV) version 3.x and gives a general background on the field of computer vision sufficient to help readers use OpenCV effectively.

## Purpose of This Book

Computer vision is a rapidly growing field largely because of four trends:

- The advent of mobile phones put millions of cameras in people's hands.
- The Internet and search engines aggregated the resulting giant flows of image and video data into huge databases.
- Computer processing power became a cheap commodity.
- Vision algorithms themselves became more mature (now with the advent of deep neural networks, which OpenCV is increasingly supporting; see *dnn* at opencv_contrib (*https://github.com/opencv/opencv_contrib*) [opencv_contrib]).

OpenCV has played a role in the growth of computer vision by enabling hundreds of thousands of people to do more productive work in vision. OpenCV 3.x now allows students, researchers, professionals, and entrepreneurs to efficiently implement projects and jump-start research by providing them with a coherent C++ computer vision architecture that is optimized over many platforms.

The purpose of this book is to:

- Comprehensively document OpenCV by detailing what function calling conventions really mean and how to use them correctly
- Give the reader an intuitive understanding of how the vision algorithms work
- Give the reader some sense of what algorithm to use and when to use it

- Give the reader a boost in implementing computer vision and machine learning algorithms by providing many working code examples to start from
- Suggest ways to fix some of the more advanced routines when something goes wrong

This book documents OpenCV in a way that allows the reader to rapidly do interesting and fun things in computer vision. It gives an intuitive understanding of how the algorithms work, which serves to guide the reader in designing and debugging vision applications and also makes the formal descriptions of computer vision and machine learning algorithms in other texts easier to comprehend and remember.

## Who This Book Is For

This book contains descriptions, working code examples, and explanations of the C++ computer vision tools contained in the OpenCV 3.x library. Thus, it should be helpful to many different kinds of users:

*Professionals and entrepreneurs*
: For practicing professionals who need to rapidly prototype or professionally implement computer vision systems, the sample code provides a quick framework with which to start. Our descriptions of the algorithms can quickly teach or remind the reader how they work. OpenCV 3.x sits on top of a *hardware acceleration layer* (HAL) so that implemented algorithms can run efficiently, seamlessly taking advantage of a variety of hardware platforms.

*Students*
: This is the text we wish had back in school. The intuitive explanations, detailed documentation, and sample code will allow you to boot up faster in computer vision, work on more interesting class projects, and ultimately contribute new research to the field.

*Teachers*
: Computer vision is a fast-moving field. We've found it effective to have students rapidly cover an accessible text while the instructor fills in formal exposition where needed and supplements that with current papers or guest lectures from experts. The students can meanwhile start class projects earlier and attempt more ambitious tasks.

*Hobbyists*
: Computer vision is fun—here's how to hack it.

We have a strong focus on giving readers enough intuition, documentation, and working code to enable rapid implementation of real-time vision applications.

## What This Book Is Not

This book is not a formal text. We do go into mathematical detail at various points,[1] but it is all in the service of developing deeper intuitions behind the algorithms or to clarify the implications of any assumptions built into those algorithms. We have not attempted a formal mathematical exposition here and might even incur some wrath along the way from those who do write formal expositions.

This book has more of an "applied" nature. It will certainly be of general help, but is not aimed at any of the specialized niches in computer vision (e.g., medical imaging or remote sensing analysis).

That said, we believe that by reading the explanations here first, a student will not only learn the theory better, but remember it longer as well. Therefore, this book would make a good adjunct text to a theoretical course and would be a great text for an introductory or project-centric course.

# About the Programs in This Book

All the program examples in this book are based on OpenCV version 3.x. The code should work under Linux, Windows, and OS X. Using references online, OpenCV 3.x has full support to run on Android and iOS. Source code for the examples in the book can be fetched from this book's website (*http://bit.ly/learningOpenCV3*); source code for OpenCV is available on GitHub (*https://github.com/opencv/opencv*); and prebuilt versions of OpenCV can be loaded from its SourceForge site (*http://sourceforge.net/projects/opencvlibrary*).

OpenCV is under ongoing development, with official releases occurring quarterly. To stay completely current, you should obtain your code updates from the aforementioned GitHub site. OpenCV maintains a website at *http://opencv.org*; for developers, there is a wiki at *https://github.com/opencv/opencv/wiki*.

# Prerequisites

For the most part, readers need only know how to program in C++. Many of the math sections in this book are optional and are labeled as such. The mathematics involve simple algebra and basic matrix algebra, and assume some familiarity with solution methods to least-squares optimization problems as well as some basic knowledge of Gaussian distributions, Bayes' law, and derivatives of simple functions.

---

1 Always with a warning to more casual users that they may skip such sections.

The math in this book is in support of developing intuition for the algorithms. The reader may skip the math and the algorithm descriptions, using only the function definitions and code examples to get vision applications up and running.

## How This Book Is Best Used

This text need not be read in order. It can serve as a kind of user manual: look up the function when you need it, and read the function's description if you want the gist of how it works "under the hood." However, the intent of this book is tutorial. It gives you a basic understanding of computer vision along with details of how and when to use selected algorithms.

This book is written to allow its use as an adjunct or primary textbook for an undergraduate or graduate course in computer vision. The basic strategy with this method is for students to read the book for a rapid overview and then supplement that reading with more formal sections in other textbooks and with papers in the field. There are exercises at the end of each chapter to help test the student's knowledge and to develop further intuitions.

You could approach this text in any of the following ways:

*Grab bag*
: Go through Chapters 1–5 in the first sitting, and then just hit the appropriate chapters or sections as you need them. This book does not have to be read in sequence, except for Chapters 18 and 19 (which cover camera calibration and stereo imaging) and Chapters 20, 21, and 22 (which cover machine learning). Entrepreneurs and students doing project-based courses might go this way.

*Good progress*
: Read just two chapters a week until you've covered Chapters 1–22 in 11 weeks (Chapter 23 will go by in an instant). Start on projects and dive into details on selected areas in the field, using additional texts and papers as appropriate.

*The sprint*
: Cruise through the book as fast as your comprehension allows, covering Chapters 1–23. Then get started on projects and go into detail on selected areas in the field using additional texts and papers. This is probably the choice for professionals, but it might also suit a more advanced computer vision course.

Chapter 20 is a brief chapter that gives general background on machine learning, which is followed by Chapters 21 and 22, which give more details on the machine learning algorithms implemented in OpenCV and how to use them. Of course, machine learning is integral to object recognition and a big part of computer vision, but it's a field worthy of its own book. Professionals should find this text a suitable launching point for further explorations of the literature—or for just getting down to business with the code in that part of the library. The machine learning interface has been substantially simplified and unified in OpenCV 3.x.

This is how we like to teach computer vision: sprint through the course content at a level where the students get the gist of how things work; then get students started on meaningful class projects while supplying depth and formal rigor in selected areas by drawing from other texts or papers in the field. This same method works for quarter, semester, or two-term classes. Students can get quickly up and running with a general understanding of their vision task and working code to match. As they begin more challenging and time-consuming projects, the instructor helps them develop and debug complex systems.

For longer courses, the projects themselves can become instructional in terms of project management. Build up working systems first; refine them with more knowledge, detail, and research later. The goal in such courses is for each project to be worthy of a conference publication and with a few project papers being published subsequent to further (post-course) work. In OpenCV 3.x, the C++ code framework, Buildbots, GitHub use, pull request reviews, unit and regression tests, and documentation are together a good example of the kind of professional software infrastructure a startup or other business should put together.

## Conventions Used in This Book

The following typographical conventions are used in this book:

*Italic*
: Indicates new terms, URLs, email addresses, filenames, file extensions, pathnames, directories, and Unix utilities.

`Constant width`
: Indicates commands, options, switches, variables, attributes, keys, functions, types, classes, namespaces, methods, modules, properties, parameters, values, objects, events, event handlers, XMLtags, HTMLtags, the contents of files, or the output from commands.

**`Constant width bold`**
: Shows commands or other text that should be typed literally by the user. Also used for emphasis in code samples.

*`Constant width italic`*
: Shows text that should be replaced with user-supplied values.

*[...]*
: Indicates a reference to the bibliography.

This icon signifies a tip, suggestion, or general note.

This icon indicates a warning or caution.

## Using Code Examples

Supplemental material (code examples, exercises, etc.) is available for download at *https://github.com/oreillymedia/Learning-OpenCV-3_examples*.

OpenCV is free for commercial or research use, and we have the same policy on the code examples in the book. Use them at will for homework, research, or for commercial products! We would very much appreciate you referencing this book when you do so, but it is not required. An attribution usually includes the title, author, publisher, and ISBN. For example: "*Learning OpenCV 3* by Adrian Kaehler and Gary Bradski (O'Reilly). Copyright 2017 Adrian Kaehler, Gary Bradski, 978-1-491-93799-0."

Other than hearing how it helped with your homework projects (which is best kept a secret), we would love to hear how you are using computer vision for academic research, teaching courses, and in commercial products when you do use OpenCV to help you. Again, it's not required, but you are always invited to drop us a line.

## O'Reilly Online Learning

For more than 40 years, *O'Reilly Media* has provided technology and business training, knowledge, and insight to help companies succeed.

Our unique network of experts and innovators share their knowledge and expertise through books, articles, and our online learning platform. O'Reilly's online learning platform gives you on-demand access to live training courses, in-depth learning paths, interactive coding environments, and a vast collection of text and video from O'Reilly and 200+ other publishers. For more information, visit *http://oreilly.com*.

## We'd Like to Hear from You

Please address comments and questions concerning this book to the publisher:

O'Reilly Media, Inc.
1005 Gravenstein Highway North
Sebastopol, CA 95472
800-998-9938 (in the United States or Canada)
707-829-0515 (international or local)
707-829-0104 (fax)

We have a web page for this book, where we list examples and any plans for future editions. You can access this information at: *http://bit.ly/learningOpenCV3*.

To comment or ask technical questions about this book, send email to *bookquestions@oreilly.com*.

For more information about our books, courses, conferences, and news, see our website at *http://www.oreilly.com*.

Find us on Facebook: *http://facebook.com/oreilly*

Follow us on Twitter: *http://twitter.com/oreillymedia*

Watch us on YouTube: *http://www.youtube.com/oreillymedia*

## Acknowledgments

A long-term open source effort sees many people come and go, each contributing in different ways. The list of contributors to this library is far too long to list here, but see the *.../opencv/docs/HTML/Contributors/doc_contributors.html* file that ships with OpenCV.

### Thanks for Help on OpenCV

Intel is where the library was born and deserves great thanks for supporting this project as it started and grew. From time to time, Intel still funds contests and contributes work to OpenCV. Intel also donated the built-in performance primitives code, which provides for seamless speedup on Intel architectures. Thank you for that.

Google has been a steady funder of development for OpenCV by sponsoring interns for OpenCV under its Google Summer of Code project; much great work has been done through this funding. Willow Garage provided several years of funding that enabled OpenCV to go from version 2.x through to version 3.0. During this time, the computer vision R&D company Itseez (recently bought by Intel Corporation) has provided extensive engineering support and web services hosting over the years. Intel has indicated verbal agreement to continue this support (thanks!).

On the software side, some individuals stand out for special mention, especially on the Russian software team. Chief among these is the Russian lead programmer Vadim Pisarevsky, who is the largest single contributor to the library. Vadim also managed and nurtured the library through the lean times when boom had turned to bust and then bust to boom; he, if anyone, is the true hero of the library. His technical insights have also been of great help during the writing of this book. Giving him managerial support has been Victor Eruhimov, a cofounder of Itseez [Itseez] and now CEO of Itseez3D.

Several people consistently help out with managing the library during weekly meetings: Grace Vesom, Vincent Rabaud, Stefano Fabri, and of course, Vadim Pisarevsky. The developer notes for these meetings can be seen at *https://github.com/opencv/opencv/wiki/Meeting_notes*.

Many people have contributed to OpenCV over time; a list of more recent ones is: Dinar Ahmatnurov, Pablo Alcantarilla, Alexander Alekhin, Daniel Angelov, Dmitriy Anisimov, Anatoly Baksheev, Cristian Balint, Alexandre Benoit, Laurent Berger, Leonid Beynenson, Alexander Bokov, Alexander Bovyrin, Hilton Bristow, Vladimir Bystritsky, Antonella Cascitelli, Manuela Chessa, Eric Christiansen, Frederic Devernay, Maria Dimashova, Roman Donchenko, Vladimir Dudnik, Victor Eruhimov, Georgios Evangelidis, Stefano Fabri, Sergio Garrido, Harris Gasparakis, Yuri Gitman, Lluis Gomez, Yury Gorbachev, Elena Gvozdeva, Philipp Hasper, Fernando J. Iglesias Garcia, Alexander Kalistratov, Andrey Kamaev, Alexander Karsakov, Rahul Kavi, Pat O'Keefe, Siddharth Kherada, Eugene Khvedchenya, Anna Kogan, Marina Kolpakova, Kirill Kornyakov, Ivan Korolev, Maxim Kostin, Evgeniy Kozhinov, Ilya Krylov, Laksono Kurnianggoro, Baisheng Lai, Ilya Lavrenov, Alex Leontiev, Gil Levi, Bo Li, Ilya Lysenkov, Vitaliy Lyudvichenko, Bence Magyar, Nikita Manovich, Juan Manuel Perez Rua, Konstantin Matskevich, Patrick Mihelich, Alexander Mordvintsev, Fedor Morozov, Gregory Morse, Marius Muja, Mircea Paul Muresan, Sergei Nosov, Daniil Osokin, Seon-Wook Park, Andrey Pavlenko, Alexander Petrikov, Philip aka Dikay900, Prasanna, Francesco Puja, Steven Puttemans, Vincent Rabaud, Edgar Riba, Cody Rigney, Pavel Rojtberg, Ethan Rublee, Alfonso Sanchez-Beato, Andrew Senin, Maksim Shabunin, Vlad Shakhuro, Adi Shavit, Alexander Shishkov, Sergey Sivolgin, Marvin Smith, Alexander Smorkalov, Fabio Solari, Adrian Stratulat, Evgeny Talanin, Manuele Tamburrano, Ozan Tonkal, Vladimir Tyan, Yannick Verdie, Pierre-Emmanuel Viel, Vladislav Vinogradov, Pavel Vlasov, Philipp Wagner, Yida Wang, Jiaolong Xu, Marian Zajko, Zoran Zivkovic.

Other contributors show up over time at *https://github.com/opencv/opencv/wiki/ChangeLog*. Finally, Arraiy [Arraiy] is now also helping maintain OpenCV.org (the free and open codebase).

## Thanks for Help on This Book

While preparing this book and the previous version of this book, we'd like to thank John Markoff, science reporter at the *New York Times*, for encouragement, key contacts, and general writing advice born of years in the trenches. We also thank our many editors at O'Reilly, especially Dawn Schanafelt, who had the patience to continue on as slips became the norm while the errant authors were off trying to found a startup. This book has been a long project that slipped from OpenCV 2.x to the current OpenCV 3.x release. Many thanks to O'Reilly for sticking with us through all that.

## Adrian Adds...

In the first edition (*Learning OpenCV*) I singled out some of the great teachers who helped me reach the point where a work like this would be possible. In the intervening years, the value of the guidance received from each of them has only grown more clear. My many thanks go out to each of them. I would like to add to this list of extraordinary mentors Tom Tombrello, to whom I owe a great debt, and in whose memory I would like to dedicate my contribution to this book. He was a man of exceptional intelligence and deep wisdom, and I am honored to have been given the opportunity to follow in his footsteps. Finally, deep thanks are due the OpenCV community, for welcoming the first edition of this book and for your patience through the many exciting, but perhaps distracting, endeavors that have transpired while this edition was being written.

This edition of the book has been a long time coming. During those intervening years, I have had the fortune to work with dozens of different companies advising, consulting, and helping them build their technology. As a board member, advisory board member, technical fellow, consultant, technical contributor, and founder, I have had the fortune to see and love every dimension of the technology development process. Many of those years were spent with Applied Minds, Inc., building and running our robotics division there, or at Applied Invention corporation, a spinout of Applied Minds, as a Fellow there. I was constantly pleased to find OpenCV at the heart of outstanding projects along the way, ranging from health care and agriculture to aviation, defense, and national security. I have been equally pleased to find the first edition of this book on people's desks in almost every institution along the way. The technology that Gary and I used to build Stanley has become integral to countless projects since, not the least of which are the many self-driving car projects now under way—any one of which, or perhaps all of which, stand ready to change and improve daily life for countless people. What a joy it is to be part of all of this! The number of incredible minds that I have encountered over the years—who have told me what benefit the first edition was to them in the classes they took, the classes they taught, the careers they built, and the great accomplishments that they completed—has been a continuous source of happiness and wonder. I am hopeful that this new edition of the book will continue to serve you all, as well as to inspire and enable a new generation of scientists, engineers, and inventors.

As the last chapter of this book closes, we start new chapters in our lives working in robotics, AI, vision, and beyond. Personally, I am deeply grateful for all of the people who have contributed the many works that have enabled this next step in my own life: teachers, mentors, and writers of books. I hope that this new edition of our book will enable others to make the next important step in their own lives, and I hope to see you there!

## Gary Adds...

I founded OpenCV in 1999 with the goal to accelerate computer vision and artificial intelligence and give everyone the infrastructure to work with that I saw at only the top labs at the time. So few goals actually work out as intended in life, and I'm thankful this goal did work out 17 (!) years later. Much of the credit for accomplishing that goal was due to the help, over the years, of many friends and contributors too numerous to mention.[2] But I will single out the original Russian group I started working with at Intel, who ran a successful computer vision company (Itseez.com) that was eventually bought back into Intel; we started out as coworkers but have since become deep friends.

With three teenagers at home, my wife, Sonya Bradski, put in more work to enable this book than I did. Many thanks and love to her. The teenagers I love, but I can't say they accelerated the book. :)

This version of the book was started back at the former startup I helped found, Industrial Perception Inc., which sold to Google in 2013. Work continued in fits and starts on random weekends and late nights ever since. Somehow it's now 2016—time flies when you are overwhelmed! Some of the speculation that I do toward the end of Chapter 23 was inspired by the nature of robot minds that I experienced with the PR2, a two-armed robot built by Willow Garage, and with the Stanley project at Stanford—the robot that won the $2 million DARPA Grand Challenge.

As we close the writing of this book, we hope to see you in startups, research labs, academic sites, conferences, workshops, VC offices, and cool company projects down the road. Feel free to say hello and chat about cool new stuff that you're doing. I started OpenCV to support and accelerate computer vision and AI for the common good; what's left is your part. We live in a creative universe where someone can create

2 We now have many contributors, as you can see by scrolling past the updates in the change logs at *https://github.com/opencv/opencv/wiki/ChangeLog*. We get so many new algorithms and apps that we now store the best in self-maintaining and self-contained modules in *opencv_contrib* (*https://github.com/opencv/opencv_contrib*)).

a pot, the next person turns that pot into a drum, and so on. Create! Use OpenCV to create something uncommonly good for us all!

CHAPTER 1

# Overview

## What Is OpenCV?

OpenCV [OpenCV] is an open source (see *http://opensource.org*) computer vision library available from *http://opencv.org*. In 1999 Gary Bradski [Bradski], working at Intel Corporation, launched OpenCV with the hopes of accelerating computer vision and artificial intelligence by providing a solid infrastructure for everyone working in the field. The library is written in C and C++ and runs under Linux, Windows, and Mac OS X. There is active development on interfaces for Python, Java, MATLAB, and other languages, including porting the library to Android and iOS for mobile applications. OpenCV has received much of its support over the years from Intel and Google, but especially from Itseez [Itseez] (recently acquired by Intel), which did the bulk of the early development work. Finally, Arraiy [Arraiy] has joined in to maintain the always open and free OpenCV.org [OpenCV].

OpenCV was designed for computational efficiency and with a strong focus on real-time applications. It is written in optimized C++ and can take advantage of multicore processors. If you desire further automatic optimization on Intel architectures [Intel], you can buy Intel's *Integrated Performance Primitives (IPP)* libraries [IPP], which consist of low-level optimized routines in many different algorithmic areas. OpenCV automatically uses the appropriate IPP library at runtime if that library is installed. Starting with OpenCV 3.0, Intel granted the OpenCV team and OpenCV community a free-of-charge subset of IPP (nicknamed IPPICV), which is built into and accelerates OpenCV by default.

One of OpenCV's goals is to provide a simple-to-use computer vision infrastructure that helps people build fairly sophisticated vision applications quickly. The OpenCV library contains over 500 functions that span many areas in vision, including factory product inspection, medical imaging, security, user interface, camera calibration, stereo vision, and robotics. Because computer vision and machine learning often go hand-in-hand, OpenCV also contains a full, general-purpose Machine Learning library (ML module). This sublibrary is focused on statistical pattern recognition and clustering. The ML module is highly useful for the vision tasks that are at the core of OpenCV's mission, but it is general enough to be used for any machine learning problem.

## Who Uses OpenCV?

Most computer scientists and practical programmers are aware of some facet of computer vision's role, but few people are aware of all the ways in which computer vision is used. For example, most people are somewhat aware of its use in surveillance, and many also know that it is increasingly being used for images and video on the Web. A few have seen some use of computer vision in game interfaces. Yet fewer people realize that most aerial and street-map images (such as in Google's Street View) make heavy use of camera calibration and image stitching techniques. Some are aware of niche applications in safety monitoring, unmanned flying vehicles, or biomedical analysis. But few are aware how pervasive machine vision has become in manufacturing: virtually everything that is mass-produced has been automatically inspected at some point using computer vision.

The open source license for OpenCV has been structured such that you can build a commercial product using all or part of OpenCV. You are under no obligation to open-source your product or to return improvements to the public domain, though we hope you will. In part because of these liberal licensing terms, there is a large user community that includes people from major companies (IBM, Microsoft, Intel, SONY, Siemens, and Google, to name only a few) and research centers (such as Stanford, MIT, CMU, Cambridge, and INRIA). There is a Yahoo Groups forum (*http://groups.yahoo.com/group/OpenCV*) where users can post questions and discussion; it has almost 50,000 members. OpenCV is popular around the world, with large user communities in China, Japan, Russia, Europe, and Israel.

Since its alpha release in January 1999, OpenCV has been used in many applications, products, and research efforts. These applications include stitching images together in satellite and web maps, image scan alignment, medical image noise reduction, object analysis, security and intrusion detection systems, automatic monitoring and safety systems, manufacturing inspection systems, camera calibration, military applications, and unmanned aerial, ground, and underwater vehicles. It has even been used in sound and music recognition, where vision recognition techniques are applied to sound spectrogram images. OpenCV was a key part of the vision system in the robot from Stanford, "Stanley," which won the $2M DARPA Grand Challenge desert robot race [Thrun06].

## What Is Computer Vision?

Computer vision[1] is the transformation of data from a still or video camera into either a decision or a new representation. All such transformations are done to achieve some particular goal. The input data may include some contextual information such as "the camera is mounted in a car" or "laser range finder indicates an object is 1 meter away." The decision might be "there is a person in this scene" or "there are 14 tumor cells on this slide." A new representation might mean turning a color image into a grayscale image or removing camera motion from an image sequence.

Because we are such visual creatures, it is easy to be fooled into thinking that computer vision tasks are easy. How hard can it be to find, say, a car when you are staring at it in an image? Your initial intuitions can be quite misleading. The human brain divides the vision signal into many channels that stream different kinds of information into your brain. Your brain has an attention system that identifies, in a task-dependent way, important parts of an image to examine while suppressing examination of other areas. There is massive feedback in the visual stream that is, as yet, little understood. There are widespread associative inputs from muscle control sensors and all of the other senses that allow the brain to draw on cross-associations made from years of living in the world. The feedback loops in the brain go back to all stages of processing, including the hardware sensors themselves (the eyes), which mechanically control lighting via the iris and tune the reception on the surface of the retina.

In a machine vision system, however, a computer receives a grid of numbers from the camera or from disk, and that's it. For the most part, there's no built-in pattern recognition, no automatic control of focus and aperture, no cross-associations with years of experience. For the most part, vision systems are still fairly naïve. Figure 1-1 shows a picture of an automobile. In that picture we see a side mirror on the driver's side of the car. What the computer "sees" is just a grid of numbers. Any given number within that grid has a rather large noise component and so by itself gives us little information, but this grid of numbers is all the computer "sees." Our task, then, becomes to turn this noisy grid of numbers into the perception "side mirror." Figure 1-2 gives some more insight into why computer vision is so hard.

1 Computer vision is a vast field. This book will give you a basic grounding in the field, but we also recommend texts by Trucco [Trucco98] for a simple introduction, Forsyth [Forsyth03] as a comprehensive reference, and Hartley [Hartley06] and Faugeras [Faugeras93] for a discussion of how 3D vision really works.

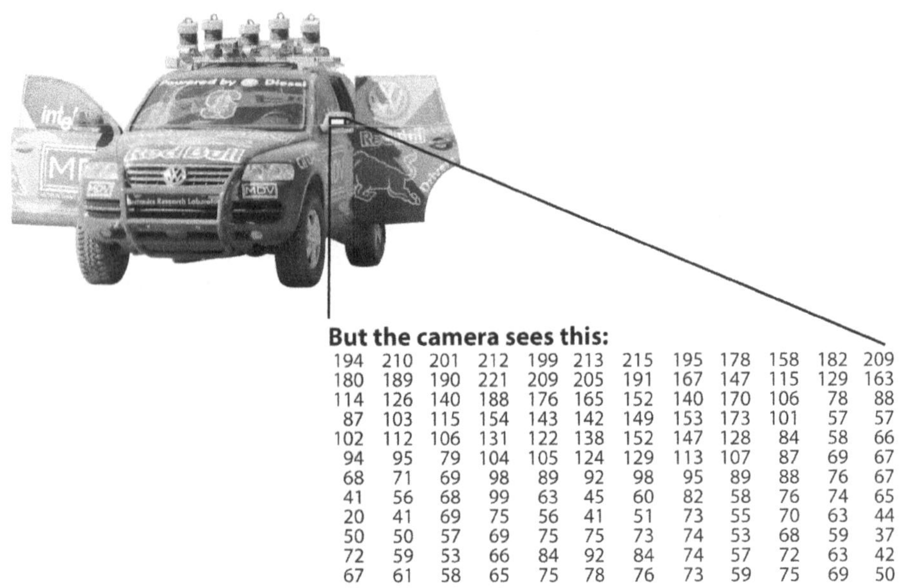


*Figure 1-1. To a computer, the car's side mirror is just a grid of numbers*

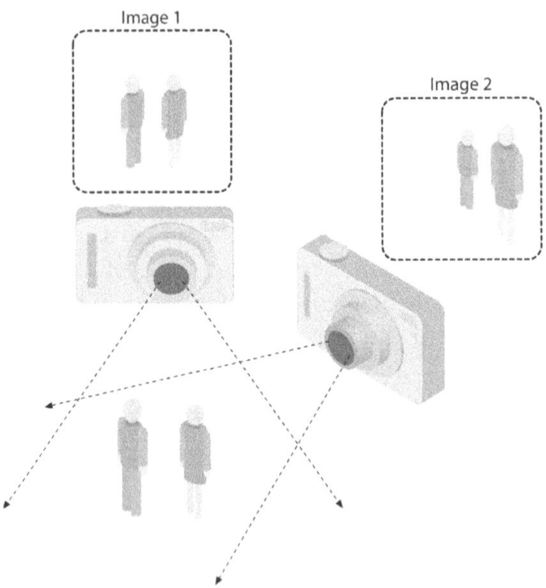


*Figure 1-2. The ill-posed nature of vision: the 2D appearance of objects can change radically with viewpoint*

In fact, the problem, as we have posed it thus far, is worse than hard: it is formally impossible to solve. Given a two-dimensional (2D) view of a 3D world, there is no unique way to reconstruct the 3D signal. Formally, such an ill-posed problem has no unique or definitive solution. The same 2D image could represent any of an infinite combination of 3D scenes, even if the data were perfect. However, as already mentioned, the data is corrupted by noise and distortions. Such corruption stems from variations in the world (weather, lighting, reflections, movements), imperfections in the lens and mechanical setup, finite integration time on the sensor (motion blur), electrical noise in the sensor or other electronics, and compression artifacts after image capture. Given these daunting challenges, how can we make any progress?

In the design of a practical system, additional contextual knowledge can often be used to work around the limitations imposed on us by visual sensors. Consider the example of a mobile robot that must find and pick up staplers in a building. The robot might use the facts that a desk is an object found inside offices and that staplers are mostly found on desks. This gives an implicit size reference; staplers must be able to fit on desks. It also helps to eliminate falsely "recognizing" staplers in impossible places (e.g., on the ceiling or a window). The robot can safely ignore a 200-foot advertising blimp shaped like a stapler because the blimp lacks the prerequisite wood-grained background of a desk. In contrast, with tasks such as image retrieval, all stapler images in a database may be of real staplers, and so large sizes and other unusual configurations may have been implicitly precluded by the assumptions of those who took the photographs; that is, the photographer perhaps took pictures only of real, normal-sized staplers. People also tend to center objects when taking pictures and tend to put them in characteristic orientations. Thus, there is often quite a bit of unintentional implicit information within photos taken by people.

Contextual information can also be modeled explicitly with machine learning techniques. Hidden variables such as size, orientation to gravity, and so on can then be correlated with their values in a labeled training set. Alternatively, one may attempt to measure hidden bias variables by using additional sensors. The use of a laser range finder to measure depth allows us to accurately measure the size of an object.

The next problem facing computer vision is noise. We typically deal with noise by using statistical methods. For example, it may be impossible to detect an edge in an image merely by comparing a point to its immediate neighbors. But if we look at the statistics over a local region, edge detection becomes much easier. A real edge should appear as a string of such immediate neighbor responses over a local region, each of whose orientation is consistent with its neighbors. It is also possible to compensate for noise by taking statistics over time. Still other techniques account for noise or distortions by building explicit models learned directly from the available data. For example, because lens distortions are well understood, one need only learn the parameters for a simple polynomial model in order to describe—and thus correct almost completely—such distortions.

The actions or decisions that computer vision attempts to make based on camera data are performed in the context of a specific purpose or task. We may want to remove noise or damage from an image so that our security system will issue an alert if someone tries to climb a fence or because we need a monitoring system that counts how many people cross through an area in an amusement park. Vision software for robots that wander through office buildings will employ different strategies than vision software for stationary security cameras because the two systems have significantly different contexts and objectives. As a general rule, the more constrained a computer vision context is, the more we can rely on those constraints to simplify the problem and the more reliable our final solution will be.

OpenCV is aimed at providing the basic tools needed to solve computer vision problems. In some cases, high-level functionalities in the library will be sufficient to solve the more complex problems in computer vision. Even when this is not the case, the basic components in the library are complete enough to enable creation of a complete solution of your own to almost any computer vision problem. In the latter case, there are several tried-and-true methods of using the library; all of them start with solving the problem using as many available library components as possible. Typically, after you've developed this first-draft solution, you can see where the solution has weaknesses and then fix those weaknesses using your own code and cleverness (better known as "solve the problem you actually have, not the one you imagine"). You can then use your draft solution as a benchmark to assess the improvements you have made. From that point, you can tackle whatever weaknesses remain by exploiting the context of the larger system in which your solution is embedded.

# The Origin of OpenCV

OpenCV grew out of an Intel research initiative to advance CPU-intensive applications. Toward this end, Intel launched many projects, including real-time ray tracing and 3D display walls. One of the authors, Gary Bradski [Bradski], working for Intel at that time, was visiting universities and noticed that some top university groups, such as the MIT Media Lab, had well-developed and internally open computer vision infrastructures—code that was passed from student to student and that gave each new student a valuable head start in developing his or her own vision application. Instead of reinventing the basic functions from scratch, a new student could begin by building on top of what came before.

Thus, OpenCV was conceived as a way to make computer vision infrastructure universally available. With the aid of Intel's Performance Library Team,[2] OpenCV started with a core of implemented code and algorithmic specifications being sent to

2 Shinn Lee was of key help.

members of Intel's Russian library team. This is the "where" of OpenCV: it started in Intel's research lab with collaboration from the Software Performance Libraries group and implementation and optimization expertise in Russia.

Chief among the Russian team members was Vadim Pisarevsky, who managed, coded, and optimized much of OpenCV and who is still at the center of much of the OpenCV effort. Along with him, Victor Eruhimov helped develop the early infrastructure, and Valery Kuriakin managed the Russian lab and greatly supported the effort. There were several goals for OpenCV at the outset:

- Advance vision research by providing not only open but also optimized code for basic vision infrastructure. No more reinventing the wheel.
- Disseminate vision knowledge by providing a common infrastructure that developers could build on, so that code would be more readily readable and transferable.
- Advance vision-based commercial applications by making portable, performance-optimized code available for free—with a license that did not require commercial applications to be open or free themselves.

Those goals constitute the "why" of OpenCV. Enabling computer vision applications would increase the need for fast processors. Driving upgrades to faster processors would generate more income for Intel than selling some extra software. Perhaps that is why this open and free code arose from a hardware vendor rather than a software company. Sometimes, there is more room to be innovative at software within a hardware company.

In any open source effort, it's important to reach a critical mass at which the project becomes self-sustaining. There have now been approximately 11 million downloads of OpenCV, and this number is growing by an average of 160,000 downloads a month. OpenCV receives many user contributions, and central development has largely moved outside of Intel.[3] OpenCV's timeline is shown in Figure 1-3. Along the way, OpenCV was affected by the dot-com boom and bust and also by numerous changes of management and direction. During these fluctuations, there were times when OpenCV had no one at Intel working on it at all. However, with the advent of multicore processors and the many new applications of computer vision, OpenCV's value began to rise. Similarly, rapid growth in the field of robotics has driven much use and development of the library. After becoming an open source library, OpenCV spent several years under active development at Willow Garage, and now is supported by the OpenCV foundation (*http://opencv.org*). Today, OpenCV is actively

3 As of this writing, Willow Garage (*http://www.willowgarage.com*), a robotics research institute and incubator, is actively supporting general OpenCV maintenance and new development in the area of robotics applications.

being developed by the foundation as well as by several public and private institutions. For more information on the future of OpenCV, see Chapter 23.

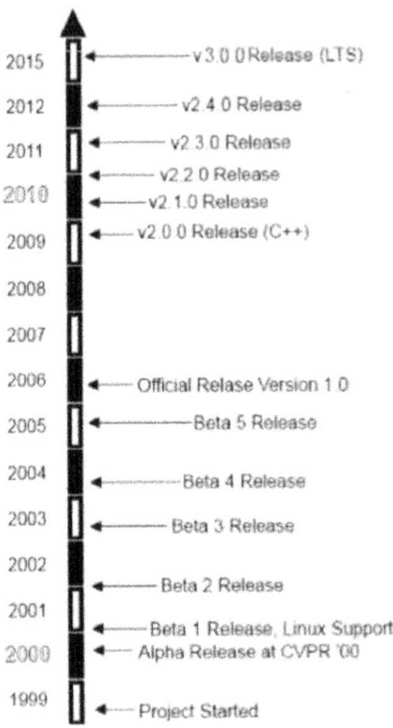


*Figure 1-3. OpenCV timeline*

## OpenCV Block Diagram

OpenCV is built in layers. At the top is the OS under which OpenCV operates. Next comes the language bindings and sample applications. Below that is the contributed code in *opencv_contrib*, which contains mostly higher-level functionality. After that is the core of OpenCV, and at the bottom are the various hardware optimizations in the *hardware acceleration layer* (HAL). Figure 1-4 shows this organization.

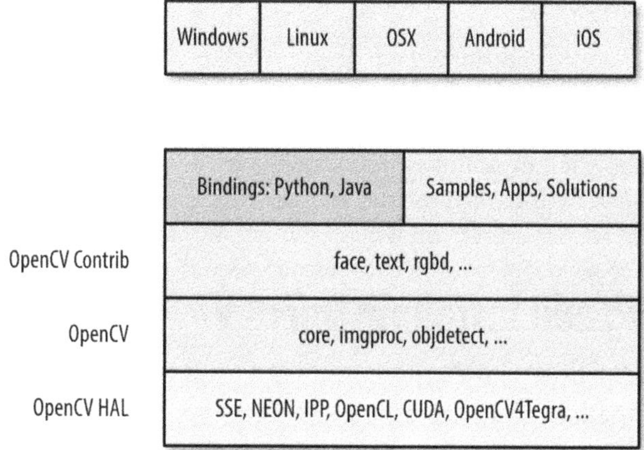


*Figure 1-4. Block diagram of OpenCV with supported operating systems*

## Speeding Up OpenCV with IPP

If available on Intel processors, OpenCV exploits a royalty-free subset of Intel's *Integrated Performance Primitives* (IPP) library, IPP 8.x (IPPICV). IPPICV can be linked into OpenCV at compile stage and if so, it replaces the corresponding low-level optimized C code (in cmake `WITH_IPP=ON/OFF`, `ON` by default). The improvement in speed from using IPP can be substantial. Figure 1-5 shows the relative speedup when IPP is used.

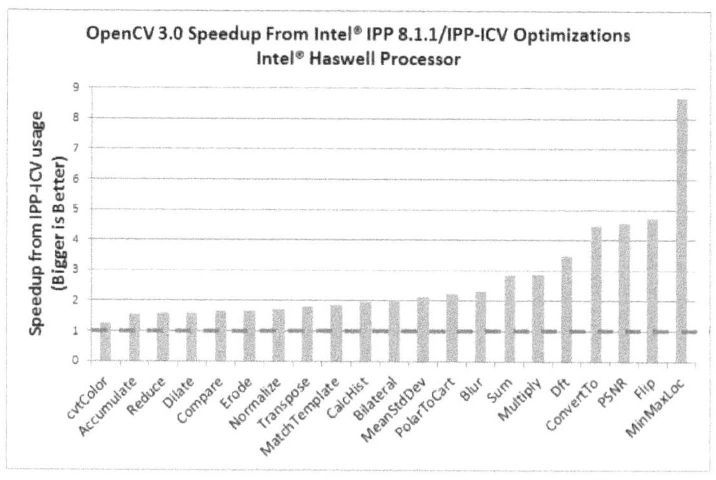


*Figure 1-5. Relative speedup when OpenCV uses IPPICV on an Intel Haswell Processor*

## Who Owns OpenCV?

Although Gary Bradski started OpenCV at Intel, the library is and always was intended to promote commercial and research use; that is its mission. It is therefore open and free, and the code itself may be used or embedded (in whole or in part) in other applications, whether commercial or research. It does not force your application code to be open or free. It does not require that you return improvements back to the library—but we hope that you will.

# Downloading and Installing OpenCV

From the main OpenCV site (*http://opencv.org*), you can download the complete source code for the latest release, as well as many recent releases. The downloads themselves are found at the downloads page (*http://opencv.org/downloads.html*). However, the most up-to-date version is always found on GitHub (*https://github.com/opencv/opencv*), where the active development branch is stored. For more recent, higher-level functionality, you can also download and build *opencv_contrib* [opencv_contrib] (*https://github.com/opencv/opencv_contrib*).

## Installation

In modern times, OpenCV uses Git as its development version control system, and CMake to build.[4] In many cases, you will not need to worry about building, as compiled libraries exist for many environments. However, as you become a more advanced user, you will inevitably want to be able to recompile the libraries with specific options tailored to your application.

### Windows

At *http://opencv.org/downloads.html*, you will see a link to download the latest version of OpenCV for Windows. This link will download an executable file, which is a self-extracting archive with prebuilt OpenCV binaries for various versions of Visual Studio. You are now almost ready to start using OpenCV.[5]

4 In olden times, OpenCV developers used Subversion for version control and automake to build. Those days, however, are long gone.

5 It is important to know that, although the Windows distribution contains binary libraries for release builds, it does not contain the debug builds of these libraries. It is therefore likely that, before developing with OpenCV, you will want to open the solution file and build these libraries yourself.

The one additional detail is that you will want to add an `OPENCV_DIR` environment variable to make it easier to tell your compiler where to find the OpenCV binaries. You can set this by going to a command prompt and typing:[6]

```
setx -m OPENCV_DIR D:\OpenCV\Build\x64\vc10
```

If you want to link OpenCV statically, this is all you will need. If you want to use OpenCV *dynamic link libraries* (DLLs), then you will also need to tell your system where to find the binary library. To do this, simply add `%OPENCV_DIR%\bin` to your library path. (For example, in Windows 10, right-click on your computer icon, select Properties, and then click on Advanced System Settings. Finally, select Environment Variables and add the OpenCV binary path to the Path variable.)

OpenCV 3 comes with IPP linked in, so you get the performance advantage of more or less modern x86 and x64 CPUs.

You can also build OpenCV from a source tarball as follows:

1. Run the CMake GUI.
2. Specify paths to the OpenCV source tree and the build directory (they must be different!).
3. Press Configure twice (choose the appropriate Visual Studio generator, or MinGW makefiles if you use MinGW), and then press Generate.
4. Open the generated solution within Visual Studio, and build it. In the case of MinGW, use the Linux instructions that follow.

### Linux

Prebuilt binaries for Linux are not included with the Linux version of OpenCV owing to the large variety of versions of GCC and GLIBC in different distributions (SuSE, Debian, Ubuntu, etc.). In many cases, however, your distribution will include OpenCV. If your distribution doesn't offer OpenCV, you will have to build it from sources. As with the Windows installation, you can start at *http://opencv.org/downloads.html*, but in this case the link will send you to SourceForge, where you can select the tarball for the current OpenCV source code bundle.

To build the libraries and demos, you'll need GTK+ 2.x or higher, including headers. You'll also need *gcc* and the essential development packages, *cmake* and *libtbb* (Intel thread building blocks), and optionally *zlib*, *libpng*, *libjpeg*, *libtiff*, and *libjasper* with development files (i.e., the versions with *-dev* at the end of their package names). You'll need Python 2.6 or later with headers installed (developer package), as well as

6 Of course, the exact path will vary depending on your installation; for example, if you are installing on a 32-bit machine, then the path will include `x86` instead of `x64`.

NumPy in order to make Python bindings work. You will also need *libavcodec* and the other *libav** libraries (including headers) from *ffmpeg*.

For the latter, install *libav/ffmpeg* packages supplied with your distribution or download *ffmpeg* from *http://www.ffmpeg.org*. The *ffmpeg* library has a *Lesser General Public License* (LGPL), but some of its components have the stricter General Public License (GPL). To use it with non-GPL software (such as OpenCV), build and use a shared *ffmpg* library:

```
$> ./configure --enable-shared
$> make
$> sudo make install
```

(When you link an LGPL library dynamically, you are not obliged to use GPL license for your code.) You will end up with */usr/local/lib/libavcodec.so.**, */usr/local/lib/libavformat.so.**, */usr/local/lib/libavutil.so.**, and include files under various */usr/local/include/libav** paths.

To actually build the library, you will need to unpack the *.tar.gz* file and go into the created source directory, and do the following:

```
mkdir release
cd release
cmake -D CMAKE_BUILD_TYPE=RELEASE -D CMAKE_INSTALL_PREFIX=/usr/local ..
make
sudo make install #  optional
```

The first and second commands create a new subdirectory and move you into it. The third command tells CMake how to configure your build. The example options we give are probably the right ones to get you started, but other options allow you to enable various options, determine what examples are built, add Python support, add CUDA GPU support, and more. By default, OpenCV's *cmake* configuration script attempts to find and use as many third-party libraries as possible. For example, if it finds CUDA SDK, it will enable GPU-accelerated OpenCV functionality. The last two commands actually build the library and install the results into the proper places. Note that you do not need to install OpenCV in order to use it in your CMake-based projects; you just need to specify the path to generate *OpenCVConfig.cmake*. In the preceding case, the file will be placed in the *release* directory. If you decided to run `sudo make install` instead, *OpenCVConfig.cmake* would be placed in */usr/local/share/OpenCV*.

Just like in the Windows case, the Linux build of OpenCV will automatically take advantage of IPP once it's installed. Starting from OpenCV 3.0, OpenCV's *cmake* configuration script will automatically download and link a free subset of IPP (IPPICV). To explicitly disable IPP if you do not want it, pass the `-D WITH_IPP=OFF` option to CMake.

### Mac OS X

Installation on OS X is very similar to Linux, except that OS X comes with its own development environment, Xcode, which includes almost everything you'll need except for CMake; you do not need GTK+, TBB, *libjpeg*, *ffmpeg*, and so on:

- By default, Cocoa is used instead of GTK+.
- By default, QTKit is used instead of *ffmpeg*.
- Grand Dispatch Central (GDC) is used instead of TBB and OpenMP.

The installation steps are then exactly the same. You may want to pass the `-G Xcode` option to CMake to generate an Xcode project for OpenCV (and for your applications) in order to build and debug the code conveniently within Xcode.

## Getting the Latest OpenCV via Git

OpenCV is under active development, and bugs are often fixed rapidly when bug reports contain accurate descriptions and code that demonstrates the bug. However, official OpenCV releases occur only once or twice a year. If you are seriously developing a project or product, you will probably want code fixes and updates as soon as they become available. To get these, you will need to access OpenCV's Git repository on GitHub.

This isn't the place for a tutorial in Git usage. If you've worked with other open source projects, then you're probably familiar with it already. If you haven't, check out *Version Control with Git* by Jon Loeliger (O'Reilly). A command-line Git client is available for Linux, OS X, and most UNIX-like systems. For Windows users, we recommend TortoiseGit (*http://code.google.com/p/tortoisegit/*); for OS X the SourceTree app may suit you.

On Windows, if you want the latest OpenCV from the Git repository, you'll need to access the directory at *https://github.com/opencv/opencv.git*.

On Linux, you can just use the following command:

```
git clone https://github.com/opencv/opencv.git
```

## More OpenCV Documentation

The primary documentation for OpenCV is the HTML documentation available at *http://opencv.org*. In addition to this, there are in-depth tutorials on many subjects at *http://docs.opencv.org/2.4.13/doc/tutorials/tutorials.html*, and an OpenCV wiki (currently located at *https://github.com/opencv/opencv/wiki*).

## Supplied Documentation

OpenCV 2.x comes with a complete reference manual and a bunch of tutorials, all in PDF format; check *opencv/doc*. Starting from OpenCV 3.x, there is no offline documentation anymore.

## Online Documentation and the Wiki

As we just mentioned, there is extensive documentation as well as a wiki available at *http://opencv.org*. The documentation there is divided into several major components:

*Reference (http://docs.opencv.org/)*
: This section contains the functions, their arguments, and some information on how to use them.

*Tutorials (http://docs.opencv.org/trunk/d9/df8/tutorial_root.html)*
: There is a large collection of tutorials; these tell you how to accomplish various things. There are tutorials for basic subjects, like how to install OpenCV or create OpenCV projects on various platforms, and more advanced topics like background subtraction of object detection.

*Quick Start (http://opencv.org/quickstart.html)*
: This is a tightly curated subset of the tutorials, containing just ones that help you get up and running on specific platforms.

*Cheat Sheet (http://docs.opencv.org/3.0-last-rst/opencv_cheatsheet.pdf)*
: This is actually a single *.pdf* file that contains a truly excellent compressed reference to almost the entire library. Thank Vadim Pisarevsky for this excellent reference as you pin these two beautiful pages to your cubicle wall.

*Wiki (https://github.com/opencv/opencv/wiki)*
: The wiki contains everything you could possibly want and more. This is where you'll find the roadmap, as well as news, open issues, bugs tracking, and countless deeper topics like how to become a contributor to OpenCV.

*Q&A (http://answers.opencv.org/questions)*
: This is a vast archive of literally thousands of questions people have asked and answered. You can go there to ask questions of the OpenCV community, or to help others by answering their questions.

All of these are accessible under the Documentation button on the OpenCV.org homepage. Of all of those great resources, one warrants a little more discussion here —the Reference. The Reference is divided into several sections, each of which pertains to a *module* in the library. The exact module list has evolved over time, but the modules are the primary organizational structure in the library. Every function in the library is part of one module. Here are the current modules:

`core`
: The "core" is the section of the library that contains all of the basic object types and their basic operations.

`imgproc`
: The image processing module contains basic transformations on images, including filters and similar convolutional operators.

`highgui` *(split to* `imgcodecs`, `videoio`, *and* `highgui` *in OpenCV 3.0)*
: This module contains user interface functions that can be used to display images or take simple user input. It can be thought of as a very lightweight window UI toolkit.

`video`
: The video library contains the functions you need to read and write video streams.

`calib3d`
: This module contains implementations of algorithms you will need to calibrate single cameras as well as stereo or multicamera arrays.

`features2d`
: This module contains algorithms for detecting, describing, and matching keypoint features.

`objdetect`
: This module contains algorithms for detecting specific objects, such as faces or pedestrians. You can train the detectors to detect other objects as well.

`ml`
: The Machine Learning library is actually an entire library in itself, and contains a wide array of machine learning algorithms implemented in such a way as to work with the natural data structures of OpenCV.

`flann`
: FLANN stands for "Fast Library for Approximate Nearest Neighbors." This library contains methods you will not likely use directly, but which are used by other functions in other modules for doing nearest neighbor searches in large data sets.

`gpu` *(split to multiple* `cuda*` *modules in OpenCV 3.0)*
: The GPU library contains implementations of most of the rest of the library functions optimized for operation on CUDA GPUs. There are also some functions that are implemented only for GPU operation. Some of these provide excellent results but require computational resources sufficiently high that implementation on non-GPU hardware would provide little utility.

`photo`
: This is a relatively new module that contains tools useful for computational photography.

`stitching`
: This entire module implements a sophisticated image stitching pipeline. This is new functionality in the library, but, like the photo module, it is an area where future growth is expected.

`nonfree` *(moved to opencv_contrib/xfeatures2d in OpenCV 3.0)*
: OpenCV contains some implementations of algorithms that are patented or otherwise burdened by usage restrictions (e.g., the SIFT algorithm). Those algorithms are segregated into their own module to indicate that you will need to do some kind of special work in order to use them in a commercial product.

`contrib` *(melted into a few opencv_contrib modules in OpenCV 3.0)*
: This module contains new things that have yet to be integrated into the whole of the library.

`legacy` *(disappeared in OpenCV 3.0)*
: This module contains old things that have yet to be banished from the library altogether.

`ocl` *(disappeared in OpenCV 3.0; replaced with T-API technology)*
: This is a newer module that could be considered analogous to the GPU module, except that it implements the Khronos OpenCL standard for open parallel programming. Though much less featured than the GPU module at this time, the `ocl` module aims to provide implementations that can run on any GPU or other Khronos-capable parallel device. (This is in contrast to the `gpu` module, which explicitly makes use of the NVidia CUDA toolkit and so will work only on NVidia GPU devices.)

Despite the ever-increasing quality of this online documentation, one task that is not within its scope is to provide a proper understanding of the algorithms implemented or of the exact meaning of the parameters these algorithms require. This book aims to provide this information, as well as a more in-depth understanding of all of the basic building blocks of the library.

## OpenCV Contribution Repository

In OpenCV 3.0, the previously monolithic library has been split into two parts: mature opencv and the current state of the art in larger vision functionality at *opencv_contrib* (*https://github.com/opencv/opencv_contrib*) [opencv_contrib]. The former is maintained by the core OpenCV team and contains (mostly) stable code, whereas the latter is less mature, is maintained and developed mostly by the community, may have parts under non-OpenCV license, and may include patented algorithms.

Here are some of the modules available in the *opencv_contrib* repository (see Appendix B for a full list at the time of this writing):

`Dnn`
: Deep neural networks

`face`
: Face recognition

`text`
: Text detection and recognition; may optionally use open source OCR Tesseract as backend

`rgbd`
: Processing RGB + depth maps, obtained with Kinect and other depth sensors (or simply computed with stereo correspondence algorithms)

`bioinspired`
: Biologically inspired vision

`ximgproc`, `xphoto`
: Advanced image processing and computational photography algorithms

`tracking`
: Modern object-tracking algorithms

### Downloading and Building Contributed Modules

On Linux and OS X, you can just use the following command to download *opencv_contrib*:

```
git clone https://github.com/opencv/opencv_contrib.git
```

On Windows, feed this address to TortoiseGit or another such client. Then you need to reconfigure OpenCV with CMake:

```
cmake -D CMAKE_BUILD_TYPE=Release \
  -D OPENCV_EXTRA_MODULES_PATH=../../opencv_contrib/modules ..
```

and rebuild it as usual. The built contributed modules will be put into the same directory as regular OpenCV binaries, and you may use them without any extra steps.

# Portability

OpenCV was designed to be portable. It was originally written to compile by any compliant C++ compiler. This meant that the C and C++ code had to be fairly standard in order to make cross-platform support easier. Table 1-1 shows the platforms on which OpenCV is known to run. Support for Intel and AMD 32-bit and 64-bit architectures (x86, x64) is the most mature, and the ARM support is rapidly improving too. Among operating systems, OpenCV fully supports Windows, Linux, OS X, Android, and iOS.

If an architecture or OS doesn't appear in Table 1-1, this doesn't mean there are no OpenCV ports to it. OpenCV has been ported to almost every commercial system, from Amazon Cloud and 40-core Intel Xeon Phi to Raspberry Pi and robotic dogs.

*Table 1-1. OpenCV portability guide for release 1.0*

| | x86/x64 | ARM | Other: MIPs, PPC |
|---|---|---|---|
| **Windows** | SIMD, IPP, Parallel, I/O | SIMD, Parallel (3.0), I/O | N/A |
| **Linux** | SIMD, IPP, Parallel,[a] I/O | SIMD, Parallel,[a] I/O | Parallel,[a] I/O* |
| **Android** | SIMD, IPP (3.0), Parallel,[b] I/O | SIMD, Parallel,[b] I/O | MIPS—basic support |
| **OS X/iOS** | SIMD, IPP (3.0), Parallel, I/O | SIMD, Parallel, I/O | N/A |
| **Other: BSD, QNX, ...** | SIMD | SIMD | |

[a] Parallelization in Linux is done via a third-party library or by enabling OpenMP.

[b] Parallelization in Android is done via Intel TBB.

Here is the legend for Table 1-1:

*SIMD*
: Vector instructions are used to gain the speed: SSE on x86/x64, NEON on ARM.

*IPP*
: Intel IPP is available. Starting from 3.0, there is free specialized IPP subset (IPPICV).

*Parallel*
: Some standard or third-party threading framework is used to distribute processing across multiple cores.

*I/O*
: Some standard or third-party API can be used to grab or write video.

## Summary

In this chapter we went over OpenCV's [OpenCV] history from its founding by Gary Bradski [Bradski] at Intel in 1999 to its current state of support by Arraiy [Arraiy]. We covered the motivation for OpenCV and some of its content. We discussed how the core library, OpenCV, has been separated from newer functionality in *opencv_contrib* (see Appendix B) along with an extensive set of links to the OpenCV-related content online. This chapter also covered how to download and install OpenCV, together with its performance and portability.

## Exercises

1. Download and install the latest release of OpenCV. Compile it in debug and release mode.
2. Download and build the latest trunk version of OpenCV using Git.
3. Describe at least three ambiguous aspects of converting 3D inputs into a 2D representation. How would you overcome these ambiguities?

CHAPTER 2

# Introduction to OpenCV

## Include Files

After installing the OpenCV library and setting up our programming environment, our next task is to make something interesting happen with code. In order to do this, we'll have to discuss header files. Fortunately, the headers reflect the new, modular structure of OpenCV introduced in Chapter 1. The main header file of interest is *.../include/opencv2/opencv.hpp*; it just calls the header files for each OpenCV module:

`#include "opencv2/core/core_c.h"`
: Old C data structures and arithmetic routines

`#include "opencv2/core/core.hpp"`
: New C++ data structures and arithmetic routines

`#include "opencv2/flann/miniflann.hpp"`
: Approximate nearest neighbor matching functions

`#include "opencv2/imgproc/imgproc_c.h"`
: Old C image processing functions

`#include "opencv2/imgproc/imgproc.hpp"`
: New C++ image processing functions

`#include "opencv2/video/photo.hpp"`
: Algorithms specific to handling and restoring photographs

`#include "opencv2/video/video.hpp"`
: Video tracking and background segmentation routines

`#include "opencv2/features2d/features2d.hpp"`
: Two-dimensional feature tracking support

`#include "opencv2/objdetect/objdetect.hpp"`
: Cascade face detector; latent SVM; HoG; planar patch detector

`#include "opencv2/calib3d/calib3d.hpp"`
: Calibration and stereo

`#include "opencv2/ml/ml.hpp"`
: Machine learning: clustering, pattern recognition

`#include "opencv2/highgui/highgui_c.h"`
: Old C image display, sliders, mouse interaction, I/O

`#include "opencv2/highgui/highgui.hpp"`
: New C++ image display, sliders, buttons, mouse, I/O

`#include "opencv2/contrib/contrib.hpp"`
: User-contributed code: flesh detection, fuzzy mean-shift tracking, spin images, self-similar features

You may use the include file *opencv.hpp* to include any and every possible OpenCV function, but it will slow down compile time. If you are using only, say, image processing functions, compile time will be faster if you include only *opencv2/imgproc/imgproc.hpp*. These include files are located on disk under the *.../modules* directory. For example, *imgproc.hpp* is located at *.../modules/imgproc/include/opencv2/imgproc/imgproc.hpp*. Similarly, the sources for the functions themselves are located under their corresponding *src* directory. For example, `cv::Canny()` in the `imgproc` module is located in *.../modules/improc/src/canny.cpp*.

With the preceding include files, we can start our first C++ OpenCV program.

You can include legacy code such as the older blob tracking, Hidden Markov Model (HMM) face detection, condensation tracker, and Eigen objects using *opencv2/legacy/legacy.hpp*, which is located in *.../modules/legacy/include/opencv2/legacy/legacy.hpp*.

## Resources

There are several good introductory PowerPoint files on the Web that provide overviews of OpenCV:

- A high-level overview of the whole library can be found at *http://is.gd/niZvJu*.
- Speedups are discussed at *http://is.gd/ShvMZE*.
- Modules are described at *http://is.gd/izlOrM*.

## First Program—Display a Picture

OpenCV provides utilities for reading from a wide array of image file types, as well as from video and cameras. These utilities are part of a toolkit called HighGUI, which is included in the OpenCV package. We will use some of these utilities to create a simple program that opens an image and displays it on the screen (Example 2-1).

*Example 2-1. A simple OpenCV program that loads an image from disk and displays it on the screen*

```
#include <opencv2/opencv.hpp> //Include file for every supported OpenCV function

int main( int argc, char** argv ) {
  cv::Mat img = cv::imread(argv[1],-1);
  if( img.empty() ) return -1;
  cv::namedWindow( "Example 2-1", cv::WINDOW_AUTOSIZE );
  cv::imshow( "Example 2-1", img );
  cv::waitKey( 0 );
  cv::destroyWindow( "Example 2-1" );
  return 0;
}
```

Note that OpenCV functions live within a namespace called `cv`. To call OpenCV functions, you must explicitly tell the compiler that you are talking about the `cv` namespace by prepending `cv::` to each function call. To get out of this bookkeeping chore, we can employ the `using namespace cv;` directive as shown in Example 2-2.[1] This tells the compiler to assume that functions might belong to that namespace. Note also the difference in include files between Examples 2-1 and 2-2; in the former, we used the general include *opencv.hpp*, whereas in the latter, we used only the necessary include file to improve compile time.

*Example 2-2. Same as Example 2-1 but employing the "using namespace" directive*

```
#include "opencv2/highgui/highgui.hpp"

using namespace cv;

int main( int argc, char** argv ) {

  Mat img = imread( argv[1], -1 );
```

1 Of course, once you do this, you risk conflicting names with other potential namespaces. If the function `foo()` exists, say, in the `cv` and `std` namespaces, you must specify which function you are talking about using either `cv::foo()` or `std::foo()` as you intend. In this book, other than in our specific example of Example 2-2, we will use the explicit form `cv::` for objects in the OpenCV namespace, as this is generally considered to be better programming style.

```
    if( img.empty() ) return -1;

    namedWindow( "Example 2-2", cv::WINDOW_AUTOSIZE );
    imshow( "Example 2-2", img );
    waitKey( 0 );

    destroyWindow( "Example 2-2" );

}
```

When compiled[2] and run from the command line with a single argument, Example 2-1 loads an image into memory and displays it on the screen. It then waits until the user presses a key, at which time it closes the window and exits. Let's go through the program line by line and take a moment to understand what each command is doing.

```
cv::Mat img = cv::imread( argv[1], -1 );
```

This line loads the image.[3] The function `cv::imread()` is a high-level routine that determines the file format to be loaded based on the filename; it also automatically allocates the memory needed for the image data structure. Note that `cv::imread()` can read a wide variety of image formats, including BMP, DIB, JPEG, JPE, PNG, PBM, PGM, PPM, SR, RAS, and TIFF. A `cv::Mat` structure is returned. This structure is the OpenCV construct with which you will deal the most. OpenCV uses this structure to handle all kinds of images: single-channel, multichannel, integer-valued, floating-point-valued, and so on. The line immediately following:

```
if( img.empty() ) return -1;
```

checks to see if an image was in fact read. Another high-level function, `cv::namedWindow()`, opens a window on the screen that can contain and display an image.

```
cv::namedWindow( "Example 2-2", cv::WINDOW_AUTOSIZE );
```

This function, provided by the HighGUI library, also assigns a name to the window (in this case, `"Example 2-2"`). Future HighGUI calls that interact with this window will refer to it by this name.

---

2 Clearly, build instructions are highly platform dependent. In this book we do not generally cover platform-specific details, but here is an example of what a build instruction might look like in a UNIX-like environment: `gcc -v example2_2.cpp -I/usr/local/include/ -L/usr/lib/ -lstdc++ -L/usr/local/lib -lopencv_highgui -lopencv_core - -o example2_2`. Note that the various components of the library are usually linked separately. In the upcoming Example 2-3 where we will include video, it would be necessary to add: `-lopencv_imgcodecs -lopencv_imgproc -lopencv_videoio -lopencv_video -lopencv_videostab`.

3 A proper program would check for the existence of `argv[1]` and, in its absence, deliver an instructional error message to the user. We will abbreviate such necessities in this book and assume that the reader is cultured enough to understand the importance of error-handling code.

The second argument to cv::namedWindow() defines window properties. It may be set either to 0 (the default value) or to cv::WINDOW_AUTOSIZE. In the former case, the size of the window will be the same regardless of the image size, and the image will be scaled to fit within the window. In the latter case, the window will expand or contract automatically when an image is loaded so as to accommodate the image's true size, but may be resized by the user.

```
cv::imshow( "Example 2-2", img );
```

Whenever we have an image in a cv::Mat structure, we can display it in an existing window with cv::imshow(). The cv::imshow() function creates a window if one does not exist (created by cv::namedWindow()). On the call to cv::imshow(), the window will be redrawn with the appropriate image in it, and the window will resize itself as appropriate if it was created with the cv::WINDOW_AUTOSIZE flag.

```
cv::waitKey( 0 );
```

The cv::waitKey() function asks the program to stop and wait for a keystroke. If a positive argument is given, the program will wait for that number of milliseconds and then continue even if nothing is pressed. If the argument is set to 0 or to a negative number, the program will wait indefinitely for a key-press.

With cv::Mat, images are automatically deallocated when they go out of scope, similar to the Standard Template Library (STL)-style container classes. This automatic deallocation is controlled by an internal reference counter. For the most part, this means we no longer need to worry about the allocation and deallocation of images, which relieves the programmer from much of the tedious bookkeeping that the OpenCV 1.0 IplImage imposed.

```
cv::destroyWindow( "Example 2-2" );
```

Finally, we can destroy the window itself. The function cv::destroyWindow() will close the window and deallocate any associated memory usage. For short programs, we will skip this step. For longer, more complex programs, the programmer should make sure to tidy up the windows before they go out of scope to avoid memory leaks.

Our next task is to construct a very simple—almost as simple as this one—program to read in and display a video file. After that, we will start to tinker a little more with the actual images.

# Second Program—Video

Playing a video with OpenCV is almost as easy as displaying a single picture. The only new issue we face is that we need some kind of loop to read each frame in sequence; we may also need some way to get out of that loop if the movie is too boring. See Example 2-3.

*Example 2-3. A simple OpenCV program for playing a video file from disk*

```
#include "opencv2/highgui/highgui.hpp"
#include "opencv2/imgproc/imgproc.hpp"

using namespace std;

int main( int argc, char** argv ) {

  cv::namedWindow( "Example 2-3", cv::WINDOW_AUTOSIZE );
  cv::VideoCapture cap;
  cap.open( string(argv[1]) );

  cv::Mat frame;
  for(;;) {
    cap >> frame;
    if( frame.empty() ) break;              // Ran out of film
    cv::imshow( "Example 2-3", frame );
    if( (char) cv::waitKey(33) >= 0 ) break;
  }

  return 0;
}
```

Here we begin the function `main()` with the usual creation of a named window (in this case, named `"Example 2-3"`). The video capture object `cv::VideoCapture` is then instantiated. This object can open and close video files of as many types as *ffmpeg* supports.

```
cap.open(string(argv[1]));
cv::Mat frame;
```

The capture object is given a string containing the path and filename of the video to be opened. Once opened, the capture object will contain all of the information about the video file being read, including state information. When created in this way, the `cv::VideoCapture` object is initialized to the beginning of the video. In the program, `cv::Mat frame` instantiates a data object to hold video frames.

```
cap >> frame;
if( frame.empty() ) break;
cv::imshow( "Example 2-3", frame );
```

Once inside of the main loop, the video file is read frame by frame from the capture object stream. The program checks to see if data was actually read from the video file —`if(frame.empty())`—and, if not, quits. If a video frame was successfully read in, it is displayed through `cv::imshow()`.

```
if( cv::waitKey(33) >= 0 ) break;
```

Once we have displayed the frame, we then wait 33 ms.[4] If the user hits a key during that time, we will exit the read loop. Otherwise, 33 ms will pass and we will execute the loop again. On exit, all the allocated data is automatically released when it goes out of scope.

## Moving Around

Now it's time to tinker around, enhance our toy programs, and explore a little more of the available functionality. The first thing we might notice about the video player in Example 2-3 is that users have no way to move around quickly within the video. So our next task is to add a slider trackbar, which will give users this ability. For more control, we will also allow the user to single-step the video by pressing the S key and to go into run mode by pressing the R key, and whenever the user jumps to a new location in the video with the trackbar, we'll pause there in single-step mode.

The HighGUI toolkit provides a number of simple instruments for working with images and video beyond the simple display functions we have just demonstrated. One especially useful mechanism is the aforementioned trackbar, which enables users to jump easily from one part of a video to another. To create a trackbar, we call `cv::createTrackbar()` and indicate which window we would like the trackbar to appear in. In order to obtain the desired functionality, we need a callback that will perform the relocation. Example 2-4 gives the details.

*Example 2-4. Adding a trackbar slider to the basic viewer window for moving around within the video file*

```
#include "opencv2/highgui/highgui.hpp"
#include "opencv2/imgproc/imgproc.hpp"
#include <iostream>
#include <fstream>

using namespace std;

int g_slider_position = 0;
int g_run = 1, g_dontset = 0; //start out in single step mode
cv::VideoCapture g_cap;

void onTrackbarSlide( int pos, void *) {

  g_cap.set( cv::CAP_PROP_POS_FRAMES, pos );
```

4 You can wait any amount of time you like. In this case, we are simply assuming that it is correct to play the video at 30 frames per second and allow user input to interrupt between each frame (thus we pause for input 33 ms between each frame). In practice, it is better to check the `cv::VideoCapture` structure in order to determine the actual frame rate of the video (more on this in Chapter 8).

```
  if( !g_dontset )
    g_run = 1;
  g_dontset = 0;

}

int main( int argc, char** argv ) {

  cv::namedWindow( "Example 2-4", cv::WINDOW_AUTOSIZE );
  g_cap.open( string(argv[1]) );
  int frames = (int) g_cap.get(CV_CAP_PROP_FRAME_COUNT);
  int tmpw   = (int) g_cap.get(CV_CAP_PROP_FRAME_WIDTH);
  int tmph   = (int) g_cap.get(CV_CAP_PROP_FRAME_HEIGHT);
  cout << "Video has " << frames << " frames of dimensions("
       << tmpw << ", " << tmph << ")." << endl;

  cv::createTrackbar("Position", "Example 2-4", &g_slider_position, frames,
                 onTrackbarSlide);

  cv::Mat frame;
  for(;;) {

    if( g_run != 0 ) {

      g_cap >> frame; if(frame.empty()) break;
      int current_pos = (int)g_cap.get(CV_CAP_PROP_POS_FRAMES);
      g_dontset = 1;

      cv::setTrackbarPos("Position", "Example 2-4", current_pos);
      cv::imshow( "Example 2-4", frame );

      g_run-=1;

    }

    char c = (char) cv::waitKey(10);
    if( c == 's' ) // single step
      {g_run = 1; cout << "Single step, run = " << g_run << endl;}
    if( c == 'r' ) // run mode
      {g_run = -1; cout << "Run mode, run = " << g_run <<endl;}
    if( c == 27 )
      break;

  }
  return(0);

}
```

In essence, the strategy is to add a global variable to represent the trackbar position and then add a callback that updates this variable and relocates the read position in the video. One call creates the trackbar and attaches the callback, and we are off and running.[5] Let's look at the details starting with the global variables.

```
int g_slider_position = 0;
int g_run             = 1;
int g_dontset         = 0;      // start out in single-step mode
VideoCapture g_cap;
```

First we define a global variable, g_slider_position, to keep the trackbar slider position state. The callback will need access to the capture object g_cap, so we promote that to a global variable as well. Because we are considerate developers and like our code to be readable and easy to understand, we adopt the convention of adding a leading g_ to any global variable. We also instantiate another global variable, g_run, which displays new frames as long it is different from zero. A positive number indicates how many frames are displayed before stopping; a negative number means the system runs in continuous video mode.

To avoid confusion, when the user clicks on the trackbar to jump to a new location in the video, we'll leave the video paused there in the single-step state by setting g_run = 1. This, however, brings up a subtle problem: as the video advances, we'd like the slider trackbar's position in the display window to advance according to our location in the video. We do this by having the main program call the trackbar callback function to update the slider's position each time we get a new video frame. However, we don't want these programmatic calls to the trackbar callback to put us into single-step mode. To avoid this, we introduce a final global variable, g_dontset, to allow us to update the trackbar's position without triggering single-step mode.

```
void onTrackbarSlide(int pos, void *) {

  g_cap.set(CV_CAP_PROP_POS_FRAMES, pos);

  if( !g_dontset )
    g_run = 1;
  g_dontset = 0;

}
```

Now we define a callback routine to be used when the user slides the trackbar. This routine will be passed a 32-bit integer, pos, which will be the new trackbar position. Inside this callback, we use the new requested position in g_cap.set() to actually advance the video playback to the new position. The if() statement sets the program to go into single-step mode after the next new frame comes in, but only if the callback was triggered by a user click, not if it was called from the main function (which sets g_dontset).

5 Note that some AVI and mpeg encodings do not allow you to move backward in the video.

The call to g_cap.set() is one we will see often in the future, along with its counterpart g_cap.get(). These routines allow us to configure (or query, in the latter case) various properties of the cv::VideoCapture object. In this case, we pass the argument CV_CAP_PROP_POS_FRAMES, which indicates that we would like to set the read position in units of frames.[6]

```
int frames = (int) g_cap.get(CV_CAP_PROP_FRAME_COUNT);
int tmpw   = (int) g_cap.get(CV_CAP_PROP_FRAME_WIDTH);
int tmph   = (int) g_cap.get(CV_CAP_PROP_FRAME_HEIGHT);
cout << "Video has " << frames << " frames of dimensions("
     << tmpw << ", " << tmph   << ")." << endl;
```

The core of the main program is the same as in Example 2-3, so we'll focus on what we've added. The first difference after opening the video is that we use g_cap.get() to determine the number of frames in the video and the width and height of the video images. These numbers are printed out. We'll need the number of frames in the video to calibrate the slider trackbar (in the next step).

```
createTrackbar("Position", "Example 2-4", &g_slider_position, frames,
               onTrackbarSlide);
```

Next we create the trackbar itself. The function cv::createTrackbar() allows us to give the trackbar a label[7] (in this case, Position) and to specify a window in which to put the trackbar. We then provide a variable that will be bound to the trackbar, the maximum value of the trackbar (the number of frames in the video), and a callback (or NULL if we don't want one) for when the slider is moved.

```
if( g_run != 0 ) {

  g_cap >> frame; if(!frame.data) break;
  int current_pos = (int)g_cap.get(CV_CAP_PROP_POS_FRAMES);
  g_dontset = 1;

  cv::setTrackbarPos("Position", "Example 2-4", current_pos);
  cv::imshow( "Example 2-4", frame );

  g_run-=1;

}
```

6 Because HighGUI is highly civilized, when a new video position is requested, it will automatically handle issues such as the possibility that the frame we have requested is not a keyframe; it will start at the previous keyframe and fast-forward up to the requested frame without us having to fuss with such details.

7 Because HighGUI is a lightweight, easy-to-use toolkit, cv::createTrackbar() does not distinguish between the name of the trackbar and the label that actually appears on the screen next to the trackbar. You may already have noticed that cv::namedWindow() likewise does not distinguish between the name of the window and the label that appears on the window in the GUI.

In the main loop, in addition to reading and displaying the video frame, we also get our current position in the video, set the `g_dontset` so that the next trackbar callback will not put us into single-step mode, and then invoke the trackbar callback to update the position of the slider trackbar displayed to the user. The global `g_run` is decremented, which has the effect of either keeping us in single-step mode or letting the video run depending on its prior state set by a user keypress, as we'll see next.

```
char c = (char) cv::waitKey(10);
if( c == 's' ) // single step
  {g_run = 1; cout << "Single step, run = " << g_run << endl;}
if( c == 'r' ) // run mode
  {g_run = -1; cout << "Run mode, run = " << g_run <<endl;}
if( c == 27 )
  break;
```

At the bottom of the main loop, we look for keyboard input from the user. If S has been pressed, we go into single-step mode (`g_run` is set to `1`, which allows reading of a single frame). If R is pressed, we go into continuous video mode (`g_run` is set to `-1` and further decrementing leaves it negative for any conceivable video size). Finally, if Esc is pressed, the program will terminate. Note again, for short programs, we've omitted the step of cleaning up the window storage using `cv::destroyWindow()`.

# A Simple Transformation

Great, so now you can use OpenCV to create your own video player, which will not be much different from countless video players out there already. But we are interested in computer vision, so we want to do some of that. Many basic vision tasks involve the application of filters to a video stream. We will modify the program we already have to do a simple operation on every frame of the video as it plays.

One particularly simple operation is *smoothing* an image, which effectively reduces the information content of the image by convolving it with a Gaussian or other similar kernel function. OpenCV makes such convolutions exceptionally easy to do. We can start by creating a new window called `"Example 2-5-out"`, where we can display the results of the processing. Then, after we have called `cv::imshow()` to display the newly captured frame in the input window, we can compute and display the smoothed image in the output window. See Example 2-5.

*Example 2-5. Loading and then smoothing an image before it is displayed on the screen*

```
#include <opencv2/opencv.hpp>

int main( int argc, char** argv ) {

  // Load an image specified on the command line.
  //
```

```
    cv::Mat image = cv::imread(argv[1],-1);

    // Create some windows to show the input
    // and output images in.
    //
    cv::namedWindow( "Example 2-5-in", cv::WINDOW_AUTOSIZE );
    cv::namedWindow( "Example 2-5-out", cv::WINDOW_AUTOSIZE );

    // Create a window to show our input image
    //
    cv::imshow( "Example 2-5-in", image );

    // Create an image to hold the smoothed output
    //
    cv::Mat out;

    // Do the smoothing
    // ( Note: Could use GaussianBlur(), blur(), medianBlur() or bilateralFilter(). )
    //
    cv::GaussianBlur( image, out, cv::Size(5,5), 3, 3);
    cv::GaussianBlur(   out, out, cv::Size(5,5), 3, 3);

    // Show the smoothed image in the output window
    //
    cv::imshow( "Example 2-5-out", out );

    // Wait for the user to hit a key, windows will self destruct
    //
    cv::waitKey( 0 );

}
```

The first call to `cv::imshow()` is no different than in our previous example. In the next call, we allocate another image structure. Next, the C++ object `cv::Mat` makes life simpler for us; we just instantiate an output matrix, `out`, and it will automatically resize/reallocate and deallocate itself as necessary as it is used. To make this point clear, we use it in two consecutive calls to `cv::GaussianBlur()`. In the first call, the input image is blurred by a 5 × 5 Gaussian convolution filter and written to `out`. The size of the Gaussian kernel should always be given in odd numbers since the Gaussian kernel (specified here by `cv::Size(5,5)`) is computed at the center pixel in that area. In the next call to `cv::GaussianBlur()`, `out` is used as both the input and output since temporary storage is assigned for us in this case. The resulting double-blurred image is displayed, and the routine then waits for any user keyboard input before terminating and cleaning up allocated data as it goes out of scope.

## A Not-So-Simple Transformation

That was pretty good, and we are learning to do more interesting things. In Example 2-5, we used Gaussian blurring for no particular purpose. We will now use a function that uses Gaussian blurring to *downsample* an image by a factor of 2 [Rosenfeld80]. If we downsample the image several times, we form a *scale space* (also known as an *image pyramid*) that is commonly used in computer vision to handle the changing scales in which a scene or object is observed.

For those who know some signal processing and the Nyquist-Shannon Sampling Theorem [Shannon49], downsampling a signal (in this case, creating an image where we are sampling every other pixel) is equivalent to convolving with a series of delta functions (think of these as "spikes"). Such sampling introduces high frequencies into the resulting signal (image). To avoid this, we want to first run a high-pass filter over the signal to band-limit its frequencies so that they are all below the sampling frequency. In OpenCV, this Gaussian blurring and downsampling is accomplished by the function `cv::pyrDown()`, which we implement in Example 2-6.

*Example 2-6. Using cv::pyrDown() to create a new image that is half the width and height of the input image*

```
#include <opencv2/opencv.hpp>

int main( int argc, char** argv ) {

  cv::Mat img1,img2;

  cv::namedWindow( "Example 2-6-in", cv::WINDOW_AUTOSIZE );
  cv::namedWindow( "Example 2-6-out", cv::WINDOW_AUTOSIZE );

  img1 = cv::imread( argv[1] );
  cv::imshow( "Example 2-6-in", img1 );

  cv::pyrDown( img1, img2);
  cv::imshow( "Example 2-6-out", img2 );

  cv::waitKey(0);

  return 0;

}
```

Let's now look at a similar but slightly more complex example involving the *Canny edge detector* [Canny86] `cv::Canny()`; see Example 2-7. In this case, the edge detector generates an image that is the full size of the input image but needs only a single-channel image to write to, so we convert to a grayscale, single-channel image first using `cv::cvtColor()` with the flag to convert blue, green, red (BGR) images to grayscale, `cv::COLOR_BGR2GRAY`.

*Example 2-7. The Canny edge detector writes its output to a single-channel (grayscale) image*

```
#include <opencv2/opencv.hpp>

int main( int argc, char** argv ) {

    cv::Mat img_rgb, img_gry, img_cny;

    cv::namedWindow( "Example Gray",  cv::WINDOW_AUTOSIZE );
    cv::namedWindow( "Example Canny", cv::WINDOW_AUTOSIZE );

    img_rgb = cv::imread( argv[1] );

    cv::cvtColor( img_rgb, img_gry, cv::COLOR_BGR2GRAY);
    cv::imshow( "Example Gray", img_gry );

    cv::Canny( img_gry, img_cny, 10, 100, 3, true );
    cv::imshow( "Example Canny", img_cny );

    cv::waitKey(0);

}
```

This allows us to string together various operators quite easily. For example, if we wanted to shrink the image twice and then look for lines that were present in the twice-reduced image, we could proceed as in Example 2-8.

*Example 2-8. Combining the pyramid down operator (twice) and the Canny subroutine in a simple image pipeline*

```
cv::cvtColor( img_rgb, img_gry, cv::COLOR_BGR2GRAY );
cv::pyrDown( img_gry, img_pyr );
cv::pyrDown( img_pyr, img_pyr2 );
cv::Canny( img_pyr2, img_cny, 10, 100, 3, true );
// do whatever with 'img_cny'
//
...
```

In Example 2-9, we show a simple way to read and write pixel values from Example 2-8.

*Example 2-9. Getting and setting pixels in Example 2-8*

```
int x = 16, y = 32;
cv::Vec3b intensity = img_rgb.at< cv::Vec3b >(y, x);

// ( Note: We could write img_rgb.at< cv::Vec3b >(y,x)[0] )
//
uchar blue  = intensity[0];
uchar green = intensity[1];
uchar red   = intensity[2];

std::cout << "At (x,y) = (" << x << ", " << y <<
          "): (blue, green, red) = (" <<
          (unsigned int)blue <<
          ", " << (unsigned int)green << ", " <<
          (unsigned int)red << ")" << std::endl;

std::cout << "Gray pixel there is: " <<
             (unsigned int)img_gry.at<uchar>(y, x) << std::endl;

x /= 4; y /= 4;
std::cout << "Pyramid2 pixel there is: " <<
             (unsigned int)img_pyr2.at<uchar>(y, x) << std::endl;

img_cny.at<uchar>(x, y) = 128; // Set the Canny pixel there to 128
```

# Input from a Camera

"Vision" can mean many things in the world of computers. In some cases, we are analyzing still frames loaded from elsewhere. In other cases, we are analyzing video that is being read from disk. In still other cases, we want to work with real-time data streaming in from some kind of camera device.

OpenCV—or more specifically, the HighGUI portion of the OpenCV library—provides us with an easy way to handle this situation. The method is analogous to how we read videos from disk since the `cv::VideoCapture` object works the same for files on disk or from a camera. For the former, you give it a path/filename, and for the latter, you give it a camera ID number (typically `0` if only one camera is connected to the system). The default value is `-1`, which means "just pick one"; naturally, this works quite well when there is only one camera to pick (see Chapter 8 for more details). Video capture from a file or from a camera is demonstrated in Example 2-10.

*Example 2-10. The same object can load videos from a camera or a file*

```
#include <opencv2/opencv.hpp>
#include <iostream>

int main( int argc, char** argv ) {
```

```
  cv::namedWindow( "Example 2-10", cv::WINDOW_AUTOSIZE );

  cv::VideoCapture cap;
  if (argc==1) {
    cap.open(0);           // open the first camera
  } else {
    cap.open(argv[1]);
  }
  if( !cap.isOpened() ) {  // check if we succeeded
    std::cerr << "Couldn't open capture." << std::endl;
    return -1;
  }

  // The rest of program proceeds as in Example 2-3
  ...
```

In Example 2-10, if a filename is supplied, OpenCV opens that file just like in Example 2-3, and if no filename is given, it attempts to open camera zero (`0`). We have added a check to confirm that something actually opened; if it didn't, an error is reported.

# Writing to an AVI File

In many applications, we will want to record streaming input or even disparate captured images to an output video stream, and OpenCV provides a straightforward method for doing this. Just as we are able to create a capture device that allows us to grab frames one at a time from a video stream, we are able to create a writer device that allows us to place frames one by one into a video file. The object that allows us to do this is `cv::VideoWriter`.

Once this call has been made, we may stream each frame to the `cv::VideoWriter` object, and finally call its `cv::VideoWriter.release()` method when we are done. Just to make things more interesting, Example 2-11 describes a program that opens a video file, reads the contents, converts them to a log-polar format (something like what your eye actually sees, as described in Chapter 11), and writes out the log-polar image to a new video file.

*Example 2-11. A complete program to read in a color video and write out the log-polar-transformed video*

```
#include <opencv2/opencv.hpp>
#include <iostream>

int main( int argc, char* argv[] ) {

  cv::namedWindow( "Example 2-11", cv::WINDOW_AUTOSIZE );
```

```
  cv::namedWindow( "Log_Polar",   cv::WINDOW_AUTOSIZE );

  // ( Note: could capture from a camera by giving a camera id as an int.)
  //
  cv::VideoCapture capture( argv[1] );

  double fps = capture.get( CV_CAP_PROP_FRAME_FPS );
  cv::Size size(
    (int)capture.get( CV_CAP_PROP_FRAME_WIDTH ),
    (int)capture.get( CV_CAP_PROP_FRAME_HEIGHT )
  );

  cv::VideoWriter writer;
  writer.open( argv[2], CV_FOURCC('M','J','P','G'), fps, size );

  cv::Mat logpolar_frame, bgr_frame;
  for(;;) {

    capture >> bgr_frame;
    if( bgr_frame.empty() ) break; // end if done

    cv::imshow( "Example 2-11", bgr_frame );

    cv::logPolar(
      bgr_frame,                      // Input color frame
      logpolar_frame,                 // Output log-polar frame
      cv::Point2f(                    // Centerpoint for log-polar transformation
        bgr_frame.cols/2,             //  x
        bgr_frame.rows/2              //  y
      ),
      40,                             // Magnitude (scale parameter)
      CV_WARP_FILL_OUTLIERS           // Fill outliers with 'zero'
    );

    cv::imshow( "Log_Polar", logpolar_frame );
    writer << logpolar_frame;

    (char) cv::waitKey(33);
    if( c == 27 ) break;        // allow the user to break out
  }

  capture.release();
}
```

Looking over this program reveals mostly familiar elements. We open one video and read some properties (frames per second, image width and height) that we'll need to open a file for the `cv::VideoWriter` object. We then read the video frame by frame from the `cv::VideoReader` object, convert the frame to log-polar format, and write the log-polar frames to this new video file one at a time until there are none left or until the user quits by pressing Esc. Then we close up.

The call to the `cv::VideoWriter` object contains several parameters that we should understand. The first is just the filename for the new file. The second is the *video codec* with which the video stream will be compressed. There are countless such codecs in circulation, but whichever codec you choose must be available on your machine (codecs are installed separately from OpenCV). In our case, we choose the relatively popular MJPG codec; we indicate this choice to OpenCV by using the macro `CV_FOURCC()`, which takes four characters as arguments. These characters constitute the "four-character code" of the codec, and every codec has such a code. The four-character code for *motion jpeg* is "MJPG," so we specify that as `CV_FOURCC('M','J','P','G')`. The next two arguments are the replay frame rate and the size of the images we will be using. In our case, we set these to the values we got from the original (color) video.

## Summary

Before moving on to the next chapter, we should take a moment to take stock of where we are and look ahead to what is coming. We have seen that the OpenCV API provides us with a variety of easy-to-use tools for reading and writing still images and videos from and to files along with capturing video from cameras. We have also seen that the library contains primitive functions for manipulating these images. What we have not yet seen are the powerful elements of the library, which allow for more sophisticated manipulation of the entire set of abstract data types that are important in solving practical vision problems.

In the next few chapters, we will delve more deeply into the basics and come to understand in greater detail both the interface-related functions and the image data types. We will investigate the primitive image manipulation operators and, later, some much more advanced ones. Thereafter, we will be ready to explore the many specialized services that the API provides for tasks as diverse as camera calibration, tracking, and recognition. Ready? Let's go!

## Exercises

Download and install OpenCV if you have not already done so. Systematically go through the directory structure. Note in particular the *docs* directory, where you can load *index.htm*, which links to the main documentation of the library. Further explore the main areas of the library. The `core` module contains the basic data structures and algorithms, `imgproc` contains the image processing and vision algorithms, `ml` includes algorithms for machine learning and clustering, and `highgui` contains the I/O functions. Check out the *.../samples/cpp* directory, where many useful examples are stored.

1. Using the install and build instructions in this book or at *http://opencv.org*, build the library in both the debug and the release versions. This may take some time, but you will need the resulting library and *dll* files. Make sure you set the *cmake* file to build the samples *.../opencv/samples/* directory.
2. Go to where you built the *.../opencv/samples/* directory (we build in *.../trunk/eclipse_build/bin*) and look for *lkdemo.cpp* (this is an example motion-tracking program). Attach a camera to your system and run the code. With the display window selected, type `r` to initialize tracking. You can add points by clicking on video positions with the mouse. You can also switch to watching only the points (and not the image) by typing `n`. Typing `n` again will toggle between "night" and "day" views.
3. Use the capture and store code in Example 2-11 together with the `PyrDown()` code of Example 2-6 to create a program that reads from a camera and stores downsampled color images to disk.
4. Modify the code in Exercise 3 and combine it with the window display code in Example 2-2 to display the frames as they are processed.
5. Modify the program of Exercise 4 with a trackbar slider control from Example 2-4 so that the user can dynamically vary the pyramid downsampling reduction level by factors of between 2 and 8. You may skip writing this to disk, but you should display the results.

CHAPTER 3

# Getting to Know OpenCV Data Types

## The Basics

In the next few chapters, we will see all of the basic data types of OpenCV, from the primitives to the larger structures that are used to handle arrays such as images and large matrices. Along the way, we will also cover the vast menagerie of functions that allow us to manipulate this data in a host of useful ways. In this chapter, we will start out by learning about the basic data types and will cover some useful utility functions that the library provides.

## OpenCV Data Types

OpenCV has many data types, which are designed to make the representation and handling of important computer vision concepts relatively easy and intuitive. At the same time, many algorithm developers require a set of relatively powerful primitives that can be generalized or extended for their particular needs. This library attempts to address both of these needs through the use of templates for fundamental data types, and specializations of those templates that make everyday operations easier.

From an organizational perspective, it is convenient to divide the data types into three major categories. First, the *basic data types* are those that are assembled directly from C++ primitives (`int`, `float`, etc.). These types include simple vectors and matrices, as well as representations of simple geometric concepts like points, rectangles, sizes, and the like. The second category contains *helper objects*. These objects represent more abstract concepts such as the garbage-collecting pointer class, range objects used for slicing, and abstractions such as termination criteria. The third category is what might be called *large array types*. These are objects whose fundamental purpose is to contain arrays or other assemblies of primitives or, more often, the basic data types. The star example of this category is the `cv::Mat` class, which is used to represent arbitrary-dimensional arrays containing arbitrary basic elements. Objects such as images are specialized uses of the `cv::Mat` class, but—unlike in earlier versions of OpenCV (i.e., before version 2.1)—such specific use does not require a different class or type. In addition to `cv::Mat`, this category contains related objects such as the sparse matrix `cv::SparseMat` class, which is more naturally suited to nondense data such as histograms. The `cv::Mat` and `cv::SparseMat` classes will be the subjects of the next chapter.

In addition to these types, OpenCV also makes heavy use of the *Standard Template Library* (STL). OpenCV particularly relies on the vector class, and many OpenCV library functions now have vector template objects in their argument lists. We will not cover STL in this book,[1] other than as necessary to explain relevant functionality. If you are already comfortable with STL, many of the template mechanisms used "under the hood" in OpenCV will be familiar to you.

## Overview of the Basic Types

The most straightforward of the basic data types is the template class `cv::Vec<>`, a container class for primitives,[2] which we will refer to as the *fixed vector classes*. Why not just use STL classes? The key difference is that the fixed vector classes are intended for small vectors whose dimensions are known at compile time. This allows for particularly efficient code to handle small common operations. What "small" means in practice is that if you have more than just a few elements, you are probably using the wrong class. (In fact, as of version 2.2, this number cannot exceed nine in any case.) In the next chapter, we will look at the `cv::Mat` class, which is the right way to handle big arrays of any number of dimensions, but for now, think of the fixed vector classes as being handy and speedy for little guys.

Even though `cv::Vec<>` is a template, you will not tend to see or use it in that form most of the time. Instead, there are aliases (`typedef`s) for common instantiations of the `cv::Vec<>` template. They have names like `cv::Vec2i`, `cv::Vec3i`, and `cv::Vec4d` (for a two-element integer vector, a three-element integer vector, or a four-element double-precision floating-point vector, respectively). In general, anything of the form

1 Readers unfamiliar with the Standard Template Library can find many excellent references online. In addition, the authors highly recommend Nicolai M. Josuttis's classic *The C++ Standard Library, Second Edition: A Tutorial and Reference* (Addison-Wesley, 2012) or Scott Meyers' excellent *Effective STL: 50 Specific Ways to Improve Your Use of the Standard Template Library* (Addison-Wesley, 2001).

2 Actually, this is an oversimplification that we will clear up a little later in the chapter. In fact, `cv::Vec<>` is a vector container for anything, and uses templating to create this functionality. As a result, `cv::Vec<>` can contain other class objects, either from OpenCV or elsewhere. In most usages, however, `cv::Vec` is used as a container for C primitive types like `int` or `float`.

cv::Vec{2,3,4,6}{b,w,s,i,f,d} is valid for any combination of two to four dimensions and the six data types.[3]

In addition to the fixed vector classes, there are also *fixed matrix classes*. They are associated with the template cv::Matx<>. Just like the fixed vector classes, cv::Matx<> is not intended to be used for large arrays, but rather is designed for the handling of certain specific small matrix operations. In computer vision, there are a lot of 2 × 2 or 3 × 3 matrices around, and a few 4 × 4, which are used for various transformations. cv::Matx<> is designed to hold these sorts of objects. As with cv::Vec<>, cv::Matx<> is normally accessed through aliases of the form cv::Matx{1,2,3,4,6}{1,2,3,4,6}{f,d}. It is important to notice that with the fixed matrix classes (like the fixed vector classes, but unlike next chapter's cv::Mat), the dimensionality of the fixed matrix classes must be known at compile time. Of course, it is precisely this knowledge that makes operations with the fixed matrix classes highly efficient and eliminates many dynamic memory allocation operations.

Closely related to the fixed vector classes are the *point classes*, which are containers for two or three values of one of the primitive types. The point classes are derived from their own template, so they are not directly descended from the fixed vector classes, but they can be cast to and from them. The main difference between the point classes and the fixed vector classes is that their members are accessed by named variables (mypoint.x, mypoint.y, etc.) rather than by a vector index (myvec[0], myvec[1], etc.). As with cv::Vec<>, the point classes are typically invoked via aliases for the instantiation of an appropriate template. Those aliases have names like cv::Point2i, cv::Point2f, and cv::Point2d, or cv::Point3i, cv::Point3f, and cv::Point3d.

The class cv::Scalar is essentially a four-dimensional point. As with the point classes, cv::Scalar is actually associated with a template that can generate an arbitrary four-component vector, but the keyword cv::Scalar specifically is aliased to a four-component vector with double-precision components. Unlike the point classes, the elements of a cv::Scalar object are accessed with an integer index, the same as cv::Vec<>. This is because cv::Scalar is directly derived from an instantiation of cv::Vec<> (specifically, from cv::Vec<double,4>).

Next on our tour are cv::Size and cv::Rect. As with the point classes, these two are derived from their own templates. cv::Size is mainly distinguished by having data members width and height rather than x and y, while cv::Rect has all four. The class cv::Size is actually an alias for cv::Size2i, which is itself an alias of a more

3 The six data types referred to here have the following conventional abbreviation in the library: b = unsigned char, w = unsigned short, s = short, i = int, f = float, d = double.

general template in the case of `width` and `height` being integers. For floating-point values of `width` and `height`, use the alias `cv::Size2f`. Similarly, `cv::Rect` is an alias for the integer form of rectangle. There is also a class to represent a rectangle that is not axis-aligned. It is called `cv::RotatedRect` and contains a `cv::Point2f` called `center`, a `cv::Size2f` called `size`, and one additional `float` called `angle`.

## Basic Types: Getting Down to Details

Each of the basic types is actually a relatively complicated object, supporting its own interface functions, overloaded operators, and the like. In this section, we will take a somewhat more encyclopedic look at what each type offers, and how some of the otherwise seemingly similar types differ from one another.

As we go over these classes, we will try to hit the high points of their interfaces, but not get into every gory detail. Instead, we will provide examples that should convey what you can and can't do with these objects. For the low-level details, you should consult *.../opencv2/core/core.hpp*.

### The point classes

Of the OpenCV basic types, the point classes are probably the simplest. As we mentioned earlier, these are implemented based on a template structure, such that there can be points of any type: integer, floating-point, and so on. There are actually two such templates, one for two-dimensional and one for three-dimensional points. The big advantage of the point classes is that they are simple and have very little overhead. Natively, they do not have a lot of operations defined on them, but they can be cast to somewhat more generalized types, such as the fixed vector classes or the fixed matrix classes (discussed later), when needed.

In most programs, the point classes are instantiated via aliases that take forms like `cv::Point2i` or `cv::Point3f`, with the last letter indicating the desired primitive from which the point is to be constructed. (Here, `b` is an unsigned character, `i` is a 32-bit integer, `f` is a 32-bit floating-point number, and `d` is a 64-bit floating-point number.)

Table 3-1 is the (relatively short) list of functions natively supported by the point classes. Note that there are several very important operations that are supported, but they are supported indirectly through implicit casting to the fixed vector classes (described in "The fixed vector classes" on page 51). These operations notably contain all of the vector and singleton[4] overloaded algebraic operators and comparison operators.

*Table 3-1. Operations supported directly by the point classes*

| Operation | Example |
|---|---|
| Default constructors | `cv::Point2i p;`<br>`cv::Point3f p;` |
| Copy constructor | `cv::Point3f p2( p1 );` |
| Value constructors | `cv::Point2i p( x0, x1 );`<br>`cv::Point3d p( x0, x1, x2 );` |
| Cast to the fixed vector classes | `(cv::Vec3f) p;` |
| Member access | `p.x; p.y; // and for three-dimensional`<br>`// point classes: p.z` |
| Dot product | `float x = p1.dot( p2 )` |
| Double-precision dot product | `double x = p1.ddot( p2 )` |
| Cross product | `p1.cross( p2 ) // (for three-dimensional point`<br>`// classes only)` |
| Query if point *p* is inside rectangle *r* | `p.inside( r ) // (for two-dimensional point`<br>`// classes only)` |

These types can be cast to and from the old C interface types `CvPoint` and `CvPoint2D32f`. In cases in which a floating-point-valued instance of one of the point classes is cast to `CvPoint`, the values will automatically be rounded.

### The cv::Scalar class

`cv::Scalar` is really a four-dimensional point class. Like the others, it is actually associated with a template class, but the alias for accessing it returns an instantiation of that template in which all of the members are double-precision floating-point numbers. The `cv::Scalar` class also has some special member functions associated with uses of four-component vectors in computer vision. Table 3-2 lists the operations supported by `cv::Scalar`.

4 You might have expected us to use the word *scalar* here, but we avoided doing so because `cv::Scalar` is an existing class in the library. As you will see shortly, a `cv::Scalar` in OpenCV is (somewhat confusingly) an array of four numbers, approximately equivalent to a `cv::Vec` with four elements! In this context, the word *singleton* can be understood to mean "a single object of whatever type the vector is an array of."

*Table 3-2. Operations supported directly by cv::Scalar*

| Operation | Example |
|---|---|
| Default constructors | `cv::Scalar s;` |
| Copy constructor | `cv::Scalar s2( s1 );` |
| Value constructors | `cv::Scalar s( x0 );`<br>`cv::Scalar s( x0, x1, x2, x3 );` |
| Element-wise multiplication | `s1.mul( s2 );` |
| (Quaternion) conjugation | `s.conj(); // (returns cv::Scalar(s0,-s1,-s2,-s2))` |
| (Quaternion) real test | `s.isReal(); // (returns true iff s1==s2==s3==0)` |

You will notice that for `cv::Scalar`, the operation "cast to the fixed vector classes" does not appear in Table 3-2 (as it did in Table 3-1). This is because, unlike the point classes, `cv::Scalar` inherits directly from an instantiation of the fixed vector class template. As a result, it inherits all of the vector algebra operations, member access functions (i.e., `operator[]`), and other properties from the fixed vector classes. We will get to that class later, but for now, just keep in mind that `cv::Scalar` is shorthand for a four-dimensional double-precision vector that has a few special member functions attached that are useful for various kinds of four-vectors.

The `cv::Scalar` class can be freely cast to and from the old C interface `CvScalar` type.

#### The size classes

The size classes are, in practice, similar to the corresponding point classes, and can be cast to and from them. The primary difference between the two is that the point classes' data members are named `x` and `y`, while the corresponding data members in the size classes are named `width` and `height`. The three aliases for the size classes are `cv::Size`, `cv::Size2i`, and `cv::Size2f`. The first two of these are equivalent and imply integer size, while the last is for 32-bit floating-point sizes. As with the point classes, the size classes can be cast to and from the corresponding old-style OpenCV classes (in this case, `CvSize` and `CvSize2D32f`). Table 3-3 lists the operations supported by the size classes.

*Table 3-3. Operations supported directly by the size classes*

| Operation | Example |
|---|---|
| Default constructors | `cv::Size sz;`<br>`cv::Size2i sz;`<br>`cv::Size2f sz;` |
| Copy constructor | `cv::Size sz2( sz1 );` |

| Operation | Example |
|---|---|
| Value constructors | `cv::Size2f sz( w, h );` |
| Member access | `sz.width; sz.height;` |
| Compute area | `sz.area();` |

Unlike the point classes, the size classes do not support casting to the fixed vector classes. This means that the size classes have more restricted utility. On the other hand, the point classes and the fixed vector classes can be cast to the size classes without any problem.

### The cv::Rect class

The rectangle classes include the members `x` and `y` of the point class (representing the upper-left corner of the rectangle) and the members `width` and `height` of the size class (representing the rectangle's size). The rectangle classes, however, do not inherit from the point or size classes, and so in general they do not inherit operators from them (see Table 3-4).

*Table 3-4. Operations supported directly by cv::Rect*

| Operation | Example |
|---|---|
| Default constructors | `cv::Rect r;` |
| Copy constructor | `cv::Rect r2( r1 );` |
| Value constructors | `cv::Rect( x, y, w, h );` |
| Construct from origin and size | `cv::Rect( p, sz );` |
| Construct from two corners | `cv::Rect( p1, p2 );` |
| Member access | `r.x; r.y; r.width; r.height;` |
| Compute area | `r.area();` |
| Extract upper-left corner | `r.tl();` |
| Extract bottom-right corner | `r.br();` |
| Determine if point *p* is inside rectangle *r* | `r.contains( p );` |

Cast operators and copy constructors exist to allow `cv::Rect` to be computed from or cast to the old-style `cv::CvRect` type as well. `cv::Rect` is actually an alias for a rectangle template instantiated with integer members.

As Table 3-5 shows, `cv::Rect` also supports a variety of overloaded operators that can be used for the computation of various geometrical properties of two rectangles or a rectangle and another object.

*Table 3-5. Overloaded operators that take objects of type cv::Rect*

| Operation | Example |
|---|---|
| Intersection of rectangles *r1* and *r2* | `cv::Rect r3 = r1 & r2;`<br>`r1 &= r2;` |
| Minimum area rectangle containing rectangles *r1* and *r2* | `cv::Rect r3 = r1 \| r2;`<br>`r1 \|= r2;` |
| Translate rectangle *r* by an amount *x* | `cv::Rect rx = r + x;`<br>`r += x;` |
| Enlarge a rectangle *r* by an amount given by size *s* | `cv::Rect rs = r + s;`<br>`r += s;` |
| Compare rectangles *r1* and *r2* for exact equality | `bool eq = (r1 == r2);` |
| Compare rectangles *r1* and *r2* for inequality | `bool ne = (r1 != r2);` |

### The cv::RotatedRect class

The `cv::RotatedRect` class is one of the few classes in the C++ OpenCV interface that is not a template underneath. Instead, it is a container that holds a `cv::Point2f` called `center`, a `cv::Size2f` called `size`, and one additional `float` called `angle`, with the latter representing the rotation of the rectangle around `center`. One very important difference between `cv::RotatedRect` and `cv::Rect` is the convention that a `cv::RotatedRect` is located in "space" relative to its center, while the `cv::Rect` is located relative to its upper-left corner. Table 3-6 lists the operations that are supported directly by `cv::RotatedRect`.

*Table 3-6. Operations supported directly by cv::RotatedRect*

| Operation | Example |
|---|---|
| Default constructors | `cv::RotatedRect rr();` |
| Copy constructor | `cv::RotatedRect rr2( rr1 );` |
| Construct from two corners | `cv::RotatedRect( p1, p2 );` |
| Value constructors; takes a point, a size, and an angle | `cv::RotatedRect rr( p, sz, theta ) ;` |
| Member access | `rr.center; rr.size; rr.angle;` |
| Return a list of the corners | `rr.points( pts[4] );` |

### The fixed matrix classes

The fixed matrix classes are for matrices whose dimensions are known at compile time (hence "fixed"). As a result, all memory for their data is allocated on the stack, which means that they allocate and clean up quickly. Operations on them are fast, and there are specially optimized implementations for small matrices (2 × 2, 3 × 3, etc.). The fixed matrix classes are also central to many of the other basic types in the C++ interface to OpenCV. The fixed vector class derives from the fixed matrix classes, and other classes either derive from the fixed vector class (like `cv::Scalar`) or they rely on casting to the fixed vector class for many important operations. As usual, the fixed matrix classes are really a template. The template is called `cv::Matx<>`, but individual matrices are usually allocated through aliases. The basic form of such an alias is `cv::Matx{1,2,...}{1,2,...}{f,d}`, where the numbers can be any number from one to six, and the trailing letter has the same meaning as with the previous types.[5]

In general, you should use the fixed matrix classes when you are representing something that is really a matrix with which you are going to do matrix algebra. If your object is really a big data array, like an image or a huge list of points, the fixed matrix classes are not the correct solution; you should be using `cv::Mat` (which we will get to in the next chapter). Fixed matrix classes are for small matrices where you know the size at compile time (e.g., a camera matrix). Table 3-7 lists the operations supported by `cv::Matx`.

*Table 3-7. Operations supported by cv::Matx*

| Operation | Example |
|---|---|
| Default constructor | `cv::Matx33f m33f; cv::Matx43d m43d;` |
| Copy constructor | `cv::Matx22d m22d( n22d );` |
| Value constructors | `cv::Matx21f m(x0,x1); cv::Matx44d m(x0,x1,x2,x3,x4,x5,x6,x7,x8,x9,x10,x11,x12,x13,x14,x15);` |
| Matrix of identical elements | `m33f = cv::Matx33f::all( x );` |
| Matrix of zeros | `m23d = cv::Matx23d::zeros();` |
| Matrix of ones | `m16f = cv::Matx16f::ones();` |
| Create a unit matrix | `m33f = cv::Matx33f::eye();` |

5 At the time of writing, the relevant header file called *core.hpp* does not actually contain every possible combination of these integers. For example, there is no 1 × 1 matrix alias, nor is there a 5 × 5. This may or may not change in later releases, but you will pretty much never want the missing ones anyway. If you really do want one of the odd ones, you can just instantiate the template yourself (e.g., `cv::Matx<5,5,float>`).

| Operation | Example |
|---|---|
| Create a matrix that can hold the diagonal of another | `m31f = cv::Matx33f::diag(); // Create a matrix of`<br>`// size 3-by-1 of floats` |
| Create a matrix with uniformly distributed entries | `m33f = cv::Matx33f::randu( min, max );` |
| Create a matrix with normally distributed entries | `m33f = cv::Matx33f::nrandn( mean, variance );` |
| Member access | `m( i, j ), m( i ); // one argument for`<br>`// one-dimensional matrices only` |
| Matrix algebra | `m1 = m0; m0 * m1; m0 + m1; m0 - m1;` |
| Singleton algebra | `m * a; a * m; m / a;` |
| Comparison | `m1 == m2; m1 != m2;` |
| Dot product | `m1.dot( m2 ); // (sum of element-wise`<br>`// multiplications, precision of m)` |
| Dot product | `m1.ddot( m2 ); // (sum of element-wise multiplications,`<br>`// double precision)` |
| Reshape a matrix | `m91f = m33f.reshape<9,1>();` |
| Cast operators | `m44f = (Matx44f) m44d` |
| Extract 2 × 2 submatrix at (*i*, *j*) | `m44f.get_minor<2, 2>( i, j );` |
| Extract row *i* | `m14f = m44f.row( i );` |
| Extract column *j* | `m41f = m44f.col( j );` |
| Extract matrix diagonal | `m41f = m44f.diag();` |
| Compute transpose | `n44f = m44f.t();` |
| Invert matrix | `n44f = m44f.inv( method ); // (default method is`<br>`// cv::DECOMP_LU)` |
| Solve linear system | `m31f = m33f.solve( rhs31f, method )`<br>`m32f = m33f.solve<2>( rhs32f, method ); // (template form[a]);`<br>`// default method is DECOMP_LU)` |
| Per-element multiplication | `m1.mul( m2 );` |

[a] The template form is used when the righthand side of the implied matrix equation has multiple columns. In this case, we are essentially solving for *k* different systems at once. This value of *k* must be supplied as the template argument to `solve<>()`. It will also determine the number of columns in the result matrix.

Note that many of the fixed matrix functions are static relative to the class (i.e., you access them directly as members of the class rather than as members of a particular object). For example, if you would like to construct a 3 × 3 identity matrix, you have a handy class function for it: `cv::Mat33f::eye()`. Note that, in this example, `eye()`

does not need any arguments because it is a member of the class, and the class is already a specialization of the `cv::Matx<>` template to 3 × 3.

### The fixed vector classes

The fixed vector classes are derived from the fixed matrix classes. They are really just convenience functions for `cv::Matx<>`. In the proper sense of C++ inheritance, it is correct to say that the fixed vector template `cv::Vec<>` is a `cv::Matx<>` whose number of columns is one. The readily available aliases for specific instantiations of `cv::Vec<>` are of the form `cv::Vec{2,3,4,6}{b,s,w,i,f,d}`, where the last character has the usual meanings (with the addition of `w`, which indicates an unsigned short). Table 3-8 shows the operations `cv::Vec` supports.

*Table 3-8. Operations supported by cv::Vec*

| Operation | Example |
|---|---|
| Default constructor | `Vec2s v2s; Vec6f v6f; // etc...` |
| Copy constructor | `Vec3f u3f( v3f );` |
| Value constructors | `Vec2f v2f(x0,x1); Vec6d v6d(x0,x1,x2,x3,x4,x5);` |
| Member access | `v4f[ i ]; v3w( j ); // (operator() and operator[]`<br>`// both work)` |
| Vector cross-product | `v3f.cross( u3f );` |

The primary conveniences of the fixed vector classes are the ability to access elements with a single ordinal, and a few specific additional functions that would not make sense for a general matrix (e.g., cross product). We can see this in Table 3-8 by the relatively small number of novel methods added to the large number of methods inherited from the fixed matrix classes.

### The complex number classes

One more class type should be included in the basic types: the complex number classes. The OpenCV complex number classes are not identical to, but are compatible with—and can be cast to and from—the classes associated with the STL complex number class template `complex<>`. The most substantial difference between the OpenCV and STL complex number classes is in member access. In the STL classes, the real and imaginary parts are accessed through the member functions `real()` and `imag()`, while in the OpenCV class, they are directly accessible as (public) member variables `re` and `im`. Table 3-9 lists the operations supported by the complex number classes.

*Table 3-9. Operations supported by the OpenCV complex number classes*

| Operation | Example |
|---|---|
| Default constructor | `cv::Complexf z1; cv::Complexd z2;` |
| Copy constructor | `cv::Complexf z2( z1 );` |
| Value constructors | `cv::Complexd z1(re0); cv::Complexd(re0,im1) ;` |
| Copy constructor | `cv::Complexf u2f( v2f );` |
| Member access | `z1.re; z1.im;` |
| Complex conjugate | `z2 = z1.conj();` |

Like many basic types, the complex classes are aliases for underlying templates. `cv::Complexf` and `cv::Complexd` are aliases for single- and double-precision complex numbers, respectively.

## Helper Objects

In addition to the basic types and the big containers (which we will get to in the next section), there is a family of helper objects that are important for controlling various algorithms (such as termination criteria) or for doing various operations on the containers (such as "ranges" or "slices"). There is also one very important object, the "smart" pointer object `cv::Ptr`. Looking into `cv::Ptr`, we will examine the garbage-collecting system, which is integral to the C++ interface to OpenCV. This system frees us from worrying about the details of object allocation and deallocation in the manner that was so onerous in the earlier C-based OpenCV interface (i.e., before version 2.1).

#### The cv::TermCriteria class

Many algorithms require a stopping condition to know when to quit. Generally, stopping criteria take the form of either some finite number of iterations that are allowed (called `COUNT` or `MAX_ITER`) or some kind of error parameter that basically says, "if you are this close, you can quit" (called `EPS`—short for *epsilon*, everyone's favorite tiny number). In many cases, it is desirable to have both of these at once so that if the algorithm never gets "close enough," it will still quit at some point.

The `cv::TermCriteria` objects encapsulate one or both of the stopping criteria so that they can be passed conveniently to an OpenCV algorithm function. They have three member variables—`type`, `maxCount`, and `epsilon`—which can be set directly (they are public) or, more often, are just set by the constructor with the form `TermCriteria( int type, int maxCount, double epsilon )`. The variable `type` is set to either `cv::TermCriteria::COUNT` or `TermCriteria::EPS`. You can also "or" (i.e., `|`) the two together. The value `cv::TermCriteria::COUNT` is a synonym for `cv::Term`

Criteria::MAX_ITER, so you can use that if you prefer. If the termination criterion includes cv::TermCriteria::COUNT, then you are telling the algorithm to terminate after maxCount iterations. If the termination criterion includes cv::TermCriteria::EPS, then you are telling the algorithm to terminate after some metric associated with the algorithm's convergence falls below epsilon.[6] The type argument has to be set accordingly for maxCount or epsilon to be used.

### The cv::Range class

The cv::Range class is used to specify a continuous sequence of integers. cv::Range objects have two elements, start and end, which—similar to cv::TermCriteria—are often set with the constructor cv::Range( int start, int end ). Ranges are inclusive of their start value, but not inclusive of their end value, so cv::Range rng( 0, 4 ) includes the values 0, 1, 2, and 3, but not 4.

Using size(), you can find the number of elements in a range. In the preceding example, rng.size() would be equal to 4. There is also a member, empty(), that tests if a range has no elements. Finally, cv::Range::all() can be used anywhere a range is required to indicate whatever range the object has available.

### The cv::Ptr template and Garbage Collection 101

One very useful object type in C++ is the "smart" pointer.[7] This pointer allows us to create a reference to something, and then pass it around. You can create more references to that thing, and then *all* of those references will be counted. As references go out of scope, the reference count for the smart pointer is decremented. Once all of the references (instances of the pointer) are gone, the "thing" will automatically be cleaned up (deallocated). You, the programmer, don't have to do this bookkeeping anymore.

Here's how this all works. First, you define an instance of the pointer template for the class object that you want to "wrap." You do this with a call like cv::Ptr<Matx33f> p( new cv::Matx33f ), or cv::Ptr<Matx33f> p = makePtr<cv::Matx33f>(). The constructor for the template object takes a pointer to the object to be pointed to. Once you do this, you have your smart pointer p, which is a sort of pointer-like object that you can pass around and use just like a normal pointer (i.e., it supports operators such as operator*() and operator->()). Once you have p, you can create other

6 The exact termination criteria are clearly algorithm dependent, but the documentation will always be clear as to how a particular algorithm interprets epsilon.

7 If you are familiar with some of the more recent additions to the C++ standard, you will recognize a similarity between the OpenCV cv::Ptr<> template and the shared_ptr<> template. Similarly, there is a smart pointer shared_ptr<> in the Boost library. Ultimately, they all function more or less the same.

objects of the same type without passing them a pointer to a new object. For example, you could create Ptr<Mat33f> q, and when you assign the value of p to q, somewhere behind the scenes, the "smart" action of the smart pointer comes into play. You see, just like a usual pointer, there is still only one actual cv::Mat33f object out there that p and q both point to. The difference is that both p and q *know* that they are each one of two pointers. Should p disappear (such as by falling out of scope), q knows that it is the only remaining reference to the original matrix. If q should then disappear and its destructor is called (implicitly), q will know that is the last one left, and that it should deallocate the original matrix. You can think of this like the last person out of a building being responsible for turning out the lights and locking the door (and in this case, burning the building to the ground as well).

The cv::Ptr<> template class supports several additional functions in its interface related to the reference-counting functionality of the smart pointer. Specifically, the functions addref() and release() increment and decrement the internal reference counter of the pointer. These are relatively dangerous functions to use, but are available in case you need to micromanage the reference counters yourself.

There is also a function called empty(), which you can use to determine if a smart pointer is pointing to an object that has been deallocated. This could happen if you called release() on the object one or more times. In this case, you would still have a smart pointer around, but the object pointed to might already have been destroyed. There is a second application of empty(), which is to determine if the internal object pointer inside the smart pointer object happens to be NULL for some other reason. For example, this might occur if you assigned the smart pointer by calling a function that might just return NULL in the first place (cvLoadImage(), fopen(), etc.).[8]

The final member of Ptr<> that you will want to know about is delete_obj(). This is a function that gets called automatically when the reference count gets to zero. By default, this function is defined but does nothing. It is there so that you can overload it in the case of instantiation of cv::Ptr<>, which points to a class that requires some specific operation in order to clean up the class to which it points. For example, let's say that you are working with an old-style (pre–version 2.1) IplImage.[9] In the old days, you might, for example, have called cvLoadImage() to load that image from disk. In the C interface, that would have looked like this:

8 For the purposes of this example, we will make reference to IplImage and cvLoadImge(), both constructs from the ancient pre–version 2.1 interface that are now deprecated. We won't really cover them in detail in this book, but all you need to know for this example is that IplImage is the old data structure for images, and cvLoadImage() was the old function to get an image from disk and return a pointer to the resulting image structure.

9 This example might seem a bit artificial, but in fact, if you have a large body of pre-v2.1 code you are trying to modernize, you will likely find yourself doing an operation like this quite often.

```
IplImage* img_p = cvLoadImage( ... );
```

The modern version of this (while still using `IplImage` rather than `cv::Mat`, which we are still working our way up to) would look like this:

```
cv::Ptr<IplImage> img_p = cvLoadImage( "an_image" );
```

or (if you prefer) this:

```
cv::Ptr<IplImage> img_p( cvLoadImage("an_image" ) );
```

Now you can use `img_p` in exactly the same way as a pointer (which is to say, for readers experienced with the pre–version 2.1 interface, "exactly as you would have back then"). Conveniently, this particular template instantiation is actually already defined for you somewhere in the vast sea of header files that make up OpenCV. If you were to go search it out, you would find the following template function defined:

```
template<> inline void cv::Ptr<IplImage>::delete_obj() {
    cvReleaseImage(&obj);
}
```

(The variable `obj` is the name of the class member variable inside `Ptr<>` that actually holds the pointer to the allocated object.) As a result of this definition, you will not need to deallocate the `IplImage*` pointer you got from `cvLoadImage()`. Instead, it will be automatically deallocated for you when `img_p` falls out of scope.

This example was a somewhat special (though highly relevant) situation, in that the case of a smart pointer to `IplImage` is sufficiently salient that it was defined for you by the library. In a somewhat more typical case, when the clean-up function does not exist for what you want, you will have to define it yourself. Consider the example of creating a file handle using a smart pointer to `FILE`.[10] In this case, we define our own overloaded version of `delete_obj()` for the `cv::Ptr<FILE>` template:

```
template<> inline void cv::Ptr<FILE>::delete_obj() {
    fclose(obj);
}
```

Then you could go ahead and use that pointer to open a file, do whatever with it, and later, just let the pointer fall out of scope (at which time the file handle would automatically be closed for you):

```
{
  cv::Ptr<FILE> f(fopen("myfile.txt", "r"));
  if(f.empty())
    throw ...;        // Throw an exception, we will get to this later on...
  fprintf(f, ...);
  ...
}
```

10 In this case, by `FILE` we mean `struct FILE`, as defined in the C standard library.

At the final brace, `f` falls out of scope, the internal reference count in `f` goes to zero, `delete_obj()` is called by `f`'s destructor, and (thus) `fclose()` is called on the file handle pointer (stored in `obj`).

A tip for gurus: a serious programmer might worry that the incrementing and decrementing of the reference count might not be sufficiently atomic for the `Ptr<>` template to be safe in multithreaded applications. This, however, is not the case, and `Ptr<>` is thread safe. Similarly, the other reference-counting objects in OpenCV are all thread safe in this same sense.

### The cv::Exception class and exception handling

OpenCV uses exceptions to handle errors. OpenCV defines its own exception type, `cv::Exception`, which is derived from the STL exception class `std::exception`. Really, this exception type has nothing special about it, other than being in the `cv::` namespace and thus distinguishable from other objects that are also derived from `std::exception`.

The type `cv::Exception` has members `code`, `err`, `func`, `file`, and `line`, which are (respectively) a numerical error code, a string indicating the nature of the error that generated the exception, the name of the function in which the error occurred, the file in which the error occurred, and an integer indicating the line on which the error occurred in that file. `err`, `func`, and `file` are all STL strings.

There are several built-in macros for generating exceptions yourself. `CV_Error( errorcode, description )` will generate and throw an exception with a fixed text description. `CV_Error_( errorcode, printf_fmt_str, [printf-args] )` works the same, but allows you to replace the fixed description with a `printf`-like format string and arguments. Finally, `CV_Assert( condition )` and `CV_DbgAssert( condition )` will both test your condition and throw an exception if the condition is not met. The latter version, however, will only operate in debug builds. These macros are the strongly preferred method of throwing exceptions, as they will automatically take care of the fields `func`, `file`, and `line` for you.

### The cv::DataType<> template

When OpenCV library functions need to communicate the concept of a particular data type, they do so by creating an object of type `cv::DataType<>`. `cv::DataType<>` itself is a template, and so the actual objects passed around are specializations of this template. This is an example of what in C++ are generally called *traits*. This allows the `cv::DataType<>` object to contain both runtime information about the type, as well as typedef statements in its own definition that allow it to refer to the same type at compile time.

This might sound a bit confusing, and it is, but that is an inevitable consequence of trying to mix runtime information and compile-time information in C++.[11] An example will help clarify. Here's the template class definition for DataType:

```
template<typename _Tp> class DataType
{
  typedef _Tp         value_type;
  typedef value_type work_type;
  typedef value_type channel_type;
  typedef value_type vec_type;

  enum {
    generic_type = 1,
    depth        = -1,
    channels     = 1,
    fmt          = 0,
    type         = CV_MAKETYPE(depth, channels)
  };
};
```

Let's try to understand what this means, and then follow it with an example. First, we can see that cv::DataType<> is a template, and expects to be specialized to a class called _Tp. It then has four typedef statements that allow the type of the cv::DataType<>, as well as some other related types, to be extracted from the cv::DataType<> instantiated object at compile time. In the template definition, these are all the same, but we will see in our example of a template specialization that they do not have to be (and often should not be). The next section is an enum that contains several components.[12] These are the generic_type, the depth, the number of channels, the format fmt, and the type. To see what all of these components mean, we'll look at two example specializations of cv::DataType<>, from *core.hpp*. The first is the cv::DataType<> definition for float:

```
template<> class DataType<float>
{
public:
  typedef float      value_type;
  typedef value_type work_type;
  typedef value_type channel_type;
```

11 You don't have this sort of problem in languages that support variable introspection and have an intrinsic runtime concept of data types.

12 If this construct is awkward to you, remember that you can always assign integer values to the "options" in an enum declaration. In effect, this is a way of stashing a bunch of integer constants that will be fixed at compile time.

```
    typedef value_type vec_type;

    enum {
      generic_type = 0,
      depth        = DataDepth<channel_type>::value,
      channels     = 1,
      fmt          = DataDepth<channel_type>::fmt,
      type         = CV_MAKETYPE(depth, channels)
    };
  };
```

The first thing to notice is that this is a definition for a C++ built-in type. It is useful to have such definitions for the built-in types, but we can also make them for more complicated objects. In this case, the `value_type` is of course `float`, and the `work_type`, `channel_type`, and `vec_type` are all the same. We will see more clearly what these are for in the next example. For the constants in the `enum`, this example will do just fine. The first variable, `generic_type`, is set to `0`, as it is zero for all types defined in *core.hpp*. The `depth` variable is the data type identifier used by OpenCV. For example, `cv::DataDepth<float>::value` resolves to the constant `CV_32F`. The entry `channels` is `1` because `float` is just a single number; we will see an alternative to this in the next example. The variable `fmt` gives a single-character representation of the format. In this case, `cv::DataDepth<float>::fmt` resolves to `f`. The last entry is `type`, which is a representation similar to depth, but includes the number of channels (in this case, one). `CV_MAKETYPE(CV_32F,1)` resolves to `CV_32FC1`.

The important thing about `DataType<>`, however, is to communicate the nature of more complicated constructs. This is essential, for example, for allowing algorithms to be implemented in a manner that is agnostic to the incoming data type (i.e., algorithms that use introspection to determine how to proceed with incoming data).

Consider the example of an instantiation of `cv::DataType<>` for a `cv::Rect<>` (itself containing an as-yet-unspecialized type `_Tp`):

```
template<typename _Tp> class DataType<Rect_<_Tp> >
{
public:
  typedef Rect_<_Tp>                                 value_type;
  typedef Rect_<typename DataType<_Tp>::work_type> work_type;
  typedef _Tp                                        channel_type;
  typedef Vec<channel_type, channels>                vec_type;

  enum {
    generic_type = 0,
    depth        = DataDepth<channel_type>::value,
    channels     = 4,
    fmt          = ((channels-1)<<8) + DataDepth<channel_type>::fmt,
    type         = CV_MAKETYPE(depth, channels)
  };
};
```

This is a much more complicated example. First, notice that cv::Rect itself does not appear. You will recall that earlier we mentioned that cv::Rect was actually an alias for a template, and that template is called cv::Rect_<>. So this template could be specialized as cv::DataType<Rect> or, for example, cv::DataType< Rect_<float> >. For the case cv::DataType<Rect>, recall that all of the elements are integers, so if we consider that case, all of the instantiations of the template parameter _Tp resolve to int.

We can see that the value_type is just the compile-time name of the thing that the cv::DataType<> is describing (namely Rect). The work_type, however, is defined to be the work_type of cv::DataType<int> (which, not surprisingly, is int). What we see is that the work_type is telling us what kind of variables the cv::DataType<> is made of (i.e., what we "do work" on). The channel type is also int. This means that if we want to represent this variable as a multichannel object, it should be represented as some number of int objects. Finally, just as channel_type tells us how to represent this cv::DataType<> as a multichannel object, vec_type tells us how to represent it as an object of type cv::Vec<>. The alias cv::DataType<Rect>::vec_type will resolve to cv::Vec<int,4>.

Moving on to the runtime constants: generic_type is again 0, depth is CV_32S, channels is 4 (because there are actually four values, the same reason the vec_type instantiated to a cv::Vec<> of size 4), fmt resolves to 0x3069 (since i is 0x69), and type resolves to CV_32SC4.

#### The cv::InputArray and cv::OutputArray classes

Many OpenCV functions take arrays as arguments and return arrays as return values, but in OpenCV, there are many kinds of arrays. We have already seen that OpenCV supports some small array types (cv::Scalar, cv::Vec, cv::Matx) and STL's std::vector<> in addition to the large array types discussed in the next chapter (cv::Mat and cv::SparseMat). In order to keep the interface from becoming onerously complicated (and repetitive), OpenCV defines the types cv::InputArray and cv::OutputArray. In effect, these types mean "any of the above" with respect to the many array forms supported by the library. There is even a cv::InputOutputArray, specifying an array for in-place computation.

The primary difference between cv::InputArray and cv::OutputArray is that the former is assumed to be const (i.e., read only). You will typically see these two types used in the definitions of library routines. You will not tend to use them yourself, but when you are being introduced to library functions, their presence means that you can use any array type, including a single cv::Scalar, and the result should be what you expect.

Related to `cv::InputArray` is the special function `cv::noArray()` that returns a `cv::InputArray`. The returned object can be passed to any input requiring `cv::InputArray` to indicate that this input is not being used. Certain functions also have optional output arrays, where you may pass `cv::noArray()` when you do not need the corresponding output.

## Utility Functions

In addition to providing the specialized primitive data types that we have seen so far in this chapter, the OpenCV library also provides some specialized functions that can be used to more efficiently handle mathematical and other operations which arise commonly in computer vision applications. In the context of the library, these are known as the *utility functions*. The utility functions include tools for mathematical operations, tests, error generations, memory and thread handling, optimization, and more. Table 3-10 lists these functions and summarizes their functionalities; detailed descriptions then follow.

*Table 3-10. Utility and system functions*

| Function | Description |
|---|---|
| `cv::alignPtr()` | Align pointer to given number of bytes |
| `cv::alignSize()` | Align buffer size to given number of bytes |
| `cv::allocate()` | Allocate a C-style array of objects |
| `cvCeil()` [a] | Round float number *x* to nearest integer not smaller than *x* |
| `cv::cubeRoot()` | Compute the cube root of a number |
| `cv::CV_Assert()` | Throw an exception if a given condition is not true |
| `CV_Error()` | Macro to build a `cv::Exception` (from a fixed string) and throw it |
| `CV_Error_()` | Macro to build a `cv::Exception` (from a formatted string) and throw it |
| `cv::deallocate()` | Deallocate a C-style array of objects |
| `cv::error()` | Indicate an error and throw an exception |
| `cv::fastAtan2()` | Calculate two-dimensional angle of a vector in degrees |
| `cv::fastFree()` | Deallocate a memory buffer |
| `cv::fastMalloc()` | Allocate an aligned memory buffer |
| `cvFloor()` | Round float number *x* to nearest integer not larger than *x* |
| `cv::format()` | Create an STL string using `sprintf`-like formatting |
| `cv::getCPUTickCount()` | Get tick count from internal CPU timer |
| `cv::getNumThreads()` | Count number of threads currently used by OpenCV |
| `cv::getOptimalDFTSize()` | Compute the best size for an array that you plan to pass to `cv::DFT()` |
| `cv::getThreadNum()` | Get index of the current thread |
| `cv::getTickCount()` | Get tick count from system |
| `cv::getTickFrequency()` | Get number or ticks per second (see `cv::getTickCount()`) |
| `cvIsInf()` | Check if a floating-point number *x* is infinity |

| Function | Description |
| --- | --- |
| cvIsNaN() | Check if a floating-point number *x* is "Not a Number" |
| cvRound() | Round float number *x* to the nearest integer |
| cv::setNumThreads() | Set number of threads used by OpenCV |
| cv::setUseOptimized() | Enables or disables the use of optimized code (SSE2, etc.) |
| cv::useOptimized() | Indicates status of optimized code enabling (see cv::setUseOptimized()) |

[a] This function has something of a legacy interface. It is a C definition, not C++ (see core *.../types_c.h*) where it is defined as an inline function. There are several others with a similar interface.

### cv::alignPtr()

```
template<T> T* cv::alignPtr(                // Return aligned pointer of type T*
  T*  ptr,                                  // pointer, unaligned
  int n    = sizeof(T)                      // align to block size, a power of 2
);
```

Given a pointer of any type, this function computes an aligned pointer of the same type according to the following computation:

```
(T*)(((size_t)ptr + n+1) & -n)
```

On some architectures, it is not even possible to read a multibyte object from an address that is not evenly divisible by the size of the object (i.e., by 4 for a 32-bit integer). On architectures such as x86, the CPU handles this for you automatically by using multiple reads and assembling your value from those reads at the cost of a substantial penalty in performance.

### cv::alignSize()

```
size_t cv::alignSize(                       // minimum size >='sz' divisible by 'n'
  size_t sz,                                // size of buffer
  int n    = sizeof(T)                      // align to block size, a power of 2
);
```

Given a number `n` (typically a return value from `sizeof()`), and a size for a buffer `sz`, `cv::alignSize()` computes the size that this buffer should be in order to contain an integer number of objects of size `n`—that is, the minimum number that is greater or equal to `sz` yet divisible by `n`. The following formula is used:

```
(sz + n-1) & -n
```

### cv::allocate()

```
template<T> T* cv::allocate(                // Return pointer to allocated buffer
  size_t sz                                 // buffer size, multiples of sizeof(T)
);
```

The function `cv::allocate()` functions similarly to the array form of `new`, in that it allocates a C-style array of `n` objects of type `T`, calls the default constructor for each object, and returns a pointer to the first object in the array.

### cv::deallocate()

```
template<T> void cv::deallocate(
  T*     ptr,                          // Pointer to buffer to free
  size_t sz                            // size of buffer, multiples of sizeof(T)
);
```

The function `cv::deallocate()` functions similarly to the array form of `delete`, in that it deallocates a C-style array of `n` objects of type `T`, and calls the destructor for each object. `cv::deallocate()` is used to deallocate objects allocated with `cv::allocate()`. The number of elements `n` passed to `cv::deallocate()` must be the same as the number of objects originally allocated with `cv::allocate()`.

### cv::fastAtan2()

```
float cv::fastAtan2(                   // Return value is 32-bit float
  float y,                             // y input value (32-bit float)
  float x                              // x input value (32-bit float)
);
```

This function computes the arctangent of an `x,y` pair and returns the angle from the origin to the indicated point. The result is reported in degrees ranging from `0.0` to `360.0`, inclusive of `0.0` but not inclusive of `360.0`.

### cvCeil()

```
int cvCeil(                            // Return the smallest int >= x
  float x                              // input value (32-bit float)
);
```

Given a floating-point number `x`, `cvCeil()` computes the smallest integer not smaller than `x`. If the input value is outside of the range representable by a 32-bit integer, the result is undefined.

### cv::cubeRoot()

```
float cv::cubeRoot(                    // Return value is 32-bit float
  float x                              // input value (32-bit float)
);
```

This function computes the cubed root of the argument `x`. Negative values of `x` are handled correctly (i.e., the return value is negative).

### cv::CV_Assert() and CV_DbgAssert()

```
// example
CV_Assert( x!=0 )
```

CV_Assert() is a macro that will test the expression passed to it and, if that expression evaluates to False (or 0), it will throw an exception. The CV_Assert() macro is always tested. Alternatively, you can use CV_DbgAssert(), which will be tested only in debug compilations.

### cv::CV_Error() and CV_Error_()

```
// example
CV_Error( ecode, estring )
CV_Error_( ecode, fmt, ... )
```

The macro CV_Error() allows you to pass in an error code ecode and a fixed C-style character string estring, which it then packages up into a cv::Exception that it then passes to cv::error() to be handled. The variant macro CV_Error_() is used if you need to construct the message string on the fly. CV_Error_() accepts the same ecode as CV_Error(), but then expects a sprintf()-style format string followed by a variable number of arguments, as would be expected by sprintf().

### cv::error()

```
void cv::error(
  const cv::Exception& ex                     // Exception to be thrown
);
```

This function is mostly called from CV_Error() and CV_Error_(). If your code is compiled in a nondebug build, it will throw the exception ex. If your code is compiled in a debug build, it will deliberately provoke a memory access violation so that the execution stack and all of the parameters will be available for whatever debugger you are running.

You will probably not call cv::error() directly, but rather rely on the macros CV_Error() and CV_Error_() to throw the error for you. These macros take the information you want displayed in the exception, package it up for you, and pass the resulting exception to cv::error().

### cv::fastFree()

```
void cv::fastFree(
  void* ptr                                   // Pointer to buffer to be freed
);
```

This routine deallocates buffers that were allocated with cv::fastMalloc() (covered next).

### cv::fastMalloc()

```
void* cv::fastMalloc(                       // Pointer to allocated buffer
  size_t size                               // Size of buffer to allocate
);
```

cv::FastMalloc() works just like the malloc() you are familiar with, except that it is often faster, and it does buffer size alignment for you. This means that if the buffer size passed is more than 16 bytes, the returned buffer will be aligned to a 16-byte boundary.

### cvFloor()

```
int cvFloor(                                // Return the largest int <= x
  float x                                   // input value (32-bit float)
};
```

Given a floating-point number x, cv::Floor() computes the largest integer not larger than x. If the input value is outside of the range representable by a 32-bit integer, the result is undefined.

### cv::format()

```
string cv::format(                          // Return STL-string
  const char* fmt,                          // formatting string, as sprintf()
  ...                                       // vargs, as sprintf()
);
```

This function is essentially the same as sprintf() from the standard library, but rather than requiring a character buffer from the caller, it constructs an STL string object and returns that. It is particularly handy for formatting error messages for the Exception() constructor (which expects STL strings in its arguments).

### cv::getCPUTickCount()

```
int64 cv::getCPUTickCount( void );          // long int CPU for tick count
```

This function reports the number of CPU ticks on those architectures that have such a construct (including, but not limited to, x86 architectures). It is important to know, however, that the return value of this function can be very difficult to interpret on many architectures. In particular, because on a multicore system a thread can be put to sleep on one core and wake up on another, the difference between the results to two subsequent calls to cv::getCPUTickCount() can be misleading or completely meaningless. Therefore, unless you are certain you know what you are doing, it is best to use cv::getTickCount() for timing measurements.[13] This function is best for tasks like initializing random number generators.

### cv::getNumThreads()

```
int cv::getNumThreads( void );              // total threads allocated to OpenCV
```

Return the current number of threads being used by OpenCV.

### cv::getOptimalDFTSize()

```
int cv::getOptimalDFTSize( int n );         // best size array to use for dft, >= n
```

When you are making calls to cv::dft(), the algorithm used by OpenCV to compute the transform is extremely sensitive to the size of the array passed to cv::dft(). The preferred sizes do obey a rule for their generation, but that rule is sufficiently complicated that it is (at best) an annoyance to compute the correct size to which to pad your array every time. The function cv::getOptimalDFTSize() takes as an argument the size of the array you would have passed to cv::dft(), and returns the size of the array you should pass to cv::dft(). OpenCV uses this information to create a larger array into which you can copy your data and pad out the rest with zeros.

### cv::getThreadNum()

```
int cv::getThreadNum( void );               // int, id of this particular thread
```

If your OpenCV library was compiled with OpenMP support, it will return the index (starting from zero) of the currently executing thread.

### cv::getTickCount()

```
int64 cv::getTickCount( void );             // long int CPU for tick count
```

This function returns a tick count relative to some architecture-dependent time. The rate of ticks is also architecture and operating system dependent, however; the time per tick can be computed by cv::getTickFrequency() (described next). This function is preferable to cv::getCPUTickCount() for most timing applications, as it is not affected by low-level issues such as which core your thread is running on and automatic throttling of CPU frequency (which most modern processors do for power-management reasons).

13 Of course, if you *really do know* what you are doing, then there is no more accurate way to get detailed timing information than from the CPU timers themselves.

### cv::getTickFrequency()

```
double cv::getTickFrequency( void );        // Tick frequency in seconds as 64-bit
```

When `cv::getTickCount()` is used for timing analysis, the exact meaning of a tick is, in general, architecture dependent. The function `cv::getTickFrequency()` computes the conversion between clock time (i.e., seconds) and abstract "ticks."

> To compute the time required for some specific thing to happen (such as a function to execute), you need only call `cv::getTickCount()` before and after the function call, subtract the results, and divide by the value of `cv::getTickFrequency()`.

### cvIsInf()

```
int cvIsInf( double x );                  // return 1 if x is IEEE754 "infinity"
```

The return value of `cvIsInf()` is `1` if `x` is plus or minus infinity and `0` otherwise. The infinity test is the test implied by the IEEE754 standard.

### cvIsNaN()

```
int cvIsNan( double x );                  // return 1 if x is IEEE754 "Not a number"
```

The return value of `cvIsNaN()` is `1` if `x` is "not a number" and `0` otherwise. The NaN test is the test implied by the IEEE754 standard.

### cvRound()

```
int cvRound( double x );                  // Return integer nearest to 'x'
```

Given a floating-point number `x`, `cvRound()` computes the integer closest to `x`. If the input value is outside of the range representable by a 32-bit integer, the result is undefined. In OpenCV 3.0 there is overloaded `cvRound( float x )` (as well as `cvFloor` and `cvCeil`), which is faster on ARM.

### cv::setNumThreads()

```
void cv::setNumThreads( int nthreads );    // Set number of threads OpenCV can use
```

When OpenCV is compiled with OpenMP support, this function sets the number of threads that OpenCV will use in parallel OpenMP regions. The default value for the number of threads is the number of logical cores on the CPU (i.e., if we have four cores each with two hyperthreads, there will be eight threads by default). If `nthreads` is set to `0`, the number of threads will be returned to this default value.

### cv::setUseOptimized()

```
void cv::setUseOptimized( bool on_off ); // If false, turn off optimized routines
```

Though early versions of OpenCV relied on outside libraries (such as IPP, the Intel Performance Primitives library) for access to high-performance optimizations such as SSE2 instructions, later versions have increasingly moved to containing that code in the OpenCV itself. By default, the use of these optimized routines is enabled, unless you specifically disabled it when you built your installation of the library. However, you can turn the use of these optimizations on and off at any time with `cv::setUseOptimized()`.

The test of the global flag for optimizations usage is done at a relatively high level inside the OpenCV library functions. The implication is that you should not call `cv::setUseOptimized()` while any other routines might be running (on any threads). You should make sure to call this routine only when you can be certain you know what is and what is not running, preferably from the very top level of your application.

### cv::useOptimized()

```
bool cv::useOptimized( void );    // return true if optimizations are enabled
```

At any time, you can check the state of the global flag, which enables the use of high-performance optimizations (see `cv::setUseOptimized()`) by calling `cv::useOptimized()`. `True` will be returned only if these optimizations are currently enabled; otherwise, this function will return `False`.

## The Template Structures

Thus far in this chapter, we have regularly alluded to the existence of template forms for almost all of the basic types. In fact, most programmers can get quite far into OpenCV without ever digging down into the templates.[14]

OpenCV versions 2.1 and later are built on a template metaprogramming style similar to STL, Boost, and similar libraries. This sort of library design can be extremely powerful, both in terms of the quality and speed of the final code, as well as the flexibility it allows the developer. In particular, template structures of the kind used in OpenCV allow for algorithms to be implemented in an abstracted way that does not specifically rely on the primitive types that are native to C++ or even native to OpenCV.

14 In fact, if your C++ programming skills are not entirely up to par, you can probably just skim or skip over this little section entirely.

In this chapter, we started with the `cv::Point` class. Though the class was introduced as a primitive, in fact when you instantiate an object of type `cv::Point`, you are actually instantiating an even more fundamental template object of type `cv::Point_<int>`.[15] This template could have been instantiated with a different type than `int`, obviously. In fact, it could have been instantiated with any type that supports the same basic set of operators as `int` (i.e., addition, subtraction, multiplication, etc.). For example, OpenCV provides a type `cv::Complex` that you could have used. You also could have used the STL complex type `std::complex`, which has nothing to do with OpenCV at all. The same is true for some other types of your own construction. This same concept generalizes to other type templates such as `cv::Scalar_<>` and `cv::Rect_<>`, as well as `cv::Matx_<>` and `cv::Vec_<>`.

When instantiating these templates on your own, you must provide the unitary type that is to be used to compose the template, as well as (where relevant) the dimensions of the template. The arguments to the common templates are shown in Table 3-11.

*Table 3-11. Common fixed length templates*

| Function | Description |
|---|---|
| `cv::Point_<Type T>` | A point consisting of a pair of objects of type `T`. |
| `cv::Rect_<Type T>` | A location, width, and height, all of type `T`. |
| `cv::Vec<Type T, int H>` | A set of `H` objects of type `T`. |
| `cv::Matx<Type T, int H, int W>` | A set of `H*W` objects of type `T`. |
| `cv::Scalar_<Type T>` | A set of four objects of type `T` (identical to `cv::Vec<T, 4>`). |

In the next chapter, we will see that the large array types, `cv::Mat` and `cv::SparseMat`, also have corresponding template types `cv::Mat<>` and `cv::SparseMat_<>`, which are similar but differ in a few important ways.

# Summary

In this chapter, we covered in detail the basic data types that are used by the OpenCV library to handle compact collections. These collections include points, but also small vectors and matrices that are often used to represent things like color (channel) vectors or coordinate vectors, as well as small matrices that operate in these spaces. We covered both the template representations used, mostly internally, by the library for

15 Note the trailing underscore—this is a common, but not universal, convention in the library used to indicate a template. In the 2.x version of the library, it was essentially universal. Since 3.x, the underscore was dropped where not specifically necessary. Thus `cv::Point_<>` still has the underscore to distinguish from the nontemplate class `cv::Point`, while `cv::Vec<>` does not have an underscore. (It was `cv::Vec_<>` in the 2.x version of the library.)

such objects, as well as the classes that are specializations of those templates. These specialization classes make up the majority of what you will use on a daily basis.

In addition to these data classes, we also covered the helper objects that allow us to express concepts such as termination criteria and value ranges. Finally, we concluded the chapter by surveying the utility functions that the library provides. These functions provide optimized implementations of important tasks that computer vision applications often encounter. Important examples of operations include special arithmetic and memory management tools.

# Exercises

1. Find and open *.../opencv/modules/core/include/opencv2/core/types_c.h*. Read through and find the many conversion helper functions.
   a. Choose a negative floating-point number.
   b. Take its absolute value, round it, and then take its ceiling and floor.
   c. Generate some random numbers.
   d. Create a floating-point `cv::Point2f` and convert it to an integer `cv::Point`. Convert a `cv::Point` to a `cv::Point2f`.
2. Compact matrix and vector types:
   a. Using the `cv::Mat33f` and `cv::Vec3f` objects (respectively), create a 3 × 3 matrix and 3-row vector.
   b. Can you multiply them together directly? If not, why not?
3. Compact matrix and vector template types:
   a. Using the `cv::Mat<>` and `cv::Vec<>` templates (respectively), create a 3 × 3 matrix and 3-row vector.
   b. Can you multiply them together directly? If not, why not?
   c. Try type-casting the vector object to a 3 × 1 matrix, using the `cv::Mat<>` template. What happens now?

CHAPTER 4

# Images and Large Array Types

## Dynamic and Variable Storage

The next stop on our journey brings us to the *large array types*. Chief among these is `cv::Mat`, which could be considered the epicenter of the entire C++ implementation of the OpenCV library. The overwhelming majority of functions in the OpenCV library are members of the `cv::Mat` class, take a `cv::Mat` as an argument, or return `cv::Mat` as a return value; quite a few are or do all three.

The `cv::Mat` class is used to represent *dense* arrays of any number of dimensions. In this context, dense means that for every entry in the array, there is a data value stored in memory corresponding to that entry, even if that entry is zero. Most images, for example, are stored as dense arrays. The alternative would be a *sparse* array. In the case of a sparse array, only nonzero entries are typically stored. This can result in a great savings of storage space if many of the entries are in fact zero, but can be very wasteful if the array is relatively dense. A common case for using a sparse array rather than a dense array would be a histogram. For many histograms, most of the entries are zero, and storing all those zeros is not necessary. For the case of sparse arrays, OpenCV has the alternative data structure, `cv::SparseMat`.

If you are familiar with the C interface (pre-version 2.1 implementation) of the OpenCV library, you will remember the data types `IplImage` and `CvMat`. You might also recall `CvArr`. In the C++ implementation, these are all gone, replaced with `cv::Mat`. This means no more dubious casting of `void*` pointers in function arguments, and in general is a tremendous enhancement in the internal cleanliness of the library.

## The cv::Mat Class: N-Dimensional Dense Arrays

The `cv::Mat` class can be used for arrays of any number of dimensions. The data is stored in the array in what can be thought of as an *n*-dimensional analog of "raster scan order." This means that in a one-dimensional array, the elements are sequential. In a two-dimensional array, the data is organized into rows, and each row appears one after the other. For three-dimensional arrays, each plane is filled out row by row, and then the planes are packed one after the other.

Each matrix contains a `flags` element signaling the contents of the array, a `dims` element indicating the number of dimensions, `rows` and `cols` elements indicating the number of rows and columns (these are not valid for `dims>2`), a `data` pointer to where the array data is stored, and a `refcount` reference counter analogous to the reference counter used by `cv::Ptr<>`. This latter member allows `cv::Mat` to behave very much like a smart pointer for the data contained in `data`. The memory layout in `data` is described by the array `step[]`. The data array is laid out such that the address of an element whose indices are given by $(i_0, i_1, \ldots, i_{N_d-1})$ is:

$$\&\left(mtx_{i_0, i_1, \ldots, i_{N_d - 1N_d}}\right) = mtx.data + mtx.step[0]^* i_0 + mtx.step[1]^* i_1 + \ldots$$

$$+ \, mtx.step\left[N_d - 1\right]^* i_{N_d - 1}$$

In the simple case of a two-dimensional array, this reduces to:

$$\&(mtx_{i,j}) = mtx.data + mtx.step[0]^* i + mtx.step[1]^* j$$

The data contained in `cv::Mat` is not required to be simple primitives. Each element of the data in a `cv::Mat` can itself be either a single number or multiple numbers. In the case of multiple numbers, this is what the library refers to as a *multichannel* array. In fact, an *n*-dimensional array and an (*n*–1)-dimensional multichannel array are actually very similar objects, but because there are many occasions in which it is useful to think of an array as a *vector-valued* array, the library contains special provisions for such structures.[1]

One reason for this distinction is memory access. By definition, an *element* of an array is the part that may be vector-valued. For example, an array might be said to be

1 Pre-2.1 OpenCV array types had an explicit element `IplImage::nChannels`, which indicated the number of channels. Because of the more general way in which such concepts are captured in the `cv::Mat` object, this information is no longer directly stored in a class variable. Rather, it is returned by a member function, `cv::channels()`.

a two-dimensional three-channel array of 32-bit floats; in this case, the element of the array is the three 32-bit floats with a size of 12 bytes. When laid out in memory, rows of an array may not be absolutely sequential; there may be small gaps that buffer each row before the next.[2] The difference between an *n*-dimensional single-channel array and an (*n*–1)-dimensional multichannel array is that this padding will always occur at the end of full rows (i.e., the channels in an element will always be sequential).

## Creating an Array

You can create an array simply by instantiating a variable of type `cv::Mat`. An array created in this manner has no size and no data type. You can, however, later ask it to allocate data by using a member function such as `create()`. One variation of `create()` takes as arguments a number of rows, a number of columns, and a type, and configures the array to represent a two-dimensional object. The type of an array determines what kind of elements it has. Valid types in this context specify both the fundamental type of element as well as the number of channels. All such types are defined in the library header, and have the form `CV_{8U,16S,16U,32S,32F,64F}C{1,2,3}`.[3] For example, `CV_32FC3` would imply a 32-bit floating-point three-channel array.

If you prefer, you can also specify these things when you first allocate the matrix. There are many constructors for `cv::Mat`, one of which takes the same arguments as `create()` (and an optional fourth argument with which to initialize all of the elements in your new array). For example:

```
cv::Mat m;

// Create data area for 3 rows and 10 columns of 3-channel 32-bit floats
m.create( 3, 10, CV_32FC3 );

// Set the values in the 1st channel to 1.0, the 2nd to 0.0, and the 3rd to 1.0
m.setTo( cv::Scalar( 1.0f, 0.0f, 1.0f ) );
```

is equivalent to:

```
cv::Mat m( 3, 10, CV_32FC3, cv::Scalar( 1.0f, 0.0f, 1.0f ) );
```

2 The purpose of this padding is to improve memory access speed.

3 OpenCV allows for arrays with more than three channels, but to construct one of these, you will have to call one of the functions `CV_{8U,16S,16U,32S,32F,64F}C()`. These functions take a single argument, which is the number of channels. So `CV_8UC(3)` is equivalent to `CV_8UC3`, but since there is no macro for `CV_8UC7`, to get this you would have to call `CV_8UC(7)`.

### The Most Important Paragraph in the Book

It is critical to understand that the data in an array is not *attached* rigidly to the array object. The `cv::Mat` object is really a header for a data area, which—in principle—is an entirely separate thing. For example, it is possible to assign one matrix `n` to another matrix `m` (i.e., `m=n`). In this case, the data pointer inside of `m` will be changed to point to the same data as `n`. The data pointed to previously by the data element of `m` (if any) will be deallocated.[4] At the same time, the reference counter for the data area that they both now share will be incremented. Last but not least, the members of `m` that characterize its data (such as `rows`, `cols`, and `flags`) will be updated to accurately describe the data now pointed to by `data` in `m`. This all results in a very convenient behavior, in which arrays can be assigned to one another, and the work necessary to do this takes place automatically behind the scenes to give the correct result.

Table 4-1 is a complete list of the constructors available for `cv::Mat`. The list appears rather unwieldy, but in fact you will use only a small fraction of these most of the time. Having said that, when you need one of the more obscure ones, you will probably be glad it is there.

*Table 4-1. cv::Mat constructors that do not copy data*

| Constructor | Description |
|---|---|
| `cv::Mat;` | Default constructor |
| `cv::Mat( int rows, int cols, int type );` | Two-dimensional arrays by type |
| `cv::Mat( int rows, int cols, int type, const Scalar& s );` | Two-dimensional arrays by type with initialization value |
| `cv::Mat( int rows, int cols, int type, void* data, size_t step=AUTO_STEP );` | Two-dimensional arrays by type with preexisting data |
| `cv::Mat( cv::Size sz, int type );` | Two-dimensional arrays by type (size in `sz`) |
| `cv::Mat( cv::Size sz, int type, const Scalar& s );` | Two-dimensional arrays by type with initialization value (size in `sz`) |

4 Technically, it will only be deallocated if `m` was the last `cv::Mat` that pointed to that particular data.

| Constructor | Description |
|---|---|
| `cv::Mat( cv::Size sz, int type, void* data, size_t step=AUTO_STEP );` | Two-dimensional arrays by type with preexisting data (size in `sz`) |
| `cv::Mat( int ndims, const int* sizes, int type );` | Multidimensional arrays by type |
| `cv::Mat( int ndims, const int* sizes, int type, const Scalar& s );` | Multidimensional arrays by type with initialization value |
| `cv::Mat( int ndims, const int* sizes, int type, void* data, size_t step=AUTO_STEP );` | Multidimensional arrays by type with preexisting data |

Table 4-1 lists the basic constructors for the `cv::Mat` object. Other than the default constructor, these fall into three basic categories: those that take a number of rows and a number of columns to create a two-dimensional array, those that use a `cv::Size` object to create a two-dimensional array, and those that construct *n*-dimensional arrays and require you to specify the number of dimensions and pass in an array of integers specifying the size of each of the dimensions.

In addition, some of these allow you to initialize the data, either by providing a `cv::Scalar` (in which case, the entire array will be initialized to that value), or by providing a pointer to an appropriate data block that can be used by the array. In this latter case, you are essentially just creating a header to the existing data (i.e., no data is copied; the data member is set to point to the data indicated by the `data` argument).

The copy constructors (Table 4-2) show how to create an array from another array. In addition to the basic copy constructor, there are three methods for constructing an array from a subregion of an existing array and one constructor that initializes the new matrix using the result of some matrix expression.

*Table 4-2. cv::Mat constructors that copy data from other cv::Mats*

| Constructor | Description |
|---|---|
| `cv::Mat( const Mat& mat );` | Copy constructor |
| `cv::Mat( const Mat& mat, const cv::Range& rows, const cv::Range& cols );` | Copy constructor that copies only a subset of rows and columns |
| `cv::Mat( const Mat& mat, const cv::Rect& roi );` | Copy constructor that copies only a subset of rows and columns specified by a region of interest |
| `cv::Mat( const Mat& mat, const cv::Range* ranges );` | Generalized region of interest copy constructor that uses an array of ranges to select from an *n*-dimensional array |
| `cv::Mat( const cv::MatExpr& expr );` | Copy constructor that initializes `m` with the result of an algebraic expression of other matrices |

The subregion (also known as "region of interest") constructors also come in three flavors: one that takes a range of rows and a range of columns (this works only on a two-dimensional matrix), one that uses a `cv::Rect` to specify a rectangular subregion (which also works only on a two-dimensional matrix), and a final one that takes an array of ranges. In this latter case, the number of valid ranges pointed to by the pointer argument `ranges` must be equal to the number of dimensions of the array `mat`. It is this third option that you must use if `mat` is a multidimensional array with `ndim` greater than 2.

If you are modernizing or maintaining pre–version 2.1 code that still contains the C-style data structures, you may want to create a new C++-style `cv::Mat` structure from an existing `CvMat` or `IplImage` structure. In this case, you have two options (Table 4-3): you can construct a header `m` on the existing data (by setting `copyData` to `false`) or you can set `copyData` to `true` (in which case, new memory will be allocated for `m` and all of the data from `old` will be copied into `m`).

*Table 4-3. cv::Mat constructors for pre–version 2.1 data types*

| Constructor | Description |
|---|---|
| `cv::Mat( const CvMat* old, bool copyData=false );` | Constructor for a new object `m` that creates `m` from an old-style `CvMat`, with optional data copy |
| `cv::Mat( const IplImage* old, bool copyData=false );` | Constructor for a new object `m` that creates `m` from an old-style `IplImage`, with optional data copy |

These constructors do a lot more for you than you might realize at first. In particular, they allow for expressions that mix the C++ and C data types by functioning as implicit constructors for the C++ data types on demand. Thus, it is possible to simply use a pointer to one of the C structures wherever a `cv::Mat` is expected and have a reasonable expectation that things will work out correctly. (This is why the `copyData` member defaults to `false`.)

In addition to these constructors, there are corresponding cast operators that will convert a `cv::Mat` into `CvMat` or `IplImage` on demand. These also do not copy data.

The last set of constructors is the template constructors (Table 4-4). These are called *template constructors* not because they construct a template form of `cv::Mat`, but because they construct an instance of `cv::Mat` from something that is itself a template. These constructors allow either an arbitrary `cv::Vec<>` or `cv::Matx<>` to be used to create a `cv::Mat` array, of corresponding dimension and type, or to use an STL `vector<>` object of arbitrary type to construct an array of that same type.

*Table 4-4. cv::Mat template constructors*

| Constructor | Description |
|---|---|
| `cv::Mat( const cv::Vec<T,n>& vec, bool copyData=true );` | Construct a one-dimensional array of type `T` and size `n` from a `cv::Vec` of the same type |
| `cv::Mat( const cv::Matx<T,m,n>& vec, bool copyData=true );` | Construct a two-dimensional array of type `T` and size `m` × `n` from a `cv::Matx` of the same type |

| Constructor | Description |
|---|---|
| `cv::Mat( const std::vector<T>& vec, bool copyData=true );` | Construct a one-dimensional array of type `T` from an STL vector containing elements of the same type |

The class `cv::Mat` also provides a number of static member functions to create certain kinds of commonly used arrays (Table 4-5). These include functions like `zeros()`, `ones()`, and `eye()`, which construct a matrix full of zeros, a matrix full of ones, or an identity matrix, respectively.[5]

*Table 4-5. Static functions that create cv::Mat*

| Function | Description |
|---|---|
| `cv::Mat::zeros( rows, cols, type );` | Create a `cv::Mat` of size `rows` × `cols`, which is full of zeros, with type `type` (`CV_32F`, etc.) |
| `cv::Mat::ones( rows, cols, type );` | Create a `cv::Mat` of size `rows` × `cols`, which is full of ones, with type `type` (`CV_32F`, etc.) |
| `cv::Mat::eye( rows, cols, type );` | Create a `cv::Mat` of size `rows` × `cols`, which is an identity matrix, with type `type` (`CV_32F`, etc.) |

## Accessing Array Elements Individually

There are several ways to access a matrix, all of which are designed to be convenient in different contexts. In recent versions of OpenCV, however, a great deal of effort has been invested to make them all comparably, if not identically, efficient. The two primary options for accessing individual elements are to access them by location or through iteration.

The basic means of direct access is the (template) member function `at<>()`. There are many variations of this function that take different arguments for arrays of different numbers of dimensions. The way this function works is that you specialize the `at<>()` template to the type of data that the matrix contains, then access that element using the row and column locations of the data you want. Here is a simple example:

```
cv::Mat m = cv::Mat::eye( 10, 10, 32FC1 );

printf(
  "Element (3,3) is %f\n",
  m.at<float>(3,3)
);
```

5 In the case of `cv::Mat::eye()` and `cv::Mat::ones()`, if the array created is multichannel, only the first channel will be set to `1.0`, while the other channels will be `0.0`.

For a multichannel array, the analogous example would look like this:

```
cv::Mat m = cv::Mat::eye( 10, 10, 32FC2 );

printf(
  "Element (3,3) is (%f,%f)\n",
  m.at<cv::Vec2f>(3,3)[0],
  m.at<cv::Vec2f>(3,3)[1]
);
```

Note that when you want to specify a template function like `at<>()` to operate on a multichannel array, the best way to do this is to use a `cv::Vec<>` object (either using a premade alias or the template form).

Similar to the vector case, you can create an array made of a more sophisticated type, such as complex numbers:

```
cv::Mat m = cv::Mat::eye( 10, 10, cv::DataType<cv::Complexf>::type );

printf(
  "Element (3,3) is %f + i%f\n",
  m.at<cv::Complexf>(3,3).re,
  m.at<cv::Complexf>(3,3).im,
);
```

It is also worth noting the use of the `cv::DataType<>` template here. The matrix constructor requires a runtime value that is a variable of type `int` that happens to take on some "magic" values that the constructor understands. By contrast, `cv::Complexf` is an actual object type, a purely compile-time construct. The need to generate one of these representations (runtime) from the other (compile time) is precisely why the `cv::DataType<>` template exists. Table 4-6 lists the available variations of the `at<>()` template.

*Table 4-6. Variations of the at<>() accessor function*

| Example | Description |
|---|---|
| `M.at<int>( i );` | Element `i` from integer array `M` |
| `M.at<float>( i, j );` | Element `( i, j )` from float array `M` |
| `M.at<int>( pt );` | Element at location `(pt.x, pt.y)` in integer matrix `M` |
| `M.at<float>( i, j, k );` | Element at location `( i, j, k )` in three-dimensional float array `M` |
| `M.at<uchar>( idx );` | Element at *n*-dimensional location indicated by `idx[]` in array `M` of unsigned characters |

To access a two-dimensional array, you can also extract a C-style pointer to a specific row of the array. This is done with the `ptr<>()` template member function of `cv::Mat`. (Recall that the data in the array is contiguous by row, thus accessing a specific column in this way would not make sense; we will see the right way to do that shortly.) As with at<>(), ptr<>() is a template function instantiated with a type name. It takes an integer argument indicating the row you wish to get a pointer to. The function returns a pointer to the primitive type of which the array is constructed (i.e., if the array type is CV_32FC3, the return value will be of type float*). Thus, given a three-channel matrix mtx of type float, the construction mtx.ptr<Vec3f>(3) would return a pointer to the first (floating-point) channel of the first element in row 3 of mtx. This is generally the fastest way to access elements of an array,[6] because once you have the pointer, you are right down there with the data.

There are thus two ways to get a pointer to the data in a matrix mtx. One is to use the ptr<>() member function. The other is to directly use the member pointer data, and to use the member array step[] to compute addresses. The latter option is similar to what one tended to do in the C interface, but is generally no longer preferred over access methods such as at<>(), ptr<>(), and the iterators. Having said this, direct address computation may still be most efficient, particularly when you are dealing with arrays of greater than two dimensions.

There is one last important point to keep in mind about C-style pointer access. If you want to access everything in an array, you will likely want to iterate one row at a time; this is because the rows may or may not be packed continuously in the array. However, the member function isContinuous() will tell you if the members are continuously packed. If they are, you can just grab the pointer to the very first element of the first row and cruise through the entire array as if it were a giant one-dimensional array.

The other form of sequential access is to use the iterator mechanism built into cv::Mat. This mechanism is based on, and works more or less identically to, the analogous mechanism provided by the STL containers. The basic idea is that OpenCV provides a pair of iterator templates, one for const and one for non-const arrays. These iterators are named cv::MatIterator<> and cv::MatConstIterator<>, respectively. The cv::Mat methods begin() and end() return objects of this type. This method of iteration is convenient because the iterators are smart enough to han-

6 The difference in performance between using at<>() and direct pointer access depends on compiler optimization. Access through at<>() tends to be comparable to (though slightly slower than) direct pointer access in code with a good optimizer, but may be more than an order of magnitude slower if that optimizer is turned off (e.g., when you do a debug build). Access through iterators is almost always slower than either of these. In almost all cases, however, using built-in OpenCV functions will be faster than any loop you write regardless of the direct access methods described here, so avoid that kind of construct wherever possible.

dle the continuous packing and noncontinuous packing cases automatically, as well as handling any number of dimensions in the array.

Each iterator must be declared and specified to the type of object from which the array is constructed. Here is a simple example of the iterators being used to compute the "longest" element in a three-dimensional array of three-channel elements (a three-dimensional vector field):

```
int sz[3] = { 4, 4, 4 };
cv::Mat m( 3, sz, CV_32FC3 );  // A three-dimensional array of size 4-by-4-by-4
cv::randu( m, -1.0f, 1.0f );   // fill with random numbers from -1.0 to 1.0

float max = 0.0f;              // minimum possible value of L2 norm
cv::MatConstIterator<cv::Vec3f> it = m.begin();
while( it != m.end() ) {

  len2 = (*it)[0]*(*it)[0]+(*it)[1]*(*it)[1]+(*it)[2]*(*it)[2];
  if( len2 > max ) max = len2;
  it++;

}
```

You would typically use iterator-based access when doing operations over an entire array, or element-wise across multiple arrays. Consider the case of adding two arrays, or converting an array from the RGB color space to the HSV color space. In such cases, the same exact operation will be done at every pixel location.

## The N-ary Array Iterator: NAryMatIterator

There is another form of iteration that, though it does not handle discontinuities in the packing of the arrays in the manner of `cv::MatIterator<>`, allows us to handle iteration over many arrays at once. This iterator is called `cv::NAryMatIterator`, and requires only that all of the arrays being iterated over be of the same geometry (number of dimensions and extent in each dimension).

Instead of returning single elements of the arrays being iterated over, the *N*-ary iterator operates by returning chunks of those arrays, called *planes*. A plane is a portion (typically a one- or two-dimensional slice) of the input array in which the data is guaranteed to be contiguous in memory.[7] This is how discontinuity is handled: you are given the contiguous chunks one by one. For each such plane, you can either operate on it using array operations, or iterate trivially over it yourself. (In this case, "trivially" means to iterate over it in a way that does not need to check for discontinuities inside of the chunk.)

7 In fact the dimensionality of the "plane" is not limited to two; it can be larger. What is always the case is that the planes will be contiguous in memory.

The concept of the plane is entirely separate from the concept of multiple arrays being iterated over simultaneously. Consider Example 4-1, in which we sum just a single multidimensional array plane by plane.

*Example 4-1. Summation of a multidimensional array, done plane by plane*

```
int main( int argc, char** argv ) {

  const int n_mat_size = 5;
  const int n_mat_sz[] = { n_mat_size, n_mat_size, n_mat_size };
  cv::Mat n_mat( 3, n_mat_sz, CV_32FC1 );

  cv::RNG rng;
  rng.fill( n_mat, cv::RNG::UNIFORM, 0.f, 1.f );

}

const cv::Mat* arrays[] = { &n_mat, 0 };
cv::Mat my_planes[1];
cv::NAryMatIterator it( arrays, my_planes );
```

At this point, you have your *N*-ary iterator. Continuing our example, we will compute the sum of `m0` and `m1`, and place the result in `m2`. We will do this plane by plane, however:

```
// On each iteration, it.planes[i] will be the current plane of the
// i-th array from 'arrays'.
//
float s = 0.f;                              // Total sum over all planes
int   n = 0;                                // Total number of planes
for (int p = 0; p < it.nplanes; p++, ++it) {
  s += cv::sum(it.planes[0])[0];
  n++;
}
```

In this example, we first create the three-dimensional array `n_mat` and fill it with 125 random floating-point numbers between `0.0` and `1.0`. To initialize the `cv::NAryMatIterator` object, we need to have two things. First, we need a C-style array containing pointers to all of the `cv::Mat`s we wish to iterate over (in this example, there is just one). This array must always be terminated with a `0` or `NULL`. Next, we need another C-style array of `cv::Mat`s that can be used to refer to the individual planes as we iterate over them (in this case, there is also just one).

Once we have created the *N*-ary iterator, we can iterate over it. Recall that this iteration is over the planes that make up the arrays we gave to the iterator. The number of planes (the same for each array, because they have the same geometry) will always be given by `it.nplanes`. The *N*-ary iterator contains a C-style array called `planes` that holds headers for the current plane in each input array. In our example, there is only one array being iterated over, so we need only refer to `it.planes[0]` (the current plane in the one and only array). In this example, we then call `cv::sum()` on each plane and accumulate the final result.

To see the real utility of the *N*-ary iterator, consider a slightly expanded version of this example in which there are two arrays we would like to sum over (Example 4-2).

*Example 4-2. Summation of two arrays using the N-ary operator*

```
int main( int argc, char** argv ) {

  const int n_mat_size = 5;
  const int n_mat_sz[] = { n_mat_size, n_mat_size, n_mat_size };
  cv::Mat n_mat0( 3, n_mat_sz, CV_32FC1 );
  cv::Mat n_mat1( 3, n_mat_sz, CV_32FC1 );

  cv::RNG rng;
  rng.fill( n_mat0, cv::RNG::UNIFORM, 0.f, 1.f );
  rng.fill( n_mat1, cv::RNG::UNIFORM, 0.f, 1.f );

  const cv::Mat* arrays[] = { &n_mat0, &n_mat1, 0 };
  cv::Mat my_planes[2];
  cv::NAryMatIterator it( arrays, my_planes );

  float s = 0.f;                    // Total sum over all planes in both arrays
  int   n = 0;                      // Total number of planes
  for(int p = 0; p < it.nplanes; p++, ++it) {
    s += cv::sum(it.planes[0])[0];
    s += cv::sum(it.planes[1])[0];
    n++;
  }
}
```

In this second example, you can see that the C-style array called `arrays` is given pointers to both input arrays, and two matrices are supplied in the `my_planes` array. When it is time to iterate over the planes, at each step, `planes[0]` contains a plane in `n_mat0`, and `planes[1]` contains the corresponding plane in `n_mat1`. In this simple example, we just sum the two planes and add them to our accumulator. In an only slightly extended case, we could use element-wise addition to sum these two planes and place the result into the corresponding plane in a third array.

Not shown in the preceding example, but also important, is the member `it.size`, which indicates the size of each plane. The size reported is the number of elements in the plane, so it does not include a factor for the number of channels. In our previous example, if `it.nplanes` were 4, then `it.size` would have been 16:

```
/////////// compute dst[*] = pow(src1[*], src2[*]) //////////////
const Mat* arrays[] = { src1, src2, dst, 0 };
float* ptrs[3];
```

```
NAryMatIterator it(arrays, (uchar**)ptrs);
for( size_t i = 0; i < it.nplanes; i++, ++it )
{
  for( size_t j = 0; j < it.size; j++ )
  {
    ptrs[2][j] = std::pow(ptrs[0][j], ptrs[1][j]);
  }
}
```

## Accessing Array Elements by Block

In the previous section, we saw ways to access individual elements of an array, either singularly or by iterating sequentially through them all. Another common situation is when you need to access a subset of an array as another array. This might be to select out a row or a column, or any subregion of the original array.

There are many methods that do this for us in one way or another, as shown in Table 4-7; all of them are member functions of the `cv::Mat` class and return a subsection of the array on which they are called. The simplest of these methods are `row()` and `col()`, which take a single integer and return the indicated row or column of the array whose member we are calling. Clearly these make sense only for a two-dimensional array; we will get to the more complicated case momentarily.

When you use `m.row()` or `m.col()` (for some array `m`), or any of the other functions we are about to discuss, it is important to understand that the data in `m` is not copied to the new arrays. Consider an expression like `m2 = m.row(3)`. This expression means to create a new array header `m2`, and to arrange its `data` pointer, `step` array, and so on, such that it will access the data in row `3` in `m`. If you modify the data in `m2`, you will be modifying the data in `m`. Later, we will visit the `copyTo()` method, which actually will copy data. The main advantage of the way this is handled in OpenCV is that the amount of time required to create a new array that accesses part of an existing array is not only very small, but also independent of the size of either the old or the new array.

Closely related to `row()` and `col()` are `rowRange()` and `colRange()`. These functions do essentially the same thing as their simpler cousins, except that they will extract an array with multiple contiguous rows (or columns). You can call both functions in one of two ways, either by specifying an integer start and end row (or column), or by passing a `cv::Range` object that indicates the desired rows (or columns). In the case of the two-integer method, the range is inclusive of the start index but exclusive of the end index (you may recall that `cv::Range` uses a similar convention).

The member function `diag()` works the same as `row()` or `col()`, except that the array returned from `m.diag()` references the diagonal elements of a matrix. `m.diag()` expects an integer argument that indicates which diagonal is to be extracted. If that argument is zero, then it will be the main diagonal. If it is positive, it will be offset from the main diagonal by that distance in the upper half of the array. If it is negative, then it will be from the lower half of the array.

The last way to extract a submatrix is with `operator()`. Using this operator, you can pass either a pair of ranges (a `cv::Range` for rows and a `cv::Range` for columns) or a `cv::Rect` to specify the region you want. This is the only method of access that will allow you to extract a subvolume from a higher-dimensional array. In this case, a pointer to a C-style array of ranges is expected, and that array must have as many elements as the number of dimensions of the array.

*Table 4-7. Block access methods of cv::Mat*

| Example | Description |
|---|---|
| `m.row( i );` | Array corresponding to row `i` of `m` |
| `m.col( j );` | Array corresponding to column `j` of `m` |
| `m.rowRange( i0, i1 );` | Array corresponding to rows `i0` through `i1-1` of matrix `m` |
| `m.rowRange( cv::Range( i0, i1 ) );` | Array corresponding to rows `i0` through `i1-1` of matrix `m` |
| `m.colRange( j0, j1 );` | Array corresponding to columns `j0` through `j1-1` of matrix `m` |
| `m.colRange( cv::Range( j0, j1 ) );` | Array corresponding to columns `j0` through `j1-1` of matrix `m` |
| `m.diag( d );` | Array corresponding to the d-offset diagonal of matrix `m` |
| `m( cv::Range(i0,i1), cv::Range(j0,j1) );` | Array corresponding to the subrectangle of matrix `m` with one corner at `i0`, `j0` and the opposite corner at (`i1-1`, `j1-1`) |
| `m( cv::Rect(i0,i1,w,h) );` | Array corresponding to the subrectangle of matrix `m` with one corner at `i0`, `j0` and the opposite corner at (`i0+w-1`, `j0+h-1`) |
| `m( ranges );` | Array extracted from `m` corresponding to the subvolume that is the intersection of the ranges given by `ranges[0]`-`ranges[ndim-1]` |

## Matrix Expressions: Algebra and cv::Mat

One of the capabilities enabled by the move to C++ in version 2.1 is the overloading of operators and the ability to create algebraic expressions consisting of matrix arrays[8] and singletons. The primary advantage of this is code clarity, as many operations can be combined into one expression that is both more compact and often more meaningful.

8 For clarity, we use the word *array* when referring to a general object of type `cv::Mat`, and use the word *matrix* for those situations where the manner in which the array is being used indicates that it is representing a mathematical object that would be called a matrix. The distinction is a purely semantic one, and not manifest in the actual design of OpenCV.

In the background, many important features of OpenCV's array class are being used to make these operations work. For example, matrix headers are created automatically as needed and workspace data areas are allocated (only) as required. When no longer needed, data areas are deallocated invisibly and automatically. The result of the computation is finally placed in the destination array by `operator=()`. However, one important distinction is that this form of `operator=()` is not assigning a `cv::Mat` or a `cv::Mat` (as it might appear), but rather a `cv::MatExpr` (the expression itself[9]) to a `cv::Mat`. This distinction is important because data is always copied into the result (lefthand) array. Recall that though `m2=m1` is legal, it means something slightly different. In this latter case, `m2` would be another reference to the data in `m1`. By contrast, `m2=m1+m0` means something different again. Because `m1+m0` is a *matrix expression*, it will be evaluated and a pointer to the results will be assigned in `m2`. The results will reside in a newly allocated data area.[10]

Table 4-8 lists examples of the algebraic operations available. Note that in addition to simple algebra, there are comparison operators, operators for constructing matrices (such as `cv::Mat::eye()`, which we encountered earlier), and higher-level operations for computing transposes and inversions. The key idea here is that you should be able to take the sorts of relatively nontrivial matrix expressions that occur when doing computer vision and express them on a single line in a clear and concise way.

*Table 4-8. Operations available for matrix expressions*

| Example | Description |
|---|---|
| `m0 + m1, m0 - m1;` | Addition or subtraction of matrices |
| `m0 + s; m0 - s; s + m0, s - m1;` | Addition or subtraction between a matrix and a singleton |
| `-m0;` | Negation of a matrix |
| `s * m0; m0 * s;` | Scaling of a matrix by a singleton |
| `m0.mul( m1 ); m0/m1;` | Per element multiplication of `m0` and `m1`, per-element division of `m0` by `m1` |
| `m0 * m1;` | Matrix multiplication of `m0` and `m1` |

9 The underlying machinery of `cv::MatExpr` is more detail than we need here, but you can think of `cv::MatExpr` as being a symbolic representation of the algebraic form of the righthand side. The great advantage of `cv::MatExpr` is that when it is time to evaluate an expression, it is often clear that some operations can be removed or simplified without evaluation (such as computing the transpose of the transpose of a matrix, adding zero, or multiplying a matrix by its own inverse).

10 If you are a real expert, this will not surprise you. Clearly a temporary array must be created to store the result of `m1+m0`. Then `m2` really is just another reference, but it is another reference to that temporary array. When `operator+()` exits, its reference to the temporary array is discarded, but the reference count is not zero. `m2` is left holding the one and only reference to that array.

<table>
<tr><th>Example</th><th>Description</th></tr>
<tr><td>m0.inv( method );</td><td>Matrix inversion of m0 (default value of method is DECOMP_LU)</td></tr>
<tr><td>m0.t();</td><td>Matrix transpose of m0 (no copy is done)</td></tr>
<tr><td>m0>m1; m0>=m1; m0==m1; m0<=m1; m0<m1;</td><td>Per element comparison, returns uchar matrix with elements 0 or 255</td></tr>
<tr><td>m0&m1; m0|m1; m0^m1; ~m0;<br>m0&s; s&m0; m0|s; s|m0; m0^s; s^m0;</td><td>Bitwise logical operators between matrices or matrix and a singleton</td></tr>
<tr><td>min(m0,m1); max(m0,m1); min(m0,s);<br>min(s,m0); max(m0,s); max(s,m0);</td><td>Per element minimum and maximum between two matrices or a matrix and a singleton</td></tr>
<tr><td>cv::abs( m0 );</td><td>Per element absolute value of m0</td></tr>
<tr><td>m0.cross( m1 ); m0.dot( m1 );</td><td>Vector cross and dot product (vector cross product is defined only for $3 \times 1$ matrices)</td></tr>
<tr><td>cv::Mat::eye( Nr, Nc, type );<br>cv::Mat::zeros( Nr, Nc, type );<br>cv::Mat::ones( Nr, Nc, type );</td><td>Class static matrix initializers that return fixed $N_r \times N_c$ matrices of type type</td></tr>
</table>

The matrix inversion operator inv() is actually a frontend to a variety of algorithms for matrix inversion. There are currently three options. The first is cv::DECOMP_LU, which means LU decomposition and works for any nonsingular matrix. The second option is cv::DECOMP_CHOLESKY, which solves the inversion by Cholesky decomposition. Cholesky decomposition works only for symmetric, positive definite matrices, but is much faster than LU decomposition for large matrices. The last option is cv::DECOMP_SVD, which solves the inversion by singular value decomposition (SVD). SVD is the only workable option for matrices that are singular or nonsquare (in which case the pseudo-inverse is computed).

Not included in Table 4-8 are all of the functions like cv::norm(), cv::mean(), cv::sum(), and so on (some of which we have not gotten to yet, but you can probably guess what they do) that convert matrices to other matrices or to scalars. Any such object can still be used in a matrix expression.

## Saturation Casting

In OpenCV, you will often do operations that risk overflowing or underflowing the available values in the destination of some computation. This is particularly common when you are doing operations on unsigned types that involve subtraction, but it can happen anywhere. To deal with this problem, OpenCV relies on a construct called *saturation casting*.

What this means is that OpenCV arithmetic and other operations that act on arrays will check for underflows and overflows automatically; in these cases, the library functions will replace the resulting value of an operation with the lowest or highest available value, respectively. Note that this is not what C language operations normally and natively do.

You may want to implement this particular behavior in your own functions as well. OpenCV provides some handy templated casting operators to make this easy for you. These are implemented as a template function called `cv::saturate_cast<>()`, which allows you to specify the type to which you would like to cast the argument. Here is an example:

```
uchar& Vxy = m0.at<uchar>( y, x );
Vxy = cv::saturate_cast<uchar>((Vxy-128)*2 + 128);}
```

In this example code, we first assign the variable `Vxy` to be a reference to an element of an 8-bit array, `m0`. We then subtract 128 from this array, multiply that by two (scale that up), and add 128 (so the result is twice as far from 128 as the original). The usual C arithmetic rules would assign `Vxy-128` to a (32-bit) signed integer; followed by integer multiplication by 2 and integer addition of 128. Notice, however, that if the original value of `Vxy` were (for example) `10`, then `Vxy-128` would be `-118`. The value of the expression would then be `-108`. This number will not fit into the 8-bit unsigned variable `Vxy`. This is where `cv::saturation_cast<uchar>()` comes to the rescue. It takes the value of `-108` and, recognizing that it is too low for an unsigned char, converts it to `0`.

## More Things an Array Can Do

At this point, we have touched on most of the members of the `cv::Mat` class. Of course, there are a few things that were missed, as they did not fall into any specific category that was discussed so far. Table 4-9 lists the "leftovers" that you will need in your daily OpenCV programming.

*Table 4-9. More class member functions of cv::Mat*

| Example | Description |
|---|---|
| `m1 = m0.clone();` | Make a complete copy of `m0`, copying all data elements as well; cloned array will be continuous |
| `m0.copyTo( m1 );` | Copy contents of `m0` onto `m1`, reallocating `m1` if necessary (equivalent to `m1=m0.clone()`) |
| `m0.copyTo( m1, mask );` | Same as `m0.copyTo(m1)`, except only entries indicated in the array mask are copied |
| `m0.convertTo( m1, type, scale, offset );` | Convert elements of `m0` to `type` (e.g., `CV_32F`) and write to `m1` after scaling by scale (default `1.0`) and adding offset (default `0.0`) |
| `m0.assignTo( m1, type );` | Internal use only (resembles `convertTo`) |
| `m0.setTo( s, mask );` | Set all entries in `m0` to singleton value `s`; if mask is present, set only those values corresponding to nonzero elements in `mask` |

| Example | Description |
|---|---|
| `m0.reshape( chan, rows );` | Changes effective shape of a two-dimensional matrix; `chan` or `rows` may be zero, which implies "no change"; data is not copied |
| `m0.push_back( s );` | Extend an `m` × 1 matrix and insert the singleton `s` at the end |
| `m0.push_back( m1 );` | Extend an `m` × `n` by `k` rows and copy `m1` into those rows; `m1` must be `k` × `n` |
| `m0.pop_back( n );` | Remove `n` rows from the end of an `m` × `n` (default value of `n` is `1`)[a] |
| `m0.locateROI( size, offset );` | Write whole size of `m0` to `cv::Size size`; if `m0` is a "view" of a larger matrix, write location of starting corner to `Point& offset` |
| `m0.adjustROI( t, b, l, r );` | Increase the size of a view by `t` pixels above, `b` pixels below, `l` pixels to the left, and `r` pixels to the right |
| `m0.total();` | Compute the total number of array elements (does not include channels) |
| `m0.isContinuous();` | Return `true` only if the rows in `m0` are packed without space between them in memory |
| `m0.elemSize();` | Return the size of the elements of `m0` in bytes (e.g., a three-channel float matrix would return 12 bytes) |
| `m0.elemSize1();` | Return the size of the subelements of `m0` in bytes (e.g., a three-channel float matrix would return 4 bytes) |
| `m0.type();` | Return a valid type identifier for the elements of `m0` (e.g., `CV_32FC3`) |
| `m0.depth();` | Return a valid type identifier for the individial channels of `m0` (e.g., `CV_32F`) |
| `m0.channels();` | Return the number of channels in the elements of `m0` |
| `m0.size();` | Return the size of the `m0` as a `cv::Size` object |
| `m0.empty();` | Return `true` only if the array has no elements (i.e., `m0.total==0` or `m0.data==NULL`) |

[a] Many implementations of "pop" functionality return the popped element. This one does not; its return type is void.

## The cv::SparseMat Class: Sparse Arrays

The `cv::SparseMat` class is used when an array is likely to be very large compared to the number of nonzero entries. This situation often arises in linear algebra with sparse matrices, but it also comes up when one wishes to represent data, particularly histograms, in higher-dimensional arrays, since most of the space will be empty in many practical applications. A sparse representation stores only data that is actually present and so can save a great deal of memory. In practice, many sparse objects would be too huge to represent at all in a dense format. The disadvantage of sparse representations is that computation with them is slower (on a per-element basis). This last point is important, in that computation with sparse matrices is not categorically slower, as there can be a great economy in knowing in advance that many operations need not be done at all.

The OpenCV sparse matrix class `cv::SparseMat` functions analogously to the dense matrix class `cv::Mat` in most ways. It is defined similarly, supports most of the same operations, and can contain the same data types. Internally, the way data is organized is quite different. While `cv::Mat` uses a data array closely related to a C data array (one in which the data is sequentially packed and addresses are directly computable from the indices of the element), `cv::SparseMat` uses a hash table to store just the nonzero elements.[11] That hash table is maintained automatically, so when the number of (nonzero) elements in the array becomes too large for efficient lookup, the table grows automatically.

## Accessing Sparse Array Elements

The most important difference between sparse and dense arrays is how elements are accessed. Sparse arrays provide four different access mechanisms: `cv::SparseMat::ptr()`, `cv::SparseMat::ref()`, `cv::SparseMat::value()`, and `cv::SparseMat::find()`.

The `cv::SparseMat::ptr()` method has several variations, the simplest of which has the template:

```
uchar* cv::SparseMat::ptr( int i0, bool createMissing, size_t* hashval=0 );
```

This particular version is for accessing a one-dimensional array. The first argument, `i0`, is the index of the requested element. The next argument, `createMissing`, indicates whether the element should be created if it is not already present in the array. When `cv::SparseMat::ptr()` is called, it will return a pointer to the element if that element is already defined in the array, but `NULL` if that element is not defined. However, if the `createMissing` argument is `true`, that element will be created and a valid non-`NULL` pointer will be returned to that new element. To understand the final argument, `hashval`, it is necessary to recall that the underlying data representation of a `cv::SparseMat` is a hash table. Looking up objects in a hash table requires two steps: first, computing the hash key (in this case, from the indices), and second, searching a list associated with that key. Normally, that list will be short (ideally only one element), so the primary computational cost in a lookup is the computation of the hash key. If this key has already been computed (as with `cv::SparseMat::hash()`, which will be covered next), then time can be saved by not recomputing it. In the case of `cv::SparseMat::ptr()`, if the argument `hashval` is left with its default argument of `NULL`, the hash key will be computed. If, however, a key is provided, it will be used.

11 Actually, zero elements may be stored if those elements have become zero as a result of computation on the array. If you want to clean up such elements, you must do so yourself. This is the function of the method `SparseMat::erase()`, which we will visit shortly.

There are also variations of `cv::SparseMat::ptr()` that allow for two or three indices, as well as a version whose first argument is a pointer to an array of integers (i.e., `const int* idx`), which is required to have the same number of entries as the dimension of the array being accessed.

In all cases, the function `cv::SparseMat::ptr()` returns a pointer to an unsigned character (i.e., `uchar*`), which will typically need to be recast to the correct type for the array.

The accessor template function `SparseMat::ref<>()` is used to return a reference to a particular element of the array. This function, like `SparseMat::ptr()`, can take one, two, or three indices, or a pointer to an array of indices, and also supports an optional pointer to the hash value to use in the lookup. Because it is a template function, you must specify the type of object being referenced. So, for example, if your array were of type `CV_32F`, then you might call `SparseMat::ref<>()` like this:

```
a_sparse_mat.ref<float>( i0, i1 ) += 1.0f;
```

The template method `cv::SparseMat::value<>()` is identical to `SparseMat::ref<>()`, except that it returns the value and not a reference to the value. Thus, this method is itself a "`const` method."[12]

The final accessor function is `cv::SparseMat::find<>()`, which works similarly to `cv::SparseMat::ref<>()` and `cv::SparseMat::value<>()`, but returns a pointer to the requested object. Unlike `cv::SparseMat::ptr()`, however, this pointer is of the type specified by the template instantiation of `cv::SparseMat::find<>()`, and so does not need to be recast. For purposes of code clarity, `cv::SparseMat::find<>()` is preferred over `cv::SparseMat::ptr()` wherever possible. `cv::SparseMat::find<>()`, however, is a `const` method and returns a `const` pointer, so the two are not always interchangeable.

In addition to direct access through the four functions just outlined, it is also possible to access the elements of a sparse matrix through iterators. As with the dense array types, the iterators are normally templated. The templated iterators are `cv::SparseMatIterator_<>` and `cv::SparseMatConstIterator_<>`, together with their corresponding `cv::SparseMat::begin<>()` and `cv::SparseMat::end<>()` routines. (The `const` forms of the `begin()` and `end()` routines return the `const` iterators.) There are also the nontemplate iterators `cv::SparseMatIterator` and `cv::SparseMatConstIt`

12 For those not familiar with "`const` correctness," it means that the method is declared in its prototype such that the `this` pointer passed to `SparseMat::value<>()` is guaranteed to be a constant pointer, and thus `SparseMat::value<>()` can be called on `const` objects, while functions like `SparseMat::ref<>()` cannot. The next function, `SparseMat::find<>()`, is also a `const` function.

erator, which are returned by the nontemplate `SparseMat::begin()` and `SparseMat::end()` routines.

In Example 4-3 we print out all of the nonzero elements of a sparse array.

*Example 4-3. Printing all of the nonzero elements of a sparse array*

```
int main( int argc, char** argv ) {

  // Create a 10x10 sparse matrix with a few nonzero elements
  //
  int size[] = {10,10};
  cv::SparseMat sm( 2, size, CV_32F );

  for( int i=0; i<10; i++ ) {           // Fill the array
    int idx[2];
    idx[0] = size[0] * rand();
    idx[1] = size[1] * rand();

    sm.ref<float>( idx ) += 1.0f;
  }

  // Print out the nonzero elements
  //
  cv::SparseMatConstIterator_<float> it     = sm.begin<float>();
  cv::SparseMatConstIterator_<float> it_end = sm.end<float>();

  for(; it != it_end; ++it) {
    const cv::SparseMat::Node* node = it.node();
    printf(" (%3d,%3d) %f\n", node->idx[0], node->idx[1], *it );
  }

}
```

In this example, we also slipped in the method `node()`, which is defined for the iterators. `node()` returns a pointer to the internal data node in the sparse matrix that is indicated by the iterator. The returned object of type `cv::SparseMat::Node` has the following definition:

```
struct Node
{
  size_t hashval;
  size_t next;
  int idx[cv::MAX_DIM];
};
```

This structure contains both the index of the associated element (note that element `idx` is of type `int[]`), as well as the hash value associated with that node (the `hashval` element is the same hash value as can be used with `SparseMat::ptr()`, `SparseMat::ref()`, `SparseMat::value()`, and `SparseMat::find()`).

## Functions Unique to Sparse Arrays

As stated earlier, sparse matrices support many of the same operations as dense matrices. In addition, there are several methods that are unique to sparse matrices. These are listed in Table 4-10, and include the functions mentioned in the previous sections.

*Table 4-10. Additional class member functions of cv::SparseMat*

| Example | Description |
|---|---|
| `cv::SparseMat sm;` | Create a sparse matrix without initialization |
| `cv::SparseMat sm( 3, sz, CV_32F );` | Create a three-dimensional sparse matrix with dimensions given by the array `sz` of type `float` |
| `cv::SparseMat sm( sm0 );` | Create a new sparse matrix that is a copy of the existing sparse matrix `sm0` |
| `cv::SparseMat( m0, try1d );` | Create a sparse matrix from an existing dense matrix `m0`; if the bool `try1d` is `true`, convert `m0` to a one-dimensional sparse matrix if the dense matrix was n × 1 or 1 × n |
| `cv::SparseMat( &old_sparse_mat );` | Create a new sparse matrix from a pointer to a pre–version 2.1 C-style sparse matrix of type `CvSparseMat` |
| `CvSparseMat* old_sm =`<br>`   (cv::SparseMat*) sm;` | Cast operator creates a pointer to a pre–version 2.1 C-style sparse matrix; that `CvSparseMat` object is created and all data is copied into it, and then its pointer is returned |
| `size_t n = sm.nzcount();` | Return the number of nonzero elements in `sm` |
| `size_t h = sm.hash( i0 );`<br>`size_t h = sm.hash( i0, i1 );`<br>`size_t h = sm.hash( i0, i1, i2 );`<br>`size_t h = sm.hash( idx );` | Return the hash value for element `i0` in a one-dimensional sparse matrix; `i0`, `i1` in a two-dimensional sparse matrix; `i0`, `i1`, `i2` in a three-dimensional sparse matrix; or the element indicated by the array of integers `idx` in an *n*-dimensional sparse matrix |
| `sm.ref<float>( i0 )         = f0;`<br>`sm.ref<float>( i0, i1 )     = f0;`<br>`sm.ref<float>( i0, i1, i2 ) = f0;`<br>`sm.ref<float>( idx )        = f0;` | Assign the value `f0` to element `i0` in a one-dimensional sparse matrix; `i0`, `i1` in a two-dimensional sparse matrix; `i0`, `i1`, `i2` in a three-dimensional sparse matrix; or the element indicated by the array of integers `idx` in an *n*-dimensional sparse matrix |
| `f0 = sm.value<float>( i0 );`<br>`f0 = sm.value<float>( i0, i1 );`<br>`f0 = sm.value<float>( i0, i1, i2 );`<br>`f0 = sm.value<float>( idx );` | Assign the value to `f0` from element `i0` in a one-dimensional sparse matrix; `i0`, `i1` in a two-dimensional sparse matrix; `i0`, `i1`, `i2` in a three-dimensional sparse matrix; or the element indicated by the array of integers `idx` in an *n*-dimensional sparse matrix |
| `p0 = sm.find<float>( i0 );`<br>`p0 = sm.find<float>( i0, i1 );`<br>`p0 = sm.find<float>( i0, i1, i2 );`<br>`p0 = sm.find<float>( idx );` | Assign to `p0` the address of element `i0` in a one-dimensional sparse matrix; `i0`, `i1` in a two-dimensional sparse matrix; `i0`, `i1`, `i2` in a three-dimensional sparse matrix; or the element indicated by the array of integers `idx` in an *n*-dimensional sparse matrix |
| `sm.erase( i0, i1, &hashval );`<br>`sm.erase( i0, i1, i2, &hashval );`<br>`sm.erase( idx, &hashval );` | Remove the element at (`i0`, `i1`) in a two-dimensional sparse matrix; at (`i0`, `i1`, `i2`) in a three-dimensional sparse matrix; or the element indicated by the array of integers `idx` in an *n*-dimensional sparse matrix. If `hashval` is not `NULL`, use the provided value instead of computing it |

| Example | Description |
| --- | --- |
| `cv::SparseMatIterator<float> it = sm.begin<float>();` | Create a sparse matrix iterator `it` and point it at the first value of the floating-point array `sm` |
| `cv::SparseMatIterator<uchar> it_end = sm.end<uchar>();` | Create a sparse matrix iterator `it_end` and initialize it to the value succeeding the final value in the byte array `sm` |

## Template Structures for Large Array Types

The concept we saw in the previous chapter, by which common library classes are related to template classes, also generalizes to `cv::Mat` and `cv::SparseMat`, and the templates `cv::Mat_<>` and `cv::SparseMat_<>`, but in a somewhat nontrivial way. When you use `cv::Point2i`, recall that this is nothing more or less than an alias (`typedef`) for `cv::Point_<int>`. In the case of the template `cv::Mat` and `cv::Mat_<>`, their relationship is not so simple. Recall that `cv::Mat` already has the capability of representing essentially any type, but it does so at construction time by explicitly specifying the base type. In the case of `cv::Mat_<>`, the instantiated template is actually *derived from* the `cv::Mat` class, and in effect specializes that class. This simplifies access and other member functions that would otherwise need to be templated.

This is worth reiterating. The purpose of using the template forms `cv::Mat_<>` and `cv::SparseMat_<>` are so you don't have to use the template forms of their member functions. Consider this example, where we have a matrix defined by:

```
cv::Mat m( 10, 10, CV_32FC2 );
```

Individual element accesses to this matrix would need to specify the type of the matrix, as in the following:

```
m.at< Vec2f >( i0, i1 ) = cv::Vec2f( x, y );
```

Alternatively, if you had defined the matrix `m` using the template class, you could use `at()` without specialization, or even just use `operator()`:

```
cv::Mat_<Vec2f> m( 10, 10 );

m.at( i0, i1 ) = cv::Vec2f( x, y );

// or...
m( i0, i1 ) = cv::Vec2f( x, y );
```

There is a great deal of simplification in your code that results from using these template definitions.

These two ways of declaring a matrix, and their associated .at methods are equivalent in efficiency. The second method, however, is considered more "correct" because it allows the compiler to detect type mismatches when m is passed into a function that requires a certain type of matrix. If:

```
cv::Mat m(10, 10, CV_32FC2 );
```

is passed into:

```
void foo((cv::Mat_<char> *)myMat);
```

failure would occur during runtime in perhaps nonobvious ways. If you instead used:

```
cv::Mat_<Vec2f> m( 10, 10 );
```

failure would be detected at compile time.

Template forms can be used to create template functions that operate on an array of a particular type. Consider our example from the previous section, where we created a small sparse matrix and then printed out its nonzero elements. We might try writing a function to achieve this as follows:

```
void print_matrix( const cv::SparseMat* sm ) {

  cv::SparseMatConstIterator_<float> it     = sm.begin<float>();
  cv::SparseMatConstIterator_<float> it_end = sm.end<float>();

  for(; it != it_end; ++it) {
    const cv::SparseMat::Node* node = it.node();
    printf(" (%3d,%3d) %f\n", node->idx[0], node->idx[1], *it );
  }
}
```

Though this function would compile and work when it is passed a two-dimensional matrix of type CV_32F, it would fail when a matrix of unexpected type was passed in. Let's look at how we could make this function more general.

The first thing we would want to address is the issue of the underlying data type. We could explicitly use the cv::SparseMat_<float> template, but it would be better still to make the function a template function. We would also need to get rid of the use of printf(), as it makes an explicit assumption that *it is a float. A better function might look like Example 4-4.

*Example 4-4. A better way to print a matrix*

```
template <class T> void print_matrix( const cv::SparseMat_<T>* sm ) {

  cv::SparseMatConstIterator_<T> it     = sm->begin();
```

```
  cv::SparseMatConstIterator_<T> it_end = sm->end();

  for(; it != it_end; ++it) {
    const typename cv::SparseMat_<T>::Node* node = it.node();
    cout <<"( " <<node->idx[0] <<", " <<node->idx[1]
      <<" ) = " <<*it <<endl;
  }
}

void calling_function1( void ) {
  ...
  cv::SparseMat_<float> sm( ndim, size );
  ...
  print_matrix<float>( &sm );
}

void calling_function2( void ) {
  ...
  cv::SparseMat sm( ndim, size, CV_32F );
  ...
  print_matrix<float>( (cv::SparseMat_<float>*) &sm );
}
```

It is worth picking apart these changes. First, though, before looking at changes, notice that the template for our function takes a pointer of type `const cv::SparseMat_<t>*`, a *pointer to* a sparse matrix template object. There is a good reason to use a pointer and not a reference here, because the caller may have a `cv::Mat` object (as used in `calling_function2()`) and not a `cv::Mat_<>` template object (as used in `calling_function1()`). The `cv::Mat` can be dereferenced and then explicitly cast to a pointer to the sparse matrix template object type.

In the templated prototype, we have promoted the function to a template of class `T`, and now expect a `cv::SparseMat_<T>*` pointer as argument. In the next two lines, we declare our iterators using the template type, but `begin()` and `end()` no longer have templated instantiations. The reason for this is that `sm` is now an instantiated template, and because of that explicit instantiation, `sm` "knows" what sort of matrix it is, and thus specialization of `begin()` and `end()` is unnecessary. The declaration of the `Node` is similarly changed so that the `Node` we are using is explicitly taken from the `cv::SparseMat_<T>` instantiated template class.[13] Finally, we change the `printf()` statement to use stream output to `cout`. This has the advantage that the printing is now agnostic to the type of `*it`.

13 The appearance of the `typename` keyword here is probably somewhat mysterious to most readers. It is a result of the dependent scoping rules in C++. If you should forget it, however, most modern compilers (e.g., g++) will throw you a friendly message reminding you to add it.

# Summary

In this chapter, we introduced the all-important OpenCV array structure `cv::Mat`, which is used to represent matrices, images, and multidimensional arrays. We saw that the `cv::Mat` class can contain any sort of primitive type, such as numbers, vectors, and others. In the case of images, they were just `cv::Mat` class objects built to contain fixed-length vectors such as `Vec3b`. This class has a wide variety of member functions that simplify the expression of many simple operations. For other common operations on arrays, a wide variety of functions exist, which we also covered in this chapter. Along the way we learned about sparse matrices, and saw that they could be used in almost any place regular `cv::Mat` structures can be used, just as we could use STL vector objects with most functions. Finally, we took a moment to dig a little deeper into the exact functioning of the template classes for large array types. There we learned that, while the primitive types are derived from their templates, the large array templates are instead derived from the basic class.

# Exercises

1. Create a 500 × 500 single channel `uchar` image with every pixel equal to zero.
   a. Create an ASCII numeric typewriter where you can type numbers into your computer and have the number show up in a 20-pixel-high by 10-pixel-wide block. As you type, the numbers will display from left to right until you hit the end of the image. Then just stop.
   b. Allow for carriage return and backspace.
   c. Allow for arrow keys to edit each number.
   d. Create a key that will convert the resulting image into a color image, each number taking on a different color.
2. We want to create a function that makes it efficient to sum up rectangular regions in an image by creating a statistics image where each "pixel" holds the sum of the rectangle from that point to the image origin. These are called *integral images* and by using just 4 points from the integral image, you can determine the sum of any rectangle in the image.
   a. Create a 100 × 200 single-channel `uchar` image with random numbers. Create a 100 × 200 single-channel `float` "integral image" with all members equal to zero.
   b. Fill in each element of the integral image with the corresponding sum of the rectangle from that pixel to the origin in the original `uchar` image.

c. How can you do part b) very efficiently in one pass using the integral numbers you've already calculated in the integral image plus the new number being added in the original image? Implement this efficient method.

d. Use the integral image to rapidly calculate the sum of pixels in any rectangle in the original image.

e. How can you modify the integral image so that you can compute the sum of a 45-degree rotated rectangle in the original image very efficiently? Describe the algorithm.

CHAPTER 5

# Array Operations

## More Things You Can Do with Arrays

As we saw in the previous chapter, there are many basic operations on arrays that are now handled by member functions of the array classes. In addition to those, however, there are many more operations that are most naturally represented as "friend" functions that either take array types as arguments, have array types as return values, or both. The functions, together with their arguments, will be covered in more detail after Table 5-1.

*Table 5-1. Basic matrix and image operators*

| Function | Description |
|---|---|
| `cv::abs()` | Return absolute value of all elements in an array |
| `cv::absdiff()` | Return absolute value of differences between two arrays |
| `cv::add()` | Perform element-wise addition of two arrays |
| `cv::addWeighted()` | Perform element-wise weighted addition of two arrays (alpha blending) |
| `cv::bitwise_and()` | Compute element-wise bit-level AND of two arrays |
| `cv::bitwise_not()` | Compute element-wise bit-level NOT of two arrays |
| `cv::bitwise_or()` | Compute element-wise bit-level OR of two arrays |
| `cv::bitwise_xor()` | Compute element-wise bit-level XOR of two arrays |
| `cv::calcCovarMatrix()` | Compute covariance of a set of *n*-dimensional vectors |
| `cv::cartToPolar()` | Compute angle and magnitude from a two-dimensional vector field |
| `cv::checkRange()` | Check array for invalid values |
| `cv::compare()` | Apply selected comparison operator to all elements in two arrays |
| `cv::completeSymm()` | Symmetrize matrix by copying elements from one half to the other |
| `cv::convertScaleAbs()` | Scale array, take absolute value, then convert to 8-bit unsigned |
| `cv::countNonZero()` | Count nonzero elements in an array |

| Function | Description |
| --- | --- |
| `cv::arrToMat()` | Convert pre–version 2.1 array types to `cv::Mat` |
| `cv::dct()` | Compute discrete cosine transform of array |
| `cv::determinant()` | Compute determinant of a square matrix |
| `cv::dft()` | Compute discrete Fourier transform of array |
| `cv::divide()` | Perform element-wise division of one array by another |
| `cv::eigen()` | Compute eigenvalues and eigenvectors of a square matrix |
| `cv::exp()` | Perform element-wise exponentiation of array |
| `cv::extractImageCOI()` | Extract single channel from pre–version 2.1 array type |
| `cv::flip()` | Flip an array about a selected axis |
| `cv::gemm()` | Perform generalized matrix multiplication |
| `cv::getConvertElem()` | Get a single-pixel type conversion function |
| `cv::getConvertScaleElem()` | Get a single-pixel type conversion and scale function |
| `cv::idct()` | Compute inverse discrete cosine transform of array |
| `cv::idft()` | Compute inverse discrete Fourier transform of array |
| `cv::inRange()` | Test if elements of an array are within values of two other arrays |
| `cv::invert()` | Invert a square matrix |
| `cv::log()` | Compute element-wise natural log of array |
| `cv::magnitude()` | Compute magnitudes from a two-dimensional vector field |
| `cv::LUT()` | Convert array to indices of a lookup table |
| `cv::Mahalanobis()` | Compute Mahalanobis distance between two vectors |
| `cv::max()` | Compute element-wise maxima between two arrays |
| `cv::mean()` | Compute the average of the array elements |
| `cv::meanStdDev()` | Compute the average and standard deviation of the array elements |
| `cv::merge()` | Merge several single-channel arrays into one multichannel array |
| `cv::min()` | Compute element-wise minima between two arrays |
| `cv::minMaxLoc()` | Find minimum and maximum values in an array |
| `cv::mixChannels()` | Shuffle channels from input arrays to output arrays |
| `cv::mulSpectrums()` | Compute element-wise multiplication of two Fourier spectra |
| `cv::multiply()` | Perform element-wise multiplication of two arrays |
| `cv::mulTransposed()` | Calculate matrix product of one array |
| `cv::norm()` | Compute normalized correlations between two arrays |
| `cv::normalize()` | Normalize elements in an array to some value |
| `cv::perspectiveTransform()` | Perform perspective matrix transform of a list of vectors |
| `cv::phase()` | Compute orientations from a two-dimensional vector field |
| `cv::polarToCart()` | Compute two-dimensional vector field from angles and magnitudes |
| `cv::pow()` | Raise every element of an array to a given power |
| `cv::randu()` | Fill a given array with uniformly distributed random numbers |
| `cv::randn()` | Fill a given array with normally distributed random numbers |
| `cv::randShuffle()` | Randomly shuffle array elements |

| Function | Description |
|---|---|
| `cv::reduce()` | Reduce a two-dimensional array to a vector by a given operation |
| `cv::repeat()` | Tile the contents of one array into another |
| `cv::saturate_cast<>()` | Convert primitive types (template function) |
| `cv::scaleAdd()` | Compute element-wise sum of two arrays with optional scaling of the first |
| `cv::setIdentity()` | Set all elements of an array to `1` for the diagonal and `0` otherwise |
| `cv::solve()` | Solve a system of linear equations |
| `cv::solveCubic()` | Find the (only) real roots of a cubic equation |
| `cv::solvePoly()` | Find the complex roots of a polynomial equation |
| `cv::sort()` | Sort elements in either the rows or columns in an array |
| `cv::sortIdx()` | Serve same purpose as `cv::sort()`, except array is unmodified and indices are returned |
| `cv::split()` | Split a multichannel array into multiple single-channel arrays |
| `cv::sqrt()` | Compute element-wise square root of an array |
| `cv::subtract()` | Perform element-wise subtraction of one array from another |
| `cv::sum()` | Sum all elements of an array |
| `cv::theRNG()` | Return a random number generator |
| `cv::trace()` | Compute the trace of an array |
| `cv::transform()` | Apply matrix transformation on every element of an array |
| `cv::transpose()` | Transpose all elements of an array across the diagonal |

In these functions, some general rules are followed. To the extent that any exceptions exist, they are noted in the function descriptions. Because one or more of these rules applies to just about every function described in this chapter, they are listed here for convenience:

*Saturation*
: Outputs of calculations are saturation-casted to the type of the output array.

*Output*
: The output array will be created with `cv::Mat::create()` if its type and size do not match the required type or size. Usually the required output type and size are the same as inputs, but for some functions size may be different (e.g., `cv::transpose`) or type may be different (e.g., `cv::split`).

*Scalars*
: Many functions such as `cv::add()` allow for the addition of two arrays or an array and a scalar. Where the prototypes make the option clear, the result of providing a scalar argument is the same as if a second array had been provided with the same scalar value in every element.

*Masks*
: Whenever a mask argument is present for a function, the output will be computed only for those elements where the mask value corresponding to that element in the output array is nonzero.

`dtype`
: Many arithmetic and similar functions do not require the types of the input arrays to be the same, and even if they are the same, the output array may be of a different type than the inputs. In these cases, the output array must have its depth explicitly specified. This is done with the `dtype` argument. When present, `dtype` can be set to any of the basic types (e.g., `CV_32F`) and the result array will be of that type. If the input arrays have the same type, then `dtype` can be set to its default value of `-1`, and the resulting type will be the same as the types of the input arrays.

*In-place operation*
: Unless otherwise specified, any operation with an array input and an array output that are of the same size and type can use the same array for both (i.e., it is allowable to write the output on top of an input).

*Multichannel*
: For those operations that do not naturally make use of multiple channels, if given multichannel arguments, each channel is processed separately.

## cv::abs()

```
cv::MatExpr cv::abs( cv::InputArray src );
cv::MatExpr cv::abs( const cv::MatExpr& src );          // Matrix expression
```

These functions compute the absolute value of an array or of some expression of arrays. The most common usage computes the absolute value of every element in an array. Because `cv::abs()` can take a matrix expression, it is able to recognize certain special cases and handle them appropriately. In fact, calls to `cv::abs()` are actually converted to calls to `cv::absDiff()` or other functions, and handled by those functions. In particular, the following special cases are implemented:

- `m2 = cv::abs( m0 - m1 )` is converted to `cv::absdiff( m0, m1, m2 )`
- `m2 = cv::abs( m0 )` is converted to `m2 = cv::absdiff( m0, cv::Scalar::all(0), m2 )`
- `m2 = cv::Mat_<Vec<uchar,n> >( cv::abs( alpha*m0 + beta ) )` (for alpha, beta real numbers) is converted to `cv::convertScaleAbs( m0, m2, alpha, beta )`

The third case might seem obscure, but this is just the case of computing a scale and offset (either of which could be trivial) to an *n*-channel array. This is typical of what you might do when computing a contrast correction for an image, for example.

In the cases that are implemented by `cv::absdiff()`, the result array will have the same size and type as the input array. In the case implemented by `cv::convertScaleAbs()`, however, the result type of the return array will always be `CV_8U`.

## cv::absdiff()

```
void cv::absdiff(
  cv::InputArray  src1,                     // First input array
  cv::InputArray  src2,                     // Second input array
  cv::OutputArray dst                       // Result array
)
```

$$dst_i = saturate(\,|\, src1_i - src2_i \,|\,)$$

Given two arrays, `cv::absdiff()` computes the difference between each pair of corresponding elements in those arrays, and places the absolute value of that difference into the corresponding element of the destination array.

## cv::add()

```
void cv::add(
  cv::InputArray  src1,                     // First input array
  cv::InputArray  src2,                     // Second input array
  cv::OutputArray dst,                      // Result array
  cv::InputArray  mask  = cv::noArray(),    // Optional, do only where nonzero
  int             dtype = -1                // Output type for result array
);
```

$$dst_i = saturate(src1_i + src2_i)$$

`cv::add()` is a simple addition function: it adds all of the elements in `src1` to the corresponding elements in `src2` and puts the results in `dst`.

For simple cases, the same result can be achieved with the matrix operation:

```
dst = src1 + src2;
```

Accumulation is also supported:

```
dst += src1;
```

## cv::addWeighted()

```
void cv::addWeighted(
  cv::InputArray  src1,               // First input array
  double          alpha,              // Weight for first input array
  cv::InputArray  src2,               // Second input array
  double          beta,               // Weight for second input array
  double          gamma,              // Offset added to weighted sum
  cv::OutputArray dst,                // Result array
  int             dtype = -1          // Output type for result array
);
```

The function `cv::addWeighted()` is similar to `cvAdd()` except that the result written to `dst` is computed according to the following formula:

$$dst_i = saturate(src1_i * \alpha + src2_i * \beta + \gamma)$$

The two source images, `src1` and `src2`, may be of any pixel type as long as both are of the same type. They may also have any number of channels (grayscale, color, etc.) as long as they agree.

This function can be used to implement *alpha blending* [Smith79; Porter84]; that is, it can be used to blend one image with another. In this case, the parameter `alpha` is the blending strength of `src1`, and `beta` is the blending strength of `src2`. You can convert to the standard alpha blend equation by setting $\alpha$ between `0` and `1`, setting $\beta$ to `1` $- \alpha$, and setting $\gamma$ to `0`; this yields:

$$dst_i = saturate(src1_i * \alpha + src2_i * (1 - \alpha))$$

However, `cv::addWeighted()` gives us a little more flexibility—both in how we weight the blended images and in the additional parameter $\gamma$, which allows for an additive offset to the resulting destination image. For the general form, you will probably want to keep `alpha` and `beta` at `0` or above, and their sum at no more than `1`; `gamma` may be set depending on average or max image value to scale the pixels up. A program that uses alpha blending is shown in Example 5-1.

*Example 5-1. Complete program to alpha-blend the ROI starting at (0,0) in src2 with the ROI starting at (x,y) in src1*

```
// alphablend <imageA> <image B> <x> <y> <width> <height> <alpha> <beta>
//
#include <cv.h>
#include <highgui.h>

int main(int argc, char** argv) {

  cv::Mat src1 = cv::imread(argv[1],1);
  cv::Mat src2 = cv::imread(argv[2],1);

  if( argc==9 && !src1.empty() && !src2.empty() ) {

    int    x     = atoi(argv[3]);
    int    y     = atoi(argv[4]);
    int    w     = atoi(argv[5]);
    int    h     = atoi(argv[6]);
    double alpha = (double)atof(argv[7]);
    double beta  = (double)atof(argv[8]);

    cv::Mat roi1( src1, cv::Rect(x,y,w,h) );
    cv::Mat roi2( src2, cv::Rect(0,0,w,h) );

    cv::addWeighted( roi1, alpha, roi2, beta, 0.0, roi1 );

    cv::namedWindow( "Alpha Blend", 1 );
    cv::imshow( "Alpha Blend", src2 );
    cv::waitKey( 0 );
  }

  return 0;
}
```

The code in Example 5-1 takes two source images: the primary one (`src1`) and the one to blend (`src2`). It reads in a rectangle ROI for `src1` and applies an ROI of the same size to `src2`, but located at the origin. It reads in `alpha` and `beta` levels but sets `gamma` to `0`. Alpha blending is applied using `cv::addWeighted()`, and the results are put into `src1` and displayed. Example output is shown in Figure 5-1, where the face of a child is blended onto a cat. Note that the code took the same ROI as in the ROI addition in Example 5-1. This time we used the ROI as the target blending region.

*Figure 5-1. The face of a child, alpha-blended onto the face of a cat*

## cv::bitwise_and()

```
void cv::bitwise_and(
  cv::InputArray  src1,                    // First input array
  cv::InputArray  src2,                    // Second input array
  cv::OutputArray dst,                     // Result array
  cv::InputArray  mask  = cv::noArray(),   // Optional, do only where nonzero
);
```

$$dst_i = src1_i \wedge src2_i$$

`cv::bitwise_and()` is a per-element bitwise conjunction operation. For every element in `src1`, the bitwise AND is computed with the corresponding element in `src2` and put into the corresponding element of `dst`.

If you are not using a mask, the same result can be achieved with the matrix operation:

```
dst = src1 & src2;
```

## cv::bitwise_not()

```
void cv::bitwise_not(
  cv::InputArray  src,                    // Input array
  cv::OutputArray dst,                    // Result array
  cv::InputArray  mask  = cv::noArray(),  // Optional, do only where nonzero
);
```

$$dst_i = \sim src1_i$$

`cv::bitwise_not()` is a per-element bitwise inversion operation. For every element in `src1`, the logical inversion is computed and placed into the corresponding element of `dst`.

If you are not using a mask, the same result can be achieved with the matrix operation:

```
dst = !src1;
```

## cv::bitwise_or()

```
void cv::bitwise_and(
  cv::InputArray  src1,                   // First input array
  cv::InputArray  src2,                   // Second input array
  cv::OutputArray dst,                    // Result array
  cv::InputArray  mask  = cv::noArray(),  // Optional, do only where nonzero
);
```

$$dst_i = src1_i \vee src2_i$$

`cv::bitwise_or()` is a per-element bitwise disjunction operation. For every element in `src1`, the bitwise OR is computed with the corresponding element in `src2` and put into the corresponding element of `dst`.

If you are not using a mask, the same result can be achieved with the matrix operation:

```
dst = src1 | src2;
```

## cv::bitwise_xor()

```
void cv::bitwise_and(
  cv::InputArray  src1,                   // First input array
  cv::InputArray  src2,                   // Second input array
  cv::OutputArray dst,                    // Result array
  cv::InputArray  mask  = cv::noArray(),  // Optional, do only where nonzero
);
```

$$dst_i = src1_i \oplus src2_i$$

cv::bitwise_and() is a per-element bitwise "exclusive or" (XOR) operation. For every element in src1, the bitwise XOR is computed with the corresponding element in src2 and put into the corresponding element of dst.

If you are not using a mask, the same result can be achieved with the matrix operation:

```
dst = src1 ^ src2;
```

## cv::calcCovarMatrix()

```
void cv::calcCovarMatrix(
  const cv::Mat* samples,          // C-array of n-by-1 or 1-by-n matrices
  int            nsamples,         // num matrices pointed to by 'samples'
  cv::Mat&       covar,            // ref to return array for covariance
  cv::Mat&       mean,             // ref to return array for mean
  int            flags,            // special variations, see Table 5-2.
  int            ctype = cv::F64   // output matrix type for covar
);

void cv::calcCovarMatrix(
  cv::InputArray samples,          // n-by-m matrix, but see 'flags' below
  cv::Mat&       covar,            // ref to return array for covariance
  cv::Mat&       mean,             // ref to return array for mean
  int            flags,            // special variations, see Table 5-2.
  int            ctype = cv::F64   // output matrix type for covar
);
```

Given any number of vectors, cv::calcCovarMatrix() will compute the mean and *covariance matrix* for the Gaussian approximation to the distribution of those points. This can be used in many ways, of course, and OpenCV has some additional flags that will help in particular contexts (see Table 5-2). These flags may be combined by the standard use of the Boolean OR operator.

*Table 5-2. Possible components of flags argument to cv::calcCovarMatrix()*

| Flag in flags argument | Meaning |
|---|---|
| `cv::COVAR_NORMAL` | Compute mean and covariance |
| `cv::COVAR_SCRAMBLED` | Fast PCA "scrambled" covariance |
| `cv::COVAR_USE_AVERAGE` | Use mean as input instead of computing it |
| `cv::COVAR_SCALE` | Rescale output covariance matrix |
| `cv::COVAR_ROWS` | Use rows of `samples` for input vectors |
| `cv::COVAR_COLS` | Use columns of `samples` for input vectors |

There are two basic calling conventions for `cv::calcCovarMatrix()`. In the first, a pointer to an array of `cv::Mat` objects is passed along with `nsamples`, the number of matrices in that array. In this case, the arrays may be $n \times 1$ or $1 \times n$. The second calling convention is to pass a single array that is $n \times m$. In this case, either the flag `cv::COVAR_ROWS` should be supplied, to indicate that there are $n$ (row) vectors of length $m$, or `cv::COVAR_COLS` should be supplied, to indicate that there are $m$ (column) vectors of length $n$.

The results will be placed in `covar` in all cases, but the exact meaning of `avg` depends on the flag values (see Table 5-2).

The flags `cv::COVAR_NORMAL` and `cv::COVAR_SCRAMBLED` are mutually exclusive; you should use one or the other but not both. In the case of `cv::COVAR_NORMAL`, the function will simply compute the mean and covariance of the points provided:

$$\Sigma^2_{normal} = z \begin{bmatrix} v_{0,0} - \bar{v}_o & \cdots & v_{m,0} - \bar{v}_o \\ \vdots & \ddots & \vdots \\ v_{0,n} - \bar{v}_n & \cdots & v_{m,n} - \bar{v}_n \end{bmatrix} \begin{bmatrix} v_{0,0} - \bar{v}_o & \cdots & v_{m,0} - \bar{v}_o \\ \vdots & \ddots & \vdots \\ v_{0,n} - \bar{v}_n & \cdots & v_{m,n} - \bar{v}_n \end{bmatrix}^T.$$

Thus the normal covariance $\Sigma^2_{normal}$ is computed from the $m$ vectors of length $n$, where $\bar{v}_n$ is defined as the $n$th element of the average vector: $\bar{v}$. The resulting covariance matrix is an $n \times n$ matrix. The factor $z$ is an optional scale factor; it will be set to `1` unless the `cv::COVAR_SCALE` flag is used.

In the case of `cv::COVAR_SCRAMBLED`, `cv::calcCovarMatrix()` will compute the following:

$$\Sigma^2_{scrambled} = z \begin{bmatrix} v_{0,0} - \bar{v}_o & \cdots & v_{m,0} - \bar{v}_o \\ \vdots & \ddots & \vdots \\ v_{0,n} - \bar{v}_n & \cdots & v_{m,n} - \bar{v}_n \end{bmatrix}^T \begin{bmatrix} v_{0,0} - \bar{v}_o & \cdots & v_{m,0} - \bar{v}_o \\ \vdots & \ddots & \vdots \\ v_{0,n} - \bar{v}_n & \cdots & v_{m,n} - \bar{v}_n \end{bmatrix}.$$

This matrix is not the usual covariance matrix (note the location of the transpose operator). This matrix is computed from the same $m$ vectors of length $n$, but the resulting *scrambled covariance* matrix is an $m \times m$ matrix. This matrix is used in some specific algorithms such as fast PCA for very large vectors (as in the *eigenfaces* technique for face recognition).

The flag `cv::COVAR_USE_AVG` is used when the mean of the input vectors is already known. In this case, the argument `avg` is used as an input rather than an output, which reduces computation time.

Finally, the flag `cv::COVAR_SCALE` is used to apply a uniform scale to the covariance matrix calculated. This is the factor $z$ in the preceding equations. When used in conjunction with the `cv::COVAR_NORMAL` flag, the applied scale factor will be `1.0/`$m$ (or, equivalently, `1.0/nsamples`). If instead `cv::COVAR_SCRAMBLED` is used, then the value of $z$ will be `1.0/`$n$ (the inverse of the length of the vectors).

The input and output arrays to `cv::calcCovarMatrix()` should all be of the same floating-point type. The size of the resulting matrix `covar` will be either $n \times n$ or $m \times m$ depending on whether the standard or scrambled covariance is being computed. It should be noted that when you are using the `cv::Mat*` form, the "vectors" input in `samples` do not actually have to be one-dimensional; they can be two-dimensional objects (e.g., images) as well.

## cv::cartToPolar()

```
void cv::cartToPolar(
  cv::InputArray  x,
  cv::InputArray  y,
  cv::OutputArray magnitude,
  cv::OutputArray angle,
  bool            angleInDegrees = false
);
```

$$magnitude_i = \sqrt{x_i^{\,2} + y_i^{\,2}}$$

$$angle_i = atan2(y_i, x_i)$$

This function `cv::cartToPolar()` takes two input arrays, `x` and `y`, which are taken to be the x- and y-components of a vector field (note that this is not a single two-channel array, but two separate arrays). The arrays `x` and `y` must be of the same size. `cv::cartToPolar()` then computes the polar representation of each of those vectors. The magnitude of each vector is placed into the corresponding location in `magnitude`, and the orientation of each vector is placed into the corresponding location in angle. The returned angles are in radians unless the Boolean variable angleInDegrees is set to true.

## cv::checkRange()

```
bool cv::checkRange(
  cv::InputArray src,
  bool           quiet  = true,
  cv::Point*     pos    = 0,           // if non-Null, location of first outlier
  double         minVal = -DBL_MAX,    // Lower check bound (inclusive)
  double         maxVal =  DBL_MAX     // Upper check bound (exclusive)
);
```

This function cv::checkRange() tests every element of the input array src and determines if that element is in a given range. The range is set by minVal and maxVal, but any NaN or inf value is also considered out of range. If an out-of-range value is found, an exception will be thrown unless quiet is set to true, in which case the return value of cv::checkRange() will be true if all values are in range and false if any value is out of range. If the pointer pos is not NULL, then the location of the first outlier will be stored in pos.

## cv::compare()

```
bool cv::compare(
  cv::InputArray  src1,              // First input array
  cv::InputArray  src2,              // Second input array
  cv::OutputArray dst,               // Result array
  int             cmpop              // Comparison operator, see Table 5-3.
);
```

This function makes element-wise comparisons between corresponding pixels in two arrays, src1 and src2, and places the results in the image dst. cv::compare() takes as its last argument a comparison operator, which may be any of the types listed in Table 5-3. In each case, the result dst will be an 8-bit array where pixels that match are marked with 255 and mismatches are set to 0.

*Table 5-3. Values of cmpop used by cv::compare() and the resulting comparison operation performed*

| Value of cmp_op | Comparison |
|---|---|
| cv::CMP_EQ | (src1i == src2i) |
| cv::CMP_GT | (src1i > src2i) |
| cv::CMP_GE | (src1i >= src2i) |
| cv::CMP_LT | (src1i < src2i) |
| cv::CMP_LE | (src1i <= src2i) |
| cv::CMP_NE | (src1i != src2i) |

All the listed comparisons are done with the same functions; you just pass in the appropriate argument to indicate what you would like done.

These comparison functions are useful, for example, in background subtraction to create a mask of the changed pixels (e.g., from a security camera) such that only novel information is pulled out of the image.

These same results can be achieved with the matrix operations:

```
dst = src1 == src2;
dst = src1  > src2;
dst = src1 >= src2;
dst = src1  < src2;
dst = src1 <= src2;
dst = src1 != src2;
```

## cv::completeSymm()

```
bool cv::completeSymm(
  cv::InputArray mtx,
  bool           lowerToUpper = false
);
```

$$mtx_{ij} = mtx_{ji} \forall\, i > j \; (\texttt{lowerToUpper = false})$$

$$mtx_{ij} = mtx_{ji} \forall\, j < i \; (\texttt{lowerToUpper = true})$$

Given a matrix (an array of dimension two) `mtx`, `cv::completeSymm()` symmetrizes the matrix by copying.[1] Specifically, all of the elements from the lower triangle are copied to their transpose position on the upper triangle of the matrix. The diagonal elements of the `mtx` are left unchanged. If the flag `lowerToUpper` is set to `true`, then the elements from the lower triangle are copied into the upper triangle instead.

## cv::convertScaleAbs()

```
void cv::convertScaleAbs(
  cv::InputArray  src,                  // Input array
  cv::OutputArray dst,                  // Result array
  double          alpha = 1.0,          // Multiplicative scale factor
  double          beta  = 0.0           // Additive offset factor
);
```

1 Mathematically inclined readers will realize that there are other symmetrizing processes for matrices that are more "natural" than this operation, but this particular operation is useful in its own right—for example, to complete a symmetric matrix when only half of it was computed—and so is exposed in the library.

$$dst_i = saturate_{uchar}(\ |\ \alpha^* src_i + \beta\ |\ )$$

The `cv::convertScaleAbs()` function is actually several functions rolled into one; it will perform four operations in sequence. The first operation is to rescale the source image by the factor `a`, the second is to offset by (add) the factor `b`, the third is to compute the absolute value of that sum, and the fourth is to cast that result (with saturation) to an unsigned char (8-bit).

When you simply pass the default values (`alpha = 1.0` or `beta = 0.0`), you need not have performance fears; OpenCV is smart enough to recognize these cases and not waste processor time on useless operations.

A similar result can be achieved, with somewhat greater generality, through the following loop:

```
for( int i = 0; i < src.rows; i++ )
    for( int j = 0; j < src.cols*src.channels(); j++ )
        dst.at<dst_type>(i, j) = satuarate_cast<dst_type>(
            (double)src.at<src_type>(i, j) * alpha + beta
        );
```

## cv::countNonZero()

```
int cv::countNonZero(                  // Return number of nonzero elements in mtx
  cv::InputArray mtx,                  // Input array
);
```

$$count = \sum_{mtx_i \neq 0} 1$$

`cv::countNonZero()` returns the number of nonzero pixels in the array `mtx`.

## cv::cvarrToMat()

```
cv::Mat cv::cvarrToMat(
  const CvArr* src,                 // Input array: CvMat, IplImage, or CvMatND
  bool         copyData = false, // if false just make new header, else copy data
  bool         allowND  = true,  // if true and possible, convert CvMatND to Mat
  int          coiMode  = 0      // if 0: error if COI set, if 1: ignore COI
);
```

`cv::cvarrToMat()` is used when you have an "old-style" (pre–version 2.1) image or matrix type and you want to convert it to a "new-style" (version 2.1 or later, which uses C++) `cv::Mat` object. By default, only the header for the new array is constructed without copying the data. Instead, the data pointers in the new header point to the existing data array (so do not deallocate it while the `cv::Mat` header is in use). If you want the data copied, just set copyData to true, and then you can freely do away with the original data object.

cv::cvarrToMat() can also take CvMatND structures, but it cannot handle all cases. The key requirement for the conversion is that the matrix should be continuous, or at least representable as a sequence of continuous matrices. Specifically, A.dim[i].size*A.dim.step[i] should be equal to A.dim.step[i-1] for all i, or at worst all but one. If allowND is set to true (default), cv::cvarrToMat() will attempt the conversion when it encounters a CvMatND structure, and throw an exception if that conversion is not possible (the preceding condition). If allowND is set to false, then an exception will be thrown whenever a CvMatND structure is encountered.

Because the concept of COI[2] is handled differently in the post–version 2.1 library (which is to say, it no longer exists), COI has to be handled during the conversion. If the argument coiMode is 0, then an exception will be thrown when src contains an active COI. If coiMode is nonzero, then no error will be reported, and instead a cv::Mat header corresponding to the entire image will be returned, ignoring the COI. (If you want to handle COI properly, you will have to check whether the image has the COI set, and if so, use cv::extractImageCOI() to create a header for just that channel.)

Most of the time, this function is used to help migrate old-style code to the new. In such cases, you will probably need to both convert old-style CvArr* structures to cv::Mat, as well as doing the reverse operation. The reverse operation is done using cast operators. If, for example, you have a matrix you defined as cv::Mat A, you can convert that to an IplImage* pointer simply with:

```
Cv::Mat A( 640, 480, cv::8UC3 );
// casting is implicit on assignment
IplImage. my_img = A;
iplImage* img        = &my_img;
```

## cv::dct()

```
void cv::dct(
  cv::InputArray  src,                  // Input array
  cv::OutputArray dst,                  // Result array
  int             flags,                // for inverse transform or row-by-row
);
```

2 "COI" is an old concept from the pre-v2 library that meant "channel of interest." In the old IplImage class, this COI was analogous to ROI (region of interest), and could be set to cause certain functions to act only on the indicated channel.

This function performs both the discrete cosine transform and the inverse transform depending on the `flags` argument. The source array `src` must be either one- or two-dimensional, and the size should be an even number (you can pad the array if necessary). The result array `dst` will have the same type and size as `src`. The argument `flags` is a bit field and can be set to one or both of `DCT_INVERSE` or `DCT_ROWS`. If `DCT_INVERSE` is set, then the inverse transform is done instead of the forward transform. If the flag `DCT_ROWS` is set, then a two-dimensional $n \times m$ input is treated as $n$ distinct one-dimensional vectors of length $m$. In this case, each such vector will be transformed independently.

The performance of `cv::dct()` depends strongly on the exact size of the arrays passed to it, and this relationship is not monotonic. There are just some sizes that work better than others. It is recommended that when passing an array to `cv::dct()`, you first determine the most optimal size that is larger than your array, and extend your array to that size. OpenCV provides a convenient routine to compute such values for you, called `cv::getOptimalDFTSize()`.

As implemented, the discrete cosine transform of a vector of length $n$ is computed via the discrete Fourier transform (`cv::dft()`) on a vector of length $n/2$. This means that to get the optimal size for a call to `cv::dct()`, you should compute it like this:

```
size_t opt_dft_size = 2 * cv::getOptimalDFTSize((N+1)/2);
```

This function (and discrete transforms in general) is covered in much greater detail in Chapter 11. In that chapter, we will discuss the details of how to pack and unpack the input and output, as well as when and why you might want to use the discrete cosine transform.

## cv::dft()

```
void cv::dft(
  cv::InputArray  src,                    // Input array
  cv::OutputArray dst,                    // Result array
  int             flags       = 0,        // for inverse transform or row-by-row
  int             nonzeroRows = 0         // only this many entries are nonzero
);
```

The `cv::dft()` function performs both the discrete Fourier transform as well as the inverse transform (depending on the `flags` argument). The source array `src` must be either one- or two-dimensional. The result array `dst` will have the same type and size as `src`. The argument `flags` is a bit field and can be set to one or more of `DFT_INVERSE`, `DFT_ROWS`, `DFT_SCALE`, `DFT_COMPLEX_OUTPUT`, or `DFT_REAL_OUTPUT`. If

DFT_INVERSE is set, then the inverse transform is done. If the flag DFT_ROWS is set, then a two-dimensional $n \times m$ input is treated as $n$ distinct one-dimensional vectors of length $m$ and each such vector will be transformed independently. The flag DFT_SCALE normalizes the results by the number of elements in the array. This is typically done for DFT_INVERSE, as it guarantees that the inverse of the inverse will have the correct normalization.

The flags DFT_COMPLEX_OUTPUT and DFT_REAL_OUTPUT are useful because when the Fourier transform of a real array is computed, the result will have a complex-conjugate symmetry. So, even though the result is complex, the number of array elements that result will be equal to the number of elements in the real input array rather than double that number. Such a packing is the default behavior of cv::dft(). To force the output to be in complex form, set the flag DFT_COMPLEX_OUTPUT. In the case of the inverse transform, the input is (in general) complex, and the output will be as well. However, if the input array (to the inverse transform) has complex-conjugate symmetry (for example, if it was itself the result of a Fourier transform of a real array), then the inverse transform will be a real array. If you know this to be the case and you would like the result array represented as a real array (thereby using half the amount of memory), you can set the DFT_REAL_OUTPUT flag. (Note that if you do set this flag, cv::dft() does not check that the input array has the necessary symmetry; it simply assumes that it does.)

The last parameter to cv::dft() is nonzeroRows. This defaults to 0, but if set to any nonzero value, will cause cv::dft() to assume that only the first nonzeroRows of the input array are actually meaningful. (If DFT_INVERSE is set, then it is only the first nonzeroRows of the output array that are assumed to be nonzero.) This flag is particularly handy when you are computing cross-correlations of convolutions using cv::dft().

The performance of cv::dft() depends strongly on the exact size of the arrays passed to it, and this relationship is not monotonic. There are just some sizes that work better than others. It is recommended that when passing an array to cv::dft(), you first determine the most optimal size larger than your array, and extend your array to that size. OpenCV provides a convenient routine to compute such values for you, called cv::getOptimalDFTSize().

Again, this function (and discrete transforms in general) is covered in much greater detail in Chapter 11. In that chapter, we will discuss the details of how to pack and unpack the input and output, as well as when and why you might want to use the discrete Fourier transform.

## cv::cvtColor()

```
void cv::cvtColor(
  cv::InputArray  src,                    // Input array
  cv::OutputArray dst,                    // Result array
  int             code,                   // color mapping code, see Table 5-4.
  int             dstCn = 0               // channels in output (0='automatic')
);
```

`cv::cvtColor()` is used to convert from one color space (number of channels) to another [Wharton71] while retaining the same data type. The input array `src` can be an 8-bit array, a 16-bit unsigned array, or a 32-bit floating-point array. The output array `dst` will have the same size and depth as the input array. The conversion operation to be done is specified by the `code` argument, with possible values shown in Table 5-4.[3] The final parameter, `dstCn`, is the desired number of channels in the destination image. If the default value of `0` is given, then the number of channels is determined by the number of channels in `src` and the conversion code.

*Table 5-4. Conversions available by means of cv::cvtColor()*

| Conversion code | Meaning |
|---|---|
| `cv::COLOR_BGR2RGB`<br>`cv::COLOR_RGB2BGR`<br>`cv::COLOR_RGBA2BGRA`<br>`cv::COLOR_BGRA2RGBA` | Convert between RGB and BGR color spaces (with or without alpha channel) |
| `cv::COLOR_RGB2RGBA`<br>`cv::COLOR_BGR2BGRA` | Add alpha channel to RGB or BGR image |
| `cv::COLOR_RGBA2RGB`<br>`cv::COLOR_BGRA2BGR` | Remove alpha channel from RGB or BGR image |
| `cv::COLOR_RGB2BGRA`<br>`cv::COLOR_RGBA2BGR`<br>`cv::COLOR_BGRA2RGB`<br>`cv::COLOR_BGR2RGBA` | Convert RGB to BGR color spaces while adding or removing alpha channel |
| `cv::COLOR_RGB2GRAY`<br>`cv::COLOR_BGR2GRAY` | Convert RGB or BGR color spaces to grayscale |
| `cv::COLOR_GRAY2RGB`<br>`cv::COLOR_GRAY2BGR`<br>`cv::COLOR_RGBA2GRAY`<br>`cv::COLOR_BGRA2GRAY` | Convert grayscale to RGB or BGR color spaces (optionally removing alpha channel in the process) |
| `cv::COLOR_GRAY2RGBA`<br>`cv::COLOR_GRAY2BGRA` | Convert grayscale to RGB or BGR color spaces and add alpha channel |

3 Long-time users of IPL should note that the function `cvCvtColor()` ignores the `colorModel` and `channelSeq` fields of the `IplImage` header. The conversions are done exactly as implied by the `code` argument.

| Conversion code | Meaning |
|---|---|
| cv::COLOR_RGB2BGR565<br>cv::COLOR_BGR2BGR565<br>cv::COLOR_BGR5652RGB<br>cv::COLOR_BGR5652BGR<br>cv::COLOR_RGBA2BGR565<br>cv::COLOR_BGRA2BGR565<br>cv::COLOR_BGR5652RGBA<br>cv::COLOR_BGR5652BGRA | Convert from RGB or BGR color space to BGR565 color representation with optional addition or removal of alpha channel (16-bit images) |
| cv::COLOR_GRAY2BGR565<br>cv::COLOR_BGR5652GRAY | Convert grayscale to BGR565 color representation or vice versa (16-bit images) |
| cv::COLOR_RGB2BGR555<br>cv::COLOR_BGR2BGR555<br>cv::COLOR_BGR5552RGB<br>cv::COLOR_BGR5552BGR<br>cv::COLOR_RGBA2BGR555<br>cv::COLOR_BGRA2BGR555<br>cv::COLOR_BGR5552RGBA<br>cv::COLOR_BGR5552BGRA | Convert from RGB or BGR color space to BGR555 color representation with optional addition or removal of alpha channel (16-bit images) |
| cv::COLOR_GRAY2BGR555<br>cv::COLOR_BGR5552GRAY | Convert grayscale to BGR555 color representation or vice versa (16-bit images) |
| cv::COLOR_RGB2XYZ<br>cv::COLOR_BGR2XYZ<br>cv::COLOR_XYZ2RGB<br>cv::COLOR_XYZ2BGR | Convert RGB or BGR image to CIE XYZ representation or vice versa (Rec 709 with D65 white point) |
| cv::COLOR_RGB2YCrCb<br>cv::COLOR_BGR2YCrCb<br>cv::COLOR_YCrCb2RGB<br>cv::COLOR_YCrCb2BGR | Convert RGB or BGR image to luma-chroma (a.k.a. YCC) color representation or vice versa |
| cv::COLOR_RGB2HSV<br>cv::COLOR_BGR2HSV<br>cv::COLOR_HSV2RGB<br>cv::COLOR_HSV2BGR | Convert RGB or BGR image to HSV (hue saturation value) color representation or vice versa |
| cv::COLOR_RGB2HLS<br>cv::COLOR_BGR2HLS<br>cv::COLOR_HLS2RGB<br>cv::COLOR_HLS2BGR | Convert RGB or BGR image to HLS (hue lightness saturation) color representation or vice versa |
| cv::COLOR_RGB2Lab<br>cv::COLOR_BGR2Lab<br>cv::COLOR_Lab2RGB<br>cv::COLOR_Lab2BGR | Convert RGB or BGR image to CIE Lab color representation or vice versa |
| cv::COLOR_RGB2Luv<br>cv::COLOR_BGR2Luv<br>cv::COLOR_Luv2RGB<br>cv::COLOR_Luv2BGR | Convert RGB or BGR image to CIE Luv color representation or vice versa |

| Conversion code | Meaning |
|---|---|
| cv::COLOR_BayerBG2RGB<br>cv::COLOR_BayerGB2RGB<br>cv::COLOR_BayerRG2RGB<br>cv::COLOR_BayerGR2RGB<br>cv::COLOR_BayerBG2BGR<br>cv::COLOR_BayerGB2BGR<br>cv::COLOR_BayerRG2BGR<br>cv::COLOR_BayerGR2BGR | Convert from Bayer pattern (single-channel) to RGB or BGR image |

We will not go into the details of these conversions nor the subtleties of some of the representations (particularly the Bayer and the CIE color spaces) here. Instead, we will just note that OpenCV contains tools to convert to and from these various color spaces, which are important to various classes of users.

The color-space conversions all use the following conventions: 8-bit images are in the range 0 to 255, 16-bit images are in the range 0 to 65,536, and floating-point numbers are in the range 0.0 to 1.0. When grayscale images are converted to color images, all components of the resulting image are taken to be equal; but for the reverse transformation (e.g., RGB or BGR to grayscale), the gray value is computed through the perceptually weighted formula:

$$Y = (0.299)R + (0.587)G + (0.114)B$$

In the case of HSV or HLS representations, hue is normally represented as a value from 0 to 360.[4] This can cause trouble in 8-bit representations and so, when you are converting to HSV, the hue is divided by 2 when the output image is an 8-bit image.

## cv::determinant()

```
double cv::determinant(
  cv::InputArray mat
);
```

$$d = \det(mat)$$

`cv::determinant()` computes the determinant of a square array. The array must be of one of the floating-point data types and must be single-channel. If the matrix is small, then the determinant is directly computed by the standard formula. For large matrices, this is not efficient, so the determinant is computed by *Gaussian elimination*.

4 Excluding 360, of course.

If you know that a matrix has a symmetric and positive determinant, you can use the trick of solving via *singular value decomposition* (SVD). For more information, see the section "Singular Value Decomposition (cv::SVD)" on page 173 in Chapter 7, but the trick is to set both `U` and `V` to `NULL`, and then just take the products of the matrix `W` to obtain the determinant.

## cv::divide()

```
void cv::divide(
  cv::InputArray  src1,                  // First input array (numerators)
  cv::InputArray  src2,                  // Second input array (denominators)
  cv::OutputArray dst,                   // Results array (scale*src1/src2)
  double          scale = 1.0,           // Multiplicative scale factor
  int             dtype = -1             // dst data type, -1 to get from src2
);

void cv::divide(
  double          scale,                 // Numerator for all divisions
  cv::InputArray  src2,                  // Input array (denominators)
  cv::OutputArray dst,                   // Results array (scale/src2)
  int             dtype = -1             // dst data type, -1 to get from src2
);
```

$$dst_i = saturate\left(scale * \frac{src1_i}{src2_i}\right)$$

$$dst_i = saturate(scale / src2_i)$$

`cv::divide()` is a simple division function; it divides all of the elements in `src1` by the corresponding elements in `src2` and puts the results in `dst`.

## cv::eigen()

```
bool cv::eigen(
  cv::InputArray  src,
  cv::OutputArray eigenvalues,
  int             lowindex    = -1,
  int             highindex   = -1
);

bool cv::eigen(
  cv::InputArray  src,
  cv::OutputArray eigenvalues,
  cv::OutputArray eigenvectors,
  int             lowindex    = -1,
  int             highindex   = -1
);
```

Given a symmetric matrix `mat`, `cv::eigen()` will compute the *eigenvectors* and *eigenvalues* of that matrix. The matrix must be of one of the floating-point types. The results array `eigenvalues` will contain the eigenvalues of `mat` in descending order. If the array `eigenvectors` was provided, the eigenvectors will be stored as the rows of that array in the same order as the corresponding eigenvalues in `eigenvalues`. The additional parameters `lowindex` and `highindex` allow you to request only some of the eigenvalues to be computed (both must be used together). For example, if `lowindex=0` and `highindex=1`, then only the largest two eigenvectors will be computed.

## cv::exp()

```
void cv::exp(
  cv::InputArray  src,
  cv::OutputArray dst
);
```

$$dst_i = e^{src_i}$$

`cv::exp()` exponentiates all of the elements in `src` and puts the results in `dst`.

## cv::extractImageCOI()

```
bool cv::extractImageCOI(
  const CvArr*    arr,
  cv::OutputArray dst,
  int             coi = -1
);
```

The function `cv::extractImageCOI()` extracts the indicated COI from a legacy-style (pre–version 2.1) array, such as an `IplImage` or `CvMat` given by `arr`, and puts the result in `dst`. If the argument `coi` is provided, then that particular COI will be extracted. If not, then the COI field in `src` will be checked to determine which channel to extract.

The `cv::extractImageCOI()` method described here is specifically for use with legacy arrays. If you need to extract a single channel from a modern `cv::Mat` object, use `cv::mixChannels()` or `cv::split()`.

## cv::flip()

```
void cv::flip(
  cv::InputArray  src,                    // Input array
  cv::OutputArray dst,                    // Result array, size and type of 'src'
  int             flipCode = 0            // >0: y-flip, 0: x-flip, <0: both
);
```

This function flips an image around the *x*-axis, the *y*-axis, or both. By default, `flipCode` is set to `0`, which flips around the *x*-axis.

If `flipCode` is set greater than zero (e.g., `+1`), the image will be flipped around the *y*-axis, and if set to a negative value (e.g., `-1`), the image will be flipped about both axes.

When doing video processing on Win32 systems, you will find yourself using this function often to switch between image formats with their origins at the upper left and lower left of the image.

## cv::gemm()

```
void cv::gemm(
  cv::InputArray  src1,                   // First input array
  cv::InputArray  src2,                   // Second input array
  double          alpha,                  // Weight for 'src1' * 'src2' product
  cv::InputArray  src3,                   // Third (offset) input array
  double          beta,                   // Weight for 'src3' array
  cv::OutputArray dst,                    // Results array
  int             flags = 0               // Use to transpose source arrays
);
```

*Generalized matrix multiplication* (GEMM) in OpenCV is performed by `cv::gemm()`, which performs matrix multiplication, multiplication by a transpose, scaled multiplication, and so on. In its most general form, `cv::gemm()` computes the following:

$$D = \alpha \cdot op(\mathrm{src}_1)^* \ \mathrm{op}(src_2) + \beta \cdot \mathrm{op}(\mathrm{src}_3)$$

where `src1`, `src2`, and `src3` are matrices, $\alpha$ and $\beta$ are numerical coefficients, and `op()` is an optional transposition of the matrix enclosed. The transpositions are controlled by the optional argument `flags`, which may be `0` or any combination (by means of Boolean OR) of `cv::GEMM_1_T`, `cv::GEMM_2_T`, and `cv::GEMM_3_T` (with each flag indicating a transposition of the corresponding matrix).

All matrices must be of the appropriate size for the (matrix) multiplication, and all should be of floating-point types. The `cv::gemm()` function also supports two-channel matrices that will be treated as two components of a single complex number.

You can achieve a similar result using the matrix algebra operators. For example:

```
cv::gemm(
  src1, src2, alpha, src3, bets, dst,cv::GEMM_1_T | cv::GEMM_3_T
);
```

would be equivalent to:

```
dst = alpha * src1.T() * src2 + beta * src3.T()
```

## cv::getConvertElem() and cv::getConvertScaleElem()

```
cv::convertData cv::getConvertElem(       // Returns a conversion function (below)
  int fromType,                           // Input pixel type (e.g., cv::8U)
  int toType                              // Output pixel type (e.g., CV_32F)
);

cv::convertScaleData cv::getConvertScaleElem(  // Returns a conversion function
  int fromType,                           // Input pixel type (e.g., cv::8U)
  int toType                              // Output pixel type (e.g., CV_32F)
);

// Conversion functions are of these forms:
//
typedef void (*ConvertData)(
  const void* from,                       // Pointer to the input pixel location
  void*       to,                         // Pointer to the result pixel location
  int         cn                          // number of channels
);

typedef void (*ConvertScaleData)(
  const void* from,                       // Pointer to the input pixel location
  void*       to,                         // Pointer to the result pixel location
  int         cn,                         // number of channels
  double      alpha,                      // scale factor
  double      beta                        // offset factor
);
```

The functions `cv::getConvertElem()` and `cv::getConvertScaleElem()` return function pointers to the functions that are used for specific type conversions in OpenCV. The function returned by `cv::getConvertElem()` is defined (via `typedef`) to the type `cv::ConvertData`, which can be passed a pointer to two data areas and a number of "channels." The number of channels is given by the argument `cn` of the conversion function, which is really the number of contiguous-in-memory objects of `fromType` to convert. This means that you could convert an entire (contiguous-in-memory) array by simply setting the number of channels equal to the total number of elements in the array.

Both `cv::getConvertElem()` and `cv::getConvertScaleElem()` take as arguments two types: `fromType` and `toType`. These types are specified with the integer constants (e.g., `CV_32F`).

In the case of `cv::getConvertScaleElem()`, the returned function takes two additional arguments, `alpha` and `beta`. These values are used by the converter function to rescale (`alpha`) and offset (`beta`) the input value before conversion to the desired type.

## cv::idct()

```
void cv::idct(
  cv::InputArray  src,                     // Input array
  cv::OutputArray dst,                     // Result array
  int             flags,                   // for row-by-row
);
```

`cv::idct()` is just a convenient shorthand for the inverse discrete cosine transform. A call to `cv::idct()` is exactly equivalent to a call to `cv::dct()` with the arguments:

```
cv::dct( src, dst, flags | cv::DCT_INVERSE );
```

## cv::idft()

```
void cv::idft(
  cv::InputArray  src,                     // Input array
  cv::OutputArray dst,                     // Result array
  int             flags       = 0,         // for row-by-row, etc.
  int             nonzeroRows = 0          // only this many entries are nonzero
);
```

`cv::idft()` is just a convenient shorthand for the inverse discrete Fourier transform. A call to `cv::idft()` is exactly equivalent to a call to `cv::dft()` with the arguments:

```
cv::dft( src, dst, flags | cv::DCT_INVERSE, outputRows );
```

Neither `cv::dft()` nor `cv::idft()` scales the output by default. So you will probably want to call `cv::idft()` with the `cv::DFT_SCALE` argument; that way, the transform and its "inverse" will be true inverse operations.

## cv::inRange()

```
void cv::inRange(
  cv::InputArray  src,                     // Input array
  cv::InputArray  upperb,                  // Array of upper bounds (inclusive)
  cv::InputArray  lowerb,                  // Array of lower bounds (inclusive)
```

```
  cv::OutputArray dst                          // Result array, cv::8UC1 type
);
```

$$dst_i = lowerb_i \le src_i \le upperb_i$$

When applied to a one-dimensional array, each element of `src` is checked against the corresponding elements of `upperb` and `lowerb`. The corresponding element of `dst` is set to `255` if the element in `src` is between the values given by `upperb` and `lowerb`; otherwise, it is set to `0`.

However, in the case of multichannel arrays for `src`, `upperb`, and `lowerb`, the output is still a single channel. The output value for element *i* will be set to `255` if and only if the values for the corresponding entry in `src` all lay inside the intervals implied for the corresponding channel in `upperb` and `lowerb`. In this sense, `upperb` and `lowerb` define an *n*-dimensional hypercube for each pixel and the corresponding value in `dst` is only set to `true` (`255`) if the pixel in `src` lies inside that hypercube.

## cv::insertImageCOI()

```
void cv::insertImageCOI(
  cv::InputArray  img,                         // Input array, single channel
  CvArr*          arr,                         // Legacy (pre v2.1) output array
  int             coi = -1                     // Target channel
);
```

Like the `cv::extractImageCOI()` function we encountered earlier in this chapter, the `cv::insertImageCOI()` function is designed to help us work with legacy (pre-v2.1) arrays like `IplImage` and `CvMat`. Its purpose is to allow us to take the data from a new-style C++ `cv::Mat` object and write that data onto one particular channel of a legacy array. The input `img` is expected to be a single-channel `cv::Mat` object, while the input `arr` is expected to be a multichannel legacy object. Both must be of the same size. The data in `img` will be copied onto the `coi` channel of `arr`.

The function `cv::extractImageCOI()` is used when you want to extract a particular COI from a legacy array into a single-channel C++ array. On the other hand, `cv::insertImageCOI()` is used to write the contents of a single-channel C++ array onto a particular channel of a legacy array. If you aren't dealing with legacy arrays and you just want to insert one C++ array into another, you can do that with `cv::merge()`.

## cv::invert()

```
double cv::invert(                              // Return 0 if 'src' is singular
  cv::InputArray  src,                          // Input Array, m-by-n
  cv::OutputArray dst                           // Result array, n-by-m
  int             method = cv::DECOMP_LU  // Method for (pseudo) inverse
);
```

`cv::invert()` inverts the matrix in `src` and places the result in `dst`. The input array must be a floating-point type, and the result array will be of the same type. Because `cv::invert()` includes the possibility of computing pseudo-inverses, the input array need not be square. If the input array is $n \times m$, then the result array will be $m \times n$. This function supports several methods of computing the inverse matrix (see Table 5-5), but the default is Gaussian elimination. The return value depends on the method used.

*Table 5-5. Possible values of method argument to cv::invert()*

| Value of method argument | Meaning |
|---|---|
| `cv::DECOMP_LU` | Gaussian elimination (LU decomposition) |
| `cv::DECOMP_SVD` | Singular value decomposition (SVD) |
| `cv::DECOMP_CHOLESKY` | Only for symmetric positive matrices |

In the case of Gaussian elimination (`cv::DECOMP_LU`), the determinant of `src` is returned when the function is complete. If the determinant is `0`, inversion failed and the array `dst` is set to all `0`s.

In the case of `cv::DECOMP_SVD`, the return value is the inverse condition number for the matrix (the ratio of the smallest to the largest eigenvalues). If the matrix `src` is singular, then `cv::invert()` in SVD mode will compute the pseudo-inverse. The other two methods (LU and Cholesky decomposition) require the source matrix to be square, nonsingular, and positive.

## cv::log()

```
void cv::log(
  cv::InputArray  src,
  cv::OutputArray dst
);
```

$$dst_i = \begin{cases} \log src_i & src_i \neq 0 \\ -C & else \end{cases}$$

`cv::log()` computes the natural log of the elements in `src1` and puts the results in `dst`. Source pixels that are less than or equal to zero are marked with destination pixels set to a large negative value.

## cv::LUT()

```
void cv::LUT(
  cv::InputArray  src,
  cv::InputArray  lut,
  cv::OutputArray dst
);
```

$$dst_i = \text{lut}(src_i)$$

The function `cv::LUT()` performs a "lookup table transform" on the input in `src`. `cv::LUT()` requires the source array `src` to be 8-bit index values. The `lut` array holds the lookup table. This lookup table array should have exactly 256 elements, and may have either a single channel or, in the case of a multichannel `src` array, the same number of channels as the source array. The function `cv::LUT()` then fills the destination array `dst` with values taken from the lookup table `lut` using the corresponding value from `src` as an index into that table.

In the case where the values in `src` are signed 8-bit numbers, they are automatically offset by +128 so that their range will index the lookup table in a meaningful way. If the lookup table is multichannel (and the indices are as well), then the value in `src` is used as a multidimensional index into `lut`, and the result array `dst` will be single channel. If `lut` is one-dimensional, then the result array will be multichannel, with each channel being separately computed from the corresponding index from `src` and the one-dimensional lookup table.

## cv::magnitude()

```
void cv::magnitude(
  cv::InputArray  x,
  cv::InputArray  y,
  cv::OutputArray dst
);
```

$$dst_i = \sqrt{x_i^2 + y_i^2}$$

`cv::magnitude()` essentially computes the radial part of a Cartesian-to-polar conversion on a two-dimensional vector field. In the case of `cv::magnitude()`, this vector field is expected to be in the form of two separate single channel arrays. These two input arrays should have the same size. (If you have a single two-channel array, cv::split() will give you separate channels.) Each element in dst is computed from the corresponding elements of x and y as the Euclidean norm of the two (i.e., the square root of the sum of the squares of the corresponding values).

## cv::Mahalanobis()

```
cv::Size cv::mahalanobis(
    cv::InputArray  vec1,
    cv::InputArray  vec2,
    cv::OutputArray icovar
);
```

cv::Mahalanobis() computes the value:

$$r_{mahalonobis} = \sqrt{(\vec{x} - \vec{\mu})^T \Sigma^{-1} (\vec{x} - \vec{\mu})}$$

The *Mahalanobis distance* is defined as the vector distance measured between a point and the center of a Gaussian distribution; it is computed using the inverse covariance of that distribution as a metric (see Figure 5-2). Intuitively, this is analogous to the *z*-score in basic statistics, where the distance from the center of a distribution is measured in units of the variance of that distribution. The Mahalanobis distance is just a multivariable generalization of the same idea.

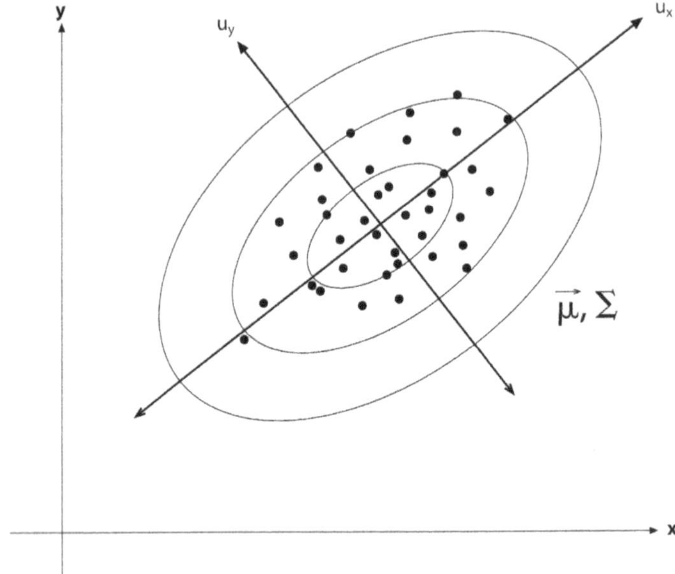


*Figure 5-2. A distribution of points in two dimensions with superimposed ellipsoids representing Mahalanobis distances of 1.0, 2.0, and 3.0 from the distribution's mean*

The vector vec1 is presumed to be the point $x$, and the vector vec2 is taken to be the distribution's mean.[5] That matrix icovar is the inverse covariance.

This covariance matrix will usually have been computed with cv::calcCovarMatrix() (described previously) and then inverted with cv::invert(). It is good programming practice to use the cv::DECOMP_SVD method for this inversion because someday you will encounter a distribution for which one of the eigenvalues is 0!

## cv::max()

```
cv::MatExpr cv::max(
  const cv::Mat& src1,                    // First input array (first position)
  const cv::Mat& src2                     // Second input array
);

MatExpr cv::max(                          // A matrix expression, not a matrix
  const cv::Mat&  src1,                   // First input array (first position)
  double          value                   // Scalar in second position
);

MatExpr cv::max(                          // A matrix expression, not a matrix
  double          value,                  // Scalar in first position
  const cv::Mat&  src1                    // Input array (second position)
);

void cv::max(
  cv::InputArray  src1,                   // First input array
  cv::InputArray  src2,                   // Second input array
  cv::OutputArray dst                     // Result array
);

void cv::max(
  const Mat&      src1,                   // First input array
  const Mat&      src2,                   // Second input array
  Mat&            dst                     // Result array
);

void cv::max(
  const Mat&      src1,                   // Input array
  double          value,                  // Scalar input
  Mat&            dst                     // Result array
);
```

5 Actually, the Mahalanobis distance is more generally defined as the distance between any two vectors; in any case, the vector vec2 is subtracted from the vector vec1. Neither is there any fundamental connection between mat in cvMahalanobiscv::Mahalanobis() and the inverse covariance; any metric can be imposed here as appropriate.

$$dst_i = \max(src_{1,i}, src_{2,i})$$

`cv::max()` computes the maximum value of each corresponding pair of pixels in the arrays `src1` and `src2`. It has two basic forms: those that return a matrix expression and those that compute a result and put it someplace you have indicated. In the three-argument form, in the case where one of the operands is a `cv::Scalar`, comparison with a multichannel array is done on a per-channel basis with the appropriate component of the `cv::Scalar`.

## cv::mean()

```
cv::Scalar cv::mean(
  cv::InputArray  src,
  cv::InputArray  mask = cv::noArray(),    // Optional, do only where nonzero
);
```

$$N = \sum_{i, mask_i \neq 0} 1$$

$$mean_c = \frac{1}{N} \sum_{i, mask_i \neq 0} src_i$$

The function `cv::mean()` computes the average value of all of the pixels in the input array `src` that are not masked out. The result is computed on a per-channel basis if `src` is multichannel.

## cv::meanStdDev()

```
void cv::meanStdDev(
  cv::InputArray  src,
  cv::OutputArray mean,
  cv::OutputArray stddev,
  cv::InputArray  mask  = cv::noArray(),   // Optional, do only where nonzero
);
```

$$N = \sum_{i, mask_i \neq 0} 1$$

$$mean_c = \frac{1}{N} \sum_{i, mask_i \neq 0} src_i$$

$$stddev_c = \sqrt{\sum_{i, mask \neq 0} (src_{c,i} - mean_c)^2}$$

The function `cv::meanStdDev()` computes the average value of the pixels in the input array `src` that are not masked out, as well as their standard deviation. The mean and standard deviation are computed on a per-channel basis if `src` is multichannel.

The standard deviation computed here is not the same as the covariance matrix. In fact, the standard deviation computed here is only the diagonal elements of the full covariance matrix. If you want to compute the full covariance matrix, you will have to use `cv::calcCovarMatrix()`.

## cv::merge()

```
void cv::merge(
  const cv::Mat*  mv,                 // C-style array of arrays
  size_t          count,              // Number of arrays pointed to by 'mv'
  cv::OutputArray dst                 // Contains all channels in 'mv'
);

void merge(
  const vector<cv::Mat>& mv,          // STL-style array of arrays
  cv::OutputArray dst                 // Contains all channels in 'mv'
);
```

`cv::merge()` is the inverse operation of `cv::split()`. The arrays contained in `mv` are combined into the output array `dst`. In the case in which `mv` is a pointer to a C-style array of `cv::Mat` objects, the additional size parameter `count` must also be supplied.

## cv::min()

```
cv::MatExpr cv::min(                  // A matrix expression, not a matrix
  const cv::Mat&  src1,               // First input array
  const cv::Mat&  src2                // Second input array
);

MatExpr cv::min(                      // A matrix expression, not a matrix
  const cv::Mat&  src1,               // First input array (first position)
  double          value               // Scalar in second position
);

MatExpr cv::min(                      // A matrix expression, not a matrix
  double          value,              // Scalar in first position
  const cv::Mat&  src1                // Input array (second position)
);

void cv::min(
  cv::InputArray  src1,               // First input array
  cv::InputArray  src2,               // Second input array
  cv::OutputArray dst                 // Result array
);
```

```
void cv::min(
  const Mat&      src1,                     // First input array
  const Mat&      src2,                     // Second input array
  Mat&            dst                       // Result array
);

void cv::min(
  const Mat&      src1,                     // Input array
  double          value,                    // Scalar input
  Mat&            dst                       // Result array
);
```

$$dst_i = \min(\text{src}_{1,i}, src_{2,i})$$

`cv::min()` computes the minimum value of each corresponding pair of pixels in the arrays `src1` and `src2` (or one source matrix and a single value). Note that the variants of `cv::min()` that return a value or a matrix expression can then be manipulated by OpenCV's matrix expression machinery.

In the three-argument form, in the case where one of the operands is a `cv::Scalar`, comparison with a multichannel array is done on a per-channel basis with the appropriate component of the `cv::Scalar`.

## cv::minMaxIdx()

```
void cv::minMaxIdx(
  cv::InputArray src,                       // Input array, single channel only
  double*        minVal,                    // min value goes here (in not NULL)
  double*        maxVal,                    // min value goes here (in not NULL)
  int*           minIdx,                    // loc of min goes here (if not NULL)
  int*           maxIdx,                    // loc of max goes here (if not NULL)
  cv::InputArray mask = cv::noArray()       // search only nonzero values
);

void cv::minMaxIdx(
  const cv::SparseMat& src,                 // Input sparse array
  double*        minVal,                    // min value goes here (in not NULL)
  double*        maxVal,                    // min value goes here (in not NULL)
  int*           minIdx,                    // C-style array, indices of min locs
  int*           maxIdx,                    // C-style array, indices of max locs
);
```

These routines find the minimal and maximal values in the array `src` and (optionally) returns their locations. The computed minimum and maximum values are placed in `minVal` and `maxVal`. Optionally, the locations of those extrema can also be returned; this will work for arrays of any number of dimensions. These locations will be written to the addresses given by `minIdx` and `maxIdx` (provided that these arguments are non-`NULL`).

`cv::minMaxIdx()` can also be called with a `cv::SparseMat` for the `src` array. In this case, the array can be of any number of dimensions and the minimum and maximum will be computed and their location returned. In this case, the locations of the extrema will be returned and placed in the C-style arrays `minLoc` and `maxLoc`. Both of those arrays, if provided, should have the same number of elements as the number of dimensions in the `src` array. In the case of `cv::SparseMat`, the minimum and maximum are computed only for what are generally referred to as *nonzero elements* in the source code; however, this terminology is slightly misleading, as what is really meant is *elements that exist* in the sparse matrix representation in memory. In fact, there may, as a result of how the sparse matrix came into being and what has been done with it in the past, be elements that *exist* and are also zero. Such elements will be included in the computation of the minimum and maximum.

When a single-dimensional array is supplied, the arrays for the locations must still have memory allocated for two integers. This is because `cv::minMaxLoc()` uses the convention that even a single-dimensional array is, in essence, an $N \times 1$ matrix. The second value returned will always be 0 for each location in this case.

## cv::minMaxLoc()

```
void cv::minMaxLoc(
  cv::InputArray src,                   // Input array
  double*        minVal,                // min value goes here (in not NULL)
  double*        maxVal,                // min value goes here (in not NULL)
  cv::Point*     minLoc,                // loc of min goes here (if not NULL)
  cv::Point*     maxLoc,                // loc of max goes here (if not NULL)
  cv::InputArray mask = cv::noArray()   // search only nonzero values
);

void cv::minMaxLoc(
  const cv::SparseMat& src,             // Input sparse array
  double*        minVal,                // min value goes here (in not NULL)
  double*        maxVal,                // min value goes here (in not NULL)
  cv::Point*     minLoc,                // C-style array, indices of min locs
  cv::Point*     maxLoc,                // C-style array, indices of max locs
);
```

This routine finds the minimal and maximal values in the array `src` and (optionally) returns their locations. The computed minimum and maximum values are placed in `minVal` and `maxVal`, respectively. Optionally, the locations of those extrema can also be returned. These locations will be written to the addresses given by `minLoc` and `maxLoc` (provided that these arguments are non-`NULL`). Because these locations are of type cv::Point, this form of the function should be used only on two-dimensional arrays (i.e., matrices or images).

As with cv::MinMaxIdx(), in the case of sparse matrices, only active entries will be considered when searching for the minimum or the maximum.

When working with multichannel arrays, you have several options. Natively, cv::minMaxLoc() does not support multichannel input. Primarily this is because this operation is ambiguous.

If you want the minimum and maximum across all channels, you can use cv::reshape() to reshape the multichannel array into one giant single-channel array. If you would like the minimum and maximum for each channel separately, you can use cv::split() or cv::mixChannels() to separate the channels out and analyze them separately.

In both forms of cv::minMaxLoc, the arguments for the minimum or maximum value or location may be set to NULL, which turns off the computation for that argument.

## cv::mixChannels()

```
void cv::mixChannels(
  const cv::Mat*         srcv,        // C-style array of matrices
  int                    nsrc,        // Number of elements in 'srcv'
  cv::Mat*               dstv,        // C-style array of target matrices
  int                    ndst,        // Number of elements in 'dstv'
  const int*             fromTo,      // C-style array of pairs, ...from,to...
  size_t                 n_pairs      // Number of pairs in 'fromTo'
);

void cv::mixChannels(
  const vector<cv::Mat>& srcv,        // STL-style vector of matrices
  vector<cv::Mat>&       dstv,        // STL-style vector of target matrices
  const int*             fromTo,      // C-style array of pairs, ...from,to...
  size_t                 n_pairs      // Number of pairs in 'fromTo'
);
```

There are many operations in OpenCV that are special cases of the general problem of rearranging channels from one or more images in the input, and sorting them into particular channels in one or more images in the output. Functions like cv::split(), cv::merge(), and (at least some cases of) cv::cvtColor() all make use of such functionality. Those methods do what they need to do by calling the much more general cv::mixChannels(). This function allows you to supply multiple arrays, each with potentially multiple channels, for the input, and the same for the output, and to map the channels from the input arrays into the channels in the output arrays in any manner you choose.

The input and output arrays can either be specified as C-style arrays of `cv::Mat` objects with an accompanying integer indicating the number of `cv::Mats`, or as an STL `vector<>` of `cv::Mat` objects. Output arrays must be preallocated with their size and number of dimensions matching those of the input arrays.

The mapping is controlled by the C-style integer array `fromTo`. This array can contain any number of integer pairs in sequence, with each pair indicating with its first value the source channel and with its second value the destination channel to which that should be copied. The channels are sequentially numbered starting at zero for the first image, then through the second image, and so on (see Figure 5-3). The total number of pairs is supplied by the argument `n_pairs`.

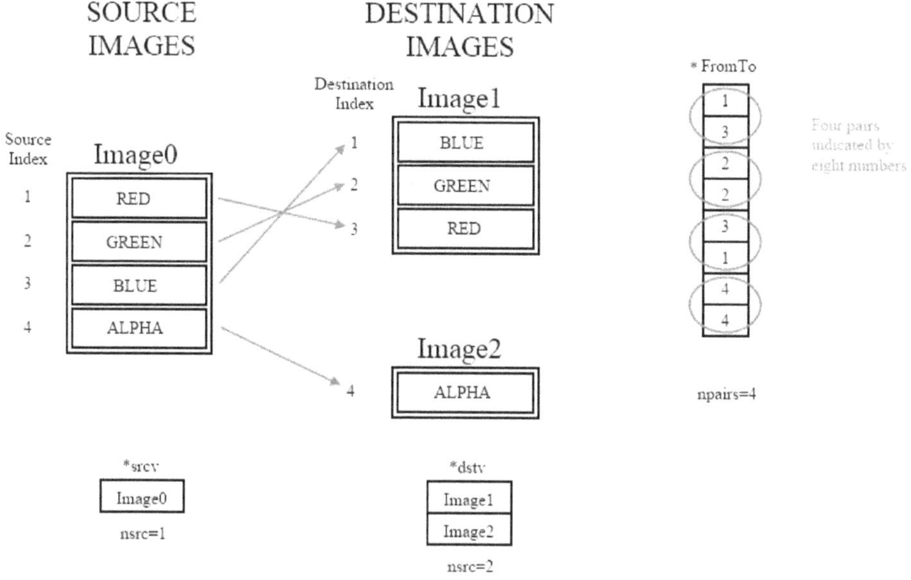


*Figure 5-3. A single four-channel RGBA image is converted to one BGR and one alpha-only image*

Unlike most other functions in the post-version 2.1 library, `cv::mixChannels()` *does not* allocate the output arrays. They must be preallocated and have the same size and dimensionality as the input arrays.

## cv::mulSpectrums()

```
doublevoid cv::mulSpectrums(
    cv::InputArray  arr1,              // First input array
    cv::InputArray  arr2,              // Second input array, same size as 'arr1'
    cv::OutputArray dst,               // Result array, same size as 'arr1'
    int             flags,             // used to indicate independent rows
    bool            conj  = false      // If true, conjugate arr2 first
);
```

In many operations involving spectra (i.e., the results from `cv::dft()` or `cv::idft()`), one wishes to do a per-element multiplication that respects the packing of the spectra (real arrays), or their nature as complex variables. (See the description of `cv::dft()` for more details.) The input arrays may be one- or two-dimensional, with the second the same size and type as the first. If the input array is two-dimensional, it may either be taken to be a true two-dimensional spectrum, or an array of one-dimensional spectra (one per row). In the latter case, `flags` should be set to `cv::DFT_ROWS`; otherwise it can be set to `0`.

When the two arrays are complex, they are simply multiplied on an element-wise basis, but `cv::mulSpectrums()` provides an option to conjugate the second array elements before multiplication. For example, you would use this option to perform correlation (using the Fourier transform), but for convolution, you would use `conj=false`.

## cv::multiply()

```
void cv::multiply(
  cv::InputArray  src1,                  // First input array
  cv::InputArray  src2,                  // Second input array
  cv::OutputArray dst,                   // Result array
  double          scale = 1.0,           // overall scale factor
  int             dtype = -1             // Output type for result array
);
```

$$dst_i = saturate(scale * src1_i * src2_i)$$

`cv::multiply()` is a simple multiplication function; it multiplies the elements in `src1` by the corresponding elements in `src2` and puts the results in `dst`.

## cv::mulTransposed()

```
void cv::mulTransposed(
  cv::InputArray  src1,                  // Input matrix
  cv::OutputArray dst,                   // Result array
  bool            aTa,                   // If true, transpose then multiply
  cv::InputArray  delta = cv::noArray(), // subtract from 'src1' before multiply
```

```
  double          scale = 1.0,           // overall scale factor
  int             dtype = -1             // Output type for result array
);
```

$$dst = \begin{cases} scale^*(src - delta)^T(src - delta) & aTa = true \\ scale^*(src - delta)(src - delta)^T & aTa = false \end{cases}$$

`cv::mulTransposed()` is used to compute the product of a matrix and its own transpose—useful, for example, in computing covariance. The matrix `src` should be two-dimensional and single-channel, but unlike `cv::GEMM()`, it is not restricted to the floating-point types. The result matrix will be the same type as the source matrix unless specified otherwise by `dtype`. If `dtype` is not negative (default), it should be either `CV_32F` or `cv::F64`; the output array `dst` will then be of the indicated type.

If a second input matrix `delta` is provided, that matrix will be subtracted from `src` before the multiplication. If no matrix is provided (i.e., `delta=cv::noArray()`), then no subtraction is done. The array `delta` need not be the same size as `src`; if `delta` is smaller than `src`, `delta` is repeated (also called *tiling*; see `cv::repeat()`) in order to produce an array whose size matches the size of `src`. The argument `scale` is applied to the matrix after the multiplication is done. Finally, the argument `aTa` is used to select the multiplication in which the transposed version of `src` is multiplied either from the left (`aTa=true`) or from the right (`aTa=false`).

## cv::norm()

```
double cv::norm(                                 // Return norm in double precision
  cv::InputArray src1,                           // Input matrix
  int            normType = cv::NORM_L2,         // Type of norm to compute
  cv::InputArray mask     = cv::noArray()        // do for nonzero values (if present)
);

double cv::norm(                                 // Return computed norm of difference
  cv::InputArray src1,                           // Input matrix
  cv::InputArray src2,                           // Second input matrix
  int            normType = cv::NORM_L2,         // Type of norm to compute
  cv::InputArray mask     = cv::noArray()        // do for nonzero values (if present)
);

double cv::norm(
  const cv::SparseMat& src,                      // Input sparse matrix
  int                  normType = cv::NORM_L2,   // Type of norm to compute
);
```

$$\|src1\|_{\infty, L1, L2}$$

$$\|src1 - src2\|_{\infty, L1, L2}$$

The `cv::norm()` function is used to compute the norm of an array (see Table 5-6) or a variety of distance norms between two arrays if two arrays are provided (see Table 5-7). The norm of a `cv::SparseMat` can also be computed, in which case zero-entries are ignored in the computation of the norm.

*Table 5-6. Norm computed by cv::norm() for different values of normType when arr2=NULL*

| normType | Result |
|---|---|
| `cv::NORM_INF` | $\|src1\|_{\infty} = \max_i abs(src1_i)$ |
| `cv::NORM_L1` | $\|src1\|_{L1} = \sum_i abs(src1_i)$ |
| `cv::NORM_L2` | $\|src1\|_{L2} = \sqrt{\sum_i src1_i^2}$ |

If the second array argument `src2` is non-`NULL`, then the norm computed is a difference norm—that is, something like the distance between the two arrays.[6] In the first three cases (shown in Table 5-7), the norm is absolute; in the latter three cases, it is rescaled by the magnitude of the second array, `src2`.

*Table 5-7. Norm computed by cv::norm() for different values of normType when arr2 is non-NULL*

| normType | Result |
|---|---|
| `cv::NORM_INF` | $\|src1 - src2\|_{\infty} = \max_i abs(src1_i - src2_i)$ |
| `cv::NORM_L1` | $\|src1 - src2\|_{L1} = \sum_i abs(src1_i - src2_i)$ |
| `cv::NORM_L2` | $\|src1 - src2\|_{L2} = \sum_i (src1_i - src2_i)^2$ |
| `cv::NORM_RELATIVE_INF` | $\frac{\|src1 - src2\|_{\infty}}{\|src2\|_{\infty}}$ |

6 At least in the case of the L2 norm, there is an intuitive interpretation of the difference norm as a Euclidean distance in a space of dimension equal to the number of pixels in the images.

| normType | Result |
|---|---|
| cv::NORM_RELATIVE_L1 | $\frac{\lVert src1 - src2 \rVert_{L1}}{\lVert src2 \rVert_{L1}}$ |
| cv::NORM_RELATIVE_L2 | $\frac{\lVert src1 - src2 \rVert_{L2}}{src2_{L2}}$ |

In all cases, `src1` and `src2` must have the same size and number of channels. When there is more than one channel, the norm is computed over all of the channels together (i.e., the sums in Tables 5-6 and 5-7 are not only over $x$ and $y$ but also over the channels).

## cv::normalize()

```
void cv::normalize(
  cv::InputArray  src1,                       // Input matrix
  cv::OutputArray dst,                        // Result matrix
  double          alpha    = 1,               // first parameter (see Table 5-8)
  double          beta     = 0,               // second parameter (see Table 5-8)
  int             normType = cv::NORM_L2,     // Type of norm to compute
  int             dtype = -1                  // Output type for result array
  cv::InputArray  mask  = cv::noArray()       // do for nonzero values (if present)
);

void cv::normalize(
  const cv::SparseMat& src,                   // Input sparse matrix
  cv::SparseMat&       dst,                   // Result sparse matrix
  double               alpha    = 1,          // first parameter (see Table 5-8)
  int                  normType = cv::NORM_L2, // Type of norm to compute
);
```

$$\lVert dst \rVert_{\infty, L1, L2} = \alpha$$

$$\min(dst) = \alpha, \ \max(dst) = \beta$$

As with so many OpenCV functions, `cv::normalize()` does more than it might at first appear. Depending on the value of `normType`, image `src` is normalized or otherwise mapped into a particular range in `dst`. The array `dst` will be the same size as `src`, and will have the same data type, unless the `dtype` argument is used. Optionally, `dtype` can be set to one of the OpenCV fundamental types (e.g., `CV_32F`) and the output array will be of that type. The exact meaning of this operation is dependent on the `normType` argument. The possible values of `normType` are shown in Table 5-8.

*Table 5-8. Possible values of normType argument to cv::normalize()*

| norm_type | Result |
|---|---|
| cv::NORM_INF | $\|\| dst \|\|_{\infty} = \max_i abs(dst_i) = \alpha$ |
| cv::NORM_L1 | $\|\| dst \|\|_{L1} = \sum_i abs(dst_i) = \alpha$ |
| cv::NORM_L2 | $\|\| dst \|\|_{L2} = \sqrt{\sum_i dst_i^2} = \alpha$ |
| cv::NORM_MINMAX | Map into range $[\alpha, \beta]$ |

In the case of the infinity norm, the array `src` is rescaled such that the magnitude of the absolute value of the largest entry is equal to `alpha`. In the case of the L1 or L2 norm, the array is rescaled so that the norm equals the value of `alpha`. If `normType` is set to `cv::MINMAX`, then the values of the array are rescaled and translated so that they are linearly mapped into the interval between `alpha` and `beta` (inclusive).

As before, if `mask` is non-`NULL` then only those pixels corresponding to nonzero values of the mask image will contribute to the computation of the norm—and only those pixels will be altered by `cv::normalize()`. Note that if the operation `dtype=cv::MINMAX` is used, the source array may not be `cv::SparseMat`. The reason for this is that the `cv::MIN_MAX` operation can apply an overall offset, and this would affect the sparsity of the array (specifically, a sparse array would become nonsparse as all of the zero elements became nonzero as a result of this operation).

## cv::perspectiveTransform()

```
void cv::perspectiveTransform(
  cv::InputArray  src,             // Input array, 2 or 3 channels
  cv::OutputArray dst,             // Result array, size, type, as src1
  cv::InputArray  mtx              // 3-by-3 or 4-by-4 transoform matrix
);
```

$$\begin{bmatrix} x \\ y \\ z \end{bmatrix} \rightarrow \begin{bmatrix} x'/w' \\ y'/w' \\ z'/w' \end{bmatrix}$$

$$\begin{bmatrix} x' \\ y' \\ z' \\ w' \end{bmatrix} \rightarrow [mtx] \begin{bmatrix} x \\ y \\ z \\ 1 \end{bmatrix}$$

The `cv::perspectiveTransform()` function performs a plane-plane projective transform of a list of points (not pixels). The input array should be a two- or three-channel array, and the matrix `mtx` should be 3 × 3 or 4 × 4, respectively, in the two cases.

`cv::perspectiveTransform()` thus transforms each element of `src` by first regarding it as a vector of length `src.channels() + 1`, with the additional dimension (the projective dimension) set initially to `1.0`. This is also known as *homogeneous coordinates*. Each extended vector is then multiplied by `mtx` and the result is rescaled by the value of the (new) projective coordinate[7] (which is then thrown away, as it is always `1.0` after this operation).

Note again that this routine is for transforming a list of points, not an image as such. If you want to apply a perspective transform to an image, you are actually asking not to transform the individual pixels, but rather to move them from one place in the image to another. This is the job of `cv::warpPerspective()`.

If you want to solve the inverse problem to find the most probable perspective transformation given many pairs of corresponding points, use `cv::getPerspectiveTransform()` or `cv::findHomography()`.

## cv::phase()

```
void cv::phase(
  cv::InputArray  x,                       // Input array of x-components
  cv::InputArray  y,                       // Input array of y-components
  cv::OutputArray dst,                     // Output array of angles (radians)
  bool            angleInDegrees = false   // degrees (if true), radians (if false)
);
```

$$dst_i = \text{atan2}(y_i, x_i)$$

`cv::phase()` computes the azimuthal (angle) part of a Cartesian-to-polar conversion on a two-dimensional vector field. This vector field is expected to be in the form of two separate single-channel arrays. These two input arrays should, of course, be of the same size. (If you happen to have a single two-channel array, a quick call to `cv::split()` will do just what you need.) Each element in `dst` is then computed from the corresponding elements of `x` and `y` as the arctangent of the ratio of the two.

7 Technically, it is possible that after multiplying by `mtx`, the value of $w'$ will be zero, corresponding to points projected to infinity. In this case, rather than dividing by zero, a value of `0` is assigned to the ratio.

## cv::polarToCart()

```
void cv::polarToCart(
  cv::InputArray  magnitude,            // Input array of magnitudes
  cv::InputArray  angle,                // Input array of angles
  cv::OutputArray x,                    // Output array of x-components
  cv::OutputArray y,                    // Output array of y-components
  bool            angleInDegrees = false // degrees (if true) radians (if false)
);
```

$x_i = magnitude_i * \cos(angle_i)$

$y_i = magnitude_i * \sin(angle_i)$

`cv::polarToCart()` computes a vector field in Cartesian $(x, y)$ coordinates from polar coordinates. The input is in two arrays, `magnitude` and `angle`, of the same size and type, specifying the magnitude and angle of the field at every point. The output is similarly two arrays that will be of the same size and type as the inputs, and which will contain the $x$ and $y$ projections of the vector at each point. The additional flag `angleInDegrees` will cause the angle array to be interpreted as angles in degrees rather than in radians.

## cv::pow()

```
void cv::pow(
  cv::InputArray  src,                  // Input array
  double          p,                    // power for exponentiation
  cv::OutputArray dst                   // Result array
);
```

$$dst_i = \begin{cases} src_i^{\,p} & p \in \mathbb{Z} \\ |\,src_i\,|^{\,p} & else \end{cases}$$

The function `cv::pow()` computes the element-wise exponentiation of an array by a given power `p`. In the case in which `p` is an integer, the power is computed directly. For noninteger `p`, the absolute value of the source value is computed first, and then raised to the power `p` (so only real values are returned). For some special values of `p`, such as integer values, or ±½, special algorithms are used, resulting in faster computation.

## cv::randu()

```
template<typename _Tp> _Tp randu(); // Return random number of specific type

void cv::randu(
  cv::InputOutArray mtx,         // All values will be randomized
  cv::InputArray    low,         // minimum, 1-by-1 (Nc=1,4), or 1-by-4 (Nc=1)
  cv::InputArray    high         // maximum, 1-by-1 (Nc=1,4), or 1-by-4 (Nc=1)
);
```

$$mtx_i \in [low_i, high_i)$$

There are two ways to call `cv::randu()`. The first method is to call the template form of `randu<>()`, which will return a random value of the appropriate type. Random numbers generated in this way are uniformly distributed[8] in the range from zero to the maximum value available for that type (for integers), and in the interval from 0.0 to 1.0 (not inclusive of 1.0) for floating-point types. This template form generates only single numbers.[9]

The second way to call `cv::randu()` is to provide a matrix `mtx` that you wish to have filled with values, and two additional arrays that specify the minimum and maximum values for the range from which you would like a random number drawn for each particular array element. These two additional values, `low` and `high`, should be $1 \times 1$ with 1 or 4 channels, or $1 \times 4$ with a single channel; they may also be of type `cv::Scalar`. In any case, they are not the size of `mtx`, but rather the size of individual entries in `mtx`.

The array `mtx` is both an input and an output, in the sense that you must allocate the matrix so that `cv::randu()` will know the number of random values you need and how they are to be arranged in terms of rows, columns, and channels.

## cv::randn()

```
void cv::randn(
  cv::InputOutArray mtx,          // All values will be randomized
  cv::InputArray    mean,         // mean values, array is in channel space
  cv::InputArray    stddev        // standard deviations, channel space
);
```

8 Uniform-distribution random numbers are generated using the Multiply-With-Carry algorithm [Goresky03].

9 In particular, this means that if you call the template form with a vector argument, such as: `cv::randu<Vec4f>`, the return value, though it will be of the vector type, will be all zeros except for the first element.

$$mtx_i \sim N(mean_i, stddev_i)$$

The function `cv::randn()` fills a matrix `mtx` with random normally distributed values.[10] The parameters from which these values are drawn are taken from two additional arrays (`mean` and `stddev`) that specify the mean and standard deviation for the distribution from which you would like a random number drawn for each particular array element.

As with the array form of `cv::randu()`, every element of `mtx` is computed separately, and the arrays `mean` and `stddev` are in the channel space for individual entries of `mtx`. Thus, if `mtx` were four channels, then `mean` and `stddev` would be $1 \times 4$ or $1 \times 1$ with four channels (or equivalently of type `cv::Scalar`).[11]

## cv::randShuffle()

```
void cv::randShuffle(
  cv::InputOutArray mtx,                    // All values will be shuffled
  double            iterFactor = 1,         // Number of times to repeat shuffle
  cv::RNG*          rng        = NULL       // your own generator, if you like
);
```

`cv::randShuffle()` attempts to randomize the entries in a one-dimensional array by selecting random pairs of elements and interchanging their position. The number of such *swaps* is equal to the size of the array `mtx` multiplied by the optional factor `iterFactor`. Optionally, a random number generator can be supplied (for more on this, see "Random Number Generator (cv::RNG)" on page 176 in Chapter 7). If none is supplied, the default random number generator `theRNG()` will be used automatically.

## cv::reduce()

```
void cv::reduce(
  cv::InputArray  src,                        // Input, n-by-m, 2-dimensional
  cv::OutputArray vec,                        // Output, 1-by-m or n-by-1
  int             dim,                        // Reduction direction 0=row, 1=col
  int             reduceOp = cv::REDUCE_SUM,  // Reduce operation (see Table 5-9)
  int             dtype = -1                  // Output type for result array
);
```

With *reduction*, you systematically transform the input matrix `src` into a vector `vec` by applying some combination rule `reduceOp` on each row (or column) and its neigh-

10 Gaussian-distribution random numbers are generated using the Ziggurat algorithm [Marsaglia00].

11 Note that `stddev` is not a square matrix; correlated number generation is not supported by `cv::randn()`.

bor until only one row (or column) remains (see Table 5-9).[12] The argument `dim` controls how the reduction is done, as summarized in Table 5-10.

*Table 5-9. The reduceOp argument in cv::reduce() selects the reduction operator*

| Value of op | Result |
|---|---|
| `cv::REDUCE_SUM` | Compute sum across vectors |
| `cv::REDUCE_AVG` | Compute average across vectors |
| `cv::REDUCE_MAX` | Compute maximum across vectors |
| `cv::REDUCE_MIN` | Compute minimum across vectors |

*Table 5-10. The dim argument in cv::reduce() controls the direction of the reduction*

| Value of dim | Result |
|---|---|
| `0` | Collapse to a single row |
| `1` | Collapse to a single column |

`cv::reduce()` supports multichannel arrays of any type. Using `dtype`, you can specify an alternative type for `dst`.

Using the `dtype` argument to specify a higher-precision format for `dst` is particularly important for `cv::REDUCE_SUM` and `cv::REDUCE_AVG`, where overflows and summation problems are possible.

## cv::repeat()

```
void cv::repeat(
  cv::InputArray  src,                  // Input 2-dimensional array
  int             ny,                   // Copies in y-direction
  int             nx,                   // Copies in x-direction
  cv::OutputArray dst                   // Result array
);

cv::Mat cv::repeat(                     // Return result array
  cv::InputArray  src,                  // Input 2-dimensional array
  int             ny,                   // Copies in y-direction
  int             nx                    // Copies in x-direction
);
```

$$dst_{i,j} = src_{i\%src.rows,\, j\%src.cols}$$

12 Purists will note that averaging is not technically a proper *fold* in the sense implied here. OpenCV has a more practical view of reductions and so includes this useful operation in `cvReduce`.

This function copies the contents of src into dst, repeating as many times as necessary to fill dst. In particular, dst can be of any size relative to src. It may be larger or smaller, and it need not have an integer relationship between any of its dimensions and the corresponding dimensions of src.

cv::repeat() has two calling conventions. The first is the old-style convention in which the output array is passed as a reference to cv::repeat(). The second actually creates and returns a cv::Mat, and is much more convenient when you are working with matrix expressions.

## cv::scaleAdd()

```
void cv::scaleAdd(
  cv::InputArray  src1,                    // First input array
  double          scale,                   // Scale factor applied to first array
  cv::InputArray  src2,                    // Second input array
  cv::OutputArray dst,                     // Result array
);
```

$$dst_i = scale * src1_i + src2_i$$

cv::scaleAdd() is used to compute the sum of two arrays, src1 and src2, with a scale factor scale applied to the first before the sum is done. The results are placed in the array dst.

The same result can be achieved with the matrix algebra operation:

```
dst = scale * src1 + src2;
```

## cv::setIdentity()

```
void cv::setIdentity(
  cv::InputOutputArray dst,                          // Array to reset values
  const cv::Scalar&    value = cv::Scalar(1.0)  // Apply to diagonal elements
);
```

$$dst_{i,j} = \begin{cases} value & i = j \\ 0 & else \end{cases}$$

cv::setIdentity() sets all elements of the array to 0 except for elements whose row and column are equal; those elements are set to 1 (or to value if provided). cv::setIdentity() supports all data types and does not require the array to be square.

This can also be done through the `eye()` member function of the `cv::Mat` class. Using `eye()` is often more convenient when you are working with matrix expressions.

```
cv::Mat A( 3, 3, CV_32F );
cv::setIdentity( A, s );
C = A + B;
```

For some other arrays `B` and `C`, and some scalar `s`, this is equivalent to:

```
C = s * cv::Mat::eye( 3, 3, CV_32F ) + B;
```

## cv::solve()

```
int cv::solve(
  cv::InputArray  lhs,                        // Lefthand side of system, n-by-n
  cv::InputArray  rhs,                        // Righthand side of system, n-by-1
  cv::OutputArray dst,                        // Results array, will be n-by-1
  int             method = cv::DECOMP_LU  // Method for solver
);
```

The function `cv::solve()` provides a fast way to solve linear systems based on `cv::invert()`. It computes the solution to:

$$C = \text{argmin}_X \| \; A \cdot X \; - \; B \|$$

where $A$ is a square matrix given by `lhs`, $B$ is the vector `rhs`, and $C$ is the solution computed by `cv::solve()` for the best vector $X$ it could find. That best vector $X$ is returned in `dst`. The actual method used to solve this system is determined by the value of the `method` argument (see Table 5-11). Only floating-point data types are supported. The function returns an integer value where a nonzero return indicates that it could find a solution.

*Table 5-11. Possible values of method argument to cv::solve()*

| Value of method argument | Meaning |
|---|---|
| `cv::DECOMP_LU` | Gaussian elimination (LU decomposition) |
| `cv::DECOMP_SVD` | Singular value decomposition (SVD) |
| `cv::DECOMP_CHOLESKY` | For symmetric positive matrices |
| `cv::DECOMP_EIG` | Eigenvalue decomposition, symmetric matrices only |
| `cv::DECOMP_QR` | QR factorization |
| `cv::DECOMP_NORMAL` | Optional additional flag; indicates that the normal equations are to be solved instead |

The methods `cv::DECOMP_LU` and `cv::DECOMP_CHOLESKY` cannot be used on singular matrices. If a singular `lhs` argument is provided, both methods will exit and return `0` (a 1 will be returned if `lhs` is nonsingular). You can use `cv::solve()` to solve overdetermined linear systems using either QR decomposition (`cv::DECOMP_QR`) or singular value decomposition (`cv::DECOMP_SVD`) methods to find the least-squares solution for the given system of equations. Both of these methods can be used in case the matrix `lhs` is singular.

Though the first five arguments in Table 5-11 are mutually exclusive, the last option, `cv::DECOMP_NORMAL`, may be combined with any of the first five (e.g., by logical OR: `cv_DECOMP_LU | cv::DECOMP_NORMAL`). If provided, then `cv::solve()` will attempt to solve the *normal equations*: $lhs^T \cdot lhs \cdot dst = lhs^T \cdot rhs$ instead of the usual system $lhs \cdot dst = rhs$.

## cv::solveCubic()

```
int cv::solveCubic(
  cv::InputArray  coeffs,
  cv::OutputArray roots
);
```

Given a cubic polynomial in the form of a three- or four-element vector `coeffs`, `cv::solveCubic()` will compute the real roots of that polynomial. If `coeffs` has four elements, the roots of the following polynomial are computed:

$$\text{coeffs}_0 x^3 + \text{coeffs}_1 x^2 + \text{coeffs}_2 x + \text{coeffs}_3 = 0$$

If `coeffs` has only three elements, the roots of the following polynomial are computed:

$$x^3 + \text{coeffs}_0 x^2 + \text{coeffs}_1 x + \text{coeffs}_2 = 0$$

The results are stored in the array `roots`, which will have either one or three elements, depending on how many real roots the polynomial has.

A word of warning about `cv::solveCubic()` and `cv::solvePoly()`: the order of the coefficients in the seemingly analogous input arrays `coeffs` is opposite in the two routines. In `cv::solveCubic()`, the highest-order coefficients come last, while in `cv::solvePoly()` the highest-order coefficients come first.

## cv::solvePoly()

```
int cv::solvePoly (
  cv::InputArray  coeffs,
  cv::OutputArray roots                  // n complex roots (2-channels)
  int             maxIters = 300         // maximum iterations for solver
);
```

Given a polynomial of any order in the form of a vector of coefficients `coeffs`, `cv::solvePoly()` will attempt to compute the roots of that polynomial. Given the array of coefficients `coeffs`, the roots of the following polynomial are computed:

$$\text{coeffs}_n x^n + \text{coeffs}_{n-1} x^{n-1} + \ \ldots \ + \text{coeffs}_1 x + \text{coeffs}_0 = 0$$

Unlike `cv::solveCubic()`, these roots are not guaranteed to be real. For an order-*n* polynomial (i.e., `coeffs` having `n+1` elements), there will be `n` roots. As a result, the array `roots` will be returned in a two-channel (real, imaginary) matrix of doubles.

## cv::sort()

```
void cv::sort(
  cv::InputArray  src,
  cv::OutputArray dst,
  int             flags
);
```

The OpenCV sort function is used for two-dimensional arrays. Only single-channel source arrays are supported. You should not think of this like sorting rows or columns in a spreadsheet; `cv::sort()` sorts every row or column *separately*. The result of the sort operation will be a new array, `dst`, which is of the same size and type as the source array.

You can sort on every row or on every column by supplying either the `cv::SORT_EVERY_ROW` or `cv::SORT_EVERY_COLUMN` flag. Sort can be in ascending or descending order, which is indicated by the `cv::SORT_ASCENDING` or `cv::SORT_DESCENDING` flag, respectively. One flag from each of the two groups is required.

## cv::sortIdx()

```
void cv::sortIdx(
  cv::InputArray  src,
  cv::OutputArray dst,
  int             flags
);
```

Similar to cv::sort(), cv::sortIdx() is used only for single-channel two-dimensional arrays. cv::sortIdx() sorts every row or column *separately*. The result of the sort operation is a new array, dst, of the same size as the source array but containing the integer indices of the sorted elements. For example, given an array A, a call to cv::sortIdx( A, B, cv::SORT_EVERY_ROW | cv::SORT_DESCENDING ) would produce:

$$A = \left\| \begin{bmatrix} 0.0 & 0.1 & 0.2 \\ 1.0 & 1.1 & 1.2 \\ 2.0 & 2.1 & 2.2 \end{bmatrix} \right\| \quad B = \left\| \begin{bmatrix} 2 & 1 & 0 \\ 2 & 1 & 0 \\ 2 & 1 & 0 \end{bmatrix} \right\|$$

In this toy case, every row was previously ordered from lowest to highest, and sorting has indicated that this should be reversed.

## cv::split()

```
void cv::split(
  const cv::Mat&    mtx,
  cv::Mat*          mv
);

void cv::split(
  const cv::Mat&    mtx,
  vector<Mat>&      mv                    // STL-style vector of n 1-channel cv::Mat's
);
```

The function cv::split() is a special, simpler case of cv::mixChannels(). Using cv::split(), you separate the channels in a multichannel array into multiple single-channel arrays. There are two ways of doing this. In the first, you supply a pointer to a C-style array of pointers to cv::Mat objects that cv::split() will use for the results of the split operation. In the second option, you supply an STL vector full of cv::Mat objects. If you use the C-style array, you need to make sure that the number of cv::Mat objects available is (at least) equal to the number of channels in mtx. If you use the STL vector form, cv::split() will handle the allocation of the result arrays for you.

## cv::sqrt()

```
void cv::sqrt(
  cv::InputArray  src,
  cv::OutputArray dst
);
```

As a special case of cv::pow(), cv::sqrt() will compute the element-wise square root of an array. Multiple channels are processed separately.

There is also (sometimes) such thing as a square root of a matrix; that is, a matrix $B$ whose relationship with some matrix $A$ is that $BB = A$. If $A$ is square and positive definite, then if $B$ exists, it is unique.

If $A$ can be diagonalized, then there is a matrix $V$ (made from the eigenvectors of $A$ as columns) such that $A = VDV^{-1}$, where $D$ is a diagonal matrix. The square root of a diagonal matrix $D$ is just the square roots of the elements of $D$. So to compute $A^{1/2}$, we simply use the matrix $V$ and get:

$$A^{1/2} = VD^{1/2}V^{-1}$$

Math fans can easily verify that this expression is correct by explicitly squaring it:

$$\left(A^{1/2}\right)^2 = \left(VD^{1/2}V^{-1}\right)\left(VD^{1/2}V^{-1}\right) = VD^{1/2}V^{-1}VD^{1/2}V^{-1}$$
$$= VDV^{-1} = A$$

In code, this would look something like:[13]

```
void matrix_square_root( const cv::Mat& A, cv::Mat& sqrtA ) {
  cv::Mat U, V, Vi, E;
  cv::eigen( A, E, U );
  V = U.T();
  cv::transpose( V, Vi ); // inverse of the orthogonal V
  cv::sqrt(E, E);         // assume that A is positively
                          // defined, otherwise its
                          // square root will be
                          // complex-valued
  sqrtA = V * Mat::diag(E) * Vi;
}
```

13 Here "something like" means that if you were really writing a responsible piece of code, you would do a lot of checking to make sure that the matrix you were handed was in fact what you thought it was (i.e., square). You would also probably want to check the return values of `cv::eigen()` and `cv::invert()`, think more carefully about the actual methods used for the decomposition and inversion, and make sure the eigenvalues were positive before blindly calling `sqrt()` on them.

## cv::subtract()

```
void cv::subtract(
  cv::InputArray  src1,                     // First input array
  cv::InputArray  src2,                     // Second input array
  cv::OutputArray dst,                      // Result array
  cv::InputArray  mask  = cv::noArray(),    // Optional, do only where nonzero
  int             dtype = -1                // Output type for result array
);
```

$$dst_i = saturate(src1_i - src2_i)$$

`cv::subtract()` is a simple subtraction function: it subtracts all of the elements in `src2` from the corresponding elements in `src1` and puts the results in `dst`.

For simple cases, the same result can be achieved with the matrix operation:

```
dst = src1 - src2;
```

Accumulation is also supported:

```
dst -= src1;
```

## cv::sum()

```
cv::Scalar cv::sum(
  cv::InputArray  arr
);
```

$$sum_c = \sum_{i,j} arr_{c\,i,j}$$

`cv::sum()` sums all of the pixels in each channel of the array `arr`. The return value is of type `cv::Scalar`, so `cv::sum()` can accommodate multichannel arrays, but only up to four channels. The sum for each channel is placed in the corresponding component of the `cv::scalar` return value.

## cv::trace()

```
cv::Scalar cv::trace(
  cv::InputArray  mat
);
```

$$Tr(mat)_c = \sum_i mat_{c\,i,i}$$

The trace of a matrix is the sum of all of the diagonal elements. The trace in OpenCV is implemented on top of `cv::diag()`, so it does not require the array passed in to be square. Multichannel arrays are supported, but the trace is computed as a scalar so each component of the scalar will be the sum over each corresponding channel for up to four channels.

## cv::transform()

```
void cv::transform(
  cv::InputArray  src,
  cv::OutputArray dst,
  cv::InputArray  mtx
);
```

$$dst_{c,i,j} = \sum_{c'} mtx_{c,c'} src_{c',i,j}$$

The function `cv::transform()` can be used to compute arbitrary linear image transforms. It treats a multichannel input array `src` as a collection of vectors in what you could think of as "channel space." Those vectors are then each multiplied by the "small" matrix `mtx` to affect a transformation in this channel space.

The `mtx` matrix must have as many rows as there are channels in `src`, or that number plus one. In the second case, the channel space vectors in `src` are automatically extended by one and the value `1.0` is assigned to the extended element.

The exact meaning of this transformation depends on what you are using the different channels for. If you are using them as color channels, then this transformation can be thought of as a linear color space transformation. Transformation between RGB and YUV color spaces is an example of such a transformation. If you are using the channels to represent the *x,y* or *x,y,z* coordinates of points, then these transformations can be thought of as rotations (or other geometrical transformations) of those points.

## cv::transpose()

```
void cv::transpose(
  cv::InputArray  src,                    // Input array, 2-dimensional, n-by-m
  cv::OutputArray dst,                    // Result array, 2-dimensional, m-by-n
);
```

`cv::transpose()` copies every element of `src` into the location in `dst` indicated by reversing the row and column indices. This function does support multichannel arrays; however, if you are using multiple channels to represent complex numbers, remember that `cv::transpose()` does not perform complex conjugation.

This same result can be achieved with the matrix member function `cv::Mat::t()`. The member function has the advantage that it can be used in matrix expressions like:

```
A = B + B.t();
```

# Summary

In this chapter, we looked at a vast array of basic operations that can be done with the all-important OpenCV array structure `cv::Mat`, which can contain matrices, images, and multidimensional arrays. We saw that the library provides operations ranging from very simple algebraic manipulations up through some relatively complicated features. Some of these operations are designed to help us manipulate arrays as images, while others are useful when the arrays represent other kinds of data. In the coming chapters, we will look at more sophisticated algorithms that implement meaningful computer vision algorithms. Relative to those algorithms, and many that you will write, the operations in this chapter will form the basic building blocks for just about anything you want to do.

# Exercises

In the following exercises, you may need to refer to the reference manual (*http://docs.opencv.org*) for details of the functions outlined in this chapter.

1. This exercise will accustom you to the idea of many functions taking matrix types. Create a two-dimensional matrix with three channels of type `byte` with data size 100 × 100. Set all the values to `0`.
    a. Draw a circle in the matrix using `void cv::circle(InputOutputArray img, cv::point center, intradius, cv::Scalar color, int thickness=1, int line_type=8, int shift=0)`.
    b. Display this image using methods described in Chapter 2.
2. Create a two-dimensional matrix with three channels of type `byte` with data size 100 × 100, and set all the values to `0`. Use the `cv::Mat` element access functions to modify the pixels. Draw a green rectangle between (20, 5) and (40, 20).
3. Create a three-channel RGB image of size 100 × 100. Clear it. Use pointer arithmetic to draw a green square between (20, 5) and (40, 20).
4. Practice using region of interest (ROI). Create a 210 × 210 single-channel `byte` image and zero it. Within the image, build a pyramid of increasing values using

ROI and `cv::Mat::setTo()`. That is: the outer border should be `0`, the next inner border should be `20`, the next inner border should be `40`, and so on until the final innermost square is set to value `200`; all borders should be 10 pixels wide. Display the image.

5. Use multiple headers for one image. Load an image that is at least 100 × 100. Create two additional headers that are ROIs where `width` = `20` and the `height` = `30`. Their origins should be at (5, 10) and (50, 60), respectively. Pass these new image subheaders to `cv::bitwise_not()`. Display the loaded image, which should have two inverted rectangles within the larger image.

6. Create a mask using `cv::compare()`. Load a real image. Use `cv::split()` to split the image into red, green, and blue images.

   a. Find and display the green image.

   b. Clone this green plane image twice (call these `clone1` and `clone2`).

   c. Find the green plane's minimum and maximum value.

   d. Set `clone1`'s values to `thresh = (unsigned char)((maximum - minimum)/ 2.0)`.

   e. Set `clone2` to `0` and use `cv::compare (green_image, clone1, clone2, cv::CMP_GE)`. Now `clone2` will have a mask of where the value exceeds `thresh` in the green image.

   f. Finally, use `cv::subtract (green_image,thresh/2, green_image, clone2)` and display the results.

CHAPTER 6

# Drawing and Annotating

## Drawing Things

We often want to draw some kind of picture, or to draw something on top of an image obtained from somewhere else. Toward this end, OpenCV provides a menagerie of functions that will allow us to make lines, squares, circles, and the like.

OpenCV's drawing functions work with images of any depth, but most of them affect only the first three channels—defaulting to only the first channel in the case of single-channel images. Most of the drawing functions support a color, a thickness, a line type (which really refers to whether to anti-alias lines), and subpixel alignment of objects.

When you specify colors, the convention is to use the `cv::Scalar` object, even though only the first three values are used most of the time. (It is sometimes convenient to be able to use the fourth value in a `cv::Scalar` to represent an alpha channel, but the drawing functions do not currently support alpha blending.) Also, by convention, OpenCV uses BGR ordering[1] for converting multichannel images to color renderings (this is what is used by the draw function `imshow()`, which actually paints images onto your screen for viewing). Of course, you don't have to use this convention, and it might not be ideal if you are using data from some other library with OpenCV headers on top of it. In any case, the core functions of the library are always agnostic to any "meaning" you might assign to a channel.

1 There is a slightly confusing point here, which is mostly due to legacy in origin. The macro `CV_RGB(r,g,b)` produces a `cv::Scalar s` with value `s.val[] = { b, g, r, 0 }`. This is as it should be, as general OpenCV functions know what is red, green, or blue only by the order, and the ordering convention for image data is BGR as stated in the text.

## Line Art and Filled Polygons

Functions that draw lines of one kind or another (segments, circles, rectangles, etc.) will usually accept a `thickness` and `lineType` parameter. Both are integers, but the only accepted values for the latter are `4`, `8`, or `cv::LINE_AA`. `thickness` is the thickness of the line measured in pixels. For circles, rectangles, and all of the other closed shapes, the `thickness` argument can also be set to `cv::FILLED` (which is an alias for `-1`). In that case, the result is that the drawn figure will be filled in the same color as the edges. The `lineType` argument indicates whether the lines should be "4-connected," "8-connected," or anti-aliased. For the first two examples in Figure 6-1, the Bresenham algorithm is used, while for the anti-aliased lines, Gaussian filtering is used. Wide lines are always drawn with rounded ends.

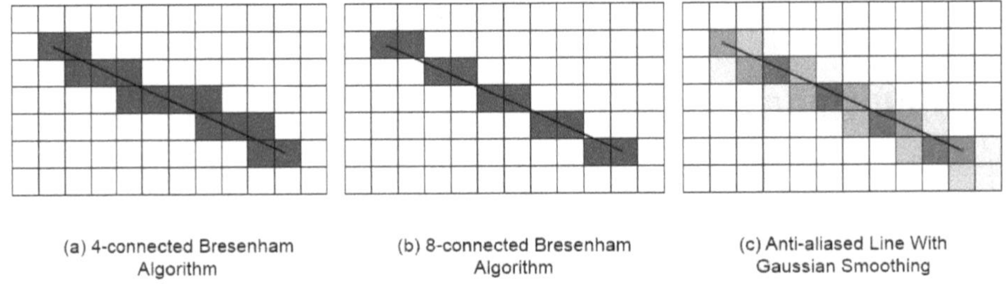


*Figure 6-1. The same line as it would be rendered using the 4-connected (a), 8-connected (b), and anti-aliased (c) line types*

For the drawing algorithms listed in Table 6-1, endpoints (lines), center points (circles), corners (rectangles), and so on are typically specified as integers. However, these algorithms support subpixel alignment through the `shift` argument. Where `shift` is available, it is interpreted as the number of bits in the integer arguments to treat as fractional bits. For example, if you say you want a circle centered at (5, 5), but set `shift` to `1`, then the circle will be drawn at (2.5, 2.5). The effect of this will typically be quite subtle, and depend on the line type used. The effect is most noticeable for anti-aliased lines.

*Table 6-1. Drawing functions*

| Function | Description |
|---|---|
| `cv::circle()` | Draw a simple circle |
| `cv::clipLine()` | Determine if a line is inside a given box |
| `cv::ellipse()` | Draw an ellipse, which may be tilted or an elliptical arc |
| `cv::ellipse2Poly()` | Compute a polygon approximation to an elliptical arc |
| `cv::fillConvexPoly()` | Draw filled versions of simple polygons |
| `cv::fillPoly()` | Draw filled versions of arbitrary polygons |

| Function | Description |
|---|---|
| `cv::line()` | Draw a simple line |
| `cv::rectangle()` | Draw a simple rectangle |
| `cv::polyLines()` | Draw multiple polygonal curves |

The following sections describe the details of each function in Table 6-1.

### cv::circle()

```
void circle(
  cv::Mat&          img,                // Image to be drawn on
  cv::Point         center,             // Location of circle center
  int               radius,             // Radius of circle
  const cv::Scalar& color,              // Color, RGB form
  int               thickness = 1,      // Thickness of line
  int               lineType  = 8,      // Connectedness, 4 or 8
  int               shift     = 0       // Bits of radius to treat as fraction
);
```

The first argument to `cv::circle()` is just your image, `img`. Next are the `center`, a two-dimensional point, and the `radius`. The remaining arguments are the standard `color`, `thickness`, `lineType`, and `shift`. The shift is applied to both the radius and the center location.

### cv::clipLine()

```
bool clipLine(                        // True if any part of line in 'imgRect'
  cv::Rect          imgRect,          // Rectangle to clip to
  cv::Point&        pt1,              // First endpoint of line, overwritten
  cv::Point&        pt2               // Second endpoint of line, overwritten
);

bool clipLine(                        // True if any part of line in image size
  cv::Size          imgSize,          // Size of image, implies rectangle at 0,0
  cv::Point&        pt1,              // First endpoint of line, overwritten
  cv::Point&        pt2               // Second endpoint of line, overwritten
);
```

This function is used to determine if a line specified by the two points `pt1` and `pt2` lies inside a rectangular boundary. In the first version, a `cv::Rect` is supplied and the line is compared to that rectangle. `cv::clipLine()` will return `False` only if the line is entirely outside of the specified rectangular region. The second version is the same except it takes a `cv::Size` argument. Calling this second version is equivalent to calling the first version with a rectangle whose (*x*, *y*) location is (`0`, `0`).

### cv::ellipse()

```
bool ellipse(
  cv::Mat&                 img,              // Image to be drawn on
  cv::Point                center,           // Location of ellipse center
  cv::Size                 axes,             // Length of major and minor axes
  double                   angle,            // Tilt angle of major axis
  double                   startAngle,       // Start angle for arc drawing
  double                   endAngle,         // End angle for arc drawing
  const cv::Scalar&        color,            // Color, BGR form
  int                      thickness = 1,    // Thickness of line
  int                      lineType  = 8,    // Connectedness, 4 or 8
  int                      shift     = 0     // Bits of radius to treat as fraction
);

bool ellipse(
  cv::Mat&                 img,              // Image to be drawn on
  const cv::RotatedRect&   rect,             // Rotated rectangle bounds ellipse
  const cv::Scalar&        color,            // Color, BGR form
  int                      thickness = 1,    // Thickness of line
  int                      lineType  = 8,    // Connectedness, 4 or 8
  int                      shift     = 0     // Bits of radius to treat as fraction
);
```

The `cv::ellipse()` function is very similar to the `cv::circle()` function, with the primary difference being the `axes` argument, which is of type `cv::Size`. In this case, the `height` and `width` arguments represent the length of the ellipse's major and minor axes. The `angle` is the angle (in degrees) of the major axis, which is measured counterclockwise from horizontal (i.e., from the *x*-axis). Similarly, the `startAngle` and `endAngle` indicate (also in degrees) the angle for the arc to start and for it to finish. Thus, for a complete ellipse, you must set these values to 0 and 360, respectively.

The alternate way to specify the drawing of an ellipse is to use a bounding box. In this case, the argument `box` of type `cv::RotatedRect` completely specifies both the size and the orientation of the ellipse. Both methods of specifying an ellipse are illustrated in Figure 6-2.

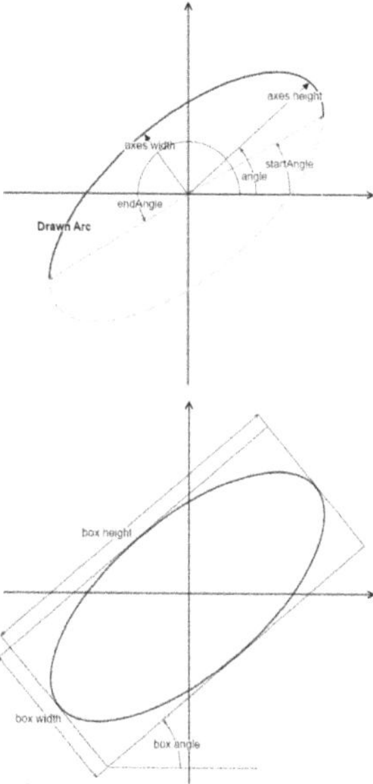


*Figure 6-2. An elliptical arc specified by the major and minor axes with tilt angle (left); a similar ellipse specified using a cv::RotatedRect (right)*

### cv::ellipse2Poly()

```
void ellipse2Poly(
  cv::Point              center,       // Location of ellipse center
  cv::Size               axes,         // Length of major and minor axes
  double                 angle,        // Tilt angle of major axis
  double                 startAngle,   // Start angle for arc drawing
  double                 endAngle,     // End angle for arc drawing
  int                    delta,        // Angle between sequential vertices
  vector<cv::Point>&     pts           // Result, STL-vector of points
);
```

The `cv::ellipse2Poly()` function is used internally by `cv::ellipse()` to compute elliptical arcs, but you can call it yourself as well. Given information about an elliptical arc (`center`, `axes`, `angle`, `startAngle`, and `endAngle`—all as defined in `cv::ellipse()`) and a parameter `delta`, which specifies the angle between subsequent points you want to sample, `cv::ellipse2Poly()` computes a sequence of points that form a polygonal approximation to the elliptical arc you specified. The computed points are returned in the `vector<> pts`.

### cv::fillConvexPoly()

```
void fillConvexPoly(
  cv::Mat&           img,                    // Image to be drawn on
  const cv::Point*   pts,                    // C-style array of points
  int                npts,                   // Number of points in 'pts'
  const cv::Scalar&  color,                  // Color, BGR form
  int                lineType   = 8,         // Connectedness, 4 or 8
  int                shift      = 0          // Bits of radius to treat as fraction
);
```

This function draws a filled polygon. It is much faster than `cv::fillPoly()` (described next) because it uses a much simpler algorithm. The algorithm used by `cv::fillConvexPoly()`, however, will not work correctly if the polygon you pass to it has self-intersections.[2] The points in `pts` are treated as sequential, and a segment from the last point in `pts` and the first point is implied (i.e., the polygon is assumed to be closed).

### cv::fillPoly()

```
void fillPoly(
  cv::Mat&           img,                    // Image to be drawn on
  const cv::Point*   pts,                    // C-style array of arrays of points
  int                npts,                   // Number of points in 'pts[i]'
  int                ncontours,              // Number of arrays in 'pts'
  const cv::Scalar&  color,                  // Color, BGR form
  int                lineType = 8,           // Connectedness, 4 or 8
  int                shift    = 0,           // Bits of radius to treat as fraction
  cv::Point          offset   = Point()      // Uniform offset applied to all points
);
```

This function draws any number of filled polygons. Unlike `cv::fillConvexPoly()`, it can handle polygons with self-intersections. The `ncontours` argument specifies how many different polygon contours there will be, and the `npts` argument is a C-style array that indicates how many points there are in each contour (i.e., `npts[i]` indicates how many points there are in polygon `i`). `pts` is a C-style array of C-style arrays containing all of the points in those polygons (i.e., `pts[i][j]` contains the `j`th point in the `i`th polygon). `cv::fillPoly()` also has one additional argument, `offset`, which is a pixel offset that will be applied to all vertex locations when the polygons are drawn. The polygons are assumed to be closed (i.e., a segment from the last element of `pts[i][]` to the first element will be assumed).

---

2 The algorithm used by `cv::fillComvexPoly()` is actually somewhat more general than implied here. It will correctly draw any polygon whose contour intersects every horizontal line at most twice (though it is allowed for the top or bottom of the polygon to be flat with respect to the horizontal). Such a polygon is said to be "monotone with respect to the horizontal."

### cv::line()

```
void line(
  cv::Mat&          img,                 // Image to be drawn on
  cv::Point         pt1,                 // First endpoint of line
  cv::Point         pt2                  // Second endpoint of line
  const cv::Scalar& color,               // Color, BGR form
  int               lineType = 8,        // Connectedness, 4 or 8
  int               shift    = 0         // Bits of radius to treat as fraction
);
```

The function cv::line() draws a straight line from pt1 to pt2 in the image img. Lines are automatically clipped by the image boundaries.

### cv::rectangle()

```
void rectangle(
  cv::Mat&          img,                 // Image to be drawn on
  cv::Point         pt1,                 // First corner of rectangle
  cv::Point         pt2                  // Opposite corner of rectangle
  const cv::Scalar& color,               // Color, BGR form
  int               lineType = 8,        // Connectedness, 4 or 8
  int               shift    = 0         // Bits of radius to treat as fraction
);

void rectangle(
  cv::Mat&          img,                 // Image to be drawn on
  cv::Rect          r,                   // Rectangle to draw
  const cv::Scalar& color,               // Color, BGR form
  int               lineType = 8,        // Connectedness, 4 or 8
  int               shift    = 0         // Bits of radius to treat as fraction
);
```

The cv::rectangle() function draws a rectangle with corners pt1 to pt2 in the image img. An alternate form of this function allows the rectangle's location and size to be specified by a single cv::Rect argument, r.

### cv::polyLines()

```
void polyLines(
  cv::Mat&          img,                 // Image to be drawn on
  const cv::Point*  pts,                 // C-style array of arrays of points
  int               npts,                // Number of points in 'pts[i]'
  int               ncontours,           // Number of arrays in 'pts'
  bool              isClosed,            // If true, connect last and first pts
  const cv::Scalar& color,               // Color, BGR form
  int               lineType = 8,        // Connectedness, 4 or 8
  int               shift    = 0         // Bits of radius to treat as fraction
);
```

This function draws any number of unfilled polygons. It can handle general polygons, including polygons with self-intersections. The ncontours argument specifies how many different polygon contours there will be, and the npts argument is a C-style array that indicates how many points there are in each contour (i.e., npts[i] indicates how many points there are in polygon i). pts is a C-style array of C-style arrays containing all of the points in those polygons (i.e., pts[i][j] contains the jth point in the ith polygon). Polygons are not assumed to be closed. If the argument isClosed is true, then a segment from the last element of pts[i][] to the first element will be assumed. Otherwise, the contour is taken to be an open contour containing only npts[i]-1 segments between the npts[i] points listed.

### cv::LineIterator

```
LineIterator::LineIterator(
  cv::Mat&        img,                    // Image to be drawn on
  cv::Point       pt1,                    // First endpoint of line
  cv::Point       pt2                     // Second endpoint of line
  int             lineType = 8,           // Connectedness, 4 or 8
  bool            leftToRight = false     // If true, always start steps on the left
);
```

The cv::LineIterator object is an iterator that is used to get each pixel of a raster line in sequence. The line iterator is our first example of a *functor* in OpenCV. We will see several more of these "objects that do stuff" in the next chapter. The constructor for the line iterator takes the two endpoints for the line as well as a line-type specifier and an additional Boolean that indicates which direction the line should be traversed.

Once initialized, the number of pixels in the line is stored in the member integer cv::LineIterator::count. The overloaded dereferencing operator cv::LineIterator::operator*() returns a pointer of type uchar*, which points to the "current" pixel. The current pixel starts at one end of the line and is incremented by means of the overloaded increment operator cv::LineIterator::operator++(). The actual traversal is done according to the Bresenham algorithm mentioned earlier.

The purpose of the cv::LineIterator is to make it possible for you to take some specific action on each pixel along the line. This is particularly handy when you are creating special effects such as switching the color of a pixel from black to white and white to black (i.e., an XOR operation on a binary image).

When accessing an individual "pixel," remember that this pixel may have one or many channels and it might be any kind of image depth. The return value from the dereferencing operator is always uchar*, so you are responsible for casting that pointer to the correct type. For example, if your image were a three-channel image of 32-bit floating-point numbers and your iterator were called iter, then you would want to cast the return (pointer) value of the dereferencing operator like this: (Vec3f*)*iter.

The style of the overloaded dereferencing operator `cv::LineIterator::operator*()` is slightly different than what you are probably used to from libraries like STL. The difference is that the return value from the iterator is itself a pointer, so the iterator behaves not like a pointer, but like a pointer to a pointer.

## Fonts and Text

One additional form of drawing is to draw text. Of course, text creates its own set of complexities, but—as always—OpenCV is more concerned with providing a simple "down and dirty" solution that will work for simple cases than a robust, complex solution (which would be redundant anyway given the capabilities of other libraries). Table 6-2 lists OpenCV's two text drawing functions.

*Table 6-2. Text drawing functions*

| Function | Description |
|---|---|
| `cv::putText()` | Draw the specified text in an image |
| `cv::getTextSize()` | Determine the width and height of a text string |

### cv::putText()

```
void cv::putText(
  cv::Mat&      img,                      // Image to be drawn on
  const string& text,                     // write this (often from cv::format)
  cv::Point     origin,                   // Upper-left corner of text box
  int           fontFace,                 // Font (e.g., cv::FONT_HERSHEY_PLAIN)
  double        fontScale,                // size (a multiplier, not "points"!)
  cv::Scalar    color,                    // Color, RGB form
  int           thickness = 1,            // Thickness of line
  int           lineType  = 8,            // Connectedness, 4 or 8
  bool          bottomLeftOrigin = false  // true='origin at lower left'
);
```

This function is OpenCV's one main text drawing routine; it simply throws some text onto an image. The text indicated by `text` is printed with its upper-left corner of the text box at `origin` and in the color indicated by `color`, unless the `bottomLeftOrigin` flag is `true`, in which case the lower-left corner of the text box is located at `origin`. The font used is selected by the `fontFace` argument, which can be any of those listed in Table 6-3.

*Table 6-3. Available fonts (all are variations of Hershey)*

| Identifier | Description |
|---|---|
| `cv::FONT_HERSHEY_SIMPLEX` | Normal size sans-serif |
| `cv::FONT_HERSHEY_PLAIN` | Small size sans-serif |
| `cv::FONT_HERSHEY_DUPLEX` | Normal size sans-serif; more complex than `cv::FONT_HERSHEY_SIMPLEX` |
| `cv::FONT_HERSHEY_COMPLEX` | Normal size serif; more complex than `cv::FONT_HERSHEY_DUPLEX` |
| `cv::FONT_HERSHEY_TRIPLEX` | Normal size serif; more complex than `cv::FONT_HERSHEY_COMPLEX` |
| `cv::FONT_HERSHEY_COMPLEX_SMALL` | Smaller version of `cv::FONT_HERSHEY_COMPLEX` |
| `cv::FONT_HERSHEY_SCRIPT_SIMPLEX` | Handwriting style |
| `cv::FONT_HERSHEY_SCRIPT_COMPLEX` | More complex variant of `cv::FONT_HERSHEY_SCRIPT_SIMPLEX` |

Any of the font names listed in Table 6-3 can also be combined (via an OR operator) with `cv::FONT_HERSHEY_ITALIC` to render the indicated font in italics. Each font has a "natural" size. When `fontScale` is not `1.0`, then the font size is scaled by this number before the text is drawn. Figure 6-3 shows a sample of each font.

*Figure 6-3. The eight fonts of Table 6-3, with the origin of each line separated from the vertical by 30 pixels*

### cv::getTextSize()

```
cv::Size cv::getTextSize(
  const string& text,
  cv::Point     origin,
  int           fontFace,
  double        fontScale,
  int           thickness,
  int*          baseLine
);
```

The `cv::getTextSize()` function answers the question of how big some text would be if you were to draw it (with some set of parameters) without actually drawing it on an image. The only novel argument to `cv::getTextSize()` is **`baseLine`**, which is actually an output parameter. **`baseLine`** is the y-coordinate of the text baseline relative to the bottommost point in the text.[3]

# Summary

In this short chapter we learned a few new functions that we can use to draw on and annotate images. These functions all operate on the same `cv::Mat` image types we have been using in the prior chapters. Most of these functions have very similar interfaces, allowing us to draw lines and curves of various thicknesses and colors. In addition to lines and curves, we also saw how OpenCV handles writing text onto an image. All of these functions are extremely useful, in practice, when we are debugging code, as well as for displaying results of our computations on top of the images they are using for input.

# Exercises

For the following exercises, modify the code of Example 2-1 to get an image displayed, or modify the code from Example 2-3 to load and display video or camera images.

1. Drawing practice: load or create and display a color image. Draw one example of every shape and line that OpenCV can draw.
2. Grayscale: load and display a color image.
   a. Turn it into three-channel grayscale (it is still an BGR image, but it looks gray to the user).

3 The "baseline" is the line on which the bottoms of characters such as *a* and *b* are aligned. Characters such as *y* and *g* hang below the baseline.

b. Draw color text onto the image.

3. Dynamic text: load and display video from a video or from a camera.

   a. Draw a *frame per second* (FPS) counter somewhere on the image.

4. Make a drawing program. Load or create an image and display it.

   a. Allow a user to draw a basic face.

   b. Make the components of the face editable (you will have to maintain a list of what was drawn, and when it is altered you might erase and redraw it the new size).

5. Use `cv::LineIterator` to count pixels on different line segments in, say, a 300 × 300 image.

   a. At what angles do you get the same number of pixels for 4-connected and 8-connected lines?

   b. For line segment angles other than the above, which counts more pixels: 4-connected or 8-connected lines?

   c. For a given line segment, explain the difference in the length of the line compared to the number of pixels you count iterating along the line for both 4-connected and 8-connected? Which connectedness is closer to the true line length?

CHAPTER 7

# Functors in OpenCV

## Objects That "Do Stuff"

As the OpenCV library has evolved, it has become increasingly common to introduce new objects that encapsulate functionality that is too complicated to be associated with a single function and which, if implemented as a set of functions, would cause the overall function space of the library to become too cluttered.[1]

As a result, new functionality is often represented by an associated new object type, which can be thought of as a "machine" that does whatever this functionality is. Most of these machines have an overloaded `operator()`, which officially makes them *function objects* or *functors*. If you are not familiar with this programming idiom, the important idea is that unlike "normal" functions, function objects are created and can maintain state information inside them. As a result, they can be set up with whatever data or configuration they need, and they are "asked" to perform services through either common member functions, or by being called as functions themselves (usually via the overloaded `operator()`[2]).

## Principal Component Analysis (cv::PCA)

*Principal component analysis*, illustrated in Figure 7-1, is the process of analyzing a distribution in many dimensions and extracting from that distribution the particular

1 We encountered one of these objects briefly in the previous chapter with the `cv::LineIterator` object.

2 Here the word *usually* means "usually when people program function objects," but does not turn out to mean "usually for the OpenCV library." There is a competing convention in the OpenCV library that uses the overloaded `operator()` to load the configuration, and a named member to provide the fundamental service of the object. This convention is substantially less canonical in general, but quite common in the OpenCV library.

subset of dimensions that carry the most information. The dimensions computed by PCA are not necessarily the basis dimensions in which the distribution was originally specified. Indeed, one of the most important aspects of PCA is the ability to generate a new basis whose axes can be ordered by their importance.[3] These basis vectors will turn out to be the eigenvectors of the covariance matrix for the distribution as a whole, and the corresponding eigenvalues will tell us about the extent of the distribution in that dimension.

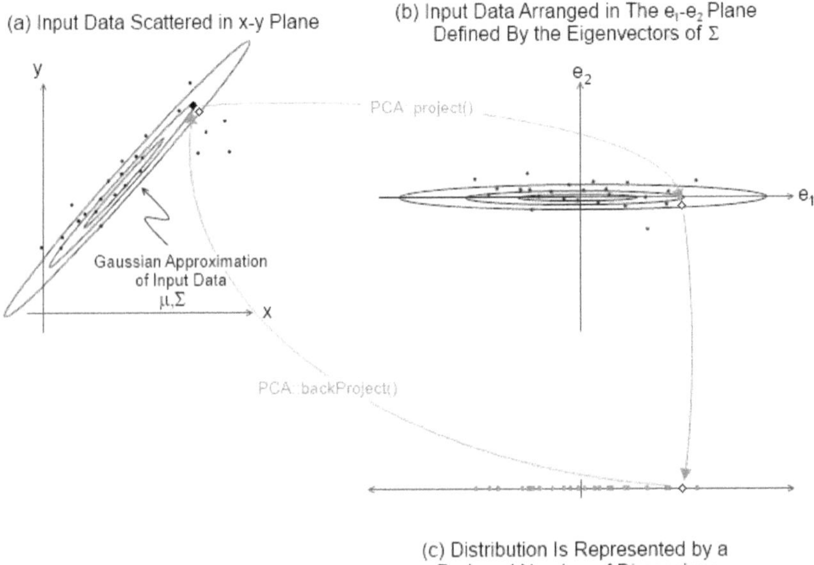


*Figure 7-1. (a) Input data is characterized by a Gaussian approximation; (b) the data is projected into the space implied by the eigenvectors of the covariance of that approximation; (c) the data is projected by the KLT projection to a space defined only by the most "useful" of the eigenvectors; superimposed: a new data point (the white diamond) is projected to the reduced dimension space by cv::PCA::project(); that same point is brought back to the original space (the black diamond) by cv::PCA::backProject()*

We are now in a position to explain why PCA is handled by one of these function objects. Given a distribution once, the PCA object can compute and retain this new basis. The big advantage of the new basis is that the basis vectors that correspond to the large eigenvalues carry most of the information about the objects in the distribu-

3 You might be thinking to yourself, "Hey, this sounds like machine learning—what is it doing in this chapter?" This is not a bad question. In modern computer vision, machine learning is becoming intrinsic to an ever-growing list of algorithms. For this reason, component capabilities, such as PCA and SVD, are increasingly considered "building blocks."

tion. Thus, without losing much accuracy, we can throw away the less informative dimensions. This dimension reduction is called a *KLT transform*.[4] Once you have loaded a sample distribution and the principal components are computed, you might want to use that information to do various things, such as apply the KLT transform to new vectors. When you make the PCA functionality a function object, it can "remember" what it needs to know about the distribution you gave it, and thereafter use that information to provide the "service" of transforming new vectors on demand.

### cv::PCA::PCA()

```
PCA::PCA();
PCA::PCA(
  cv::InputArray data,                // Data, as rows or cols in 2d array
  cv::InputArray mean,                // average, if known, 1-by-n or n-by-1
  int            flags,               // Are vectors rows or cols of 'data'
  int            maxComponents = 0    // Max dimensions to retain
);
```

The PCA object has a default constructor, `cv::PCA()`, which simply builds the PCA object and initializes the empty structure. The second form executes the default construction, then immediately proceeds to pass its arguments to `PCA::operator()()` (discussed next).

### cv::PCA::operator()()

```
PCA::operator()(
  cv::InputArray data,                // Data, as rows or cols in 2d array
  cv::InputArray mean,                // average, if known, 1-by-n or n-by-1
  int            flags,               // Are vectors rows or cols of 'data'
  int            maxComponents = 0    // Max dimensions to retain
);
```

The overloaded `operator()()` for `PCA` builds the model of the distribution inside of the PCA object. The `data` argument is an array containing all of the samples that constitute the distribution. Optionally, `mean`, a second array that contains the mean value in each dimension, can be supplied (`mean` can either be $n \times 1$ or $1 \times n$). The data can be arranged as an $n \times D$ ($n$ rows of samples, each of $D$ dimensions) or $D \times n$ array ($n$ columns of samples, each of $D$ dimensions). The `flags` argument is currently used only to specify the arrangement of the data in `data` and `mean`. In particular, `flags` can be set to either `cv::PCA_DATA_AS_ROW` or `cv::PCA_DATA_AS_COL`, to indicate that either `data` is $n \times D$ and `mean` is $n \times 1$ or `data` is $D \times n$ and `mean` is $1 \times n$, respectively.

4 KLT stands for "Karhunen-Loeve Transform," so the phrase *KLT transform* is a bit of a malapropism. It is, however, at least as often said one way as the other.

The final argument, maxComponents, specifies the maximum number of components (dimensions) that PCA should retain. By default, all of the components are retained.

Any subsequent call to cv::PCA::operator()() will overwrite the internal representations of the eigenvectors and eigenvalues, so you can recycle a PCA object whenever you need to (i.e., you don't have to reallocate a new one for each new distribution you want to handle if you no longer need the information about the previous distribution).

### cv::PCA::project()

```
cv::Mat PCA::project(                        // Return results, as a 2d matrix
  cv::InputArray  vec                        // points to project, rows or cols, 2d
) const;

void PCA::project(
  cv::InputArray  vec                        // points to project, rows or cols, 2d
  cv::OutputArray result                     // Result of projection, reduced space
) const;
```

Once you have loaded your reference distribution with cv::PCA::operator()(), you can start asking the PCA object to do useful things for you like compute the KLT projection of some set of vectors onto the basis vectors computed by the principal component analysis. The cv::PCA::project() function has two forms; the first returns a matrix containing the results of the projections, while the second writes the results to a matrix you provide. The advantage of the first form is that you can use it in matrix expressions.

The vec argument contains the input vectors. vec is required to have the same number of dimensions and the same "orientation" as the data array that was passed to PCA when the distribution was first analyzed (i.e., if your data was columns when you called cv::PCA::operator()(), vec should also have the data arranged into columns).

The returned array will have the same number of objects as vec with the same orientation, but the dimensionality of each object will be whatever was passed to maxComponents when the PCA object was first configured with cv::PCA::operator()().

### cv::PCA::backProject()

```
cv::Mat PCA::backProject(                   // Return results, as a 2d matrix
  cv::InputArray  vec                       // Result of projection, reduced space
} const;

void PCA::backProject(
  cv::InputArray  vec                       // Result of projection, reduced space
```

```
    cv::OutputArray result              // "reconstructed" vectors, full dimension
  ) const;
```

The `cv::PCA::backProject()` function performs the reverse operation of `cv::PCA::project()`, with the analogous restrictions on the input and output arrays. The `vec` argument contains the input vectors, which this time are from the projected space. They will have the same number of dimensions as you specified with `maxComponents` when you configured the PCA object and the same "orientation" as the `data` array that was passed to PCA when the distribution was first analyzed (i.e., if your data was columns when you called `cv::PCA::operator()`, `vec` should also have the data arranged into columns).

The returned array will have the same number of objects as `vec` with the same orientation, but the dimensionality of each object will be the dimensionality of the original data you gave to the PCA object when the PCA object was first configured with `cv::PCA::operator()()`.

If you did not retain all of the dimensions when you configured the PCA object in the beginning, the result of back-projecting vectors —which are themselves projections of some vector $\vec{x}$ from the original data space—will not be equal to $\vec{x}$. Of course, the difference should be small, even if the number of components retained was much smaller than the original dimension of $\vec{x}$, as this is the point of using PCA in the first place.

## Singular Value Decomposition (cv::SVD)

The `cv::SVD` class is similar to `cv::PCA` in that it is the same kind of function object. Its purpose, however, is quite different. The singular value decomposition is essentially a tool for working with nonsquare, ill-conditioned, or otherwise poorly behaved matrices such those you encounter when solving underdetermined linear systems.

Mathematically, the *singular value decomposition* (SVD) is the decomposition of an $m \times n$ matrix $A$ into the form:

$$A = U \cdot W \cdot V^T$$

where $W$ is a diagonal matrix and $U$ and $V$ are $m \times m$ and $n \times n$ (unitary) matrices, respectively. Of course, the matrix $W$ is also an $m \times n$ matrix, so here "diagonal" means that any element whose row and column numbers are not equal is necessarily `0`.

### cv::SVD()

```
SVD::SVD();
SVD::SVD(
  cv::InputArray A,                    // Linear system, array to be decomposed
  int            flags = 0             // what to construct, can A can scratch
);
```

The SVD object has a default constructor, `cv::SVD()`, which simply builds the SVD object and initializes the empty structure. The second form basically executes the default construction, then immediately proceeds to pass its arguments to `cv::SVD::operator()()` (discussed next).

### cv::SVD::operator()()

```
SVD::& SVD::operator() (
  cv::InputArray A,                    // Linear system, array to be decomposed
  int            flags = 0             // what to construct, can A be scratch
);
```

The operator `cv::SVD::operator()()` passes to the `cv::SVD` object the matrix to be decomposed. The matrix *A*, as described earlier, is decomposed into a matrix *U*, a matrix *V* (actually the transpose of *V*, which we will call *Vt*), and a set of singular values (which are the diagonal elements of the matrix *W*).

The flags can be any one of `cv::SVD::MODIFY_A`, `cv::SVD::NO_UV`, or `cv::SVD::FULL_UV`. The latter two are mutually exclusive, but either can be combined with the first. The flag `cv::SVD::MODIFY_A` indicates that it is OK to modify the matrix *A* when computing. This speeds up computation a bit and saves some memory. It is more important when the input matrix is already very large. The flag `cv::SVD::NO_UV` tells `cv::SVD` to not explicitly compute the matrices *U* and *Vt*, while the flag `cv::SVD::FULL_UV` indicates that not only would you like *U* and *Vt* computed, but that you would also like them to be represented as full-size square orthogonal matrices.

### cv::SVD::compute()

```
void SVD::compute(
  cv::InputArray  A,                // Linear system, array to be decomposed
  cv::OutputArray W,                // Output array 'W', singular values
  cv::OutputArray U,                // Output array 'U', left singular vectors
  cv::OutputArray Vt,               // Output array 'Vt', right singular vectors
  int             flags = 0         // what to construct, and if A can be scratch
);
```

This function is an alternative to using `cv::SVD::operator()()` to decompose the matrix *A*. The primary difference is that the matrices *W*, *U*, and *Vt* are stored in the user-supplied arrays, rather than being kept internally. The flags supported are exactly those supported by `cv::SVD::operator()()`.

### cv::SVD::solveZ()

```
void SVD::solveZ(
  cv::InputArray  A,                      // Linear system, array to be decomposed
  cv::OutputArray z                       // One possible solution (unit length)
);
```

$$\vec{z} = \operatorname{argmin}_{\vec{x}:\|\vec{x}\|=1} \| A \cdot \vec{x} \|$$

Given an underdetermined (singular) linear system, `cv::SVD::solveZ()` will (attempt to) find a unit length solution of $A \cdot \vec{x} = 0$ and place the solution in the array `z`. Because the linear system is singular, however, it may have no solution, or it may have an infinite family of solutions. `cv::SVD::solveZ()` will find a solution, if one exists. If no solution exists, then the return value $\vec{z}$ will be a vector that minimizes $A \cdot \vec{x}$, even if this is not, in fact, zero.

### cv::SVD::backSubst()

```
void SVD::backSubst(
  cv::InputArray  b,                      // Righthand side of linear system
  cv::OutputArray x                       // Found solution to linear system
);

void SVD::backSubst(
  cv::InputArray  W,                      // Output array 'W', singular values
  cv::InputArray  U,                      // Output array 'U', left singular vectors
  cv::InputArray  Vt,                     // Output array 'Vt', right singular vectors
  cv::InputArray  b,                      // Righthand side of linear system
  cv::OutputArray x                       // Found solution to linear system
);
```

Assuming that the matrix *A* has been previously passed to the `cv::SVD` object (and thus decomposed into *U*, *W*, and *Vt*), the first form of `cv::SVD::backSubst()` attempts to solve the system:

$$(UWV^T) \cdot \vec{x} = \vec{b}$$

The second form does the same thing, but expects the matrices *W*, *U*, and *Vt* to be passed to it as arguments. The actual method of computing `dst` is to evaluate the following expression:

$$\vec{x} = V_t^T \cdot diag(W)^{-1} \cdot U^T \cdot \vec{b} \sim A^{-1} \cdot \vec{b}$$

This method produces a *pseudosolution* for an overdetermined system, which is the best solution in the sense of minimizing the least-squares error.[5] Of course, it will also exactly solve a correctly determined linear system.

In practice, it is relatively rare that you would want to use `cv::SVD::backSubst()` directly. This is because you can do precisely the same thing by calling `cv::solve()` and passing the `cv::DECOMP_SVD` method flag—which is a lot easier. Only in the less common case in which you need to solve many different systems with the same *lefthand* side ($x$) would you be better off calling `cv::SVD::backSubst()` directly—as opposed to solving the *same* system many times with different *righthand* sides ($b$), which you might as well do with `cv::solve()`.

## Random Number Generator (cv::RNG)

A random number tor (RNG) object holds the state of a pseudorandom sequence that generates random numbers. The benefit of using it is that you can conveniently maintain multiple streams of pseudorandom numbers.

When programming large systems, it is a good practice to use separate random number streams in different modules of the code. This way, removing one module does not change the behavior of the streams in the other modules.

Once created, the random number generator provides the "service" of generating random numbers on demand, drawn from either a uniform or a Gaussian distribution. The generator uses the *Multiply with Carry* (MWC) algorithm [Goresky03] for uniform distributions and the *Ziggurat* algorithm [Marsaglia00] for the generation of numbers from a Gaussian distribution.

### cv::theRNG()

```
cv::RNG& theRNG( void );                    // Return a random number generator
```

The `cv::theRNG()` function returns the default random number generator for the thread from which it was called. OpenCV automatically creates one instance of `cv::RNG` for each thread in execution. This is the same random number generator

5 The object $diag(W)^{-1}$ is a matrix whose diagonal elements $\lambda_i^*$ are defined in terms of the diagonal elements $\lambda_I$ of $W$ by $\lambda_i^* = \lambda_i^{-1}$ for $\lambda_i \geq \varepsilon$. This value $\varepsilon$ is the *singularity threshold*, a very small number that is typically proportional to the sum of the diagonal elements of $W$ (i.e., $\varepsilon_0 \Sigma_i \lambda_i$).

that is implicitly accessed by functions like `cv::randu()` or `cv::randn()`. Those functions are convenient if you want a single number or to initialize a single array. However, if you have a loop of your own that needs to generate a lot of random numbers, you are better off grabbing a reference to a random number generator—in this case, the default generator, but you could use your own instead—and using `RNG::operator T()` to get your random numbers (more on that operator shortly).

### cv::RNG()

```
cv::RNG::RNG( void );
cv::RNG::RNG( uint64 state );                 // create using the seed 'state'
```

You can create an RNG object with either the default constructor, or by passing it a 64-bit unsigned integer that it will use as the seed of the random number sequence. If you call the default constructor (or pass `0` to the second variation) the generator will initialize with a standardized value.[6]

### cv::RNG::operator T(), where T is your favorite type

```
cv::RNG::operator uchar();
cv::RNG::operator schar();
cv::RNG::operator ushort();
cv::RNG::operator short int();
cv::RNG::operator int();
cv::RNG::operator unsigned();
cv::RNG::operator float();
cv::RNG::operator double();
```

`cv::RNG::operator T()` is really a set of different methods that return a new random number from `cv::RNG` of some specific type. Each of these is an overloaded cast operator, so in effect you cast the RNG object to whatever type you want, as shown in Example 7-1. (The style of the cast operation is up to you; this example shows both the `int(x)` and the `(int)x` forms.)

*Example 7-1. Using the default random number generator to generate a pair of integers and a pair of floating-point numbers*

```
cv::RNG rng = cv::theRNG();
cout << "An integer:      " << (int)rng   << endl;
cout << "Another integer: " << int(rng)   << endl;
cout << "A float:         " << (float)rng << endl;
cout << "Another float:   " << float(rng) << endl;
```

6 This "standard value" is not zero because, for that value, many random number generators (including the ones used by RNG) will return nothing but zeros thereafter. Currently, this standard value is $2^{32} - 1$.

When integer types are generated, they will be generated (using the MWC algorithm described earlier and thus uniformly) across the entire range of available values. When floating-point types are generated, they will always be in the range from the interval `[0.0, 1.0)`.[7]

### cv::RNG::operator()

```
unsigned int cv::RNG::operator()();         // Return random value from 0-UINT_MAX
unsigned int cv::RNG::operator()( unsigned int N );  // Return value from 0-(N-1)
```

When you are generating integer random numbers, the overloaded `operator()()` allows a convenient way to just grab another one. In essence, calling `my_rng()` is equivalent to calling `(unsigned int)my_rng`. The somewhat-more-interesting form of `cv::RNG::operator()()` takes an integer argument `N`. This form returns (using the MWC algorithm described earlier and thus uniformly) a random unsigned integer *modulo* `N`. Thus, the range of integers returned by `my_rng( N )` is then the range of integers from `0` to `N-1`.

### cv::RNG::uniform()

```
int    cv::RNG::uniform( int a,    int b    );      // Return value from a-(b-1)
float  cv::RNG::uniform( float a,  float b  );      // Return value in range [a,b)
double cv::RNG::uniform( double a, double b );      // Return value in range [a,b)
```

This function allows you to generate a random number uniformly (using the MWC algorithm) in the interval `[a, b)`.

The C++ compiler does not consider the return value of a function when determining which of multiple similar forms to use, only the arguments. As a result, if you call `float x = my_rng.uniform(0,1)` you will get `0.f`, because `0` and `1` are integers and the only integer in the interval `[0, 1)` is `0`. If you want a floating-point number, you should use something like `my_rng.uniform(0.f, 1.f)`, and for a double, use `my_rng.uniform(0.,1.)`. Of course, explicitly casting the arguments also works.

### cv::RNG::gaussian()

```
double  cv::RNG::gaussian( double sigma ); // Gaussian number, zero mean,
                                           // std-dev='sigma'
```

7 In case this notation is unfamiliar, in the designation of an interval using square brackets, `[` indicates that this limit is inclusive, and in the designation using parentheses, `(` indicates that this limit is noninclusive. Thus the notation `[0.0,1.0)` means an interval from 0.0 to 1.0 inclusive of 0.0 but not inclusive of 1.0.

This function allows you to generate a random number from a zero-mean Gaussian distribution (using the Ziggurat algorithm) with standard deviation sigma.

### cv::RNG::fill()

```
void  cv::RNG::fill(
  InputOutputArray mat,               // Input array, values will be overwritten
  int              distType,          // Type of distribution (Gaussian or uniform)
  InputArray       a,                 // min (uniform) or mean (Gaussian)
  InputArray       b                  // max (uniform) or std-deviation (Gaussian)
 );
```

The cv::RNG::fill() algorithm fills a matrix mat of up to four channels with random numbers drawn from a specific distribution. That distribution is selected by the distType argument, which can be either cv::RNG::UNIFORM or cv::RNG::NORMAL. In the case of the uniform distribution, each element of mat will be filled with a random value generated from the interval $mat_{c,i} \in [a_c, b_c)$. In the case of the Gaussian (cv::RNG::NORMAL) distribution, each element is generated from a distribution with the mean taken from a and the standard deviation taken from b:$mat_{c,i} \in N(a_c, b_c)$. It is important to note that the arrays a and b are not of the dimension of mat; instead, they are $n_c \times 1$ or $1 \times n_c$ where $n_c$ is the number of channels in mat (i.e., there is not a separate distribution for each element of mat; a and b specify one distribution, not one distribution for every element of mat.)

If you have a multichannel array, then you can generate individual entries in "channel space" from a multivariate distribution simply by giving the appropriate mean and standard deviation for each channel in the input arrays a and b. This distribution, however, will be drawn from a distribution with only zero entries in the off-diagonal elements of its covariance matrix. (This is because each element is generated completely independently of the others.) If you need to draw from a more general distribution, the easiest method is to generate the values from an identity covariation matrix with zero mean using cv::RNG::fill(), and then rotate back to your original basis using cv::transform().

## Summary

In this chapter we introduced the concept of functors, as well as the manner in which they are used in the OpenCV library. We looked at a few such objects that are of general utility and saw how they worked. These included the PCA and SVD objects, as well as the very useful random number generator RNG. Later on, as we delve into the more advanced algorithms the library provides, we will see that this same concept is used in many of the more modern additions to the library.

## Exercises

1. Using the `cv::RNG` random number generator:
   a. Generate and print three floating-point numbers, each drawn from a uniform distribution from 0.0 to 1.0.
   b. Generate and print three double-precision numbers, each drawn from a Gaussian distribution centered at 0.0 and with a standard deviation of 1.0.
   c. Generate and print three unsigned bytes, each drawn from a uniform distribution from 0 to 255.
2. Using the `fill()` method of the `cv::RNG` random number generator, create an array of:
   a. 20 floating-point numbers with a uniform distribution from 0.0 to 1.0.
   b. 20 floating-point numbers with a Gaussian distribution centered at 0.0 and with a standard deviation of 1.0.
   c. 20 unsigned bytes with a uniform distribution from 0 to 255.
   d. 20 color triplets, each of three bytes with a uniform distribution from 0 to 255.
3. Using the `cv::RNG` random number generator, create an array of 100 three-byte objects such that:
   a. The first and second dimensions have a Gaussian distribution, centered at 64 and 192, respectively, each with a variance of 10.
   b. The third dimension has a Gaussian distribution, centered at 128 and with a variance of 2.
   c. Using the `cv::PCA` object, compute a projection for which `maxComponents=2`.
   d. Compute the mean in both dimensions of the projection; explain the result.
4. Beginning with the following matrix:

$$A = \begin{matrix} 1 & 1 \\ 0 & 1 \\ -1 & 1 \end{matrix}$$

   a. First compute, by hand, the matrix $A^T A$. Find the eigenvalues $(e_1, e_2)$, and eigenvectors $(\vec{v}_1, \vec{v}_2)$, of $A^T A$. From the eigenvalues, compute the singular values $(\sigma_1, \sigma_2) = (\sqrt{e_1}, \sqrt{e_2})$.
   b. Compute the matrices $V = [\vec{v}_1, \vec{v}_2]$, and $U = [\vec{u}_1, \vec{u}_2, \vec{u}_3]$. Recall that $\vec{u}_1 = \frac{1}{\sigma_1} A\vec{v}_1$, $\vec{u}_2 = \frac{1}{\sigma_2} A\vec{v}_2$, and $\vec{u}_3$ is a vector orthogonal to both $\vec{u}_1$ and $\vec{u}_2$. Hint: recall that the

cross product of two vectors is always orthogonal to both terms in the cross product.

c. The matrix $\Sigma$ is defined (given this particular value of $A$) to be:

$$\Sigma = \begin{matrix} \sigma_1 & 0 \\ 0 & \sigma_2 \\ 0 & 0 \end{matrix}$$

Using this definition of $\Sigma$, and the preceding results for $V$ and $U$, verify $A = U \Sigma V^T$ by direct multiplication.

d. Using the `cv::SVD` object, compute the preceding matrices $\Sigma$, $V$, and $U$ and verify that the results you computed by hand are correct. Do you get exactly what you expected? If not, explain.

CHAPTER 8

# Image, Video, and Data Files

## HighGUI: Portable Graphics Toolkit

The OpenCV functions that allow us to interact with the operating system, the filesystem, and hardware such as cameras are mostly found in the module called HighGUI (which stands for "high-level graphical user interface"). HighGUI allows us to read and write graphics-related files (both images and video), to open and manage windows, to display images, and to handle simple mouse, pointer, and keyboard events. We can also use it to create other useful doodads—like sliders, for example—and then add them to our windows. If you are a GUI guru in your window environment of choice, then you might find that much of what HighGUI offers is redundant. Even so, you might find that the benefit of cross-platform portability is itself a tempting morsel.

In this chapter, we will investigate the portion of HighGUI that deals with capture and storage for both still and video images. In the following chapter, we will learn how we can display images in windows using the cross-platform tools supplied by HighGUI, as well as other native and cross-platform window toolkits.

From our initial perspective, the HighGUI library in OpenCV can be divided into three parts: the hardware part, the filesystem part, and the GUI part. We will take a moment to give an overview of what is in each part before we really dive in.

In OpenCV 3.0, HighGUI has been split into three modules: `imgcodecs` (image encoding/decoding), `videoio` (capturing and encoding video), and a portion now called `highgui` (which is the UI part). For better backward compatibility, the `highgui.hpp` header includes `videoio.hpp` and `imgcodecs.hpp` headers, so most of the OpenCV 2.x code will be compatible with 3.x. Going forward, we will be using the name *HighGUI* to refer to all of the image I/O, video I/O, and UI functionality described in this chapter, and the code examples should be compatible with both OpenCV 2.x and 3.x. But keep in mind that if you use OpenCV 3.0 and you need just video capturing capabilities or you just read and write images, you may use `videoio`/`imgcodecs` separately from other HighGUI components.

The hardware part is primarily concerned with the operation of cameras. In most operating systems, interaction with a camera is a tedious and painful task. HighGUI provides an easy way to query a camera and retrieve its latest image. It hides all of the nasty stuff, and tries to keep everyone happy.

The filesystem part is concerned primarily with loading and saving images. One nice feature of the library is that it allows us to read video using the same methods we would use to read a camera. We can therefore abstract ourselves away from the particular device we're using and get on with writing the interesting code. In a similar spirit, HighGUI provides us with a (relatively) universal pair of functions to load and save still images. These functions simply use the filename extension to determine the file type and automatically handle all of the decoding or encoding that is necessary. In addition to these image-specific functions, OpenCV provides a set of XML/YML-based functions that make it convenient to load and store all sorts of other data in a simple, human-readable, text-based format.

The third part of HighGUI is the window system (or GUI). The library provides some simple functions that allow us to open a window and throw an image into that window. It also allows us to register and respond to mouse and keyboard events on that window. These features are most useful when we're trying to get off the ground with a simple application. By tossing in some slider bars, we are able to prototype a surprising variety of applications using only the HighGUI library. If we want to link to Qt, we can even get a little more functionality.[1] We will cover all of this in the next chapter on window toolkits.

1 This is Qt, the cross-platform widget toolkit. We will talk more about how it works later in this chapter.

## Working with Image Files

OpenCV provides special functions for loading and saving images. These functions deal, either explicitly or implicitly, with the complexities associated with compressing and decompressing that image data. They differ in several respects from the more universal XML/YAML-based functions that we will learn about in "Data Persistence" on page 198. The primary distinction is that because the former functions are designed for images specifically, as opposed to general arrays of data, they rely heavily on existing backends for compression and decompression. In this way, they are able to handle each of the common file formats in the special manner it requires. Some of these compression and decompression schemes have been developed with the idea that it is possible to lose some information without degrading the visual experience of the image. Clearly such lossy compression schemes are not a good idea for arrays of non-image data.

Artifacts introduced by lossy compression schemes can also cause headaches for computer vision algorithms. In many cases, algorithms will find and respond to compression artifacts that are completely invisible to humans.

The key difference to remember is that the loading and saving functions we will discuss first are really interfaces to the resources for handling image files that are already present in your operating system or its available libraries. In contrast, the XML/YAML data persistence system mentioned earlier, which we will get to later in this chapter, is entirely internal to OpenCV.

## Loading and Saving Images

The most common tasks we will need to accomplish are loading and saving files from disk. The easiest way to do this is with the high-level functions `cv::imread()` and `cv::imwrite()`. These functions handle the complete task of decompression and compression as well as the actual interaction with the file system.

### Reading files with cv::imread()

The first thing to learn is how to get an image out of the filesystem and into our program. The function that does this is `cv::imread()`:

```
cv::Mat cv::imread(
  const string& filename,                  // Input filename
  int           flags    = cv::IMREAD_COLOR  // Flags set how to interpret file
);
```

When opening an image, cv::imread() doesn't look at the file extension. Instead, it analyzes the first few bytes of the file (a.k.a. its *signature* or "magic sequence") and determines the appropriate codec using that. The second argument, flags, can be set to one of several values as described in Table 8-1. By default, flags is set to cv::IMREAD_COLOR. This value indicates that images are to be loaded as three-channel images with 8 bits per channel. In this case, even if the image is actually grayscale in the file, the resulting image in memory will still have three channels, with all of the channels containing identical information. Alternatively, if flags is set to cv::IMREAD_GRAYSCALE, then the image will be loaded as grayscale, regardless of the number of channels in the file. The final option is to set flags to cv::IMREAD_ANYCOLOR. In that case, the image will be loaded "as is," with the result being three-channel if the file is color, and one-channel if the file is grayscale.[2]

In addition to the color-related flags, cv::imread() supports the flag cv::IMREAD_ANYDEPTH, which indicates that if an input image's channels have more than 8 bits, then it should be loaded without conversion (i.e., the allocated array will be of the type indicated in the file).

*Table 8-1. Flags accepted by cv::imread()*

| Parameter ID | Meaning | Default |
|---|---|---|
| cv::IMREAD_COLOR | Always load to three-channel array. | yes |
| cv::IMREAD_GRAYSCALE | Always load to single-channel array. | no |
| cv::IMREAD_ANYCOLOR | Channels as indicated by file (up to three). | no |
| cv::IMREAD_ANYDEPTH | Allow loading of more than 8-bit depth. | no |
| cv::IMREAD_UNCHANGED | Equivalent to combining: cv::IMREAD_ANYCOLOR \| cv::IMREAD_ANYDEPTH[a] | no |

[a] This is not precisely true. IMREAD_UNCHANGED has another unique effect: it will preserve the alpha channel in an image when that image is loaded. Note that even IMREAD_ANYCOLOR will still effectively crop the depth to three channels.

cv::imread() does not give a runtime error when it fails to load an image; it simply returns an empty cv::Mat (i.e., cv::Mat::empty()==true).

2 At the time of writing, "as is" does not support the loading of a fourth channel for those file types that support alpha channels. In such cases, the fourth channel will be ignored and the file will be treated as if it had only three channels.

### Writing files with cv::imwrite()

The obvious complementary function to `cv::imread()` is `cv::imwrite()`, which takes three arguments:

```
bool cv::imwrite(
  const string&      filename,              // Input filename
  cv::InputArray     image,                 // Image to write to file
  const vector<int>& params  = vector<int>() // (Optional) for parameterized fmts
);
```

The first argument gives the filename, whose extension is used to determine the format in which the file will be stored. Here are some popular extensions supported by OpenCV:

- *.jpg* or *.jpeg*: baseline JPEG; 8-bit; one- or three-channel input
- *.jp2*: JPEG 2000; 8-bit or 16-bit; one- or three-channel input
- *.tif* or *.tiff*: TIFF; 8- or 16-bit; one-, three-, or four-channel input
- *.png*: PNG; 8- or 16-bit; one-, three-, or four-channel input
- *.bmp*: BMP; 8-bit; one-, three-, or four-channel input
- *.ppm*, *.pgm*: NetPBM; 8-bit; one-channel (PGM) or three-channel (PPM)

The second argument is the image to be stored. The third argument is used for parameters that are accepted by the particular file type being used for the write operation. The `params` argument expects an STL vector of integers, with those integers being a sequence of parameter IDs followed by the value to be assigned to that parameter (i.e., alternating between the parameter ID and the parameter value). For the parameter IDs, there are aliases provided by OpenCV, as listed in Table 8-2.

*Table 8-2. Parameters accepted by cv::imwrite()*

| Parameter ID | Meaning | Range | Default |
|---|---|---|---|
| `cv::IMWRITE_JPG_QUALITY` | JPEG quality | `0-100` | 95 |
| `cv::IMWRITE_PNG_COMPRESSION` | PNG compression (higher values mean more compression) | `0-9` | 3 |
| `cv::IMWRITE_PXM_BINARY` | Use binary format for PPM, PGM, or PBM files | `0 or 1` | 1 |

The `cv::imwrite()` function will store only 8-bit, single-, or three-channel images for most file formats. Backends for flexible image formats like PNG, TIFF, or JPEG 2000 allow storing 16-bit or even float formats and some allow four-channel images

(BGR plus alpha) as well. The return value will be `true` if the save was successful and should be `false` if the save was not.[3]

## A Note About Codecs

Remember, however, that `cv::imwrite()` is intended for images, and relies heavily on software libraries intended for handling image file types. These libraries are generically referred to as *codecs* ("*co*-mpression and *dec*-ompression librarie-*s*"). Your operating system will likely have many codecs available, with (at least) one being available for each of the different common file types.

OpenCV comes with the codecs you will need for some file formats (JPEG, PNG, TIFF, etc.). For each of the codecs, there are three possibilities: a) to not include support for this codec, b) to use the codec supplied with OpenCV (build it together with other OpenCV modules), or c) use the external library (libjpeg, libpng, etc.) correspondingly. On Windows the default option is b. On OS X/Linux the default option is c; if CMake did not find the codec, it will use b. You may explicitly override this setting if you want. In Linux, if you want option c, install the codec together with the development files (e.g., *libjpeg* and *libjpeg-dev*).

## Compression and Decompression

As already mentioned, the `cv::imread()` and `cv::imwrite` functions are high-level tools that handle a number of separate things necessary to ultimately get your image written to or read from the disk. In practice, it is often useful to be able to handle some of those subcomponents individually and, in particular, to be able to compress or decompress an image in memory (using the codecs we just reviewed).

### Compressing files with cv::imencode()

Images can be compressed directly from OpenCV's array types. In this case, the result will not be an array type, but rather a simple character buffer. This should not be surprising, as the resulting object is now in some format that is meaningful only to the codec that compressed it, and will (by construction) not be the same size as the original image.

```
void cv::imencode(
  const string&      ext,                        // Extension specifies codec
  cv::InputArray     img,                        // Image to be encoded
  vector<uchar>&     buf,                        // Encoded file bytes go here
```

3 The reason we say "should" is that, in some OS environments, it is possible to issue save commands that will actually cause the operating system to throw an exception. Normally, however, a `0` value will be returned to indicate failure.

```
    const vector<int>& params  = vector<int>() // (Optional) for parameterized fmts
  );
```

The first argument to `cv::imencode()` is `ext`, the file extension represented as a string, which is associated with the desired compression. Of course, no file is actually written, but the extension is not only an intuitive way to refer to the desired format. Also, the extension is the actual key used by most operating systems to index the available codecs. The next argument, `img`, is the image to be compressed, and the argument following that, `buf`, is the character array into which the compressed image will be placed. This buffer will be automatically resized by `cv::imencode()` to the correct size for the compressed image. The final argument, `params`, is used to specify any parameters that may be required (or desired) for the specific compression codec to be used. The possible values for `params` are those listed in Table 8-2 for `cv::imwrite()`.

#### Uncompressing files with cv::imdecode()

```
cv::Mat cv::imdecode(
  cv::InputArray buf,                        // Encoded file bytes are here
  int            flags = cv::IMREAD_COLOR // Flags set how to interpret file
);
```

Just as `cv::imencode()` allows us to compress an image into a character buffer, `cv::imdecode()` allows us to decompress from a character buffer into an image array. `cv::imdecode()` takes just two arguments, the first being the buffer[4] `buf` (which normally has `std::vector<uchar>` type) and the second being the `flags` argument, which takes the same options as the flags used by `cv::imread()` (see Table 8-1). As was the case with `cv::imread()`, `cv::imdecode()` does not need a file extension (as `cv::imencode()` did) because it can deduce the correct codec to use from the first bytes of the compressed image in the buffer.

Just as `cv::imread()` returns an empty (`cv::Mat::empty()==true`) array if it cannot read the file it is given, `cv::imdecode()` returns an empty array if the buffer it is given is empty, contains invalid or unusable data, and so on.

## Working with Video

When working with video, we must consider several issues, including (of course) how to read and write video files to and from disk. Of course, once we can do that, we will want to know how to actually play back such files onscreen, either for debugging

4 You should not be surprised that the input `buf` is not type `vector<uchar>&`, as it was with `cv::imencode()`. Recall that the type `cv::InputArray` covers many possibilities and that `vector<>` is one of them.

or as the final output of our program; we'll start with disk IO and get to playback next.

## Reading Video with the cv::VideoCapture Object

The first thing we need is the `cv::VideoCapture` object. This is another one of those "objects that do stuff" we encountered in the previous chapter. This object contains the information needed for reading frames from a camera or video file. Depending on the source, we use one of three different calls to create a `cv::VideoCapture` object:

```
cv::VideoCapture::VideoCapture(
  const string&      filename,           // Input filename
);
cv::VideoCapture::VideoCapture(
  int device                             // Video capture device id
);
cv::VideoCapture::VideoCapture();
```

In the case of the first constructor, we can simply give a filename for a video file (*.MPG*, *.AVI*, etc.) and OpenCV will open the file and prepare to read it. If the open is successful and we are able to start reading frames, the member function `cv::VideoCapture::isOpened()` will return `true`.

A lot of people don't always check these sorts of things, assuming that nothing will go wrong. Don't do that here. The returned value of `cv::VideoCapture::isOpened()` will be `false` if for some reason the file could not be opened (e.g., if the file does not exist), but that is not the only possible cause. The constructed object will also not be ready to be used if the codec with which the video is compressed is not known. Because of the many issues surrounding codecs (legal as well as technical), this is not as rare of an occurrence as one might hope.

As with the image codecs, you will need to have the appropriate library already residing on your computer in order to successfully read the video file. This is why it is always important for your code to check the return value of `cv::VideoCapture::isOpened()`, because even if it works on one machine (where the needed codec DLL or shared library is available), it might not work on another machine (where that codec is missing). Once we have an open `cv::VideoCapture` object, we can begin reading frames and do a number of other things. But before we get into that, let's take a look at how to capture images from a camera.

The variant of `cv::VideoCapture::VideoCapture()` that takes an integer `device` argument works very much like the `string` version we just discussed, except without the headache from the codecs.[5] In this case, we give an *identifier* that indicates a camera we would like to access and how we expect the operating system to talk to that camera. For the camera, this is just an identification number—it is zero (`0`) when we have only one camera and increments upward when there are multiple cameras on the same system. The other part of the identifier is called the *domain* of the camera and indicates (in essence) what type of camera we have. The domain can be any of the predefined constants shown in Table 8-3.

*Table 8-3. Camera "domain" indicating where HighGUI should look for your camera*

| Camera capture constant | Numerical value |
|---|---|
| `cv::CAP_ANY` | `0` |
| `cv::CAP_MIL` | `100` |
| `cv::CAP_VFW` | `200` |
| `cv::CAP_V4L` | `200` |
| `cv::CAP_V4L2` | `200` |
| `cv::CAP_FIREWIRE` | `300` |
| `cv::CAP_IEEE1394` | `300` |
| `cv::CAP_DC1394` | `300` |
| `cv::CAP_CMU1394` | `300` |
| `cv::CAP_QT` | `500` |
| `cv::CAP_DSHOW` | `700` |
| `cv::CAP_PVAPI` | `800` |
| `cv::CAP_OPENNI` | `900` |
| `cv::CAP_ANDROID` | `1000` |
| `...` | |

When we construct the `device` argument for `cv::VideoCapture::VideoCapture()`, we pass in an identifier that is the sum of the domain index and the camera index. For example:

```
cv::VideoCapture capture( cv::CAP_FIREWIRE );
```

In this example, `cv::VideoCapture::VideoCapture()` will attempt to open the first (i.e., number `0`) FireWire camera. In most cases, the domain is unnecessary when we have only one camera; it is sufficient to use `cv::CAP_ANY` (which is conveniently equal to `0`, so we don't even have to type that in). One last useful hint before we move on:

5 Of course, to be completely fair, we should confess that the headache caused by different codecs has been replaced by the analogous headache of determining which cameras are supported on our system and/or the camera driver software.

on some platforms, you can pass -1 to cv::VideoCapture::VideoCapture(), which will cause OpenCV to open a window that allows you to select the desired camera.

Your last option is to create the capture object without providing any information about what is to be opened.

```
cv::VideoCapture cap;

cap.open( "my_video.avi" );
```

In this case, the capture object will be there, but not ready for use until you explicitly open the source you want to read from. You do this with the cv::VideoCapture::open() method, which, like the cv::VideoCapture constructor, can take either an STL string or a device ID as arguments. In either case, cv::VideoCapture::open() will have exactly the same effect as calling the cv::VideoCapture constructor with the same argument.

#### Reading frames with cv::VideoCapture::read()

```
bool cv::VideoCapture::read(
  cv::OutputArray image                    // Image into which to read data
);
```

Once you have a cv::VideoCapture object, you can start reading frames. The easiest way to do this is to call cv::VideoCapture::read(), which will simply go to the open file represented by cv::VideoCapture and get the next frame, inserting that frame into the provided array image. This action will automatically "advance" the video capture object such that a subsequent call to cv::VideoCapture::read() will return the next frame, and so on.

If the read was not successful (e.g., if you have reached the end of your file), then this function call will return false (otherwise, it will return true). Similarly, the array object you supplied to the function will also be empty.

#### Reading frames with cv::VideoCapture::operator>>()

```
cv::VideoCapture& cv::VideoCapture::operator>>(
  cv::Mat& image                           // Image into which to read data
);
```

In addition to using the read method of cv::VideoCapture, you can use the overloaded function cv::VideoCapture::operator>>() (i.e., the "stream read" operator) to read the next frame from your video capture object. In this case, cv::VideoCapture::operator>>() behaves exactly the same as cv::VideoCapture::read(), except that because it is a stream operator, it returns a reference to the original cv::VideoCapture object regardless of whether it succeeded. In this case, you must check that the return array is not empty.

### Reading frames with cv::VideoCapture::grab() and cv::VideoCapture::retrieve()

Instead of taking images one at a time from your camera or video source and decoding them as you read them, you can break this process down into a *grab* phase, which is little more than a memory copy, and a *retrieve* phase, which handles the actual decoding of the grabbed data.

```
bool cv::VideoCapture::grab( void );
bool cv::VideoCapture::retrieve(
  cv::OutputArray image,                 // Image into which to read data
  int             channel = 0            // Used for multihead devices
);
```

The `cv::VideoCapture::grab()` function copies the currently available image to an internal buffer that is invisible to the user. Why would you want OpenCV to put the frame somewhere you can't access it? The answer is that this grabbed frame is unprocessed, and `grab()` is designed simply to get it onto the computer (typically from a camera) as quickly as possible.

> There are many reasons to grab and retrieve separately rather than together as would be the case in calling `cv::VideoCapture::read()`. The most common situation arises when there are multiple cameras (e.g., with stereo imaging). In this case, it is important to have frames that are separated in time by the minimum amount possible (ideally they would be simultaneous for stereo imaging). Therefore, it makes the most sense to first grab all the frames and then come back and decode them after you have them all safely in your buffers.

As was the case with `cv::VideoCapture::read()`, `cv::VideoCapture::grab()` returns `true` only if the grab was successful.

Once you have grabbed your frame, you can call `cv::VideoCapture::retrieve()`, which handles the de coding as well as the allocation and copying necessary to return the frame to you as a `cv::Mat` array. `cv::VideoCapture::retrieve()` functions analogously to `cv::VideoCapture::read()` except that it operates from the internal buffer to which `cv::VideoCapture::grab()` copies frames. The other important difference between `cv::VideoCapture::read()` and `cv::VideoCapture::retrieve()` is the additional argument `channel`. The `channel` argument is used when the device being accessed natively has multiple "heads" (i.e., multiple imagers). This is typically the case for devices designed specifically to be stereo imagers, as well as slightly more exotic devices such as the Kinect.[6] In these cases, the value of `channel` will indicate

6 The currently supported multihead cameras are Kinect and Videre; others may be added later.

which image from the device is to be retrieved. In these cases, you would call `cv::VideoCapture::grab()` just once and then call `cv::VideoCapture::retrieve()` as many times as needed to retrieve all of the images in the camera (each time with a different value for `channel`).

### Camera properties: cv::VideoCapture::get() and cv::VideoCapture::set()

Video files contain not only the video frames themselves, but also important metadata, which can be essential for handling the files correctly. When a video file is opened, that information is copied into the `cv::VideoCapture` object's internal data area. It is very common to want to read that information from the `cv::VideoCapture` object, and sometimes also useful to write to that data area ourselves. The `cv::VideoCapture::get()` and `cv::VideoCapture::set()` functions allow us to perform these two operations:

```
double cv::VideoCapture::get(
  int    propid                        // Property identifier (see Table 8-4)
);

bool cv::VideoCapture::set(
  int    propid                        // Property identifier (see Table 8-4)
  double value                         // Value to which to set the property
);
```

The routine `cv::VideoCapture::get()` accepts any of the property IDs shown in Table 8-4.[7]

*Table 8-4. Video capture properties used by cv::VideoCapture::get() and cv::VideoCapture::set()*

| Video capture property | Camera only | Meaning |
|---|---|---|
| `cv::CAP_PROP_POS_MSEC` | | Current position in video file (milliseconds) or video capture timestamp |
| `cv::CAP_PROP_POS_FRAMES` | | Zero-based index of next frame |
| `cv::CAP_PROP_POS_AVI_RATIO` | | Relative position in the video (range is `0.0` to `1.0`) |
| `cv::CAP_PROP_FRAME_WIDTH` | | Width of frames in the video |
| `cv::CAP_PROP_FRAME_HEIGHT` | | Height of frames in the video |
| `cv::CAP_PROP_FPS` | | Frame rate at which the video was recorded |
| `cv::CAP_PROP_FOURCC` | | Four character code indicating codec |

7 It should be understood that not all of the properties recognized by OpenCV will be recognized or handled by the "backend" behind the capture. For example, the capture mechanisms operating behind the scenes on Android, Firewire on Linux (via dc1394), QuickTime, or a Kinect (via OpenNI) are all going to be very different, and not all of them will offer all of the services implied by this long list of options. And expect this list to grow as new system types make new options possible.

| Video capture property | Camera only | Meaning |
|---|---|---|
| cv::CAP_PROP_FRAME_COUNT | | Total number of frames in a video file |
| cv::CAP_PROP_FORMAT | | Format of the Mat objects returned (e.g., CV_8UC3) |
| cv::CAP_PROP_MODE | | Indicates capture mode; values are specific to video backend being used (e.g., DC1394) |
| cv::CAP_PROP_BRIGHTNESS | ✓ | Brightness setting for camera (when supported) |
| cv::CAP_PROP_CONTRAST | ✓ | Contrast setting for camera (when supported) |
| cv::CAP_PROP_SATURATION | ✓ | Saturation setting for camera (when supported) |
| cv::CAP_PROP_HUE | ✓ | Hue setting for camera (when supported) |
| cv::CAP_PROP_GAIN | ✓ | Gain setting for camera (when supported) |
| cv::CAP_PROP_EXPOSURE | ✓ | Exposure setting for camera (when supported) |
| cv::CAP_PROP_CONVERT_RGB | ✓ | If nonzero, captured images will be converted to have three channels |
| cv::CAP_PROP_WHITE_BALANCE | ✓ | White balance setting for camera (when supported) |
| cv::CAP_PROP_RECTIFICATION | ✓ | Rectification flag for stereo cameras (DC1394-2.x only) |

Most of these properties are self-explanatory. POS_MSEC is the current position in a video file, measured in milliseconds. POS_FRAME is the current position in frame number. POS_AVI_RATIO is the position given as a number between 0.0 and 1.0. (This is actually quite useful when you want to position a trackbar to allow folks to navigate around your video.) FRAME_WIDTH and FRAME_HEIGHT are the dimensions of the individual frames of the video to be read (or to be captured at the camera's current settings). FPS is specific to video files and indicates the number of frames per second at which the video was captured; you will need to know this if you want to play back your video and have it come out at the right speed. FOURCC is the *four-character code* for the compression codec to be used for the video you are currently reading (more on these shortly). FRAME_COUNT should be the total number of frames in the video, but this figure is not entirely reliable.

All of these values are returned as type double, which is perfectly reasonable except for the case of FOURCC (FourCC) [FourCC85]. Here you will have to recast the result in order to interpret it, as shown in Example 8-1.

*Example 8-1. Unpacking a four-character code to identify a video codec*

```
cv::VideoCapture cap( "my_video.avi" );

unsigned f = (unsigned)cap.get( cv::CAP_PROP_FOURCC );
char fourcc[] = {
                (char) f,       // First character is lowest bits
                (char)(f >> 8), // Next character is bits 8-15
                (char)(f >> 16), // Next character is bits 16-23
                (char)(f >> 24), // Last character is bits 24-31
```

```
        '\0'              // and don't forget to terminate
    };
```

For each of these video capture properties, there is a corresponding `cv::VideoCapture::set()` function that will attempt to set the property. These are not all meaningful things to do; for example, you should not be setting the `FOURCC` of a video you are currently reading. Attempting to move around the video by setting one of the position properties will work, but only for some video codecs (we'll have more to say about video codecs in the next section).

## Writing Video with the cv::VideoWriter Object

The other thing we might want to do with video is writing it out to disk. OpenCV makes this easy; it is essentially the same as reading video but with a few extra details.

Just as we did with the `cv::VideoCapture` device for reading, we must first create a `cv::VideoWriter` device before we can write out our video. The video writer has two constructors; one is a simple default constructor that just creates an uninitialized video object that we will have to open later, and the other has all of the arguments necessary to actually set up the writer.

```
cv::VideoWriter::VideoWriter(
  const string& filename,                // Input filename
  int           fourcc,                  // codec, use CV_FOURCC() macro
  double        fps,                     // Frame rate (stored in output file)
  cv::Size      frame_size,              // Size of individual images
  bool          is_color  = true         // if false, you can pass gray frames
);
```

You will notice that the video writer requires a few extra arguments relative to the video reader. In addition to the filename, we have to tell the writer what codec to use, what the frame rate is, and how big the frames will be. Optionally, we can tell OpenCV that the frames are already in color (i.e., three-channel). If you set `isColor` to `false`, you can pass grayscale frames and the video writer will handle them correctly.

As with the video reader, you can also create the video writer with a default constructor, and then configure the writer with the `cv::VideoWriter::open()` method, which takes the same arguments as the full constructor. For example:

```
cv::VideoWriter out;

out.open(
  "my_video.mpg",
  CV_FOURCC('D','I','V','X'),    // MPEG-4 codec
  30.0,                          // Frame rate (FPS)
  cv::Size( 640, 480 ),          // Write out frames at 640x480 resolution
  true                           // Expect only color frames
);
```

Here, the codec is indicated by its four-character code. (For those of you who are not experts in compression codecs, they all have a unique four-character identifier associated with them.) In this case, the `int` that is named `fourcc` in the argument list for `cv::VideoWriter::VideoWriter()` is actually the four characters of the `fourcc` packed together. Since this comes up relatively often, OpenCV provides a convenient macro, `CV_FOURCC(c0,c1,c2,c3)`, that will do the bit packing for you. Don't forget that you are providing characters (single quotes), not strings (double quotes).

Once you have given your video writer all the information it needs to set itself up, it is always a good idea to ask it if it is ready to go. You do this with the `cv::VideoWriter::isOpened()` method, which will return `true` if you are good to go. If it returns `false`, that could mean that you don't have write permission to the directory for the file you indicated, or (most often) that the codec you specified is not available.

The codecs available to you depend on your operating system installation and the additional libraries you have installed. For portable code, it is especially important to be able to gracefully handle the case in which the desired codec is not available on any particular machine.

#### Writing frames with cv::VideoWriter::write()

Once you have verified that your video writer is ready to write, you can write frames by simply passing your array to the writer:

```
cv::VideoWriter::write(
  const Mat& image                // Image to write as next frame
);
```

This image must have the same size as the size you gave to the writer when you configured it in the first place. If you told the writer that the frames would be in color, this must also be a three-channel image. If you indicated to the writer (via `isColor`) that the images may not be in color, then you can supply a single-channel (grayscale) image.

#### Writing frames with cv::VideoWriter::operator<<()

The video writer also supports the idiom of the overloaded output stream operator (`operator<<()`). In this case, once you have your video writer open, you can write images to the video stream in the same manner you would write to `cout` or a file `ofstream` object:

```
my_video_writer << my_frame;
```

In practice, it is primarily a matter of style which method you will choose.

# Data Persistence

In addition to standard video compression, OpenCV provides a mechanism for serializing and deserializing its various data types to and from disk in either YAML or XML format. These methods can be used to load or store any number of OpenCV data objects (including basic types like `int`, `float`, etc.) in a single file. Such functions are separate from the specialized functions we saw earlier in this chapter that handle the more particular situation of loading and saving compressed image files and video data. In this section, we will focus on general object persistence: reading and writing matrices, OpenCV structures, configuration, and logfiles.

The basic mechanism for reading and writing files is the `cv::FileStorage` object. This object essentially represents a file on disk, but does so in a manner that makes accessing the data represented in the file easy and natural.

## Writing to a cv::FileStorage

```
cv::FileStorage::FileStorage;
cv::FileStorage::FileStorage( string fileName, int flag );
```

The `cv::FileStorage` object is a representation of an XML or YAML data file. You can create it and pass a filename to the constructor, or you can just create an unopened storage object with the default constructor and open the file later with `cv::FileStorage::open()`, where the `flag` argument should be either `cv::FileStorage::WRITE` or `cv::FileStorage::APPEND`.

```
cv::FileStorage::open( string fileName, int flag );
```

Once you have opened the file you want to write to, you can write using the operator `cv::FileStorage::operator<<()` in the same manner you might write to `stdout` with an STL stream. Internally, however, there is quite a bit more going on when you write in this manner.

Data inside `cv::FileStorage` is stored in one of two forms, either as a "mapping" (i.e., key/value pairs) or a "sequence" (a series of unnamed entries). At the top level, the data you write to the file storage is all a mapping, and inside of that mapping you can place other mappings or sequences, and mappings or sequences inside of those as deep as you like.

```
myFileStorage << "someInteger" << 27;                      // save an integer
myFileStorage << "anArray" << cv::Mat::eye(3,3,CV_32F);  // save an array
```

To create a sequence entry, you first provide the string name for the entry, and then the entry itself. The entry can be a number (integer, float, etc.), a string, or any OpenCV data type.

If you would like to create a new mapping or sequence, you can do so with the special characters { (for a mapping) or [ (for a sequence). Once you have started the mapping or sequence, you can add new elements and then finally close the mapping or sequence with } or ] (respectively).

```
myFileStorage << "theCat" << "{";
myFileStorage << "fur" << "gray" << "eyes" << "green" << "weightLbs" << 16;
myFileStorage << "}";
```

Once you have created a mapping, you enter each element with a name and the data following, just as you did for the top-level mapping. If you create a sequence, you simply enter the new data one item after another until you close the sequence.

```
myFileStorage << "theTeam" << "[";
myFileStorage << "eddie" << "tom" << "scott";
myFileStorage << "]";
```

Once you are completely done writing, you close the file with the `cv::FileStorage::release()` member function.

Example 8-2 is an explicit code sample from the OpenCV documentation.

*Example 8-2. Using cv::FileStorage to create a .yml data file*

```
#include "opencv2/opencv.hpp"
#include <time.h>

int main(int, char** argv)
{

    cv::FileStorage fs("test.yml", cv::FileStorage::WRITE);

    fs << "frameCount" << 5;

    time_t rawtime; time(&rawtime);
    fs << "calibrationDate" << asctime(localtime(&rawtime));

    cv::Mat cameraMatrix = (
      cv::Mat_<double>(3,3)
      << 1000, 0, 320, 0, 1000, 240, 0, 0, 1
    );
    cv::Mat distCoeffs = (
      cv::Mat_<double>(5,1)
      << 0.1, 0.01, -0.001, 0, 0
    );
    fs << "cameraMatrix" << cameraMatrix << "distCoeffs" << distCoeffs;

    fs << "features" << "[";
    for( int i = 0; i < 3; i++ )
    {
        int x = rand() % 640;
```

```
        int y = rand() % 480;
        uchar lbp = rand() % 256;

        fs << "{:" << "x" << x << "y" << y << "lbp" << "[:";
        for( int j = 0; j < 8; j++ )
            fs << ((lbp >> j) & 1);
        fs << "]" << "}";
    }
    fs << "]";
    fs.release();

    return 0;
}
```

The result of running this program would be a YAML file with the following contents:


```
%YAML:1.0
frameCount: 5
calibrationDate: "Fri Jun 17 14:09:29 2011\n"
cameraMatrix: !!opencv-matrix
   rows: 3
   cols: 3
   dt: d
   data: [ 1000., 0., 320., 0., 1000., 240., 0., 0., 1. ]
distCoeffs: !!opencv-matrix
   rows: 5
   cols: 1
   dt: d
   data: [ 1.0000000000000001e-01, 1.0000000000000000e-02,
       -1.0000000000000000e-03, 0., 0. ]
features:
   - { x:167, y:49,  lbp:[ 1, 0, 0, 1, 1, 0, 1, 1 ] }
   - { x:298, y:130, lbp:[ 0, 0, 0, 1, 0, 0, 1, 1 ] }
   - { x:344, y:158, lbp:[ 1, 1, 0, 0, 0, 0, 1, 0 ] }
```


In the example code, you will notice that sometimes all of the data in a mapping or sequence is stored on a single line and other times it is stored with one element per line. This is not an automatic formatting behavior. Instead, it is created by a variant of the mapping and sequence creation strings: `"{:"` and `"}"` mappings, and `"[:"` and `"]"` for sequences. This feature is meaningful only for YAML output; if the output file is XML, this nuance is ignored and the mapping or sequence is stored as it would have been without the variant.

## Reading from a cv::FileStorage

```
FileStorage::FileStorage( string fileName, int flag );
```

The `cv::FileStorage` object can be opened for reading the same way it is opened for writing, except that the `flag` argument should be set to `cv::FileStorage::READ`. As with writing, you can also create an unopened file storage object with the default constructor and open it later with `cv::FileStorage::open()`.

```
FileStorage::open( string fileName, int flag );
```

Once the file has been opened, the data can be read with either the overloaded array operator `cv::FileStorage::operator[]()` or with the iterator `cv::FileNodeIterator`. Once you are completely done reading, you then close the file with the `cv::FileStorage::release()` member function.

To read from a mapping, the `cv::FileStorage::operator[]()` is passed the string key associated with the desired object. To read from a sequence, the same operator can be called with an integer argument instead. The return value of this operator is not the desired object, however; it is an object of type `cv::FileNode`, which represents the value that goes with the given key in an abstract form. The `cv::FileNode` object can be manipulated in a variety of ways, which we will now investigate.

## cv::FileNode

Once you have a `cv::FileNode` object, you can do one of several things with it. If it represents an object (or a number or a string), you can just load it into a variable of the appropriate type with the overloaded extraction operator `cv::FileNode::operator>>()`.

```
cv::Mat anArray;
myFileStorage["calibrationMatrix"] >> anArray;
```

The `cv::FileNode` object also supports direct casting to any of the basic data types.

```
int aNumber;
myFileStorage["someInteger"] >> aNumber;
```

This is equivalent to:

```
int aNumber;
aNumber = (int)myFileStorage["someInteger"];
```

As mentioned earlier, there is also an iterator for moving through file nodes that can be used as well. Given a `cv::FileNode` object, the member functions `cv::FileNode::begin()` and `cv::FileNode::end()` have their usual interpretations as providing the first and "after last" iterator for either a mapping or a sequence. The iterator, on dereferencing with the usual overloaded dereferencing operator `cv::FileNodeIterator::operator*()`, will return another `cv::FileNode` object. Such iterators support the usual incrementing and decrementing operators. If the iterator was iterating through a mapping, then the returned `cv::FileNode` object will have a name that can be retrieved with `cv::FileNode::name()`.

Of the methods listed in Table 8-5, one requires special clarification: `cv::FileNode::type()`. The returned value is an enumerated type defined in the class `cv::FileNode`. The possible values are given in Table 8-6.

*Table 8-5. Member functions of cv::FileNode*

| Example | Description |
|---|---|
| `cv::FileNode fn` | File node object default constructor |
| `cv::FileNode fn1( fn0 )` | File node object copy constructor; creates a node `fn1` from a node `fn0` |
| `cv::FileNode fn1( fs, node )` | File node constructor; creates a C++ style `cv::FileNode` object from a C-style `CvFileStorage*` pointer `fs` and a C-style `CvFileNode*` pointer node |
| `fn[ (string)key ]`<br>`fn[ (const char*)key ]` | STL string or C-string accessor for named child (of mapping node); converts key to the appropriate child node |
| `fn[ (int)id ]` | Accessor for numbered child (of sequence node); converts ID to the appropriate child node |
| `fn.type()` | Returns node type `enum` |
| `fn.empty()` | Determines if node is empty |
| `fn.isNone()` | Determines if node has value `None` |
| `fn.isSeq()` | Determines if node is a sequence |
| `fn.isMap()` | Determines if node is a mapping |
| `fn.isInt()`<br>`fn.isReal()`<br>`fn.isString()` | Determines if node is an integer, a floating-point number, or a string (respectively) |
| `fn.name()` | Returns nodes name if node is a child of a mapping |
| `size_t sz=fn.size()` | Returns a number of elements in a sequence or mapping |
| `(int)fn`<br>`(float)fn`<br>`(double)fn`<br>`(string)fn` | Extracts the value from a node containing an integer, 32-bit float, 64-bit float, or string (respectively) |

*Table 8-6. Possible return values for cv::FileNode::type()*

| Example | Description |
|---|---|
| `cv::FileNode::NONE = 0` | Node is of type `None` |
| `cv::FileNode::INT = 1` | Node contains an integer |
| `cv::FileNode::REAL = 2`<br>`cv::FileNode::FLOAT = 2` | Node contains a floating-point number[a] |
| `cv::FileNode::STR = 3`<br>`cv::FileNode::STRING = 3` | Node contains a string |

| Example | Description |
|---|---|
| `cv::FileNode::REF = 4` | Node contains a reference (i.e., a compound object) |
| `cv::FileNode::SEQ = 5` | Node is itself a sequence of other nodes |
| `cv::FileNode::MAP = 6` | Node is itself a mapping of other nodes |
| `cv::FileNode::FLOW = 8` | Node is a compact representation of a sequence or mapping |
| `cv::FileNode::USER = 16` | Node is a registered object (e.g., a matrix) |
| `cv::FileNode::EMPTY = 32` | Node has no value assigned to it |
| `cv::FileNode::NAMED = 64` | Node is a child of a mapping (i.e., it has a name) |

[a] Note that the floating-point types are not distinguished. This is a somewhat subtle point. Recall that XML and YAML are ASCII text file formats. As a result, all floating-point numbers are of no specific precision until cast to an internal machine variable type. So, at the time of parsing, all floating-point numbers are represented only as an abstract floating-point type.

Note that the last four `enum` values are powers of two starting at 8. This is because a node may have any or all of these properties in addition to one of the first eight listed types.

Example 8-3 (also from the OpenCV documentation) shows how we could read the file we wrote previously.

*Example 8-3. Using cv::FileStorage to read a .yml file*

```
cv::FileStorage fs2("test.yml", cv::FileStorage::READ);

// first method: use (type) operator on FileNode.
int frameCount = (int)fs2["frameCount"];

// second method: use cv::FileNode::operator >>
//
std::string date;
fs2["calibrationDate"] >> date;

cv::Mat cameraMatrix2, distCoeffs2;
fs2["cameraMatrix"] >> cameraMatrix2;
fs2["distCoeffs"] >> distCoeffs2;

cout << "frameCount: "        << frameCount    << endl
     << "calibration date: "  << date          << endl
     << "camera matrix: "     << cameraMatrix2 << endl
     << "distortion coeffs: " << distCoeffs2   << endl;

cv::FileNode features   = fs2["features"];
cv::FileNodeIterator it = features.begin(), it_end = features.end();
int idx                 = 0;
std::vector<uchar> lbpval;
```

```
// iterate through a sequence using FileNodeIterator
for( ; it != it_end; ++it, idx++ )
{
    cout << "feature #" << idx << ": ";
    cout << "x=" << (int)(*it)["x"] << ", y=" << (int)(*it)["y"] << ", lbp: (";

    // ( Note: easily read numerical arrays using FileNode >> std::vector. )
    //
    (*it)["lbp"] >> lbpval;
    for( int i = 0; i < (int)lbpval.size(); i++ )
        cout << " " << (int)lbpval[i];
    cout << ")" << endl;

}
fs.release();
```

# Summary

In this chapter we saw several different ways to interact with disk storage and physical devices. We learned that the HighGUI module in the library provides simple tools for reading and writing common image file formats from disk and that these tools would automatically handle the compression and decompression of these file formats. We also learned that video is handled in a similar way and that the same tools that are used for reading video information from disk could also be used to capture video from a camera. Finally, we saw that OpenCV provides a powerful tool for storing and restoring its own native types into XML and YML files. These files allowed the data to be organized into key/value maps for easy retrieval once read into memory.

# Exercises

This chapter completes our introduction to basic I/O programming and data structures in OpenCV. The following exercises build on this knowledge and create useful utilities for later use.

1. Create a program that (1) reads frames from a video, (2) turns the result to grayscale, and (3) performs Canny edge detection on the image. Display all three stages of processing in three different windows, with each window appropriately named for its function.

    a. Display all three stages of processing in one image. (Hint: create another image of the same height but three times the width as the video frame. Copy the images into this, either by using pointers or (more cleverly) by creating three new image headers that point to the beginning of and to one-third and two-thirds of the way into the `imageData`. Then use `Mat::copyTo()`.)

b. Write appropriate text labels describing the processing in each of the three slots.

2. Create a program that reads in and displays an image. When the user's mouse clicks on the image, read in the corresponding pixel values (blue, green, red) and write those values as text to the screen at the mouse location.

   a. For the program of Exercise 2, display the mouse coordinates of the individual image when clicking anywhere within the three-image display.

3. Create a program that reads in and displays an image.

   a. Allow the user to select a rectangular region in the image by drawing a rectangle with the mouse button held down, and highlight the region when the mouse button is released. Be careful to save an image copy in memory so that your drawing into the image does not destroy the original values there. The next mouse click should start the process all over again from the original image.

   b. In a separate window, use the drawing functions to draw a graph in blue, green, and red that represents how many pixels of each value were found in the selected box. This is the *color histogram* of that color region. The *x*-axis should be eight bins that represent pixel values falling within the ranges 0–31, 32–63,..., 223–255. The *y*-axis should be counts of the number of pixels that were found in that bin range. Do this for each color channel, BGR.

4. Make an application that reads and displays a video and is controlled by sliders. One slider will control the position within the video from start to end in 10 increments; another binary slider should control pause/unpause. Label both sliders appropriately.

5. Create your own simple paint program.

   a. Write a program that creates an image, sets it to `0`, and then displays it. Allow the user to draw lines, circles, ellipses, and polygons on the image using the left mouse button. Create an eraser function when the right mouse button is held down.

   b. Enable "logical drawing" by allowing the user to set a slider setting to AND, OR, and XOR. That is, if the setting is AND then the drawing will appear only when it crosses pixels greater than 0 (and so on for the other logical functions).

6. Write a program that creates an image, sets it to `0`, and then displays it. When the user clicks on a location, he or she can type in a label there. Allow Backspace to edit and provide for an abort key. Hitting Enter should fix the label at the spot it was typed.

7. Perspective transform:

   a. Write a program that reads in an image and uses the numbers 1–9 on the keypad to control a perspective transformation matrix (refer to our discussion of the `cv::warpPerspective()` in the section "cv::warpPerspective(): Dense perspective transform" on page 313 in Chapter 11). Tapping any number should increment the corresponding cell in the perspective transform matrix; tapping with the Shift key depressed should decrement the number associated with that cell (stopping at `0`). Each time a number is changed, display the results in two images: the raw image and the transformed image.

   b. Add functionality to zoom in or out.

   c. Add functionality to rotate the image.

8. Face fun. Go to the */samples/cpp/* directory and build the *facedetect.cpp* code. Draw a skull image (or find one on the Web) and store it to disk. Modify the *facedetect* program to load in the image of the skull.

   a. When a face rectangle is detected, draw the skull in that rectangle. Hint: you could look up the `cv::resize()` function. You may then set the ROI to the rectangle and use `Mat::copyTo()` to copy the properly resized image there.

   b. Add a slider with 10 settings corresponding to `0.0` to `1.0`. Use this slider to alpha-blend the skull over the face rectangle using the `cv::addWeighted()` function.

9. Image stabilization. Go to the */samples/cpp/* directory and build the *lkdemo* code (the motion tracking or *optical flow* code). Create and display a video image in a much larger window image. Move the camera slightly but use the optical flow vectors to display the image in the same place within the larger window. This is a rudimentary image stabilization technique.

10. Create a structure of an integer, a `cv::Point` and a `cv::Rect`; call it "my_struct."

    a. Write two functions: `void write_my_struct(FileStorage& fs, const string& name, const my_struct& ms)` and `void read_my_struct(const FileStorage& fs, const FileNode& ms_node, my_struct& ms)`. Use them to write and read `my_struct`.

    b. Write and read an array of 10 `my_struct` structures.

CHAPTER 9

# Cross-Platform and Native Windows

## Working with Windows

The previous chapter introduced the *HighGUI* toolkit. There, we looked at how that toolkit could help us with file- and device-related tasks. In addition to those features, the HighGUI library also provides basic built-in features for creating windows, displaying images in those windows, and making some user interaction possible with those windows. The native OpenCV graphical user interface (GUI) functions have been part of the library for a long time, and have the advantages of being stable, portable,[1] and easy to use.

Though convenient, the UI features of the HighGUI library have the disadvantage of being not particularly complete. As a result, there has been an effort to modernize the UI portion of HighGUI, and to add a number of useful new features, by converting from "native" interfaces to the use of Qt. Qt is a cross-platform toolkit, and so new features need be implemented only a single time in the library, rather than once for each supported platform. Needless to say, this has made development of the Qt interface more attractive, so it has more capabilities and will probably grow in the future, leaving the features of the native interface to become static legacy code.

In this section, we will first take a look at the native functions, and then move on to the differences, and particularly the new features, offered by the Qt-based interface.

1 They are "portable" because they make use of native window GUI tools on various platforms. This means X11 on Linux, Cocoa on Mac OS X, and the raw Win32 API on Microsoft Windows machines. However, this portability extends only to those platforms for which there is an implementation in the library. There are platforms on which OpenCV can be used for which there are no available implementations of the HighGUI library (e.g., Android).

Finally, we will look at how you would integrate OpenCV data types with some existing platform-specific toolkits for some popular operating systems.

## HighGUI Native Graphical User Interface

This section describes the core interface functions that are part of OpenCV and require no external toolkit support. If you compile OpenCV with Qt support, some of these functions will behave somewhat differently or have some additional options; we will cover that case in the following section. For the moment, however, we will focus on the bare-bones HighGUI UI tools.

The HighGUI user input tools support only three basic interactions—specifically, keypresses, mouse clicks on the image area, and the use of simple trackbars. These basic functions are usually sufficient for simple mockups and debugging, but hardly ideal for end-user-facing applications. For that, you will (at least) want to use the Qt-based interface or some other more full-featured UI toolkit.

The main advantages of the native tools are that they are fast and easy to use, and don't require you to install any additional libraries.

### Creating a window with cv::namedWindow()

First, we want to be able to create a window and show an image on the screen using HighGUI. The function that does the first part for us is `cv::namedWindow()`, and it expects a name for the new window and one flag. The name appears at the top of the window, and is also used as a handle for the window that can be passed to other HighGUI functions.[2] The `flag` argument indicates whether the window should autosize itself to fit an image we put into it. Here is the full prototype:

```
int cv::namedWindow(
  const string&  name,                      // Handle used to identify window
  int            flags = 0                  // Used to tell window to autosize
);
```

For now, the only valid options available for `flags` are to set it to `0` (the default value), which indicates that users are able (and required) to resize the window, or to set it to `cv::WINDOW_AUTOSIZE`.[3] If `cv::WINDOW_AUTOSIZE` is set, then HighGUI resizes the window to fit automatically whenever a new image is loaded, but users cannot resize the window.

2 In OpenCV, windows are referenced by name instead of by some unfriendly (and invariably OS-dependent) "handle." Conversion between handles and names happens under HighGUI's hood, so you needn't worry about it.

3 Later in this chapter, we will look at the (optional) Qt-based backend for HighGUI. If you are using that backend, there are more options available for `cv::namedWindow()` and other functions.

Once we create a window, we usually want to put something inside it. But before we do that, let's see how to get rid of the window when it is no longer needed. For this, we use `cv::destroyWindow()`, a function whose only argument is a string: the name given to the window when it was created.

```
int cv::destroyWindow(
  const string&  name,                       // Handle used to identify window
);
```

### Drawing an image with cv::imshow()

Now we are ready for what we really want to do: load an image and put it into the window where we can view it and appreciate its profundity. We do this via one simple function, `cv::imshow()`:

```
void cv::imshow(
  const string&  name,                       // Handle used to identify window
  cv::InputArray image                       // Image to display in window
);
```

The first argument is the name of the window within which we intend to draw. The second argument is the image to be drawn. Note that the window will keep a copy of the drawn image and will repaint it as needed from that buffer, so a subsequent modification of the input `image` will not change the contents of the window unless a subsequent call to `cv::imshow()` is made.

### Updating a window and cv::waitKey()

The function `cv::waitKey()` is to wait for some specified (possibly indefinite) amount of time for a keypress on the keyboard, and to return that key value when it is received. `cv::waitKey()` accepts a keypress from any open OpenCV window (but will not function if no such window exists).

```
int cv::waitKey(
  int delay = 0                       // Milliseconds until giving up (0='never')
);
```

`cv::waitKey()` takes a single argument, `delay`, which is the amount of time (in milliseconds) for which it will wait for a keypress before returning automatically. If the delay is set to `0`, `cv::waitKey()` will wait indefinitely for a keypress. If no keypress comes before `delay` milliseconds has passed, `cv::waitKey()` will return `-1`.

There is a second, less obvious function of `cv::waitKey()`, which is to provide an opportunity for any open OpenCV window to be updated. This means that if you do not call cv::waitKey(), your image may never be drawn in your window, or your window may behave strangely (and badly) when moved, resized, or uncovered.[4]

### An example displaying an image

Let's now put together a simple program that will display an image on the screen (see Example 9-1). We can read a filename from the command line, create a window, and put our image in the window in 15 lines (including comments!). This program will display our image as long as we want to sit and appreciate it, and then exit when the Esc key (ASCII value of 27) is pressed.

*Example 9-1. Creating a window and displaying an image in that window*

```
int main( int argc, char** argv ) {

  // Create a named window with the name of the file
  //
  cv::namedWindow( argv[1], 1 );

  // Load the image from the given filename
  //
  cv::Mat = cv::imread( argv[1] );

  // Show the image in the named window
  //
  cv::imshow( argv[1], img );

  // Idle until the user hits the Esc key
  //
  while( true ) {
    if( cv::waitKey( 100 /* milliseconds */ ) == 27 ) break;
  }

  // Clean up and don't be piggies
  //
  cv::destroyWindow( argv[1] );

  exit(0);
}
```

4 What this sentence really means is that cv::waitKey() is the *only* function in HighGUI that can fetch and handle events. This means that if it is not called periodically, no normal event processing will take place. As a corollary to this, if HighGUI is being used within an environment that takes care of event processing, then you may not need to call cv::waitKey(). For more information on this detail, see cv::startWindowThread() in a few pages.

For convenience, we have used the filename as the window name. This is nice because OpenCV automatically puts the window name at the top of the window, so we can tell which file we are viewing (see Figure 9-1). Easy as cake.

*Figure 9-1. A simple image displayed with cv::imshow()*

Before we move on, there are a few other window-related functions you ought to know about. They are:

```
void cv::moveWindow( const char* name, int x, int y );
void cv::destroyAllWindows( void );
int  cv::startWindowThread( void );
```

`cv::moveWindow()` simply moves a window on the screen so that its upper-left corner is positioned at pixel location: *x*, *y*. `cv::destroyAllWindows()` is a useful cleanup function that closes all of the windows and deallocates the associated memory.

On Linux and Mac OS X, `cv::startWindowThread()` tries to start a thread that updates the window automatically, and handles resizing and so forth. A return value of `0` indicates that no thread could be started—for example, because there is no support for this feature in the version of OpenCV that you are using. Note that, if you do not start a separate window thread, OpenCV can react to user interface actions only when it is explicitly given time to do so (this happens when your program invokes `cv::waitKey()`).

### Mouse events

Now that we can display an image to a user, we might also want to allow the user to interact with the image we have created. Since we are working in a window environment and since we already learned how to capture single keystrokes with `cv::waitKey()`, the next logical thing to consider is how to "listen to" and respond to mouse events.

Unlike keyboard events, mouse events are handled by a more traditional callback mechanism. This means that, to enable response to mouse clicks, we must first write a callback routine that OpenCV can call whenever a mouse event occurs. Once we have done that, we must register the callback with OpenCV, thereby informing OpenCV that this is the correct function to use whenever the user does something with the mouse over a particular window.

Let's start with the callback. For those of you who are a little rusty on your event-driven program lingo, the *callback* can be any function that takes the correct set of arguments and returns the correct type. Here, we must be able to tell the function to be used as a callback exactly what kind of event occurred and where it occurred. The function must also be told if the user was pressing such keys as Shift or Alt when the mouse event occurred. A pointer to such a function is called a `cv::MouseCallback`. Here is the exact prototype that your callback function must match:

```
void your_mouse_callback(
  int   event,                         // Event type (see Table 9-1)
  int   x,                             // x-location of mouse event
  int   y,                             // y-location of mouse event
  int   flags,                         // More details on event (see Table 9-1)
  void* param                          // Parameters from cv::setMouseCallback()
);
```

Now, whenever your function is called, OpenCV will fill in the arguments with their appropriate values. The first argument, called the `event`, will have one of the values shown in Table 9-1.

*Table 9-1. Mouse event types*

| Event | Numerical value |
|---|---|
| cv::EVENT_MOUSEMOVE | 0 |
| cv::EVENT_LBUTTONDOWN | 1 |
| cv::EVENT_RBUTTONDOWN | 2 |
| cv::EVENT_MBUTTONDOWN | 3 |
| cv::EVENT_LBUTTONUP | 4 |
| cv::EVENT_RBUTTONUP | 5 |
| cv::EVENT_MBUTTONUP | 6 |
| cv::EVENT_LBUTTONDBLCLK | 7 |

| Event | Numerical value |
|---|---|
| cv::EVENT_RBUTTONDBLCLK | 8 |
| cv::EVENT_MBUTTONDBLCLK | 9 |

The second and third arguments will be set to the x- and y-coordinates of the mouse event. Note that these coordinates represent the pixel coordinates in the image independent of the other details of the window.[5]

The fourth argument, called `flags`, is a bit field in which individual bits indicate special conditions present at the time of the event. For example, `cv::EVENT_FLAG_SHIFTKEY` has a numerical value of `16` (i.e., the fifth bit, or `1<<4`); so, if we wanted to test whether the Shift key were down, we could simply compute the bitwise AND of `flags & cv::EVENT_FLAG_SHIFTKEY`. Table 9-2 shows a complete list of the flags.

*Table 9-2. Mouse event flags*

| Flag | Numerical value |
|---|---|
| cv::EVENT_FLAG_LBUTTON | 1 |
| cv::EVENT_FLAG_RBUTTON | 2 |
| cv::EVENT_FLAG_MBUTTON | 4 |
| cv::EVENT_FLAG_CTRLKEY | 8 |
| cv::EVENT_FLAG_SHIFTKEY | 16 |
| cv::EVENT_FLAG_ALTKEY | 32 |

The final argument is a void pointer that can be used to have OpenCV pass additional information, in the form of a pointer, to whatever kind of structure you need.[6]

Next, we need the function that registers the callback. That function is called `cv::setMouseCallback()`, and it requires three arguments.

```
void cv::setMouseCallback(
  const string&     windowName,          // Handle used to identify window
  cv::MouseCallback on_mouse,            // Callback function
  void*             param     = NULL     // Additional parameters for callback fn.
);
```

The first argument is the name of the window to which the callback will be attached; only events in that particular window will trigger this specific callback. The second

5 In general, this is not the same as the pixel coordinates of the event that would be returned by the OS. This is because OpenCV is concerned with telling you where *in the image* the event happened, not *in the window* (to which the OS typically references mouse event coordinates).

6 A common situation in which the `param` argument is used is when the callback itself is a static member function of a class. In this case, you will probably want to pass the `this` pointer to indicate which class object instance the callback is intended to affect.

argument is your callback function. The third argument, param, allows us to specify the param information that should be given to the callback whenever it is executed. This is, of course, the same param we were just discussing with the callback prototype.

In Example 9-2, we write a small program to draw boxes on the screen with the mouse. The function my_mouse_callback() responds to mouse events, and it uses the events to determine what to do when it is called.

*Example 9-2. Toy program for using a mouse to draw boxes on the screen*

```
#include <opencv2/opencv.hpp>

// Define our callback which we will install for
// mouse events
//
void my_mouse_callback(
   int event, int x, int y, int flags, void* param
);

Rect box;
bool drawing_box = false;

// A little subroutine to draw a box onto an image
//
void draw_box( cv::Mat& img, cv::Rect box ) {
  cv::rectangle(
    img,
    box.tl(),
    box.br(),
    cv::Scalar(0x00,0x00,0xff)    /* red */
  );
}

void help() {
  std::cout << "Call: ./ch4_ex4_1\n" <<
    " shows how to use a mouse to draw regions in an image." << std::endl;
}

int main( int argc, char** argv ) {

  help();
  box = cv::Rect(-1,-1,0,0);
  cv::Mat image(200, 200, CV_8UC3), temp;
  image.copyTo(temp);

  box   = cv::Rect(-1,-1,0,0);
  image = cv::Scalar::all(0);

  cv::namedWindow( "Box Example" );
```

```
  // Here is the crucial moment where we actually install
  // the callback. Note that we set the value of 'params' to
  // be the image we are working with so that the callback
  // will have the image to edit.
  //
  cv::setMouseCallback(
    "Box Example",
    my_mouse_callback,
    (void*)&image
  );

  // The main program loop. Here we copy the working image
  // to the temp image, and if the user is drawing, then
  // put the currently contemplated box onto that temp image.
  // Display the temp image, and wait 15ms for a keystroke,
  // then repeat.
  //
  for(;;) {

    image.copyTo(temp);
    if( drawing_box ) draw_box( temp, box );
    cv::imshow( "Box Example", temp );

    if( cv::waitKey( 15 ) == 27 ) break;
  }

  return 0;
}

// This is our mouse callback. If the user
// presses the left button, we start a box.
// When the user releases that button, then we
// add the box to the current image. When the
// mouse is dragged (with the button down) we
// resize the box.
//
void my_mouse_callback(
   int event, int x, int y, int flags, void* param
) {

  cv::Mat& image = *(cv::Mat*) param;

  switch( event ) {

    case cv::EVENT_MOUSEMOVE: {
      if( drawing_box ) {
        box.width  = x-box.x;
        box.height = y-box.y;
      }
    }
    break;
```

```
    case cv::EVENT_LBUTTONDOWN: {
      drawing_box = true;
      box = cv::Rect( x, y, 0, 0 );
    }
    break;

    case cv::EVENT_LBUTTONUP: {
      drawing_box = false;
      if( box.width < 0  ) {
        box.x += box.width;
        box.width *= -1;
      }
      if( box.height < 0 ) {
        box.y += box.height;
        box.height *= -1;
      }
      draw_box( image, box );
    }
    break;
  }

}
```

### Sliders, trackbars, and switches

HighGUI provides a convenient slider element. In HighGUI, sliders are called *trackbars*. This is because their o riginal (historical) intent was for selecting a particular frame in the playback of a video. Of course, once trackbars were added to HighGUI, people began to use them for all of the usual things one might do with sliders, as well as many unusual ones (we'll discuss some of these in "Surviving without buttons" on page 218).

As with the parent window, the slider is given a unique name (in the form of a character string) and is thereafter always referred to by that name. The HighGUI routine for creating a trackbar is:

```
int cv::createTrackbar(
  const string&        trackbarName,      // Handle used to identify trackbar
  const string&        windowName,        // Handle used to identify window
  int*                 value,             // Slider position gets put here
  int                  count,             // Total counts for slider at far right
  cv::TrackbarCallback onChange     = NULL,// Callback function (optional)
  void*                param        = NULL // Additional params for callback fn.
);
```

The first two arguments are the name for the trackbar itself and the name of the parent window to which the trackbar will be attached. When the trackbar is created, it is added to either the top or the bottom of the parent window.[7] The trackbar will not occlude any image that is already in the window; rather, it will make the window slightly bigger. The name of the trackbar will appear as a "label" for the trackbar. As with the location of the trackbar itself, the exact location of this label depends on the operating system, but most often it is immediately to the left, as in Figure 9-2.

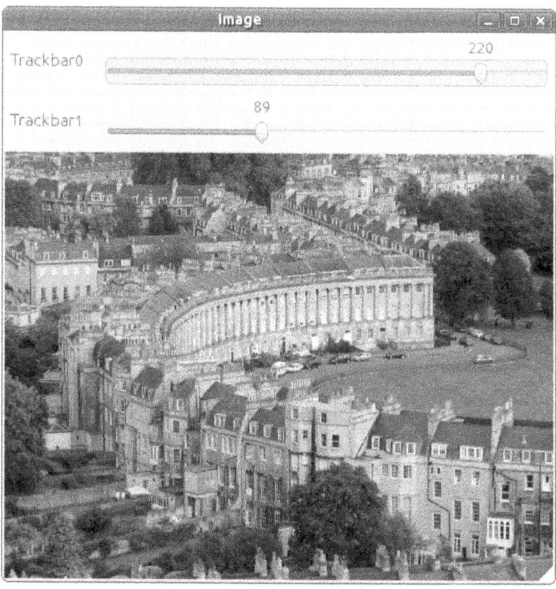


*Figure 9-2. A simple application displaying an image; this window has two trackbars attached: Trackbar0 and Trackbar1*

The next two arguments are `value`, a pointer to an integer that will automatically be set to the value to which the slider has been moved, and `count`, a numerical value for the maximum value of the slider.

The last argument is a pointer to a callback function that will be automatically called whenever the slider is moved. This is exactly analogous to the callback for mouse events. If used, the callback function must have the form specified by `cv::TrackbarCallback`, which means that it must match the following prototype:

```
void your_trackbar_callback(
  int   pos,                        // Trackbar slider position
  void* param = NULL                // Parameters from cv::setTrackbarCallback()
);
```

7 Whether it is added to the top or bottom depends on the operating system, but it will always appear in the same place on a given platform.

This callback is not actually required, so if you don't want a callback, then you can simply set this value to `NULL`. Without a callback, the only effect of the user moving the slider will be the value of `*value` being updated. (Of course, if you don't have a callback, you will be responsible for polling this value if you are going to respond to it being changed.)

The final argument to `cv::createTrackbar()` is `params`, which can be any pointer. This pointer will be passed to your callback as its `params` argument whenever the callback is executed. This is very helpful for, among other things, allowing you to handle trackbar events without having to introduce global variables.

Finally, here are two more routines that will allow you to programmatically read or set the value of a trackbar just by using its name:

```
int cv::getTrackbarPos(
  const string& trackbarName,          // Handle used to identify trackbar, label
  const string& windowName,            // Handle used to identify window
);

void cv::setTrackbarPos(
  const string& trackbarName,          // Handle used to identify trackbar, label
  const string& windowName,            // Handle used to identify window
  int   pos                            // Trackbar slider position
);
```

These functions allow you to read or set the value of a trackbar from anywhere in your program.

#### Surviving without buttons

Unfortunately, the native interface in HighGUI does not provide any explicit support for buttons. It is thus common practice, among the particularly lazy, to instead use sliders with only two positions.[8] Another option that occurs often in the OpenCV samples in *.../opencv/samples/c/* is to use keyboard shortcuts instead of buttons (see, e.g., the *floodfill* demo in the OpenCV source-code bundle).

*Switches* are just sliders (trackbars) that have only two positions, "on" (`1`) and "off" (`0`) (i.e., `count` has been set to `1`). You can see how this is an easy way to obtain the functionality of a button using only the available trackbar tools. Depending on exactly how you want the switch to behave, you can use the trackbar callback to auto-

8 For the less lazy, the common practice is to compose the image you are displaying with a "control panel" you have drawn and then use the mouse event callback to test for the mouse's location when the event occurs. When the (*x*, *y*) location is within the area of a button you have drawn on your control panel, the callback is set to perform the button action. In this way, all "buttons" are internal to the mouse event callback routine associated with the parent window. But really, if you need this kind of functionality now, it is probably best just to use the Qt backend.

matically reset the button back to 0 (as in Example 9-3; this is something like the standard behavior of most GUI "buttons") or to automatically set other switches to 0 (which gives the effect of a "checkbox" or, with a little more work, a "radio button").

*Example 9-3. Using a trackbar to create a "switch" that the user can turn on and off; this program plays a video and uses the switch to create a pause functionality*

```
// An example program in which the user can draw boxes on the screen.
//
#include <opencv2/opencv.hpp>
#include <iostream>

using namespace std;

//
// Using a trackbar to create a "switch" that the user can turn on and off.
// We make this value global so everyone can see it.
//
int g_switch_value = 1;
void switch_off_function() { cout << "Pause\n"; }; //YOU COULD DO MORE
void switch_on_function()  { cout << "Run\n"; };

// This will be the callback that we give to the trackbar.
//
void switch_callback( int position, void* ) {
  if( position == 0 ) {
    switch_off_function();
  } else {
    switch_on_function();
  }
}

void help() {
    cout << "Call: my.avi" << endl;
    cout << "Shows putting a pause button in a video." << endl;
}

int main( int argc, char** argv ) {

  cv::Mat frame; // To hold movie images
  cv::VideoCapture g_capture;
  help();
  if( argc < 2 || !g_capture.open( argv[1] ) ){
    cout << "Failed to open " << argv[1] << " video file\n" << endl;
    return -1;
  }

  // Name the main window
  //
  cv::namedWindow( "Example", 1 );
```

```
  // Create the trackbar. We give it a name,
  // and tell it the name of the parent window.
  //
  cv::createTrackbar(
    "Switch",
    "Example",
    &g_switch_value,
    1,
    switch_callback
  );

  // This will cause OpenCV to idle until
  // someone hits the Esc key.
  //
  for(;;) {
    if( g_switch_value ) {
          g_capture >> frame;
          if( frame.empty() ) break;
          cv::imshow( "Example", frame);
      }
      if( cv::waitKey(10)==27 ) break;
  }

  return 0;
}
```

You can see that this will turn on and off just like a light switch. In our example, whenever the trackbar "switch" is set to `0`, the callback executes the function `switch_off_function()`, and whenever it is switched on, the `switch_on_function()` is called.

## Working with the Qt Backend

As we described earlier, the thinking in the development of the HighGUI portion of OpenCV has been to rely on separate libraries for any serious GUI functionality. This makes sense, as it is not OpenCV's purpose to reinvent that particular wheel and there are plenty of great GUI toolkits out there that are well maintained and which evolve and adapt to the changing times under their own excellent development teams.

The basic GUI tools we have seen so far provide what rudimentary functionality they do by means of native libraries in various platforms that are wrapped up in such a way that you, the developer, don't see those native libraries. This has worked reasonably well, but, even before the addition of mobile platform support, it was becoming difficult to maintain. For this reason, there is an increasing shift to relying on a single cross-platform toolkit as the means of delivering even these basic functionalities. That cross-platform toolkit is Qt.

From the perspective of the OpenCV library, there is a great advantage to using such an outside toolkit. Functionality is gained, and at the same time, development time is reduced (which would otherwise be taken away from the library's core goal).

In this section we will learn how to use the newer Qt-based HighGUI interface. It is very likely that all future evolution of HighGUI functionality will happen in this component, as it is much more efficient to work here than in the multiple legacy native interfaces.

It is important to note, however, that using HighGUI with the Qt backend is not the same as using Qt directly (we will explore that possibility briefly at the end of this chapter). The HighGUI interface is still the HighGUI interface; it simply uses Qt behind the scenes in place of the various native libraries. One side effect of this is that it isn't that convenient to extend the Qt interface. If you want more than HighGUI gives you, you are still pretty stuck with writing your own window layer. On the bright side, the Qt interface gives you a lot more to work with, and so perhaps you will not find that extra level of complexity necessary as often, or perhaps ever.[9]

### Getting started

If you have built your OpenCV installation with Qt support on,[10] then when you open a window, it will automatically have two new features. These are the *toolbar* and the *status bar* (see Figure 9-3). These objects come up complete, with all of the contents you see in the figure. In particular, the toolbar contains buttons that allow you to pan (the first four arrow buttons), zoom (the next four buttons), save the current image (the ninth button), and pop up a properties window (more on this last one a little later).

9 You will see that the Qt-based HighGUI interface is still mainly intended for developers doing scientific work or debugging systems. If you are doing end-user-facing commercial code, you will still almost certainly want a more powerful and expressive UI toolkit.

10 This means that when you configured the build with `cmake`, you used the `-D WITH_QT=ON` option.

*Figure 9-3. This image is displayed with the Qt interface enabled; it shows the toolbar, the status bar, and a text overlay that, in this case, contains the name of the image*

The status bar in Figure 9-3 contains information about what is under your mouse at any given moment. The *x*, *y* location is displayed, as well as the RGB value of the pixel at which the mouse currently points.

All of this you get "for free" just for compiling your code with the Qt interface enabled. If you have compiled with Qt and you *don't* want these decorations, then you can simply add the `CV_GUI_NORMAL` flag when you call `cv::namedWindow()`, and they will disappear.[11]

### The actions menu

As you can see, when you create the window with `CV_GUI_EXPANDED`, you will see a range of buttons in the toolbar. An alternative to the toolbar, which will always be available whether you use `CV_GUI_EXPANDED` or not, is the *pop-up menu*. This pop-up menu (shown in Figure 9-4) contains the same options as the toolbar, and you can display it at any time by right-clicking on the image.

11 There is also a `CV_GUI_EXTENDED` flag which, in theory, creates these decorations but its numerical value is 0x00, so it is the default behavior anyhow.

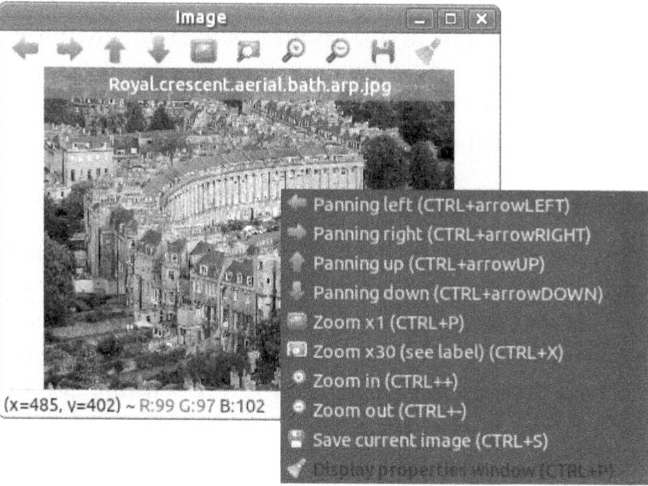


*Figure 9-4. Here the Qt extended UI window is shown with the pop-up menu, which provides the same options as the toolbar (along with an explanation of the buttons and their keyboard shortcuts)*

### The text overlay

Another option provided by the Qt GUI is the ability to put a short-lived banner across the top of the image you are displaying. This banner is called the *overlay*, and appears with a shaded box around it for easy reading. This is an exceptionally handy feature if you just want to throw some simple information like the frame number or frame rate on a video, or a filename of the image you are viewing. You can display an overlay on any window, even if you are using `CV_GUI_NORMAL`.

```
int cv::displayOverlay(
  const string& name,             // Handle used to identify window
  const string& text,             // Text you want to display
  int           delay             // Milliseconds to show text (0='forever')
);
```

The `cv::displayOverlay()` function takes three arguments. The first is the name of the window on which you want the overlay to appear. The second argument is whatever text you would like to appear in the window. (One word of warning here: the text has a fixed size, so if you try to cram too much stuff in there, it will just overflow.[12] By default, the text is always center justified.) The third and final argument, `delay`, is the amount of time (in milliseconds) that the overlay will stay in place. If

12 You can, however, insert new lines. So, for example, if you were to give the `text` string `"Hello\nWorld"`, then the word *Hello* would appear on the first (top) line, and the word *World* would appear on a second line right below it.

delay is set to 0, then the overlay will stay in place indefinitely—or at least until you write over it with another call to cv::displayOverlay(). In general, if you call cv::displayOverlay() before the delay timer for a previous call is expired, the previous text is removed and replaced with the new text, and the timer is reset to the new delay value regardless of what is left in the timer before the new call.

### Writing your own text into the status bar

In addition to the overlay, you can also write text into the status bar. By default, the status bar contains information about the pixel over which your mouse pointer is located (if any). You can see in Figure 9-3 that the status bar contains an *x*, *y* location and the RGB color value of the pixel that was under the pointer when the figure was made. You can replace this text with your own text with the cv::displayStatusBar() method:

```
int cv::displayStatusBar(
  const string& name,              // Handle used to identify window
  const string& text,              // Text you want to display
  int           delay              // Milliseconds to show text (0='forever')
);
```

Unlike cv::displayOverlay(), cv::displayStatusBar() can be used only on windows that were created with the CV_GUI_EXPANDED flag (i.e., ones that have a status bar in the first place). When the delay timer is expired (if you didn't set it to 0), then the default *x*, *y* and RGB text will reappear in the status bar.

### The properties window

So far, we haven't really discussed the last button on the toolbar (which corresponds to the last option on the pop-up menu, the one that is "darkened" in Figure 9-5). This option opens up an entirely new window called the *properties window*. The properties window is a convenient place to put trackbars and buttons (yes, the Qt GUI supports buttons) that you don't want in your face all the time. It's important to remember, however, that there is just one properties window *per application*, so you don't really *create* it, you just *configure* it.

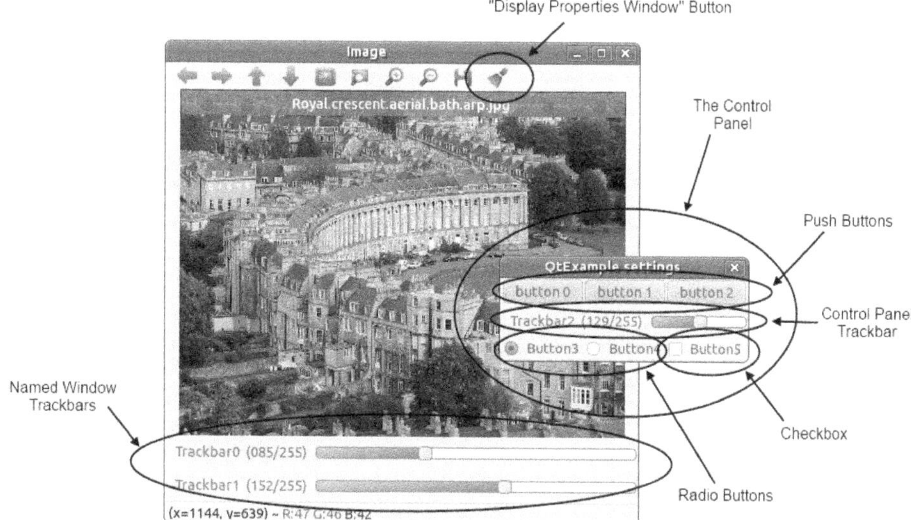


*Figure 9-5. This window contains two trackbars; you can also see the control panel, which contains three push buttons, a trackbar, two radio buttons, and a checkbox*

The properties window will not be available unless you have assigned some trackbars or buttons to it (more on how to do this momentarily). If it is available, then you can display it by clicking the Display Properties Window button on the toolbar (the one on the far right), clicking the identical button on the action menu, or pressing Ctrl-P while your mouse is over any window.

### Trackbars revisited

In the previous section on the HighGUI native interface, we saw that it was possible to add trackbars to windows. The trackbars in Figure 9-5 were created using the same `cv::createTrackbar()` command we saw earlier. The only real difference is that the trackbars in Figure 9-5 are prettier than the ones we created using the non-Qt interface (recall Figure 9-2).

The important new concept in the Qt interface, however, is that we can also create trackbars in the properties window. We do so simply by creating the trackbar as we normally would, but by specifying an empty string as the window name to which the trackbar will be attached.

```
int contrast = 128;
cv::createTrackbar( "Contrast:", "", &contrast, 255, on_change_contrast );
```

For example, this fragment would create a trackbar in the properties window that would be labeled "Contrast:", and whose value would start out at 128, with a maximum value of 255. Whenever the slider is adjusted, the callback on_change_contrast() will be called.

### Creating buttons with cv::createButton()

One of the most helpful new capabilities provided by the Qt interface is the ability to create buttons. This includes normal push buttons, radio-style (mutually exclusive) buttons, and checkboxes. All buttons created are always located in the control panel.

All three styles of buttons are created with the same method:

```
int cv::createButton(
  const string&       buttonName,             // Handle used to identify trackbar
  cv::ButtonCallback onChange     = NULL, // Callback for button event
  void*               params,                 // (Optional) params for button event
  int                 buttonType   = cv::PUSH_BUTTON,  // PUSH_BUTTON or RADIOBOX
  int                 initialState = 0        // Start state of the button
);
```

The button expects a name, buttonName, that will appear on or next to the button. If you like, you can neglect this argument and simply provide an empty string, in which case the button name will be automatically generated in a serialized manner (e.g., "button 0," "button 1," etc.). The second argument, onChange, is a callback that will be called whenever the button is clicked. The prototype for such a callback must match the declaration for cv::ButtonCallback, which is:

```
void your_button_callback(
  int   state,                                // Identifies button event type
  void* params                                // Parameters from cv::createButton()
);
```

When your callback is called as a result of someone clicking a button, it will be given the value state, which is derived from what just happened to the button. The pointer param that you gave to cv::createButton() will also be passed to your callback, filling its param argument.

The buttonType argument can take one of three values: cv::PUSH_BUTTON, cv::RADIOBOX, or cv::CHECKBOX. The first corresponds to your standard button—you click it, it calls your callback. In the case of the checkbox, the value will be 1 or 0 depending on whether the box was checked or unchecked. The same is true for a radio button, except that when you click a radio button, the callback is called both for the button you just clicked and for the button that is now unclicked (as a result of the mutex nature of radio buttons). All buttons in the same row—*button bars*, described next—are assumed to be part of the same mutex group.

When buttons are created, they are automatically organized into *button bars*. A button bar is a group of buttons that occupies a "row" in the properties window. Consider the following code, which generated the control panel you saw in Figure 9-5.

```
cv::namedWindow( "Image", cv::GUI_EXPANDED );
cv::displayOverlay( "Image", file_name, 0 );
cv::createTrackbar( "Trackbar0", "Image", &mybar0, 255 );
cv::createTrackbar( "Trackbar1", "Image", &mybar1, 255 );

cv::createButton( "", NULL, NULL, cv::PUSH_BUTTON );
cv::createButton( "", NULL, NULL, cv::PUSH_BUTTON );
cv::createButton( "", NULL, NULL, cv::PUSH_BUTTON );
cv::createTrackbar( "Trackbar2", "", &mybar1, 255 );
cv::createButton( "Button3", NULL, NULL, cv::RADIOBOX, 1 );
cv::createButton( "Button4", NULL, NULL, cv::RADIOBOX, 0 );
cv::createButton( "Button5", NULL, NULL, cv::CHECKBOX, 0 );
```

You will notice that `Trackbar0` and `Trackbar1` are created in the window called `"Image"`, while `Trackbar2` is created in an unnamed window (the properties window). The first three `cv::createButton()` calls are not given a name for the button, and you can see in Figure 9-5 the automatically assigned names are placed on the buttons. You will also notice in Figure 9-5 that the first three buttons are in one row, while the second group of three is on another. This is because of the trackbar.

Buttons are created one after another, each to the right of its predecessor, until (unless) a trackbar is created. Because a trackbar consumes an entire row, it is given its own row below the buttons. If more buttons are created, they will appear on a new row thereafter.[13]

### Text and fonts

Just as the Qt interface allowed for much prettier trackbars and other elements, Qt also allows for much prettier and more versatile text. To write text using the Qt interface, you must first create a `CvFont` object,[14] which you then use whenever you want to put some text on the screen. Fonts are created via the `cv::fontQt()` function:

```
CvFont fontQt(                                 // Return font characterization struct
  const string& fontName,                      // e.g., "Times"
  int           pointSize,                     // Size of font, using "point" system.
  cv::Scalar    color   = cv::Scalar::all(0), // BGR color as scalar (no alpha)
  int           weight  = cv::FONT_NORMAL,     // Font weight, 1-100 (Table 9-3)
  int           spacing = 0                    // Space between individual characters
);
```

The first argument to `cv::fontQt()` is the system font name. This might be something like "Times." If your system does not have an available font with this name, then a default font will be chosen for you. The second argument, `pointSize`, is the

13 Unfortunately, there is no "carriage return" for button placement.

14 You will notice that the name of this object is `CvFont` rather than what you might expect: `cv::Font`. This is a legacy to the old pre-C++ interface. `CvFont` is a `struct`, and is not in the `cv::` namespace.

size of the font (i.e., `12` = "12 point," `14` = "14 point," etc.) You may set this to `0`, in which case a default font size (typically 12 point) will be selected for you.

The argument color can be set to any `cv::Scalar` and will set the color in which the font is drawn; the default value is black. `weight` can take one of several predefined values, or any integer between `1` and `100`. The predefined aliases and their values are shown in Table 9-3.

*Table 9-3. Predefined aliases for Qt-font weight and their associated values*

| Camera capture constant | Numerical value |
|---|---|
| `cv::FONT_LIGHT` | 25 |
| `cv::FONT_NORMAL` | 50 |
| `cv::FONT_DEMIBOLD` | 63 |
| `cv::FONT_BOLD` | 75 |
| `cv::FONT_BLACK` | 87 |

The final argument is `spacing`, which controls the spacing between individual characters. It can be negative or positive.

Once you have your font, you can put text on an image (and thus on the screen)[15] with `cv::addText()`.

```
void cv::addText(
  cv::Mat&      image,                      // Image onto which to write
  const string& text,                       // Text to write
  cv::Point     location,                   // Location of lower-left corner of text
  CvFont*       font                        // OpenCV font characerization struct
);
```

The arguments to `cv::addText()` are just what you would expect: the `image` to write on, the `text` to write, where to write it, and the `font` to use—with the latter being a font you defined using `cv::fontQt`. The `location` argument corresponds to the lower-left corner of the first character in `text` (or, more precisely, the beginning of the *baseline* for that character).

15 It is important to notice here that `cv::addText()` is somewhat unlike all of the rest of the functions in the Qt interface—though not inconsistent with the behavior of its non-Qt analog `cv::putText()`. Specifically, `cv::addText()` does not put text in or on a *window*, but rather in an *image*. This means that you are actually changing the pixel values of the image—which is different than what would happen if, for example, you were to use `cv::displayOverlay()`.

### Setting and getting window properties

Many of the state properties of a window set at creation are queryable when you are using the Qt backend, and many of those can be changed (set) even after the window's creation.

```
void   cv::setWindowProperty(
  const string& name,              // Handle used to identify window
  int           prop_id,           // Identifies window property (Table 9-4)
  double        prop_value         // Value to which to set property
);

double cv::getWindowProperty(
  const string& name,              // Handle used to identify window
  int           prop_id            // Identifies window property (Table 9-4)
);
```

To get a window property, you need only call `cv::getWindowProperty()` and supply the `name` of the window and the property ID (`prop_id` argument) of the property you are interested in (see Table 9-4). Similarly, you can use `cv::setWindowProperty()` to set window properties with the same property IDs.

*Table 9-4. Gettable and settable window properties*

| Property name | Description |
|---|---|
| `cv::WIND_PROP_FULL_SIZE` | Set to either `cv::WINDOW_FULLSCREEN` for fullscreen window, or to `cv::WINDOW_NORMAL` for regular window. |
| `cv::WIND_PROP_AUTOSIZE` | Set to either `cv::WINDOW_AUTOSIZE` to have the window automatically size to the displayed image, or `cv::WINDOW_NORMAL` to have the image size to the window. |
| `cv::WIND_PROP_ASPECTRATIO` | Set to either `cv::WINDOW_FREERATIO` to allow the window to have any aspect ratio (as a result of user resizing) or `cv::WINDOW_KEEPRATIO` to allow user resizing to affect only absolute size (and not aspect ratio). |

### Saving and recovering window state

The Qt interface also allows the state of windows to be saved and restored. This can be very convenient, as it includes not only the location and size of your windows, but also the state of all of the trackbars and buttons. The interface state is saved with the `cv::saveWindowParameters()` function, which takes a single argument indicating the window to be saved:

```
void cv::saveWindowParameters(
  const string& name                    // Handle used to identify window
);
```

Once the state of the window is saved, it can be restored with the complementary `cv::loadWindowParameters()` function:

```
void cv::loadWindowParameters(
  const string& name                    // Handle used to identify window
);
```

The real magic here is that the load command will work correctly even if you have quit and restarted your program. The nature of how this works is not important to us here, but one detail you should know is that the state information, wherever it is saved, is saved under a key that is constructed from the executable name. So if you change the name of the executable, the state will not restore (though you can change the executable's *location* without having this problem).

## Interacting with OpenGL

One of the most exciting things that the Qt interface allows you to do is to generate imagery with OpenGL and overlay that imagery on top of your own image.[16] This can be extremely effective for visualizing and debugging robotic or augmented-reality applications, or anywhere in which you are trying to generate a three-dimensional model from your image and want to see the result on top of the original. Figure 9-6 shows a very simple example of what you can do.

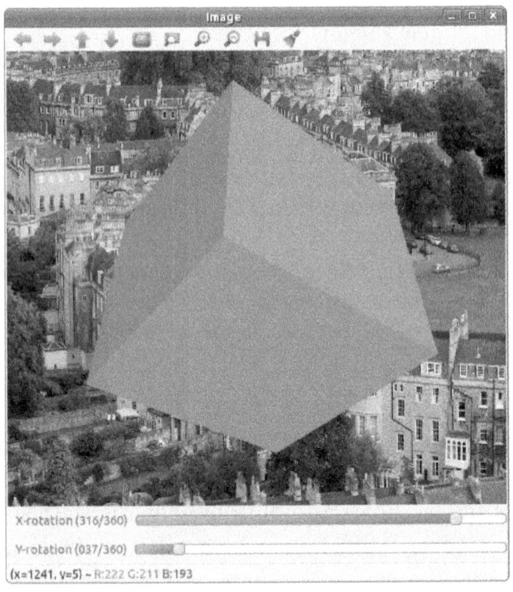


*Figure 9-6. Here OpenGL is used to render a cube on top of our previous image*

16 In order to use these commands, you will need to have built OpenCV with the CMake flag `-D WITH_QT_OPENGL=ON`.

The basic concept is very simple: you create a callback that is an OpenGL draw function and then register it with the interface. From there, OpenCV takes care of the rest. The callback is then called every time the window is drawn (which includes whenever you call `cv::imshow()`, as you would with successive frames of a video stream). Your callback should match the prototype for `cv::OpenGLCallback()`, which means that it should be something like the following:

```
void your_opengl_callback(
  void* params              // (Optional) Params from cv::setOpenGLCallback()
);
```

Once you have your callback, you can configure the OpenGL interface with `cv::setOpenGLCallback()`:

```
void cv::setOpenGLCallback(
  const string&      windowName,          // Handle used to identify window
  cv::OpenGLCallback callback,            // OpenGL callback routine
  void*              params    = NULL     // (Optional) parameters for callback
);
```

As you can see, there is not much you need to do in order to set things up. In addition to specifying the name of the window on which the drawing will be done and supplying the `callback` function, you have a third argument, `params`, which allows you to specify a pointer that will be passed to `callback` whenever it is called.

It is probably worth calling out here explicitly that none of this sets up the camera, lighting, or other aspects of your OpenGL activities. Internally there is a wrapper around your OpenGL callback that will set up the projection matrix using a call to `gluPerspective()`. If you want anything different (which you almost certainly will), you will have to clear and configure the projection matrix at the beginning of your callback.

In Example 9-4, we have taken a simple example from the OpenCV documentation that draws a cube in OpenGL, but we have replaced the fixed rotation angles in that cube with variables (`rotx` and `roty`), which we have made the values of the two sliders in our earlier examples. Now the user can rotate the cube with the sliders while enjoying the beautiful scenery behind it.

*Example 9-4. Slightly modified code from the OpenCV documentation that draws a cube every frame; this modified version uses the global variables rotx and roty that are connected to the sliders in Figure 9-6*

```
void on_opengl( void* param ) {

  glMatrixModel( GL_MODELVIEW );
```

```
  glLoadIdentity();

  glTranslated( 0.0, 0.0, -1.0 );

  glRotatef( rotx, 1, 0, 0 );
  glRotatef( roty, 0, 1, 0 );
  glRotatef( 0, 0, 0, 1 );

  static const int coords[6][4][3] = {
    { { +1, -1, -1 }, { -1, -1, -1 }, { -1, +1, -1 }, { +1, +1, -1 } },
    { { +1, +1, -1 }, { -1, +1, -1 }, { -1, +1, +1 }, { +1, +1, +1 } },
    { { +1, -1, +1 }, { +1, -1, -1 }, { +1, +1, -1 }, { +1, +1, +1 } },
    { { -1, -1, -1 }, { -1, -1, +1 }, { -1, +1, +1 }, { -1, +1, -1 } },
    { { +1, -1, +1 }, { -1, -1, +1 }, { -1, -1, -1 }, { +1, -1, -1 } },
    { { -1, -1, +1 }, { +1, -1, +1 }, { +1, +1, +1 }, { -1, +1, +1 } }
  };

  for (int i = 0; i < 6; ++i) {
    glColor3ub( i*20, 100+i*10, i*42 );
    glBegin(GL_QUADS);
    for (int j = 0; j < 4; ++j) {
      glVertex3d(
        0.2 * coords[i][j][0],
        0.2 * coords[i][j][1],
        0.2 * coords[i][j][2]
      );
    }
    glEnd();
  }
}
```

## Integrating OpenCV with Full GUI Toolkits

Even OpenCV's built-in Qt interface is still just a handy way of accomplishing some simple tasks that come up often while we are developing code or exploring algorithms. When it comes time to actually build an end-user-facing application, neither the native UI nor the Qt-based interface is going to do it. In this section, we will briefly explore some of the issues and techniques for working with OpenCV and three existing toolkits: Qt, wxWidgets, and the Windows Template Library (WTL).

There are countless UI toolkits out there, and we would not want to waste time digging into each of them. Having said that, it is useful to explore how to handle the issues that will arise if you want to use OpenCV with a more fully featured toolkit. These few that we actually do explore here should give you enough insight about the recurring issues that you should have no trouble figuring out what to do in some other similar environment.

The primary issue is how to convert OpenCV images to the form that the toolkit expects for graphics, and to know which widget or component in the toolkit is going to do that display work for you. From there, you don't need much else that is specific to OpenCV. Notably, you will not need or want the features of the UI toolkits we have covered in this chapter.

### An example of OpenCV and Qt

Here we will show an example of using the actual Qt toolkit to write a program that reads a video file and displays it on the screen. There are several subtleties, some of which have to do with how one uses Qt, and others to do with OpenCV. Of course, we will focus on the latter, but it is worth taking a moment to notice how the former affects our current goal.

Example 9-5 shows the top-level code for our program; it just creates a Qt application and adds our `QMoviePlayer` widget. Everything interesting will happen inside that object.

*Example 9-5. An example program ch4_qt.cpp, which takes a single argument indicating a video file; that video file will be replayed inside of a Qt object that we will define, called QMoviePlayer*

```
#include <QApplication>
#include <QLabel>
#include <QMoviePlayer.hpp>
int main( int argc, char* argv[] ) {

  QApplication app( argc, argv );

  QMoviePlayer mp;
  mp.open( argv[1] );
  mp.show();

  return app.exec();
}
```

The interesting stuff is in the `QMoviePlayer` object. Let's take a look at the header file which defines that object in Example 9-6.

*Example 9-6. The QMoviePlayer object header file QMoviePlayer.hpp*

```
#include "ui_QMoviePlayer.h"
#include <opencv2/opencv.hpp>
#include <string>

using namespace std;

class QMoviePlayer : public QWidget {

  Q_OBJECT;
```

```
public:
QMoviePlayer( QWidget *parent = NULL );
virtual ~QMoviePlayer() {;}

bool open( string file );

private:
Ui::QMoviePlayer  ui;
cv::VideoCapture m_cap;

QImage  m_qt_img;
cv::Mat m_cv_img;
QTimer* m_timer;

void paintEvent( QPaintEvent* q );
void _copyImage( void );

public slots:
void nextFrame();

};
```

There is a lot going on here. The first thing that happens is the inclusion of the file *ui_QMoviePlayer.h*. This file was automatically generated by the Qt Designer. What matters here is that it is just a `QWidget` that contains nothing but a `QFrame` called `frame`. The member `ui::QMoviePlayer` is that interface object that is defined in *ui_QMoviePlayer.h*.

In this file, there is also a `QImage` called `m_qt_img` and a `cv::Mat` called `m_cv_img`. These will contain the Qt and OpenCV representations of the image we are getting from our video. Finally, there is a `QTimer`, which is what will take the place of `cv::waitKey()`, allowing us to replay the video frames at the correct rate. The remaining functions will become clear as we look at their actual definitions in *QMoviePlayer.cpp* (Example 9-7).

*Example 9-7. The QMoviePlayer object source file: QMoviePlayer.cpp*

```
#include "QMoviePlayer.hpp"
#include <QTimer>
#include <QPainter>

QMoviePlayer::QMoviePlayer( QWidget *parent )
 : QWidget( parent )
{
  ui.setupUi( this );
}
```

The top-level constructor for `QMoviePlayer` just calls the setup function, which was automatically built for us for the UI member.

```
bool QMoviePlayer::open( string file ) {

  if( !m_cap.open( file ) ) return false;

  // If we opened the file, set up everything now:
  //
  m_cap.read( m_cv_img );
  m_qt_img = QImage(
    QSize( m_cv_img.cols, m_cv_img.rows ),
    QImage::Format_RGB888
  );
  ui.frame->setMinimumSize( m_qt_img.width(), m_qt_img.height() );
  ui.frame->setMaximumSize( m_qt_img.width(), m_qt_img.height() );
  _copyImage();

  m_timer = new QTimer( this );
  connect(
    m_timer,
    SIGNAL( timeout() ),
    this,
    SLOT( nextFrame() )
  );
  m_timer->start( 1000. / m_cap.get( cv::CAP_PROP_FPS ) );

  return true;

}
```

When an `open()` call is made on the `QMoviePlayer`, several things have to happen. The first is that the `cv::VideoCapture` member object `m_cap` needs to be opened. If that fails, we just return. Next we read the first frame into our OpenCV image member `m_cv_img`. Once we have this, we can set up the Qt image object `m_qt_img`, giving it the same size as the OpenCV image. After that, we resize the frame object in the UI element to be the same size as the incoming images as well.

We will look at the call to `QMoviePlayer::_copyImage()` in a moment; this is going to handle the very important process of converting the image we have already captured into `m_cv_img` onto the Qt image `m_qt_img`, which we are actually going to have Qt paint onto the screen for us.

The last thing we do in `QMoviePlayer::open()` is set up a `QTimer` such that, when it "goes off," it will call the function `QMoviePlayer::nextFrame()` (which will, not surprisingly, get the next frame). The call to `m_timer->start()` is how we both start the timer running and indicate that it should go off at the correct rate implied by `cv::CAP_PROP_FPS` (i.e., 1,000 milliseconds divided by the frame rate).

```
void QMoviePlayer::_copyImage( void ) {

  // Copy the image data into the Qt QImage
  //
```

```
    cv::Mat cv_header_to_qt_image(
      cv::Size(
        m_qt_img.width(),
        m_qt_img.height()
      ),
      CV_8UC3,
      m_qt_img.bits()
    );
    cv::cvtColor( m_cv_img, cv_header_to_qt_image, cv::BGR2RGB );

}
```

The `QMoviePlayer::_copyImage()` function is responsible for copying the image from the buffer `m_cv_img` into the Qt image buffer `m_qt_img`. The way we do this shows off a nice feature of the `cv::Mat` object. First, we define a `cv::Mat` object called `cv_header_to_qt_image`. When we define that object, we actually tell it what area to use for its data area, and hand it the data area for the Qt `QImage` object `m_qt_img.bits()`. We then call `cv::cvtColor` to do the copying, which handles the subtlety that OpenCV prefers BGR ordering, while Qt prefers RGB.

```
void QMoviePlayer::nextFrame() {

  // Nothing to do if capture object is not open
  //
  if( !m_cap.isOpened() ) return;

  m_cap.read( m_cv_img );
  _copyImage();

  this->update();

}
```

The `QMoviePlayer::nextFrame()` function actually handles the reading of subsequent frames. Recall that this routine is called whenever the `QTimer` expires. It reads the new image into the OpenCV buffer, calls `QMoviePlayer::_copyImage()` to copy it into the Qt buffer, and then makes an update call on the `QWidget` that this is all part of (so that Qt knows that something has changed).

```
void QMoviePlayer::paintEvent( QPaintEvent* e ) {

  QPainter painter( this );

  painter.drawImage( QPoint( ui.frame->x(), ui.frame->y()), m_qt_img );

}
```

Last but not least is the `QMoviePlayer::paintEvent()` function. This is called by Qt whenever it is necessary to actually draw the `QMoviePlayer` widget. This function just creates a `QPainter` and tells it to draw the current Qt image `m_qt_img` (starting at the corner of the screen).

### An example of OpenCV and wxWidgets

In Example 9-8, we will use a different cross-platform toolkit, wxWidgets. The wxWidgets toolkit is similar in many ways to Qt in terms of its GUI components but, naturally, it is in the details that things tend to become difficult. As with the Qt example, we will have one top-level file that basically puts everything in place and a code and header file pair that defines an object that encapsulates our example task of playing a video. This time our object will be called `WxMoviePlayer` and we will build it based on the UI classes provided by wxWidgets.

*Example 9-8. An example program ch4_wx.cpp, which takes a single argument indicating a video file; that video file will be replayed inside of a wxWidgets object that we will define, called WxMoviePlayer*

```
#include "wx/wx.h"
#include "WxMoviePlayer.hpp"

// Application class, the top level object in wxWidgets
//
class MyApp : public wxApp {
  public:
    virtual bool OnInit();
};

// Behind the scenes stuff to create a main() function and attach MyApp
//
DECLARE_APP( MyApp );
IMPLEMENT_APP( MyApp );

// When MyApp is initialized, do these things.
//
bool MyApp::OnInit() {

  wxFrame* frame = new wxFrame( NULL, wxID_ANY, wxT("ch4_wx") );
  frame->Show( true );

  WxMoviePlayer* mp = new WxMoviePlayer(
    frame,
    wxPoint( -1, -1 ),
    wxSize( 640, 480 )
  );
  mp->open( wxString(argv[1]) );
  mp->Show( true );

  return true;
```

```
}
```

The structure here looks a little more complicated than the Qt example, but the content is very similar. The first thing we do is create a class definition for our application, which we derive from the library class `wxApp`. The only thing different about our class is that it will overload the `MyApp::OnInit()` function with our own content. After declaring `class MyApp`, we call two macros: `DECLARE_APP()` and `IMPLEMENT_APP()`. In short, these are creating the `main()` function and installing an instance of `MyApp` as "the application." The last thing we do in our main program is to actually fill out the function `MyApp::OnInit()` that will be called when our program starts. When `MyApp::OnInit()` is called, it creates the window (called a *frame* in wxWidgets), and an instance of our `WxMoviePlayer` object in that frame. It then calls the open method on the `WxMoviePlayer` and hands it the name of the movie file we want to open.

Of course, all of the interesting stuff is happening inside of the `WxMoviePlayer` object. Example 9-9 shows the header file for that object.

*Example 9-9. The WxMoviePlayer object header file WxMoviePlayer.hpp*

```
#include "opencv2/opencv.hpp"

#include "wx/wx.h"
#include <string>

#define TIMER_ID 0

using namespace std;

class WxMoviePlayer : public wxWindow {

  public:
    WxMoviePlayer(
      wxWindow*       parent,
      const wxPoint& pos,
      const wxSize&   size
    );
    virtual ~WxMoviePlayer() {};
    bool open( wxString file );

  private:

    cv::VideoCapture m_cap;
    cv::Mat          m_cv_img;
    wxImage          m_wx_img;
    wxBitmap         m_wx_bmp;
    wxTimer*         m_timer;
```

```
    wxWindow*        m_parent;

    void _copyImage( void );

    void OnPaint( wxPaintEvent& e );
    void OnTimer( wxTimerEvent& e );
    void OnKey(   wxKeyEvent&   e );

  protected:
    DECLARE_EVENT_TABLE();
};
```

There are several important things to notice in this declaration. The `WxMoviePlayer` object is derived from `wxWindow`, which is the generic class used by wxWidgets for just about anything that will be visible on the screen. We have three event-handling methods, `OnPaint()`, `onTimer()`, and `OnKey()`; these will handle drawing, getting a new image from the video, and closing the file with the Esc key, respectively. Finally, you will notice that there is an object of type `wxImage` and an object of type `wxBitmap`, in addition to the OpenCV `cv:Mat` type image. In wxWidgets, *bitmaps* (which are operating system dependent) are distinguished from *images* (which are device-independent representations of image data). The exact role of these two will be clear shortly as we look at the code file *WxMoviePlayer.cpp* (see Example 9-10).

*Example 9-10. The WxMoviePlayer object source file WxMoviePlayer.cpp*

```
#include "WxMoviePlayer.hpp"

BEGIN_EVENT_TABLE( WxMoviePlayer, wxWindow )
  EVT_PAINT( WxMoviePlayer::OnPaint )
  EVT_TIMER( TIMER_ID, WxMoviePlayer::OnTimer )
  EVT_CHAR( WxMoviePlayer::OnKey )
END_EVENT_TABLE()
```

The first thing we do is to set up the callbacks that will be associated with individual events. We do this through macros provided by the wxWidgets framework.[17]

```
    WxMoviePlayer::WxMoviePlayer(
      wxWindow*      parent,
      const wxPoint& pos,
      const wxSize&  size
    ) : wxWindow( parent, -1, pos, size, wxSIMPLE_BORDER ) {
```

17 The astute reader will notice that the keyboard event is "hooked up" to the `WxMoviePlayer` widget and not to the top-level application or the frame (as was the case for the Qt example, and as is the case for HighGUI). There are various ways to accomplish this, but wxWidgets really prefers your keyboard events to be bound locally to visible objects in your UI, rather than globally. Since this is a simple example, we chose to just do the easiest thing and bind the keyboard events directly to the movie player.

```
    m_timer          = NULL;
    m_parent         = parent;
}
```

When the movie player is created, its timer element is NULL (we will set that up when we actually have a video open). We do take note of the parent of the player, however. (In this case, that parent will be the wxFrame we created to put it in.) We will need to know which frame is the parent when it comes time to close the application in response to the Esc key.

```
void WxMoviePlayer::OnPaint( wxPaintEvent& event ) {
  wxPaintDC dc( this );

  if( !dc.Ok() ) return;

  int x,y,w,h;
  dc.BeginDrawing();
    dc.GetClippingBox( &x, &y, &w, &h );
    dc.DrawBitmap( m_wx_bmp, x, y );
  dc.EndDrawing();

  return;
}
```

The WxMoviePlayer::OnPaint() routine is called whenever the window needs to be repainted on screen. Notice that when we execute WxMoviePlayer::OnPaint(), the information we need to actually do the painting is assumed to be in m_wx_bpm, the wxBitmap object. Because the wxBitmap is the system-dependent representation, it is already prepared to be copied to the screen. The next two methods, WxMovie Player::_copyImage() and WxMoviePlayer::open(), will show how it got created in the first place.

```
void WxMoviePlayer::_copyImage( void ) {

  m_wx_bmp = wxBitmap( m_wx_img );

  Refresh( FALSE ); // indicate that the object is dirty
  Update();

}
```

The WxMoviePlayer::_copyImage() method will get called whenever a new image is read from the cv::VideoCapture object. It doesn't appear to do much, but actually a lot is going on in its short body. First and foremost is the construction of the wxBit map m_wx_bmp from the wxImage m_wx_img. The constructor is handling the conversion from the abstract representation used by wxImage (which, we will see, looks very much like the representation used by OpenCV) to the device- and system-specific representation used by your particular machine. Once that copy is done, a call to

Refresh() indicates that the widget is "dirty" and needs redrawing, and the subsequent call to Update() indicates that the time for that redrawing is now.

```
bool WxMoviePlayer::open( wxString file ) {

  if( !m_cap.open( std::string( file.mb_str() ) )) {
    return false;
  }

  // If we opened the file, set up everything now:
  //
  m_cap.read( m_cv_img );

  m_wx_img = wxImage(
    m_cv_img.cols,
    m_cv_img.rows,
    m_cv_img.data,
    TRUE  // static data, do not free on delete()
  );

  _copyImage();

  m_timer = new wxTimer( this, TIMER_ID );
  m_timer->Start( 1000. / m_cap.get( cv::CAP_PROP_FPS ) );

  return true;

}
```

The WxMoviePlayer::open() method also does several important things. The first is to actually open the cv::VideoCapture object, but there is a lot more to be done. Next, an image is read off of the player and is used to create a wxImage object that "points at" the OpenCV cv::Mat image. This is the opposite philosophy to the one we used in the Qt example: in this case, it turns out to be a little more convenient to create the cv::Mat first and have it own the data, and then to create the GUI toolkit's image object second and have it be just a header to that existing data. Next, we call WxMoviePlayer::_copyImage(), and that function converts the OpenCV image m_cv_img into the native bitmap for us.

Finally, we create a wxTimer object and tell it to wake us up every few milliseconds—with that number being computed from the FPS reported by the cv::VideoCapture object. Whenever that timer expires, a wxTimerEvent is generated and passed to WxMoviePlayer::OnTimer(), which you will recall is the handler of such events.

```
void WxMoviePlayer::OnTimer( wxTimerEvent& event ) {

  if( !m_cap.isOpened() ) return;

  m_cap.read( m_cv_img );
  cv::cvtColor( m_cv_img, m_cv_img, cv::BGR2RGB );
```

```
    _copyImage();

}
```

That handler doesn't do too much; primarily it just reads a new frame from the video, converts that frame from BGR to RGB for display, and then calls our `WxMoviePlayer::_copyImage()`, which makes the next bitmap for us.

```
void WxMoviePlayer::OnKey( wxKeyEvent& e ) {

  if( e.GetKeyCode() == WXK_ESCAPE ) m_parent->Close();

}
```

Finally, we have our handler for any keypresses. It simply checks to see if that key was the Esc key, and if so, closes the program. Note that we do not close the `WxMoviePlayer` object, but rather the parent frame. Closing the frame is the same as closing the window any other way; it shuts down the application.

### An example of OpenCV and the Windows Template Library

In this example, we will use the native windows GUI API.[18] The Windows Template Library (WTL) is a very thin C++ wrapper around the raw Win32 API. WTL applications are structured similarly to MFC, in that there is an application/document-view structure. For the purposes of this sample, we will start by running the WTL Application Wizard from within Visual Studio (Figure 9-7), creating a new SDI Application, and ensuring that Use a View Window is selected under User Interface Features (it should be, by default).

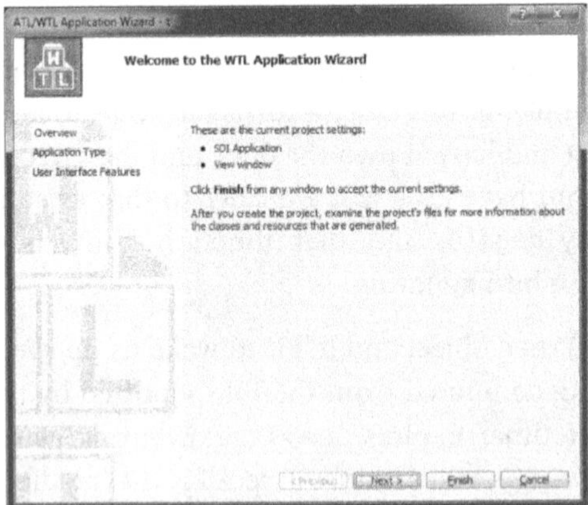


*Figure 9-7. The WTL Application Wizard*

18 Special thanks to Sam Leventer, who is the original author of this WTL example code.

The exact filenames generated by the wizard will depend on the name you give your project. For Example 9-11, the project is named *OpenCVTest*, and we will mostly be working in the `COpenCVTestView` class.

*Example 9-11. An example header file for our custom View class*

```
class COpenCVTestView : public CWindowImpl<COpenCVTestView> {

public:
  DECLARE_WND_CLASS(NULL)

  bool OpenFile(std::string file);
  void _copyImage();


  BOOL PreTranslateMessage(MSG* pMsg);

  BEGIN_MSG_MAP(COpenCVTestView)
    MESSAGE_HANDLER(WM_ERASEBKGND, OnEraseBkgnd)
    MESSAGE_HANDLER(WM_PAINT, OnPaint)
    MESSAGE_HANDLER(WM_TIMER, OnTimer)
  END_MSG_MAP()

// Handler prototypes (uncomment arguments if needed):
//  LRESULT MessageHandler(
//    UINT    /*uMsg*/,
//    WPARAM  /*wParam*/,
//    LPARAM  /*lParam*/,
//    BOOL&   /*bHandled*/
//  );
//  LRESULT CommandHandler(
//    WORD    /*wNotifyCode*/,
//    WORD    /*wID*/,
//    HWND    /*hWndCtl*/,
//    BOOL&   /*bHandled*/
//  );
//  LRESULT NotifyHandler(
//    int     /*idCtrl*/,
//    LPNMHDR /*pnmh*/,
//    BOOL&   /*bHandled*/
//  );

  LRESULT OnPaint(
    UINT    /*uMsg*/,
    WPARAM  /*wParam*/,
    LPARAM  /*lParam*/,
    BOOL&   /*bHandled*/
  );
  LRESULT OnTimer(
    UINT    /*uMsg*/,
    WPARAM  /*wParam*/,
```

```
    LPARAM  /*lParam*/,
    BOOL&   /*bHandled*/
  );
  LRESULT OnEraseBkgnd(
    UINT    /*uMsg*/,
    WPARAM  /*wParam*/,
    LPARAM  /*lParam*/,
    BOOL&   /*bHandled*/
  );

private:
  cv::VideoCapture m_cap;
  cv::Mat          m_cv_img;

  RGBTRIPLE*       m_bitmapBits;
};
```

The structure here is very similar to the preceding wxWidgets example. The only change outside of the view code is for the `Open` menu item handler, which will be in your `CMainFrame` class. It will need to call into the view class to open the video:

```
LRESULT CMainFrame::OnFileOpen(
  WORD /*wNotifyCode*/,
  WORD /*wID*/,
  HWND /*hWndCtl*/,
  BOOL& /*bHandled*/
) {
  WTL::CFileDialog dlg(TRUE);
  if (IDOK == dlg.DoModal(m_hWnd)) {
    m_view.OpenFile(dlg.m_szFileName);
  }
  return 0;
}

bool COpenCVTestView::OpenFile(std::string file) {

  if( !m_cap.open( file ) ) return false;

  // If we opened the file, set up everything now:
  //
  m_cap.read( m_cv_img );

  // could create a DIBSection here, but let's just allocate memory for raw bits
  //
  m_bitmapBits = new RGBTRIPLE[m_cv_img.cols * m_cv_img.rows];

  _copyImage();

  SetTimer(0, 1000.0f / m_cap.get( cv::CAP_PROP_FPS ) );

  return true;
}
```

```
void COpenCVTestView::_copyImage() {

  // Copy the image data into the bitmap
  //
  cv::Mat cv_header_to_qt_image(
    cv::Size(
      m_cv_img.cols,
      m_cv_img.rows
    ),
    CV_8UC3,
    m_bitmapBits
  );
  cv::cvtColor( m_cv_img, cv_header_to_qt_image, cv::BGR2RGB );
}

LRESULT COpenCVTestView::OnPaint(
  UINT   /* uMsg     */,
  WPARAM /* wParam   */,
  LPARAM /* lParam   */,
  BOOL&  /* bHandled */
) {
  CPaintDC dc(m_hWnd);

  WTL::CRect rect;
  GetClientRect(&rect);

  if( m_cap.isOpened() ) {

    BITMAPINFO bmi = {0};
    bmi.bmiHeader.biSize = sizeof(bmi.bmiHeader);
    bmi.bmiHeader.biCompression = BI_RGB;
    bmi.bmiHeader.biWidth       = m_cv_img.cols;

    // note that bitmaps default to bottom-up, use negative height to
    // represent top-down
    //
    bmi.bmiHeader.biHeight = m_cv_img.rows * -1;

    bmi.bmiHeader.biPlanes = 1;
    bmi.bmiHeader.biBitCount = 24;  // 32 if you use RGBQUADs for the bits

    dc.StretchDIBits(
      0,                     0,
      rect.Width(),          rect.Height(),
      0,                     0,
      bmi.bmiHeader.biWidth, abs(bmi.bmiHeader.biHeight),
      m_bitmapBits,
      &bmi,
      DIB_RGB_COLORS,
      SRCCOPY
```

```
    );

  } else {

    dc.FillRect(rect, COLOR_WINDOW);

  }

  return 0;
}

LRESULT COpenCVTestView::OnTimer(
  UINT   /* uMsg     */,
  WPARAM /* wParam   */,
  LPARAM /* lParam   */,
  BOOL&  /* bHandled */
) {
  // Nothing to do if capture object is not open
  //
  if( !m_cap.isOpened() ) return 0;

  m_cap.read( m_cv_img );
  _copyImage();

  Invalidate();

  return 0;
}

LRESULT COpenCVTestView::OnEraseBkgnd(
  UINT   /* uMsg     */,
  WPARAM /* wParam   */,
  LPARAM /* lParam   */,
  BOOL&  /* bHandled */
) {
  // since we completely paint our window in the OnPaint handler, use
  // an empty background handler
  return 0;
}
```

This code illustrates how to use bitmap-based drawing in a C++ application in Windows. This method is simpler but less efficient than using DirectShow to handle the video stream.

If you are using the .NET Runtime (either through C#, VB.NET, or Managed C++), then you may want to look into a package that completely wraps OpenCV, such as Emgu (*http://emgu.com*).

## Summary

We have seen that OpenCV provides a number of ways to bring computer vision programs to the screen. The native HighGUI tools are convenient and easy to use, but not so great for functionality or final polish.

For a little more capability, the Qt-based HighGUI tools add buttons and some nice gadgets for manipulating your image on the screen—which is very helpful for debugging, parameter tuning, and studying the subtle effects of changes in your program. Because those methods lack extensibility and are likely unsuitable for the production of professional applications, we went on to look at a few examples of how you might combine OpenCV with existing fully featured GUI toolkits.

## Exercises

1. Using HighGui only, create a window into which you can load and view four images at once, each of size at least 300 × 300. You should be able to click on each of the images and print out the correct $(x, y)$ location of the click relative to the image, not the larger window. The printout should be text written on the image you clicked on.
2. Using QT, create a window into which you can load and view four images at once. Implement the box drawing code of Example 9-2 such that you can draw boxes within each window, but do not allow a box to draw over the image boundary that you are drawing in.
3. Using QT, create a window sufficient to contain a 500 × 500 image. When a button is pushed for that window, a smaller 100 × 100 window appears that magnifies the area in the first image that the mouse is over. A slider should allow magnifications of 1×, 2×, 3×, and 4×. Handle the case where the magnification around the mouse will step over the boundary of the 500 × 500 image. Black pixels should be shown in the magnification window. When the button is pushed again, the small window vanishes and magnification doesn't work. The button toggles magnification on and off.
4. Using QT, create a 1,000 × 1,000 window. When a button is pushed on, you can click in the window and type and edit text. Do not allow the text to go beyond the boundary of the window. Allow for typing and backspacing.
5. Build and run the rotating cube described in Example 9-12. Modify it so that you have buttons: rotate right, left, up, and down. When you press the buttons, the cube should rotate.

# CHAPTER 10
# Filters and Convolution

## Overview

At this point, we have all of the basics at our disposal. We understand the structure of the library as well as the basic data structures it uses to represent images. We understand the HighGUI interface and can actually run a program and display our results on the screen. Now that we understand these primitive methods required to manipulate image structures, we are ready to learn some more sophisticated operations.

We will now move on to higher-level methods that treat the images as images, and not just as arrays of colored (or grayscale) values. When we say "image processing" in this chapter, we mean just that: using higher-level operators that are defined on image structures in order to accomplish tasks whose meaning is naturally defined in the context of graphical, visual images.

## Before We Begin

There are a couple of important concepts we will need throughout this chapter, so it is worth taking a moment to review them before we dig into the specific image-processing functions that make up the bulk of this chapter. First, we'll need to understand filters (also called kernels) and how they are handled in OpenCV. Next, we'll take a look at how boundary areas are handled when OpenCV needs to apply a filter, or another function of the area around a pixel, when that area spills off the edge of the image.

### Filters, Kernels, and Convolution

Most of the functions we will discuss in this chapter are special cases of a general concept called *image filtering*. A filter is any algorithm that starts with some image $I(x, y)$

and computes a new image $I'(x, y)$ by computing for each pixel location $x, y$ in $I'$ some function of the pixels in $I$ that are in some small area around that same $x, y$ location. The template that defines both this small area's shape, as well as how the elements of that small area are combined, is called a *filter* or a *kernel*.[1] In this chapter, many of the important kernels we encounter will be *linear kernels*. This means that the value assigned to point $x, y$ in $I'$ can be expressed as a weighted sum of the points around (and usually including) $x, y$ in $I$.[2] If you like equations, this can be written as:

$$I'(x, y) = \sum_{i,j \in kernel} k_{i,j} \cdot I(x + i, y + j)$$

This basically says that for some kernel of whatever size (e.g., 5 × 5), we should sum over the area of the kernel, and for each pair $i, j$ (representing one point in the kernel), we should add a contribution equal to some value $k_{i,j}$ multiplied by the value of the pixel in $I$ that is offset from $x, y$ by $i, j$. The size of the array $I$ is called the *support* of the kernel.[3] Any filter that can be expressed in this way (i.e., with a linear kernel) is also known as a *convolution*, though the term is often used somewhat casually in the computer vision community to include the application of any filter (linear or otherwise) over an entire image.

It is often convenient (and more intuitive) to represent the kernel graphically as an array of the values of $k_{i,j}$ (see Figure 10-1). We will typically use this representation throughout the book when it is necessary to represent a kernel.

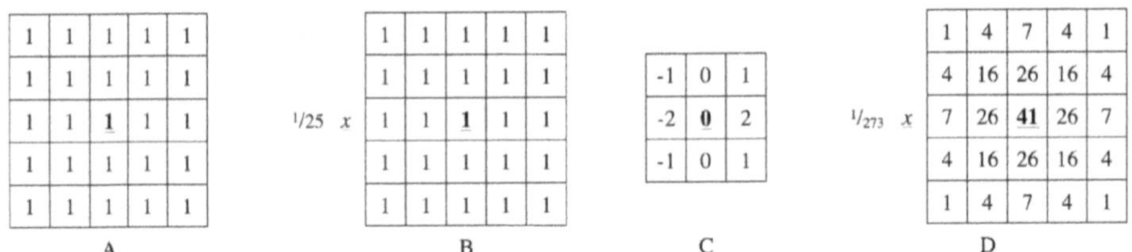


*Figure 10-1. (A) A 5 × 5 box kernel, (B) a normalized 5 × 5 box kernel, (C) a 3 × 3 Sobel "x-derivative" kernel, and (D) a 5 × 5 normalized Gaussian kernel; in each case, the "anchor" is represented in bold*

1 These two terms can be considered essentially interchangeable for our purposes. The signal processing community typically prefers the word *filter*, while the mathematical community tends to prefer *kernel*.

2 An example of a nonlinear kernel that comes up relatively often is the *median filter*, which replaces the pixel at $x, y$ with the median value inside of the kernel area.

3 For technical purists, the "support" of the kernel actually consists of only the nonzero portion of the kernel array.

### Anchor points

Each kernel shown in Figure 10-1 has one value depicted in bold. This is the *anchor point* of the kernel. This indicates how the kernel is to be aligned with the source image. For example, in Figure 10-1(D), the number 41 appears in bold. This means that in the summation used to compute $I'(x, y)$, it is $I(x, y)$ that is multiplied by $^{41}/_{273}$ (and similarly, the terms corresponding to $I(x - 1, y)$ and $I(x + 1, y)$ are multiplied by $^{26}/_{273}$).

## Border Extrapolation and Boundary Conditions

An issue that will come up with some frequency as we look at how images are processed in OpenCV is how borders are handled. Unlike some other image-handling libraries,[4] the filtering operations in OpenCV (`cv::blur()`, `cv::erode()`, `cv::dilate()`, etc.) produce output images of the same size as the input. To achieve that result, OpenCV creates "virtual" pixels outside of the image at the borders. You can see this would be necessary for an operation like `cv::blur()`, which is going to take all of the pixels in a neighborhood of some point and average them to determine a new value for that point. How could a meaningful result be computed for an edge pixel that does not have the correct number of neighbors? In fact, it will turn out that in the absence of any clearly "right" way of handling this, we will often find ourselves explicitly asserting how this issue is to be resolved in any given context.

### Making borders yourself

Most of the library functions you will use will create these virtual pixels for you. In that context, you will only need to tell the particular function how you would like those pixels created.[5] Just the same, in order to know what your options mean, it is best to take a look at the function that allows you to explicitly create "padded" images that use one method or another.

The function that does this is `cv::copyMakeBorder()`. Given an image you want to pad out, and a second image that is somewhat larger, you can ask `cv::copyMakeBorder()` to fill all of the pixels in the larger image in one way or another.

4 For example, MATLAB.

5 Actually, the pixels are usually not even really created; rather, they are just "effectively created" by the generation of the correct boundary conditions in the evaluation of the particular function in question.

```
void cv::copyMakeBorder(
  cv::InputArray    src,                         // Input image
  cv::OutputArray   dst,                         // Result image
  int               top,                         // Top side padding (pixels)
  int               bottom,                      // Bottom side padding (pixels)
  int               left,                        // Left side padding (pixels)
  int               right,                       // Right side padding (pixels)
  int               borderType,                  // Pixel extrapolation method
  const cv::Scalar& value = cv::Scalar()         // Used for constant borders
);
```

The first two arguments to `cv::copyMakeBorder()` are the smaller source image and the larger destination image. The next four arguments specify how many pixels of padding are to be added to the source image on the `top`, `bottom`, `left`, and `right` edges. The next argument, `borderType`, actually tells `cv::copyMakeBorder()` how to determine the correct values to assign to the padded pixels (as shown in Figure 10-2).

*Figure 10-2. The same image is shown padded using each of the six different borderType options available to cv::copyMakeBorder() (the "NO BORDER" image in the upper left is the original for comparison)*

To understand what each option does in detail, consider an extremely zoomed-in section at the edge of each image (Figure 10-3).

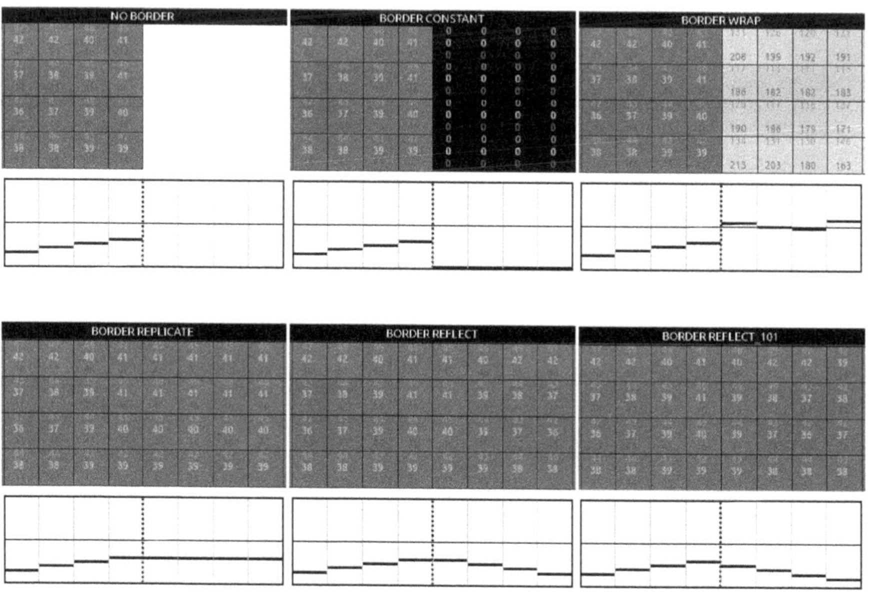


*Figure 10-3. An extreme zoom in at the left side of each image—for each case, the actual pixel values are shown, as well as a schematic representation; the vertical dotted line in the schematic represents the edge of the original image*

As you can see by inspecting the figures, some of the available options are quite different. The first option, a constant border (`cv::BORDER_CONSTANT`) sets all of the pixels in the border region to some fixed value. This value is set by the `value` argument to `cv::copyMakeBorder()`. (In Figures 10-2 and 10-3, this value happens to be `cv::Scalar(0,0,0)`.) The next option is to wrap around (`cv::BORDER_WRAP`), assigning each pixel that is a distance *n* off the edge of the image the value of the pixel that is a distance *n* in from the opposite edge. The replicate option, `cv::BORDER_REPLICATE`, assigns every pixel off the edge the same value as the pixel on that edge. Finally, there are two slightly different forms of reflection available: `cv::BORDER_REFLECT` and `cv::BORDER_REFLECT_101`. The first assigns each pixel that is a distance *n* off the edge of the image the value of the pixel that is a distance *n* in from that same edge. In contrast, `cv::BORDER_REFLECT_101` assigns each pixel that is a distance *n* off the edge of the image the value of the pixel that is a distance *n* + 1 in from that same edge (with the result that the very edge pixel is not replicated). In most cases, `cv::BORDER_REFLECT_101` is the default behavior for OpenCV methods. The value of `cv::BORDER_DEFAULT` resolves to `cv::BORDER_REFLECT_101`. Table 10-1 summarizes these options.

*Table 10-1. borderType options available to cv::copyMakeBorder(), as well as many other functions that need to implicitly create boundary conditions*

| Border type | Effect |
|---|---|
| cv::BORDER_CONSTANT | Extend pixels by using a supplied (constant) value |
| cv::BORDER_WRAP | Extend pixels by replicating from opposite side |
| cv::BORDER_REPLICATE | Extend pixels by copying edge pixel |
| cv::BORDER_REFLECT | Extend pixels by reflection |
| cv::BORDER_REFLECT_101 | Extend pixels by reflection, edge pixel is not "doubled" |
| cv::BORDER_DEFAULT | Alias for cv::BORDER_REFLECT_101 |

### Manual extrapolation

On some occasions, you will want to compute the location of the reference pixel to which a particular off-the-edge pixel is referred. For example, given an image of width $w$ and height $h$, you might want to know what pixel in that image is being used to assign a value to virtual pixel $(w + dx, h + dy)$. Though this operation is essentially extrapolation, the function that computes such a result for you is (somewhat confusingly) called `cv::borderInterpolate()`:

```
int cv::borderInterpolate(          // Returns coordinate of "donor" pixel
  int p,                            // 0-based coordinate of extrapolated pixel
  int len,                          // Length of array (on relevant axis)
  int borderType                    // Pixel extrapolation method
);
```

The `cv::borderInterpolate()` function computes the extrapolation for one dimension at a time. It takes a coordinate `p`, a length `len` (which is the actual size of the image in the associated direction), and a `borderType` value. So, for example, you could compute the value of a particular pixel in an image under a mixed set of boundary conditions, using `BORDER_REFLECT_101` in one dimension, and `BORDER_WRAP` in another:

```
float val = img.at<float>(
  cv::borderInterpolate( 100, img.rows, BORDER_REFLECT_101 ),
  cv::borderInterpolate(  -5, img.cols, BORDER_WRAP )
);
```

This function is typically used internally to OpenCV (for example, inside of `cv::copyMakeBorder`) but it can come in handy in your own algorithms as well. The possible values for `borderType` are exactly the same as those used by `cv::copyMakeBorder`. Throughout this chapter, we will encounter functions that take a `borderType` argument; in all of those cases, they take the same list of argument.

# Threshold Operations

You'll often run into situations where you have done many layers of processing steps and want either to make a final decision about the pixels in an image or to categorically reject those pixels below or above some value while keeping the others. The OpenCV function `cv::threshold()` accomplishes these tasks (see survey [Sezgin04]). The basic idea is that an array is given, along with a threshold, and then something happens to every element of the array depending on whether it is below or above the threshold. If you like, you can think of threshold as a very simple convolution operation that uses a 1 × 1 kernel and then performs one of several nonlinear operations on that one pixel:[6]

```
double cv::threshold(
  cv::InputArray    src,            // Input image
  cv::OutputArray   dst,            // Result image
  double            thresh,         // Threshold value
  double            maxValue,       // Max value for upward operations
  int               thresholdType   // Threshold type to use (Example 10-3)
);
```

As shown in Table 10-2, each threshold type corresponds to a particular comparison operation between the *i*th source pixel ($src_i$) and the threshold `thresh`. Depending on the relationship between the source pixel and the threshold, the destination pixel $dst_i$ may be set to `0`, to $src_i$, or the given maximum value `maxValue`.

*Table 10-2. thresholdType options for cv::threshold()*

| Threshold type | Operation |
|---|---|
| `cv::THRESH_BINARY` | $DST_I = (SRC_I > thresh)\ ?\ MAXVALUE : 0$ |
| `cv::THRESH_BINARY_INV` | $DST_I = (SRC_I > thresh)\ ?\ 0 : MAXVALUE$ |
| `cv::THRESH_TRUNC` | $DST_I = (SRC_I > thresh)\ ?\ THRESH : SRC_I$ |
| `cv::THRESH_TOZERO` | $DST_I = (SRC_I > thresh)\ ?\ SRC_I : 0$ |
| `cv::THRESH_TOZERO_INV` | $DST_I = (SRC_I > thresh)\ ?\ 0 : SRC_I$ |

Figure 10-4 should help to clarify the exact implications of each available value for `thresholdType`, the thresholding operation.

6 The utility of this point of view will become clearer as we proceed through this chapter and look at other, more complex convolution operations. Many useful operations in computer vision can be expressed as a sequence of common convolutions, and more often than not, the last one of those convolutions is a threshold operation.

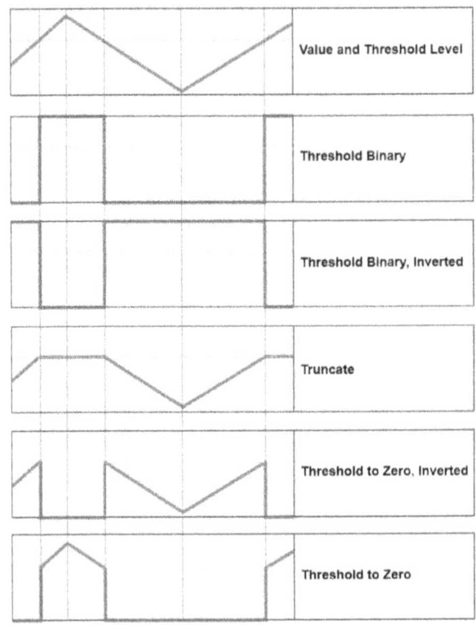


*Figure 10-4. Results of varying the threshold type in cv::threshold(); the horizontal line through each chart represents a particular threshold level applied to the top chart and its effect for each of the five types of threshold operations below*

Let's look at a simple example. In Example 10-1, we sum all three channels of an image and then clip the result at `100`.

*Example 10-1. Using cv::threshold() to sum three channels of an image*

```
#include <opencv2/opencv.hpp>
#include <iostream>
using namespace std;

void sum_rgb( const cv::Mat& src, cv::Mat& dst ) {

  // Split image onto the color planes.
  //
  vector< cv::Mat> planes;
  cv::split(src, planes);

  cv::Mat b = planes[0], g = planes[1], r = planes[2], s;

  // Add equally weighted rgb values.
  //
  cv::addWeighted( r, 1./3., g, 1./3., 0.0, s );
  cv::addWeighted( s, 1., b, 1./3., 0.0, s );
```

```
    // Truncate values above 100.
    //
    cv::threshold( s, dst, 100, 100, cv::THRESH_TRUNC );

}

void help() {
    cout << "Call: ./ch10_ex10_1 faceScene.jpg" << endl;
    cout << "Shows use of alpha blending (addWeighted) and threshold" << endl;
}

int main(int argc, char** argv) {

    help();

    if(argc < 2) { cout << "specify input image" << endl; return -1; }

    // Load the image from the given file name.
    //
    cv::Mat src = cv::imread( argv[1] ), dst;
    if( src.empty() ) { cout << "can not load " << argv[1] << endl; return -1; }
    sum_rgb( src, dst);

    // Create a named window with the name of the file and
    // show the image in the window
    //
    cv::imshow( argv[1], dst );

    // Idle until the user hits any key.
    //
    cv::waitKey(0);

    return 0;
}
```

Some important ideas are shown here. One is that we don't want to add directly into an 8-bit array (with the idea of normalizing next) because the higher bits will overflow. Instead, we use equally weighted addition of the three color channels (`cv::addWeighted()`); then the sum is truncated to saturate at the value of `100` for the return. Had we used a floating-point temporary image for `s` in Example 10-1, we could have substituted the code shown in Example 10-2 instead. Note that `cv::accumulate()` can accumulate 8-bit integer image types into a floating-point image.

*Example 10-2. Alternative method to combine and threshold image planes*

```
void sum_rgb( const cv::Mat& src, cv::Mat& dst ) {

  // Split image onto the color planes.
  //
  vector<cv::Mat> planes;
  cv::split(src, planes);

  cv::Mat b = planes[0], g = planes[1], r = planes[2];

  // Accumulate separate planes, combine and threshold.
  //
  cv::Mat s = cv::Mat::zeros(b.size(), CV_32F);
  cv::accumulate(b, s);
  cv::accumulate(g, s);
  cv::accumulate(r, s);

  // Truncate values above 100 and rescale into dst.
  //
  cv::threshold( s, s, 100, 100, cv::THRESH_TRUNC );
  s.convertTo(dst, b.type());
}
```

## Otsu's Algorithm

It is also possible to have `cv::threshold()` attempt to determine the optimal value of the threshold for you. You do this by passing the special value `cv::THRESH_OTSU` as the value of `thresh`.

Briefly, Otsu's algorithm is to consider all possible thresholds, and to compute the variance $\sigma_i^2$ for each of the two classes of pixels (i.e., the class below the threshold and the class above it). Otsu's algorithm minimizes:

$$\sigma_w^2 \equiv w_1(t) \cdot \sigma_1^2 + w_2(t) \cdot \sigma_2^2$$

where $w_1(t)$ and $w_2(t)$ are the relative weights for the two classes given by the number of pixels in each class, and $\sigma_1^2$ and $\sigma_2^2$ are the variances in each class. It turns out that minimizing the variance of the two classes in this way is the same as maximizing the variance between the two classes. Because an exhaustive search of the space of possible thresholds is required, this is not a particularly fast process.

## Adaptive Threshold

There is a modified threshold technique in which the threshold level is itself variable (across the image). In OpenCV, this method is implemented in the `cv::adaptiveThreshold()` [Jain86] function:

```
void cv::adaptiveThreshold(
  cv::InputArray    src,              // Input image
  cv::OutputArray   dst,              // Result image
  double            maxValue,         // Max value for upward operations
  int               adaptiveMethod,   // mean or Gaussian
  int               thresholdType     // Threshold type to use (Example 10-3)
  int               blockSize,        // Block size
  double            C                 // Constant
);
```

`cv::adaptiveThreshold()` allows for two different adaptive threshold types depending on the settings of `adaptiveMethod`. In both cases, we set the *adaptive threshold* $T(x, y)$ on a pixel-by-pixel basis by computing a weighted average of the $b \times b$ region around each pixel location minus a constant, where $b$ is given by `blockSize` and the constant is given by `C`. If the method is set to `cv::ADAPTIVE_THRESH_MEAN_C`, then all pixels in the area are weighted equally. If it is set to `cv::ADAPTIVE_THRESH_GAUSSIAN_C`, then the pixels in the region around $(x, y)$ are weighted according to a Gaussian function of their distance from that center point.

Finally, the parameter `thresholdType` is the same as for `cv::threshold()` shown in Table 10-2.

The adaptive threshold technique is useful when there are strong illumination or reflectance gradients that you need to threshold relative to the general intensity gradient. This function handles only single-channel 8-bit or floating-point images, and it requires that the source and destination images be distinct.

Example 10-3 shows source code for comparing `cv::adaptiveThreshold()` and `cv::threshold()`. Figure 10-5 illustrates the result of processing an image that has a strong lighting gradient across it with both functions. The lower-left portion of the figure shows the result of using a single global threshold as in `cv::threshold()`; the lower-right portion shows the result of adaptive local threshold using `cv::adaptiveThreshold()`. We get the whole checkerboard via adaptive threshold, a result that is impossible to achieve when using a single threshold. Note the calling-convention messages at the top of the code in Example 10-3; the parameters used for Figure 10-5 were:

```
./adaptThresh 15 1 1 71 15 ../Data/cal3-L.bmp
```

Source image

Binary threshold

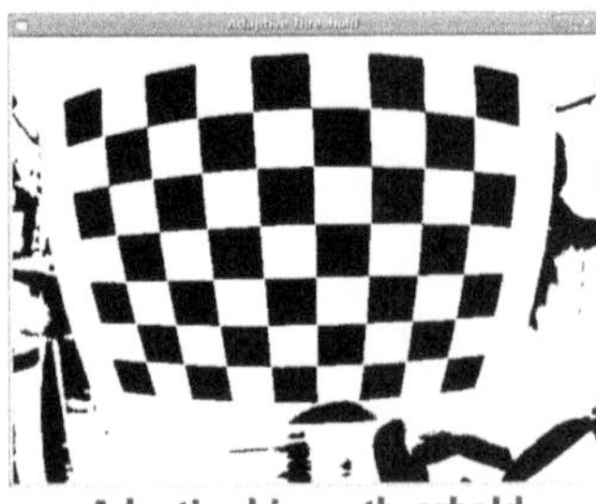

Adaptive binary threshold

*Figure 10-5. Binary threshold versus adaptive binary threshold: the input image (top) was turned into a Boolean image using a global threshold (lower left) and an adaptive threshold (lower right); raw image courtesy of Kurt Konolige*

*Example 10-3. Threshold versus adaptive threshold*

```
#include <opencv2/opencv.hpp>
#include <cstdlib>
#include <iostream>

using namespace std;

int main( int argc, char** argv )
{
  if(argc != 7) { cout <<
   "Usage: " <<argv[0] <<" fixed_threshold invert(0=off|1=on) "
   "adaptive_type(0=mean|1=gaussian) block_size offset image\n"
   "Example: " <<argv[0] <<" 100 1 0 15 10 fruits.jpg\n"; return -1; }

  // Command line.
  //
  double fixed_threshold = (double)atof(argv[1]);
  int threshold_type  = atoi(argv[2]) ? cv::THRESH_BINARY : cv::THRESH_BINARY_INV;
  int adaptive_method = atoi(argv[3]) ? cv::ADAPTIVE_THRESH_MEAN_C
                                      : cv::ADAPTIVE_THRESH_GAUSSIAN_C;
  int block_size = atoi(argv[4]);
  double offset  = (double)atof(argv[5]);
  cv::Mat Igray = cv::imread(argv[6], cv::IMREAD_GRAYSCALE);
```

```
  // Read in gray image.
  //
  if( Igray.empty() ){ cout << "Can not load " << argv[6] << endl; return -1; }

  // Declare the output images.
  //
  cv::Mat It, Iat;

  // Thresholds.
  //
  cv::threshold(
    Igray,
    It,
    fixed_threshold,
    255,
    threshold_type);
  cv::adaptiveThreshold(
    Igray,
    Iat,
    255,
    adaptive_method,
    threshold_type,
    block_size,
    offset
  );

  // Show the results.
  //
  cv::imshow("Raw",Igray);
  cv::imshow("Threshold",It);
  cv::imshow("Adaptive Threshold",Iat);
  cv::waitKey(0);

  return 0;
}
```

# Smoothing

*Smoothing*, also called *blurring* as depicted in Figure 10-6, is a simple and frequently used image-processing operation. There are many reasons for smoothing, but it is often done to reduce noise or camera artifacts. Smoothing is also important when we wish to reduce the resolution of an image in a principled way (we will discuss this in more detail in "Image Pyramids" on page 302 in Chapter 11).

OpenCV offers five different smoothing operations, each with its own associated library function, which each accomplish slightly different kinds of smoothing. The `src` and `dst` arguments in all of these functions are the usual source and destination arrays. After that, each smoothing operation has parameters that are specific to the associated operation. Of these, the only common parameter is the last, `borderType`.

This argument tells the smoothing operation how to handle pixels at the edge of the image.

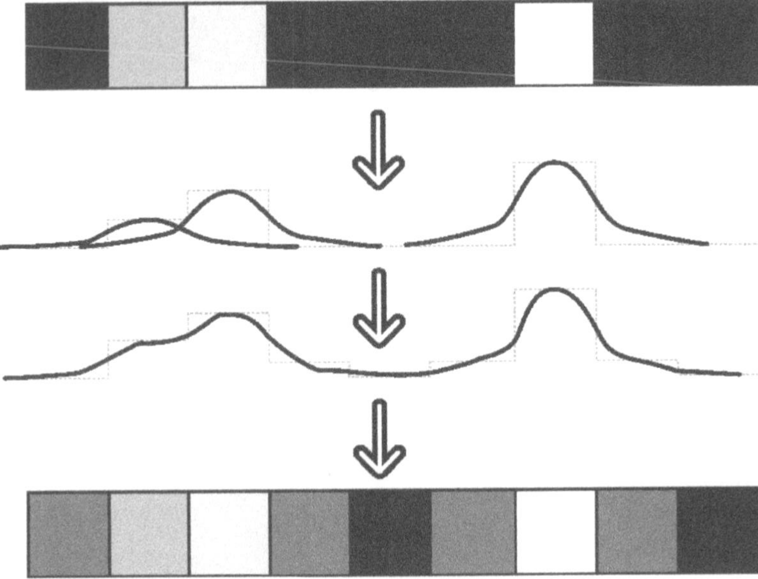

*Figure 10-6. Gaussian blur on 1D-pixel array*

## Simple Blur and the Box Filter

```
void cv::blur(
  cv::InputArray  src,                              // Input image
  cv::OutputArray dst,                              // Result image
  cv::Size        ksize,                            // Kernel size
  cv::Point       anchor     = cv::Point(-1,-1),    // Location of anchor point
  int             borderType = cv::BORDER_DEFAULT   // Border extrapolation to use
);
```

The *simple blur* operation is provided by `cv::blur()`. Each pixel in the output is the simple mean of all of the pixels in a window (i.e., the *kernel*), around the corresponding pixel in the input. The size of this window is specified by the argument `ksize`. The argument `anchor` can be used to specify how the kernel is aligned with the pixel being computed. By default, the value of `anchor` is `cv::Point(-1,-1)`, which indicates that the kernel should be centered relative to the filter. In the case of multichannel images, each channel will be computed separately.

The simple blur is a specialized version of the *box filter*, as shown in Figure 10-7. A box filter is any filter that has a rectangular profile and for which the values $k_{i,j}$ are all equal. In most cases, $k_{i,j} = 1$ for all $i, j$, or $k_{i,j} = 1/A$, where $A$ is the area of the filter. The latter case is called a *normalized box filter*, the output of which is shown in Figure 10-8.

```
void cv::boxFilter(
  cv::InputArray  src,                                  // Input image
  cv::OutputArray dst,                                  // Result image
  int             ddepth,                               // Output depth (e.g., CV_8U)
  cv::Size        ksize,                                // Kernel size
  cv::Point       anchor     = cv::Point(-1,-1),  // Location of anchor point
  bool            normalize  = true,              // If true, divide by box area
  int             borderType = cv::BORDER_DEFAULT // Border extrapolation to use
);
```

The OpenCV function `cv::boxFilter()` is the somewhat more general form of which `cv::blur()` is essentially a special case. The main difference between `cv::boxFilter()` and `cv::blur()` is that the former can be run in an unnormalized mode (`normalize = false`), and that the depth of the output image `dst` can be controlled. (In the case of `cv::blur()`, the depth of `dst` will always equal the depth of `src`.) If the value of `ddepth` is set to `-1`, then the destination image will have the same depth as the source; otherwise, you can use any of the usual aliases (e.g., `CV_32F`).

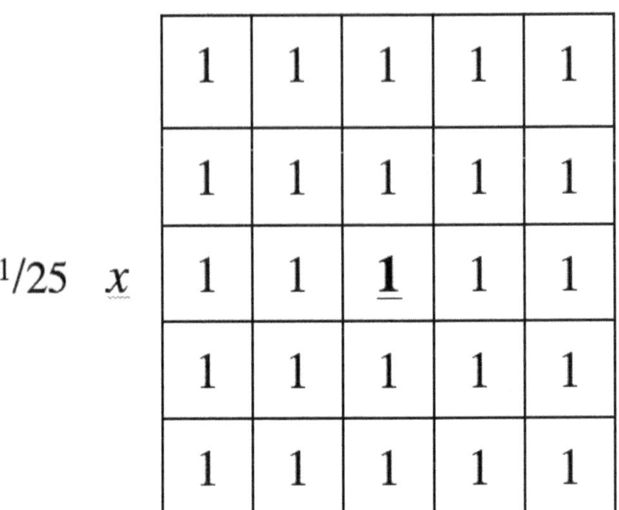


*Figure 10-7. A 5 × 5 blur filter, also called a normalized box filter*

*Figure 10-8. Image smoothing by block averaging: on the left are the input images; on the right, the output images*

## Median Filter

The *median filter* [Bardyn84] replaces each pixel by the median or "middle-valued" pixel (as opposed to the mean pixel) in a rectangular neighborhood around the center pixel.[7] Results of median filtering are shown in Figure 10-9. Simple blurring by averaging can be sensitive to noisy images, especially images with large isolated outlier values (e.g., *shot noise* in digital photography). Large differences in even a small number of points can cause a noticeable movement in the average value. Median filtering is able to ignore the outliers by selecting the middle points.

```
void cv::medianBlur(
  cv::InputArray  src,                  // Input image
  cv::OutputArray dst,                  // Result image
  cv::Size        ksize                 // Kernel size
);
```

The arguments to `cv::medianBlur` are essentially the same as for the filters you've learned about in this chapter so far: the source array `src`, the destination array `dst`, and the kernel size `ksize`. For `cv::medianBlur()`, the anchor point is always assumed to be at the center of the kernel.

*Figure 10-9. Blurring an image by taking the median of surrounding pixels*

7 Note that the median filter is an example of a nonlinear kernel, which cannot be represented in the pictorial style shown back in Figure 10-1.

## Gaussian Filter

The next smoothing filter, the *Gaussian filter*, is probably the most useful. Gaussian filtering involves convolving each point in the input array with a (normalized) Gaussian kernel and then summing to produce the output array:

```
void cv::GaussianBlur(
  cv::InputArray  src,                              // Input image
  cv::OutputArray dst,                              // Result image
  cv::Size        ksize,                            // Kernel size
  double          sigmaX,                           // Gaussian half-width in x-direction
  double          sigmaY     = 0.0,                 // Gaussian half-width in y-direction
  int             borderType = cv::BORDER_DEFAULT // Border extrapolation to use
);
```

For the Gaussian blur (an example kernel is shown in Figure 10-10), the parameter `ksize` gives the width and height of the filter window. The next parameter indicates the sigma value (half width at half max) of the Gaussian kernel in the x-dimension. The fourth parameter similarly indicates the sigma value in the y-dimension. If you specify only the $x$ value, and set the $y$ value to `0` (its default value), then the $y$ and $x$ values will be taken to be equal. If you set them both to `0`, then the Gaussian's parameters will be automatically determined from the window size through the following formulae:

$$\sigma_x = \left(\frac{n_x - 1}{2}\right) \cdot 0.30 + 0.80, \; n_x = ksize.width - 1$$

$$\sigma_y = \left(\frac{n_y - 1}{2}\right) \cdot 0.30 + 0.80, \; n_y = ksize.height - 1$$

Finally, `cv::GaussianBlur()` takes the usual `borderType` argument.

$^1/_{141}$ x

| | | | | |
|---|---|---|---|---|
| 1 | 4 | 7 | 4 | 1 |
| 7 | 26 | **41** | 26 | 7 |
| 1 | 4 | 7 | 4 | 1 |

*Figure 10-10. An example Gaussian kernel where ksize = (5,3), sigmaX = 1, and sigmaY = 0.5*

The OpenCV implementation of Gaussian smoothing also provides a higher performance optimization for several common kernels. 3 × 3, 5 × 5, and 7 × 7 kernels with the "standard" sigma (i.e., `sigmaX = 0.0`) give better performance than other kernels. Gaussian blur supports single- or three-channel images in either 8-bit or 32-bit floating-point formats, and it can be done in place. Results of Gaussian blurring are shown in Figure 10-11.

*Figure 10-11. Gaussian filtering (blurring)*

## Bilateral Filter

```
void cv::bilateralFilter(
  cv::InputArray  src,                    // Input image
  cv::OutputArray dst,                    // Result image
  int             d,                      // Pixel neighborhood size (max distance)
  double          sigmaColor,             // Width param for color weight function
  double          sigmaSpace,             // Width param for spatial weight function
  int             borderType = cv::BORDER_DEFAULT  // Border extrapolation to use
);
```

The fifth and final form of smoothing supported by OpenCV is called *bilateral filtering* [Tomasi98], an example of which is shown in Figure 10-12. Bilateral filtering is one operation from a somewhat larger class of image analysis operators known as *edge-preserving smoothing*. Bilateral filtering is most easily understood when contrasted to Gaussian smoothing. A typical motivation for Gaussian smoothing is that pixels in a real image should vary slowly over space and thus be correlated to their neighbors, whereas random noise can be expected to vary greatly from one pixel to the next (i.e., noise is not spatially correlated). It is in this sense that Gaussian smoothing reduces noise while preserving signal. Unfortunately, this method breaks down near edges, where you do expect pixels to be uncorrelated with their neighbors across the edge. As a result, Gaussian smoothing blurs away edges. At the cost of what is unfortunately substantially more processing time, bilateral filtering provides a means of smoothing an image without smoothing away its edges.

*Figure 10-12. Results of bilateral smoothing*

Like Gaussian smoothing, bilateral filtering constructs a weighted average of each pixel and its neighboring components. The weighting has two components, the first of which is the same weighting used by Gaussian smoothing. The second component is also a Gaussian weighting but is based not on the spatial distance from the center pixel but rather on the difference in intensity[8] from the center pixel.[9] You can think of bilateral filtering as Gaussian smoothing that weighs similar pixels more highly than less similar ones, keeping high-contrast edges sharp. The effect of this filter is typically to turn an image into what appears to be a watercolor painting of the same scene.[10] This can be useful as an aid to segmenting the image.

8 In the case of multichannel (i.e., color) images, the difference in intensity is replaced with a weighted sum over colors. This weighting is chosen to enforce a Euclidean distance in the CIE Lab color space.

9 Technically, the use of Gaussian distribution functions is not a necessary feature of bilateral filtering. The implementation in OpenCV uses Gaussian weighting even though the method allows many possible weighting functions.

10 This effect is particularly pronounced after multiple iterations of bilateral filtering.

Bilateral filtering takes three parameters (other than the source and destination). The first is the diameter `d` of the pixel neighborhood that is considered during filtering. The second is the width of the Gaussian kernel used in the color domain called `sigmaColor`, which is analogous to the sigma parameters in the Gaussian filter. The third is the width of the Gaussian kernel in the spatial domain called `sigmaSpace`. The larger the second parameter, the broader the range of intensities (or colors) that will be included in the smoothing (and thus the more extreme a discontinuity must be in order to be preserved).

The filter size `d` has a strong effect (as you might expect) on the speed of the algorithm. Typical values are less than or equal to `5` for video processing, but might be as high as `9` for non-real-time applications. As an alternative to specifying `d` explicitly, you can set it to `-1`, in which case, it will be automatically computed from `sigmaSpace`.

In practice, small values of `sigmaSpace` (e.g., `10`) give a very light but noticeable effect, while large values (e.g., `150`) have a very strong effect and tend to give the image a somewhat "cartoonish" appearance.

# Derivatives and Gradients

One of the most basic and important convolutions is computing derivatives (or approximations to them). There are many ways to do this, but only a few are well suited to a given situation.

## The Sobel Derivative

In general, the most common operator used to represent differentiation is the *Sobel derivative* [Sobel73] operator (see Figures 10-13 and 10-14). Sobel operators exist for any order of derivative as well as for mixed partial derivatives (e.g., $\partial^2/\partial x\partial y$).

*Figure 10-13. The effect of the Sobel operator when used to approximate a first derivative in the x-dimension*

```
void cv::Sobel(
  cv::InputArray  src,                   // Input image
  cv::OutputArray dst,                   // Result image
  int             ddepth,                // Pixel depth of output (e.g., CV_8U)
  int             xorder,                // order of corresponding derivative in x
  int             yorder,                // order of corresponding derivative in y
  cv::Size        ksize      = 3,        // Kernel size
  double          scale      = 1,        // Scale (applied before assignment)
  double          delta      = 0,        // Offset (applied before assignment)
  int             borderType = cv::BORDER_DEFAULT  // Border extrapolation
);
```

Here, `src` and `dst` are your image input and output. The argument `ddepth` allows you to select the depth (type) of the generated output (e.g., `CV_32F`). As a good example of how to use `ddepth`, if `src` is an 8-bit image, then the `dst` should have a depth of at least `CV_16S` to avoid overflow. `xorder` and `yorder` are the orders of the derivative. Typically, you'll use `0`, `1`, or at most `2`; a `0` value indicates no derivative in that direction.[11] In all cases except one, the `ksize` × `ksize` separable kernel is used to calculate the derivative. When `ksize` = 1, the 3×1 or 1×3 kernel is used (that is, no Gaussian smoothing is done). `ksize` = 1 can only be used for the first or the second x- or y-

11 Either `xorder` or `yorder` must be nonzero.

derivatives. The `ksize` parameter should be odd and is the width (and the height) of the filter to be used. Currently, kernel sizes up to `31` are supported.[12] The `scale` factor and `delta` are applied to the derivative before storing in `dst`. This can be useful when you want to actually visualize a derivative in an 8-bit image you can show on the screen:

$$dst_i = scale \cdot \left\{ \sum_{i,j \in sobel\_kernel} k_{i,j} * I(x+i,\ y+j) \right\} + delta$$

The `borderType` argument functions exactly as described for other convolution operations.

*Figure 10-14. The effect of the Sobel operator when used to approximate a first derivative in the y-dimension*

Sobel operators have the nice property that they can be defined for kernels of any size, and those kernels can be constructed quickly and iteratively. The larger kernels give a better approximation to the derivative because they are less sensitive to noise. However, if the derivative is not expected to be constant over space, clearly a kernel that is too large will no longer give a useful result.

12 In practice, it really only makes sense to set the kernel size to 3 or greater. If you set `ksize` to 1, then the kernel size will automatically be adjusted up to 3.

To understand this more exactly, we must realize that a Sobel operator is not really a derivative as it is defined on a discrete space. What the Sobel operator actually represents is a fit to a polynomial. That is, the Sobel operator of second order in the x-direction is not really a second derivative; it is a local fit to a parabolic function. This explains why one might want to use a larger kernel: that larger kernel is computing the fit over a larger number of pixels.

## Scharr Filter

In fact, there are many ways to approximate a derivative in the case of a discrete grid. The downside of the approximation used for the Sobel operator is that it is less accurate for small kernels. For large kernels, where more points are used in the approximation, this problem is less significant. This inaccuracy does not show up directly for the $X$ and $Y$ filters used in `cv::Sobel()`, because they are exactly aligned with the x- and y-axes. The difficulty arises when you want to make image measurements that are approximations of *directional derivatives* (i.e., direction of the image gradient by using the arctangent of the ratio $y/x$ of two directional filter responses).[13]

To put this in context, a concrete example of where you may want such image measurements is in the process of collecting shape information from an object by assembling a histogram of gradient angles around the object. Such a histogram is the basis on which many common shape classifiers are trained and operated. In this case, inaccurate measures of gradient angle will decrease the recognition performance of the classifier.

For a 3 × 3 Sobel filter, the inaccuracies are more apparent the farther the gradient angle is from horizontal or vertical. OpenCV addresses this inaccuracy for small (but fast) 3 × 3 Sobel derivative filters by a somewhat obscure use of the special `ksize` value `cv::SCHARR` in the `cv::Sobel()` function. The Scharr filter is just as fast but more accurate than the Sobel filter, so it should always be used if you want to make image measurements using a 3 × 3 filter. The filter coefficients for the Scharr filter are shown in Figure 10-15 [Scharr00].

13 As you might recall, there are functions `cv::cartToPolar()` and `cv::polarToCart()` that implement exactly this transformation. If you find yourself wanting to call `cv::cartToPolar()` on a pair of x- and y-derivative images, you should probably be using `CV_SCHARR` to compute those images.

| -3 | 0 | +3 |
|---|---|---|
| -10 | 0 | +10 |
| -3 | 0 | +3 |

A

| -3 | -10 | -3 |
|---|---|---|
| 0 | 0 | 0 |
| +3 | +10 | +3 |

B

*Figure 10-15. The 3 × 3 Scharr filter using flag cv::SCHARR*

## The Laplacian

OpenCV *Laplacian function* (first used in vision by Marr [Marr82]) implements a discrete approximation to the Laplacian operator:[14]

$$Laplace\ (f) = \frac{\partial^2 f}{\partial x^2} + \frac{\partial^2 f}{\partial y^2}$$

Because the Laplacian operator can be defined in terms of second derivatives, you might well suppose that the discrete implementation works something like the second-order Sobel derivative. Indeed it does, and in fact, the OpenCV implementation of the Laplacian operator uses the Sobel operators directly in its computation:

```
void cv::Laplacian(
  cv::InputArray  src,                      // Input image
  cv::OutputArray dst,                      // Result image
  int             ddepth,                   // Depth of output image (e.g., CV_8U)
  int             ksize     = 3,            // Kernel size
  double          scale     = 1,            // Scale applied before assignment to dst
  double          delta     = 0,            // Offset applied before assignment to dst
  int             borderType = cv::BORDER_DEFAULT  // Border extrapolation to use
);
```

The `cv::Laplacian()` function takes the same arguments as the `cv::Sobel()` function, with the exception that the orders of the derivatives are not needed. This aperture `ksize` is precisely the same as the aperture appearing in the Sobel derivatives and, in effect, gives the size of the region over which the pixels are sampled in the computation of the second derivatives. In the actual implementation, for `ksize` anything other than `1`, the Laplacian operator is computed directly from the sum of the

14 Note that the Laplacian *operator* is distinct from the Laplacian *pyramid*, which we will discuss in Chapter 11.

corresponding Sobel operators. In the special case of `ksize=1`, the Laplacian operator is computed by convolution with the single kernel shown in Figure 10-16.

| | | |
|---|---|---|
| 0 | 1 | 0 |
| 1 | -4 | 1 |
| 0 | 1 | 0 |

*Figure 10-16. The single kernel used by cv::Laplacian() when ksize = 1*

The Laplacian operator can be used in a variety of contexts. A common application is to detect "blobs." Recall that the form of the Laplacian operator is a sum of second derivatives along the x-axis and y-axis. This means that a single point or any small blob (smaller than the aperture) that is surrounded by higher values will tend to maximize this function. Conversely, a point or small blob that is surrounded by lower values will tend to maximize the negative of this function.

With this in mind, the Laplacian operator can also be used as a kind of edge detector. To see how this is done, consider the first derivative of a function, which will (of course) be large wherever the function is changing rapidly. Equally important, it will grow rapidly as we approach an edge-like discontinuity and shrink rapidly as we move past the discontinuity. Hence, the derivative will be at a local maximum somewhere within this range. Therefore, we can look to the 0s of the second derivative for locations of such local maxima. Edges in the original image will be 0s of the Laplacian operator. Unfortunately, both substantial and less meaningful edges will be 0s of the Laplacian, but this is not a problem because we can simply filter out those pixels that also have larger values of the first (Sobel) derivative. Figure 10-17 shows an example of using a Laplacian operator on an image together with details of the first and second derivatives and their zero crossings.

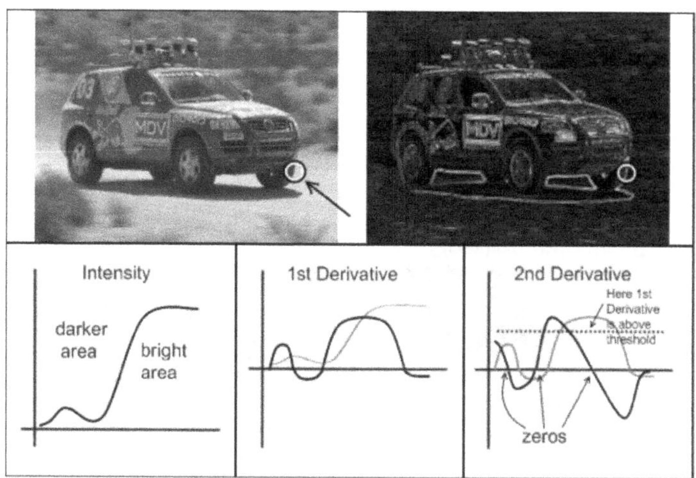


*Figure 10-17. Laplace transform (upper right) of the racecar image: zooming in on the tire (circled) and considering only the x-dimension, we show a (qualitative) representation of the brightness as well as the first and second derivatives (lower three cells); the 0s in the second derivative correspond to edges, and the 0 corresponding to a large first derivative is a strong edge*

# Image Morphology

OpenCV also provides a fast, convenient interface for doing *morphological transformations* [Serra83] on an image. Figure 10-18 shows the most popular morphological transformations. Image morphology is its own topic and, especially in the early years of computer vision, a great number of morphological operations were developed. Most were developed for one specific purpose or another, and some of those found broader utility over the years. Essentially, all morphology operations are based on just two primitive operations. We will start with those, and then move on to the more complex operations, each of which is typically defined in terms of its simpler predecessors.

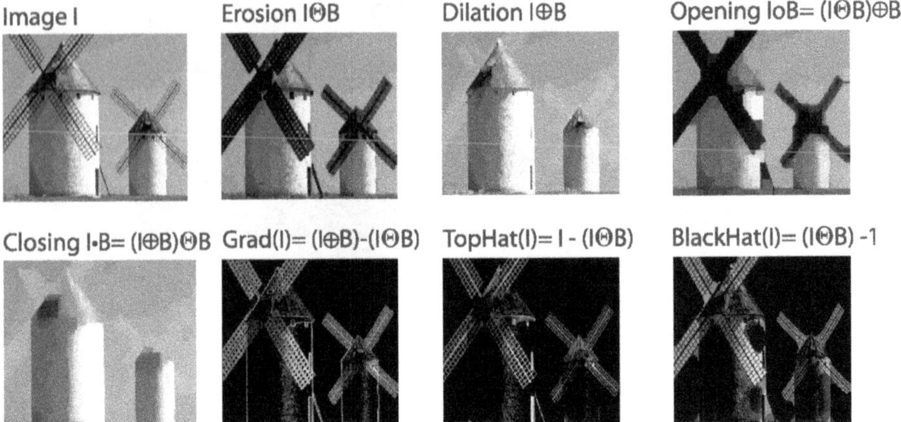


*Figure 10-18. Summary results for all morphology operators*

## Dilation and Erosion

The basic morphological transformations are called *dilation* and *erosion*, and they arise in a wide variety of contexts such as removing noise, isolating individual elements, and joining disparate elements in an image. More sophisticated morphology operations, based on these two basic operations, can also be used to find intensity peaks (or holes) in an image, and to define (yet another) particular form of an image gradient.

Dilation is a convolution of some image with a kernel in which any given pixel is replaced with the *local maximum* of all of the pixel values covered by the kernel. As we mentioned earlier, this is an example of a nonlinear operation, so the kernel cannot be expressed in the form shown back in Figure 10-1. Most often, the kernel used for dilation is a "solid" square kernel, or sometimes a disk, with the anchor point at the center. The effect of dilation is to cause filled[15] regions within an image to grow as diagrammed in Figure 10-19.

15 Here the term *filled* means those pixels whose value is nonzero. You could read this as "bright," since the local maximum actually takes the pixel with the highest intensity value under the template (kernel). It is worth mentioning that the diagrams that appear in this chapter to illustrate morphological operators are in this sense inverted relative to what would happen on your screen (because books write with dark ink on light paper instead of light pixels on a dark screen).

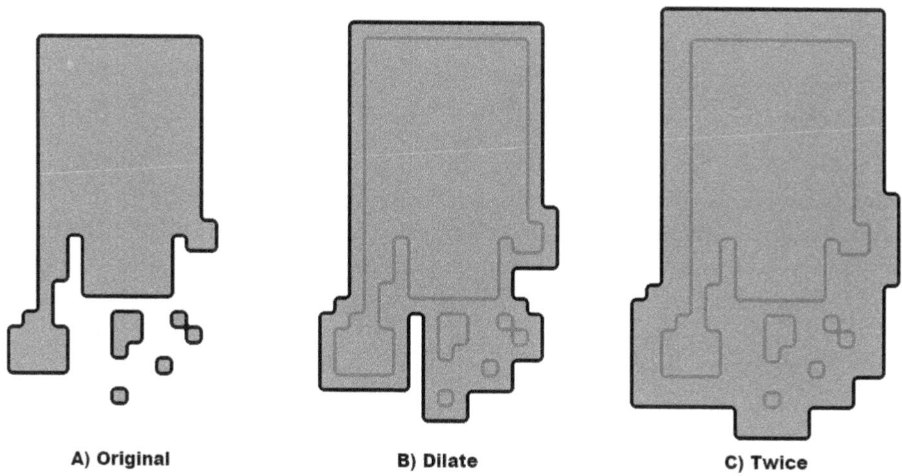


*Figure 10-19. Morphological dilation: take the maximum under a square kernel*

Erosion is the converse operation. The action of the erosion operator is equivalent to computing a *local minimum* over the area of the kernel.[16] Erosion is diagrammed in Figure 10-20.

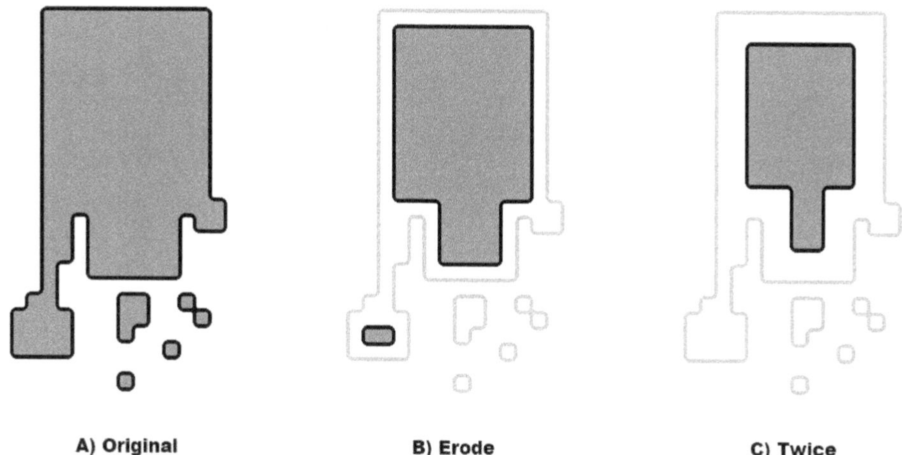


*Figure 10-20. Morphological erosion: take the minimum under a square kernel*

16 To be precise, the pixel in the destination image is set to the value equal to the minimal value of the pixels under the kernel in the source image.

Image morphology is often done on Boolean[17] images that result from a threshold operation. However, because dilation is just a max operator and erosion is just a min operator, morphology may be used on intensity images as well.

In general, whereas dilation expands a bright region, erosion reduces such a bright region. Moreover, dilation will tend to fill concavities and erosion will tend to remove protrusions. Of course, the exact result will depend on the kernel, but these statements are generally true so long as the kernel is both convex and filled.

In OpenCV, we effect these transformations using the `cv::erode()` and `cv::dilate()` functions:

```
void cv::erode(
  cv::InputArray    src,                                  // Input image
  cv::OutputArray   dst,                                  // Result image
  cv::InputArray    element,                              // Structuring, a cv::Mat()
  cv::Point         anchor     = cv::Point(-1,-1),        // Location of anchor point
  int               iterations = 1,                       // Number of times to apply
  int               borderType = cv::BORDER_CONSTANT      // Border extrapolation
  const cv::Scalar& borderValue = cv::morphologyDefaultBorderValue()
);
void cv::dilate(
  cv::InputArray    src,                   // Input image
  cv::OutputArray   dst,                   // Result image
  cv::InputArray    element,                              // Structuring, a cv::Mat()
  cv::Point         anchor     = cv::Point(-1,-1),        // Location of anchor point
  int               iterations = 1,                       // Number of times to apply
  int               borderType = cv::BORDER_CONSTANT      // Border extrapolation
  const cv::Scalar& borderValue = cv::morphologyDefaultBorderValue()
);
```

Both `cv::erode()` and `cv::dilate()` take a source and destination image, and both support "in place" calls (in which the source and destination are the same image). The third argument is the kernel, to which you may pass an uninitialized array `cv::Mat()`, which will cause it to default to using a 3 × 3 kernel with the anchor at its center (we will discuss how to create your own kernels later). The fourth argument is the number of iterations. If not set to the default value of `1`, the operation will be applied multiple times during the single call to the function. The `borderType` argument is the usual border type, and the `borderValue` is the value that will be used for off-the-edge pixels when the `borderType` is set to `cv::BORDER_CONSTANT`.

17 It should be noted that OpenCV does not actually have a Boolean image data type. The minimum size representation is 8-bit characters. Those functions that *interpret* an image as Boolean do so by classifying all pixels as either zero (`False` or `0`) or nonzero (`True` or `1`).

The results of an erode operation on a sample image are shown in Figure 10-21, and those of a dilation operation on the same image are shown in Figure 10-22. The erode operation is often used to eliminate "speckle" noise in an image. The idea here is that the speckles are eroded to nothing while larger regions that contain visually significant content are not affected. The dilate operation is often used to try to find *connected components* (i.e., large discrete regions of similar pixel color or intensity). The utility of dilation arises because in many cases a large region might otherwise be broken apart into multiple components as a result of noise, shadows, or some other similar effect. A small dilation will cause such components to "melt" together into one.

*Figure 10-21. Results of the erosion, or "min," operator: bright regions are isolated and shrunk*

*Figure 10-22. Results of the dilation, or "max," operator: bright regions are expanded and often joined*

To recap: when OpenCV processes the `cv::erode()` function, what happens beneath the hood is that the value of some point $p$ is set to the minimum value of all of the points covered by the kernel when aligned at $p$; for the `cv::dilate()` operator, the equation is the same except that max is considered rather than min:

$$erode(x, y) = \min_{(i,j) \in kernel} src(x + i,\ y + j)$$

$$dilate(x, y) = \max_{(i,j) \in kernel} src(x + i,\ y + j)$$

You might be wondering why we need a complicated formula when the earlier heuristic description was perfectly sufficient. Some users actually prefer such formulas but, more importantly, the formulas capture some generality that isn't apparent in the qualitative description. Observe that if the image is not Boolean, then the min and max operators play a less trivial role. Take another look at Figures 10-21 and 10-22, which show the erosion and dilation operators (respectively) applied to two real images.

## The General Morphology Function

When you are working with Boolean images and image masks where the pixels are either on (>0) or off (=0), the basic erode and dilate operations are usually sufficient. When you're working with grayscale or color images, however, a number of additional operations are often helpful. Several of the more useful operations can be handled by the multipurpose `cv::morphologyEx()` function.

```
void cv::morphologyEx(
  cv::InputArray    src,                          // Input image
  cv::OutputArray   dst,                          // Result image
  int               op,                           // Operator (e.g. cv::MOP_OPEN)
  cv::InputArray    element,                      // Structuring element, cv::Mat()
  cv::Point         anchor      = cv::Point(-1,-1),  // Location of anchor point
  int               iterations  = 1,                 // Number of times to apply
  int               borderType  = cv::BORDER_DEFAULT // Border extrapolation
  const cv::Scalar& borderValue = cv::morphologyDefaultBorderValue()

);
```

In addition to the arguments that we saw with the `cv::dilate()` and `cv::erode()` functions, `cv::morphologyEx()` has one new—and very important—parameter. This new argument, called `op`, is the specific operation to be done. The possible values of this argument are listed in Table 10-3.

*Table 10-3. cv::morphologyEx() operation options*

| Value of operation | Morphological operator | Requires temp image? |
|---|---|---|
| `cv::MOP_OPEN` | Opening | No |
| `cv::MOP_CLOSE` | Closing | No |
| `cv::MOP_GRADIENT` | Morphological gradient | Always |
| `cv::MOP_TOPHAT` | Top Hat | For in-place only (`src = dst`) |
| `cv::MOP_BLACKHAT` | Black Hat | For in-place only (`src = dst`) |

## Opening and Closing

The first two operations, *opening* and *closing*, are actually simple combinations of the erosion and dilation operators. In the case of opening, we erode first and then dilate (Figure 10-23). Opening is often used to count regions in a Boolean image. For example, if we have thresholded an image of cells on a microscope slide, we might use opening to separate out cells that are near each other before counting the regions.

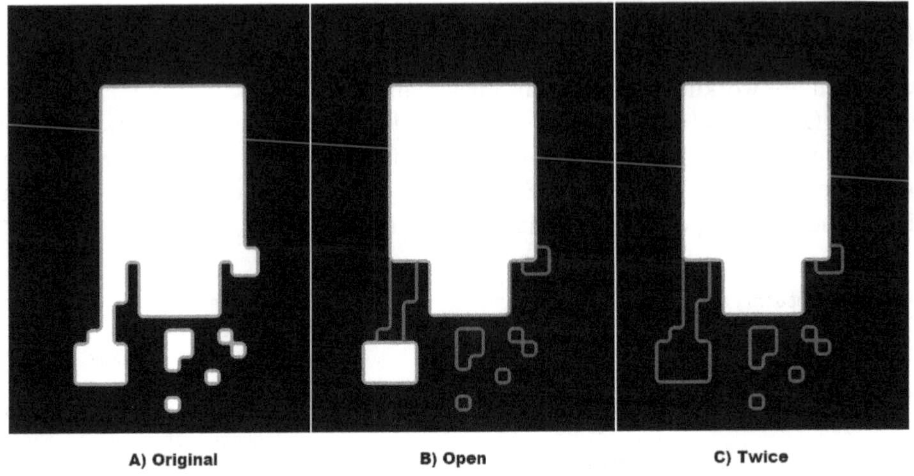


*Figure 10-23. Morphological opening applied to a simple Boolean image*

In the case of closing, we dilate first and then erode (Figure 10-24). Closing is used in most of the more sophisticated connected-component algorithms to reduce unwanted or noise-driven segments. For connected components, usually an erosion or closing operation is performed first to eliminate elements that arise purely from noise, and then an opening operation is used to connect nearby large regions. (Notice that, although the end result of using opening or closing is similar to using erosion or dilation, these new operations tend to preserve the area of connected regions more accurately.)

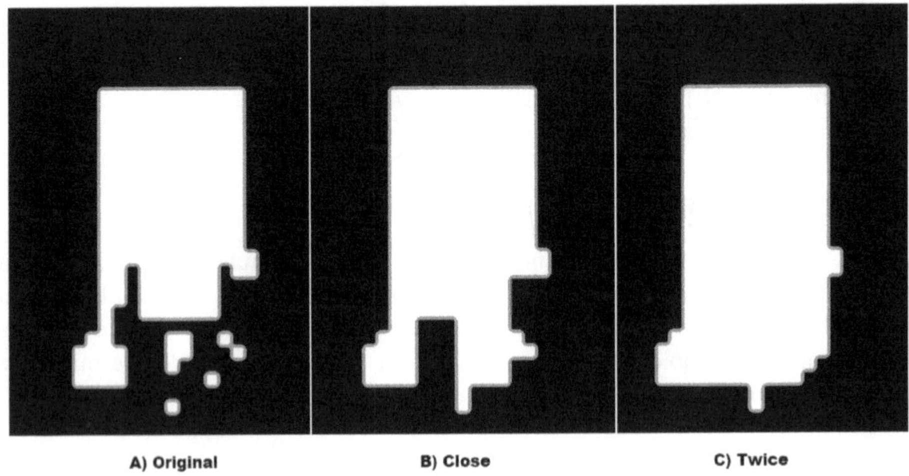


*Figure 10-24. Morphological closing applied to a simple Boolean image*

When used on non-Boolean images, the most prominent effect of closing is to eliminate lone outliers that are lower in value than their neighbors, whereas the effect of opening is to eliminate lone outliers that are higher than their neighbors. Results of using the opening operator are shown in Figures 10-25 and 10-26, and results of the closing operator are shown in Figures 10-27 and 10-28.

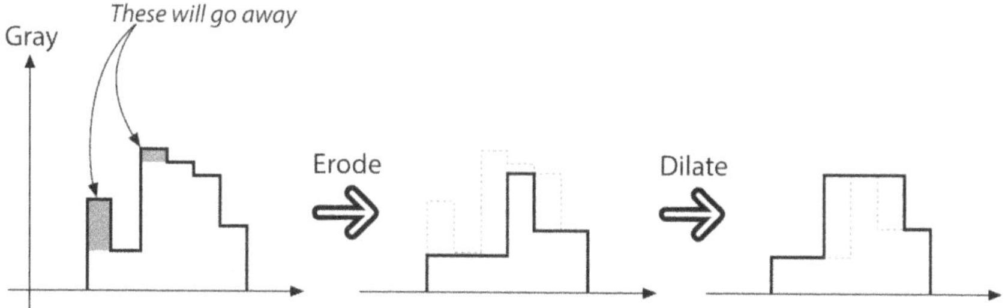


*Figure 10-25. Morphological opening operation applied to a (one-dimensional) non-Boolean image: the upward outliers are eliminated*

*Figure 10-26. Results of morphological opening on an image: small bright regions are removed, and the remaining bright regions are isolated but retain their size*

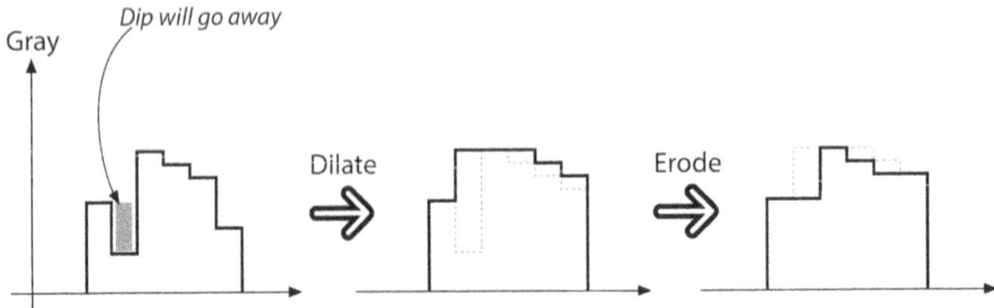


*Figure 10-27. Morphological closing operation applied to a (one-dimensional) non-Boolean image: the downward outliers are eliminated*

*Figure 10-28. Results of morphological closing on an image: bright regions are joined but retain their basic size*

One last note on the opening and closing operators concerns how the `iterations` argument is interpreted. You might expect that asking for two iterations of closing would yield something like dilate-erode-dilate-erode. It turns out that this would not be particularly useful. What you usually want (and what you get) is dilate-dilate-erode-erode. In this way, not only the single outliers but also neighboring pairs of outliers will disappear. Figures 10-23(C) and 10-24(C) illustrate the effect of calling open and close (respectively) with an iteration count of two.

## Morphological Gradient

Our next available operator is the *morphological gradient*. For this one, it is probably easier to start with a formula and then figure out what it means:

$$gradient(\text{src}) = dilate(\text{src}) - erode(\text{src})$$

As we can see in Figure 10-29, the effect of subtracting the eroded (slightly reduced) image from the dilated (slightly enlarged) image is to leave behind a representation of the edges of objects in the original image.

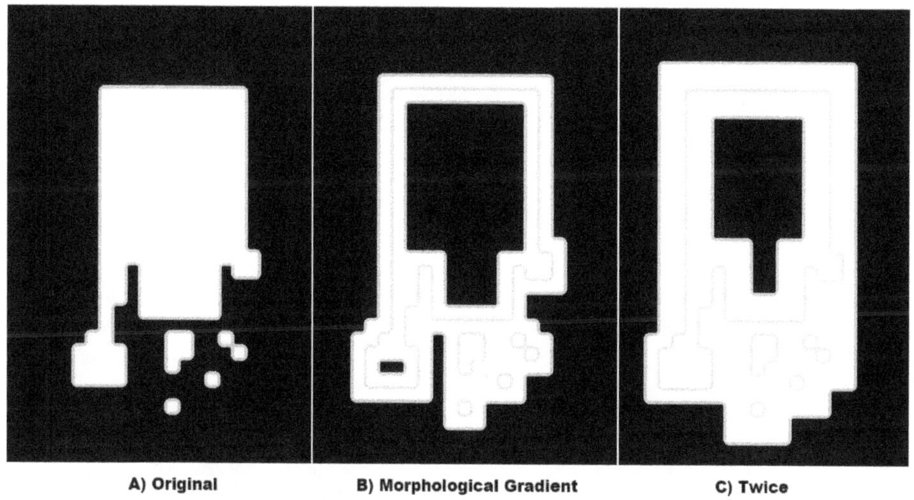


*Figure 10-29. Morphological gradient applied to a simple Boolean image*

With a grayscale image (Figure 10-30), we see that the value of the operator is telling us something about how fast the image brightness is changing; this is why the name "morphological gradient" is justified. Morphological gradient is often used when we want to isolate the perimeters of bright regions so we can treat them as whole objects (or as whole parts of objects). The complete perimeter of a region tends to be found because a contracted version is subtracted from an expanded version of the region, leaving a complete perimeter edge. This differs from calculating a gradient, which is much less likely to work around the full perimeter of an object. Figure 10-31 shows the result of the morphological gradient operator.

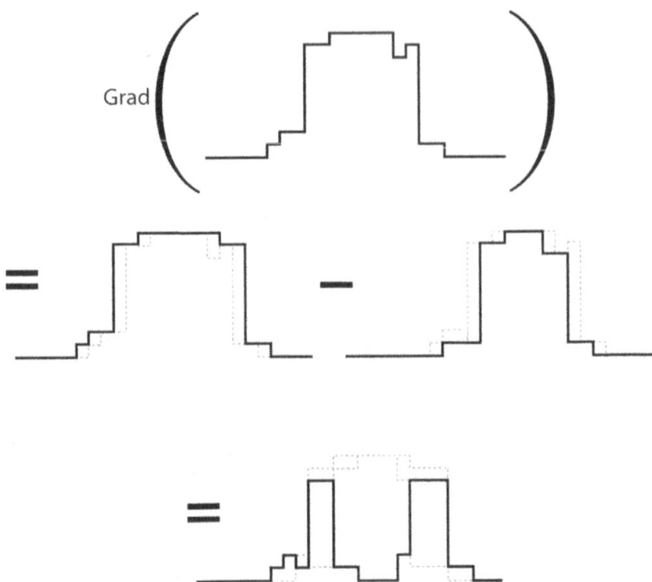


*Figure 10-30. Morphological gradient applied to (one-dimensional) non-Boolean image: as expected, the operator has its highest values where the grayscale image is changing most rapidly*

*Figure 10-31. Results of the morphological gradient operator: bright perimeter edges are identified*

## Top Hat and Black Hat

The last two operators are called *Top Hat* and *Black Hat* [Meyer78]. These operators are used to isolate patches that are, respectively, brighter or dimmer than their immediate neighbors. You would use these when trying to isolate parts of an object that exhibit brightness changes relative only to the object to which they are attached. This often occurs with microscope images of organisms or cells, for example. Both operations are defined in terms of the more primitive operators, as follows:

> *TopHat*(src) = src – *open*(src) // Isolate brighter
> *BlackHat*(src) = *close*(src) – src // Isolate dimmer

As you can see, the Top Hat operator subtracts the opened form of A from A. Recall that the effect of the open operation was to exaggerate small cracks or local drops. Thus, subtracting open(A) from A should reveal areas that are lighter than the surrounding region of A, relative to the size of the kernel (see Figures 10-32 and 10-33); conversely, the Black Hat operator reveals areas that are darker than the surrounding region of A (Figures 10-34 and 10-35). Summary results for all the morphological operators discussed in this chapter are shown back in Figure 10-18.[18]

18 Both of these operations (Top Hat and Black Hat) are most useful in grayscale morphology, where the structuring element is a matrix of real numbers (not just a Boolean mask) and the matrix is added to the current pixel neighborhood before taking a minimum or maximum. As of this writing, however, this is not yet implemented in OpenCV.

*Figure 10-32. Results of morphological Top Hat operation: bright local peaks are isolated*

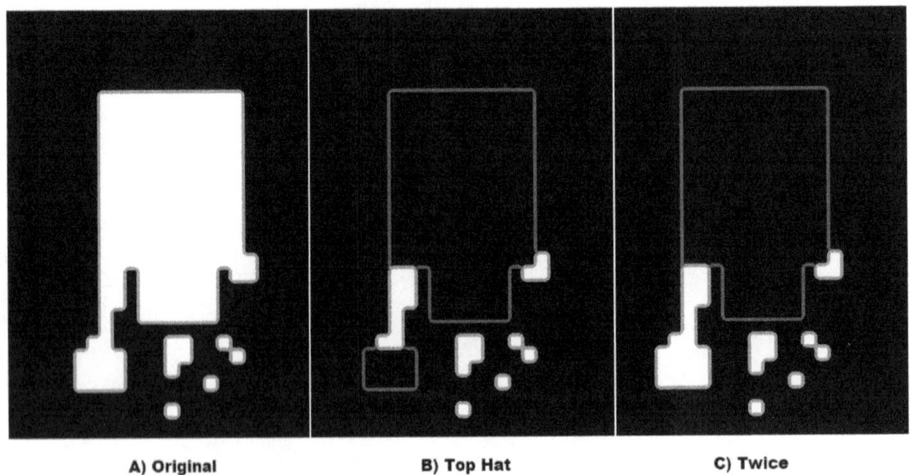


*Figure 10-33. Results of morphological Top Hat operation applied to a simple Boolean image*

*Figure 10-34. Results of morphological Black Hat operation: dark holes are isolated*

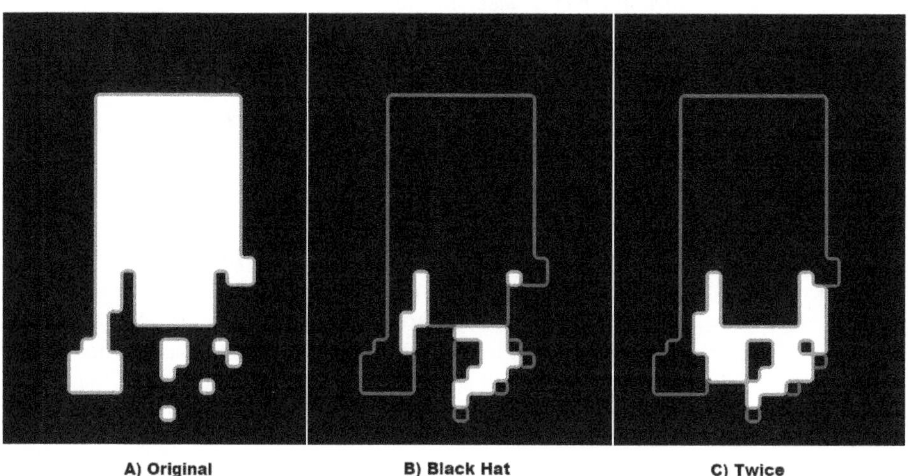


*Figure 10-35. Results of morphological Black Hat operation applied to a simple Boolean image*

## Making Your Own Kernel

In the morphological operations we have looked at so far, the kernels considered were always square and 3 × 3. If you need something a little more general than that,

OpenCV allows you to create your own kernel. In the case of morphology, the kernel is often called a *structuring element*, so the routine that allows you to create your own morphology kernels is called `cv::getStructuringElement()`.

In fact, you can just create any array you like and use it as a structuring element in functions like `cv::dilate()`, `cv::erode()`, or `cv::morphologyEx()`, but this is often more work than is necessary. Often what you need is a nonsquare kernel of an otherwise common shape. This is what `cv::getStructuringElement()` is for:

```
cv::Mat cv::getStructuringElement(
  int       shape,                    // Element shape, e.g., cv::MORPH_RECT
  cv::Size  ksize,                    // Size of structuring element (odd num!)
  cv::Point anchor = cv::Point(-1,-1) // Location of anchor point
);
```

The first argument, `shape`, controls which basic shape will be used to create the element (Table 10-4), while `ksize` and `anchor` specify the size of the element and the location of the anchor point, respectively. As usual, if the `anchor` argument is left with its default value of `cv::Point(-1,-1)`, then `cv::getStructuringElement()` will take this to mean that the anchor should automatically be placed at the center of the element.

*Table 10-4. cv::getStructuringElement() element shapes*

| Value of shape | Element | Description |
|---|---|---|
| `cv::MORPH_RECT` | Rectangular | $E_{i,j} = 1, \forall\, i, j$ |
| `cv::MORPH_ELLIPSE` | Elliptic | Ellipse with axes `ksize.width` and `ksize.height`. |
| `cv::MORPH_CROSS` | Cross-shaped | $E_{i,j} = 1$, iff $i == anchor.y$ or $j == anchor.x$ |

Of the options for the shapes shown in Table 10-4, the last is there only for legacy compatibility. In the old C API (v1.x), there was a separate struct used for the purpose of expressing convolution kernels. There is no need to use this functionality now, as you can simply pass any `cv::Mat` to the morphological operators as a structuring element if you need something more complicated than the basic shape-based elements created by `cv::getStructuringElement()`.

# Convolution with an Arbitrary Linear Filter

In the functions we have seen so far, the basic mechanics of the convolution were happening deep down below the level of the OpenCV API. We took some time to understand the basics of convolution, and then went on to look at a long list of functions that implemented different kinds of useful convolutions. In essentially every case, there was a kernel that was implied by the function we chose, and we just passed that function a little extra information that parameterized that particular filter type. For linear filters, however, it is possible to just provide the entire kernel and let OpenCV handle the convolution for us.

From an abstract point of view, this is very straightforward: we just need a function that takes an array argument to describe the kernel and we are done. At a practical level, there is an important subtlety that strongly affects performance. That subtlety is that some kernels are *separable*, and others are not.

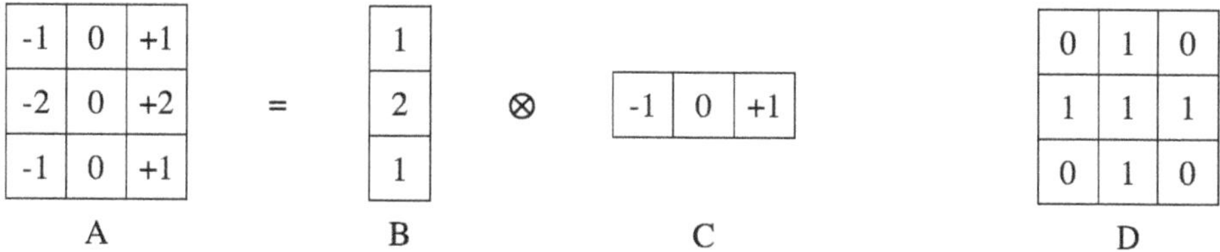


*Figure 10-36. The Sobel kernel (A) is separable; it can be expressed as two one-dimensional convolutions (B and C); D is an example of a nonseparable kernel*

A separable kernel is one that can be thought of as two one-dimensional kernels, which we apply by first convolving with the x-kernel and then with the y-kernel. The benefit of this decomposition is that the computational cost of a kernel convolution is approximately the image area multiplied by the kernel area.[19] This means that convolving your image of area $A$ by an $n \times n$ kernel takes time proportional to $An^2$, while convolving your image once by an $n \times 1$ kernel and then by a $1 \times n$ kernel takes time proportional to $An + An = 2An$. For even $n$ as small as 3 there is a benefit, and the benefit grows with $n$.

## Applying a General Filter with cv::filter2D()

Given that the number of operations required for an image convolution, at least at first glance,[20] seems to be the number of pixels in the image multiplied by the number of pixels in the kernel, this can be a lot of computation and so is not something you want to do with some `for` loop and a lot of pointer dereferencing. In situations like this, it is better to let OpenCV do the work for you and take advantage of the internal optimizations. The OpenCV way to do all of this is with `cv::filter2D()`:

19 This statement is only exactly true for convolution in the spatial domain, which is how OpenCV handles only small kernels.

20 We say "at first glance" because it is also possible to perform convolutions in the frequency domain. In this case, for an $n \times n$ image and an $m \times m$ kernel with $n >> m$, the computational time will be proportional to $n^2\log(n)$ and not to the $n^2m^2$ that is expected for computations in the spatial domain. Because the frequency domain computation is independent of the size of the kernel, it is more efficient for large kernels. OpenCV automatically decides whether to do the convolution in the frequency domain based on the size of the kernel.

```
cv::filter2D(
  cv::InputArray  src,                                // Input image
  cv::OutputArray dst,                                // Result image
  int             ddepth,                             // Output depth (e.g., CV_8U)
  cv::InputArray  kernel,                             // Your own kernel
  cv::Point       anchor     = cv::Point(-1,-1),      // Location of anchor point
  double          delta      = 0,                     // Offset before assignment
  int             borderType = cv::BORDER_DEFAULT     // Border extrapolation to use
);
```

Here we create an array of the appropriate size, fill it with the coefficients of our linear filter, and then pass it together with the source and destination images into `cv::filter2D()`. As usual, we can specify the depth of the resulting image with `ddepth`, the anchor point for the filter with `anchor`, and the border extrapolation method with `borderType`. The kernel can be of even size if its anchor point is defined; otherwise, it should be of odd size. If you want an overall offset applied to the result after the linear filter is applied, you can use the argument `delta`.

## Applying a General Separable Filter with cv::sepFilter2D

In the case where your kernel is separable, you will get the best performance from OpenCV by expressing it in its separated form and passing those one-dimensional kernels to OpenCV (e.g., passing the kernels shown back in Figures 10-36(B) and 10-36(C) instead of the one shown in Figure 10-36(A). The OpenCV function `cv::sepFilter2D()` is like `cv::filter2D()`, except that it expects these two one-dimensional kernels instead of one two-dimensional kernel.

```
cv::sepFilter2D(
  cv::InputArray  src,                                // Input image
  cv::OutputArray dst,                                // Result image
  int             ddepth,                             // Output depth (e.g., CV_8U)
  cv::InputArray  rowKernel,                          // 1-by-N row kernel
  cv::InputArray  columnKernel,                       // M-by-1 column kernel
  cv::Point       anchor     = cv::Point(-1,-1),      // Location of anchor point
  double          delta      = 0,                     // Offset before assignment
  int             borderType = cv::BORDER_DEFAULT     // Border extrapolation to use
);
```

All the arguments of `cv::sepFilter2D()` are the same as those of `cv::filter2D()`, with the exception of the replacement of the `kernel` argument with the `rowKernel` and `columnKernel` arguments. The latter two are expected to be $n_1 \times 1$ and $1 \times n_2$ arrays (with $n_1$ not necessarily equal to $n_2$).

## Kernel Builders

The following functions can be used to obtain popular kernels: `cv::getDerivKernel()`, which constructs the Sobel and Scharr kernels, and `cv::getGaussianKernel()`, which constructs Gaussian kernels.

### cv::getDerivKernel()

The actual kernel array for a derivative filter is generated by `cv::getDerivKernel()`.

```
void cv::getDerivKernels(
  cv::OutputArray kx,
  cv::OutputArray ky,
  int             dx,                  // order of corresponding derivative in x
  int             dy,                  // order of corresponding derivative in y
  int             ksize,               // Kernel size
  bool            normalize = true,    // If true, divide by box area
  int             ktype     = CV_32F   // Type for filter coefficients
);
```

The result of `cv::getDerivKernel()` is placed in the `kx` and `ky` array arguments. You might recall that the derivative type kernels (Sobel and Scharr) are separable kernels. For this reason, you will get back two arrays, one that is 1 × `ksize` (row coefficients, `kx`) and another that is `ksize` × 1 (column coefficients, `ky`). These are computed from the x- and y-derivative orders `dx` and `dy`. The derivative kernels are always square, so the size argument `ksize` is an integer. `ksize` can be any of `1`, `3`, `5`, `7`, or `cv::SCHARR`. The `normalize` argument tells `cv::getDerivKernels()` if it should normalize the kernel elements "correctly." For situations where you are operating on floating-point images, there is no reason not to set `normalize` to `true`, but when you are doing operations on integer arrays, it is often more sensible to not normalize the arrays until some later point in your processing, so you won't throw away precision that you will later need.[21] The final argument, `ktype`, indicates the type of the filter coefficients (or, equivalently, the type of the arrays `kx` and `ky`). The value of `ktype` can be either `CV_32F` or `CV_64F`.

### cv::getGaussianKernel()

The actual kernel array for a Gaussian filter is generated by `cv::getGaussianKernel()`.

```
cv::Mat cv::getGaussianKernel(
  int             ksize,               // Kernel size
  double          sigma,               // Gaussian half-width
  int             ktype = CV_32F       // Type for filter coefficients
);
```

As with the derivative kernel, the Gaussian kernel is separable. For this reason, `cv::getGaussianKernel()` computes only a `ksize` × 1 array of coefficients. The value of `ksize` can be any odd positive number. The argument `sigma` sets the stan-

21 If you are, in fact, going to do this, you will need that normalization coefficient at some point. The normalization coefficient you will need is: $2^{ksize*2-dx-dy-2}$.

dard deviation of the approximated Gaussian distribution. The coefficients are computed from sigma according to the following function:

$$k_i = \alpha \cdot e^{-\frac{(i-(ksize-1)2)^2}{(2\sigma)^2}}$$

That is, the coefficient alpha is computed such that the filter overall is normalized. `sigma` may be set to `-1`, in which case the value of sigma will be automatically computed from the size `ksize`.[22]

## Summary

In this chapter, we learned about general image convolution, including the importance of how boundaries are handled in convolutions. We also learned about image kernels, and the difference between linear and nonlinear kernels. Finally, we learned how OpenCV implements a number of common image filters, and what those filters do to different kinds of input data.

## Exercises

1. Load an image with interesting textures. Smooth the image in several ways using `cv::smooth()` with `smoothtype=cv::GAUSSIAN`.
   a. Use a symmetric 3 × 3, 5 × 5, 9 × 9, and 11 × 11 smoothing window size and display the results.
   b. Are the output results nearly the same by smoothing the image twice with a 5 × 5 Gaussian filter as when you smooth once with two 11 × 11 filters? Why or why not?
2. Create a 100 × 100 single-channel image. Set all pixels to `0`. Finally, set the center pixel equal to `255`.
   a. Smooth this image with a 5 × 5 Gaussian filter and display the results. What did you find?
   b. Do this again but with a 9 × 9 Gaussian filter.
   c. What does it look like if you start over and smooth the image twice with the 5 × 5 filter? Compare this with the 9 × 9 results. Are they nearly the same? Why or why not?

22 In this case, $\sigma = 0.3 \cdot \left(\frac{ksize - 1}{2} - 1\right) + 0.8$.

3. Load an interesting image, and then blur it with `cv::smooth()` using a Gaussian filter.

   a. Set `param1=param2=9`. Try several settings of `param3` (e.g., `1`, `4`, and `6`). Display the results.

   b. Set `param1=param2=0` before setting `param3` to `1`, `4`, and `6`. Display the results. Are they different? Why?

   c. Use `param1=param2=0` again, but this time set `param3=1` and `param4=9`. Smooth the picture and display the results.

   d. Repeat Exercise 3c but with `param3=9` and `param4=1`. Display the results.

   e. Now smooth the image once with the settings of Exercise 3c and once with the settings of Exercise 3d. Display the results.

   f. Compare the results in Exercise 3e with smoothings that use `param3=param4=9` and `param3=param4=0` (i.e., a 9 × 9 filter). Are the results the same? Why or why not?

4. Use a camera to take two pictures of the same scene while moving the camera as little as possible. Load these images into the computer as `src1` and `src1`.

   a. Take the absolute value of `src1` minus `src1` (subtract the images); call it `diff12` and display. If this were done perfectly, `diff12` would be black. Why isn't it?

   b. Create `cleandiff` by using `cv::erode()` and then `cv::dilate()` on `diff12`. Display the results.

   c. Create `dirtydiff` by using `cv::dilate()` and then `cv::erode()` on `diff12` and then display.

   d. Explain the difference between `cleandiff` and `dirtydiff`.

5. Create an outline of an object. Take a picture of a scene. Then, without moving the camera, put a coffee cup in the scene and take a second picture. Load these images and convert both to 8-bit grayscale images.

   a. Take the absolute value of their difference. Display the result, which should look like a noisy mask of a coffee mug.

   b. Do a binary threshold of the resulting image using a level that preserves most of the coffee mug but removes some of the noise. Display the result. The "on" values should be set to `255`.

   c. Do a `cv::MOP_OPEN` on the image to further clean up noise.

   d. Using the erosion operator and logical XOR function, turn the mask of the coffee cup image into an outline of the coffee cup (only the edge pixels remaining).

6. High dynamic range: go into a room with strong overhead lighting and tables that shade the light. Take a picture. With most cameras, either the lighted parts of the scene are well exposed and the parts in shadow are too dark, or the lighted parts are overexposed and the shadowed parts are OK. Create an adaptive filter to help balance out such an image; that is, in regions that are dark on average, boost the pixels up some, and in regions that are very light on average, decrease the pixels somewhat.
7. Sky filter: create an adaptive "sky" filter that smooths only bluish regions of a scene so that only the sky or lake regions of a scene are smoothed, not ground regions.
8. Create a clean mask from noise. After completing Exercise 5, continue by keeping only the largest remaining shape in the image. Set a pointer to the upper left of the image and then traverse the image. When you find a pixel of value `255` ("on"), store the location and then flood-fill it using a value of `100`. Read the connected component returned from flood fill and record the area of filled region. If there is another larger region in the image, then flood-fill the smaller region using a value of `0` and delete its recorded area. If the new region is larger than the previous region, then flood-fill the previous region using the value `0` and delete its location. Finally, fill the remaining largest region with `255`. Display the results. We now have a single, solid mask for the coffee mug.
9. Use the mask created in Exercise 8 or create another mask of your own (perhaps by drawing a digital picture, or simply use a square). Load an outdoor scene. Now use this mask with `copyTo()` to copy an image of a mug into the scene.
10. Create a low-variance random image (use a random number call such that the numbers don't differ by much more than three and most numbers are near zero). Load the image into a drawing program such as PowerPoint, and then draw a wheel of lines meeting at a single point. Use bilateral filtering on the resulting image and explain the results.
11. Load an image of a scene and convert it to grayscale.
    a. Run the morphological Top Hat operation on your image and display the results.
    b. Convert the resulting image into an 8-bit mask.
    c. Copy a grayscale value into the original image where the Top Hat mask (from Part b of this exercise) is nonzero. Display the results.
12. Load an image with many details.
    a. Use `resize()` to reduce the image by a factor of 2 in each dimension (hence the image will be reduced by a factor of 4). Do this three times and display the results.

b. Now take the original image and use `cv::pyrDown()` to reduce it three times, and then display the results.

c. How are the two results different? Why are the approaches different?

13. Load an image of an interesting or sufficiently "rich" scene. Using `cv::threshold()`, set the threshold to `128`. Use each setting type in Figure 10-4 on the image and display the results. You should familiarize yourself with thresholding functions because they will prove quite useful.

    a. Repeat the exercise but use `cv::adaptiveThreshold()` instead. Set `param1=5`.

    b. Repeat part a of this exercise using `param1=0` and then `param1=-5`.

14. Approximate a bilateral (edge preserving) smoothing filter. Find the major edges in an image and hold these aside. Then use `cv::pyrMeanShiftFiltering()` to segment the image into regions. Smooth each of these regions separately and then alpha-blend these smooth regions together with the edge image into one whole image that smooths regions but preserves the edges.

15. Use `cv::filter2D()` to create a filter that detects only 60-degree lines in an image. Display the results on a sufficiently interesting image scene.

16. Separable kernels: create a 3 × 3 Gaussian kernel using rows [(1/16, 2/16, 1/16), (2/16, 4/16, 2/16), (1/16, 2/16, 1/16)] and with anchor point in the middle.

    a. Run this kernel on an image and display the results.

    b. Now create two one-dimensional kernels with anchors in the center: one going "across" (1/4, 2/4, 1/4), and one going down (1/4, 2/4, 1/4). Load the same original image and use `cv::filter2D()` to convolve the image twice, once with the first 1D kernel and once with the second 1D kernel. Describe the results.

    c. Describe the order of complexity (number of operations) for the kernel in part a and for the kernels in part b. The difference is the advantage of being able to use separable kernels and the entire Gaussian class of filters—or any linearly decomposable filter that is separable, since convolution is a linear operation.

17. Can you make a separable kernel from the Scharr filter shown in Figure 10-15? If so, show what it looks like.

18. In a drawing program such as PowerPoint, draw a series of concentric circles forming a bull's-eye.

    a. Make a series of lines going into the bull's-eye. Save the image.

    b. Using a 3 × 3 aperture size, take and display the first-order x- and y-derivatives of your picture. Then increase the aperture size to 5 × 5, 9 × 9, and 13 × 13. Describe the results.

19. Create a new image that is just a 45-degree line, white on black. For a given series of aperture sizes, we will take the image's first-order x-derivative ($dx$) and first-order y-derivative ($dy$). We will then take measurements of this line as follows. The ($dx$) and ($dy$) images constitute the gradient of the input image. The magnitude at location $(i, j)$ is $mag(i, j) = \sqrt{dx^2(i, j) + dy^2(i, j)}$ and the angle is $\Theta(i, j) = atan2(dy(i, j), dx(i, j))$. Scan across the image and find places where the magnitude is at or near maximum. Record the angle at these places. Average the angles and report that as the measured line angle.

    a. Do this for a 3 × 3 aperture Sobel filter.

    b. Do this for a 5 × 5 filter.

    c. Do this for a 9 × 9 filter.

    d. Do the results change? If so, why?

CHAPTER 11

# General Image Transforms

## Overview

In the previous chapters, we covered the class of image transformations that can be understood specifically in terms of convolution. Of course, there are a lot of useful operations that cannot be expressed in this way (i.e., as a little window scanning over the image doing one thing or another). In general, transformations that can be expressed as convolutions are local, meaning that even though they may change the entire image, the effect on any particular pixel is determined by only a small number of pixels around it. The transforms we will look at in this chapter generally will not have this property.

Some very useful *image transforms* are simple, and you will use them all the time—*resize*, for example. Others are for more specialized purposes. The image transforms we will look at in this chapter convert one image into another. The output image will often be a different size as the input, or will differ in other ways, but it will still be in essence "a picture" in the same sense as the input. In Chapter 12, we will consider operations that render images into some potentially entirely different representation.

There are a number of useful transforms that arise repeatedly in computer vision. OpenCV provides complete implementations of some of the more common ones as well as building blocks to help you implement your own, more complex, transformations.

## Stretch, Shrink, Warp, and Rotate

The simplest image transforms we will encounter are those that resize an image, either to make it larger or smaller. These operations are a little less trivial than you might think, because resizing immediately implies questions about how pixels are interpolated (for enlargement) or merged (for reduction).

## Uniform Resize

We often encounter an image of some size that we would like to convert to some other size. We may want to upsize or downsize the image; both of these tasks are accomplished by the same function.

### cv::resize()

The `cv::resize()` function handles all of these resizing needs. We provide our input image and the size we would like it converted to, and it will generate a new image of the desired size.

```
void cv::resize(
  cv::InputArray  src,                                  // Input image
  cv::OutputArray dst,                                  // Result image
  cv::Size        dsize,                                // New size
  double          fx            = 0,                    // x-rescale
  double          fy            = 0,                    // y-rescale
  int             interpolation = CV::INTER_LINEAR // interpolation method
);
```

We can specify the size of the output image in two ways. One way is to use *absolute sizing*; in this case, the `dsize` argument directly sets the size we would like the result image `dst` to be. The other option is to use *relative sizing*; in this case, we set `dsize` to `cv::Size(0,0)`, and set `fx` and `fy` to the scale factors we would like to apply to the x- and y-axes, respectively.[1] The last argument is the `interpolation` method, which defaults to linear interpolation. The other available options are shown in Table 11-1.

*Table 11-1. cv::resize() interpolation options*

| Interpolation | Meaning |
|---|---|
| `cv::INTER_NEAREST` | Nearest neighbor |
| `cv::INTER_LINEAR` | Bilinear |
| `cv::INTER_AREA` | Pixel area resampling |
| `cv::INTER_CUBIC` | Bicubic interpolation |
| `cv::INTER_LANCZOS4` | Lanczos interpolation over 8 × 8 neighborhood. |

Interpolation is an important issue here. Pixels in the source image sit on an integer grid; for example, we can refer to a pixel at location (20, 17). When these integer locations are mapped to a new image, there can be gaps—either because the integer

1 Either `dsize` must be `cv::Size(0,0)` or `fx` and `fy` must both be `0`.

source pixel locations are mapped to float locations in the destination image and must be rounded to the nearest integer pixel location, or because there are some locations to which no pixels are mapped (think about doubling the image size by stretching it; then every other destination pixel would be left blank). These problems are generally referred to as *forward projection* problems. To deal with such rounding problems and destination gaps, we actually solve the problem backward: we step through each pixel of the destination image and ask, "Which pixels in the source are needed to fill in this destination pixel?" These source pixels will almost always be on fractional pixel locations, so we must interpolate the source pixels to derive the correct value for our destination value. The default method is bilinear interpolation, but you may choose other methods (as shown in Table 11-1).

The easiest approach is to take the resized pixel's value from its closest pixel in the source image; this is the effect of choosing the `interpolation` value `cv::INTER_NEAREST`. Alternatively, we can linearly weight the 2 × 2 surrounding source pixel values according to how close they are to the destination pixel, which is what `cv::INTER_LINEAR` does. We can also virtually place the new, resized pixel over the old pixels and then average the covered pixel values, as done with `cv::INTER_AREA`.[2] For yet smoother interpolation, we have the option of fitting a cubic spline between the 4 × 4 surrounding pixels in the source image and then reading off the corresponding destination value from the fitted spline; this is the result of choosing the `cv::INTER_CUBIC` interpolation method. Finally, we have the Lanczos interpolation, which is similar to the cubic method, but uses information from an 8 × 8 area around the pixel.[3]

It is important to notice the difference between `cv::resize()` and the similarly named `cv::Mat::resize()` member function of the `cv::Mat` class. `cv::resize()` creates a new image of a different size, over which the original pixels are mapped. The `cv::Mat::resize()` member function resizes the image whose member you are calling, and crops that image to the new size. Pixels are not interpolated (or extrapolated) in the case of `cv::Mat::resize()`.

2 At least that's what happens when `cv::resize()` shrinks an image. When it expands an image, `cv::INTER_AREA` amounts to the same thing as `cv::INTER_NEAREST`.

3 The subtleties of the Lanczos filter are beyond the scope of this book, but this filter is commonly used in processing digital images because it has the effect of increasing the *perceived* sharpness of the image.

## Image Pyramids

Image pyramids [Adelson84] are heavily used in a wide variety of vision applications. An image pyramid is a collection of images—all arising from a single original image —that are successively downsampled until some desired stopping point is reached. (This stopping point could be a single-pixel image!)

There are two kinds of image pyramids that arise often in the literature and in applications: the Gaussian [Rosenfeld80] and Laplacian [Burt83] pyramids [Adelson84]. The Gaussian pyramid is used to downsample images, and the Laplacian pyramid (discussed shortly) is required when we want to reconstruct an upsampled image from an image lower in the pyramid.

### cv::pyrDown()

Normally, we produce layer $(i + 1)$ in the Gaussian pyramid (we denote this layer $G_{i+1}$) from layer $G_i$ of the pyramid, by first convolving $G_i$ with a Gaussian kernel and then removing every even-numbered row and column. Of course, in this case, it follows that each image is exactly one-quarter the area of its predecessor. Iterating this process on the input image $G_0$ produces the entire pyramid. OpenCV provides us with a method for generating each pyramid stage from its predecessor:

```
void cv::pyrDown(
  cv::InputArray  src,                       // Input image
  cv::OutputArray dst,                       // Result image
  const cv::Size& dstsize = cv::Size()       // Output image size
);
```

The `cv::pyrDown()` method will do exactly this for us if we leave the destination size argument `dstsize` set to its default value of `cv::Size()`. To be a little more specific, the default size of the output image is `( (src.cols+1)/2, (src.rows+1)/2 )`.[4] Alternatively, we can supply a `dstsize`, which will indicate the size we would like for the output image; `dstsize`, however, must obey some very strict constraints. Specifically:

$$| \ dstsize.width*2 - src.cols \ | \ \leq 2$$

$$| \ dstsize.height*2 - src.rows \ | \ \leq 2$$

This restriction means that the destination image is *very close to* half the size of the source image. The `dstsize` argument is used only for handling somewhat esoteric cases in which very tight control is needed on how the pyramid is constructed.

4 The +1s are there to make sure odd-sized images are handled correctly. They have no effect if the image was even sized to begin with.

### cv::buildPyramid()

It is a relatively common situation that you have an image and wish to build a sequence of new images that are each downscaled from their predecessor. The function `cv::buildPyramid()` creates such a stack of images for you in a single call.

```
void cv::buildPyramid(
  cv::InputArray            src,           // Input image
  cv::OutputArrayOfArrays dst,             // Output images from pyramid
  int                       maxlevel       // Number of pyramid levels
);
```

The argument `src` is the source image. The argument `dst` is of a somewhat unusual-looking type `cv::OutputArrayOfArrays`, but you can think of this as just being an STL `vector<>` of objects of type `cv::OutputArray`. The most common example of this would be `vector<cv::Mat>`. The argument `maxlevel` indicates how many pyramid levels are to be constructed.

The argument `maxlevel` is any integer greater than or equal to `0`, and indicates the number of pyramid images to be generated. When `cv::buildPyramid()` runs, it will return a vector in `dst` that is of length `maxlevel+1`. The first entry in `dst` will be identical to `src`. The second will be half as large—that is, as would result from calling `cv::pyrDown()`. The third will be half the size of the second, and so on (see the left-hand image in of Figure 11-1).

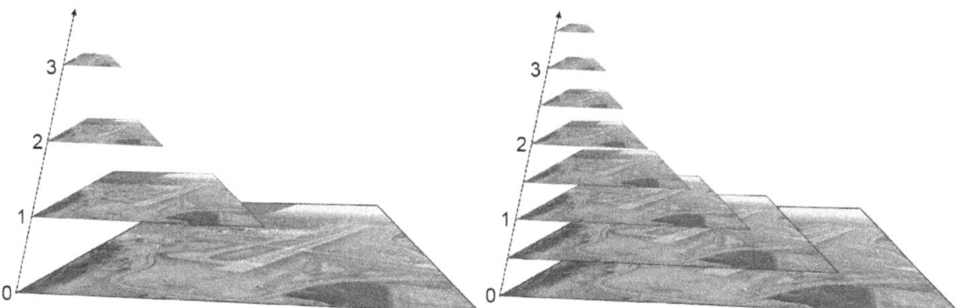


*Figure 11-1. An image pyramid generated with maxlevel=3 (left); two pyramids interleaved together to create a $\sqrt{2}$ pyramid (right)*

In practice, you will often want a pyramid with a finer logarithmic scaling than factors of two. One way to achieve this is to simply call `cv::resize()` yourself as many times as needed for whatever scale factor you want to use—but this can be quite slow. An alternative (for some common scale factors) is to call `cv::resize()` only once for each interleaved set of images you want, and then call `cv::buildPyramid()` on each of those resized "bases." You can then interleave these results together for one large, finer-grained pyramid. Figure 11-1 (right) shows an example in which two pyramids are generated. The original image is first rescaled by a factor of $\sqrt{2}$, and then `cv::buildPyramid()` is called on that one image to make a second pyramid of four intermediate images. Once combined with the original pyramid, the result is a finer pyramid with a scale factor of $\sqrt{2}$ across the entire pyramid.

### cv::pyrUp()

Similarly, we can convert an existing image to an image that is twice as large in each direction by the following analogous (but not inverse!) operation:

```
void cv::pyrUp(
  cv::InputArray  src,                        // Input image
  cv::OutputArray dst,                        // Result image
  const cv::Size& dstsize = cv::Size()        // Output image size
);
```

In this case, the image is first upsized to twice the original in each dimension, with the new (even) rows filled with 0s. Thereafter, a convolution is performed with the Gaussian filter[5] to approximate the values of the "missing" pixels.

Analogous to `cv::PyrDown()`, if `dstsize` is set to its default value of `cv::Size()`, the resulting image will be exactly twice the size (in each dimension) as `src`. Again, we can supply a `dstsize` that will indicate the size we would like for the output image `dstsize`, but it must again obey some very strict constraints. Specifically:

$$|\, dstsize.width*2 - src.cols \,| \leq (dstsize.width\%2)$$
$$|\, dstsize.height*2 - src.rows \,| \leq (dstsize.height\%2)$$

5 This filter is also normalized to four, rather than to one. This is appropriate because the inserted rows have 0s in all of their pixels before the convolution. (Normally, the sum of Gaussian kernel elements would be 1, but in case of 2x pyramid upsampling—in the 2D case—all the kernel elements are multiplied by 4 to recover the average brightness after the zero rows and columns are inserted.)

This restriction means that the destination image is *very close to* double the size of the source image. As before, the `dstsize` argument is used only for handling somewhat esoteric cases in which very tight control is needed over how the pyramid is constructed.

### The Laplacian pyramid

We noted previously that the operator `cv::pyrUp()` is not the inverse of `cv::pyrDown()`. This should be evident because `cv::pyrDown()` is an operator that loses information. In order to restore the original (higher-resolution) image, we would require access to the information that was discarded by the downsampling process. This data forms the *Laplacian pyramid*. The *i*th layer of the Laplacian pyramid is defined by the relation:

$$L_i = G_i - UP(G_{i+1}) \otimes g_{5x5}$$

Here the operator `UP()` upsizes by mapping each pixel in location $(x, y)$ in the original image to pixel $(2x + 1, 2y + 1)$ in the destination image; the $\otimes$ symbol denotes convolution; and $g_{5\times5}$ is a $5 \times 5$ Gaussian kernel. Of course, $UP(G_{i+1}) \otimes g_{5\times5}$ is the definition of the `cv::pyrUp()` operator provided by OpenCV. Hence, we can use OpenCV to compute the Laplacian operator directly as:

$$L_i = G_i - pyrUp(G_{i+1})$$

The Gaussian and Laplacian pyramids are shown diagrammatically in Figure 11-2, which also shows the inverse process for recovering the original image from the subimages. Note how the Laplacian is really an approximation that uses the difference of Gaussians, as revealed in the preceding equation and diagrammed in the figure.

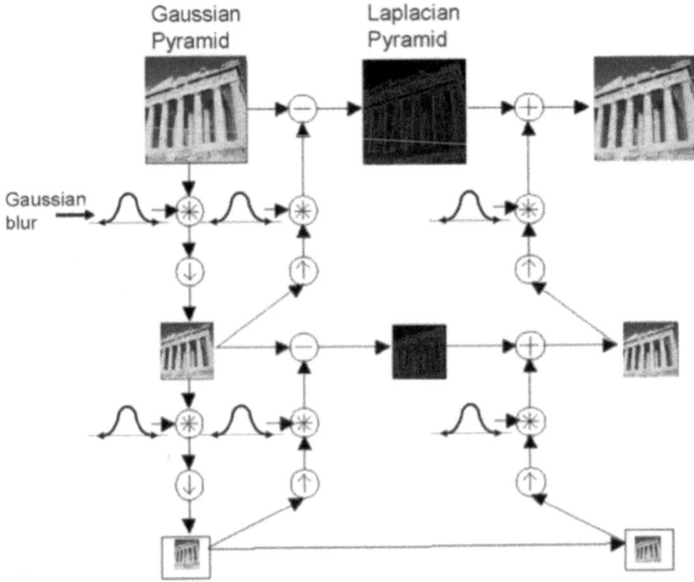


*Figure 11-2. The Gaussian pyramid and its inverse, the Laplacian pyramid*

## Nonuniform Mappings

In this section, we turn to *geometric* manipulations of images—that is, those transformations that have their origin at the intersection of three-dimensional geometry and projective geometry.[6] Such manipulations include both uniform and nonuniform resizing (the latter is known as *warping*). There are many reasons to perform these operations: for example, warping and rotating an image so that it can be superimposed on a wall in an existing scene or artificially enlarging a set of training images used for object recognition.[7] The functions that can stretch, shrink, warp, and/or rotate an image are called *geometric transforms* (for an early exposition, see [Semple79]). For planar areas, there are two flavors of geometric transforms: transforms that use a 2 × 3 matrix, which are called *affine transforms*; and transforms based on a 3 × 3 matrix, which are called *perspective transforms* or *homographies*. You can think of the latter transformation as a method for computing the way in which a plane in three dimensions is perceived by a particular observer, who might not be looking at that plane straight on.

6 We will cover these transformations in detail here, and will return to them in Chapter 19 when we discuss how they can be used in the context of three-dimensional vision techniques.

7 This activity might seem a bit dodgy; after all, wouldn't it be better to just use a recognition method that's invariant to local affine distortions? Nonetheless, this method has a long history and is quite useful in practice.

An affine transformation is any transformation that can be expressed in the form of a matrix multiplication followed by a vector addition. In OpenCV, the standard style of representing such a transformation is as a 2 × 3 matrix. We define:

$$A \equiv \begin{bmatrix} a_{00} & a_{01} \\ a_{10} & a_{11} \end{bmatrix} B \equiv \begin{bmatrix} b_0 \\ b_1 \end{bmatrix} T \equiv \begin{bmatrix} A & B \end{bmatrix} X \equiv \begin{bmatrix} x \\ y \end{bmatrix} X' \equiv \begin{bmatrix} x \\ y \\ 1 \end{bmatrix}$$

You can easily see that the effect of the affine transformation $A \cdot X + B$ is exactly equivalent to extending the vector $X$ into the vector $X'$ and simply left-multiplying $X'$ by $T$.

Affine transformations can be visualized as follows: Any parallelogram $ABCD$ in a plane can be mapped to any other parallelogram $A'B'C'D'$ by some affine transformation. If the areas of these parallelograms are nonzero, then the implied affine transformation is defined uniquely by (three vertices of) the two parallelograms. If you like, you can think of an affine transformation as drawing your image into a big rubber sheet and then deforming the sheet by pushing or pulling[8] on the corners to make different kinds of parallelograms.

When we have multiple images that we know to be slightly different views of the same object, we might want to compute the actual transforms that relate the different views. In this case, affine transformations are often used, instead of perspective transforms, to model the views because they have fewer parameters and so are easier to solve for. The downside is that true perspective distortions can be modeled only by a *homography*,[9] so affine transforms yield a representation that cannot accommodate all possible relationships between the views. On the other hand, for small changes in viewpoint the resulting distortion is affine, so in some circumstances, an affine transformation may be sufficient.

Affine transforms can convert rectangles to parallelograms. They can squash the shape but must keep the sides parallel; they can rotate it and/or scale it. Perspective transformations offer more flexibility; a perspective transform can turn a rectangle into an arbitrary quadrangle. Figure 11-3 shows schematic examples of various affine and perspective transformations; Figure 11-4, later in this chapter, shows such examples using an image.

---

8 One can even pull in such a manner as to invert the parallelogram.

9 *Homography* is the mathematical term for mapping points on one surface to points on another. In this sense, it is a more general term than used here. In the context of computer vision, homography almost always refers to mapping between points on two image planes that correspond to the same location on a planar object in the real world. Such a mapping is representable by a single 3 × 3 orthogonal matrix (more on this in Chapter 19).

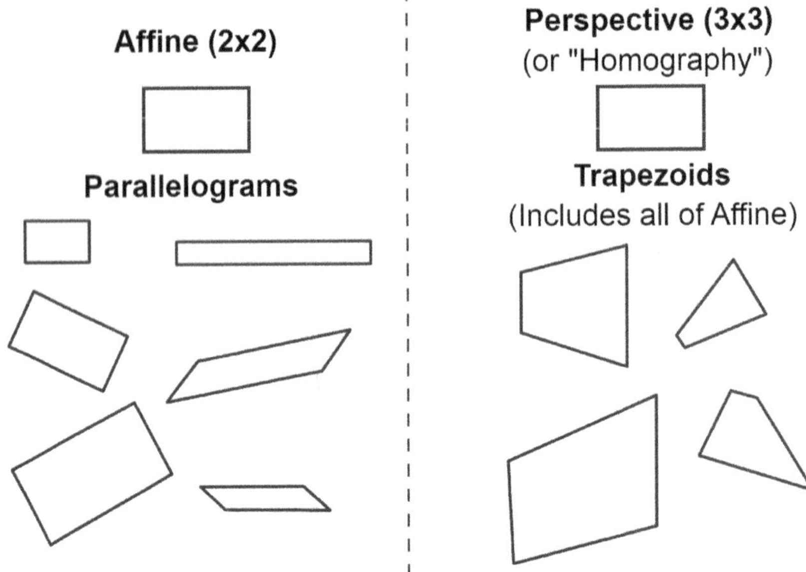


*Figure 11-3. Affine and perspective transformations*

## Affine Transformation

There are two situations that arise with affine transformations. In the first case, we have an image (or a region of interest) we'd like to transform; in the second case, we have a list of points for which we'd like to compute the result of a transformation. These cases are very similar in concept, but quite different in terms of practical implementation. As a result, OpenCV has two different functions for these situations.

### cv::warpAffine(): Dense affine transformations

In the first case, the obvious input and output formats are images, and the implicit requirement is that the warping assumes the pixels are a *dense representation* of the underlying image. This means that image warping must necessarily handle interpolations so that the output images are smooth and look natural. The affine transformation function provided by OpenCV for dense transformations is `cv::warpAffine()`:

```
void cv::warpAffine(
  cv::InputArray    src,                                 // Input image
  cv::OutputArray   dst,                                 // Result image
  cv::InputArray    M,                                   // 2-by-3 transform mtx
  cv::Size          dsize,                               // Destination image size
  int               flags       = cv::INTER_LINEAR,      // Interpolation, inverse
  int               borderMode  = cv::BORDER_CONSTANT,   // Pixel extrapolation
  const cv::Scalar& borderValue = cv::Scalar()           // For constant borders
);
```

Here `src` and `dst` are your source and destination arrays, respectively. The input `M` is the $2 \times 3$ matrix we introduced earlier that quantifies the desired transformation. Each element in the destination array is computed from the element of the source array at the location given by:

$$dst(x, y) = src(M_{00}x + M_{01}y + M_{02}, M_{10}x + M_{11}y + M_{12})$$

In general, however, the location indicated by the righthand side of this equation will not be an integer pixel. In this case, it is necessary to use interpolation to find an appropriate value for *dst*(*x*, *y*). The next argument, `flags`, selects the interpolation method. The available interpolation methods are those in Table 11-1, the same as `cv::resize()`, plus one additional option, `cv::WARP_INVERSE_MAP` (which may be added with the usual Boolean OR). This option is a convenience that allows for inverse warping from `dst` to `src` instead of from `src` to `dst`. The final two arguments are for border extrapolation, and have the same meaning as similar arguments in image convolutions (see Chapter 10).

### cv::getAffineTransform(): Computing an affine map matrix

OpenCV provides two functions to help you generate the map matrix `M`. The first is used when you already have two images that you know to be related by an affine transformation or that you'd like to approximate in that way:

```
cv::Mat cv::getAffineTransform(              // Return 2-by-3 matrix
  const cv::Point2f* src,                    // Coordinates *three* of vertices
  const cv::Point2f* dst                     // Target coords, three vertices
);
```

Here `src` and `dst` are arrays containing three two-dimensional (*x*, *y*) points. The return value is an array that is the affine transform computed from those points.

In essence, the arrays of points in `src` and `dst` in `cv::getAffineTransform()` define two parallelograms. The points in `src` will be mapped by an application `cv::warpAffine()`, using the resulting matrix `M`, to the corresponding points in `dst`; all other points will be dragged along for the ride. Once these three independent corners are mapped, the mapping of all of the other points is completely determined.

Example 11-1 shows some code that uses these functions. In the example, we obtain the `cv::warpAffine()` matrix parameters by first constructing two three-component arrays of points (the corners of our representative parallelogram) and then convert that to the actual transformation matrix using `cv::getAffineTransform()`. We then do an affine warp followed by a rotation of the image. For our array of representative points in the source image, called `srcTri[]`, we take the three points: `(0,0)`, `(0,height-1)`, and `(width-1,0)`. We then specify the locations to which these points will be mapped in the corresponding array `dstTri[]`.

*Example 11-1. An affine transformation*

```
#include <opencv2/opencv.hpp>
#include <iostream>

using namespace std;

int main(int argc, char** argv) {

  if(argc != 2) {
    cout << "Warp affine\nUsage: " <<argv[0] <<" <imagename>\n" << endl;
    return -1;
  }

  cv::Mat src = cv::imread(argv[1],1);
  if( src.empty() ) { cout << "Can not load " << argv[1] << endl; return -1; }

  cv::Point3f srcTri[] = {
     cv::Point2f(0,0),             // src Top left
     cv::Point2f(src.cols-1, 0), // src Top right
     cv::Point2f(0, src.rows-1)  // src Bottom left
  };

  cv::Point2f dstTri[] = {
     cv::Point2f(src.cols*0.f,   src.rows*0.33f), // dst Top left
     cv::Point2f(src.cols*0.85f, src.rows*0.25f), // dst Top right
     cv::Point2f(src.cols*0.15f, src.rows*0.7f)   // dst Bottom left
  };

  // COMPUTE AFFINE MATRIX
  //
  cv::Mat warp_mat = cv::getAffineTransform(srcTri, dstTri);
  cv::Mat dst, dst2;
  cv::warpAffine(
    src,
    dst,
    warp_mat,
    src.size(),
    cv::INTER_LINEAR,
    cv::BORDER_CONSTANT,
    cv::Scalar()
  );
  for( int i = 0; i < 3; ++i )
    cv::circle(dst, dstTri[i], 5, cv::Scalar(255, 0, 255), -1, cv::AA);

  cv::imshow("Affine Transform Test", dst);
  cv::waitKey();

  for(int frame=0;;++frame) {

    // COMPUTE ROTATION MATRIX
    cv::Point2f center(src.cols*0.5f, src.rows*0.5f);
```

```
    double angle = frame*3 % 360, scale = (cos((angle - 60)* cv::PI/180) + 1.05)*0.8;

    cv::Mat rot_mat = cv::getRotationMatrix2D(center, angle, scale);

    cv::warpAffine(
      src,
      dst,
      rot_mat,
      src.size(),
      cv::INTER_LINEAR,
      cv::BORDER_CONSTANT,
      cv::Scalar()
    );
    cv::imshow("Rotated Image", dst);
    if(cv::waitKey(30) >= 0 )
      break;

  }

  return 0;
}
```

The second way to compute the map matrix `M` is to use `cv::getRotationMatrix2D()`, which computes the map matrix for a rotation around some arbitrary point, combined with an optional rescaling. This is just one possible kind of affine transformation, so it is less general than is `cv::getAffineTransform()`, but it represents an important subset that has an alternative (and more intuitive) representation that's easier to work with in your head:

```
cv::Mat cv::getRotationMatrix2D(               // Return 2-by-3 matrix
  cv::Point2f  center                          // Center of rotation
  double       angle,                          // Angle of rotation
  double       scale                           // Rescale after rotation
);
```

The first argument, `center`, is the center point of the rotation. The next two arguments give the magnitude of the rotation and the overall rescaling. The function returns the map matrix `M`, which (as always) is a 2 × 3 matrix of floating-point numbers.

If we define $\alpha$ = scale * cos(angle) and $\beta$ = scale * sin(angle), then this function computes the matrix `M` to be:

$$\begin{bmatrix} \alpha & \beta & (1-\alpha)\cdot center_x - \beta \cdot center_y \\ -\beta & \alpha & \beta \cdot center_x - (1-\alpha)\cdot center_y \end{bmatrix}$$

You can combine these methods of setting the map matrix `M` to obtain, for example, an image that is rotated, scaled, *and* warped.

### cv::transform(): Sparse affine transformations

We have explained that `cv::warpAffine()` is the right way to handle dense mappings. For sparse mappings (i.e., mappings of lists of individual points), it is best to use `cv::transform()`. You will recall from Chapter 5 that the transform method has the following prototype:

```
void cv::transform(
  cv::InputArray  src,                     // Input N-by-1 array (Ds channels)
  cv::OutputArray dst,                     // Output N-by-1 array (Dd channels)
  cv::InputArray  mtx                      // Transform matrix (Ds-by-Dd)
);
```

In general, `src` is an $N \times 1$ array with $D_s$ channels, where $N$ is the number of points to be transformed and $D_s$ is the dimension of those source points. The output array `dst` will be the same size but may have a different number of channels, $D_d$. The transformation matrix `mtx` is a $D_s \times D_d$ matrix that is then applied to every element of `src`, after which the results are placed into `dst`.

Note that `cv::transform()` acts on the channel indices of every point in an array. For the current problem, we assume that the array is essentially a large vector ($N \times 1$ or $1 \times N$) of these multichannel objects. The important thing to remember is that the index that the transformation matrix is relative to is the channel index, not the "vector" index of the large array.

In the case of transformations that are simple rotations in this channel space, our transformation matrix `mtx` will be a $2 \times 2$ matrix only, and it can be applied directly to the two-channel indices of `src`. In fact this is true for rotations, stretching, and warping as well in some simple cases. Usually, however, to do a general affine transformation (including translations and rotations about arbitrary centers, and so on), it is necessary to extend the number of channels in `src` to three, so that the action of the more usual $2 \times 3$ affine transformation matrix is defined. In this case, all of the third-channel entries must be set to `1` (i.e., the points must be supplied in homogeneous coordinates). Of course, the output array will still be a two-channel array.

### cv::invertAffineTransform(): Inverting an affine transformation

Given an affine transformation represented as a $2 \times 3$ matrix, it is often desirable to be able to compute the inverse transformation, which can be used to "put back" all of the transformed points to where they came from. This is done with `cv::invertAffineTransform()`:

```
void cv::invertAffineTransform(
  cv::InputArray  M,                           // Input 2-by-3 matrix
  cv::OutputArray iM                           // Output also a 2-by-3 matrix
);
```

This function takes a 2 × 3 array `M` and returns another 2 × 3 array `iM` that inverts `M`. Note that `cv::invertAffineTransform()` does not actually act on any image, it just supplies the inverse transform. Once you have `iM`, you can use it as you would have used `M`, with either `cv::warpAffine()` or `cv::transform()`.

## Perspective Transformation

To gain the greater flexibility offered by perspective transforms (also called *homographies*), we need a new function that will allow us to express this broader class of transformations. First we remark that, even though a perspective projection is specified completely by a single matrix, the projection is not actually a linear transformation. This is because the transformation requires division by the final dimension (usually *Z*; see Chapter 19) and thus loses a dimension in the process.

As with affine transformations, image operations (dense transformations) are handled by different functions than transformations on point sets (sparse transformations).

### cv::warpPerspective(): Dense perspective transform

The dense perspective transform uses an OpenCV function that is analogous to the one provided for dense affine transformations. Specifically, `cv::warpPerspective()` has all of the same arguments as `cv::warpAffine()`, except with the small, but crucial, distinction that the map matrix must now be 3 × 3.

```
void cv::warpPerspective(
  cv::InputArray    src,                                 // Input image
  cv::OutputArray   dst,                                 // Result image
  cv::InputArray    M,                                   // 3-by-3 transform mtx
  cv::Size          dsize,                               // Destination image size
  int               flags       = cv::INTER_LINEAR,      // Interpolation, inverse
  int               borderMode  = cv::BORDER_CONSTANT,   // Extrapolation method
  const cv::Scalar& borderValue = cv::Scalar()           // For constant borders
);
```

Each element in the destination array is computed from the element of the source array at the location given by:

$$dst(x, y) = src\left(\frac{M_{00}x + M_{01}y + M_{02}}{M_{20}x + M_{21}y + M_{22}}, \frac{M_{10}x + M_{11}y + M_{12}}{M_{20}x + M_{21}y + M_{22}}\right)$$

As with the affine transformation, the location indicated by the right side of this equation will not (generally) be an integer location. Again the `flags` argument is used to select the desired interpolation method, and has the same possible values as the corresponding argument to `cv::warpAffine()`.

### cv::getPerspectiveTransform(): Computing the perspective map matrix

As with the affine transformation, for filling the map matrix in the preceding code, we have a convenience function that can compute the transformation matrix from a list of point correspondences:

```
cv::Mat cv::getPerspectiveTransform(          // Return 3-by-3 matrix
  const cv::Point2f* src,                     // Coordinates of *four* vertices
  const cv::Point2f* dst                      // Target coords, four vertices
);
```

The `src` and `dst` argument are now arrays of four (not three) points, so we can independently control how the corners of (typically) a rectangle in `src` are mapped to (generally) some rhombus in `dst`. Our transformation is completely defined by the specified destinations of the four source points. As mentioned earlier, for perspective transformations, the return value will be a $3 \times 3$ array; see Example 11-2 for sample code. Other than the $3 \times 3$ matrix and the shift from three to four control points, the perspective transformation is otherwise exactly analogous to the affine transformation we already introduced.

*Example 11-2. Code for perspective transformation*

```
#include <opencv2/opencv.hpp>
#include <iostream>

using namespace std;

int main(int argc, char** argv) {

  if(argc != 2) {
    cout << "Perspective Warp\nUsage: " <<argv[0] <<" <imagename>\n" << endl;
    return -1;
  }

  Mat src = cv::imread(argv[1],1);
  if( src.empty() ) { cout << "Can not load " << argv[1] << endl; return -1; }

  cv::Point2f srcQuad[] = {
    cv::Point2f(0,          0),          // src Top left
    cv::Point2f(src.cols-1, 0),          // src Top right
    cv::Point2f(src.cols-1, src.rows-1), // src Bottom right
    cv::Point2f(0,          src.rows-1)  // src Bottom left
  };
```

```
  cv::Point2f dstQuad[] = {
    cv::Point2f(src.cols*0.05f, src.rows*0.33f),
    cv::Point2f(src.cols*0.9f,  src.rows*0.25f),
    cv::Point2f(src.cols*0.8f,  src.rows*0.9f),
    cv::Point2f(src.cols*0.2f,  src.rows*0.7f)
  };

  // COMPUTE PERSPECTIVE MATRIX
  //
  cv::Mat warp_mat = cv::getPerspectiveTransform(srcQuad, dstQuad);
  cv::Mat dst;
  cv::warpPerspective(src, dst, warp_mat, src.size(), cv::INTER_LINEAR,
                      cv::BORDER_CONSTANT, cv::Scalar());

  for( int i = 0; i < 4; i++ )
    cv::circle(dst, dstQuad[i], 5, cv::Scalar(255, 0, 255), -1, cv::AA);

  cv::imshow("Perspective Transform Test", dst);
  cv::waitKey();
  return 0;

}
```

#### cv::perspectiveTransform(): Sparse perspective transformations

`cv::perspectiveTransform()` is a special function that performs perspective transformations on lists of points. Because `cv::transform()` is limited to linear operations, it cannot properly handle perspective transforms. This is because such transformations require division by the third coordinate of the homogeneous representation ($x = f^*X/Z, y = f^*\ Y/Z$). The special function `cv::perspectiveTransform()` takes care of this for us:

```
void cv::perspectiveTransform(
  cv::InputArray  src,                // Input N-by-1 array (2 or 3 channels)
  cv::OutputArray dst,                // Output N-by-1 array (2 or 3 channels)
  cv::InputArray  mtx                 // Transform matrix (3-by-3 or 4-by-4)
);
```

As usual, the `src` and `dst` arguments are, respectively, the array of source points to be transformed and the array of destination points resulting from the transformation. These arrays should be two- or three-channel arrays. The matrix `mtx` can be either a 3 × 3 or a 4 × 4 matrix. If it is 3 × 3, then the projection is from two dimensions to two; if the matrix is 4 × 4, then the projection is from three dimensions to three.

In the current context, we are transforming a set of points in an image to another set of points in an image, which sounds like a mapping from two dimensions to two dimensions. This is not exactly correct, however, because the perspective transformation is actually mapping points on a two-dimensional plane embedded in a three-dimensional space back down to a (different) two-dimensional subspace. Think of this as being just what a camera does. (We will return to this topic in greater detail when discussing cameras in later chapters.) The camera takes points in three dimensions and maps them to the two dimensions of the camera imager. This is essentially what is meant when the source points are taken to be in "homogeneous coordinates." We are adding a dimension to those points by introducing the $Z$ dimension and then setting all of the $Z$ values to 1. The projective transformation is then projecting back out of that space onto the two-dimensional space of our output. This is a rather long-winded way of explaining why, when mapping points in one image to points in another, you will need a $3 \times 3$ matrix.

Outputs of the code in Examples 11-1 and 11-2 are shown in Figure 11-4 for affine and perspective transformations. In these examples, we transform actual images; you can compare these with the simple diagrams back in Figure 11-3.

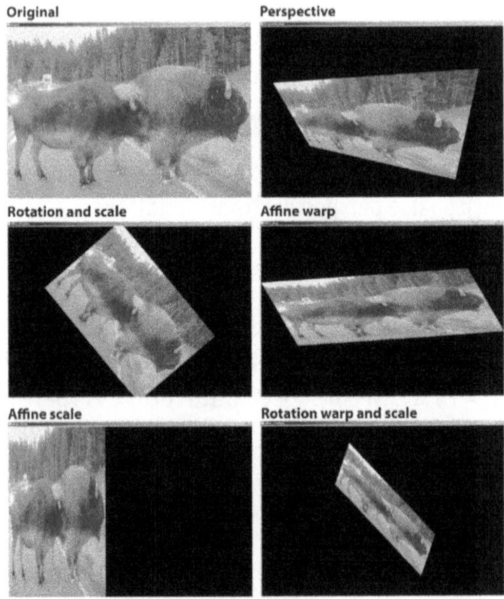


*Figure 11-4. Perspective and affine mapping of an image*

## General Remappings

The affine and perspective transformations we have seen so far are actually specific cases of a more general process. Under the hood, those two transformations both have the same basic behavior: they take pixels from one place in the source image and map them to another place in the destination image. In fact, there are other useful operations that have the same structure. In this section, we will look at another few transformations of this kind, and then look at how OpenCV makes it possible to implement your own general mapping transformations.

## Polar Mappings

In Chapter 5, we briefly encountered two functions, `cv::cartToPolar()` and `cv::polarToCart()`, which could be used to convert arrays of points in an $x$, $y$ Cartesian representation to (or from) arrays of points in an $r\text{–}\theta$ polar representation.

There is a slight style inconsistency here between the polar mapping functions and the perspective and affine transformation functions. The polar mapping functions expect pairs of single-channel arrays, rather than double-channel arrays, as their way of representing two-dimensional vectors. This difference stems from the way the two functions are traditionally used, rather than any intrinsic difference between what they are doing.

The functions `cv::cartToPolar()` and `cv::polarToCart()` are employed by more complex routines such as `cv::logPolar()` (described later) but are also useful in their own right.

### cv::cartToPolar(): Converting from Cartesian to polar coordinates

To map from Cartesian coordinates to polar coordinates, we use the function `cv::cartToPolar()`:

```
void cv::cartToPolar(
  cv::InputArray  x,                        // Input single channel x-array
  cv::InputArray  y,                        // Input single channel y-array
  cv::OutputArray magnitude,                // Output single channel mag-array
  cv::OutputArray angle,                    // Output single channel angle-array
  bool            angleInDegrees = false    // Set true for degrees, else radians
);
```

The first two arguments `x`, and `y`, are single-channel arrays. Conceptually, what is being represented here is not just a list of points, but a *vector field*[10]—with the x-component of the vector field at any given point being represented by the value of the array `x` at that point, and the y-component of the vector field at any given point being represented by the value of the array `y` at that point. Similarly, the result of this function appears in the arrays `magnitude` and `angle`, with each point in `magnitude` representing the length of the vector at that point in `x` and `y`, and each point in `angle` representing the orientation of that vector. The angles recorded in `angle` will, by default, be in radians—that is, $[0, 2\pi)$. If the argument `angleInDegrees` is set to `true`, however, then the angles array will be recorded in degrees [0, 360). Also note that the

10 If you are not familiar with the concept of a vector field, it is sufficient for our purposes to think of this as a two-component vector associated with every point in "image."

angles are computed using (approximately) `atan2(y,x)`, so an angle of 0 corresponds to a vector pointing in the $\hat{x}$ direction.

As an example of where you might use this function, suppose you have already taken the x- and y-derivatives of an image, either by using `cv::Sobel()` or by using convolution functions via `cv::DFT()` or `cv::filter2D()`. If you stored the x-derivatives in an image `dx_img` and the y-derivatives in `dy_img`, you could now create an edge-angle recognition histogram; that is, you could collect all the angles provided the magnitude or strength of the edge pixel is above some desired threshold. To calculate this, we would first create two new destination images (and call them `img_mag` and `img_angle`, for example) for the directional derivatives and then use the function `cv::cartToPolar( dx_img, dy_img, img_mag, img_angle, 1 )`. We would then fill a histogram from `img_angle` as long as the corresponding "pixel" in `img_mag` is above our desired threshold.

In Chapter 22, we will discuss image recognition and image features. This process is actually the basis of how an important image feature used in object recognition, called *HOG* (histogram of oriented gradients), is calculated.

#### cv::polarToCart(): Converting from polar to Cartesian coordinates

The function `cv::cartToPolar()` performs the reverse mapping from polar coordinates to Cartesian coordinates.

```
void cv::polarToCart(
  cv::InputArray  magnitude,              // Output single channel mag-array
  cv::InputArray  angle,                  // Output single channel angle-array
  cv::OutputArray x,                      // Input single channel x-array
  cv::OutputArray y,                      // Input single channel y-array
  bool            angleInDegrees = false  // Set true for degrees, else radians
);
```

The inverse operation is also often useful, allowing us to convert from polar back to Cartesian coordinates. It takes essentially the same arguments as `cv::cartToPolar()`, with the exception that `magnitude` and `angle` are now inputs, and `x` and `y` are now the results.

## LogPolar

For two-dimensional images, the log-polar transform [Schwartz80] is a change from Cartesian to *log-polar* coordinates: $(x, y) \leftrightarrow re^{i\theta}$, where $r = \sqrt{x^2 + y^2}$ and $\theta = atan2(y, x)$. Next, to separate out the polar coordinates into a $(\rho, \theta)$ space that is relative to some center point $(x_c, y_c)$, we take the log so that $\rho = log(\sqrt{(x - x_c)^2 + (y - y_c)^2})$ and $\theta = atan2(y - y_c, x - x_c)$. For image purposes—when we need to "fit" the interesting stuff into the available image memory—we typically apply a scaling factor $m$ to $\rho$. Figure 11-5 shows a square object on the left and its encoding in log-polar space.

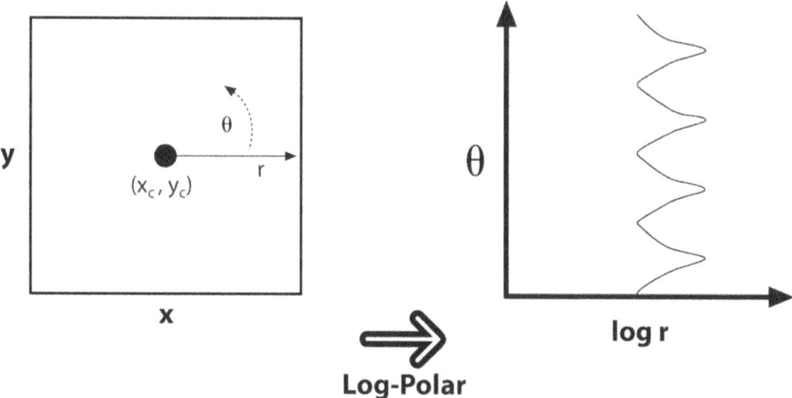


*Figure 11-5. The log-polar transform maps (x, y) into (log(r),θ); here, a square is displayed in the log-polar coordinate system*

You might be wondering why anyone would want to do this. The log-polar transform takes its inspiration from the human visual system. Your eye has a small but dense area of photoreceptors in its center (the *fovea*), and the density of receptors falls off rapidly (exponentially) from there. Try staring at a spot on the wall and holding your finger at arm's length in your line of sight. Then, keep staring at the spot while slowly moving your finger away from your face; note how the detail rapidly decreases as the image of your finger on your retina moves away from your fovea. This structure also has certain nice mathematical properties (beyond the scope of this book) that concern preserving the angles of line intersections.

More important for us is that the log-polar transform can be used to create two-dimensional invariant representations of object views by shifting the transformed image's center of mass to a fixed point in the log-polar plane; see Figure 11-6. On the left are three shapes that we want to recognize as "square." The problem is, they look very different. One is much larger than the others and another is rotated. The log-polar transform appears on the right in Figure 11-6. Observe that size differences in the $(x, y)$ plane are converted to shifts along the $\log(r)$ axis of the log-polar plane and that the rotation differences are converted to shifts along the $\theta$-axis in the log-polar plane. If we take the transformed center of each square in the log-polar plane and then recenter that point to a certain fixed position, then all the squares will show up identically in the log-polar plane. This yields a type of invariance to two-dimensional rotation and scaling.[11]

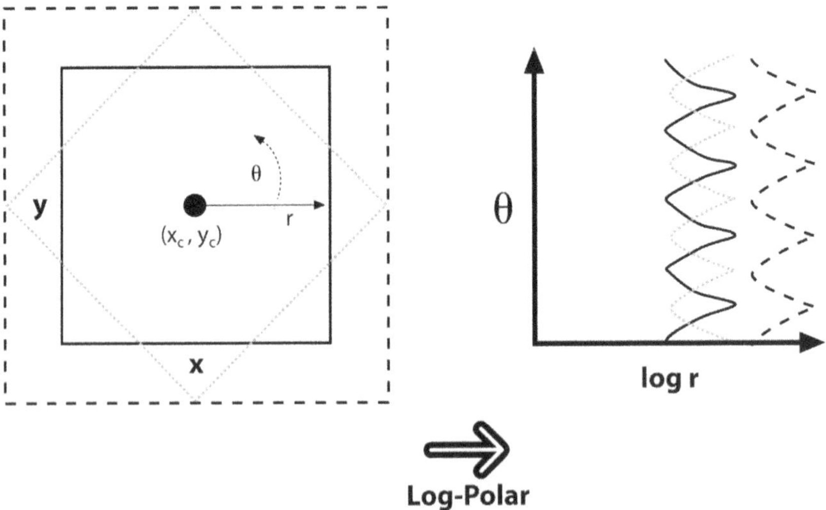


*Figure 11-6. Log-polar transform of rotated and scaled squares: size goes to a shift on the log(r) axis and rotation to a shift on the θ-axis*

## cv::logPolar()

The OpenCV function for a log-polar transform is `cv::logPolar()`:

```
void cv::logPolar(
  cv::InputArray  src,                          // Input image
  cv::OutputArray dst,                          // Output image
  cv::Point2f     center,                       // Center of transform
  double          m,                            // Scale factor
  int             flags = cv::INTER_LINEAR      // interpolation and fill modes
                        | cv::WARP_FILL_OUTLIERS
);
```

The `src` and `dst` are the usual input and output images. The parameter `center` is the center point $(x_c, y_c)$ of the log-polar transform; `m` is the scale factor, which should be set so that the features of interest dominate the available image area. The `flags` parameter allows for different interpolation methods. The interpolation methods are

11 In Chapter 22, we'll learn about recognition. For now, simply note that it wouldn't be a good idea to derive a log-polar transform for a whole object because such transforms are quite sensitive to the exact location of their center points. What is more likely to work for object recognition is to detect a collection of key points (such as corners or blob locations) around an object, truncate the extent of such views, and then use the centers of those key points as log-polar centers. These local log-polar transforms could then be used to create local features that are (partially) scale- and rotation-invariant and that can be associated with a visual object.

the same set of standard interpolations available in OpenCV (see Table 11-1). The interpolation methods can be combined with either or both of the flags `CV::WARP_FILL_OUTLIERS` (to fill points that would otherwise be undefined) or `CV::WARP_INVERSE_MAP` (to compute the reverse mapping from log-polar to Cartesian coordinates).

Sample log-polar coding is given in Example 11-3, which demonstrates the forward and backward (inverse) log-polar transform. The results on a photographic image are shown in Figure 11-7.

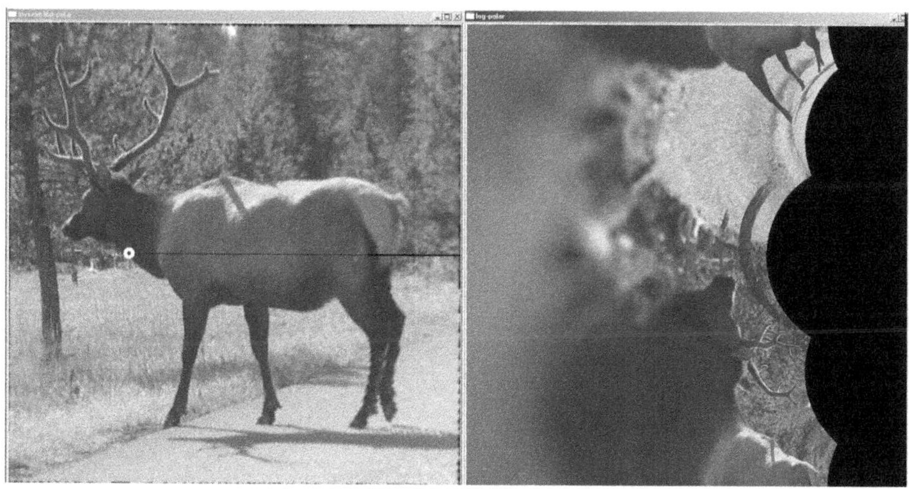

*Figure 11-7. Log-polar example on an elk with transform centered at the white circle on the left; the output is on the right*

*Example 11-3. Log-polar transform example*

```
#include <opencv2/opencv.hpp>
#include <iostream>

using namespace std;

int main(int argc, char** argv) {

  if(argc != 3) {
    cout << "LogPolar\nUsage: " <<argv[0] <<" <imagename> <M value>\n"
      <<"<M value>~30 is usually good enough\n";
    return -1;
  }

  cv::Mat src = cv::imread(argv[1],1);

  if( src.empty() ) { cout << "Can not load " << argv[1] << endl; return -1; }
```

```
  double M = atof(argv[2]);
  cv::Mat dst(src.size(), src.type()), src2(src.size(), src.type());

  cv::logPolar(
    src,
    dst,
    cv::Point2f(src.cols*0.5f, src.rows*0.5f),
    M,
    cv::INTER_LINEAR | cv::WARP_FILL_OUTLIERS
  );
  cv::logPolar(
    dst,
    src2,
    cv::Point2f(src.cols*0.5f, src.rows*0.5f),
    M,
    cv::INTER_LINEAR | cv::WARP_INVERSE_MAP
  );
  cv::imshow( "log-polar", dst );
  cv::imshow( "inverse log-polar", src2 );

  cv::waitKey();

  return 0;
}
```

## Arbitrary Mappings

We sometimes want to accomplish interpolation programmatically; that is, we'd like to apply some known algorithm that will determine the mapping. In other cases, however, we'd like to do this mapping ourselves. Before diving into some methods that will compute (and apply) these mappings for us, let's take a moment to look at the function responsible for applying the mappings that these other methods rely upon.

One common use of `cv::remap()` is to rectify (correct distortions in) calibrated and stereo images. We will see functions in Chapters 18 and 19 that convert calculated camera distortions and alignments into `mapx` and `mapy` parameters.

The OpenCV function we want is called `cv:remap()`.

### cv::remap(): General image remapping

```
void cv::remap(
  cv::InputArray    src,                                // Input image
  cv::OutputArray   dst,                                // Output image
  cv::InputArray    map1,                               // target x for src pix
  cv::InputArray    map2,                               // target y for src pix
  int               interpolation = cv::INTER_LINEAR,   // Interpolation, inverse
  int               borderMode    = cv::BORDER_CONSTANT, // Extrapolation method
```

```
    const cv::Scalar& borderValue   = cv::Scalar()          // For constant borders
  );
```

The first two arguments of `cv::remap()` are the source and destination images, respectively. The next two arguments, `map1` and `map2`, indicate the target *x* and *y* locations, respectively, where any particular pixel is to be relocated. This is how you specify your own general mapping. These should be the same size as the source and destination images, and must be one of the following data types: `CV::S16C2`, `CV::F32C1`, or `CV::F32C2`. Noninteger mappings are allowed: `cv::remap()` will do the interpolation calculations for you automatically.

The next argument, `interpolation`, contains flags that tell `cv::remap()` exactly how that interpolation is to be done. Any one of the values listed in Table 11-1 will work—except for `cv::INTER_AREA`, which is not implemented for `cv::remap()`.

# Image Repair

Images are often corrupted by noise. There may be dust or water spots on a lens, scratches on older images, or parts of an image that were vandalized. *Inpainting* [Telea04] is a method for removing such damage by taking the color and texture at the border of the damaged area and propagating and mixing it inside the damaged area. See Figure 11-8 for an application that involves the removal of writing from an image.

*Figure 11-8. An image damaged by overwritten text (left) is restored by inpainting (right)*

## Inpainting

Inpainting works provided the damaged area is not too "thick" and enough of the original texture and color remains around the boundaries of the damage. Figure 11-9 shows what happens when the damaged area is too large.

*Figure 11-9. Inpainting cannot magically restore textures that are completely removed: the navel of the orange has been completely blotted out (left); inpainting fills it back in with mostly orangelike texture (right)*

The prototype for `cv::inpaint()` is:

```
void cv::inpaint(
  cv::InputArray  src,                // Input image: 8-bit, 1 or 3 channels
  cv::InputArray  inpaintMask,        // 8-bit, 1 channel. Inpaint nonzeros
  cv::OutputArray dst,                // Result image
  double          inpaintRadius,      // Range to consider around pixel
  int             flags               // Select NS or TELEA
);
```

Here `src` is an 8-bit, single-channel, grayscale image or a three-channel color image to be repaired, and `inpaintMask` is an 8-bit, single-channel image of the same size as `src` in which the damaged areas (e.g., the writing seen in the left panel of Figure 11-8) have been marked by nonzero pixels; all other pixels are set to `0` in `inpaintMask`. The output image will be written to `dst`, which must have the same size and number of channels as `src`. The `inpaintRadius` is the area around each inpainted pixel that will be factored into the resulting output color of that pixel. As in Figure 11-9, interior pixels within a thick enough inpainted region may take their color entirely from other inpainted pixels closer to the boundaries. Almost always, one uses a small radius such as 3 because too large a radius will result in a noticeable blur. Finally, the `flags` parameter allows you to experiment with two different methods of inpainting:

`cv::INPAINT_NS` (Navier-Stokes method), and `cv::INPAINT_TELEA` (A. Telea's method).

## Denoising

Another important problem that arises is noise in the image. In many applications, the primary source of noise arises from effects of low-light conditions. In low light, the gain of a digital imager must be increased and the result is that noise is also amplified. The character of this kind of noise is typically random isolated pixels that appear either too bright or too dark, but discoloration is also possible in color images.

The denoising algorithm implemented in OpenCV is called Fast Non-Local Means Denoising (FNLMD), and is based on work by Antoni Buades, Bartomeu Coll, and Jean-Michel Morel [Buades05]. While simple denoising algorithms essentially rely on averaging individual pixels with their neighbors, the central concept of FNLMD is to look for *similar pixels* elsewhere in the image, and average among those. In this context, a pixel is considered to be a similar pixel not because it is similar in color or intensity, but because it is similar in environment. The key logic here is that many images contain repetitive structures, and so even if your pixel is corrupted by noise, there will be many other similar pixels that are not.

The identification of similar pixels proceeds based on a window $B(p, s)$ centered on pixel $p$ and of size $s$. Given such a window around the point we wish to update, we can compare that window with an analogous window around some other pixel $q$. We define the square distance between $B(p, s)$ and $B(q, s)$ to be:

$$d^2(B(p,s), B(q,s)) = \frac{1}{3(2s+1)} \sum_{c=1}^{3} \sum_{j \varepsilon B(0,s)} (I_c(p+j) - I_c(q+j))^2$$

Where $c$ is the color index, $I_c(p)$ is the intensity of the image in channel $c$ at point $p$, and the summation over $j$ is over the elements of the patch. From this square distance, a weight can be assigned to every other pixel relative to the pixel currently being updated. This weight is given by the formula:

$$w(p,q) = e^{-\frac{\max(d^2 - 2\sigma^2, 0.0)}{h^2}}$$

In this weight function, $\sigma$ is the standard deviation expected in the noise (in intensity units), and $h$ is a generic filtering parameter that determines how quickly patches will become irrelevant as their square distance grows from the patch we are updating. In general, increasing the value of $h$ will increase the noise removed but at the expense of some of the image detail. Decreasing the value of $h$ will preserve detail, but also more of the noise.

Typically, there is a decreasing return in considering patches very far away (in pixel units) from the pixel being updated, as the number of such patches increases quadratically with the distance allowed. For this reason, normally an overall area, called the *search window*, is defined and only patches in the search window contribute to the update. The update of the current pixel is then given by a simple weighted average of all other pixels in the search window using the given exponentially decaying weight function.[12] This is why the algorithm is called "non-local"; the patches that contribute to the repair of a given pixel are only loosely correlated to the location of the pixel being recomputed.

The OpenCV implementation of FNLMD contains several different functions, each of which applies to slightly different circumstances.

### Basic FNLMD with cv::fastNlMeansDenoising()

```
void cv::fastNlMeansDenoising(
  cv::InputArray  src,                        // Input image
  cv::OutputArray dst,                        // Output image
  float           h                  = 3,  // Weight decay parameter
  int             templateWindowSize = 7,  // Size of patches used for comparison
  int             searchWindowSize   = 21  // Maximum patch distance to consider
);
```

The first of these four functions, `cv::fastNlMeansDenoising()`, implements the algorithm as described exactly. We compute the result array `dst` from the input array `src` using a patch area of `templateWindowSize` and a decay parameter of `h`, and patches inside of `searchWindowSize` distance are considered. The image may be one-, two-, or three-channel, but must be or type `cv::8U`.[13] Table 11-2 lists some values, provided from the authors of the algorithm, that you can use to help set the decay parameter, `h`.

12 There is one subtlety here, which is that the weight of the contribution of the pixel $p$ in its own recalculation would be $w(p, p) = e^0 = 1$. In general this results in too high a weight relative to other similar pixels and very little change occurs in the value at $p$. For this reason, the weight at $p$ is normally chosen to be the maximum of the weights of the pixels within the area $B(p, s)$.

13 Note that though this image allows for multiple channels, it is not the best way to handle color images. For color images, it is better to use `cv::fastNlMeansDenoisingColored()`.

*Table 11-2. Recommended values for cv::fastNlMeansDenoising() for grayscale images*

| Noise: σ | Patch size: s | Search window | Decay parameter: h |
|---|---|---|---|
| $0 < \sigma \leq 15$ | $3 \times 3$ | $21 \times 21$ | $0.40 \cdot \sigma$ |
| $15 < \sigma \leq 30$ | $5 \times 5$ | $21 \times 21$ | $0.40 \cdot \sigma$ |
| $30 < \sigma \leq 45$ | $7 \times 7$ | $35 \times 35$ | $0.35 \cdot \sigma$ |
| $45 < \sigma \leq 75$ | $9 \times 9$ | $35 \times 35$ | $0.35 \cdot \sigma$ |
| $75 < \sigma \leq 100$ | $11 \times 11$ | $35 \times 35$ | $0.30 \cdot \sigma$ |

### FNLMD on color images with cv::fastNlMeansDenoisingColor()

```
void cv::fastNlMeansDenoisingColored(
  cv::InputArray  src,                        // Input image
  cv::OutputArray dst,                        // Output image
  float           h                  = 3,  // Luminosity weight decay parameter
  float           hColor             = 3,  // Color weight decay parameter
  int             templateWindowSize = 7,  // Size of patches used for comparison
  int             searchWindowSize   = 21  // Maximum patch distance to consider
);
```

The second variation of the FNLMD algorithm is used for color images. It accepts only images of type `cv::8UC3`. Though it would be possible in principle to apply the algorithm more or less directly to an RGB image, in practice it is better to convert the image to a different color space for the computation. The function `cv::fastNlMeansDenoisingColored()` first converts the image to the LAB color space, then applies the FNLMD algorithm, then converts the result back to RGB. The primary advantage of this is that in color there are, in effect, three decay parameters. In an RGB representation, however, it would be unlikely that you would want to set any of them to distinct values. But in the LAB space, it is natural to assign a different decay parameter to the luminosity component than to the color components. The function `cv::fastNlMeansDenoisingColored()` allows you to do just that. The parameter `h` is used for the luminosity decay parameter, while the new parameter `hColor` is used for the color channels. In general, the value of `hColor` will be quite a bit smaller than `h`. In most contexts, `10` is a suitable value. Table 11-3 lists some values that you can use to help set the decay parameter, `h`.

*Table 11-3. Recommended values for cv::fastNlMeansDenoising() for color images*

| Noise: σ | Patch size: s | Search window | Decay parameter: h |
|---|---|---|---|
| $0 < \sigma \leq 25$ | $3 \times 3$ | $21 \times 21$ | $0.55 \cdot \sigma$ |
| $25 < \sigma \leq 55$ | $5 \times 5$ | $35 \times 35$ | $0.40 \cdot \sigma$ |
| $55 < \sigma \leq 100$ | $7 \times 7$ | $35 \times 35$ | $0.35 \cdot \sigma$ |

#### FNLMD on video with cv::fastNlMeansDenoisingMulti() and cv::fastNlMeansDenoisingColorMulti()

```
void cv::fastNlMeansDenoisingMulti(
  cv::InputArrayOfArrays srcImgs,                // Sequence of several images
  cv::OutputArray        dst,                    // Output image
  int                    imgToDenoiseIndex,      // Index of image to denoise
  int                    temporalWindowSize,     // Num images to use (odd)
  float                  h                  = 3, // Weight decay parameter
  int                    templateWindowSize = 7, // Size of comparison patches
  int                    searchWindowSize   = 21 // Maximum patch distance
);
void cv::fastNlMeansDenoisingColoredMulti(
  cv::InputArrayOfArrays srcImgs,                // Sequence of several images
  cv::OutputArray        dst,                    // Output image
  int                    imgToDenoiseIndex,      // Index of image to denoise
  int                    temporalWindowSize,     // Num images to use (odd)
  float                  h                  = 3, // Weight decay param
  float                  hColor             = 3, // Weight decay param for color
  int                    templateWindowSize = 7, // Size of comparison patches
  int                    searchWindowSize   = 21 // Maximum patch distance
);
```

The third and fourth variations are used for sequential images, such as those that might be captured from video. In the case of sequential images, it is natural to imagine that frames other than just the current one might contain useful information for denoising a pixel. In most applications the noise will not be constant between images, while the signal will likely be similar or even identical. The functions `cv::fastNlMeansDenoisingMulti()` and `cv::fastNlMeansDenoisingColorMulti()` expect an array of images, `srcImgs`, rather than a single image. Additionally, they must be told which image in the sequence is actually to be denoised; this is done with the parameter `imgToDenoiseIndex`. Finally, a temporal window must be provided that indicates the number of images from the sequence to be used in the denoising. This parameter must be odd, and the implied window is always centered on `imgToDenoiseIndex`. (Thus, if you were to set `imgToDenoiseIndex` to 4 and `temporalWindowSize` to 5, the images that would be used in the denoising would be 2, 3, 4, 5, and 6.)

## Histogram Equalization

Cameras and image sensors must not only accommodate the naturally occurring contrast in a scene but also manage the image sensors' exposure to the available light levels. In a standard camera, the shutter and lens aperture settings are used to ensure that sensors receive neither too much nor too little light. However, the range of contrasts in a particular image is often too much for the sensor's available dynamic range. As a result, there is a trade-off between capturing the dark areas (e.g., shadows), which require a longer exposure time, and the bright areas, which require shorter exposure to avoid saturating "whiteouts." In many cases, both cannot be done effectively in the same image.

After the picture has been taken, there's nothing we can do about what the sensor recorded; however, we can still take what's there and try to expand the dynamic range of the image to increase its contrast. The most commonly used technique for this is *histogram equalization.*[14] In Figure 11-10, we can see that the image on the left is poor because there's not much variation of the range of values. This is evident from the histogram of its intensity values on the right. Because we are dealing with an 8-bit image, its intensity values can range from `0` to `255`, but the histogram shows that the actual intensity values are all clustered near the middle of the available range. Histogram equalization is a method for stretching this range out.

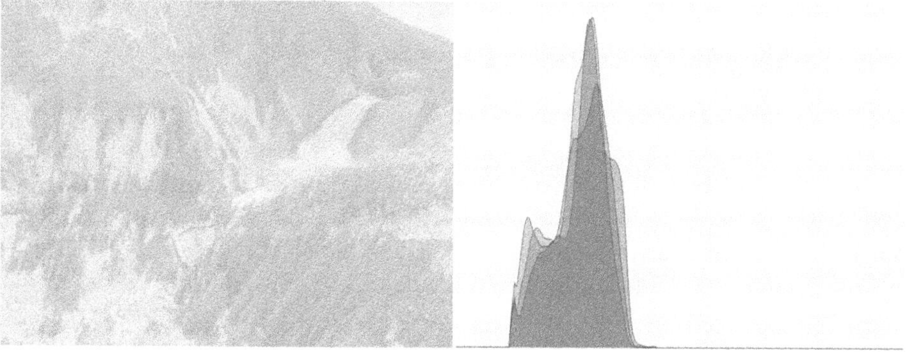

*Figure 11-10. The image on the left has poor contrast, as is confirmed by the histogram of its intensity values on the right*

The underlying math behind histogram equalization involves mapping one distribution (the given histogram of intensity values) to another distribution (a wider and, ideally, uniform distribution of intensity values). That is, we want to spread out the $y$ values of the original distribution as evenly as possible in the new distribution. It turns out that there is a good answer to the problem of spreading out distribution values: the remapping function should be the *cumulative distribution function*. An example of the cumulative distribution function is shown in Figure 11-11 for the somewhat idealized case of a density distribution that was originally pure Gaussian. However, cumulative density can be applied to any distribution; it is just the running sum of the original distribution from its negative to its positive bounds.

14 Histogram equalization is an old mathematical technique; its use in image processing is described in various textbooks [Jain86; Russ02; Acharya05], conference papers [Schwarz78], and even in biological vision [Laughlin81]. If you are wondering why histogram equalization is not in the chapter on histograms (Chapter 13), it is because histogram equalization makes no explicit use of any histogram data types. Although histograms are used internally, the function (from the user's perspective) requires no histograms at all.

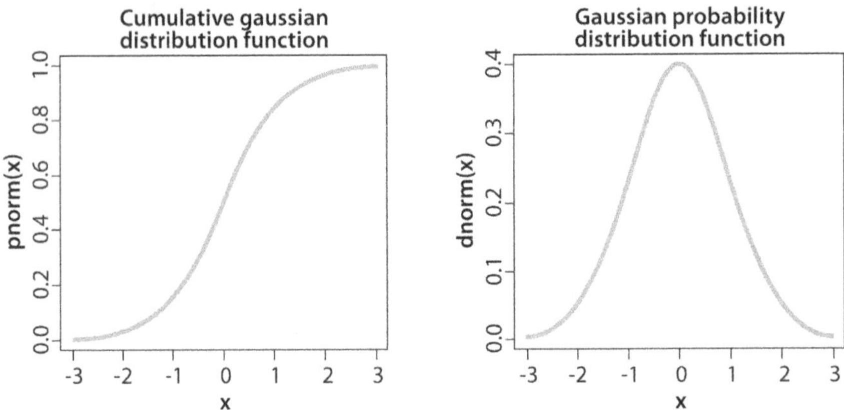


*Figure 11-11. Result of cumulative distribution function (left) computed for a Gaussian distribution (right)*

We may use the cumulative distribution function to remap the original distribution to an equally spread distribution (see Figure 11-12) simply by looking up each *y* value in the original distribution and seeing where it should go in the equalized distribution. For continuous distributions the result will be an exact equalization, but for digitized/discrete distributions the results may be far from uniform.

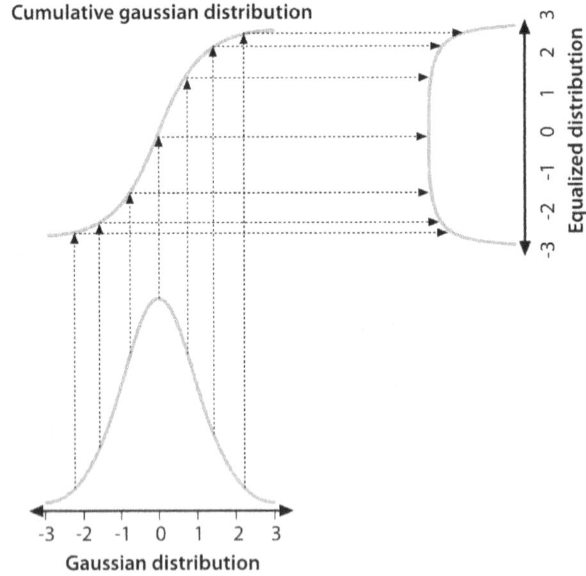


*Figure 11-12. Using the cumulative density function to equalize a Gaussian distribution*

Applying this equalization process to Figure 11-10 yields the equalized intensity distribution histogram and resulting image in Figure 11-13.

*Figure 11-13. Histogram equalized results: the spectrum has been spread out*

## cv::equalizeHist(): Contrast equalization

OpenCV wraps this whole process up in one neat function.

```
void cv::equalizeHist(
  const cv::InputArray  src,                    // Input image
  cv::OutputArray dst                           // Result image
);
```

In `cv::equalizeHist()`, the source `src` must be a single-channel, 8-bit image. The destination image `dst` will be the same. For color images, you will have to separate the channels and process them one by one.[15]

# Summary

In this chapter, we learned a variety of methods that can be used to transform images. These transformations included scale transformations, as well as affine and perspective transformations. We learned how to remap vector functions from Cartesian to polar representations. What all of these functions have in common is their conversion of one image into another through a global operation on the entire image. We saw one function that could handle even the most general remapping, relative to which many of the functions discussed earlier in the chapter could be seen as special cases.

15 In practice, separately applying histogram equalization to each channel in an RGB image is not likely to give aesthetically satisfying results. It is probably better to convert to a more suitable space, such as LAB, and then apply histogram equalization only to the luminosity channel.

We also encountered some algorithms that are useful in computational photography, such as inpainting, denoising, and histogram equalization. These algorithms are useful for handling images from camera and video streams generally, and are often handy when you want to implement other computer vision techniques on top of grainy or otherwise poor-quality video data.

# Exercises

1. Find and load a picture of a face where the face is frontal, has eyes open, and takes up most or all of the image area. Write code to find the pupils of the eyes.

A Laplacian "likes" a bright central point surrounded by dark. Pupils are just the opposite. Invert and convolve with a sufficiently large Laplacian.

2. Look at the diagrams of how the log-polar function transforms a square into a wavy line.
   a. Draw the log-polar results if the log-polar center point were sitting on one of the corners of the square.
   b. What would a circle look like in a log-polar transform if the center point were inside the circle and close to the edge?
   c. Draw what the transform would look like if the center point were sitting just outside of the circle.
3. A log-polar transform takes shapes of different rotations and sizes into a space where these correspond to shifts in the $\theta$-axis and $\log(r)$ axis. The Fourier transform is translation invariant. How can we use these facts to force shapes of different sizes and rotations to automatically give equivalent representations in the log-polar domain?
4. Draw separate pictures of large, small, large rotated, and small rotated squares. Take the log-polar transform of these each separately. Code up a two-dimensional shifter that takes the center point in the resulting log-polar domain and shifts the shapes to be as identical as possible.
5. Load an image, take a perspective transform, and then rotate it. Can this transform be done in one step?
6. Inpainting works pretty well for the repair of writing over textured regions. What would happen if the writing obscured a real object edge in a picture? Try it.

7. Practice histogram equalization on images that you load in, and report the results.
8. Explain the difference between histogram equalization of an image and denoising an image.

CHAPTER 12

# Image Analysis

## Overview

In the preceding chapter, we learned about the image transformations that OpenCV makes available to us. These transformations were essentially mappings that converted an input image into an output image such that the output remained, essentially, a picture, just like the input. In this chapter we will consider operations that render images into some potentially entirely different representation.

These new representations will usually still be arrays of values, but those values might be quite different in meaning than the intensity values in the input image. For example, the first function we will consider is the *discrete Fourier transform*, in which the output "image," though still an array, contains a frequency representation of the input image. In a few cases, the result of the transformation will be something like a list of components, and not an array at all, as would be the case for the *Hough line transform*.

Finally, we will learn about image segmentation methods that can be used to represent an image in terms of meaningfully connected regions.

## Discrete Fourier Transform

For any set of values that are indexed by a discrete (integer) parameter, it is possible to define a *discrete Fourier transform* (DFT)[1] in a manner analogous to the Fourier transform of a continuous function. For $N$ complex numbers $x_0, x_1, x_2, ..., x_{N-1}$, the one-dimensional DFT is defined by the following formula (where $i = \sqrt{-1}$):

$$g_k = \sum_{n=0}^{N-1} f_n e^{-\frac{2\pi i}{N} kn}$$

A similar transform can be defined for a two-dimensional array of numbers (of course, higher-dimensional analogs exist also):

$$g_{k_x,k_y} = \sum_{n_x=0}^{N_x-1} \sum_{n_y=0}^{N_y-1} f_{n_x,n_y} e^{-\frac{2\pi i}{N}(k_x n_x + k_y n_y)}$$

In general, one might expect that the computation of the $N$ different terms $g_k$ would require $O(N^2)$ operations. In fact, there are several *fast Fourier transform* (FFT) algorithms capable of computing these values in $O(N \log N)$ time.

### cv::dft(): The Discrete Fourier Transform

The OpenCV function `cv::dft()` implements one such FFT algorithm. The `cv::dft()` function can compute FFTs for one- and two-dimensional arrays of inputs. In the latter case, the two-dimensional transform can be computed or, if desired, only the one-dimensional transforms of each individual row can be computed (this operation is much faster than calling `cv::dft()` several times):

```
void cv::dft(
  cv::InputArray    src,                   // Input array (real or complex)
  cv::OutputArray   dst,                   // Output array
  int               flags       = 0,       // for inverse, or other options
  int               nonzeroRows = 0        // number of rows to not ignore
);
```

The input array must be of floating-point type and may be single- or double-channel. In the single-channel case, the entries are assumed to be real numbers, and the output

1 Joseph Fourier [Fourier] was the first to find that some functions can be decomposed into an infinite series of other functions, which became a field known as *Fourier analysis*. Some key text on methods of decomposing functions into their Fourier series are Morse for physics [Morse53] and Papoulis in general [Papoulis62]. The fast Fourier transform was invented by Cooley and Tukey in 1965 [Cooley65], though Carl Gauss worked out the key steps as early as 1805 [Johnson84]. Early use in computer vision is described by Ballard and Brown [Ballard82].

will be packed in a special space-saving format called *complex conjugate symmetrical* (CCS).[2] If the source and channel are two-channel matrices or images, then the two channels will be interpreted as the real and imaginary components of the input data. In this case, there will be no special packing of the results, and some space will be wasted with a lot of 0s in both the input and output arrays.[3]

The special packing of result values that is used with single-channel CCS output is as follows.

For a one-dimensional array:

$$\left[Re\,Y_0 \quad Re\,Y_1 \quad Im\,Y_1 \quad Re\,Y_2 \quad Im\,Y_2 \quad \ldots \quad Re\,Y_{\left(\frac{N}{2}-1\right)} \quad Im\,Y_{\left(\frac{N}{2}-1\right)} \quad Re\,Y_{\left(\frac{N}{2}\right)}\right]$$

For a two-dimensional array:

$$\begin{bmatrix}
Re\,Y_{00} & Re\,Y_{01} & Im\,Y_{01} & Re\,Y_{02} & Im\,Y_{02} & \cdots & Re\,Y_{0,\frac{N_x}{2}-1} & Im\,Y_{0,\frac{N_x}{2}-1} & Re\,Y_{0,\frac{N_x}{2}} \\
Re\,Y_{10} & Re\,Y_{11} & Im\,Y_{11} & Re\,Y_{12} & Im\,Y_{12} & \cdots & Re\,Y_{1,\frac{N_x}{2}-1} & Im\,Y_{1,\frac{N_x}{2}-1} & Re\,Y_{1,\frac{N_x}{2}} \\
Im\,Y_{20} & Re\,Y_{21} & Im\,Y_{21} & Re\,Y_{22} & Im\,Y_{22} & \cdots & Re\,Y_{2,\frac{N_x}{2}-1} & Im\,Y_{2,\frac{N_x}{2}-1} & Im\,Y_{2,\frac{N_x}{2}} \\
\vdots & \vdots & \vdots & \vdots & \vdots & \vdots\;\vdots & & \vdots & \vdots \\
Re\,Y_{\frac{N_y}{2}-1,0} & Re\,Y_{N_y-3,1} & Im\,Y_{N_y-3,1} & Re\,Y_{N_y-3,2} & Im\,Y_{N_y-3,2} & \cdots & Re\,Y_{N_y-3,\frac{N_x}{2}-1} & Im\,Y_{N_y-3,\frac{N_x}{2}-1} & Re\,Y_{N_y-3,\frac{N_x}{2}} \\
Im\,Y_{\frac{N_y}{2}-1,0} & Re\,Y_{N_y-2,1} & Im\,Y_{N_y-2,1} & Re\,Y_{N_y-2,2} & Im\,Y_{N_y-2,2} & \cdots & Re\,Y_{N_y-2,\frac{N_x}{2}-1} & Im\,Y_{N_y-2,\frac{N_x}{2}-1} & Im\,Y_{N_y-2,\frac{N_x}{2}} \\
Re\,Y_{\frac{N_y}{2},0} & Re\,Y_{N_y-1,1} & Im\,Y_{N_y-1,1} & Re\,Y_{N_y-1,2} & Im\,Y_{N_y-1,2} & \cdots & Re\,Y_{N_y-1,\frac{N_x}{2}-1} & Im\,Y_{N_y-1,\frac{N_x}{2}-1} & Re\,Y_{N_y-1,\frac{N_x}{2}}
\end{bmatrix}$$

It is worth taking a moment to look closely at the indices of these arrays. Certain values in the array are guaranteed to be `0` (more accurately, certain values of $f_k$ are guaranteed to be real). Also note that the last row listed in the table will be present only if $N_y$ is even and that the last column will be present only if $N_x$ is even. In the case of the two-dimensional array being treated as $N_y$ separate one-dimensional arrays rather than a full two-dimensional transform (we'll take a look at how to do this), all of the result rows will be analogous to the single row listed for the output of the one-dimensional array.

The third argument, called **`flags`**, indicates exactly what operation is to be done. As usual, **`flags`** is treated as a bit array, so you can combine any flags you need with

2 As a result of this compact representation, the size of the output array for a single-channel image is the same as the size of the input array because the elements that are provably zero are omitted. In the case of the two-channel (complex) array, the output size will, of course, also be equal to the input size.

3 When using this method, you must be sure to explicitly set the imaginary components to `0` in the two-channel representation. An easy way to do this is to create a matrix full of 0s using `cv::Mat::zeros()` for the imaginary part and then call `cv::merge()` with a real-valued matrix to form a temporary complex array on which to run `cv::dft()` (possibly in place). This procedure will result in full-size, unpacked, complex matrix of the spectrum.

Boolean OR. The transformation we started with is known as a *forward transform* and is selected by default. The inverse transform[4] is defined in exactly the same way except for a change of sign in the exponential and a scale factor. To perform the inverse transform without the scale factor, use the flag `cv::DFT_INVERSE`. The flag for the scale factor is `cv::DFT_SCALE`, which results in all of the output being scaled by a factor of $N^{-1}$ (or $(N_x N_y)^{-1}$ for a two-dimensional transform). This scaling is necessary if the sequential application of the forward transform and the inverse transform is to bring us back to where we started. Because one often wants to combine `cv::DFT_INVERSE` with `cv::DFT_SCALE`, there are several shorthand notations for this kind of operation. In addition to just combining the two operations, you can use `cv::DFT_INV_SCALE` (or `cv::DFT_INVERSE_SCALE` if you're not into brevity). The last flag you may want to have handy is `cv::DFT_ROWS`, which allows you to tell `cv::dft()` to treat a two-dimensional array as a collection of one-dimensional arrays that should each be transformed separately as if they were $N_y$ distinct vectors of length $N_x$. This can significantly reduce overhead when you're doing many transformations at a time. By using `cv::DFT_ROWS`, you can also implement three-dimensional (and higher) DFT.

Though the default behavior of the forward transform is to produce results in CCS format (which results in an output array exactly the same size as the input array), you can explicitly ask OpenCV to not do this with the flag `cv::DFT_COMPLEX_OUTPUT`. The result will be the full complex array (with all of the zeros in it). Conversely, when you're performing an inverse transformation on a complex array, the result is normally also a complex array. If the source array had complex conjugate symmetry,[5] you can ask OpenCV to produce a purely real array (which will be smaller than the input array) by passing the `cv::DFT_REAL_OUTPUT` flag.

In order to understand the last argument, `nonzero_rows`, we must digress for a moment to explain that, in general, DFT algorithms strongly prefer input vectors of some lengths over input vectors of other lengths; similarly, for arrays of some sizes over arrays of other sizes. In most DFT algorithms, the preferred sizes are powers of 2 (i.e., $2^n$ for some integer $n$). In the case of the algorithm used by OpenCV, the preference is that the vector lengths, or array dimensions, be $2^p3^q5^r$ for some integers $p$, $q$, and $r$. Hence the usual procedure is to create a somewhat larger array and then to copy your array into that somewhat roomier zero-padded array. For convenience,

4 With the inverse transform, the input is packed in the special format described previously. This makes sense because, if we first called the forward DFT and then ran the inverse DFT on the results, we would expect to wind up with the original data—that is, of course, if we remember to use the `cv::DFT_SCALE` flag!

5 This is not to say that it is in CCS format, only that it possesses the symmetry, as it would if (for example) it were the result of a forward transform of a purely real array in the first place. Also, note that you are *telling* OpenCV that the input array has this symmetry—it will trust you. It does not actually check to verify that this symmetry is present.

there is a handy utility function, cv::getOptimalDFTSize(), which takes the (integer) length of your vector and returns the first equal or larger size that can be expressed in the form given (i.e., $2^p 3^q 5^r$). Despite the need for this padding, it is possible to indicate to cv::dft() that you really do not care about the transform of those rows that you had to add down below your actual data (or, if you are doing an inverse transform, which rows in the result you do not care about). In either case, you can use nonzero_rows to indicate how many rows contain meaningful data. This will provide some savings in computation time.

## cv::idft(): The Inverse Discrete Fourier Transform

As we saw earlier, the function cv::dft() can be made to implement not only the discrete Fourier transform, but also the inverse operation (with the provision of the correct flags argument). It is often preferable, if only for code readability, to have a separate function that does this inverse operation by default.

```
void cv::idft(
  cv::InputArray  src,                          // Input array (real or complex)
  cv::OutputArray dst,                          // Output array
  int             flags       = 0,              // for variations
  int             nonzeroRows = 0               // number of rows to not ignore
);
```

Calling cv::idft() is exactly equivalent to calling cv::dft() with the cv::DFT_INVERSE flag (in addition to whatever flags you supply to cv::idft(), of course).

## cv::mulSpectrums(): Spectrum Multiplication

In many applications that involve computing DFTs, one must also compute the per-element multiplication of the two resulting spectra. Because such results are complex numbers, typically packed in their special high-density CCS format, it would be tedious to unpack them and handle the multiplication via the "usual" matrix operations. Fortunately, OpenCV provides the handy cv::mulSpectrums() routine, which performs exactly this function for us:

```
void cv::mulSpectrums(
  cv::InputArray  src1,                         // Input array (ccs or complex)
  cv::InputArray  src2,                         // Input array (ccs or complex)
  cv::OutputArray dst,                          // Result array
  int             flags,                        // for row-by-row computation
  bool            conj = false                  // true to conjugate src2
);
```

Note that the first two arguments are arrays, which must be either CCS packed single-channel spectra or two-channel complex spectra—as you would get from calls to cv::dft(). The third argument is the destination array, which will be of the same size and type as the source arrays. The final argument, `conj`, tells `cv::mulSpectrums()` exactly what you want done. In particular, it may be set to `false` for implementing the preceding pair multiplication or set to `true` if the element from the first array is to be multiplied by the complex conjugate of the corresponding element of the second array.[6]

## Convolution Using Discrete Fourier Transforms

It is possible to greatly increase the speed of a convolution by using DFT via the convolution theorem [Titchmarsh26] that relates convolution in the spatial domain to multiplication in the Fourier domain [Morse53; Bracewell65; Arfken85].[7] To accomplish this, we first compute the Fourier transform of the image and then the Fourier transform of the convolution filter. Once this is done, we can perform the convolution in the transform space in linear time with respect to the number of pixels in the image. It is worthwhile to look at the source code for computing such a convolution, as it will also provide us with many good examples of using `cv::dft()`. The code is shown in Example 12-1, which is taken directly from the OpenCV reference.

*Example 12-1. Using cv::dft() and cv::idft() to accelerate the computation of convolutions*

```
#include <opencv2/opencv.hpp>
#include <iostream>

using namespace std;

int main(int argc, char** argv) {

  if(argc != 2) {
    cout << "Fourier Transform\nUsage: " <<argv[0] <<" <imagename>" << endl;
    return -1;
  }

  cv::Mat A = cv::imread(argv[1],0);

  if( A.empty() ) { cout << "Cannot load " << argv[1] << endl; return -1; }

  cv::Size patchSize( 100, 100 );
  cv::Point topleft( A.cols/2, A.rows/2 );
  cv::Rect roi( topleft.x, topleft.y, patchSize.width, patchSize.height );
  cv::Mat B = A( roi );
```

6 The primary usage of this argument is the implementation of a correlation in Fourier space. It turns out that the only difference between convolution (which we will discuss in the next section) and correlation is the conjugation of the second array in the spectrum multiplication.

7 Recall that OpenCV's DFT algorithm implements the FFT whenever the data size makes the FFT faster.

```
    int dft_M = cv::getOptimalDFTSize( A.rows+B.rows-1 );
    int dft_N = cv::getOptimalDFTSize( A.cols+B.cols-1 );

    cv::Mat dft_A = cv::Mat::zeros( dft_M, dft_N, CV::F32 );
    cv::Mat dft_B = cv::Mat::zeros( dft_M, dft_N, CV::F32 );

    cv::Mat dft_A_part = dft_A( Rect(0, 0, A.cols,A.rows) );
    cv::Mat dft_B_part = dft_B( Rect(0, 0, B.cols,B.rows) );

    A.convertTo( dft_A_part, dft_A_part.type(), 1, -mean(A)[0] );
    B.convertTo( dft_B_part, dft_B_part.type(), 1, -mean(B)[0] );

    cv::dft( dft_A, dft_A, 0, A.rows );
    cv::dft( dft_B, dft_B, 0, B.rows );

    // set the last parameter to false to compute convolution instead of correlation
    //
    cv::mulSpectrums( dft_A, dft_B, dft_A, 0, true );
    cv::idft( dft_A, dft_A, DFT_SCALE, A.rows + B.rows - 1 );

    cv::Mat corr = dft_A( Rect(0, 0, A.cols + B.cols - 1, A.rows + B.rows - 1) );
    cv::normalize( corr, corr, 0, 1, NORM_MINMAX, corr.type() );
    cv::pow( corr, 3., corr );

    cv::B ^= cv::Scalar::all( 255 );

    cv::imshow( "Image", A );
    cv::imshow( "Correlation", corr );
    cv::waitKey();

    return 0;

}
```

In Example 12-1, we can see that the input arrays are first created and then initialized. Next, two new arrays are created whose dimensions are optimal for the DFT algorithm. The original arrays are copied into these new arrays and the transforms are then computed. Finally, the spectra are multiplied together and the inverse transform is applied to the product. The transforms are the slowest[8] part of this operation; an $N \times N$ image takes $O(N^2 \log N)$ time and so the entire process is also completed in that time (assuming that $N > M$ for an $M \times M$ convolution kernel). This time is much faster than $O(N^2M^2)$, the non-DFT convolution time required by the more naïve method.

8 By "slowest" we mean "asymptotically slowest"—in other words, that this portion of the algorithm takes the most time for very large $N$. This is an important distinction. In practice, as we saw in the earlier section on convolutions, it is not always optimal to pay the overhead for conversion to Fourier space. In general, when you are convolving with a small kernel it is not worth the trouble to make this transformation.

## cv::dct(): The Discrete Cosine Transform

For real-valued data, it is often sufficient to compute what is, in effect, only half of the discrete Fourier transform. The *discrete cosine transform* (DCT) [Ahmed74; Jain77] is defined analogously to the full DFT by the following formula:

$$c_k = \left(\frac{1}{N}\right)^{\frac{1}{2}} x_0 + \sum_{n=1}^{N-1} \left(\frac{2}{N}\right)^{\frac{1}{2}} x_n \cos\left(\left(k + \frac{1}{2}\right)\frac{n}{N}\pi\right)$$

Of course, there is a similar transform for higher dimensions. Note that, by convention, the normalization factor is applied to both the cosine transform and its inverse (which is not the convention for the discrete Fourier transform).

The basic ideas of the DFT apply also to the DCT, but now all the coefficients are real-valued.[9] The actual OpenCV call is:

```
void cv::dct(
  cv::InputArray  src,                    // Input array (even size)
  cv::OutputArray dst,                    // Output array
  int             flags = 0               // for row-by-row or inverse
);
```

The `cv::dct()` function expects arguments like those for `cv::dft()` except that, because the results are real-valued, there is no need for any special packing of the result array (or of the input array in the case of an inverse transform). Unlike `cv::dft()`, however, the input array must have an even number of elements (you can pad the last element with a zero if necessary to achieve this). The `flags` argument can be set to `cv::DCT_INVERSE` to generate the inverse transformation, and may be combined with `cv::DCT_ROWS` with the same effect as with `cv::dft()`. Because of the different normalization convention, both the forward and inverse cosine transforms always contain their respective contribution to the overall normalization of the transform; hence, there is no analog to `cv::DFT_SCALE` for `cv::dct()`.

As with `cv::dft()`, performance strongly depends on array size. In fact, deep down, the implementation of `cv::dct()` actually calls `cv::dft()` on an array exactly half the size of your input array. For this reason, the optimal size of an array to pass to `cv::dct()` is exactly double the size of the optimal array you would pass to `cv::dft()`. Putting everything together, the best way to get an optimal size for `cv::dct()` is to compute:

9 Astute readers might object that the cosine transform is being applied to a vector that is not a manifestly even function. However, with `cv::dct()`, the algorithm simply treats the vector as if it were extended to negative indices in a mirrored manner.

```
size_t optimal_dct_size = 2 * cv::getOptimalDFTSize( (N+1)/2 );
```

where `N` is the actual size of your data that you want to transform.

## cv::idct(): The Inverse Discrete Cosine Transform

Just as with `cv::idft()` and `cv::dft()`, `cv::dct()` can be asked to compute the inverse cosine transform using the `flags` argument. As before, code readability is often improved with the use of a separate function that does this inverse operation by default:

```
void cv::idct(
  cv::InputArray  src,                    // Input array
  cv::OutputArray dst,                    // Output array
  int             flags = 0,              // for row-by-row computation
);
```

Calling `cv::idct()` is exactly equivalent to calling `cv::dct()` with the `cv::DCT_INVERSE` flag (in addition to any other flags you supply to `cv::idct()`).

# Integral Images

OpenCV allows you to easily calculate an integral image with the appropriately named `cv::integral()` function. An *integral image* [Viola04] is a data structure that allows rapid summing of subregions.[10] Such summations are useful in many applications; a notable one is the computation of *Haar wavelets*, which are used in some face recognition and similar algorithms.

OpenCV supports three variations of the integral image. They are the *sum*, the *square-sum*, and the *tilted-sum*. In each case the resulting image is the same size as the original image, plus one in each direction.

A standard integral image sum has the form:

$$sum(x, y) = \sum_{y'<y} \sum_{x'<x} image(x', y')$$

The square-sum image is the sum of squares:

$$sum_{square}(x, y) = \sum_{y'<y} \sum_{x'<x} [image(x', y')]^2$$

10 The citation given is the best for more details on the method, but it was actually introduced in computer vision in 2001 in a paper titled "Robust Real-Time Object Detection" by the same authors. The method was previously used as early as 1984 in computer graphics, where the integral image is known as a *summed area table*.

The tilted-sum is like the sum except that it is for the image rotated by 45 degrees:

$$sum_{tilted}(x, y) = \sum_{y'<y} \sum_{abs(x'-x)<y} image(x', y')$$

Using these integral images, you may calculate sums, means, and standard deviations over arbitrary upright or "tilted" rectangular regions of the image. As a simple example, to sum over a simple rectangular region described by the corner points $(x^1, y^1)$ and $(x^2, y^2)$, where $x^2 > x^1$ and $y^2 > y^1$, we'd compute:

$$\sum_{y_1 \le y < y_2} \sum_{x_1 \le x < x_2} image(x, y) = [sum(x_2, y_2) - sum(x_1, y_2) - sum(x_2, y_1) + sum(x_1, y_1)]$$

In this way, it is possible to do fast blurring, approximate gradients, compute means and standard deviations, and perform fast block correlations even for variable window sizes.

To make this all a little clearer, consider the 7 × 5 image shown in Figure 12-1; the region is shown as a bar chart in which the height associated with the pixels represents the brightness of those pixel values. The same information is shown in Figure 12-2, numerically on the left and in integral form on the right. We compute standard summation integral images $I'(x, y)$ by going across rows, proceeding row by row using the previously computed integral image values together with the current raw image pixel value $I(x, y)$ to calculate the next integral image value as follows:

$$I'(x, y) = [I(x, y) - I(x - 1, y) - I(x, y - 1) + I(x - 1, y - 1)]$$

The last term is subtracted because this value is double-counted when the second and third terms are added. You can verify that this works by testing some values in Figure 12-2.

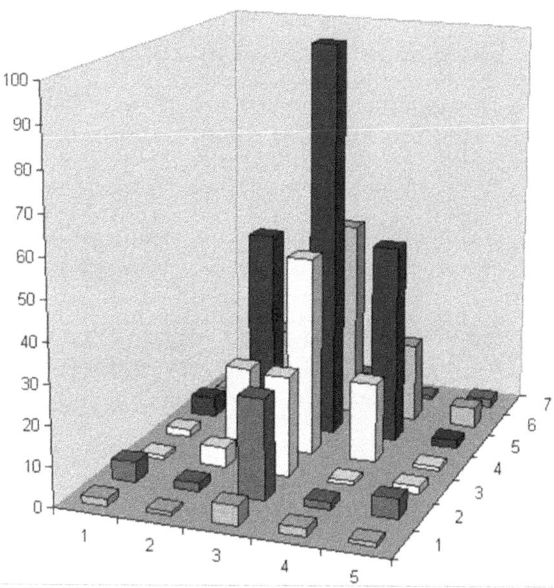


*Figure 12-1. A simple 7 × 5 image shown as a bar chart with x, y, and height equal to pixel value*

When using the integral image to compute a region, we can see in Figure 12-2 that, in order to compute the central rectangular area bounded by the 20s in the original image, we'd calculate 398 – 9 – 10 + 1 = 380. Thus, a rectangle of any size can be computed using four measurements, resulting in $O(1)$ computational complexity.

| | | | | |
|---|---|---|---|---|
| 1 | 2 | 5 | 1 | 2 |
| 2 | **20** | **50** | **20** | 5 |
| 5 | **50** | **100** | **50** | 2 |
| 2 | **20** | **50** | **20** | 1 |
| 1 | 5 | 25 | 1 | 2 |
| 5 | 2 | 25 | 2 | 5 |
| 2 | 1 | 5 | 2 | 1 |

| | | | | | |
|---|---|---|---|---|---|
| 0 | 0 | 0 | 0 | 0 | 0 |
| 0 | 1 | 3 | 8 | 9 | 11 |
| 0 | 3 | **25** | **80** | **101** | 108 |
| 0 | 8 | **80** | **235** | **306** | 315 |
| 0 | 10 | **102** | **307** | **398** | 408 |
| 0 | 11 | 108 | 338 | 430 | 442 |
| 0 | 16 | 115 | 370 | 464 | 481 |
| 0 | 18 | 118 | 378 | 474 | 492 |

*Figure 12-2. The 7 × 5 image of Figure 12-1 shown numerically at left (with the origin assumed to be the upper left) and converted to an 8 × 6 integral image at right*

## cv::integral() for Standard Summation Integral

The different forms of integral are (somewhat confusingly) distinguished in the C++ API only by their arguments. The form that computes the basic sum has only three.

```
void cv::integral(
  cv::InputArray  image,                    // Input array
  cv::OutputArray sum,                      // Output sum results
  int             sdepth = -1               // Results depth (e.g., cv::F32)
);
```

The first and second are the input and result images. If the input image is of size $W \times H$, then the output image will be of size $(W + 1) \times (H + 1)$.[11] The third argument, `sdepth`, specifies the desired depth of the sum (destination) image. `sdepth` can be any of `CV::S32`, `CV::F32`, or `CV::F64`.[12]

## cv::integral() for Squared Summation Integral

The squared sum is computed with the same function as the regular sum, except for the provision of an additional output argument for the squared sum.

```
void cv::integral(
  cv::InputArray  image,                    // Input array
  cv::OutputArray sum,                      // Output sum results
  cv::OutputArray sqsum,                    // Output sum of squares results
  int             sdepth = -1               // Results depth (e.g., cv::F32)
);
```

The `cv::OutputArray` argument `sqsum` indicates to `cv::integral()` that the square sum should be computed in addition to the regular sum. As before, `sdepth` specifies the desired depth of the resulting images. `sdepth` can be any of `CV::S32`, `CV::F32`, or `CV::F64`.

## cv::integral() for Tilted Summation Integral

Similar to the squared sum, the tilted sum integral is essentially the same function, with an additional argument for the additional result.

```
void cv::integral(
  cv::InputArray  image,                    // Input array
  cv::OutputArray sum,                      // Output sum results
```

11 This allows for the rows of zeros, which are implied by the fact that summing zero terms results in a sum of zero.

12 It is worth noting that even though `sum` and `tilted_sum` allow 32-bit float as output for input images of 32-bit float type, it is recommended to use 64-bit float, particularly for larger images. After all, a modern large image can be many millions of pixels.

```
    cv::OutputArray sqsum,                      // Output sum of squares results
    cv::OutputArray tilted,                     // Output tilted sum results
    int             sdepth = -1                 // Results depth (e.g., cv::F32)
  );
```

The additional `cv::OutputArray` argument `tilted` is computed by this form of `cv::integral()`, in addition to the other sums; thus, all of the other arguments are the same.

# The Canny Edge Detector

Though it is possible to expose edges in images with simple filters such as the Laplace filter, it is possible to improve on this method substantially. The simple Laplace filter method was refined by J. Canny in 1986 into what is now commonly called the *Canny edge detector* [Canny86]. One of the differences between the Canny algorithm and the simpler, Laplace-based algorithm described in the previous chapter is that, in the Canny algorithm, the first derivatives are computed in $x$ and $y$ and then combined into four directional derivatives. The points where these directional derivatives are local maxima are then candidates for assembling into edges. The most significant new dimension to the Canny algorithm is this phase in which it assembles the individual edge-candidate pixels into *contours*.[13]

The algorithm forms these contours by applying a *hysteresis threshold* to the pixels. This means that there are two thresholds, an upper and a lower. If a pixel has a gradient larger than the upper threshold, then it is accepted as an edge pixel; if a pixel is below the lower threshold, it is rejected. If the pixel's gradient is between the thresholds, then it will be accepted only if it is connected to a pixel that is above the high threshold. Canny recommended a ratio of high:low threshold between 2:1 and 3:1. Figures 12-3 and 12-4 show the results of applying `cv::Canny()` to a test pattern and a photograph using high:low hysteresis threshold ratios of 5:1 and 3:2, respectively.

13 We'll have much more to say about contours later. As you await those revelations, keep in mind that the `cv::Canny()` routine does not actually return objects of a contour type; if we want them, we will have to build those from the output of `cv::Canny()` by using `cv::findContours()`. Everything you ever wanted to know about contours will be covered in Chapter 14.

*Figure 12-3. Results of Canny edge detection for two different images when the high and low thresholds are set to 50 and 10, respectively*

*Figure 12-4. Results of Canny edge detection for two different images when the high and low thresholds are set to 150 and 100, respectively*

## cv::Canny()

The OpenCV implementation of the Canny edge detection algorithm converts an input image into an "edge image."

```
void cv::Canny(
  cv::InputArray  image,                // Input single channel image
  cv::OutputArray edges,                // Output edge image
  double          threshold1,           // "lower" threshold
  double          threshold2,           // "upper" threshold
  int             apertureSize = 3,     // Sobel aperture
  bool            L2gradient   = false  // true=L2-norm (more accurate)
);
```

The `cv::Canny()` function expects an input image, which must be single-channel, and an output image, which will also be grayscale (but which will actually be a Boolean image). The next two arguments are the low and high thresholds. The next-to-last argument, `apertureSize`, is the aperture used by the Sobel derivative operators that are called inside of the implementation of `cv::Canny()`. The final argument `L2gradient` is used to select between computing the directional gradient "correctly" using the proper $L_2$-norm, or using a faster, less accurate $L_1$-norm-based method. If the argument `L2gradient` is set to `true`, the more accurate form is used:

$$|\,grad(x, y)\,|_{L_2} = \sqrt{\left(\frac{dI}{dx}\right)^2 + \left(\frac{dI}{dy}\right)^2}$$

If `L2gradient` is set to `false`, the faster form is used:

$$|\,grad(x, y)\,|_{L_1} = \left|\frac{dI}{dx}\right| + \left|\frac{dI}{dy}\right|$$

# Hough Transforms

The *Hough transform*[14] is a method for finding lines, circles, or other simple forms in an image. The original Hough transform was a line transform, which is a relatively fast way of searching a binary image for straight lines. The transform can be further generalized to cases other than just simple lines.

## Hough Line Transform

The basic theory of the Hough line transform is that any point in a binary image could be part of some set of possible lines. If we parameterize each line by, for exam-

14 Hough developed the transform for use in physics experiments [Hough59]; its use in vision was introduced by Duda and Hart [Duda72].

ple, a slope $a$ and an intercept $b$, then a point in the original image is transformed to a locus of points in the $(a, b)$ plane corresponding to all of the lines passing through that point, of which it could potentially be a part (see Figure 12-5). If we convert every nonzero pixel in the input image into such a set of points in the output image and sum over all such contributions, then lines that appear in the input (i.e., $(x, y)$ plane) image will appear as local maxima in the output (i.e., $(a, b)$ plane) image. Because we are summing the contributions from each point, the $(a, b)$ plane is commonly called the *accumulator plane*.

*Figure 12-5. The Hough line transform finds many lines in each image; some of the lines found are expected, but others may not be*

It might occur to you that the slope-intercept form is not really the best way to represent all the lines passing through a point (i.e. because of the considerably different density of lines as a function of the slope, and the related fact that the interval of possible slopes goes from $-\infty$ to $+\infty$). This is why the actual parameterization of the transform image used in numerical computation is somewhat different. The preferred parameterization represents each line as a point in polar coordinates $(\rho, \theta)$, with the implied line being the line passing through the indicated point but perpendicular to the radial from the origin to that point (see Figure 12-6). The equation for such a line is:

$$\rho = x\cos\theta + y\sin\theta$$

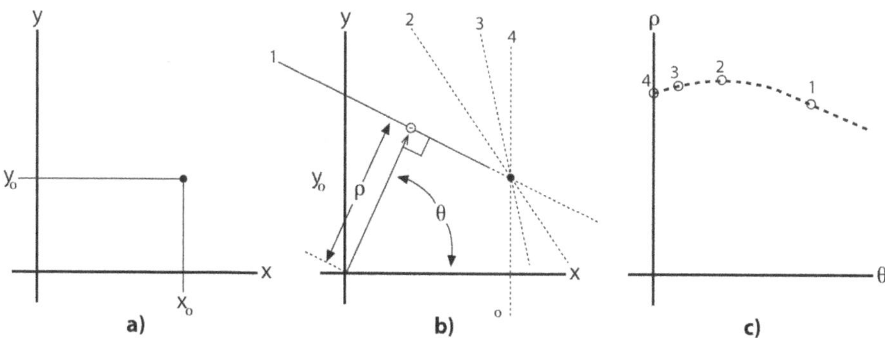


*Figure 12-6. A point ($x_0$, $y_0$) in the image plane (a) implies many lines, each parameterized by a different ρ and θ (b); these lines in the (ρ, θ) plane, taken together, form a curve of characteristic shape (c)*

The OpenCV Hough transform algorithm does not make this computation explicit to the user. Instead, it simply returns the local maxima in the ($\rho$, $\theta$) plane. However, you will need to understand this process in order to understand the arguments to the OpenCV Hough line transform function.

OpenCV supports three different kinds of Hough line transform: the *standard Hough transform* (SHT) [Duda72], the *multiscale Hough transform* (MHT), and the *progressive probabilistic Hough transform* (PPHT).[15] The SHT is the algorithm we just covered. The MHT algorithm is a slight refinement that gives more accurate values for the matched lines. The PPHT is a variation of this algorithm that, among other things, computes an extent for individual lines in addition to the orientation (as shown in Figure 12-7). It is called "probabilistic" because, rather than accumulating every possible point in the accumulator plane, it accumulates only a fraction of them. The idea is that if the peak is going to be high enough anyhow, then hitting it only a fraction of the time will be enough to find it; the result of this conjecture can be a substantial reduction in computation time.

15 The *probabilistic Hough transform* (PHT) was introduced by Kiryati, Eldar, and Bruckshtein in 1991 [Kiryati91]; the PPHT was introduced by Matas, Galambos, and Kittler in 1999 [Matas00].

*Figure 12-7. The Canny edge detector (param1=50, param2=150) is run first, with the results shown in gray, and the progressive probabilistic Hough transform (param1=50, param2=10) is run next, with the results overlaid in white; you can see that the strong lines are generally picked up by the Hough transform*

### cv::HoughLines(): The standard and multiscale Hough transforms

The standard and multiscale Hough transforms are both implemented in a single function—`cv::HoughLines()`—with the distinction being in the use (or nonuse) of two optional parameters.

```
void cv::HoughLines(
  cv::InputArray  image,                // Input single channel image
  cv::OutputArray lines,                // N-by-1 two-channel array
  double          rho,                  // rho resolution (pixels)
  double          theta,                // theta resolution (radians)
  int             threshold,            // Unnormalized accumulator threshold
  double          srn       = 0,        // rho refinement (for MHT)
  double          stn       = 0         // theta refinement (for MHT)
);
```

The first argument is the input image. It must be an 8-bit image, but the input is treated as binary information (i.e., all nonzero pixels are considered to be equivalent). The second argument is the place where the found lines will be stored. It will be an $N \times 1$ two-channel array of floating-point type (the number of columns, $N$, will be the number of lines returned).[16] The two channels will contain the rho ($\rho$) and theta ($\theta$) values for each found line.

The next two arguments, `rho` and `theta`, set the resolution desired for the lines (i.e., the resolution of the accumulator plane). The units of `rho` are pixels and the units of `theta` are radians; thus, the accumulator plane can be thought of as a two-dimensional histogram with cells of dimension `rho` pixels by `theta` radians. The `threshold` value is the value in the accumulator plane that must be reached for the algorithm to report a line. This last argument is a bit tricky in practice; it is not normalized, so you should expect to scale it up with the image size for SHT. Remember that this argument is, in effect, indicating the number of points (in the edge image) that must support the line for the line to be returned.

The parameters `srn` and `stn` are not used by the standard Hough transform; they control an extension of the SHT algorithm called the multiscale Hough transform (MHT). For MHT, these two parameters indicate higher resolutions to which the parameters for the lines should be computed. MHT first computes the locations of the lines to the accuracy given by the `rho` and `theta` parameters, and then goes on to refine those results by a factor of `srn` and `stn`, respectively (i.e., the final resolution in `rho` is `rho` divided by `srn`, and the final resolution in `theta` is `theta` divided by `stn`). Leaving these parameters set to `0` causes the SHT algorithm to be run.

### cv::HoughLinesP(): The progressive probabilistic Hough transform

```
void cv::HoughLinesP(
  cv::InputArray  image,                  // Input single channel image
  cv::OutputArray lines,                  // N-by-1 4-channel array
  double          rho,                    // rho resolution (pixels)
  double          theta,                  // theta resolution (radians)
  int             threshold,              // Unnormalized accumulator threshold
  double          minLineLength = 0,      // required line length
  double          maxLineGap    = 0       // required line separation
);
```

The `cv::HoughLinesP()` function works very much like `cv::HoughLines()`, with two important differences. The first is that the `lines` argument will be a four-channel array (or a vector of objects all of type `Vec4i`). The four channels will be the $(x_0,y_0)$ and $(x_1,y_1)$ (in that order), the $(x,y)$ locations of the two endpoints of the found line segment. The second important difference is the meaning of the two parameters. For the PPHT, the `minLineLength` and `maxLineGap` arguments set the minimum length of a line segment that will be returned, and the separation between collinear segments required for the algorithm not to join them into a single longer segment.

---

16 As usual, depending on the object type you pass to `lines`, this could be either a $1 \times N$ array with two channels, or if you like, a `std::vector<>` with $N$ entries, with each entry being of type `Vec2f`.

## Hough Circle Transform

The *Hough circle transform* [Kimme75] (see Figure 12-8) works in a manner roughly analogous to the Hough line transforms just described. The reason it is only "roughly" is that—if you were to try doing the exactly analogous thing—the accumulator *plane* would have to be replaced with an accumulator *volume* with three dimensions: one for $x$ and one for $y$ (the location of the circle center), and another for the circle radius $r$. This would mean far greater memory requirements and much slower speed. The implementation of the circle transform in OpenCV avoids this problem by using a somewhat trickier method called the *Hough gradient method.*

*Figure 12-8. The Hough circle transform finds some of the circles in the test pattern and (correctly) finds none in the photograph*

The Hough gradient method works as follows. First, the image is passed through an edge-detection phase (in this case, `cv::Canny()`). Next, for every nonzero point in the edge image, the local gradient is considered (we compute the gradient by first computing the first-order Sobel x- and y-derivatives via `cv::Sobel()`). Using this gradient, we increment every point along the line indicated by this slope—from a specified minimum to a specified maximum distance—in the accumulator. At the same time, the location of every one of these nonzero pixels in the edge image is noted. The candidate centers are then selected from those points in this (two-dimensional) accumulator that are both above some given threshold and larger than all of their immediate neighbors. These candidate centers are sorted in descending order of their accumulator values, so that the centers with the most supporting pixels appear first. Next, for each center, all of the nonzero pixels (recall that this list was built earlier) are considered. These pixels are sorted according to their distance from the center. Working out from the smallest distances to the maximum radius, we select a single radius that is best supported by the nonzero pixels. A center is kept if it has sufficient support from the nonzero pixels in the edge image *and* if it is a sufficient distance from any previously selected center.

This implementation enables the algorithm to run much faster and, perhaps more importantly, helps overcome the problem of the otherwise sparse population of a three-dimensional accumulator, which would lead to a lot of noise and render the results unstable. On the other hand, this algorithm has several shortcomings that you should be aware of.

First, the use of the Sobel derivatives to compute the local gradient—and the attendant assumption that this can be considered equivalent to a local tangent—is not a numerically stable proposition. It might be true "most of the time," but you should expect this to generate some noise in the output.

Second, the entire set of nonzero pixels in the edge image is considered for every candidate center; hence, if you make the accumulator threshold too low, the algorithm will take a long time to run. Third, because only one circle is selected for every center, if there are concentric circles then you will get only one of them.

Finally, because centers are considered in ascending order of their associated accumulator value and because new centers are not kept if they are too close to previously accepted centers, there is a bias toward keeping the larger circles when multiple circles are concentric or approximately concentric. (It is only a "bias" because of the noise arising from the Sobel derivatives; in a smooth image at infinite resolution, it would be a certainty.)

#### cv::HoughCircles(): the Hough circle transform

The Hough circle transform function `cv::HoughCircles()` has similar arguments to the line transform.

```
void cv::HoughCircles(
  cv::InputArray  image,                  // Input single channel image
  cv::OutputArray circles,                // N-by-1 3-channel or vector of Vec3f
  int             method,                 // Always cv::HOUGH_GRADIENT
  double          dp,                     // Accumulator resolution (ratio)
  double          minDist,                // Required separation (between lines)
  double          param1    = 100,        // Upper Canny threshold
  double          param2    = 100,        // Unnormalized accumulator threshold
  int             minRadius = 0,          // Smallest radius to consider
  int             maxRadius = 0           // Largest radius to consider
);
```

The input `image` is again an 8-bit image. One significant difference between `cv::HoughCircles()` and `cv::HoughLines()` is that the latter requires a binary image. The `cv::HoughCircles()` function will internally (automatically) call `cv::Sobel()`[17] for you, so you can provide a more general grayscale image.

The result array, `circles`, will be either a matrix-array or a vector, depending on what you pass to `cv::HoughCircles()`. If a matrix is used, it will be a one-dimensional array of type `CV::F32C3`; the three channels will be used to encode the location of the circle and its radius. If a vector is used, it must be of type `std::vector<Vec3f>`. The `method` argument must always be set to `cv::HOUGH_GRADIENT`.

The parameter `dp` is the resolution of the accumulator image used. This parameter allows us to create an accumulator of a lower resolution than the input image. It makes sense to do this because there is no reason to expect the circles that exist in the image to fall naturally into the same number of bins as the width or height of the image itself. If `dp` is set to `1`, then the resolutions will be the same; if set to a larger number (e.g., `2`), then the accumulator resolution will be smaller by that factor (in this case, half). The value of `dp` cannot be less than `1`.

The parameter `minDist` is the minimum distance that must exist between two circles in order for the algorithm to consider them distinct circles.

For the method set to `cv::HOUGH_GRADIENT`, the next two arguments, `param1` and `param2`, are the edge (Canny) threshold and the accumulator threshold, respectively. You may recall that the Canny edge detector actually takes two different thresholds itself. When `cv::Canny()` is called internally, the first (higher) threshold is set to the value of `param1` passed into `cv::HoughCircles()`, and the second (lower) threshold is set to exactly half that value. The parameter `param2` is the one used to threshold the accumulator and is exactly analogous to the `threshold` argument of `cv::HoughLines()`.

The final two parameters are the minimum and maximum radius of circles that can be found. This means that these are the radii of circles for which the accumulator has a representation. Example 12-2 shows an example program using `cv::HoughCircles()`.

17 The function `cv::Sobel()`, not `cv::Canny()`, is called internally. The reason is that `cv::HoughCircles()` needs to estimate the orientation of a gradient at each pixel, and this is difficult to do with a binary edge map.

*Example 12-2. Using cv::HoughCircles() to return a sequence of circles found in a grayscale image*

```
#include <opencv2/opencv.hpp>
#include <iostream>
#include <math.h>

using namespace cv;
using namespace std;

int main(int argc, char** argv) {

  if(argc != 2) {
    cout << "Hough Circle detect\nUsage: " <<argv[0] <<" <imagename>\n" << endl;
    return -1;
  }

  cv::Mat src, image;

  src  = cv::imread( argv[1], 1 );
  if( src.empty() ) { cout << "Cannot load " << argv[1] << endl; return -1; }

  cv::cvtColor(src, image, cv::BGR2GRAY);
  cv::GaussianBlur(image, image, Size(5,5), 0, 0);

  vector<cv::Vec3f> circles;
  cv::HoughCircles(image, circles, cv::HOUGH_GRADIENT, 2, image.cols/10);

  for( size_t i = 0; i < circles.size(); ++i ) {
    cv::circle(
      src,
      cv::Point(cvRound(circles[i][0]), cvRound(circles[i][1])),
      cvRound(circles[i][2]),
      cv::Scalar(0,0,255),
      2,
      cv::AA
    );
  }

  cv::imshow( "Hough Circles", src);
  cv::waitKey(0);

  return 0;

}
```

It is worth reflecting momentarily on the fact that, no matter what tricks we employ, there is no getting around the requirement that circles be described by three degrees of freedom ($x$, $y$, and $r$), in contrast to only two degrees of freedom ($\rho$ and $\theta$) for lines. The result will invariably be that any circle-finding algorithm requires more memory and computation time than the line-finding algorithms we looked at previously. With this in mind, it's a good idea to bound the radius parameter as tightly as circumstances allow in order to keep these costs under control.[18] The Hough transform was extended to arbitrary shapes by Ballard in 1981 [Ballard81] basically by considering objects as collections of gradient edges.

# Distance Transformation

The *distance transform* of an image is defined as a new image in which every output pixel is set to a value equal to the distance to the nearest zero pixel in the input image —according to some specific distance metric. It should be immediately obvious that the typical input to a distance transform should be some kind of edge image. In most applications the input to the distance transform is an output of an edge detector such as the Canny edge detector that has been inverted (so that the edges have value `0` and the nonedges are nonzero).

There are two methods available to compute the distance transform. The first method uses a mask that is typically a 3 × 3 or 5 × 5 array. Each point in the array defines the "distance" to be associated with a point in that particular position relative to the center of the mask. Larger distances are built up (and thus approximated) as sequences of "moves" defined by the entries in the mask. This means that using a larger mask will yield more accurate distances. When we use this method, given a specific distance metric, the appropriate mask is automatically selected from a set known to OpenCV. This is the "original" method developed by Borgefors (1986) [Borgefors86]. The second method computes exact distances, and is due to Felzenszwalb [Felzenszwalb04]. Both methods run in time linear in the total number of pixels, but the exact algorithm is a bit slower.

The distance metric can be any of several different types including the classic L2 (Cartesian) distance metric. Figure 12-9 shows two examples of using the distance transform on a pattern and an image.

18 Although `cv::HoughCircles()` catches centers of the circles quite well, it sometimes fails to find the correct radius. Therefore, in an application where only a center must be found (or where some different technique can be used to find the actual radius), the radius returned by `cv::HoughCircles()` can be ignored.

**User defined 3x3 mask (a=1, b=1.5)**

| | | | | | | |
|---|---|---|---|---|---|---|
| 4.5 | 4 | 3.5 | 3 | 3.5 | 4 | 4.5 |
| 4 | 3 | 2.5 | 2 | 2.5 | 3 | 4 |
| 3.5 | 2.5 | 1.5 | 1 | 1.5 | 2.5 | 3.5 |
| 3 | 2 | 1 | 0 | 1 | 2 | 3 |
| 3.5 | 2.5 | 1.5 | 1 | 1.5 | 2.5 | 3.5 |
| 4 | 3 | 2.5 | 2 | 2.5 | 3 | 4 |
| 4.5 | 4 | 3.5 | 3 | 3.5 | 4 | 4.5 |

**User defined 5x5 mask (a=1, b=1.5, c=2)**

| | | | | | | |
|---|---|---|---|---|---|---|
| 4.5 | 3.5 | 3 | 3 | 3 | 3.5 | 4.5 |
| 3.5 | 3 | 2 | 2 | 2 | 3 | 3.5 |
| 3 | 2 | 1.5 | 1 | 1.5 | 2 | 3 |
| 3 | 2 | 1 | 0 | 1 | 2 | 3 |
| 3 | 2 | 1.5 | 1 | 1.5 | 2 | 3 |
| 3.5 | 3 | 2 | 2 | 2 | 3 | 3.5 |
| 4 | 3.5 | 3 | 3 | 3 | 3.5 | 4 |

*Figure 12-9. First a Canny edge detector was run with param1=100 and param2=200; then the distance transform was run with the output scaled by a factor of 5 to increase visibility*

## cv::distanceTransform() for Unlabeled Distance Transform

When you call the OpenCV distance transform function, the output image will be a 32-bit floating-point image (i.e., `CV::F32`).

```
void cv::distanceTransform(
  cv::InputArray  src,                    // Input image
  cv::OutputArray dst,                    // Result image
  int             distanceType,           // Distance metric to use
  int             maskSize                // Mask to use (3, 5, or see below)
);
```

`cv::distanceTransform()` has two parameters. The first is `distanceType`, which indicates the distance metric to be used. Your choices here are `cv::DIST_C`, `cv::DIST_L1`, and `cv::DIST_L2`. These methods compute the distance to the nearest zero based on integer steps along a grid. The difference between the methods is that `cv::DIST_C` is the distance when the steps are counted on a four-connected grid (i.e., diagonal moves are not allowed), and `cv::DIST_L1` gives the number of steps on an eight-connected grid (i.e., diagonal moves are allowed). When `distanceType` is set to `cv::DIST_L2`, `cv::distanceTransform()` attempts to compute the exact Euclidean distances.

After the distance type is the `maskSize`, which may be 3, 5, or `cv::DIST_MASK_PRECISE`. In the case of 3 or 5, this argument indicates that a 3 × 3 or 5 × 5 mask should be used with the Borgefors method. If you are using `cv::DIST_L1` or `cv::DIST_C`, you can always use a 3 × 3 mask and you will get exact results. If you are using `cv::DIST_L2`, the Borgefors method is always approximate, and using the larger 5 × 5 mask will result in a better approximation to the L2 distance, at the cost of a slightly slower computation. Alternatively, `cv::DIST_MASK_PRECISE` can be used to indicate the Felzenszwalb algorithm (when used with `cv::DIST_L2`).

## cv::distanceTransform() for Labeled Distance Transform

It is also possible to ask the distance transform algorithm to not only calculate the distances, but to also report which object that minimum distance is to. These "objects" are called *connected components*. We will have a lot more to say about connected components in Chapter 14 but, for now, you can think of them as exactly what they sound like: structures made of continuously connected groups of zeros in the source image.

```
void cv::distanceTransform(
  cv::InputArray  src,                          // Input image
  cv::OutputArray dst,                          // Result image
  cv::OutputArray labels,                       // Connected component ids
  int             distanceType,                 // Distance metric to use
  int             maskSize,                     // (3, 5, or see below)
  int             labelType = cv::DIST_LABEL_CCOMP // How to label
);
```

If a `labels` array is provided, then as a result of running `cv::distanceTransform()` it will be of the same size as `dst`. In this case, connected components will be computed automatically, and the label associated with the nearest such component will be placed in each pixel of `labels`. The output "labels" array will basically be the discrete Voronoi diagram.

If you are wondering how to differentiate labels, consider that for any pixel that is `0` in `src`, then the corresponding distance must also be `0`. In addition, the label for that pixel must be the label of the connected component it is part of. As a result, if you want to know what label was given to any particular zero pixel, you need only look up that pixel in `labels`.

The argument `labelType` can be set either to `cv::DIST_LABEL_CCOMP` or `cv::DIST_LABEL_PIXEL`. In the former case, the function automatically finds connected components of zero pixels in the input image and gives each one a unique label. In the latter case, all zero pixels are given distinct labels.

# Segmentation

The topic of image segmentation is a large one, which we have touched on in several places already, and will return to in more sophisticated contexts later in the book. Here, we will focus on several methods of the library that specifically implement techniques that are either segmentation methods in themselves, or primitives that will be used later by more sophisticated tactics. Note that, at this time, there is no general "magic" solution for image segmentation, and it remains a very active area in computer vision research. Despite this, many good techniques have been developed that are reliable at least in some specific domain, and in practice can yield very good results.

## Flood Fill

Flood fill [Heckbert90; Shaw04; Vandevenne04] is an extremely useful function that is often used to mark or isolate portions of an image for further processing or analysis. Flood fill can also be used to derive, from an input image, masks that can be used by subsequent routines to speed or restrict processing to only those pixels indicated by the mask. The function `cv::floodFill()` itself takes an optional mask that can be further used to control where filling is done (e.g., for multiple fills of the same image).

In OpenCV, flood fill is a more general version of the sort of fill functionality that you probably already associate with typical computer painting programs. For both, a *seed point* is selected from an image and then all similar neighboring points are colored with a uniform color. The difference is that the neighboring pixels need not all be identical in color.[19] The result of a flood fill operation will always be a single contiguous region. The `cv::floodFill()` function will color a neighboring pixel if it is within a specified range (`loDiff` to `upDiff`) of either the current pixel or if (depending on the settings of `flags`) the neighboring pixel is within a specified range of the original `seed` value. Flood filling can also be constrained by an optional mask argument. There are two different prototypes for the `cv::floodFill()` routine, one that accepts an explicit `mask` parameter, and one that does not.

```
int cv::floodFill(
  cv::InputOutputArray image,                    // Input image, 1 or 3 channels
  cv::Point            seed,                     // Start point for flood
  cv::Scalar           newVal,                   // Value for painted pixels
  cv::Rect*            rect,                     // Output bounds painted domain
  cv::Scalar           lowDiff  = cv::Scalar(),// Maximum down color distance
  cv::Scalar           highDiff = cv::Scalar(),// Maximum up color distance
  int                  flags                     // Local/global, and mask-only
);

int cv::floodFill(
  cv::InputOutputArray image,                     // Input w-by-h, 1 or 3 channels
  cv::InputOutputArray mask,                      // 8-bit, w+2-by-h+2 (Nc=1)
```

19 Users of contemporary painting and drawing programs should note that most of them now employ a filling algorithm very much like `cv::floodFill()`.

```
    cv::Point            seed,                          // Start point for flood
    cv::Scalar           newVal,                        // Value for painted pixels
    cv::Rect*            rect,                          // Output bounds painted domain
    cv::Scalar           lowDiff  = cv::Scalar(),  // Maximum down color distance
    cv::Scalar           highDiff = cv::Scalar(),  // Maximum up color distance
    int                  flags                          // Local/global, and mask-only
  );
```

The parameter `image` is the input image, which can be 8-bit or a floating-point type, and must either have one or three channels. In general, this `image` array will be modified by `cv::floodFill()`. The flood-fill process begins at the location `seed`. The seed will be set to value `newVal`, as will all subsequent pixels colored by the algorithm. A pixel will be colorized if its intensity is not less than a colorized neighbor's intensity minus `loDiff` and not greater than the colorized neighbor's intensity plus `upDiff`. If the `flags` argument includes `cv::FLOODFILL_FIXED_RANGE`, then a pixel will be compared to the original seed point rather than to its neighbors. Generally, the `flags` argument controls the connectivity of the fill, what the fill is relative to, whether we are filling only a mask, and what values are used to fill the mask. Our first example of flood fill is shown in Figure 12-10.

*Figure 12-10. Results of flood fill (top image is filled with gray, bottom image with white) from the dark circle located just off center in both images; in this case, the upDiff and loDiff parameters were each set to 7.0*

The `mask` argument indicates a mask that can function both as an input to `cv::floodFill()` (in which case, it constrains the regions that can be filled) and as an output from cv::floodFill() (in which case, it will indicate the regions that actually were filled). mask must be a single-channel, 8-bit image whose size is exactly two pixels larger in width and height than the source image.[20]

In the sense that mask is an input to cv::floodFill(), the algorithm will not flood across nonzero pixels in the mask. As a result, you should zero it before use if you don't want masking to block the flooding operation.

When the mask is present, it will also be used as an output. When the algorithm runs, every "filled" pixel will be set to a nonzero value in the mask. You have the option of adding the value cv::FLOODFILL_MASK_ONLY to flags (using the usual Boolean OR operator). In this case, the input image will not be modified at all. Instead, only mask will be modified.

If the flood-fill mask is used, then the mask pixels corresponding to the repainted image pixels are set to 1. Don't be confused if you fill the mask and see nothing but black upon display; the filled values are there, but the mask image needs to be rescaled if you want to display it so you can actually see it on the screen. After all, the difference between 0 and 1 is pretty small on an intensity scale of 0 to 255.

Two possible values for flags have already been mentioned: cv::FLOODFILL_FIXED_RANGE and cv::FLOODFILL_MASK_ONLY. In addition to these, you can also add the numerical values 4 or 8.[21] In this case, you are specifying whether the flood-fill algorithm should consider the pixel array to be *four-connected* or *eight-connected*. In the former case, a four-connected array is one in which pixels are only connected to their four nearest neighbors (left, right, above, and below). In the eight-connected case, pixels are considered to be connected to diagonally neighboring pixels as well.

The flags argument is slightly tricky because it has three parts that are possibly less intuitive than they could be. The *low* 8 bits (0–7) can be set to 4 or 8, as we saw before, to control the connectivity considered by the filling algorithm. The *high* 8 bits (16–23) are the ones that contain the flags such as cv::FLOODFILL_FIXED_RANGE and cv::FLOODFILL_MASK_ONLY. The *middle* bits (8–15), however, are used a little bit dif-

20 This is done to make processing easier and faster for the internal algorithm. Note that since the mask is larger than the original image, pixel (*x*,*y*) in image corresponds to pixel (*x*+1,*y*+1) in mask. Therefore, you may find this an excellent opportunity to use cv::Mat::getSubRect().

21 The text here reads "add," but recall that flags is really a bit-field argument. Conveniently, however, 4 and 8 are single bits. So you can use "add" or "OR," whichever you prefer (e.g., flags = 8 | cv::FLOODFILL_MASK_ONLY).

ferently to actually represent a numerical value: the value with which you want the mask to be filled. If the middle bits of `flags` are 0s, the mask will be filled with 1s (the default), but any other value will be interpreted as an 8-bit unsigned integer. All these flags may be linked together via OR. For example, if you want an eight-way connectivity fill, filling only a fixed range, filling the mask (not the image), and filling using a value of 47, then the parameter to pass in would be:

```
flags = 8
      | cv::FLOODFILL_MASK_ONLY
      | cv::FLOODFILL_FIXED_RANGE
      | (47<<8);
```

Figure 12-11 shows flood fill in action on a sample image. Using `cv::FLOODFILL_FIXED_RANGE` with a wide range resulted in most of the image being filled (starting at the center). We should note that `newVal`, `loDiff`, and `upDiff` are prototyped as type `cv::Scalar` so they can be set for three channels at once. For example, `loDiff = cv::Scalar(20,30,40)` will set `loDiff` thresholds of `20` for red, `30` for green, and `40` for blue.

*Figure 12-11. Results of flood fill (top image is filled with gray, bottom image with white) from the dark circle located just off center in both images; in this case, flood fill was done with a fixed range and with a high and low difference of 25.0*

## Watershed Algorithm

In many practical contexts, we would like to segment an image but do not have the benefit of any kind of separate background mask. One technique that is often effective in this context is the *watershed algorithm* [Meyer92], which converts lines in an image into "mountains" and uniform regions into "valleys" that can be used to help segment objects. The watershed algorithm first takes the gradient of the intensity image; this has the effect of forming valleys or *basins* (the low points) where there is no texture, and of forming mountains or *ranges* (high ridges corresponding to edges) where there are dominant lines in the image. It then successively floods basins starting from caller-specified points until these regions meet. Regions that merge across the marks so generated are segmented as belonging together as the image "fills up." In this way, the basins connected to the marker point become "owned" by that marker. We then segment the image into the corresponding marked regions.

More specifically, the watershed algorithm allows a user (or another algorithm) to mark parts of an object or background that are known to be part of the object or background. Alternatively, the caller can draw a simple line or collection of lines that effectively tells the watershed algorithm to "group points like these together." The watershed algorithm then segments the image by allowing marked regions to "own" the edge-defined valleys in the gradient image that are connected with the segments. Figure 12-12 clarifies this process.

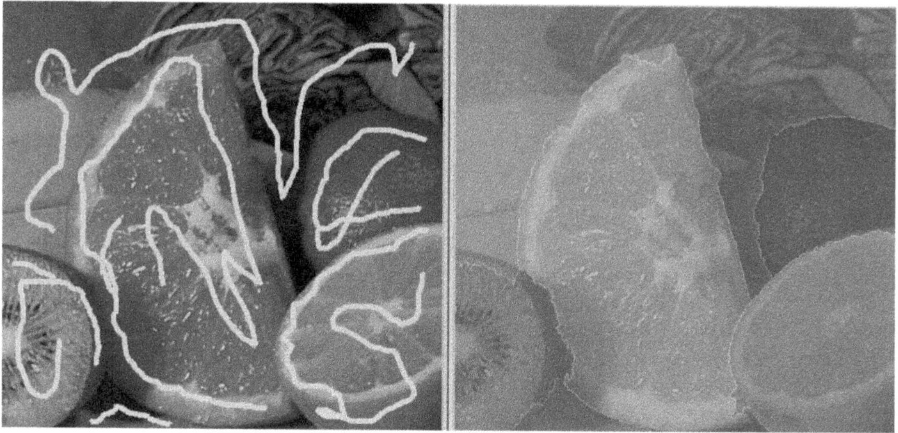

*Figure 12-12. Watershed algorithm: after a user has marked objects that belong together (left), the algorithm merges the marked area into segments (right)*

The function specification of the watershed segmentation algorithm is:

```
void cv::watershed(
  cv::InputArray       image,                  // Input 8-bit, three channels
  cv::InputOutputArray markers                 // 32-bit float, single channel
);
```

Here, `image` must be an 8-bit, three-channel (color) image and `markers` is a single-channel integer (`CV::S32`) image of the same (*x*, *y*) dimensions. On input, the value of `markers` is `0` *except* where the caller has indicated by using positive numbers that some regions belong together. For example, in the left panel of Figure 12-12, the orange might have been marked with a "1," the lemon with a "2," the lime with "3," the upper background with "4," and so on.

After the algorithm has run, all of the former zero-value pixels in `markers` will be set to one of the given markers (i.e., all of the pixels of the orange are hoped to come out with a "1" on them, the pixels of the lemon with a "2," etc.), except the boundary pixels between regions, which will be set to `-1`. Figure 12-12 (right) shows an example of such a segmentation.

It is tempting to think that all regions will be separated by pixels with marker value `-1` at their boundaries. However, this is not actually the case. Notably, if two neighboring pixels were input originally with nonzero but distinct values, they will remain touching and not separated by a `-1` pixel on output.

## Grabcuts

The Grabcuts algorithm, introduced by Rother, Kolmogorov, and Blake [Rother04], extends the Graphcuts algorithm [Boykov01] for use in user-directed image segmentation. The Grabcuts algorithm is capable of obtaining excellent segmentations, often with no more than a bounding rectangle around the foreground object to be segmented.

The original Graphcuts algorithm used user-labeled foreground and user-labeled background regions to establish distribution histograms for those two classes of image regions. It then combined the assertion that unlabeled foreground or background should conform to similar distributions with the idea that these regions tend to be smooth and connected (i.e., a bunch of blobs). These assertions were then combined into an *energy functional* that gave a low energy (i.e., cost) to solutions that conformed to these assertions and a high energy to solutions that violated them. The algorithm obtained the final result by minimizing this energy function.[22]

The Grabcuts algorithm extends Graphcuts in several important ways. The first is that it replaces the histogram models with a different (Gaussian mixture) model, enabling the algorithm to work on color images. In addition, it solves the energy functional minimization problem in an iterative manner, which provides better results overall, and allows much greater flexibility in the labeling provided by the user. Notably, this latter point makes possible even one-sided labelings, which identify either only background or only foreground pixels (where Graphcuts required both).

The implementation in OpenCV allows the caller to either just provide a rectangle around the object to be segmented, in which case the pixels "under" the rectangle's boundary (i.e., outside of it) are taken to be background and no foreground is specified. Alternatively, the caller can specify an overall mask in which pixels are categorized as being either definitely foreground, definitely background, probably foreground, or probably background.[23] In this case, the definite regions will be used to classify the other regions, with the latter being classified into the definite categories by the algorithm.

The OpenCV implementation of Grabcuts is implemented by the `cv::Grabcuts()` function:

```
void cv::grabCut(
  cv::InputArray       img,
  cv::InputOutputArray mask,
  cv::Rect             rect,
  cv::InputOutputArray bgdModel,
  cv::InputOutputArray fgdModel,
  int                  iterCount,
  int                  mode      = cv::GC_EVAL
);
```

Given an input image `img`, the resulting labeling will be computed by `cv::grabCut()` and placed in the output array `mask`. This `mask` array can also be used as an input. This is determined by the `mode` variable. If `mode` contains the flag

22 This minimization is a nontrivial problem. In practice it is performed through a technique called *Mincut*, which is how both the Graphcuts and Grabcuts algorithms get their respective names.

23 Perhaps unintuitively, the algorithm, as implemented, does not allow for a "don't know" prior labeling.

cv::GC_INIT_WITH_MASK,[24] then the values currently in mask when it is called will be used to initialize the labeling of the image. The mask is expected to be a single-channel image of cv::8U type in which every value is one of the following enumerations.

| Enumerated value | Numerical value | Meaning |
|---|---|---|
| cv::GC_BGD | 0 | Definitely background |
| cv::GC_FGD | 1 | Definitely foreground |
| cv::PR_GC_BGD | 2 | Probably background |
| cv::PR_GC_FGD | 3 | Probably foreground |

The argument rect is used only when you are not using mask initialization. When the mode contains the flag cv::GC_INIT_WITH_RECT, the entire region outside of the provided rectangle is taken to be "definitely background," while the rest is automatically set to "probably foreground."

The next two arrays are essentially temporary buffers. When you first call cv::grabCut(), they can be empty arrays. However, if for some reason you should run the Grabcuts algorithm for some number of iterations and then want to restart the algorithm for more iterations (possibly after allowing a user to provide additional "definite" pixels to guide the algorithm), you will need to pass in the same (unmodified) buffers that were filled by the previous run (in addition to using the mask you got back from the previous run as input for the next run).

Internally, the Grabcuts algorithm essentially runs the Graphcuts algorithm some number of times (with the minor extensions mentioned previously). In between each such run, the mixture models are recomputed. The itercount argument determines how many such iterations will be applied. Typical values for itercount are 10 or 12, though the number required may depend on the size and nature of the image being processed.

## Mean-Shift Segmentation

Mean-shift segmentation finds the peaks of color distributions over space [Comaniciu99]. It is related to *the mean-shift algorithm*, which we will discuss later when we

24 Actually, the way that the cv::grabCut() is implemented, you do not need to explicitly provide the cv::GC_INIT_WITH_MASK flag. This is because mask initialization is actually the default behavior. So, as long as you do not provide the cv::GC_INIT_WITH_RECT flag, you will get mask initialization. However, this is not implemented as a default argument, but rather a default in the procedural logic of the function, and is therefore not guaranteed to remain unchanged in future releases of the library. It is best to either use the cv::GC_INIT_WITH_MASK flag or the cv::GC_INIT_WITH_RECT flag explicitly; doing so not only adds future proofing, but also enhances general clarity.

talk about tracking and motion in Chapter 17. The main difference between the two is that the former looks at spatial distributions of color (and is thus related to our current topic of segmentation), while the latter tracks those distributions through time in successive frames. The function that does this segmentation based on the color distribution peaks is `cv::pyrMeanShiftFiltering()`.

Given a set of multidimensional data points whose dimensions are (*x*, *y*, blue, green, red), mean-shift can find the highest density "clumps" of data in this space by scanning a *window* over the space. Notice, however, that the spatial variables (*x*, *y*) can have very different ranges from the color-magnitude ranges (blue, green, red). Therefore, mean-shift needs to allow for different window radii in different dimensions. In this case we should have, at a minimum, one radius for the spatial variables (`spatialRadius`) and one radius for the color magnitudes (`colorRadius`). As mean-shift windows move, all the points traversed by the windows that converge at a peak in the data become connected to or "owned" by that peak. This ownership, radiating out from the densest peaks, forms the segmentation of the image. The segmentation is actually done over a scale pyramid—`cv::pyrUp()`, `cv::pyrDown()`—as described in Chapter 11, so that color clusters at a high level in the pyramid (shrunken image) have their boundaries refined at lower levels in the pyramid.

The output of the mean-shift segmentation algorithm is a new image that has been "posterized," meaning that the fine texture is removed and the gradients in color are mostly flattened. You can then further segment such images using whatever algorithm is appropriate for your needs (e.g., `cv::Canny()` combined with `cv::findContours()`, if in fact a contour segmentation is what you ultimately want).

The function prototype for `cv::pyrMeanShiftFiltering()` looks like this:

```
void cv::pyrMeanShiftFiltering(
  cv::InputArray   src,                      // 8-bit, Nc=3 image
  cv::OutputArray  dst,                      // 8-bit, Nc=3, same size as src
  cv::double       sp,                       // Spatial window radius
  cv::double       sr,                       // Color window radius
  int              maxLevel = 1,             // Max pyramid level
  cv::TermCriteria termcrit = TermCriteria(
    cv::TermCriteria::MAX_ITER | cv::TermCriteria::EPS,
    5,
    1
  )
);
```

In `cv::pyrMeanShiftFiltering()` we have an input image `src` and an output image `dst`. Both must be 8-bit, three-channel color images of the same width and height. The `spatialRadius` and `colorRadius` define how the mean-shift algorithm averages color and space together to form a segmentation. For a 640 × 480 color image, it works well to set `spatialRadius` equal to `2` and `colorRadius` equal to `40`. The next parameter of this algorithm is `max_level`, which describes how many levels of scale pyramid you want used for segmentation. A `max_level` of 2 or 3 works well for a 640 × 480 color image.

The final parameter is `cv::TermCriteria`, which we have seen in some previous algorithms. `cv::TermCriteria` is used for all iterative algorithms in OpenCV. The mean-shift segmentation function comes with good defaults if you just want to leave this parameter blank.

Figure 12-13 shows an example of mean-shift segmentation using the following values:

```
cv::pyrMeanShiftFiltering( src, dst, 20, 40, 2);
```

*Figure 12-13. Mean-shift segmentation over scale using cv::pyrMeanShiftFiltering() with parameters max_level=2, spatialRadius=20, and colorRadius=40; similar areas now have similar values and so can be treated as super pixels (larger statistically similar areas), which can speed up subsequent processing significantly*

# Summary

In this chapter, we expanded our repertoire of techniques for image analysis. Building on the general image transforms from the previous chapter, we learned new methods that we can use to better understand the images we are working with and which, as we will see, will form the foundation for many more complex algorithms. Tools such as the distance transform, integral images, and the segmentation techniques will turn out to be important building blocks for other algorithms in OpenCV as well as for your own image analysis.

# Exercises

1. In this exercise, we learn to experiment with parameters by setting good `lowThresh` and `highThresh` values in `cv::Canny()`. Load an image with suitably interesting line structures. We'll use three different high:low threshold settings of 1.5:1, 2.75:1, and 4:1.

   a. Report what you see with a high setting of less than `50`.

   b. Report what you see with high settings between `50` and `100`.

   c. Report what you see with high settings between `100` and `150`.

   d. Report what you see with high settings between `150` and `200`.

   e. Report what you see with high settings between `200` and `250`.

   f. Summarize your results and explain what happens as best you can.

2. Load an image containing clear lines and circles such as a side view of a bicycle. Use the Hough line and Hough circle calls and see how they respond to your image.

3. Can you think of a way to use the Hough transform to identify any kind of shape with a distinct perimeter? Explain how.

4. Take the Fourier transform of a small Gaussian distribution and the Fourier transform of an image. Multiply them and take the inverse Fourier transform of the results. What have you achieved? As the filters get bigger, you will find that working in the Fourier space is much faster than in the normal space.

5. Load an interesting image, and then blur it with `cv::smooth()` using a Gaussian filter.

Use long skinny rectangles; subtract and add them in place.

   a. Set `param1` = `param2` = 9. Try several settings of `param3` (e.g., 1, 4, and 6). Display the results.

   b. Set `param1` = `param2` = 0 before setting `param3` to 1, 4, and 6. Display the results. Are they different? Why?

   c. Use `param1` = `param2` = 0 again, but this time set `param3` = 1 and `param4` = 9. Smooth the picture and display the results.

   d. Repeat Exercise 5c but with `param3` = 9 and `param4` = 1. Display the results.

e. Now smooth the image once with the settings of Exercise 5c and once with the settings of Exercise 5d. Display the results.

f. Compare the results in Exercise 5e with smoothings that use `param3` = `param4` = 9 and `param3` = `param4` = 0 (i.e., a 9 × 9 filter). Are the results the same? Why or why not?

6. Write a function to compute an integral image that is rotated 45 degrees; you can then use that image to find the sum of a 45-degree rotated rectangle from four points.

7. Explain how you could use the distance transform to automatically align a known shape with a test shape when the scale is known and held fixed. How would this be done over multiple scales?

8. Write a function to blur an image using `cv::GaussianBlur()` with a kernel size of (50,50). Time it. Use the DFT of a 50 × 50 Gaussian kernel to do the same kind of blur much faster.

9. Write an image-processing function to interactively remove people from an image. Use `cv::grabCut()` to segment the person, and then use `cv::inpaint()` to fill in the resulting hole (recall that we learned about `cv::inpaint()` in the previous chapter).

10. Take a sufficiently interesting image. Use `cv::pyrMeanShiftFiltering()` to segment it. Use `cv::floodFill()` to mask off two resulting segments, and then use that mask to blur all but those segments in the image.

11. Create an image with a 20 × 20 square in it. Rotate it to an arbitrary angle. Take a distance transform of this image. Create a 20 × 20 square shape. Use the distance transform image to algorithmically overlay the shape onto the rotated square in the image you made.

12. In the 2005 DARPA Grand Challenge robot race, the authors on the Stanford team used a kind of color clustering algorithm to separate road from nonroad. The colors were sampled from a laser-defined trapezoid of road patch in front of the car. Other colors in the scene that were close in color to this patch—and whose connected component connected to the original trapezoid—were labeled as road. See Figure 12-14 where the watershed algorithm was used to segment the road after a trapezoid mark was used inside the road and an inverted "U" mark outside the road. Suppose we could automatically generate these marks. What could go wrong with this method of segmenting the road?

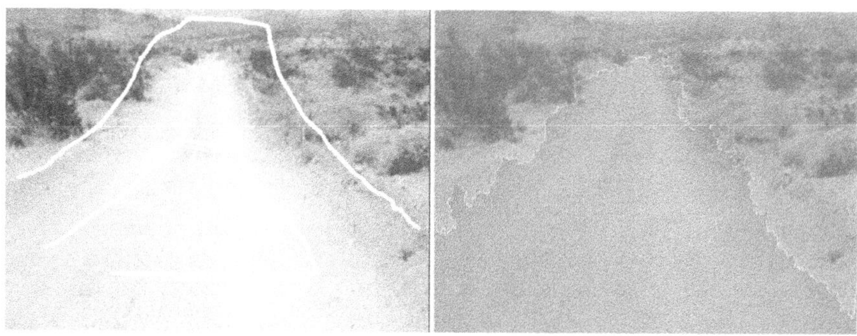

*Figure 12-14. Using the watershed algorithm to identify a road: markers are put in the original image (left), and the algorithm yields the segmented road (right)*

CHAPTER 13

# Histograms and Templates

In the course of analyzing images, objects, and video information, we frequently want to represent what we are looking at as a *histogram*. Histograms can be used to represent such diverse things as the color distribution of an object, an edge gradient template of an object [Freeman95], or the distribution of probabilities representing our current hypothesis about an object's location. Figure 13-1 shows the use of histograms for rapid gesture recognition. Edge gradients were collected from "up," "right," "left," "stop," and "OK" hand gestures. A webcam was then set up to watch a person who used these gestures to control web videos. In each frame, color interest regions were detected from the incoming video; then edge gradient directions were computed around these interest regions, and these directions were collected into orientation bins within a histogram. The histograms were then matched against the gesture models to recognize the gesture. The vertical bars in Figure 13-1 show the match levels of the different gestures. The gray horizontal line represents the threshold for acceptance of the "winning" vertical bar corresponding to a gesture model.

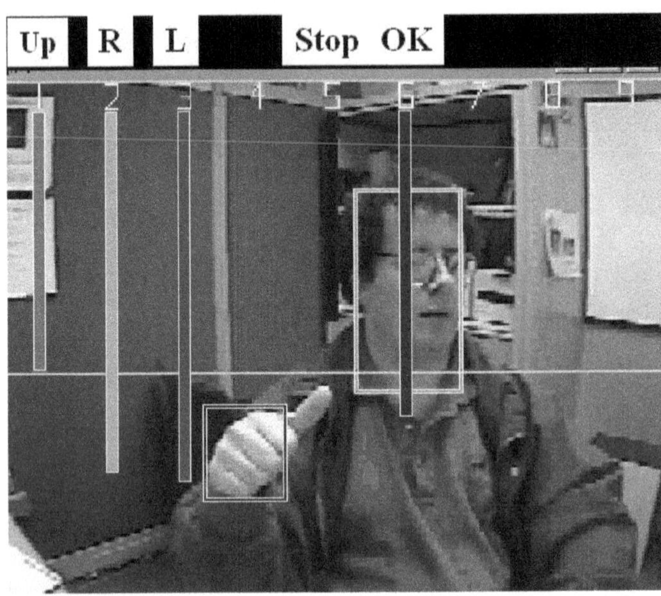


*Figure 13-1. Local histograms of gradient orientations are used to find the hand and its gesture; here the "winning" gesture (longest vertical bar) is a correct recognition of "L" (move left)*

Histograms are used in many computer vision applications. For example, they're used to detect scene transitions in videos by indicating when the edge and color statistics markedly change from frame to frame. You can use them to identify interest points in images by assigning each interest point a "tag" consisting of histograms of nearby features. Histograms of edges, colors, corners, and so on form a general feature type that is passed to classifiers for object recognition. Sequences of color or edge histograms are used to identify whether videos have been copied on the Web. And the list goes on—histograms are one of the classic tools of computer vision.

Histograms are simply collected *counts* of the underlying data organized into a set of predefined *bins*. They can be populated by counts of features computed from the data, such as gradient magnitudes and directions, color, or just about any other characteristic. In any case, they are used to obtain a statistical picture of the underlying distribution of data. The histogram usually has fewer dimensions than the source data. Figure 13-2 depicts a typical situation. The figure shows a two-dimensional distribution of points (upper left); we impose a grid (upper right) and count the data points in each *grid cell*, yielding a one-dimensional histogram (lower right). Because the raw data points can represent just about anything, the histogram is a handy way of representing whatever it is that you have learned from your image.

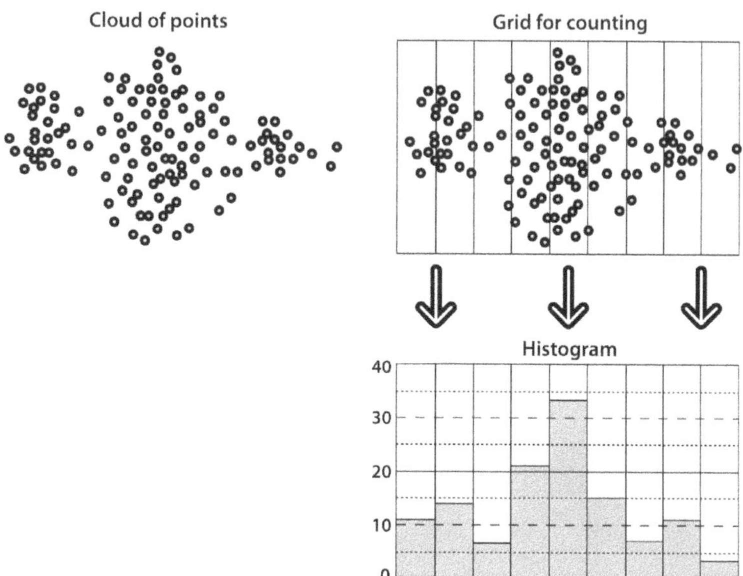


*Figure 13-2. Typical histogram example: starting with a cloud of points (upper left), a counting grid is imposed (upper right) that yields a one-dimensional histogram of point counts (lower right)*

Histograms that represent continuous distributions do so by quantizing the points into each grid cell.[1] This is where problems can arise, as shown in Figure 13-3. If the grid is too wide (upper left), then the output is too coarse and we lose the structure of the distribution. If the grid is too narrow (upper right), then there is not enough averaging to represent the distribution accurately and we get small, "spiky" cells.

1 This is also true of histograms representing information that falls naturally into discrete groups when the histogram uses fewer bins than the natural description would suggest or require. An example of this is representing 8-bit intensity values in a 10-bin histogram: each bin would then combine the points associated with approximately 25 different intensities, (erroneously) treating them all as equivalent.

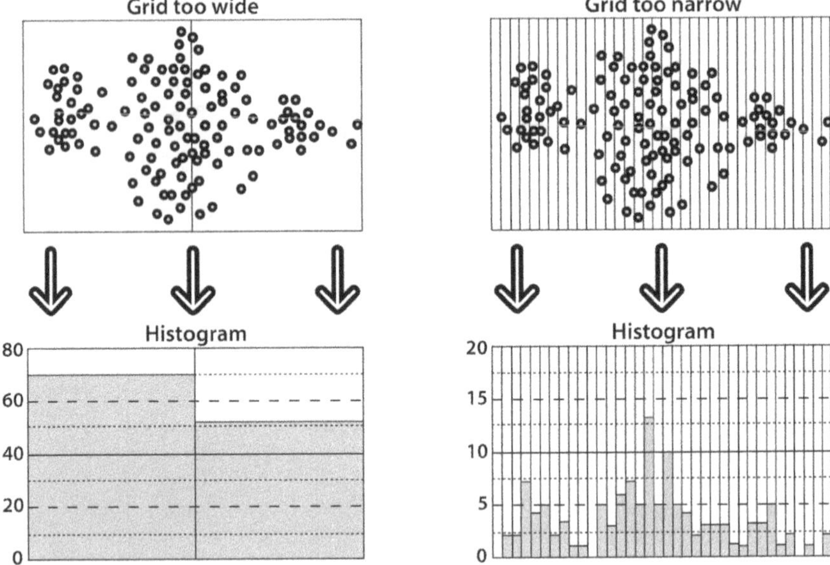


*Figure 13-3. A histogram's accuracy depends on its grid size: a grid that is too wide yields too coarse quantization in the histogram values (left); a grid that is too small yields "spiky" and singleton results from too small samples (right)*

OpenCV has a data type for representing histograms. The histogram data structure is capable of representing histograms in one or many dimensions, and it contains all the data necessary to track bins of both uniform and nonuniform sizes. And, as you might expect, it comes equipped with a variety of useful functions that allow us to easily perform common operations on our histograms.

# Histogram Representation in OpenCV

Histograms are represented in OpenCV as arrays, using the same array structures as are used for other data.[2] This means that you can use `cv::Mat` if you have a one- or two-dimensional array (with the array being $N \times 1$ or $1 \times N$ in the former case), `vector<>` types, or sparse matrices. Of course, the interpretation of the array in a histograms context is different, even though the underlying data structure is identical. For an $n$-dimensional array, the interpretation is as an $n$-dimensional array of histogram bins, in which the value for any particular element represents the number of counts

2 This is a substantial change in the C++ API relative to the C API. In the latter, there is a specific structure called a `CvHistogram` for representing histogram data. The elimination of this structure in the C++ interface creates a much simpler, more unified library.

associated with (i.e., in the range represented by) that particular bin. This distinction is important in the sense that bin numbers, being just indices into an array of some dimensionality, are simple integers. The identity of the bin—that is, what it represents—is separate from the bin's integer index. Whenever you are working with histograms, you will need to convert between measured values and histogram bin indices. For example, a histogram representing people's weights might have bins for 20–40, 40–60, 60–80, and 80–100 kilograms. In this case, these weights are the values represented by the bins, but the bin indices are still just 0, 1, 2, and 3. Many OpenCV functions will do this task (or some part of this task) for you.

Often, when you are working with higher-dimensional histograms, it will be the case that most of the entries in that histogram are zero. The `cv::SparseMat` class is very good for representing such cases. In fact, histograms are the primary reason for the existence of the `cv::SparseMat` class. Most of the basic functions that work on dense arrays will also work on sparse rays, but we will touch on a few important exceptions in the next section.

## cv::calcHist(): Creating a Histogram from Data

The function `cv::calcHist()` computes the bin values for a histogram from one or more arrays of data. Recall that the dimensions of the histogram are not related to the dimensionality of the input arrays or their size, but rather to their number. Each dimension of the histogram represents counting (and binning) across all pixels values in one of the channels of one of the input arrays. You are not required to use every channel in every image, however; you can pick whatever subset you would like of the channels of the arrays passed to `cv::calcHist()`. The function interface to `cv::calcHist()` is as follows:

```
void cv::calcHist(
  const cv::Mat*  images,                // C-style array of images, 8U or 32F
  int             nimages,               // number of images in 'images' array
  const int*      channels,              // C-style list of int's, lists channels
  cv::InputArray  mask,                  // in 'images' count, iff 'mask' nonzero
  cv::OutputArray hist,                  // output histogram array
  int             dims,                  // hist dimensionality < cv::MAX_DIMS (32)
  const int*      histSize,              // C-style array, hist sizes in each dim
  const float**   ranges,                // C-style array, 'dims' pairs set bin sizes
  bool            uniform    = true,     // true for uniform binning
  bool            accumulate = false     // true, add to 'hist' else replace
);

void cv::calcHist(
  const cv::Mat*  images,                // C-style array of images, 8U or 32F
  int             nimages,               // number of images in 'images' array
  const int*      channels,              // C-style list of int's, lists channels
  cv::InputArray  mask,                  // in 'images' count, iff 'mask' nonzero
  cv::SparseMat&  hist,                  // output histogram (sparse) array
  int             dims,                  // hist dimensionality < cv::MAX_DIMS (32)
  const int*      histSize,              // C-style array, hist sizes in each dim
  const float**   ranges,                // C-style array, 'dims' pairs set bin sizes
```

```
  bool              uniform    = true, // true for uniform binning
  bool              accumulate = false // if true, add to 'hist', else replace
);

void cv::calcHist(
  cv::InputArrayOfArrays  images,             // vector of 8U or 32F images
  const vector<int>&      channels,           // lists channels to use
  cv::InputArray          mask,               // in 'images' count, iff 'mask'
                                              // nonzero
  cv::OutputArray         hist,               // output histogram array
  const vector<int>       histSize,           // hist sizes in each dimension
  const vector<float>&    ranges,             // pairs give bin sizes in a
                                              // flat list
  bool                    accumulate = false  // if true, add to 'hist', else
                                              // replace
);
```

There are three forms of the `cv::calcHist()` function, two of which use "old-fashioned" C-style arrays, while the third uses the now preferred STL vector template type arguments. The primary distinction between the first two is whether the computed results are to be organized into a dense or a sparse array.

The first arguments are the array data, with `images` being either a pointer to a C-style list of arrays or the more modern `cv::InputArrayOfArrays`. In either case, the role of `images` is to contain one or more arrays from which the histogram will be constructed. All of these arrays must be the same size, but each can have any number of channels. These arrays may be 8-bit integers or of 32-bit floating-point type, but the type of all of the arrays must match. In the case of the C-style array input, the additional argument `nimages` indicates the number of arrays pointed to by `images`. The argument `channels` indicates which channels to consider when creating the histogram. Once again, `channels` may be a C-style array or an STL vector of integers. These integers identify which channels from the input arrays are to be used for the output histogram. The channels are numbered sequentially, so the first $N_c^{(0)}$ channels in `images[0]` are numbered 0 through $N_c^{(0)} - 1$, while the next $N_c^{(1)}$ channels in `images[1]` are numbered $N_c^{(0)}$ through $N_c^{(0)} + N_c^{(1)} - 1$, and so on. Of course, the number of entries in `channels` is equal to the number of dimensions of the histogram you are creating.

The array `mask` is an optional mask that, if present, will be used to select which pixels of the arrays in `images` will contribute to the histogram. `mask` must be an 8-bit array and the same size as the input arrays. Any pixel in `images` corresponding to a nonzero pixel in `mask` will be counted. If you do not wish to use a mask, you can pass `cv::noArray()` instead.

The `hist` argument is the output histogram you would like to fill. The `dims` argument is the number of dimensions that histogram will have. Recall that `dims` is also the number of entries in the `channels` array, indicating how each dimension is to be computed. The `histSize` argument may be a C-style array or an STL-style vector of integers and indicates the number of bins that should be allocated in each dimension of `hist`. The number of entries in `histSize` must also be equal to `dims`.

While `histSize` indicates the number of bins in each dimension, `ranges` indicates the values that correspond to each bin in each dimension. `ranges` also can be a C-style array or an STL vector. In the C-style array case, each entry `ranges[i]` is another array, and the length of ranges must be equal to `dims`. In this case, the entry `ranges[i]` indicates the bin structure of the corresponding `i`th dimension. How `ranges[i]` is interpreted depends on the value of the argument `uniform`. If `uniform` is `true`, then all of the bins in the `i`th dimension are of the same size, and all you need to do is specify the (inclusive) lower bound of the lowest bin and the (noninclusive) upper bound of the highest bin (e.g., `ranges[i] = {0,100.0}`). If, on the other hand, `uniform` is `false`, then if there are $N_i$ bins in the `i`th dimension, there must be $N_i + 1$ entries in `ranges[i]`. Thus, the `j`th entry will be interpreted as the (inclusive) lower bound of bin `j` and the (noninclusive) upper bound of bin `j – 1`. In the case in which `ranges` is of type `vector<float>`, the entries have the same meaning as C-style array values, but they are "flattened" into one single-level array (i.e., for the uniform case, there are just two entries in ranges per histogram dimension and they are in the order of the dimensions, while for the nonuniform case, there will be $N_i + 1$ entries per dimension, and they are again all in the order of the dimensions). Figure 13-4 shows these cases.

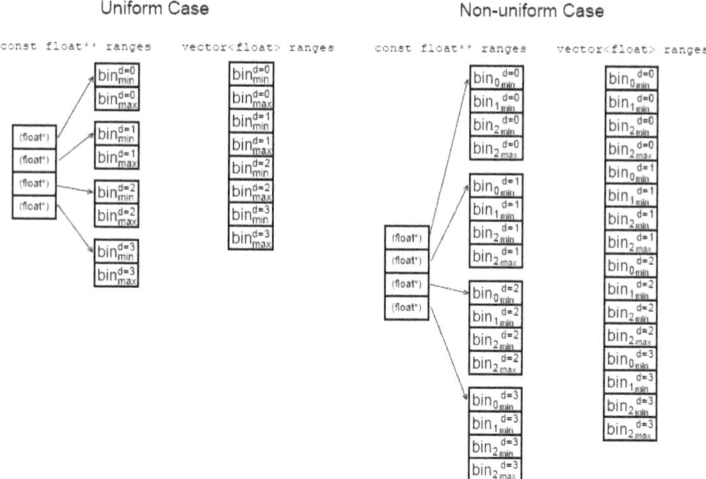


*Figure 13-4. The ranges argument may be either a C-style array of arrays, or a single STL-style vector of floating point numbers. In the case of a uniform histogram, only the minimum and maximum bin edges must be supplied. For a nonuniform histogram, the lower edge of each bin in each dimension must be supplied, as well as the maximum value*

The final argument, `accumulate`, tells OpenCV that the array `hist` is not to be deleted, reallocated, or otherwise set to `0` before new counts are added from the arrays in `images`.

# Basic Manipulations with Histograms

Even though the data structure for the histogram is the same as the data structure used for matrices and image arrays, this particular interpretation of the data structure invites new operations on these arrays that accomplish tasks specific to histograms. In this section, we will touch on some simple operations that are specific to histograms, as well as review how some important histogram manipulations can be performed with array operations that we have already discussed in prior chapters.

## Histogram Normalization

When dealing with a histogram, we first need to accumulate information into its various bins. Once we have done this, however, it is often desirable to work with the histogram in *normalized form* so that individual bins will represent the fraction of the total number of events assigned to the entire histogram. In the C++ API, we can accomplish this by simply using the array algebra operators and operations:

```
cv::Mat normalized = my_hist / sum(my_hist)[0];
```

or:

```
cv::normalize( my_hist, my_hist, 1, 0, NORM_L1 );
```

## Histogram Threshold

It is also common that you wish to threshold a histogram and (for example) throw away all bins whose entries contain less than some minimum number of counts. Like normalization, this operation can be accomplished without the use of any particular special histogram routine. Instead, you can use the standard array threshold function:

```
cv::threshold(
  my_hist,                    // input histogram
  my_thresholded_hist,        // result, all values<threshold set to zero
  threshold,                  // cutoff value
  0,                          // value does not matter in this case
  cv::THRESH_TOZERO           // threshold type
);
```

## Finding the Most Populated Bin

In some cases, you would like to find the bins that are above some threshold, and throw away the others. In other cases, you would like to simply find the one bin that has the most weight in it. This is particularly common when the histogram is being used to represent a probability distribution. In this case, the array `cv::minMaxLoc()` will give you what you want.

In the case of a two-dimensional array, you can use the `cv::InputArray` form of `cv::minMaxLoc()`:

```
void cv::minMaxLoc(
  cv::InputArray src,                    // Input array
  double*        minVal,                 // put minimum value (if not NULL)
  double*        maxVal = 0,             // put maximum value (if not NULL)
  cv::Point*     minLoc = 0,             // put minimum location (if not NULL)
  cv::Point*     maxLoc = 0,             // put maximum location (if not NULL)
  cv::InputArray mask   = cv::noArray() // ignore points for which mask is zero
);
```

The arguments `minVal` and `maxVal` are pointers to locations you provide for `cv::minMaxLoc()` to store the minimum and maximum values that have been identified. Similarly, `minLoc` and `maxLoc` are pointers to variables (of type `cv::Point`, in this case) where the actual locations of the minimum and maximum can be stored. If you do not want one or more of these four results to be computed, you can simply pass `NULL` for that (pointer) variable and that information will not be computed:

```
double    max_val;
cv::Point max_pt;

cv::minMaxLoc(
  my_hist,     // input histogram
  NULL,        // don't care about the min value
  &max_val,    // place to put the maximum value
  NULL,        // don't care about the location of the min value
  &max_pt      // place to put the maximum value location (a cv::Point)
);
```

In this example, though, the histogram would need to be two-dimensional.[3] If your histogram is of sparse array type, then there is no problem. Recall that there is an alternate form of `cv::minMaxLoc()` for sparse arrays:

```
void cv::minMaxLoc(
  const cv::SparseMat& src,                    // Input (sparse) array
  double*              minVal,                 // put min value (if not NULL)
  double*              maxVal = 0,             // put max value (if not NULL)
  int*                 minLoc = 0,             // put min location (if not NULL)
  int*                 maxLoc = 0,             // put max location (if not NULL)
  cv::InputArray       mask   = cv::noArray() // ignore points if mask is zero
);
```

3 If you have a one-dimensional `vector<>` array, you can just use `cv::Mat( vec ).reshape(1)` to make it an $N \times 1$ array in two dimensions.

Note that this form of cv::minMaxLoc() actually differs from the previous form in several ways. In addition to taking a sparse matrix as the source, it also takes type int* for the minIdx and maxIdx variables instead of cv::Point* for the analogous minLoc and maxLoc variables. This is because the sparse matrix form of cv::minMaxLoc() supports arrays of any dimensionality. Therefore, you need to allocate the location variables yourself and make sure that they have the correct amount of space available for the *n*-dimensional index associated with a point in the (*n*-dimensional) sparse histogram:

```
double maxval;
int    max_pt[CV_MAX_DIM];

cv::minMaxLoc(
  my_hist,              // input sparse histogram
  NULL,                 // don't care about the min value
  &max_val,             // place to put the maximum value
  NULL,                 // don't care about the location of the min value
  max_pt                // place to put the maximum value location (a cv::Point)
);
```

It turns out that if you want to find the minimum or maximum of an *n*-dimensional array that is not sparse, you need to use another function. This function works essentially the same as cv::minMaxLoc(), and has a similar name, but is not quite the same creature:

```
void cv::minMaxIdx(
  cv::InputArray src,
  double*        minVal,                  // put min value (if not NULL)
  double*        maxVal = 0,              // put max value (if not NULL)
  int*           minLoc = 0,              // put min location indices (if not NULL)
  int*           maxLoc = 0,              // put max location indices (if not NULL)
  cv::InputArray mask   = cv::noArray()   // ignore points if mask is zero
);
```

In this case, the arguments have the same meanings as the corresponding arguments in the two forms of cv::minMaxLoc(). You must allocate minIdx and maxIdx to C-style arrays of the correct size yourself (as before). One word of warning is in order here however: if the input array src is one-dimensional, you should allocate minIdx and maxIdx to be of dimension two. The reason for this is that cv::minMaxIdx() treats a one-dimensional array as a two-dimensional array internally. As a result, if the maximum is found at position *k*, the return value for maxIdx will be (k,0) for a single-column matrix and (0,k) for a single-row matrix.

## Comparing Two Histograms

Yet another indispensable tool for working with histograms, first introduced by Swain and Ballard [Swain91] and further generalized by Schiele and Crowley

[Schiele96], is the ability to compare two histograms in terms of some specific criteria for similarity. The function `cv::compareHist()` does just this.

```
double cv::compareHist(
  cv::InputArray        H1,        // First histogram to be compared
  cv::InputArray        H2,        // Second histogram to be compared
  int                   method     // comparison method (see options below)
);

double cv::compareHist(
  const cv::SparseMat& H1,        // First histogram to be compared
  const cv::SparseMat& H2,        // Second histogram to be compared
  int                   method     // comparison method (see options below)
);
```

The first two arguments are the histograms to be compared, which should be of the same size. The third argument is where we select our desired distance metric. We can use this method, of course, to match two whole images by taking the histogram of each and comparing the histogram via the methods described next. We can also use it to find objects in images by taking a histogram of the object and searching different subregions of an image, taking their histograms and seeing how well they match using one of the following histogram comparison methods.[4]

The four available options are as follows:

### Correlation method (cv::COMP_CORREL)

The first comparison is based on statistical correlation; it implements the Pearson correlation coefficient, and is typically appropriate when $H_1$ and $H_2$ can be interpreted as probability distributions.

$$d_{correl}(H_1, H_2) = \frac{\sum_i H_1'(i) \cdot H_2'(i)}{\sqrt{\sum_i H_1'^2(i) \cdot \sum_i H_2'^2(i)}}$$

Here, $H_k'(i) = H_k(i) - N^{-1}\sum_j h_k(j)$ and $N$ is equal to the number of bins in the histogram.

For *correlation*, a high score represents a better match than a low score. A perfect match is `1` and a maximal mismatch is `-1`; a value of `0` indicates no correlation (random association).

4 In OpenCV 2.4, there was a function that automated this subregion matching, `cv::calcBackProjectPatch()`, but it was removed from OpenCV 3.0 and onward because it was slow.

### Chi-square method (cv::COMP_CHISQR_ALT)

For *chi-square*,[5] the distance metric is based on the chi-squared test statistic, which is an alternate test of whether two distributions are correlated.

$$d_{chi\text{-}square}(H_1, H_2) = \sum_i \frac{(H_1(i) - H_2(i))^2}{H_1(i) + H_2(i)}$$

For this test, a low score represents a better match than a high score. A perfect match is `0` and a total mismatch is unbounded (depending on the size of the histogram).

### Intersection method (cv::COMP_INTERSECT)

The *histogram intersection* method is based on a simple intersection of the two histograms. This means that it asks, in effect, what do these two have in common, and sums over all of the bins of the histograms.

$$d_{intersection}(H_1, H_2) = \sum_i \min(H_1(i), H_2(i))$$

For this metric, high scores indicate good matches and low scores indicate bad matches. If both histograms are normalized to `1`, then a perfect match is `1` and a total mismatch is `0`.

### Bhattacharyya distance method (cv::COMP_BHATTACHARYYA)

The last option, called the *Bhattacharyya distance* [Bhattacharyya43], is also a measure of the overlap of two distributions.

$$d_{correl}(H_1, H_2) = \sqrt{1 - \frac{\sum_i H_1(i) \cdot H_2(i)}{\sqrt{\sum_i H_1(i) \sum_i H_2(i)}}}$$

In this case, low scores indicate good matches and high scores indicate bad matches. A perfect match is `0` and a total mismatch is `1`.

With `cv::COMP_BHATTACHARYYA`, a special factor in the code is used to normalize the input histograms. In general, however, you should normalize histograms *before* comparing them, because concepts like histogram intersection make little sense (even if allowed) without normalization.

5 The chi-square test was invented by Karl Pearson [Pearson], who founded the field of mathematical statistics.

The simple case depicted in Figure 13-5 should clarify matters. In fact, this is about the simplest case that could be imagined: a one-dimensional histogram with only two bins. The model histogram has a `1.0` value in the left bin and a `0.0` value in the right bin. The last three rows show the comparison histograms and the values generated for them by the various metrics (the EMD metric will be explained shortly).

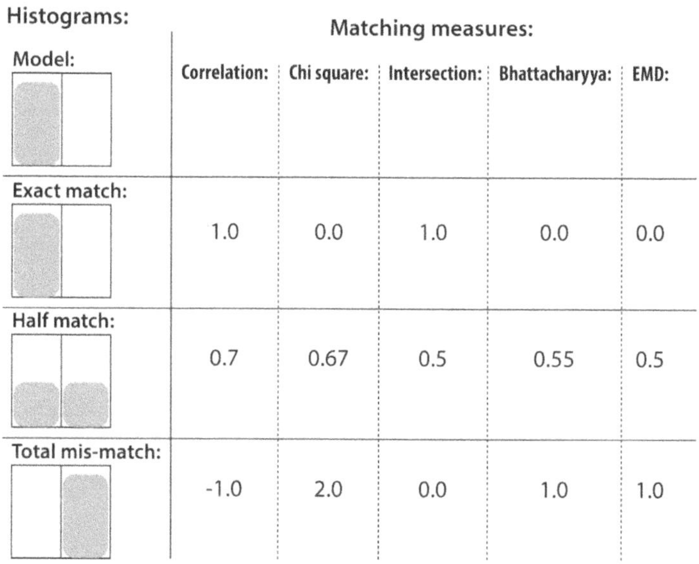

| Histograms: | Matching measures: | | | | |
|---|---|---|---|---|---|
| Model: | Correlation: | Chi square: | Intersection: | Bhattacharyya: | EMD: |
| Exact match: | 1.0 | 0.0 | 1.0 | 0.0 | 0.0 |
| Half match: | 0.7 | 0.67 | 0.5 | 0.55 | 0.5 |
| Total mis-match: | -1.0 | 2.0 | 0.0 | 1.0 | 1.0 |

*Figure 13-5. Histogram matching measures*

Figure 13-5 provides a quick reference for the behavior of different matching types. Close inspection of these matching algorithms in the figure will reveal a disconcerting observation: if histogram bins shift by just one slot, as with the chart's first and third comparison histograms, then all these matching methods (except EMD) yield a maximal mismatch even though these two histograms have a similar "shape." The rightmost column in Figure 13-5 reports values returned by EMD, also a type of distance measure. In comparing the third to the model histogram, the EMD measure quantifies the situation precisely: the third histogram has moved to the right by one unit. We will explore this measure further in the section "Earth Mover's Distance" on page 391.

In the authors' experience, intersection works well for quick-and-dirty matching, and chi-square or Bhattacharyya work best for slower but more accurate matches. The EMD measure gives the most intuitive matches but is much slower.

## Histogram Usage Examples

It's probably time for some helpful examples. The program in Example 13-1 (adapted from the OpenCV code bundle) shows how we can use some of the functions just discussed. This program computes a hue-saturation histogram from an incoming image, and then draws that histogram as an illuminated grid.

*Example 13-1. Histogram computation and display*

```
#include <opencv2/opencv.hpp>
#include <iostream>

using namespace std;

int main( int argc, char** argv ){

  if(argc != 2) {
    cout << "Computer Color Histogram\nUsage: " <<argv[0] <<" <imagename>" << endl;
    return -1;
  }

  cv::Mat src = cv::imread( argv[1],1 );
  if( src.empty() ) { cout << "Cannot load " << argv[1] << endl; return -1; }

  // Compute the HSV image, and decompose it into separate planes.
  //
  cv::Mat hsv;
  cv::cvtColor(src, hsv, cv::COLOR_BGR2HSV);

  float h_ranges[]      = {0, 180}; // hue is [0, 180]
  float s_ranges[]      = {0, 256};
  const float* ranges[] = {h_ranges, s_ranges};
  int histSize[]        = {30, 32}, ch[] = {0, 1};

  cv::Mat hist;

  // Compute the histogram
  //
  cv::calcHist(&hsv, 1, ch, cv::noArray(), hist, 2, histSize, ranges, true);
  cv::normalize(hist, hist, 0, 255, cv::NORM_MINMAX);

  int scale = 10;
  cv::Mat hist_img(histSize[0]*scale, histSize[1]*scale, CV_8UC3);

  // Draw our histogram.
  //
  for( int h = 0; h < histSize[0]; h++ ) {
    for( int s = 0; s < histSize[1]; s++ ){
      float hval = hist.at<float>(h, s);
      cv::rectangle(
        hist_img,
        cv::Rect(h*scale,s*scale,scale,scale),
        cv::Scalar::all(hval),
        -1
      );
```

```
    }
  }

  cv::imshow("image", src);
  cv::imshow("H-S histogram", hist_img);
  cv::waitKey();

  return 0;

}
```

In this example, we have spent a fair amount of time preparing the arguments for `cv::calcHist()`, which is not uncommon.

In many practical applications, it is useful to consider the color histograms associated with human skin tone. By way of example, Figure 13-6 contains histograms taken from a human hand under various lighting conditions. The left column shows images of a hand in an indoor environment, a shaded outdoor environment, and a sunlit outdoor environment. In the middle column are the blue, green, and red (BGR) histograms corresponding to the observed flesh tone of the hand. In the right column are the corresponding HSV histograms, where the vertical axis is V (value), the radius is S (saturation), and the angle is H (hue). Notice that indoors is the darkest, outdoors in shadow is a bit brighter, and outdoors in the sun is the brightest. Note also that the colors shift somewhat as a result of the changing color of the illuminating light.

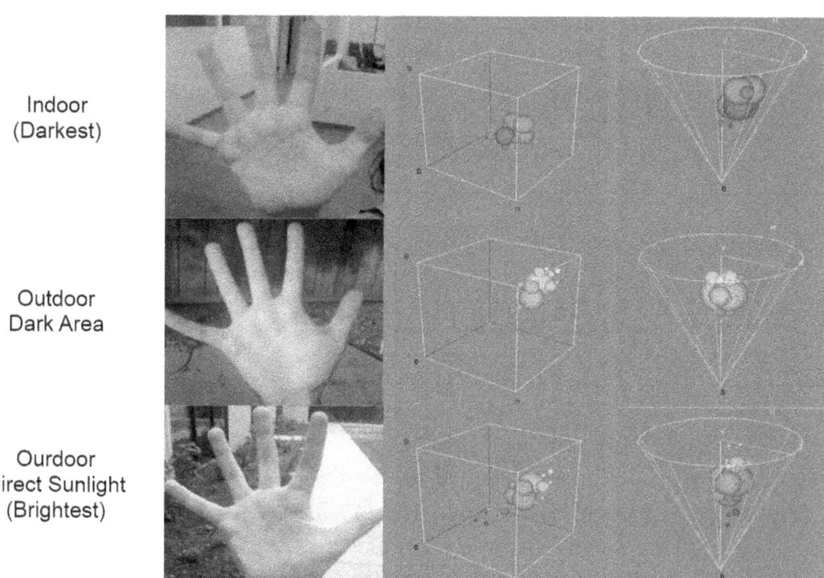


*Figure 13-6. Histogram of flesh colors under indoor (upper left), shadowed outdoor (middle left), and direct sun outdoor (lower left) lighting conditions; the center and righthand columns display the associated BGR and HSV histograms, respectively*

As a test of histogram comparison, we could take a portion of one palm (e.g., the top half of the indoor palm), and compare the histogram representation of the colors in that image either with the histogram representation of the colors in the remainder of that image or with the histogram representations of the other two hand images. Flesh tones are often easier to pick out after conversion to an HSV color space. Restricting ourselves to the hue and saturation planes is not only sufficient but also helps with recognition of flesh tones across ethnic groups.

To put this experiment into practice (see Example 13-2), we take the three images of a hand under different lighting conditions (Figure 13-6). First we construct a histogram from the hand portion of the top image (the dark one), which we will use as our reference. We then compare that histogram to a histogram taken from the hand in the bottom half of that same image, and then to the hands that appear in the next two (whole) images. The first image is an indoor image, while the latter two are outdoors. The matching results are shown in Table 13-1. Note that some of the distance metrics return a small number when the distance is small, and a larger number when it is high, while other metrics do the opposite. This is what we should have expected from the simple analysis of the matching measures shown in Figure 13-5. Also note that while the indoor, lower-half image matches well, the intensity and color of the light outdoors creates bad matches for the skin.

*Table 13-1. Histogram comparison, via four matching methods, of palm-flesh colors in upper half of indoor palm with listed variant palm-flesh color. For reference, the expected score for a perfect match is provided in the first row.*

| Comparison | CORREL | CHISQR | INTERSECT | BHATTACHARYYA |
|---|---|---|---|---|
| (Perfect match) | (1.0) | (0.0) | (1.0) | (0.0) |
| Indoor lower half | 0.96 | 0.14 | 0.82 | 0.2 |
| Outdoor shade | 0.09 | 1.57 | 0.13 | 0.8 |
| Outdoor sun | 0.0 | 1.98 | 0.01 | 0.99 |

# Some More Sophisticated Histograms Methods

Everything we've discussed so far was reasonably basic. Each of the functions provided for a relatively obvious need. Collectively, they form a good foundation for much of what you might want to do with histograms in the context of computer vision (and probably in other contexts as well). At this point we want to look at some more sophisticated methods available within OpenCV that are extremely useful in certain applications. These routines include a more advanced method of comparing two histograms as well as tools for computing and/or visualizing which portions of an image contribute to a given portion of a histogram.

## Earth Mover's Distance

We saw earlier that lighting changes can cause significant shifts in color values (see Figure 13-6), although such shifts tend not to change the shape of the histogram of color values, but instead shift the color value locations and thus cause the histogram-matching schemes we've covered to fail. The difficulty with histogram *match* measures is that they can return a large difference in the case where two histograms are similarly shaped, but only displaced relative to one another. It is often desirable to have a *distance* measure that performs like a match, but is less sensitive to such displacements. Earth mover's distance (EMD) [Rubner00] is such a metric; it essentially measures how much work it would take to "shovel" one histogram shape into another, including moving part (or all) of the histogram to a new location. It works in any number of dimensions.

Return again to Figure 13-5; we see the "earth shoveling" nature of EMD's distance measure in the rightmost column. An exact match is a distance of `0`. Half a match is half a "shovel full," the amount it would take to spread half of the left histogram into the next slot. Finally, moving the entire histogram one step to the right would require an entire unit of distance (i.e., to change the model histogram into the "totally mismatched" histogram).

The EMD algorithm itself is quite general; it allows users to set their own distance metric or their own cost-of-moving matrix. You can record where the histogram "material" flowed from one histogram to another, and employ nonlinear distance metrics derived from prior information about the data. The EMD function in OpenCV is `cv::EMD()`:

```
float cv::EMD(
  cv::InputArray  signature1,               // sz1-by-(dims+1) float array
  cv::InputArray  signature2,               // sz2-by-(dims+1) float array
  int             distType,                 // distance type (e.g., 'cv::DIST_L1')
  cv::InputArray  cost       = noArray(),   // sz1-by-sz2 array (if cv::DIST_USER)
  float*          lowerBound = 0,           // input/output low bound on distance
  cv::OutputArray flow       = noArray()    // output, sz1-by-sz2
);
```

Although we're applying the EMD to histograms, the interface prefers that we talk to it in terms of what the algorithm calls *signatures* for the first two array parameters. These signatures are arrays that are always of type `float` and consist of rows containing the histogram bin count followed by its coordinates. For the one-dimensional histogram of Figure 13-5, the signatures (listed array rows) for the lefthand column of histograms (skipping the model) are as follows: top, [[1, 0], [0, 1]]; middle, [[0.5, 0], [0.5, 1]]; bottom, [[0, 0], [1, 1]]. If we had a bin in a three-dimensional histogram with a bin count of 537 at $(x, y, z)$ index (7, 43, 11), then the signature row for that bin would be [537, 7, 43, 11]. In general, this will be a necessary step before calling cv::EMD(); you will need to convert your histograms into signatures. (We will go through this in a little more detail in Example 13-2.)

The parameter distType should be any of: *Manhattan distance* (cv::DIST_L1), *Euclidean distance* (cv::DIST_L2), *checkerboard distance* (cv::DIST_C), or a user-defined distance metric (cv::DIST_USER). In the case of the user-defined distance metric, the user supplies this information in the form of a (precalculated) cost matrix via the cost argument. (In this case, cost is an $n_1 \times n_2$ matrix, with $n_1$ and $n_2$ the sizes of signature1 and signature2, respectively.)

The argument lowerBound has two functions (one as input, the other as output). As a return value, it is a lower bound on the distance between the centers of mass of the two histograms. In order for this lower bound to be computed, one of the standard distance metrics must be in use (i.e., not cv::DIST_USER), and the total weights of the two signatures must be the same (as would be the case for normalized histograms). If you supply a lower-bound argument, you must also initialize that variable to a meaningful value. This value is used as the lower bound on separations for which the EMD will be computed at all.[6] Of course, if you want the EMD computed no matter what the distance is, you can always initialize lowerBound to 0.

The next argument, flow, is an optional $n_1 \times n_2$ matrix that can be used to record the *flow* of mass from the ith point of signature1 to the jth point of signature2. In essence, this tells you how the mass was rearranged to give the computed total EMD.

As an example, suppose we have two histograms, hist1 and hist2, which we want to convert into two signatures, sig1 and sig2. Just to make things more difficult, let's suppose that these are two-dimensional histograms (as in the preceding code examples) of dimension h_bins by s_bins. Example 13-2 shows how to convert these two histograms into two signatures.

*Example 13-2. Creating signatures from histograms for EMD; note that this code is the source of the data in Table 13-1, in which the hand histogram is compared in different lighting conditions*

```
#include <opencv2/opencv.hpp>
#include <iostream>

using namespace std;
```

6 This is important because it is typically possible to compute the lower bound for a pair of histograms much more quickly than the actual EMD. As a result, in many practical cases, if the EMD is above some bound, you probably do not care about the actual value of the EMD, only that it is "too big" (i.e., the things you are comparing are "not similar"). In this case, it is quite helpful to have cv::EMD() exit once it is known that the EMD value will be big enough that you do not care about the exact value.

```
void help( char** argv ){
  cout << "\nCall is:\n"
  << argv[0] <<" modelImage0 testImage1 testImage2 badImage3\n\n"
  << "for example: "
  << "  ./ch7_ex7_3_expanded HandIndoorColor.jpg HandOutdoorColor.jpg "
  << "HandOutdoorSunColor.jpg fruits.jpg\n"
  << "\n";
}

// Compare 3 images' histograms
int main( int argc, char** argv ) {

  if( argc != 5 ) { help( argv ); return -1; }

  vector<cv::Mat> src(5);
  cv::Mat         tmp;
  int             i;

  tmp = cv::imread( argv[1],  1);
  if( tmp.empty() ) {
    cerr << "Error on reading image 1," << argv[1] << "\n" << endl;
    help();
    return(-1);
  }

  // Parse the first image into two image halves divided halfway on y
  //
  cv::Size size  = tmp.size();
  int width      = size.width;
  int height     = size.height;
  int halfheight = height >> 1;

  cout <<"Getting size [["   <<tmp.cols <<"] [" <<tmp.rows <<"]]\n" <<endl;
  cout <<"Got size (w,h): (" <<size.width <<"," <<size.height <<")" <<endl;

  src[0] = cv::Mat(cv::Size(width,halfheight), CV_8UC3);
  src[1] = cv::Mat(cv::Size(width,halfheight), CV_8UC3);

  // Divide the first image into top and bottom halves into src[0] and src[1]
  //
  cv::Mat_<cv::Vec3b>::iterator tmpit = tmp.begin<cv::Vec3b>();

  // top half
  //
  cv::Mat_<cv::Vec3b>::iterator s0it = src[0].begin<cv::Vec3b>();
  for(i = 0; i < width*halfheight; ++i, ++tmpit, ++s0it) *s0it = *tmpit;

  // Bottom half
  //
  cv::Mat_<cv::Vec3b>::iterator s1it = src[1].begin<cv::Vec3b>();
  for(i = 0; i < width*halfheight; ++i, ++tmpit, ++s1it) *s1it = *tmpit;
```

```
// Load the other three images
//
for(i = 2; i<5; ++i){
  src[i] = cv::imread(argv[i], 1);
  if(src[i].empty()) {
    cerr << "Error on reading image " << i << ": " << argv[i] << "\n" << endl;
    help();
    return(-1);
  }
}

// Compute the HSV image, and decompose it into separate planes.
//
vector<cv::Mat> hsv(5), hist(5), hist_img(5);
int          h_bins      = 8;
int          s_bins      = 8;
int          hist_size[] = { h_bins, s_bins }, ch[] = {0, 1};
float        h_ranges[]  = { 0, 180 };                  // hue range is [0,180]
float        s_ranges[]  = { 0, 255 };
const float* ranges[]    = { h_ranges, s_ranges };
int          scale       = 10;

for(i = 0; i<5; ++i) {
  cv::cvtColor( src[i], hsv[i], cv::BGR2HSV );
  cv::calcHist( &hsv[i], 1, ch, noArray(), hist[i], 2, hist_size, ranges, true );
  cv::normalize( hist[i], hist[i], 0, 255, cv::NORM_MINMAX );
  hist_img[i] = cv::Mat::zeros( hist_size[0]*scale, hist_size[1]*scale, CV_8UC3 );

  // Draw our histogram For the 5 images
  //
  for( int h = 0; h < hist_size[0]; h++ )
    for( int s = 0; s < hist_size[1]; s++ ){
      float hval = hist[i].at<float>(h, s);
      cv::rectangle(
        hist_img[i],
        cv::Rect(h*scale, s*scale, scale, scale),
        cv::Scalar::all(hval),
        -1
      );
    }
}

// Display
//
cv::namedWindow( "Source0", 1 );cv::imshow( "Source0", src[0] );
cv::namedWindow( "HS Histogram0", 1 );cv::imshow( "HS Histogram0", hist_img[0] );

cv::namedWindow( "Source1", 1 );cv::imshow( "Source1", src[1] );
cv::namedWindow( "HS Histogram1", 1 ); cv::imshow( "HS Histogram1", hist_img[1] );

cv::namedWindow( "Source2", 1 ); cv::imshow( "Source2", src[2] );
cv::namedWindow( "HS Histogram2", 1 ); cv::imshow( "HS Histogram2", hist_img[2] );
```

```
  cv::namedWindow( "Source3", 1 ); cv::imshow( "Source3", src[3] );
  cv::namedWindow( "HS Histogram3", 1 ); cv::imshow( "HS Histogram3", hist_img[3] );

  cv::namedWindow( "Source4", 1 ); cv::imshow( "Source4", src[4] );
  cv::namedWindow( "HS Histogram4", 1 ); cv::imshow( "HS Histogram4", hist_img[4] );

  // Compare the histogram src0 vs 1, vs 2, vs 3, vs 4
  cout << "Comparison:\n"
    << "Corr                Chi                  Intersect          Bhat\n"
    << endl;

  for(i=1; i<5; ++i) {  // For each histogram
    cout << "Hist[0] vs Hist[" << i << "]: " << endl;;
    for(int j=0; j<4; ++j) { // For each comparison type
      cout << "method[" << j << "]: " << cv::compareHist(hist[0],hist[i],j) << "  ";
    }
    cout << endl;
  }

  //Do EMD and report
  //
  vector<cv::Mat> sig(5);
  cout << "\nEMD: " << endl;

  // Oi Vey, parse histograms to earth movers signatures
  //
  for( i=0; i<5; ++i) {

    vector<cv::Vec3f> sigv;

    // (re)normalize histogram to make the bin weights sum to 1.
    //
    cv::normalize(hist[i], hist[i], 1, 0, cv::NORM_L1);
    for( int h = 0; h < h_bins; h++ )
      for( int s = 0; s < s_bins; s++ ) {
        float bin_val = hist[i].at<float>(h, s);
        if( bin_val != 0 )
          sigv.push_back( cv::Vec3f(bin_val, (float)h, (float)s));
      }

    // make Nx3 32fC1 matrix, where N is the number of nonzero histogram bins
    //
    sig[i] = cv::Mat(sigv).clone().reshape(1);
    if( i > 0 )
      cout << "Hist[0] vs Hist[" << i << "]: "
           << EMD(sig[0], sig[i], cv::DIST_L2) << endl;
  }

  cv::waitKey(0);

}
```

## Back Projection

Back projection is a way of recording how well the pixels fit the distribution of pixels in a histogram model. For example, if we have a histogram of flesh color, then we can use back projection to find flesh-colored areas in an image. The function for doing this kind of lookup has two variations, one for dense arrays and one for sparse arrays.

### Basic back projection: cv::calcBackProject()

Back projection computes a vector from the selected channels of the input images just like `cv::calcHist()`, but instead of accumulating events in the output histogram it reads the input histogram and reports the bin value already present. In the context of statistics, if you think of the input histogram as a (prior) probability distribution for the particular vector (color) on some object, then back projection is computing the probability that any particular part of the image is in fact drawn from that prior distribution (e.g., part of the object).

```
void cv::calcBackProject(
  const cv::Mat*  images,              // C-style array of images, 8U or 32F
  int             nimages,             // number of images in 'images' array
  const int*      channels,            // C-style list, ints identifying channels
  cv::InputArray  hist,                // input histogram array
  cv::OutputArray backProject,         // output single channel array
  const float**   ranges,              // C-style array, 'dims' pairs of bin sizes
  double          scale     = 1,       // Optional scale factor for output
  bool            uniform   = true     // true for uniform binning
);

void cv::calcBackProject(
  const cv::Mat*  images,              // C-style array of images, 8U or 32F
  int             nimages,             // number of images in 'images' array
  const int*      channels,            // C-style list, ints identifying channels
  const cv::SparseMat& hist,           // input (sparse) histogram array
  cv::OutputArray backProject,         // output single channel array
  const float**   ranges,              // C-style array, 'dims' pairs of bin sizes
  double          scale     = 1,       // Optional scale factor for output
  bool            uniform   = true     // true for uniform binning
);

void cv::calcBackProject(
  cv::InputArrayOfArrays images,              // STL-vector of images, 8U or 32F
  const vector<int>&     channels,            // STL-vector, channels indices
  cv::InputArray         hist,                // input histogram array
  cv::OutputArray        backProject,         // output single channel array
  const vector<float>&   ranges,              // STL-style vector, range boundaries
  double                 scale     = 1,       // Optional scale factor for output
  bool                   uniform   = true     // true for uniform binning
);
```

There are three versions of `cv::calcBackProject()`. The first two use C-style arrays for their inputs. One of these supports dense histograms and one supports sparse histograms. The third version uses the newer style of template-based inputs rather than C-style pointers.[7] In both cases, the image is provided in the form of a set of single- or multichannel arrays (the `images` variable), while the histogram is precisely the form of histogram that is produced by `cv::calcHist()` (the `hist` variable). The set of single-channel arrays is exactly the same form as what you would have used when you called `cv::calcHist()` in the first place, only this time it is the image you want to compare your histogram to. If the argument `images` is a C-style array (type `cv::Mat*`), you will also need to tell `cv::calcBackProject()` how many elements it has; this is the function of the `nimages` argument.

The `channels` argument is a list of the channels that will actually be used in the back projection. Once again, the form of this argument is the same as the form of the corresponding argument used by `cv::calcHist()`. We relate each integer entry in the `channels` array to a channel in the `arrays` input by enumerating the channels in order, starting with the first array (`arrays[0]`), then for the second array (`images[1]`), and so on (e.g., if there were three matrices pointed to by `images`, with three channels each, their channels would correspond to the values `0`, `1`, and `2` for the first array; `3`, `4`, and `5` for the second array; and `6`, `7`, and `8` for the third array). As you can see, though the number of entries in `channels` must be the same as the dimensionality of the histogram `hist`, that number need not be the same as the number of arrays in (or the total number of channels represented by) `images`.

The results of the back-projection computation will be placed in the array `backProject`, which will be the same size and type as `images[0]`, and have a single channel.

Because histogram data is stored in the same matrix structures used for other data, there is no place to record the bin information that was used in the original construction of the histogram. In this sense, to really comprehend a histogram completely, the associated `cv::Mat` (or `cv::SparseMat` or whatever) is needed, as well as the original `ranges` data structure that was used when the histogram was created by `cv::calcHist()`.[8] This is why this range of information needs to be supplied to `cv::calcBackProject()` in the `ranges` argument.

Finally, there are two optional arguments, `scale` and `uniform`. `scale` is an optional scale factor that is applied to the return values placed in `backProject`. (This is partic-

7 Of these three, the third is the generally preferred form in modern code (i.e., the use of the C-style arrays for input is considered "old-fashioned" in most modern OpenCV code).

8 An alternative approach would have been to define another data type for histograms that inherited from `cv::Mat`, but which also contained this bin information. The authors of the library chose not to take this route in the 2.0 (and later) version of the library in favor of simplifying the library.

ularly useful if you want to visualize the results.) `uniform` indicates whether the input histogram is a uniform histogram (in the sense of `cv::calcHist()`). Because `uniform` defaults to `true`, this argument is needed only for nonuniform histograms.

Example 13-1 showed how to convert an image into single-channel planes and then make an array of them. As just described, the values in `backProject` are set to the values in the associated bin in `hist`. If the histogram is normalized, then this value can be associated with a conditional probability value (i.e., the probability that a pixel in `image` is a member of the type characterized by the histogram in `hist`).[9] In Figure 13-7, we use a flesh-color histogram to derive a probability of flesh image.

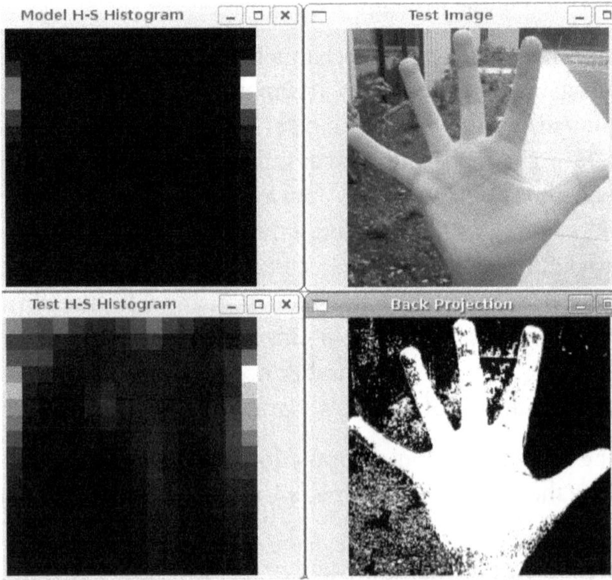


*Figure 13-7. Back projection of histogram values onto each pixel based on its color: the HS (hue and saturation planes of an HSV representation of the image) flesh-color histogram (upper left) is used to convert the hand image (upper right) into the flesh-color probability image (lower right); the lower-left panel is the histogram of the hand image*

9 Specifically, in the case of our flesh-tone HS histogram, if *C* is the color of the pixel and *F* is the probability that a pixel is flesh, then this probability map gives us *p(C|F)*, the probability of drawing that color if the pixel actually is flesh. This is not the same as *p(F|C)*, the probability that the pixel is flesh given its color. However, these two probabilities are related by Bayes' theorem [Bayes1763], so, if we know the overall probability of encountering a flesh-colored object in a scene as well as the total probability of encountering of the range of flesh colors, then we can compute *p(F|C)* from *p(C|F)*. Specifically, Bayes' theorem establishes the following relation:

$$p(F|C) = \frac{p(F)}{p(C)} p(C \mid F)$$

One method of finding an object or a desired region in a new image is as follows:

1. Create a histogram of the object or region that you want to search for.
2. To find this object or region in a new image, use the histogram you calculated together with `cv::calcBackProject()` to create a back-projected image. (In the back-projected image, areas of peak values are likely to contain the object or region matches of interest.)
3. In each of the high-valued areas in the back-projected image, take a local histogram and use `cv::compareHist()` with your object or region histogram to confirm whether that area actually contains the object or region you are looking for.

When `backProject` is a byte image rather than a float image, you should either not normalize the histogram or else scale it up before use. The reason is that the highest possible value in a normalized histogram is `1`, so anything less than that will be rounded down to `0` in the 8-bit image. You might also need to scale `backProject` in order to see the values with your eyes, depending on how high the values are in your histogram.

# Template Matching

Template matching via `cv::matchTemplate()` is not based on histograms; rather, the function matches an actual image patch against an input image by "sliding" the patch over the input image using one of the matching methods described in this section. One example is shown in Figure 13-8.

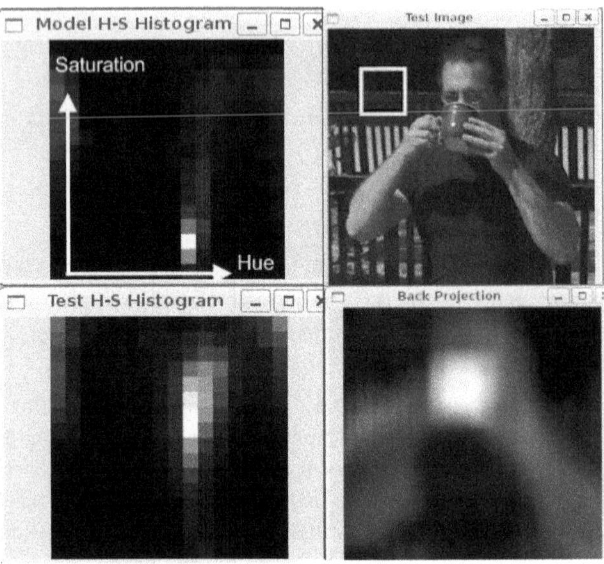


*Figure 13-8. Using cv::matchTemplate() with cv::TM_CCOEFF_NORMED to locate an object (here, a coffee cup) whose size approximately matches the patch size (white box in upper-right panel): the sought object is modeled by a hue-saturation histogram (upper left), which can be compared with an HS histogram for the image as a whole (lower left); the result of cv::matchTemplate() (lower right) is that the object is easily picked out from the scene*

If, as in Figure 13-9, we have an image patch containing a face, then we can slide that face over an input image looking for strong matches that would indicate another face is present.

```
void cv::matchTemplate(
  cv::InputArray  image,     // Input image to be searched, 8U or 32F, size W-by-H
  cv::InputArray  templ,     // Template to use, same type as 'image', size w-by-h
  cv::OutputArray result,    // Result image, type 32F, size (W-w+1)-by(H-h+1)
  int             method     // Comparison method to use
);
```

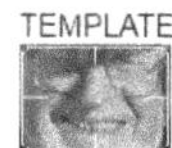

*Figure 13-9. cv::matchTemplate() sweeps a template image patch across another image looking for matches*

The input to `cv::matchTemplate()` starts with a single 8-bit or floating-point plane or color `image`. The matching model in `templ` is just a patch from another (presumably similar) image containing the object for which you are searching. The computed output will be put in the `result` image, which should be a single-channel byte or floating-point image of size `(image.width - templ.width + 1, image.height - templ.height + 1)`. The matching `method` is chosen from one of the options listed next (we use *I* to denote the input image, *T* the template, and *R* the result image in the definitions). For each of these, there is also a normalized version.[10]

## Square Difference Matching Method (cv::TM_SQDIFF)

These methods match the squared difference, so a perfect match will be `0` and bad matches will be large:

$$R_{sq\_diff} = \sum_{x',y'} [T(x', y') - I(x + x', y + y')]^2$$

10 The normalized versions were first developed by Galton [Galton] as described by Rodgers [Rodgers88]. The normalized methods are useful, as they can help reduce the effects of lighting differences between the template and the image. In each case, the normalization coefficient is the same.

## Normalized Square Difference Matching Method (cv::TM_SQDIFF_NORMED)

As with `cv::TM_SQDIFF`, a perfect match for `cv::TM_SQDIFF_NORMED` will return a `0`.

$$R_{sq\_diff\_normed} = \frac{\sum_{x',y'} \left[ T(x',y') - I(x+x',y+y') \right]^2}{\sqrt{\sum_{x',y'} T(x',y')^2 \cdot \sum_{x',y'} I(x+x',y+y')^2}}$$

## Correlation Matching Methods (cv::TM_CCORR)

These methods multiplicatively match the template against the image, so a perfect match will be large and bad matches will be small or `0`.

$$R_{ccorr} = \sum_{x',y'} T(x',y') \cdot I(x+x',y+y')$$

## Normalized Cross-Correlation Matching Method (cv::TM_CCORR_NORMED)

As with `cv::TM_CCORR`, an extreme mismatch for `cv::TM_CCORR_NORMED` will return a score near `0`.

$$R_{ccorr\_normed} = \frac{\sum_{x',y'} T(x',y') \cdot I(x+x',y+y')}{\sqrt{\sum_{x',y'} T(x',y')^2 \cdot \sum_{x',y'} I(x+x',y+y')^2}}$$

## Correlation Coefficient Matching Methods (cv::TM_CCOEFF)

These methods match a template relative to its mean against the image relative to its mean, so a perfect match will be `1` and a perfect mismatch will be `-1`; a value of `0` simply means that there is no correlation (random alignments).

$$R_{ccoeff} = \sum_{x',y'} T'(x',y') \cdot I'(x+x',y+y')$$

$$T'(x',y') = T(x',y') - \frac{\sum_{x'',y''} T(x'',y'')}{(w-h)}$$

$$I'(x+x',y+y') = I(x+x',y+y') - \frac{\sum_{x'',y''} I(x'',y'')}{(w-h)}$$

## Normalized Correlation Coefficient Matching Method (cv::TM_CCOEFF_NORMED)

As with `cv::TM_CCOEFF`, a relative match for `cv::TM_CCOEFF_NORMED` will return a positive score and a relative mismatch will return a negative score.

$$R_{ccoeff\_normed} = \frac{\sum_{x',y'} T'(x',y') \cdot I'(x+x',y+y')}{\sqrt{\sum_{x',y'} T'(x',y')^2 \cdot \sum_{x',y'} I'(x+x',y+y')^2}}$$

Here $T'$ and $I'$ are as defined for `cv::TM_CCOEFF`.

As usual, we obtain more accurate matches (at the cost of more computations) as we move from simpler measures (square difference) to more sophisticated ones (correlation coefficient). It's best to do some test trials of all these settings and then choose the one that best trades off accuracy for speed in your application.

Be careful when interpreting your results. The square-difference methods show best matches with a minimum, whereas the correlation and correlation-coefficient methods show best matches at maximum points.

Once we use `cv::matchTemplate()` to obtain a matching `result` image, we can then use `cv::minMaxLoc()` or `cv::minMaxIdx()` to find the location of the best match. Again, we want to ensure there's an area of good match around that point in order to avoid random template alignments that just happen to work well. A good match should have good matches nearby, because slight misalignments of the template shouldn't vary the results too much for real matches. You can look for the best-matching "hill" by slightly smoothing the result image before seeking the maximum (for correlation or correlation-coefficient) or minimum (for square-difference matching methods). The morphological operators (for example) can be helpful in this context.

Example 13-3 should give you a good idea of how the different template-matching techniques behave. This program first reads in a template and image to be matched, and then performs the matching via the methods we've discussed in this section.

*Example 13-3. Template matching*

```
#include <opencv2/opencv.hpp>
#include <iostream>

using namespace std;

void help( argv ){
```

```
  cout << "\n"
    <<"Example of using matchTemplate(). The call is:\n"
    <<"\n"
    <<argv[0] <<" template image_to_be_searched\n"
    <<"\n"
    <<"   This routine will search using all methods:\n"
    <<"         cv::TM_SQDIFF        0\n"
    <<"         cv::TM_SQDIFF_NORMED 1\n"
    <<"         cv::TM_CCORR         2\n"
    <<"         cv::TM_CCORR_NORMED  3\n"
    <<"         cv::TM_CCOEFF        4\n"
    <<"         cv::TM_CCOEFF_NORMED 5\n"
    <<"\n";
}

// Display the results of the matches
//
int main( int argc, char** argv ) {

  if( argc != 3) {
    help( argv );
    return -1;
  }

  cv::Mat src, templ, ftmp[6];                // ftmp is what to display on

  // Read in the template to be used for matching:
  //
  if((templ=cv::imread(argv[1], 1)).empty()) {
    cout << "Error on reading template " << argv[1] << endl;
    help( argv );return -1;
  }

  // Read in the source image to be searched:
  //
  if((src=cv::imread(argv[2], 1)).empty()) {
    cout << "Error on reading src image " << argv[2] << endl;
    help( argv );return -1;
  }

  // Do the matching of the template with the image
  for(int i=0; i<6; ++i){
    cv::matchTemplate( src, templ, ftmp[i], i);
    cv::normalize(ftmp[i],ftmp[i],1,0,cv::MINMAX);
  }

  // Display
  //
  cv::imshow( "Template", templ );
  cv::imshow( "Image", src );
```

```
  cv::imshow( "SQDIFF", ftmp[0] );
  cv::imshow( "SQDIFF_NORMED", ftmp[1] );
  cv::imshow( "CCORR", ftmp[2] );
  cv::imshow( "CCORR_NORMED", ftmp[3] );
  cv::imshow( "CCOEFF", ftmp[4] );
  cv::imshow( "CCOEFF_NORMED", ftmp[5] );

  // Let user view results:
  //
  cv::waitKey(0);
}
```

Note the use of `cv::normalize()` in this code, which allows us to display the results in a consistent manner. (Recall that some of the matching methods can return negative-valued results.) We use the `cv::NORM_MINMAX` flag when normalizing; this tells the function to shift and scale the floating-point images so that all returned values are between `0` and `1`. Figure 13-10 shows the results of sweeping the face template over the source image (shown in Figure 13-9) using each of `cv::matchTemplate()`'s available matching methods. In outdoor imagery especially, it's almost always better to use one of the normalized methods. Among those, correlation coefficient gives the most clearly delineated match—but, as expected, at a greater computational cost. For a specific application, such as automatic parts inspection or tracking features in a video, you should try all the methods and find the speed and accuracy trade-off that best serves your needs.

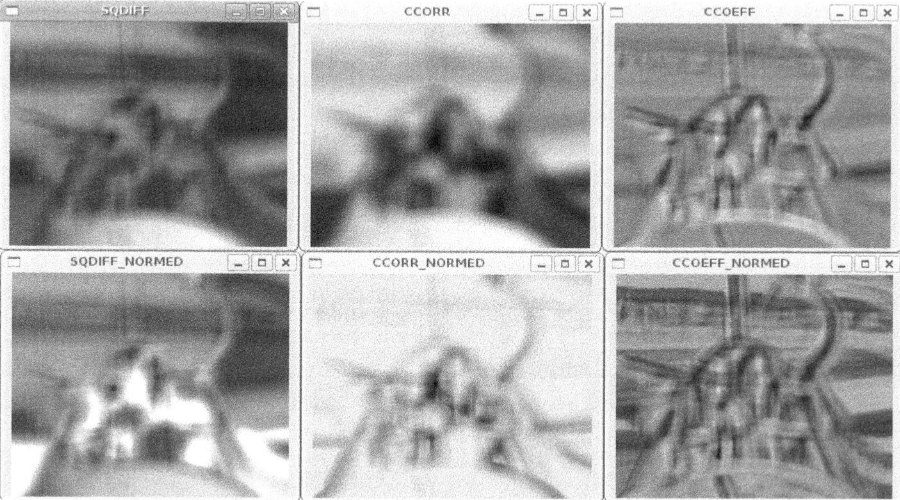


*Figure 13-10. Match results of six matching methods for the template search depicted in Figure 13-9: the best match for square difference is 0 and for the other methods it's the maximum point; thus, matches are indicated by dark areas in the left column and by bright spots in the other two columns*

## Summary

In this chapter, we learned how OpenCV represents histograms as dense or sparse matrix objects. In practice, such histograms are typically used to represent probability density functions that associate a probability amplitude to every element of an array of some number of dimensions. We also covered how this may be used for object or region recognition. We learned how to do basic operations on these arrays, which is useful when interpreting arrays as probability distributions—such as normalization and comparison with other distributions. We concluded with a discussion of template matching. Template matching can be extremely powerful in highly structured images.

## Exercises

1. Generate 1,000 random numbers $r_i$ between 0 and 1. Decide on a bin size and then take a histogram of $1/r_i$.
   a. Are there similar numbers of entries (i.e., within a factor of ±10) in each histogram bin?
   b. Propose a way of dealing with distributions that are highly nonlinear so that each bin has, within a factor of 10, the same amount of data.
2. Take three images of a hand in each of the three lighting conditions discussed in the text. Use `cv::calcHist()` to make an BGR histogram of the flesh color of one of the hands photographed indoors.
   a. Try using just a few large bins (e.g., 2 per dimension), a medium number of bins (16 per dimension), and many bins (256 per dimension). Then run a matching routine (using all histogram matching methods) against the other indoor lighting images of hands. Describe what you find.
   b. Now add 8 and then 32 bins per dimension and try matching across lighting conditions (train on indoor, test on outdoor). Describe the results.
3. As in Exercise 2, gather BGR histograms of hand flesh color. Take one of the indoor histogram samples as your model and measure EMD (earth mover's distance) against the second indoor histogram and against the first outdoor shaded and first outdoor sunlit histograms. Use these measurements to set a distance threshold.
   a. Using this EMD threshold, see how well you detect the flesh histogram of the third indoor histogram, the second outdoor shaded, and the second outdoor sunlit histograms. Report your results.

b. Take histograms of randomly chosen nonflesh background patches to see how well your EMD discriminates. Can it reject the background while matching the true flesh histograms?

4. Using your collection of hand images, design a histogram that can determine under which of the three lighting conditions a given image was captured. Toward this end, you should create features—perhaps sampling from parts of the whole scene, sampling brightness values, and/or sampling relative brightness (e.g., from top to bottom patches in the frame) or gradients from center to edges.

5. Assemble three histograms of flesh models from each of our three lighting conditions (indoor, outdoor in the shade, and outdoor in the sun).

   a. Use each histogram on images from indoor, outdoor shaded, and outdoor sunlit on each of the three conditions. Report how well each histogram works on images of its own condition and on the two other conditions.

   b. Use what you learned to create a "switching histogram" model. First use the scene detector to determine which histogram model to use: indoor, outdoor shaded, or outdoor sunlit. Then use the corresponding flesh model to accept or reject the second flesh patch under all three conditions. How well does this switching model work?

6. Create a flesh-region interest (or "attention") detector.

   a. Just indoors for now, use several samples of hand and face flesh to create a BGR histogram.

   b. On three new indoor scenes, two with flesh and one without, use `cv::calcBackProject()` to create a back-projected image.

   c. Use `cv::meanShift()` at 16 equally spaced grid points in the image with mean-shift window size equal to ¼ of the image width and height to find peaks in the back-projected image.

   d. In a ⅛ × ⅛ region around each peak, collect a histogram.

   e. Use `cv::compareHist()` with all the methods of comparison to find flesh regions in the three collected images.

   f. Report your results. Which comparison methods are most accurate?

7. Create a flesh-region interest (or "attention") detector.

   a. Just indoors for now, use several samples of hand and face flesh to create an RGB histogram.

   b. Use `cv::calcBackProject()` to find areas of flesh.

   c. Use `cv::erode()` from Chapter 10 to clean up noise and then `cv::floodFill()` (from Chapter 12) to find large areas of flesh in an image. These are your "attention" regions.

8. Try some hand-gesture recognition. Photograph a hand about two feet from the camera and create some (nonmoving) hand gestures: thumb up, thumb left, thumb right.

    a. Using your attention detector from Exercise 7, take image gradients in the area of detected flesh around the hand and create a histogram model of resultant gradient orientations for each of the three gestures. Also create a gradient histogram of the face (if there's a face in the image) so that you'll have a (nongesture) model of that large flesh region. You might also take histograms of some similar but nongesture hand positions, just so they won't be confused with the actual gestures.

    b. Test for recognition using a webcam: use the flesh interest regions to find "potential hands"; take gradients in each flesh region; use histogram matching with the preceding gradient-model histograms and set a threshold to detect the gesture. If two models are above threshold, take the better match as the winner.

    c. Move your hand one to two feet farther back and see if the gradient histogram can still recognize the gestures. Report.

9. Repeat Exercise 8 but with EMD for the matching. What happens to EMD as you move your hand back?

CHAPTER 14
# Contours

Although algorithms like the Canny edge detector can be used to find the edge pixels that separate different segments in an image, they do not tell you anything about those edges as entities in themselves. The next step is to be able to assemble those edge pixels into contours. By now you have probably come to expect that there is a convenient function in OpenCV that will do exactly this for you, and indeed there is: `cv::findContours()`. We will start out this chapter with some basics that we will need in order to use this function. With those concepts in hand, we will get into contour finding in some detail. Thereafter, we will move on to the many things we can do with contours after they've been computed.

## Contour Finding

A *contour* is a list of points that represent, in one way or another, a curve in an image. This representation can be different depending on the circumstance at hand. There are many ways to represent a curve. Contours are represented in OpenCV by STL-style `vector<>` template objects in which every entry in the vector encodes information about the location of the next point on the curve. It should be noted that though a sequence of 2D points (`vector<cv::Point>` or `vector<cv::Point2f>`) is the most common representation, there are other ways to represent contours as well. One example of such a construct is the *Freeman chain*, in which each point is represented as a particular "step" in a given direction from the prior point. We will get into such variations in more detail as we encounter them. For now, the important thing to know is that contours are almost always STL vectors, but are not necessarily limited to the obvious vectors of `cv::Point` objects.

The function `cv::findContours()` computes contours from binary images. It can take images created by `cv::Canny()`, which have edge pixels in them, or images created by functions like `cv::threshold()` or `cv::adaptiveThreshold()`, in which the edges are implicit as boundaries between positive and negative regions.[1]

## Contour Hierarchies

Before getting down to exactly how to extract contours, it is worth taking a moment to understand exactly what a contour is, and how groups of contours can be related to one another. Of particular interest is the concept of a contour tree, which is important for understanding one of the most useful ways `cv::findContours()`[2] can communicate its results to us.

Take a moment to look at Figure 14-1, which depicts the functionality of `cv::findContours()`. The upper part of the figure shows a test image containing a number of colored regions (labeled A through E) on a white background.[3] Also shown in the figure are the contours that will be located by `cv::findContours()`. Those contours are labeled cX or hX, where *c* stands for "contour," *h* stands for "hole," and *X* is some number. Some of those contours are dashed lines; they represent *exterior boundaries* of the white regions (i.e., nonzero regions). OpenCV and `cv::findContours()` distinguish between these exterior boundaries and the dotted lines, which you may think of either as *interior boundaries* or as the exterior boundaries of *holes* (i.e., zero regions).

The concept of containment here is important in many applications. For this reason, OpenCV can be asked to assemble the found contours into a *contour tree*[4] that encodes the containment relationships in its structure. A contour tree corresponding to this test image would have the contour called c0 at the root node, with the holes h00 and h01 as its children. Those would in turn have as children the contours that they directly contain, and so on.

1 There are some subtle differences between passing edge images and binary images to `cvFindContours()`; we will discuss those shortly.

2 The retrieval methods derive from Suzuki [Suzuki85].

3 For clarity, the dark areas are depicted as gray in the figure, so simply imagine that this image is thresholded such that the gray areas are set to black before passing to `cv::findContours()`.

4 Contour trees first appeared in Reeb [Reeb46] and were further developed by [Bajaj97], [Kreveld97], [Pascucci02], and [Carr04].

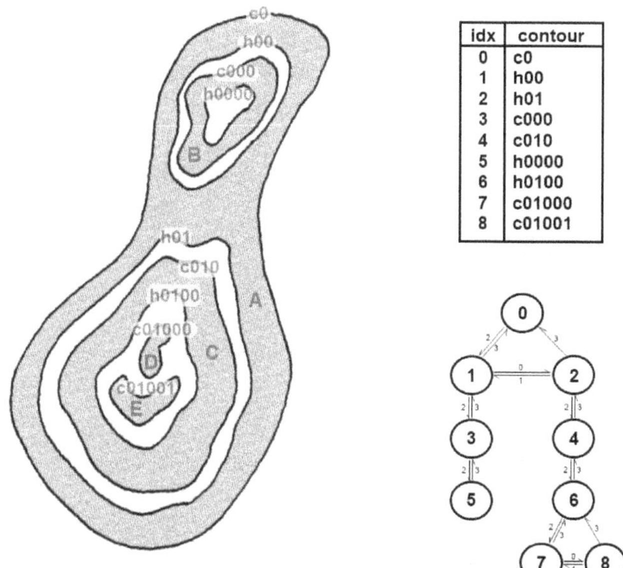


| idx | contour |
|---|---|
| 0 | c0 |
| 1 | h00 |
| 2 | h01 |
| 3 | c000 |
| 4 | c010 |
| 5 | h0000 |
| 6 | h0100 |
| 7 | c01000 |
| 8 | c01001 |

*Figure 14-1. A test image (left side) passed to cv::findContours(). There are five colored regions (labeled A, B, C, D, and E), but contours are formed from both the exterior and interior edges of each colored region. The result is nine contours in total. Each contour is identified and appears in an output list (the contours argument—upper right). Optionally, a hierarchical representation can also be generated (the hierarchy argument—lower right). In the graph shown lower right (corresponding to the constructed contour tree), each node is a contour, and the links in the graph are labeled with the index in the four-element data structure associated with each node in the hierarchy array*

There are many possible ways to represent such a tree. OpenCV represents such trees with arrays (typically of vectors) in which each entry in the array represents one particular contour. In that array, each entry contains a set of four integers (typically represented as an element of type `cv::Vec4i`, just like an entry in a four-channel array). In this case, however, there is a "special" meaning attached to each component of the node's vector representation. Each node in the hierarchy list has four integer components. Each component indicates another node in the hierarchy with a particular relationship to the current node. Where a particular relationship does not exist, that element of the data structure is set to `-1` (e.g., element 3, the parent ID for the root node, would have value `-1` because it has no parent).

By way of example, consider the contours in Figure 14-1. The five colored regions result in a total of nine contours (counting both the exterior and the interior edges of each region). If a contour tree is constructed from these nine contours, each node will have as children those contours that are contained within it. The resulting tree is visualized in the lower right of Figure 14-1. For each node, those links that are valid are also visualized, and the links are labeled with the index associated with that link in the four-element data structure for that node (see Table 14-1).

*Table 14-1. Meaning of each component in the four-element vector representation of each node in a contour hierarchy list*

| Index | Meaning |
|---|---|
| 0 | Next contour (same level) |
| 1 | Previous contour (same level) |
| 2 | First child (next level down) |
| 3 | Parent (next level up) |

It is interesting to note the consequences of using `cv::findContours()` on an image generated by `cv::canny()` or a similar edge detector relative to what happens with a binary image such as the test image shown in Figure 14-1. Deep down, `cv::findContours()` does not really know anything about edge images. This means that, to `cv::findContours()`, an "edge" is just a very thin "white" area. As a result, for every exterior contour, there will be a hole contour that almost exactly coincides with it. This hole is actually just inside of the exterior boundary. You can think of it as the white-to-black transition that marks the interior boundary of the edge.

#### Finding contours with cv::findContours()

With this concept of contour trees in hand, we can look at the `cv::findContours()` function and see exactly how we tell it what we want and how we interpret its response:

```
void cv::findContours(
  cv::InputOutputArray    image,         // Input "binary" 8-bit single channel
  cv::OutputArrayOfArrays contours,      // Vector of vectors or points
  cv::OutputArray         hierarchy,     // (optional) topology information
  int                     mode,          // Contour retrieval mode (Figure 14-2)
  int                     method,              // Approximation method
  cv::Point               offset = cv::Point() // (optional) Offset every point
);

void cv::findContours(
  cv::InputOutputArray    image,         // Input "binary" 8-bit single channel
  cv::OutputArrayOfArrays contours,      // Vector of vectors or points
  int                     mode,          // Contour retrieval mode (Figure 14-2)
  int                     method,        // Approximation method
  cv::Point               offset = cv::Point() // (optional) Offset every point
);
```

The first argument is the input image; this image should be an 8-bit, single-channel image and will be interpreted as binary (i.e., as if all nonzero pixels were equivalent to one another). When it runs, `cv::findContours()` will actually use this image as scratch space for computation, so if you need that image for anything later, you should make a copy and pass that to `cv::findContours()`. The second argument is an array of arrays, which in most practical cases will mean an STL vector of STL vectors. This will be filled with the list of contours found (i.e., it will be a vector of contours, where `contours[i]` will be a specific contour and thus `contours[i][j]` would refer to a specific vertex in `contours[i]`).

The next argument, `hierarchy`, can be either supplied or not supplied (through one of the two forms of the function just shown). If supplied, `hierarchy` is the output that describes the tree structure of the contours. The output `hierarchy` will be an array (again, typically an STL vector) with one entry for each contour in `contours`. Each such entry will contain an array of four elements, each indicating the node to which a particular link from the current node is connected (see Table 14-1).

The `mode` argument tells OpenCV how you would like the contours extracted. There are four possible values for `mode`:

`cv::RETR_EXTERNAL`
: Retrieves only the extreme outer contours. In Figure 14-1, there is only one exterior contour, so Figure 14-2 indicates that the first contour points to that outermost sequence and that there are no further connections.

`cv::RETR_LIST`
: Retrieves all the contours and puts them in the list. Figure 14-2 depicts the "hierarchy" resulting from the test image in Figure 14-1. In this case, nine contours are found and they are all connected to one another by `hierarchy[i][0]` and `hierarchy[i][1]` (`hierarchy[i][2]` and `hierarchy[i][3]` are not used here).[5]

`cv::RETR_CCOMP`
: Retrieves all the contours and organizes them into a two-level hierarchy, where the top-level boundaries are external boundaries of the components and the second-level boundaries are boundaries of the holes. Referring to Figure 14-2, we can see that there are five exterior boundaries, of which three contain holes. The holes are connected to their corresponding exterior boundaries by `hierarchy[i][2]` and `hierarchy[i][3]`. The outermost boundary, c0, contains two holes. Because `hierarchy[i][2]` can contain only one value, the node can have only

5 You are not very likely to use `cv::RETR_LIST`. This option primarily made sense in previous versions of the OpenCV library in which the contour's return value was not automatically organized into a list as the `vector<>` type now implies.

one child. All of the holes inside of c0 are connected to one another by the `hierarchy[i][0]` and `hierarchy[i][1]` pointers.

`cv::RETR_TREE`
: Retrieves all the contours and reconstructs the full hierarchy of nested contours. In our example (Figures 14-1 and 14-2), this means that the root node is the outermost contour, c0. Below c0 is the hole h00, which is connected to the other hole, h01, at the same level. Each of those holes in turn has children (the contours c000 and c010, respectively), which are connected to their parents by vertical links. This continues down to the innermost contours in the image, which become the leaf nodes in the tree.

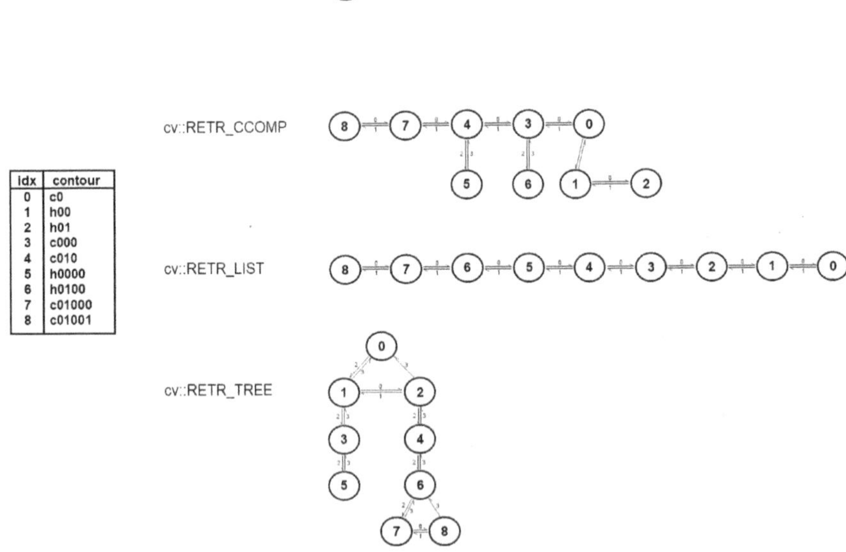


*Figure 14-2. The way in which the tree node variables are used to "hook up" all of the contours located by cv::findContours(). The contour nodes are the same as in Figure 14-1*

The next values pertain to the `method` (i.e., how the contours are represented):

`cv::CHAIN_APPROX_NONE`
: Translates all the points from the contour code into points. This operation will produce a large number of points, as each point will be one of the eight neighbors of the previous point. No attempt is made to reduce the number of vertices returned.

`cv::CHAIN_APPROX_SIMPLE`
: Compresses horizontal, vertical, and diagonal segments, leaving only their ending points. For many special cases, this can result in a substantial reduction of the number of points returned. The extreme example would be a rectangle (of any size) that is oriented along the *x*-*y* axes. In this case, only four points would be returned.

`cv::CHAIN_APPROX_TC89_L1` *or* `cv::CHAIN_APPROX_TC89_KCOS`
: Applies one of the flavors of the Teh-Chin chain approximation algorithm.[6] The Teh-Chin, or T-C, algorithm is a more sophisticated (and more compute-intensive) method for reducing the number of points returned. The T-C algorithm requires no additional parameters to run.

The final argument to `cv::findContours()` is `offset`. This argument is optional. If present, every point in the returned contour will be shifted by this amount. This is particularly useful when either the contours are extracted from a region of interest but you would like them represented in the parent image's coordinate system, or the reverse case, where you are extracting the contours in the coordinates of a larger image but would like to express them relative to some subregion of the image.

## Drawing Contours

One of the most straightforward things you might want to do with a list of contours, once you have it, is to draw the contours on the screen. For this we have `cv::drawContours()`:

```
void  cv::drawContours(
  cv::InputOutputArray   image,                     // Will draw on input image
  cv::InputArrayOfArrays contours,                  // Vector of vectors or points
  int                    contourIdx,                // Contour to draw (-1 is "all")
  const cv::Scalar&      color,                     // Color for contours
  int                    thickness = 1,             // Thickness for contour lines
  int                    lineType  = 8,             // Connectedness ('4' or '8')
  cv::InputArray         hierarchy = noArray(),     // optional (from findContours)
  int                    maxLevel  = INT_MAX,       // Max descent in hierarchy
  cv::Point              offset    = cv::Point()    // (optional) Offset all points
)
```

The first argument, `image`, is simple: it is the image on which to draw the contours. The next argument, `contour`, is the list of contours to be drawn. This is in the same form as the `contour` output of `cv::findContours()`; it is a list of lists of points. The

6 If you are interested in the details of how this algorithm works, you can consult C. H. Teh and R. T. Chin, "On the Detection of Dominant Points on Digital Curve," *PAMI* 11, no. 8 (1989): 859–872. Because the algorithm requires no tuning parameters, however, you can get quite far without knowing the deeper details of the algorithm.

contourIdx argument can be used to select either a single contour to draw or to tell cv::drawContours() to draw all of the contours on the list provided in the contours argument. If contourIdx is a positive number, that particular contour will be drawn. If contourIdx is negative (usually it is just set to -1), all contours are drawn.

The color, thickness, and lineType arguments are similar to the corresponding arguments in other draw functions such as cv::Line(). As usual, the color argument is a four-component cv::Scalar, the thickness is an integer indicating the thickness of the lines to be drawn in pixels, and the lineType may be either 4 or 8, indicating whether the line is to be drawn as a four-connected (ugly), eight-connected (not too ugly), or cv::AA (pretty) line.

The hierarchy argument corresponds to the hierarchy output from cv::findContours(). The hierarchy works with the maxLevel argument. The latter limits the depth in the hierarchy to which contours will be drawn in your image. Setting maxLevel to 0 indicates that only "level 0" (the highest level) in the hierarchy should be drawn; higher numbers indicate that number of layers down from the highest level that should be included. Looking at Figure 14-2, you can see that this is useful for contour trees; it is also potentially useful for connected components (cv::RETR_CCOMP) in case you would like only to visualize exterior contours (but not "holes"—interior contours).

Finally, we can give an offset to the draw routine so that the contour will be drawn elsewhere than at the absolute coordinates by which it was defined. This feature is particularly useful when the contour has already been converted to center-of-mass or other local coordinates. offset is particularly helpful in the case in which you have used cv::findContours() one or more times in different image subregions (ROIs) but now want to display all the results within the original large image. Conversely, we could use offset if we'd extracted a contour from a large image and then wanted to form a small mask for this contour.

## A Contour Example

Example 14-1 is drawn from the OpenCV package. Here we create a window with an image in it. A trackbar sets a simple threshold, and the contours in the thresholded image are drawn. The image is updated whenever the trackbar is adjusted.

*Example 14-1. Finding contours based on a trackbar's location; the contours are updated whenever the trackbar is moved*

```
#include <opencv2/opencv.hpp>
#include <iostream>

using namespace std;
```

```
cv::Mat g_gray, g_binary;
int g_thresh = 100;

void on_trackbar( int, void* ) {

  cv::threshold( g_gray, g_binary, g_thresh, 255, cv::THRESH_BINARY );
  vector< vector< cv::Point> > contours;
  cv::findContours(
    g_binary,
    contours,
    cv::noArray(),
    cv::RETR_LIST,
    cv::CHAIN_APPROX_SIMPLE
  );
  g_binary = cv::Scalar::all(0);

  cv::drawContours( g_binary, contours, -1, cv::Scalar::all(255));
  cv::imshow( "Contours", g_binary );

}

int main( int argc, char** argv ) {

  if( argc != 2 || ( g_gray = cv::imread(argv[1], 0)).empty() ) {
    cout << "Find threshold dependent contours\nUsage: " <<argv[0]
      <<"fruits.jpg" << endl;
    return -1;
  }
  cv::namedWindow( "Contours", 1 );

  cv::createTrackbar(
    "Threshold",
    "Contours",
    &g_thresh,
    255,
    on_trackbar
  );
  on_trackbar(0, 0);

  cv::waitKey();

  return 0;

}
```

Here, everything of interest to us is happening inside of the function `on_trackbar()`. The image `g_gray` is thresholded such that only those pixels brighter than `g_thresh` remain nonzero. The `cv::findContours()` function is then called on this thresholded image. Finally, `cv::drawContours()` is called, and the contours are drawn (in white) onto the grayscale image.

## Another Contour Example

In Example 14-2, we find contours on an input image and then proceed to draw them one by one. This is a good example to tinker with on your own to explore the effects of changing either the contour finding mode (`cv::RETR_LIST` in the code) or the `max_depth` that is used to draw the contours (`0` in the code). If you set `max_depth` to a larger number, notice that the example code steps through the contours returned by `cv::findContours()` by means of `hierarchy[i][1]`. Thus, for some topologies (`cv::RETR_TREE`, `cv::RETR_CCOMP`, etc.), you may see the same contour more than once as you step through.

*Example 14-2. Finding and drawing contours on an input image*

```
#include <opencv2/opencv.hpp>
#include <algorithm>
#include <iostream>

using namespace std;

struct AreaCmp {
    AreaCmp(const vector<float>& _areas) : areas(&_areas) {}
    bool operator()(int a, int b) const { return (*areas)[a] > (*areas)[b]; }
    const vector<float>* areas;
};


int main(int argc, char* argv[]) {

  cv::Mat img, img_edge, img_color;

  // load image or show help if no image was provided
  //
  if( argc != 2 || (img = cv::imread(argv[1],cv::LOAD_IMAGE_GRAYSCALE)).empty() ) {
    cout << "\nExample 8_2 Drawing Contours\nCall is:\n./ch8_ex8_2 image\n\n";
    return -1;
  }

  cv::threshold(img, img_edge, 128, 255, cv::THRESH_BINARY);
  cv::imshow("Image after threshold", img_edge);
  vector< vector< cv::Point > > contours;
  vector< cv::Vec4i > hierarchy;

  cv::findContours(
    img_edge,
    contours,
    hierarchy,
    cv::RETR_LIST,
    cv::CHAIN_APPROX_SIMPLE
  );
```

```
  cout << "\n\nHit any key to draw the next contour, ESC to quit\n\n";
  cout << "Total Contours Detected: " << contours.size() << endl;

  vector<int> sortIdx(contours.size());
  vector<float> areas(contours.size());
  for( int n = 0; n < (int)contours.size(); n++ ) {
    sortIdx[n] = n;
    areas[n] = contourArea(contours[n], false);
  }

  // sort contours so that the largest contours go first
  //
  std::sort( sortIdx.begin(), sortIdx.end(), AreaCmp(areas ));

  for( int n = 0; n < (int)sortIdx.size(); n++ ) {
    int idx = sortIdx[n];
    cv::cvtColor( img, img_color, cv::GRAY2BGR );
    cv::drawContours(
      img_color, contours, idx,
      cv::Scalar(0,0,255), 2, 8, hierarchy,
      0                 // Try different values of max_level, and see what happens
    );
    cout << "Contour #" << idx << ": area=" << areas[idx] <<
      ", nvertices=" << contours[idx].size() << endl;
    cv::imshow(argv[0], img_color);
    int k;
    if( (k = cv::waitKey()&255) == 27 )
      break;
  }
  cout << "Finished all contours\n";

  return 0;

}
```

## Fast Connected Component Analysis

Another approach, closely related to contour analysis, is *connected component analysis*. After segmenting an image, typically by thresholding, we can use connected component analysis to efficiently isolate and process the resulting image regions one by one. The input required by OpenCV's connected component algorithm is a binary (black-and-white) image, and the output is a labeled pixel map where nonzero pixels from the same connected component get the same unique label. For example, there are five connected components in Figure 14-1, the biggest one with two holes, two smaller ones with one hole each, and two small components without holes. Connected component analysis is quite popular in background segmentation algorithms as the post-processing filter that removes small noise patches and in problems like OCR where there is a well-defined foreground to extract. Of course, we want to run such a basic operation quickly. A slower "manual" way to do this would be to use cv::find

Contours() (where you pass the cv::RETR_CCOMP flag) and then subsequently loop over the resulting connected components where cv::drawContours() with color=component_label and thickness=-1 is called. This is slow for several reasons:

- cv::findContours() first allocates a separate STL vector for each contour, and there can be hundreds—sometimes thousands—of contours in the image.
- Then, when you want to fill a nonconvex area bounded by one or more contours, cv::drawContours() is also slow and involves building and sorting a collection of all the tiny line segments bounding the area.
- Finally, collecting some basic information about a connected component (such as an area or bounding box) requires extra, sometimes expensive, calls.

Thankfully, as of OpenCV 3 there is a great alternative to all this complex stuff—namely, the cv::connectedComponents() and cv::connectedComponentsWithStats() functions:

```
int  cv::connectedComponents (
  cv::InputArrayn image,                  // input 8-bit single-channel (binary)
  cv::OutputArray labels,                 // output label map
  int             connectivity = 8,       // 4- or 8-connected components
  int             ltype        = CV_32S // Output label type (CV_32S or CV_16U)
);

int  cv::connectedComponentsWithStats (
  cv::InputArrayn image,                  // input 8-bit single-channel (binary)
  cv::OutputArray labels,                 // output label map
  cv::OutputArray stats,                  // Nx5 matrix (CV_32S) of statistics:
                                          // [x0, y0, width0, height0, area0;
                                          //  ... ; x(N-1), y(N-1), width(N-1),
                                          // height(N-1), area(N-1)]
  cv::OutputArray centroids,              // Nx2 CV_64F matrix of centroids:
                                          // [ cx0, cy0; ... ; cx(N-1), cy(N-1)]
  int             connectivity = 8,       // 4- or 8-connected components
  int             ltype        = CV_32S // Output label type (CV_32S or CV_16U)
);
```

cv::connectedComponents() simply creates the label map. cv::connectedComponentsWithStats() does the same but also returns some important information about each connected component, such as the bounding box, area, and center of mass (also known as the centroid). If you do not need the centroids, pass cv::noArray() for the OutputArray centroids parameter. Both functions return the number of found connected components. The functions do *not* use cv::findContours() and cv::drawContours(); instead, they implement the direct and very efficient algorithm described in "Two Strategies to Speed Up Connected Component Labeling Algorithms" [Wu08].

Let's consider a short example that draws the labeled connected components while removing small ones (see Example 14-3).

*Example 14-3. Drawing labeled connected components*

```
#include <opencv2/opencv.hpp>
#include <algorithm>
#include <iostream>

using namespace std;
int main(int argc, char* argv[]) {

  cv::Mat img, img_edge, labels, img_color, stats;

  // load image or show help if no image was provided
  if( argc != 2
    || (img = cv::imread( argv[1], cv::LOAD_IMAGE_GRAYSCALE )).empty()
  ) {
    cout << "\nExample 8_3 Drawing Connected componnents\n" \
      << "Call is:\n" <<argv[0] <<" image\n\n";
    return -1;
  }

  cv::threshold(img, img_edge, 128, 255, cv::THRESH_BINARY);
  cv::imshow("Image after threshold", img_edge);

  int i, nccomps = cv::connectedComponentsWithStats (
    img_edge, labels,
    stats, cv::noArray()
  );
  cout << "Total Connected Components Detected: " << nccomps << endl;

  vector<cv::Vec3b> colors(nccomps+1);
  colors[0] = Vec3b(0,0,0); // background pixels remain black.
  for( i = 1; i <= nccomps; i++ ) {
    colors[i] = Vec3b(rand()%256, rand()%256, rand()%256);
    if( stats.at<int>(i-1, cv::CC_STAT_AREA) < 100 )
      colors[i] = Vec3b(0,0,0); // small regions are painted with black too.
  }
  img_color = Mat::zeros(img.size(), CV_8UC3);
  for( int y = 0; y < img_color.rows; y++ )
    for( int x = 0; x < img_color.cols; x++ )
    {
      int label = labels.at<int>(y, x);
      CV_Assert(0 <= label && label <= nccomps);
      img_color.at<cv::Vec3b>(y, x) = colors[label];
     }
  cv::imshow("Labeled map", img_color);
  cv::waitKey();
  return 0;
}
```

## More to Do with Contours

When analyzing an image, there are many different things we might want to do with contours. After all, most contours are—or are candidates to be—things that we are interested in identifying or manipulating. The various relevant tasks include characterizing the contours, simplifying or approximating them, matching them to templates, and so on.

In this section, we will examine some of these common tasks and visit the various functions built into OpenCV that will either do these things for us or at least make it easier for us to perform our own tasks.

### Polygon Approximations

If we are drawing a contour or are engaged in shape analysis, it is common to approximate a contour representing a polygon with another contour having fewer vertices. There are many different ways to do this; OpenCV offers implementations of two of them.

#### Polygon approximation with cv::approxPolyDP()

The routine `cv::approxPolyDP()` is an implementation of one of these two algorithms:[7]

```
void cv::approxPolyDP(
  cv::InputArray  curve,        // Array or vector of 2-dimensional points
  cv::OutputArray approxCurve,  // Result, type is same as 'curve'
  double          epsilon,      // Max distance from 'curve' to 'approxCurve'
  bool            closed        // If true, assume link from last to first vertex
);
```

The `cv::approxPolyDP()` function acts on one polygon at a time, which is given in the input `curve`. The output of `cv::approxPolyDP()` will be placed in the `approxCurve` output array. As usual, these polygons can be represented as either STL vectors of `cv::Point` objects or as OpenCV `cv::Mat` arrays of size $N \times 1$ (but having two channels). Whichever representation you choose, the input and output arrays used for `curve` and `approxCurve` should be of the same type.

The parameter `epsilon` is the accuracy of approximation you require. The meaning of the `epsilon` parameter is that this is the largest deviation you will allow between the original polygon and the final approximated polygon. `closed`, the last argument,

7 For aficionados, the method we are discussing here is the Douglas-Peucker (DP) approximation [Douglas73]. Other popular methods are the Rosenfeld-Johnson [Rosenfeld73] and Teh-Chin [Teh89] algorithms. Of those two, the Teh-Chin algorithm is not available in OpenCV as a reduction method, but is available at the time of the extraction of the polygon (see "Finding contours with cv::findContours()" on page 412).

indicates whether the sequence of points indicated by `curve` should be considered a closed polygon. If set to `true`, the curve will be assumed to be closed (i.e., the last point is to be considered connected to the first point).

#### The Douglas-Peucker algorithm explained

To better understand how to set the `epsilon` parameter, and to better understand the output of `cv::approxPolyDP()`, it is worth taking a moment to understand exactly how the algorithm works. In Figure 14-3, starting with a contour (panel b), the algorithm begins by picking two extremal points and connecting them with a line (panel c). Then the original polygon is searched to find the point farthest from the line just drawn, and that point is added to the approximation.

The process is iterated (panel d), adding the next most distant point to the accumulated approximation, until all of the points are less than the distance indicated by the precision parameter (panel f). This means that good candidates for the parameter are some fraction of the contour's length, or of the length of its bounding box, or a similar measure of the contour's overall size.

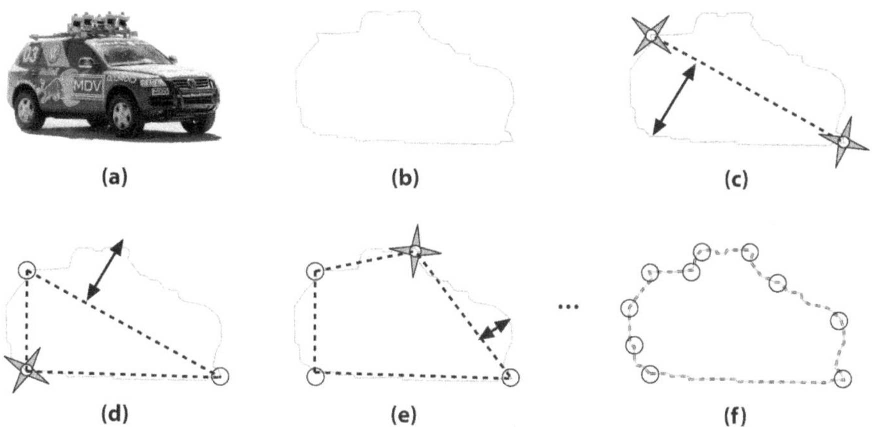


*Figure 14-3. Visualization of the DP algorithm used by cv::approxPolyDP(): the original image (a) is approximated by a contour (b) and then—starting from the first two maximally separated vertices (c)—the additional vertices are iteratively selected from that contour (d–f)*

## Geometry and Summary Characteristics

Another task that one often faces with contours is computing their various *summary characteristics*. These might include length or some other form of size measure of the overall contour. Other useful characteristics are the *contour moments*, which can be used to summarize the gross shape characteristics of a contour; we will address these in the next section. Some of the methods we will discuss work equally well for any collection of points (i.e., even those that do not imply a piecewise curve between those points). We will mention along the way which methods make sense only for curves (such as computing arc length) and which make sense for any general set of points (such as bounding boxes).

### Length using cv::arcLength()

The subroutine `cv::arcLength()` will take a contour and return its length.

```
double  cv::arcLength(
  cv::InputArray  points,       // Array or vector of 2-dimensional points
  bool            closed        // If true, assume link from last to first vertex
);
```

The first argument of `cv::arcLength()` is the contour itself, whose form may be any of the usual representations of a curve (i.e., STL vector of points or an array of two-channel elements). The second argument, `closed`, indicates whether the contour should be treated as closed. If the contour is considered closed, the distance from the last point in `points` to the first contributes to the overall arc length.

`cv::arcLength()` is an example of a case where the `points` argument is implicitly assumed to represent a curve, and so is not particularly meaningful for a general set of points.

### Upright bounding box with cv::boundingRect()

Of course, the length and area are simple characterizations of a contour. One of the simplest ways to characterize a contour is to report a bounding box for that contour. The simplest version of that would be to simply compute the upright bounding rectangle. This is what `cv::boundingRect()` does for us:

```
cv::Rect cv::boundingRect(      // Return upright rectangle bounding the points
  cv::InputArray  points,       // Array or vector of 2-dimensional points
);
```

The `cv::boundingRect()` function just takes one argument, which is the curve whose bounding box you would like computed. The function returns a value of type `cv::Rect`, which is the bounding box you are looking for.

The bounding box computation is meaningful for any set of points, regardless of whether those points represent a curve or are just some arbitrary constellation of points.

### A minimum area rectangle with cv::minAreaRect()

One problem with the bounding rectangle from `cv::boundingRect()` is that it returns a `cv::Rect` and so can represent only a rectangle whose sides are oriented horizontally and vertically. In contrast, the routine `cv::minAreaRect()` returns the minimal rectangle that will bound your contour, and this rectangle may be inclined relative to the vertical; see Figure 14-4. The arguments are otherwise similar to `cv::boundingRect()`. The OpenCV data type `cv::RotatedRect` is just what is needed to represent such a rectangle. Recall that it has the following definition:

```
class cv::RotatedRect {
  cv::Point2f center;       // Exact center point (around which to rotate)
  cv::Size2f  size;         // Size of rectangle (centered on 'center')
  float       angle;        // degrees
};
```

So, in order to get a little tighter fit, you can call `cv::minAreaRect()`:

```
cv::RotatedRect cv::minAreaRect( // Return rectangle bounding the points
  cv::InputArray  points,        // Array or vector of 2-dimensional points
);
```

As usual, `points` can be any of the standard representations for a sequence of points, and is equally meaningful for curves and for arbitrary point sets.

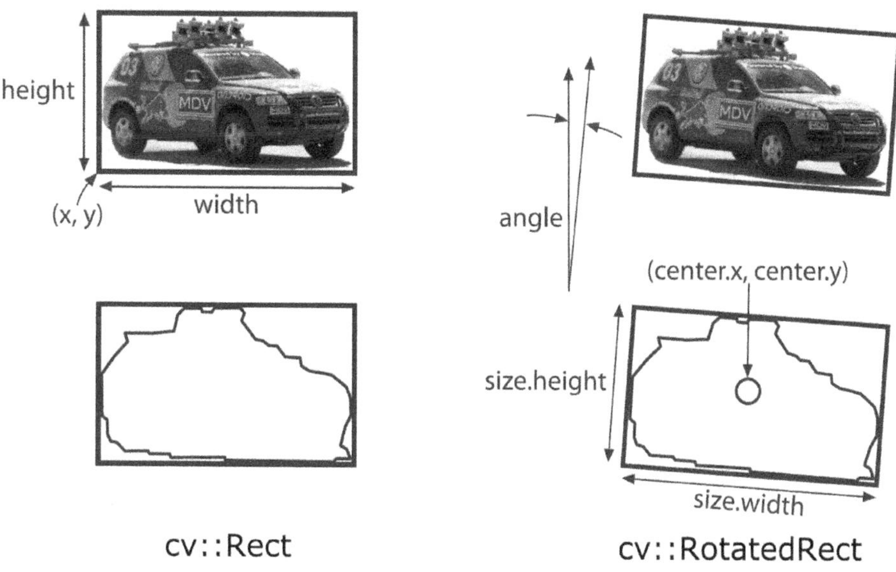


*Figure 14-4. cv::Rect can represent only upright rectangles, but cv::RotatedRect can handle rectangles of any inclination*

### A minimal enclosing circle using cv::minEnclosingCircle()

Next, we have `cv::minEnclosingCircle()`.[8] This routine works pretty much the same way as the bounding box routines, with the exception that there is no convenient data type for the return value. As a result, you must pass in references to variables you would like set by `cv::minEnclosingCircle()`:

```
void  cv::minEnclosingCircle(
   cv::InputArray points,        // Array or vector of 2-dimensional points
   cv::Point2f&   center,        // Result location of circle center
   float&         radius         // Result radius of circle
);
```

The input `points` is just the usual sequence of points representation. The `center` and `radius` variables are variables you will have to allocate and which will be set for you by `cv::minEnclosingCircle()`.

The `cv::minEnclosingCircle` function is equally meaningful for curves as for general point sets.

### Fitting an ellipse with cv::fitEllipse()

As with the minimal enclosing circle, OpenCV also provides a method for fitting an ellipse to a set of points:

```
cv::RotatedRect cv::fitEllipse(  // Return rect bounding ellipse (Figure 14-5)
  cv::InputArray  points         // Array or vector of 2-dimensional points
);
```

`cv::fitEllipse()` takes just a points array as an argument.

At first glance, it might appear that `cv::fitEllipse()` is just the elliptical analog of `cv::minEnclosingCircle()`. There is, however, a subtle difference between `cv::minEnclosingCircle()` and `cv::fitEllipse()`, which is that the former simply computes the smallest circle that completely encloses the given points, whereas the latter uses a fitting function and returns the ellipse that is the best approximation to the point set. This means that not all points in the contour will even be enclosed in the ellipse returned by `cv::fitEllipse()`.[9] The fitting is done through a least-squares fitness function.

The results of the fit are returned in a `cv::RotatedRect` structure. The indicated box exactly encloses the ellipse (see Figure 14-5).

---

8 For more information on the inner workings of these fitting techniques, see Fitzgibbon and Fisher [Fitzgibbon95] and Zhang [Zhang96].

9 Of course, if the number of points is sufficiently small (or in certain other degenerate cases—including all of the points being collinear), it is possible for all of the points to lie on the ellipse. In general, however, some points will be inside, some will be outside, and few if any will actually lie on the ellipse itself.

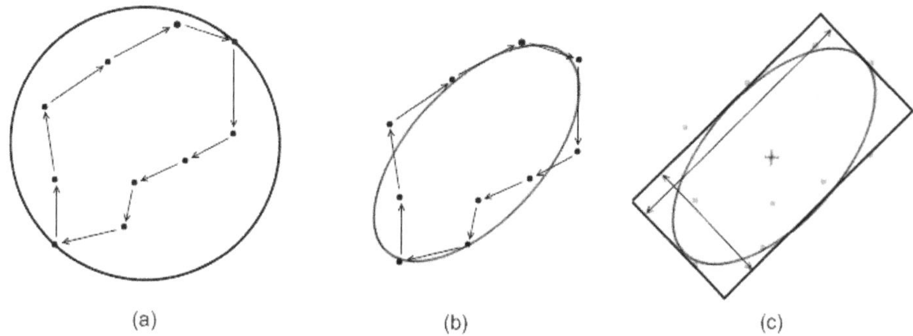


*Figure 14-5. Ten-point contour with the minimal enclosing circle superimposed (a) and with the best fitting ellipsoid (b). A "rotated rectangle" is used by OpenCV to represent that ellipsoid (c)*

### Finding the best line fit to your contour with cv::fitLine()

In many cases, your "contour" will actually be a set of points that you believe is approximately a straight line—or, more accurately, that you believe to be a noisy sample whose underlying origin is a straight line. In such a situation, the problem is to determine what line would be the best explanation for the points you observe. In fact, there are many reasons why one might want to find a best line fit, and as a result, there are many variations of how to actually do that fitting.

We do this fitting by minimizing a cost function, which is defined to be:

$$cost(\vec{\theta}) = \Sigma_{points:i}\rho(r_i) \text{ , where } r_i = r(\vec{\theta}, \vec{x}_i)$$

Here $\vec{\theta}$ is the set of parameters that define the line, $\vec{x}_i$ is the `i`th point in the contour, and $r_i$ is the distance between that point and the line defined by $\vec{\theta}$. Thus, it is the function $\rho(r_i)$ that fundamentally distinguishes the different available fitting methods. In the case of $\rho(r_i) = \frac{1}{2}r_i^2$, the cost function will become the familiar least-squares fitting procedure that is probably familiar to most readers from elementary statistics. The more complex distance functions are useful when more robust fitting methods are needed (i.e., fitting methods that handle outlier data points more gracefully). Table 14-2 shows the available forms for $\rho(r)$ and the associated OpenCV enum values used by `cv::fitLine()`.

*Table 14-2. Available distance metrics for the distType parameter in cv::fitLine()*

| distType | Distance metric | |
|---|---|---|
| cv::DIST_L2 | $\rho(r_i) = \frac{1}{2} r_i^2$ | Least-squares method |
| cv::DIST_L1 | $\rho(r_i) = r_i$ | |
| cv::DIST_L12 | $\rho(r_i) = 2 \cdot \left(\sqrt{1 + \frac{1}{2} r^2} - 1\right)$ | |
| cv::DIST_FAIR | $\rho(r_i) = C^2 \cdot \left(\frac{r}{C} - \log\left(1 + \frac{r}{C}\right)\right)$ | $C = 1.3998$ |
| cv::DIST_WELSCH | $\rho(r_i) = \frac{C^2}{2} \cdot \left(1 - \exp\left(-\left(\frac{r}{C}\right)^2\right)\right)$ | $C = 2.9846$ |
| cv::DIST_HUBER | $\rho(r_i) = \begin{cases} \frac{1}{2} r_i^2 & r < C \\ C \cdot \left(r - \frac{C}{2}\right) & r \geq C \end{cases}$ | $C = 1.345$ |

The OpenCV function `cv::fitLine()` has the following function prototype:

```
void cv::fitLine(
  cv::InputArray  points,       // Array or vector of 2-dimensional points
  cv::OutputArray line,         // Vector of Vec4f (2d), or Vec6f (3d)
  int             distType,     // Distance type (Table 14-2)
  double          param,        // Parameter for distance metric (Table 14-2)
  double          reps,         // Radius accuracy parameter
  double          aeps          // Angle accuracy parameter
);
```

The argument `points` is mostly what you have come to expect, a representation of a set of points either as a `cv::Mat` array or an STL vector. One very important difference, however, between `cv::fitLine()` and many of the other functions we are looking at in this section is that `cv::fitLine()` accepts both two- and three-dimensional points. The output `line` is a little strange. That entry should be of type `cv::Vec4f` (for a two-dimensional line), or `cv::Vec6f` (for a three-dimensional line) where the first half of the values gives the line direction and the second half a point on the line. The third argument, `distType`, allows us to select the distance metric we would like to use. The possible values for `distType` are shown in Table 14-2. The `param` argument is used to supply a value for the parameters used by some of the distance metrics (these parameters appear as the variables C in Table 14-2). This parameter can be set to `0`, in which case `cv::fitLine()` will automatically select the optimal value for the selected distance metric. The parameters `reps` and `aeps` represent your required accuracy for the origin of the fitted line (the *x*, *y*, *z* parts) and for the angle of the line (the $v_x$, $v_y$, $v_z$ parts). Typical values for these parameters are `1e-2` for both of them.

#### Finding the convex hull of a contour using cv::convexHull()

There are many situations in which we need to simplify a polygon by finding its *convex hull.* The convex hull of a polygon or contour is the polygon that completely contains the original, is made only of points from the original, and is everywhere convex (i.e., the internal angle between any three sequential points is less than 180 degrees). An example of a convex hull is shown in Figure 14-6. There are many reasons to compute convex hulls. One particularly common reason is that testing whether a point is inside a convex polygon can be very fast, and it is often worthwhile to test first whether a point is inside the convex hull of a complicated polygon before even bothering to test whether it is in the true polygon.

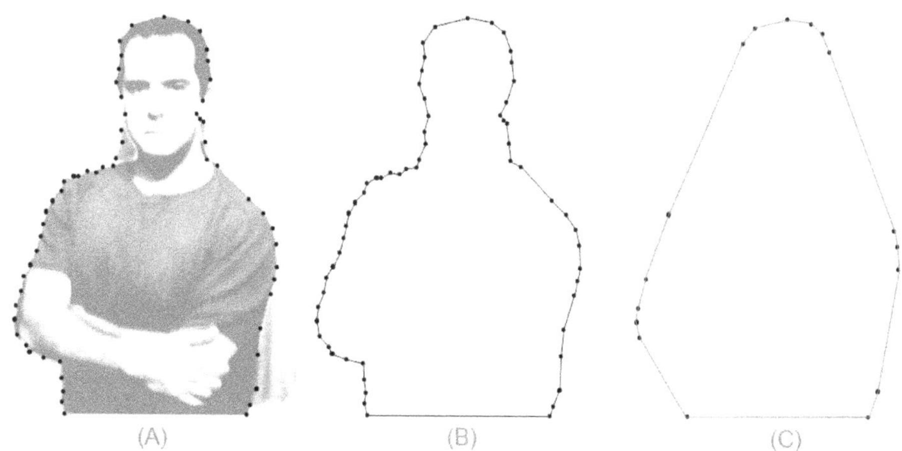


*Figure 14-6. An image (a) is converted to a contour (b). The convex hull of that contour (c) has far fewer points and is a much simpler piece of geometry*

To compute the convex hull of a contour, OpenCV provides the function `cv::convexHull()`:

```
void cv::convexHull(
  cv::InputArray  points,                 // Array or vector of 2-d points
  cv::OutputArray hull,                   // Array of points or integer indices
  bool            clockwise    = false,   // true='output points will be clockwise'
  bool            returnPoints = true     // true='points in hull', else indices
);
```

The `points` input to `cv::convexHull()` can be any of the usual representations of a contour. The argument `hull` is where the resulting convex hull will appear. For this argument, you have two options: if you like, you can provide the usual contour type structure, and `cv::convexHull()` will fill that up with the points in the resulting convex hull. The other option (see `returnPoints`, discussed momentarily) is to provide not an array of points but an array of integers. In this case, `cv::convexHull()` will associate an index with each point that is to appear in the hull and place those indices in `hull`. In this case, the indexes will begin at zero, and the index value `i` will refer to the point `points[i]`.

The clockwise argument indicates how you would like to have cv::convexHull() express the hull it computes. If clockwise is set to true, then the hull will be in clockwise order; otherwise, it will be in counterclockwise order. The final argument, returnPoints, is associated with the option to return point indices rather than point values. If points is an STL vector object, this argument is ignored, because the type of the vector template (int versus cv::Point) can be used to infer what you want. If, however, points is a cv::Mat type array, returnPoints must be set to true if you are expecting point coordinates, and false if you are expecting indices.

## Geometrical Tests

When you are dealing with bounding boxes and other summary representations of polygon contours, it is often desirable to perform such simple geometrical checks as polygon overlap or a fast overlap check between bounding boxes. OpenCV provides a small but handy set of routines for this sort of geometrical checking.

Many important tests that apply to rectangles are supported through interfaces provided by those rectangle types. For example, the contains() method of type cv::Rect can be passed a point, and it will determine whether that point is inside the rectangle.

Similarly, the minimal rectangle containing two rectangles can be computed with the logical OR operator (e.g., rect1 | rect2), while the intersection of two rectangles can be computed with the logical AND operator (e.g., rect1 & rect2). Unfortunately, at this time, there are no such operators for cv::RotatedRect.

For operations on general curves, however, there are library functions that can be used.

### Testing if a point is inside a polygon with cv::pointPolygonTest()

This first geometry toolkit function is cv::pointPolygonTest(), which allows you to test whether a point is inside of a polygon (indicated by the array contour). In particular, if the argument measureDist is set to true, then the function returns the distance to the nearest contour edge; that distance is 0 if the point is inside the contour and positive if the point is outside. If the measure_dist argument is false, then the return values are simply +1, -1, or 0 depending on whether the point is inside, outside, or on an edge (or vertex), respectively. As always, the contour itself can be either an STL vector or an $n \times 1$ two-channel array of points.

```
double cv::pointPolygonTest(    // Return distance to boundary (or just side)
  cv::InputArray contour,       // Array or vector of 2-dimensional points
  cv::Point2f    pt,            // Test point
  bool           measureDist    // true 'return distance', else {0,+1,-1} only
);
```

### Testing whether a contour is convex with cv::isContourConvex()

It is common to want to know whether or not a contour is convex. There are lots of reasons for this, but one of the most typical is that there are a lot of algorithms for working with polygons that either work only on convex polygons or that can be simplified dramatically for the case of convex polygons. To test whether a polygon is convex, simply call `cv::isContourConvex()` and pass it your contour in any of the usual representations. The contour passed will always be assumed to be a closed polygon (i.e., a link between the last and first points in the contour is presumed to be implied):

```
bool cv::isContourConvex(          // Return true if contour is convex
  cv::InputArray  contour          // Array or vector of 2-dimensional points
);
```

Because of the nature of the implementation, `cv::isContourConvex()` requires that the contour passed to it be a *simple polygon*. This means that the contour must not have any self-intersections.

# Matching Contours and Images

Now that you have a pretty good idea of what a contour is and how to work with contours as objects in OpenCV, we'll move to the topic of how to use them for some practical purposes. The most common task associated with contours is matching them in some way with one another. We may have two computed contours that we'd like to compare, or a computed contour and some abstract template with which we'd like to compare our contour. We will discuss both of these cases.

## Moments

One of the simplest ways to compare two contours is to compute *contour moments*, which represent certain high-level characteristics of a contour, an image, or a set of points. (The entire discussion that follows will apply equally well to contours, images, or point sets, so for convenience we will just refer to these options collectively as *objects*.) Numerically, the moments are defined by the following formula:

$$m_{p,q} = \sum_{i=1}^{N} I(x_i, y_i) x^p y^q$$

In this expression, the moment $m_{p,q}$ is defined as a sum over all of the pixels in the object, in which the value of the pixel at point $x, y$ is multiplied by the factor $x^p y^q$. In the case of the $m_{00}$ moment, this factor is equal to `1`—so if the image is a binary image (i.e., one in which every pixel is either `0` or `1`), then $m_{00}$ is just the area of the nonzero pixels in the image. In the case of a contour, the result is the length of the contour,[10] and in the case of a point set it is just the number of points. After a little thought, you should be able to convince yourself that for the same binary image, the $m_{10}$ and $m_{01}$ moments, divided by the $m_{00}$ moment, are the average $x$ and $y$ values across the object. The term *moments* relates to the how this term is used in statistics, and the higher-order moments can be related to what are called the moments of a statistical distribution (i.e., area, average, variance, etc.). In this sense, you can think of the moments of a nonbinary image as being the moments of a binary image in which any individual pixel can be occupied by multiple objects.

### Computing moments with cv::moments()

The function that computes these moments for us is:

```
cv::Moments cv::moments(              // Return structure contains moments
  cv::InputArray points,              // 2-dimensional points or an "image"
  bool           binaryImage = false // false='interpret image values as "mass"'
)
```

The first argument, `points`, is the contour we are interested in, and the second, `binaryImage`, tells OpenCV whether the input image should be interpreted as a binary image. The `points` argument can be either a two-dimensional array (in which case it will be understood to be an image) or a set of points represented as an $N \times 1$ or $1 \times N$ array (with two channels) or an STL vector of `cv::Point` objects. In the latter cases (the sets of points), `cv::moments` will interpret these points not as a discrete set of points, but as a contour with those points as vertices.[11] The meaning of this second argument is that if `true`, all nonzero pixels will be treated as having value `1`, rather than whatever actual value is stored there. This is particularly useful when the image is the output of a threshold operation that might, for example, have `255` as its nonzero values. The `cv::moments()` function returns an instance of the `cv::Moments` object. That object is defined as follows:

10 Mathematical purists might object that $m_{00}$ should be not the contour's length but rather its area. But because we are looking here at a contour and not a filled polygon, the length and the area are actually the same in a discrete pixel space (at least for the relevant distance measure in our pixel space). There are also functions for computing moments of `cv::Array` images; in that case, $m_{00}$ would actually be the area of nonzero pixels. Indeed, the distinction is not entirely academic, however; if a contour is actually represented as a set of vertex points, the formula used to compute the length will not give precisely the same area as we would compute by first rasterizing the contour (i.e., using `cv::drawContours()`) and then computing the area of that rasterization—though the two should converge to the same value in the limit of infinite resolution.

11 In the event that you may need to handle a set of points, rather than a contour, it is most convenient to simply create an image containing those points.

```
class Moments {
public:
    double m00;                        //   zero order moment              (x1)
    double m10, m01;                   //  first order moments             (x2)
    double m20, m11, m02;              // second order moments             (x3)
    double m30, m21, m12, m03;         //  third order moments             (x4)
    double mu20, mu11, mu02;           // second order central moments     (x3)
    double mu30, mu21, mu12, mu03;     //  third order central moments     (x4)
    double nu20, nu11, nu02;           // second order normalized central (x3)
    double nu30, nu21, nu12, nu03;     //  third order normalized central (x4)
  Moments();
  Moments(
    double m00,
    double m10, double m01,
    double m20, double m11, double m02,
    double m30, double m21, double m12, double m03
  );
  Moments( const CvMoments& moments ); // convert v1.x struct to C++ object
  operator CvMoments() const;          // convert C++ object to v1.x struct
}
```

A single call to `cv::moments()` will compute all of the moments up to third order (i.e., moments for which $p + q \leq 3$). It will also compute *central moments* and *normalized central moments*. We will discuss those next.

## More About Moments

The moment computation just described gives some rudimentary characteristics of a contour that can be used to compare two contours. However, the moments resulting from that computation are not the best parameters for such comparisons in most practical cases. In general, the moments we have discussed so far will not be the same for two otherwise identical contours that are displaced relative to each other, of different size, or rotated relative to each other.

### Central moments are invariant under translation

Given a particular contour or image, the $m_{00}$ moment of that contour will clearly be the same no matter where that contour appears in an image. The higher-order moments, however, clearly will not be. Consider the $m_{10}$ moment, which we identified earlier with the average x-position of a pixel in the object. Clearly given two otherwise identical objects in different places, the average x-position is different. It may be less obvious on casual inspection, but the second-order moments, which tell us something about the spread of the object, are also not invariant under translation.[12] This is not particularly convenient, as we would certainly like (in most cases) to be able to use these moments to compare an object that might appear anywhere in an image to a reference object that appeared somewhere (probably somewhere else) in some reference image.

The solution to this is to compute *central moments*, which are usually denoted $\mu_{p,\,q}$ and defined by the following relation:

$$\mu_{p,q} = \sum_{i=1}^{N} I(x_i, y_i)(x - \bar{x})^p(y - \bar{y})^q$$

where:

$$\bar{x} = \frac{m_{10}}{m_{00}}$$

and:

$$\bar{y} = \frac{m_{01}}{m_{00}}$$

Of course, it should be immediately clear that $\mu_{00} = m_{00}$ (because the terms involving $p$ and $q$ vanish anyhow), and that the $\mu_{10}$ and $\mu_{01}$ central moments are both equal to `0`. The higher-order moments are thus the same as the noncentral moments but measured with respect to the "center of mass" of (or in the coordinates of the center of mass of) the object as a whole. Because these measurements are relative to this center, they do not change if the object appears in any arbitrary location in the image.

You will notice that there are no elements `mu00`, `mu10`, or `mu01` in the object `cv::Moments`. This is simply because these values are "trivial" (i.e., `mu00` = `m00`, and `mu10` = `mu01` = `0`). The same is true for the normalized central moments (except that `nu00` = `1`, while `nu10` and `nu01` are both `0`). For this reason they are not included in the structure, as they would just waste memory storing redundant information.

12 For those who are not into this sort of math jargon, the phrase "invariant under translation" means that some quantity computed for some object is unchanged if that entire object is moved (i.e., "translated") from one place to another in the image. The phrase "invariant under rotation" similarly means that the quantity being computed is unchanged of the object is rotated in the image.

### Normalized central moments are also invariant under scaling

Just as the central moments allow us to compare two different objects that are in different locations in our image, it is also often important to be able to compare two different objects that are the same except for being different sizes. (This sometimes happens because we are looking for an object of a type that appears in nature of different sizes—e.g., bears—but more often it is simply because we do not necessarily know how far the object will be from the imager that generated our image in the first place.)

Just as the central moments achieve translational invariance by subtracting out the average, the *normalized central moments* achieve scale invariance by factoring out the overall size of the object. The formula for the normalized central moments is the following:

$$\nu_{p,q} = \frac{\mu_{p,q}}{m_{00}^{\left(\frac{p+q}{2}+1\right)}}$$

This marginally intimidating formula simply says that the normalized central moments are equal to the central moments up to a normalization factor that is itself just some power of the area of the object (with that power being greater for higher-order moments).

There is no specific function for computing normalized moments in OpenCV, as they are computed automatically by `cv::Moments()` when the standard and central moments are computed.

### Hu invariant moments are invariant under rotation

Finally, the *Hu invariant moments* are linear combinations of the normalized central moments. The idea here is that, by combining the different normalized central moments, we can create invariant functions representing different aspects of the image in a way that is invariant to scale, rotation, and (for all but the one called $h_1$) reflection.

For the sake of completeness, we show here the actual definitions of the Hu moments:

$$h_1 = \nu_{20} + \nu_{02}$$

$$h_2 = (\nu_{20} - \nu_{02})^2 + 4\nu_{11}^2$$

$$h_3 = (\nu_{30} - 3\nu_{12})^2 + (3\nu_{21} - \nu_{03})^2$$

$$h_4 = (\nu_{30} + \nu_{12})^2 + (\nu_{21} + \nu_{03})^2$$

$$h_5 = (\nu_{30} - 3\nu_{12})(\nu_{30} + \nu_{12})\left[(\nu_{30} + \nu_{12})^2 - 3(\nu_{21} + \nu_{03})^2\right] + (3\nu_{21} - \nu_{03})(\nu_{21} + \nu_{03})[3(\nu_{30} + \nu_{12}) - (\nu_{21} + \nu_{03})]$$

$$h_6 = (\nu_{20} - \nu_{02})[(\nu_{30} + \nu_{12})^2 - (\nu_{21} + \nu_{03})^2] + 4\nu_{11}(\nu_{30} + \nu_{12})(\nu_{21} + \nu_{03})$$

$$h_7 = (3\nu_{21} - \nu_{03})(\nu_{30} + \nu_{12})\left[(\nu_{30} + \nu_{12})^2 - 3(\nu_{21} + \nu_{03})^2\right] - (\nu_{30} - 3\nu_{12})(\nu_{21} + \nu_{03})\left[3(\nu_{30} + \nu_{12})^2 - (\nu_{21} + \nu_{03})^2\right]$$

Looking at Figure 14-7 and Table 14-3, we can gain a sense of how the Hu moments behave. Observe first that the moments tend to be smaller as we move to higher orders. This should be no surprise because, by their definition, higher Hu moments have more powers of various normalized factors. Since each of those factors is less than `1`, the products of more and more of them will tend to be smaller numbers.

*Figure 14-7. Images of five simple characters; looking at their Hu moments yields some intuition concerning their behavior*

*Table 14-3. Values of the Hu moments for the five simple characters shown in Figure 14-6*

| | $h_1$ | $h_2$ | $h_3$ | $h_4$ | $h_5$ | $h_6$ | $h_7$ |
|---|---|---|---|---|---|---|---|
| A | 2.837e–1 | 1.961e–3 | 1.484e–2 | 2.265e–4 | –4.152e–7 | 1.003e–5 | –7.941e–9 |
| I | 4.578e–1 | 1.820e–1 | 0.000 | 0.000 | 0.000 | 0.000 | 0.000 |
| O | 3.791e–1 | 2.623e–4 | 4.501e–7 | 5.858e–7 | 1.529e–13 | 7.775e–9 | –2.591e–13 |
| M | 2.465e–1 | 4.775e–4 | 7.263e–5 | 2.617e–6 | –3.607e–11 | –5.718e–8 | –7.218e–24 |
| F | 3.186e–1 | 2.914e–2 | 9.397e–3 | 8.221e–4 | 3.872e–8 | 2.019e–5 | 2.285e–6 |

Other factors of particular interest are that the *I*, which is symmetric under 180-degree rotations and reflection, has a value of exactly `0` for $h_3$ through $h_7$, and that the *O*, which has similar symmetries, has all nonzero moments (though in fact, two of these are essentially zero also). We leave it to the reader to look at the figures, compare the various moments, and build a basic intuition for what those moments represent.

### Computing Hu invariant moments with cv::HuMoments()

While the other moments were all computed with the same function `cv::moments()`, the Hu invariant moments are computed with a second function that takes the `cv::Moments` object you got from `cv::moments()` and returns a list of numbers for the seven invariant moments:

```
void cv::HuMoments(
  const cv::Moments& moments, // Input is result from cv::moments() function
  double*            hu       // Return is C-style array of 7 Hu moments
);
```

The function `cv::HuMoments()` expects a `cv::Moments` object and a pointer to a C-style array you should have already allocated with room for the seven invariant moments.

## Matching and Hu Moments

Naturally, with Hu moments we would like to compare two objects and determine whether they are similar. Of course, there are many possible definitions of "similar." To make this process somewhat easier, the OpenCV function `cv::matchShapes()` allows us to simply provide two objects and have their moments computed and compared according to a criterion that we provide.

```
double  cv::MatchShapes(
  cv::InputArray object1,       // First array of 2D points or cv:8UC1 image
  cv::InputArray object2,       // Second array of 2D points or cv:8UC1 image
  int            method,        // Comparison method (Table 14-4)
  double         parameter = 0  // Method-specific parameter
);
```

These objects can be either grayscale images or contours. In either case, `cv::matchShapes()` will compute the moments for you before proceeding with the comparison. The `method` used in `cv::matchShapes()` is one of the three listed in Table 14-4.

*Table 14-4. Matching methods used by cv::matchShapes()*

| Value of method | cv::matchShapes() return value |
|---|---|
| `cv::CONTOURS_MATCH_I1` | $\Delta_1 = \sum_{i=1..7} \left\| \frac{1}{\eta_i^A} - \frac{1}{\eta_i^B} \right\|$ |
| `cv::CONTOURS_MATCH_I2` | $\Delta_2 = \sum_{i=1..7} \left\| \eta_i^A - \eta_i^B \right\|$ |
| `cv::CONTOURS_MATCH_I3` | $\Delta_1 = \sum_{i=1..7} \left\| \frac{\eta_i^A - \eta_i^B}{\eta_i^A} \right\|$ |

In the table, $\eta_i^A$ and $\eta_i^B$ are defined as:

$$\eta_i^A = \text{sign}(h_i^A) \cdot \log(h_i^A)$$

and:

$$\eta_i^B = \text{sign}(h_i^B) \cdot \log(h_i^B)$$

In these expressions, $h_i^A$ and $h_i^B$ are the Hu invariant moments of images *A* and *B*, respectively.

Each of the three values defined in Table 14-3 has a different meaning in terms of how the comparison metric is computed. This metric determines the value ultimately returned by `cv::matchShapes()`. The final `parameter` argument is not currently used, so we can safely leave it at the default value of `0` (it is there for future comparison metrics that may require an additional user-provided parameter).

## Using Shape Context to Compare Shapes

Using moments to compare shapes is a classic technique that dates back to the 80s, but there are much better modern algorithms designed for this purpose. In OpenCV 3 there is a dedicated module called `shape` that implements a few such algorithms, in particular Shape Context [Belongie02].

The shape module is still under development, so we will cover only the high-level structure (very briefly) and some very useful parts here that are available for immediate use.

### Structure of the shape module

The shape module is built around an abstraction called a `cv::ShapeDistanceExtractor`. This abstract type is used for any functor whose purpose is to compare two (or more) shapes and return some kind of distance metric that can be used to quantify their dissimilarity. The word *distance* is chosen because, in most cases at least, the dissimilarity will have the properties expected of a distance, such as being always non-negative and being equal to zero only when the shapes are identical. Here are the important parts of the definition of `cv::ShapeDistanceExtractor`.

```
class ShapeContextDistanceExtractor : public ShapeDistanceExtractor {
  public:
  ...
  virtual float computeDistance( InputArray contour1, InputArray contour2 ) = 0;
};
```

Individual shape distance extractors are derived from this class. We will cover two of them that are currently available. Before we do, however, we need to briefly consider two other abstract functor types: `cv::ShapeTransformer` and `cv::HistogramCostExtractor`.

```
class ShapeTransformer : public Algorithm {

public:
  virtual void estimateTransformation(
    cv::InputArray      transformingShape,
    cv::InputArray      targetShape,
    vector<cv::DMatch>& matches
  ) = 0;

  virtual float applyTransformation(
    cv::InputArray      input,
    cv::OutputArray     output      = noArray()
  ) = 0;

  virtual void warpImage(
    cv::InputArray      transformingImage,
    cv::OutputArray     output,
    int                 flags       = INTER_LINEAR,
    int                 borderMode  = BORDER_CONSTANT,
    const cv::Scalar&   borderValue = cv::Scalar()
  ) const = 0;
};

class HistogramCostExtractor : public Algorithm {

public:
  virtual void  buildCostMatrix(
    cv::InputArray      descriptors1,
    cv::InputArray      descriptors2,
    cv::OutputArray     costMatrix
  )                                                  = 0;

  virtual void  setNDummies( int nDummies )          = 0;
  virtual int   getNDummies() const                  = 0;

  virtual void  setDefaultCost( float defaultCost ) = 0;
  virtual float getDefaultCost() const               = 0;
};
```

The *shape transformer* classes are used to represent any of a wide class of algorithms that can remap a set of points to another set of points, or more generally, an image to another image. Affine and perspective transformations (which we have seen earlier) can be implemented as shape transformers, as is an important transformer called the *thin plate spline transform*. The latter transform derives its name from a physical analogy with a thin metal plate and essentially solves for the mapping that would result if some number of "control" points on a thin metal plate were moved to some set of other locations. The resulting transform is the dense mapping that would result if that thin metal plate were to respond to these deformations of the control points. It turns out that this is a widely useful construction, and it has many applications in image alignment and shape matching. In OpenCV, this algorithm is implemented in the form of the functor `cv::ThinPlateSplineShapeTransformer`.

The *histogram cost extractor* generalizes the construction we saw earlier in the case of the earth mover distance, in which we wanted to associate a cost with "shoveling dirt" from one bin to another. In some cases, this cost is constant or linear in the distance "shoveled," but in other cases we would like to associate a different cost with moving counts from one bin to another. The EMD algorithm had its own (legacy) interface for such a specification, but the `cv::HistogramCostExtractor` base class and its derived classes give us a way to handle general instances of this problem. Table 14-5 shows a list of derived cost extractors.

*Table 14-5. Cost extractors derived from the cv::HistogramCostExtractor abstract class*

| Derived extractor class | Cost used by extractor |
|---|---|
| `cv::NormHistogramCostExtractor` | Cost computed from L2 or other norm |
| `cv::ChiHistogramCostExtractor` | Compare using chi-squared distance |
| `cv::EMDHistogramCostExtractor` | Cost is same as in EMD cost matrix using L2 norm |
| `cv::EMDL1HistogramCostExtractor` | Cost is same as in EMD cost matrix using L1 norm |

For each of these extractors and transformers, there is a factory method with a name like `createX()`, with `X` being the desired functor name—for example, `cv::createChiHistogramCostExtractor()`. With these two types in hand, we are now ready to look at some specializations of the shape distance extractor.

#### The shape context distance extractor

As mentioned at the beginning of this section, OpenCV 3 contains an implementation of shape context distance [Belongie02], packaged inside a functor derived from `cv:: ShapeDistanceExtractor`. This method, called `cv::ShapeContextDistanceExtractor`, uses the shape transformer and histogram cost extractor functors in its implementation.

```
namespace cv {

  class ShapeContextDistanceExtractor : public ShapeDistanceExtractor {

    public:
    ...
    virtual float computeDistance(
      InputArray contour1,
      InputArray contour2
```

```
    ) = 0;
  };

  Ptr<ShapeContextDistanceExtractor> createShapeContextDistanceExtractor(
    int   nAngularBins                            = 12,
    int   nRadialBins                             = 4,
    float innerRadius                             = 0.2f,
    float outerRadius                             = 2,
    int   iterations                              = 3,
    const Ptr<HistogramCostExtractor> &comparer
                                       = createChiHistogramCostExtractor(),
    const Ptr<ShapeTransformer>       &transformer
                                       = createThinPlateSplineShapeTransformer()
  );

}
```

In the essence, the Shape Context algorithm computes a representation of each of the two (or *N*) compared shapes. Each representation considers a subset of points on the shape boundary and, for each sampled point, it builds a certain histogram reflecting the shape appearance in polar coordinates when viewed from that point. All histograms have the same size (`nAngularBins * nRadialBins`). Histograms for a point $p_i$ in shape #1 and a point $q_j$ in shape #2 are compared using classical chi-squared distance. Then the algorithm computes the optimal 1:1 correspondence between points (p's and q's) so that the total sum of chi-squared distances is minimal. The algorithm is not the fastest—even computing the cost matrix takes `N*N*nAngularBins*nRadialBins`, where `N` is the size of the sampled subsets of boundary points—but it gives rather decent results, as you can see in Example 14-4.

*Example 14-4. Using the shape context distance extractor*

```
#include "opencv2/opencv.hpp"
#include <algorithm>
#include <iostream>
#include <string>

using namespace std;
using namespace cv;

static vector<Point> sampleContour( const Mat& image, int n=300 ) {

  vector<vector<Point> > _contours;
  vector<Point> all_points;
  findContours(image, _contours, RETR_LIST, CHAIN_APPROX_NONE);
  for (size_t i=0; i <_contours.size(); i++) {
    for (size_t j=0; j <_contours[i].size(); j++)
      all_points.push_back( _contours[i][j] );

  // If too little points, replicate them
```

```
  //
  int dummy=0;
  for (int add=(int)all_points.size(); add<n; add++)
    all_points.push_back(all_points[dummy++]);

  // Sample uniformly
  random_shuffle(all_points.begin(), all_points.end());
  vector<Point> sampled;
  for (int i=0; i<n; i++)
    sampled.push_back(all_points[i]);
  return sampled;
}

int main(int argc, char** argv) {

  string path     = "../data/shape_sample/";
  int indexQuery = 1;

  Ptr<ShapeContextDistanceExtractor> mysc = createShapeContextDistanceExtractor();

  Size sz2Sh(300,300);
  Mat img1=imread(argv[1], IMREAD_GRAYSCALE);
  Mat img2=imread(argv[2], IMREAD_GRAYSCALE);
  vector<Point> c1 = sampleContour(img1);
  vector<Point> c2 = sampleContour(img2);
  float dis = mysc->computeDistance( c1, c2 );
  cout << "shape context distance between " <<
     argv[1] << " and " << argv[2] << " is: " << dis << endl;

  return 0;

}
```

Check out the *...samples/cpp/shape_example.cpp* example in OpenCV 3 distribution, which is a more advanced variant of Example 14-4.

#### Hausdorff distance extractor

Similarly to the Shape Context distance, the Hausdorff distance is another measure of shape dissimilarity available through the `cv::ShapeDistanceExtractor` interface. We define the Hausdorff [Huttenlocher93] distance, in this context, by first taking all of the points in one image and for each one finding the distance to the nearest point in the other. The largest such distance is the *directed Hausdorff distance*. The *Hausdorff distance* is the larger of the two directed Hausdorff distances. (Note that the directed Hausdorff distance is, by itself, not symmetric, while the Hausdorff distance is manifestly symmetric by construction.) In equations, the Hausdorff distance $H()$ is defined in terms of the directed Hausdorff distance $h()$ between two sets $A$ and $B$ as follows:[13]

$$H(A, B) = maxx(h(A, B), h(B, a))$$

with:

$$h(A, B) = \max_{a \varepsilon A} \min_{b \varepsilon B} \| a - b \|$$

Here $\| \cdot \|$ is some norm relative to the points of A and B (typically the Euclidean distance).

In essence, the Hausdorff distance measures the distance between the "worst explained" pair of points on the two shapes. The Hausdorff distance extractor can be created with the factory method: `cv::createHausdorffDistanceExtractor()`.

```
cv::Ptr<cv::HausdorffDistanceExtractor> cv::createHausdorffDistanceExtractor(
  int   distanceFlag = cv::NORM_L2,
  float rankProp     = 0.6
);
```

Remember that the returned `cv::HausdorffDistanceExtractor` object will have the same interface as the Shape Context distance extractor, and so is called using its `cv::computeDistance()` method.

# Summary

In this chapter we learned about contours, sequences of points in two dimensions. These sequences could be represented as STL vectors of two-dimensional point objects (e.g., `cv::Vec2f`), as $N \times 1$ dual-channel arrays, or as $N \times 2$ single-channel arrays. Such sequences can be used to represent contours in an image plane, and there are many features built into the library to help us construct and manipulate these contours.

Contours are generally useful for representing spatial partitions or segmentations of an image. In this context, the OpenCV library provides us with tools for comparing such partitions to one another, as well as for testing properties of these partitions, such as convexity, moments, or the relationship of an arbitrary point with such a contour. Finally, OpenCV provides many ways to match contours and shapes. We

13 Please excuse the use of $H()$ for Hausdorff distance and $h()$ for directed Hausdorff distance when, only a few pages ago, we used $h$ for Hu invariant moments.

saw some of the features available for this purposes, including both the older style features and the newer features based on the Shape Distance Extractor interface.

## Exercises

1. We can find the extremal points (i.e., the two points that are farthest apart) in a closed contour of $N$ points by comparing the distance of each point to every other point.
   a. What is the complexity of such an algorithm?
   b. Explain how you can do this faster.
2. What is the maximal closed contour length that could fit into a $4 \times 4$ image? What is its contour area?
3. Describe an algorithm for determining whether a closed contour is convex—without using `cv::isContourConvex()`.
4. Describe algorithms:
   a. for determining whether a point is above a line.
   b. for determining whether a point is inside a triangle.
   c. for determining whether a point is inside a polygon—without using `cv::pointPolygonTest()`.
5. Using PowerPoint or a similar program, draw a white circle of radius `20` on a black background (the circle's circumference will thus be $2\ \pi\ 20 \approx 125.7$. Save your drawing as an image.
   a. Read the image in, turn it into grayscale, threshold, and find the contour. What is the contour length? Is it the same (within rounding) or different from the calculated length?
   b. Using 125.7 as a base length of the contour, run `cv::approxPolyDP()` using as parameters the following fractions of the base length: 90%, 66%, 33%, 10%. Find the contour length and draw the results.
6. Suppose we are building a bottle detector and wish to create a "bottle" feature. We have many images of bottles that are easy to segment and find the contours of, but the bottles are rotated and come in various sizes. We can draw the contours and then find the Hu moments to yield an invariant bottle-feature vector. So far, so good—but should we draw filled-in contours or just line contours? Explain your answer.
7. When using `cv::moments()` to extract bottle contour moments in Exercise 6, how should we set `isBinary`? Explain your answer.

8. Take the letter shapes used in the discussion of Hu moments. Produce variant images of the shapes by rotating to several different angles, scaling larger and smaller, and combining these transformations. Describe which Hu features respond to rotation, which to scale, and which to both.
9. Go to Google images and search for "ArUco markers." Choose some larger ones.
   a. Are moments good for finding ArUco images?
   b. Are moments or Hu features good for reading ArUco codes?
   c. Is `cv::matchShapes()` good for reading ArUco codes?
10. Make a shape in PowerPoint (or another drawing program) and save it as an image. Make a scaled, a rotated, and a rotated *and* scaled version of the object, and then store these as images. Compare them using `cv::matchShapes()`.
11. Modify the shape context example or *shape_example.cpp* from OpenCV 3 to use Hausdorff distance instead of a shape context.
12. Get five pictures of five hand gestures. (When taking the photos, either wear a black coat or a colored glove so that a selection algorithm can find the outline of the hand.)
   a. Try recognizing the gestures with `cv::matchShapes()`.
   b. Try recognizing the gestures with `cv::computeDistance()`.
   c. Which one works better and why?

CHAPTER 15

# Background Subtraction

## Overview of Background Subtraction

Because of its simplicity and because camera locations are fixed in many contexts, *background subtraction* (a.k.a. *background differencing*) remains a key image-processing operation for many applications, notably video security ones. Toyama, Krumm, Brumitt, and Meyers give a good overview and comparison of many techniques [Toyama99]. In order to perform background subtraction, we first must "learn" a model of the background.

Once learned, this *background model* is compared against the current image, and then the known background parts are subtracted away. The objects left after subtraction are presumably new foreground objects.

Of course, "background" is an ill-defined concept that varies by application. For example, if you are watching a highway, perhaps average traffic flow should be considered background. Normally, background is considered to be any static or periodically moving parts of a scene that remain static or periodic over the period of interest. The whole ensemble may have time-varying components, such as trees waving in morning and evening wind but standing still at noon. Two common but substantially distinct environment categories that are likely to be encountered are indoor and outdoor scenes. We are interested in tools that will help us in both of these environments.

In this chapter, we will first discuss the weaknesses of typical background models, and then will move on to discuss higher-level scene models. In that context, we present a quick method that is mostly good for indoor static background scenes whose lighting doesn't change much. We then follow this with a "codebook" method that is slightly slower but can work in both outdoor and indoor scenes; it allows for periodic movements (such as trees waving in the wind) and for lighting to change slowly or periodically. This method is also tolerant to learning the background even when there are occasional foreground objects moving by. We'll top this off with another discussion of connected components (first seen in Chapters 12 and 14) in the context of cleaning up foreground object detection. We will then compare the quick background method against the codebook background method. This chapter will conclude with a discussion of the implementations available in the OpenCV library of two modern algorithms for background subtraction. These algorithms use the principles discussed in the chapter, but also include both extensions and implementation details that make them more suitable for real-world application.

# Weaknesses of Background Subtraction

Although the background modeling methods mentioned here work fairly well for simple scenes, they suffer from an assumption that is often violated: that the behavior of all the pixels in the image is statistically independent from the behavior of all the others. Notably, the methods we describe here learn a model for the variations a pixel experiences without considering any of its neighboring pixels. In order to take surrounding pixels into account, we could learn a multipart model, a simple example of which would be an extension of our basic independent pixel model to include a rudimentary sense of the brightness of neighboring pixels. In this case, we use the brightness of neighboring pixels to distinguish when neighboring pixel values are relatively bright or dim. We then learn effectively two models for the individual pixel: one for when the surrounding pixels are bright and one for when the surrounding pixels are dim. In this way, we have a model that takes into account the surrounding *context*. But this comes at the cost of twice as much memory uses and more computation, since we now need different values for when the surrounding pixels are bright or dim. We also need twice as much data to fill out this two-state model. We can generalize the idea of "high" and "low" contexts to a multidimensional histogram of single and surrounding pixel intensities as well and perhaps make it even more complex by doing all this over a few time steps. Of course, this richer model over space and time would require still more memory, more collected data samples, and more computational resources.[1]

Because of these extra costs, the more complex models are usually avoided. We can often more efficiently invest our resources in cleaning up the *false-positive* pixels that result when the independent pixel assumption is violated. This cleanup usually takes the form of image-processing operations (`cv::erode()`, `cv::dilate()`, and

1 In cases in which a computer is expected to "learn" something from data, it is often the case that the primary practical obstacle to success turns out to be having enough data. The more complex your model becomes, the easier it is to get yourself into a situation in which the expressive power of your model vastly exceeds your capability to generate training data for that model. We will revisit this issue in more detail in Chapter 20.

`cv::floodFill()`, mostly) that eliminate stray patches of pixels. We've discussed these routines previously (Chapter 10) in the context of finding large and compact[2] *connected components* within noisy data. We will employ connected components again in this chapter and so, for now, will restrict our discussion to approaches that assume pixels vary independently.

# Scene Modeling

How do we define background and foreground? If we're watching a parking lot and a car comes in to park, then this car is a new foreground object. But should it stay foreground forever? How about a trash can that was moved? It will show up as foreground in two places: the place it was moved to and the "hole" it was moved from. How do we tell the difference? And again, how long should the trash can (and its hole) remain foreground? If we are modeling a dark room and suddenly someone turns on a light, should the whole room become foreground? To answer these questions, we need a higher-level "scene" model, in which we define multiple levels between foreground and background states, and a timing-based method of slowly relegating unmoving foreground patches to background patches. We will also have to detect and create a new model when there is a global change in a scene.

In general, a scene model might contain multiple layers, from "new foreground" to older foreground on down to background. There might also be some motion detection so that, when an object is moved, we can identify both its "positive" aspect (its new location) and its "negative" aspect (its old location, the "hole").

In this way, a new foreground object would be put in the "new foreground" object level and marked as a positive object or a hole. In areas where there was no foreground object, we could continue updating our background model. If a foreground object does not move for a given time, it is demoted to "older foreground," where its pixel statistics are provisionally learned until its learned model joins the learned background model.

For global change detection such as turning on a light in a room, we might use global frame differencing. For example, if many pixels change at once, then we could classify it as a global rather than local change and then switch to using a different model for the new situation.

## A Slice of Pixels

Before we go on to modeling pixel changes, let's get an idea of what pixels in an image can look like over time. Consider a camera looking out a window on a scene of

2 Here we are using mathematician's definition of *compact*, which has nothing to do with size.

a tree blowing in the wind. Figure 15-1 shows what the pixels in a given line segment of the image look like over 60 frames. We wish to model these kinds of fluctuations. Before doing so, however, let's take a small digression to discuss how we sampled this line because it's a generally useful trick for creating features and for debugging.

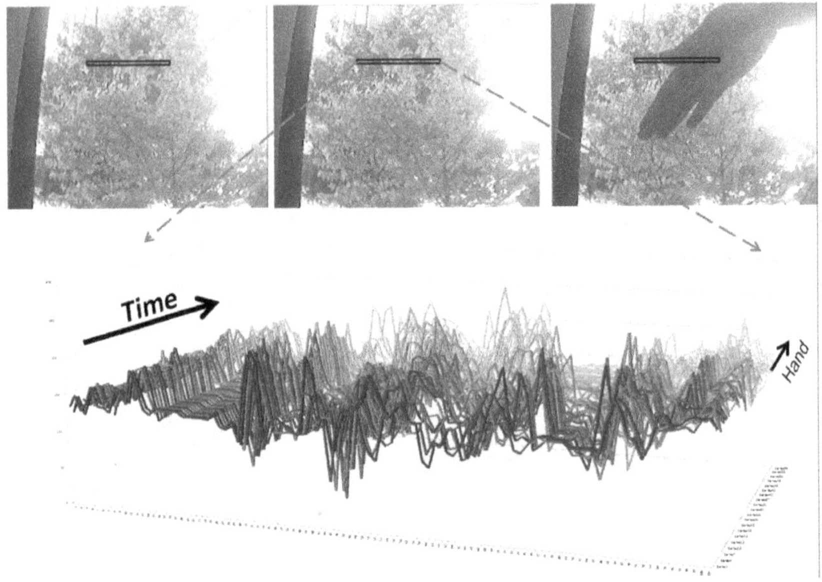


*Figure 15-1. Fluctuations of a line of pixels in a scene of a tree moving in the wind over 60 frames: some dark areas (upper left) are quite stable, whereas moving branches (upper center) can vary widely*

Because this comes up quite often in various contexts, OpenCV makes it easy to sample an arbitrary line of pixels. This is done with the object called the *line iterator*, which we encountered way back in Chapter 6. The line iterator, `cv::LineIterator`, is an object that, once instantiated, can be queried to give us information about all of the points along a line in sequence.

The first thing we need to do is to instantiate a line iterator object. We do this with the `cv::LineIterator` constructor:

```
cv::LineIterator::LineIterator(
  const cv::Mat& image,                     // Image to iterate over
  cv::Point      pt1,                       // Start point for iterator
  cv::Point      pt2,                       // End point for iterator
  int            connectivity  = 8,         // Connectivity, either 4 or 8
  bool           left_to_right = false      // true='fixed iteration direction'
);
```

Here, the input `image` may be of any type or number of channels. The points `pt1` and `pt2` are the ends of the line segment. The `connectivity` can be `4` (the line can step right, left, up, or down) or `8` (the line can additionally step along the diagonals). Finally, if `left_to_right` is set to `0` (`false`), then `line_iterator` scans from `pt1` to `pt2`; otherwise, it will go from the leftmost to the rightmost point.[3]

The iterator can then just be incremented through, pointing to each of the pixels along the line between the given endpoints. We increment the iterator with the usual `cv::LineIterator::operator++()`. All the channels are available at once. If, for example, our line iterator is called `line_iterator`, then we can access the current point by dereferencing the iterator (e.g., `*line_iterator`). One word of warning is in order here, however: the return type of `cv::LineIterator::operator*()` is not a pointer to a built-in OpenCV vector type (i.e., `cv::Vec<>` or some instantiation of it), but rather a `uchar*` pointer. This means that you will typically want to cast this value yourself to something like `cv::Vec3f*` (or whatever is appropriate for the array `image`).[4]

With this convenient tool in hand, we can extract some data from a file. The program in Example 15-1 generates from a movie file the sort of data seen in Figure 15-1.

*Example 15-1. Reading out the RGB values of all pixels in one row of a video and accumulating those values into three separate files*

```
#include <opencv2/opencv.hpp>
#include <iostream>
#include <fstream>

using namespace std;

void help(char** argv ) {
  cout << "\n"
    << "Read out RGB pixel values and store them to disk\nCall:\n"
    << argv[0] <<" avi_file\n"
    << "\n This will store to files blines.csv, glines.csv and rlines.csv\n\n"
```

3 The `left_to_right` flag was introduced because a discrete line drawn from `pt1` to `pt2` does not always match the line from `pt2` to `pt1`. Therefore, setting this flag gives the user a consistent rasterization regardless of the `pt1, pt2` order.

4 In some cases, you can get away with being a little sloppy here. Specifically, when the image is already of unsigned character type, you can just access the elements directly with constructions like `(*line_iterator)[0]`, `(*line_iterator)[1]`, and so on. On close inspection, these are actually dereferencing the iterator to get a character pointer, then using the built-in C offset dereference bracket operator, rather than casting the dereferenced iterator to an OpenCV vector type like `Vec3f` and accessing the channel through the overloaded dereferencing operator of that class. In the end, for the special case of `Vec3b` (or any number of channels), it happens to all come out the same in the end.

```
    << endl;
}

int main( int argc, char** argv) {
  // Argument handling
  //
  if(argc != 2) { help(argv); return -1; }
  cv::namedWindow( argv[0], CV_WINDOW_AUTOSIZE );
  cv::VideoCapture cap;
  if((argc < 2)|| !cap.open(argv[1]))
  {
    cerr << "Couldn't open video file" << endl;
    help(argv);
    cap.open(0);
    return -1;
  }

  //Prepare Output
  //
  cv::Point pt1(10,10), pt2(30,30);
  int max_buffer;
  cv::Mat rawImage;
  ofstream b,g,r;
  b.open("blines.csv");
  g.open("glines.csv");
  r.open("rlines.csv");

  // MAIN PROCESSING LOOP:
  //
  for(;;) {
    cap >> rawImage;
    if( !rawImage.data ) break;
    cv::LineIterator it( rawImage, pt1, pt2, 8);
    for( int j=0; j<it.count; ++j,++it ) {
      b << (int)(*it)[0] << ", ";
      g << (int)(*it)[1] << ", ";
      r << (int)(*it)[2] << ", ";
      (*it)[2] = 255;   // Mark this sample in red
    }
    cv::imshow( argv[0], rawImage );
    int c = cv::waitKey(10);
    b << "\n"; g << "\n"; r << "\n";
  }

  // CLEAN UP:
  //
  b << endl; g << endl; r << endl;
  b.close(); g.close(); r.close();
  cout << "\n"
    << "Data stored to files: blines.csv, glines.csv and rlines.csv\n\n"
    << endl;
}
```

In Example 15-1, we stepped through the points, one at a time, and processed each one. Another common and useful way to approach the problem is to create a buffer (of the appropriate type), and then copy the entire line into it before processing the buffer. In that case, the buffer copy would have looked something like the following:

```
cv::LineIterator it( rawImage, pt1, pt2, 8 );

vector<cv::Vec3b> buf( it.count );

for( int i=0; i < it.count; i++, ++it )
  buf[i] = &( (const cv::Vec3b*) it );
```

The primary advantage of this approach is that if the image `rawImage` were not of an unsigned character type, this method handles the casting of the components to the appropriate vector type in a somewhat cleaner way.

We are now ready to move on to some methods for modeling the kinds of pixel fluctuations seen in Figure 15-1. As we move from simple to increasingly complex models, we will restrict our attention to those models that will run in real time and within reasonable memory constraints.

## Frame Differencing

The very simplest background subtraction method is to subtract one frame from another (possibly several frames later) and then label any difference that is "big enough" the foreground. This process tends to catch the edges of moving objects. For simplicity, let's say we have three single-channel images: `frameTime1`, `frameTime2`, and `frameForeground`. The image `frameTime1` is filled with an older grayscale image, and `frameTime2` is filled with the current grayscale image. We could then use the following code to detect the magnitude (absolute value) of foreground differences in `frameForeground`:

```
cv::absdiff(
  frameTime1,                    // First input array
  frameTime2,                    // Second input array
  frameForeground                // Result array
);
```

Because pixel values always exhibit noise and fluctuations, we should ignore (set to `0`) small differences (say, less than 15), and mark the rest as big differences (set to `255`):

```
cv::threshold(
  frameForeground,               // Input image
  frameForeground,               // Result image
  15,                            // Threshold value
  255,                           // Max value for upward operations
  cv::THRESH_BINARY              // Threshold type to use
);
```

The image `frameForeground` then marks candidate foreground objects as `255` and background pixels as `0`. We need to clean up small noise areas as discussed earlier; we might do this with `cv::erode()` or by using connected components. For color images, we could use the same code for each color channel and then combine the channels with the `cv::max()` function. This method is much too simple for most applications other than merely indicating regions of motion. For a more effective background model, we need to keep some statistics about the means and average differences of pixels in the scene. You can look ahead to the section "A Quick Test" on page 484 to see examples of frame differencing in Figures 15-6 and 15-7.

# Averaging Background Method

The averaging method basically learns the average and standard deviation (or similarly, but computationally faster, the average difference) of each pixel as its model of the background.

Consider the pixel line from Figure 15-1. Instead of plotting one sequence of values for each frame (as we did in that figure), we can represent the variations of each pixel throughout the video in terms of an average and average differences (Figure 15-2). In the same video, a foreground object (a hand) passes in front of the camera. That foreground object is not nearly as bright as the sky and tree in the background. The brightness of the hand is also shown in the figure.

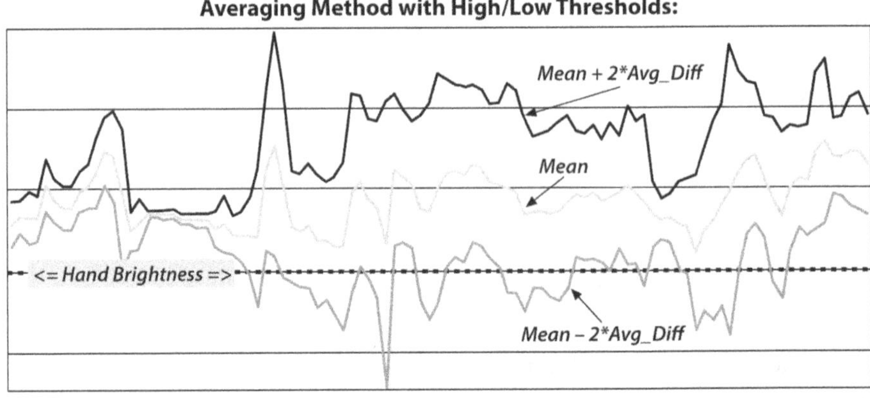


*Figure 15-2. Data from Figure 15-1 presented in terms of average differences: an object (a hand) that passes in front of the camera is somewhat darker, and the brightness of that object is reflected in the graph*

The averaging method makes use of four OpenCV routines: `cv::Mat::operator +=()`, to accumulate images over time; `cv::absdiff()`, to accumulate frame-to-frame image differences over time; `cv::inRange()`, to segment the image (once a background model has been learned) into foreground and background regions; and `cv::max()`, to compile segmentations from different color channels into a single mask image. Because this is a rather long code example, we will break it into pieces and discuss each piece in turn.

First, we create pointers for the various scratch and statistics-keeping images we will need along the way (see Example 15-2). It will prove helpful to sort these pointers according to the type of images they will later hold.

*Example 15-2. Learning a background model to identify foreground pixels*

```
#include <opencv2/opencv.hpp>
#include <iostream>
#include <cstdlib>
#include <fstream>

using namespace std;

// Global storage
//
// Float, 3-channel images
//
cv::Mat image; //, rawImage;
cv::Mat IavgF, IdiffF, IprevF, IhiF, IlowF;
cv::Mat tmp, tmp2, mask;

// Float, 1-channel images
//
vector<cv::Mat> Igray(3);
vector<cv::Mat> Ilow(3);
vector<cv::Mat> Ihi(3);

// Byte, 1-channel image
//
cv::Mat Imaskt;

// Thresholds
//
float high_thresh = 15.0;  //scaling the thresholds in backgroundDiff()
float low_thresh = 13.0;

// Counts number of images learned for averaging later
//
float Icount;
```

Next, we create a single call to allocate all the necessary intermediate images.[5] For convenience, we pass in a single image (from our video) that can be used as a reference for sizing the intermediate images:

```
// I is just a sample image for allocation purposes
// (passed in for sizing)
//
void AllocateImages( const cv::Mat& I ) {
  cv::Size sz = I.size();
  IavgF = cv::Mat::zeros(sz, CV_32FC3 );
  IdiffF = cv::Mat::zeros(sz, CV_32FC3 );
  IprevF = cv::Mat::zeros(sz, CV_32FC3 );
  IhiF = cv::Mat::zeros(sz, CV_32FC3 );
  IlowF = cv::Mat::zeros(sz, CV_32FC3 );
  Icount = 0.00001; // Protect against divide by zero
  tmp = cv::Mat::zeros( sz, CV_32FC3 );
  tmp2 = cv::Mat::zeros( sz, CV_32FC3 );
  Imaskt = cv::Mat( sz, CV_32FC1 );
}
```

In the next piece of code, we learn the accumulated background image and the accumulated absolute value of frame-to-frame image differences (a computationally quicker proxy[6] for learning the standard deviation of the image pixels). This is typically called for 30 to 1,000 frames, sometimes taking just a few frames from each second or sometimes taking all available frames. The routine will be called with a three-color-channel image of depth 8 bits:

```
// Learn the background statistics for one more frame
// I is a color sample of the background, 3-channel, 8u
//
void accumulateBackground( cv::Mat& I ){
  static int first = 1; // nb. Not thread safe
  I.convertTo( tmp, CV_32F ); // convert to float
  if( !first ){
    IavgF += tmp;
    cv::absdiff( tmp, IprevF, tmp2 );
    IdiffF += tmp2;
    Icount += 1.0;
  }
  first = 0;
```

5 In the example here, the accumulator images are of type `float-32`. This is probably fine if the number of frames is not too large, but otherwise it may be preferable to use a `float-64` type.

6 Notice our use of the word *proxy*. Average difference is not mathematically equivalent to standard deviation, but in this context it is close enough to yield results of similar quality. The advantage of average difference is that it is slightly faster to compute than standard deviation. With only a tiny modification of the code example you can use standard deviations instead and compare the quality of the final results for yourself; we'll discuss this more explicitly later in this section.

```
    IprevF = tmp;
}
```

We first use `cv::Mat::convertTo()` to turn the raw background 8-bit-per-channel, three-color-channel image into a floating-point, three-channel image. We then accumulate the raw floating-point images into `IavgF`. Next, we calculate the frame-to-frame absolute difference image using `cv::absdiff()` and accumulate that into image `IdiffF`. Each time we accumulate these images, we increment the image count `Icount`, a global variable, to use for averaging later.

Once we have accumulated enough frames, we convert them into a statistical model of the background; that is, we compute the means and deviation measures (the average absolute differences) of each pixel:

```
void createModelsfromStats() {
  IavgF *= (1.0/Icount);
  IdiffF *= (1.0/Icount);

  // Make sure diff is always something
  //
  IdiffF += cv::Scalar( 1.0, 1.0, 1.0 );
  setHighThreshold( high_thresh);
  setLowThreshold( low_thresh);
}
```

In this section, we use `cv::Mat::operator*=()` to calculate the average raw and absolute difference images by dividing by the number of input images accumulated. As a precaution, we ensure that the average difference image is at least `1`; we'll need to scale this factor when calculating a foreground-background threshold and would like to avoid the degenerate case in which these two thresholds could become equal.

The next two routines, `setHighThreshold()` and `setLowThreshold()`, are utility functions that set a threshold based on the *frame-to-frame average absolute differences* (FFAAD). The FFAAD can be thought of as the basic metric against which we compare observed changes in order to determine whether they are significant. The call `setHighThreshold(7.0)`, for example, fixes a threshold such that any value that is seven times the FFAAD above average for that pixel is considered foreground; likewise, `setLowThreshold(6.0)` sets a threshold bound that is six times the FFAAD below the average for that pixel. Within this range around the pixel's average value, objects are considered to be background. These threshold functions are:

```
void setHighThreshold( float scale ) {
  IhiF = IavgF + (IdiffF * scale);
  cv::split( IhiF, Ihi );
}

void setLowThreshold( float scale ) {
  IlowF = IavgF - (IdiffF * scale);
  cv::split( IlowF, Ilow );
```

```
}

void adjustThresholds(char** argv, cv::Mat &img) {
  int key = 1;
  while((key = cv::waitKey()) != 27 && key != 'Q' && key != 'q')  // Esc or Q or
                                                                   // q to exit
  {
    if(key == 'L') { low_thresh += 0.2;}
    if(key == 'l') { low_thresh -= 0.2;}
    if(key == 'H') { high_thresh += 0.2;}
    if(key == 'h') { high_thresh -= 0.2;}
    cout << "H or h, L or l, esq or q to quit;  high_thresh = "
                       << high_thresh << ", " << "low_thresh = "
                       << low_thresh << endl;
    setHighThreshold(high_thresh);
    setLowThreshold(low_thresh);
    backgroundDiff(img, mask);
    showForgroundInRed(argv, img);
  }
}
```

In `setLowThreshold()` and `setHighThreshold()`, we first scale the difference image (the FFAAD) prior to adding or subtracting these ranges relative to `IavgF`. This action sets the `IhiF` and `IlowF` range for each channel in the image via `cv::split()`.

Once we have our background model, complete with high and low thresholds, we use it to segment the image into foreground (things not "explained" by the background image) and the background (anything that fits within the high and low thresholds of our background model). We perform segmentation by calling:

```
// Create a binary: 0,255 mask where 255 (red) means foreground pixel
// I      Input image, 3-channel, 8u
// Imask  Mask image to be created, 1-channel 8u
//
void backgroundDiff(
    cv::Mat& I,
    cv::Mat& Imask) {

  I.convertTo( tmp, CV_32F ); // To float
  cv::split( tmp, Igray );

  // Channel 1
  //
  cv::inRange( Igray[0], Ilow[0], Ihi[0], Imask );

  // Channel 2
  //
  cv::inRange( Igray[1], Ilow[1], Ihi[1], Imaskt );
  Imask = cv::min( Imask, Imaskt );

  // Channel 3
  //
```

```
    cv::inRange( Igray[2], Ilow[2], Ihi[2], Imaskt );
    Imask = cv::min( Imask, Imaskt );

    // Finally, invert the results
    //
    Imask = 255 - Imask;
}
```

This function first converts the input image `I` (the image to be segmented) into a floating-point image by calling `cv::Mat::convertTo()`. We then convert the three-channel image into separate one-channel image planes using `cv::split()`. Next we check these color channel planes to see whether they are within the high and low range of the average background pixel via the `cv::inRange()` function, which sets the grayscale 8-bit depth image `Imaskt` to max (`255`) when it's in range and to `0` otherwise. For each color channel, we logically AND[7] the segmentation results into a mask image `Imask`, since strong differences in any color channel are considered evidence of a foreground pixel here. Finally, we invert `Imask` using `cv::operator-()`, because foreground should be the values out of range, not in range. The mask image is the output result.

By way of putting it all together, we define the function `main()` that reads in a video and builds a background model. For our example, we run the video in a training mode until the user hits the space bar, after which the video runs in a mode in which any foreground objects detected are highlighted in red:

```
void help(char** argv ) {
  cout << "\n"
  << "Train a background model on  the first <#frames to train on> frames of
  << an incoming video, then run the model\n"
  << argv[0] <<" <#frames to train on> <avi_path/filename>\n"
  << "For example:\n"
  << argv[0] << " 50 ../tree.avi\n"
  << endl;
}

void showForgroundInRed( char** argv, const cv::Mat &img) {
    cv::Mat rawImage;
    cv::split( img, Igray );
    Igray[2] = cv::max( mask, Igray[2] );
    cv::merge( Igray, rawImage );
    cv::imshow( argv[0], rawImage );
}

int main( int argc, char** argv) {
```

7 In this circumstance, you could have used the bitwise OR operator as well, because the images being ORed are unsigned character images and only the values `0x00` and `0xff` are relevant. In general, however, the `cv::max()` operation is a good way to get a "fuzzy" OR, which responds sensibly to a range of values.

```
cv::namedWindow( argv[0], cv::WINDOW_AUTOSIZE );
cv::VideoCapture cap;
if((argc < 3)|| !cap.open(argv[2])) {
  cerr << "Couldn't run the program" << endl;
  help(argv);
  cap.open(0);
  return -1;
}
int number_to_train_on = atoi( argv[1] );

// FIRST PROCESSING LOOP (TRAINING):
//
int frame_count = 0;
int key;
bool first_frame = true;
cout << "Total frames to train on = " << number_to_train_on << endl; //db
while(1) {
  cout << "frame#: " << frame_count << endl;
  cap >> image;
  if(frame_count == 0) { AllocateImages(image);}
  if( !image.data ) exit(1); // Something went wrong, abort
  accumulateBackground( image );
  cv::imshow( argv[0], image );
  frame_count++;
  if( (key = cv::waitKey(7)) == 27 || key == 'q' || key == 'Q' ||  frame_count >=
                  number_to_train_on) break;  //Allow early exit on space, esc, q
}

// We have accumulated our training, now create the models
//
cout << "Creating the background model" << endl;
createModelsfromStats();
cout << "Done!  Hit any key to continue into single step. " <<
                  " Hit 'a' or 'A' to adjust thresholds, esq, 'q' or 'Q'
                  to quit\n" << endl;

// SECOND PROCESSING LOOP (TESTING):
//
while((key = cv::waitKey()) != 27 || key == 'q' || key == 'Q'  ) {
// esc, 'q' or 'Q' to exit
  cap >> image;
  if( !image.data ) exit(0);
  cout <<  frame_count++ << endl;
  backgroundDiff( image, mask );

  // A simple visualization is to write to the red channel
  //
  showForgroundInRed( argv, image);
  if(key == 'a') {
    cout << "In adjust thresholds, 'H' or 'h' == high thresh up " <<
                                "or down; 'L' or 'l' for low thresh
                                up or down." << endl;
```

```
        cout << " esq, 'q' or 'Q' to quit " << endl;
        adjustThresholds(argv, image);
        cout << "Done with adjustThreshold, back to frame stepping, " <<
                                "esc, q or Q to quit." << endl;
      }
    }
    exit(0);
  }
```

We've just seen a simple method of learning background scenes and segmenting foreground objects. It will work well only with scenes that do not contain moving background components (it would fail with a waving curtain or waving trees, features that generate bi- or multi-modal signatures). It also assumes that the lighting remains fairly constant (as in indoor static scenes). You can look ahead to Figure 15-6 to check the performance of this averaging method.

## Accumulating Means, Variances, and Covariances

The averaging background method just described made use of the accumulation operator `cv::Mat::operator+=()` to do what was essentially the simplest possible task: sum up a bunch of data that we could then normalize into an average. The average is a convenient statistical quantity for a lot of reasons, of course, but one often-overlooked advantage it has is the fact that it can be computed incrementally in this way.[8] This means that we can do processing incrementally without needing to accumulate all of the data before analyzing. We will now consider a slightly more sophisticated model, which can also be computed online in this way.

Our next model will represent the intensity (or color) variation within a pixel by computing a *Gaussian model* for that variation. A one-dimensional Gaussian model is characterized by a single mean (or average) and a single *variance* (which tells us something about the expected spread of measured values about the mean). In the case of a $d$-dimensional model (e.g., a three-color model), there will be a $d$-dimensional vector for the mean, and a $d^2$-element matrix that represents not only the individual variances of the $d$-dimensions, but also the covariances, which represent correlations between each of the individual dimensions.

As promised, each of these quantities—the means, the variances, and the covariances—can be computed in an incremental manner. Given a stream of incoming images,

8 For purists, our implementation is not exactly a purely incremental computation, as we divide by the number of samples at the end. There does, however, exist a purely incremental method for updating the average when a new data point is introduced, but the "nearly incremental" version used is substantially more computationally efficient. We will continue throughout the chapter to refer to methods as "incremental" if they can be computed from purely cumulative functions of the data combined with factors associated with overall normalization.

we can define three functions that will accumulate the necessary data, and three functions that will actually convert those accumulations into the model parameters.

The following code assumes the existence of a few global variables:

```
cv::Mat sum;
cv::Mat sqsum;
int     image_count = 0;
```

### Computing the mean with cv::Mat::operator+=()

As we saw in our previous example, the best method to compute the pixel means is to add them all up using `cv::Mat::operator+=()` and then divide by the total number of images to obtain the mean:

```
void accumulateMean(
  cv::Mat& I
) {
  if( sum.empty ) {
    sum = cv::Mat::zeros( I.size(), CV_32FC(I.channels()) );
  }
  I.convertTo( scratch, sum.type() );
  sum += scratch;
  image_count++;
}
```

The preceding function, `accumulateMean()`, is then called on each incoming image. Once all of the images that are going to be used for the background model have been computed, you can call the next function, `computeMean()`, to get a single "image" that contains the averages for every pixel across your entire input set:

```
cv::Mat& computeMean(
  cv::Mat& mean
) {
  mean = sum / image_count;
}
```

### Computing the mean with cv::accumulate()

OpenCV provides another function, which is essentially similar to just using the `cv::Mat::operator+=()` operator, but with two important distinctions. The first is that it will automatically handle the `cv::Mat::convertTo()` functionality (and thus remove the need for a scratch image), and the second is that it allows the use of an image mask. This function is `cv::accumulate()`. The ability to use an image mask when computing a background model is very useful, as you often have some other information that some part of the image should not be included in the background model. For example, you might be building a background model of a highway or other uniformly colored area, and be able to immediately determine from color that some objects are not part of the background. This sort of thing can be very helpful in a real-world situation in which there is little or no opportunity to get access to the scene in the complete absence of foreground objects.

The accumulate function has the following prototype:

```
void accumulate(

  cv::InputArray       src,                // Input, 1 or 3 channels, 8U or F32
  cv::InputOutputArray dst,                // Result image, F32 or F64
  cv::InputArray       mask = cv::noArray() // Use src pixel if mask pixel != 0
);
```

Here the array `dst` is the array in which the accumulation is happening, and `src` is the new image that will be added. `cv::accumulate()` admits an optional mask. If present, only the pixels in `dst` that correspond to nonzero elements in `mask` will be updated.

With `cv::accumulate()`, the previous `accumulateMean()` function can be simplified to:

```
void accumulateMean(
  cv::Mat& I
) {
  if( sum.empty ) {
    sum = cv::Mat::zeros( I.size(), CV_32FC(I.channels()) );
  }
  cv::accumulate( I, sum );
  image_count++;
}
```

#### Variation: Computing the mean with cv::accumulateWeighted()

Another alternative that is often useful is to use a *running average*. The running average is given by the following formula:

$$acc(x,y) = (1 - \alpha) \cdot acc(x, y) + \alpha \cdot image(x, y)$$

For a constant value of $\alpha$, running averages are not equivalent to the result of summing with `cv::Mat::operator+=()` or `cv::accumulate()`. To see this, simply consider adding three numbers (2, 3, and 4) with $\alpha$ set to `0.5`. If we were to accumulate them with `cv::accumulate()`, then the sum would be 9 and the average 3. If we were to accumulate them with `cv::accumulateWeighted()`, the first sum would give $0.5 \cdot 2 + 0.5 \cdot 3 = 2.5$, and then adding the third term would give $0.5 \cdot 2.5 + 0.5 \cdot 4 = 3.25$. The reason the second number is larger is that the most recent contributions are given more weight than those further in the past. Such a running average is also called a *tracker* for just this reason. You can think of the parameter $\alpha$ as setting the time scale necessary for the influence of previous frames to fade—the smaller it is, the faster the influence of past frames fades away.

To accumulate running averages across entire images, we use the OpenCV function `cv::accumulateWeighted()`:

```
void accumulateWeighted(
  cv::InputArray       src,                  // Input, 1 or 3 channels, 8U or F32
  cv::InputOutputArray dst,                  // Result image, F32 or F64
  double               alpha,                // Weight factor applied to src
  cv::InputArray       mask = cv::noArray() // Use src pixel if mask pixel != 0
);
```

Here `dst` is the array in which the accumulation is happening, and `src` is the new image that will be added. The `alpha` value is the weighting parameter. Like `cv::accumulate()`, `cv::accumulateWeighted()` admits an optional mask. If present, only the pixels in `dst` that correspond to nonzero elements in `mask` will be updated.

#### Finding the variance with the help of cv::accumulateSquare()

We can also accumulate squared images, which will allow us to compute quickly the variance of individual pixels. You may recall from your last class in statistics that the variance of a finite population is defined by the formula:

$$\sigma^2 = \frac{1}{N}\sum_{i=0}^{N-1}(x_i - \bar{x})^2$$

where $\bar{x}$ is the mean of $x$ for all $N$ samples. The problem with this formula is that it entails making one pass through the images to compute $\bar{x}$ and then a second pass to compute $\sigma^2$. A little algebra should convince you that the following formula will work just as well:

$$\sigma^2 = \left(\frac{1}{N}\sum_{i=0}^{N-1} x_i^2\right) - \left(\frac{1}{N}\sum_{i=0}^{N-1} x_i\right)^2$$

Using this form, we can accumulate both the pixel values and their squares in a single pass. Then, the variance of a single pixel is just the average of the square minus the square of the average. With this in mind, we can define an accumulation function and a computation function as we did with the mean. As with the mean, one option would be to first do an element-by-element squaring of the incoming image, and then to accumulate that with something like `sqsum += I.mul(I)`. This, however, has several disadvantages, the most significant of which is that `I.mul(I)` does not do any kind of implicit type conversion (as we saw that the `cv::Mat::operator+=()` operator did not do either). As a result, elements of (for example) an 8-bit array, when squared, will almost inevitably cause overflows. As with `cv::accumulate()`, however, OpenCV provides us with a function that does what we need all in a single convenient package—`cv::accumulateSquare()`:

```
void accumulateSquare(
  cv::InputArray        src,                  // Input, 1 or 3 channels, 8U or F32
  cv::InputOutputArray dst,                  // Result image, F32 or F64
  cv::InputArray        mask = cv::noArray() // Use src pixel if mask pixel != 0
);
```

With the help of `cv::accumulateSquare()`, we can write a function to accumulate the information we need for our variance computation:

```
void accumulateVariance(
  cv::Mat& I
) {
  if( sum.empty ) {
    sum   = cv::Mat::zeros( I.size(), CV_32FC(I.channels()) );
    sqsum = cv::Mat::zeros( I.size(), CV_32FC(I.channels()) );
  }
  cv::accumulate( I, sum );
  cv::accumulateSquare( I, sqsum );
  image_count++;
}
```

The associated computation function would then be:

```
// note that 'variance' is sigma^2
//
void computeVariance(
  cv::Mat& variance
) {
  double one_by_N = 1.0 / image_count;
  variance        = one_by_N  * sqsum - (one_by_N * one_by_N) * sum.mul(sum);
}
```

### Finding the covariance with cv::accumulateWeighted()

The variance of the individual channels in a multichannel image captures some important information about how similar we *expect* background pixels in future images to be to our observed average. This, however, is still a very simplistic model for both the background and our "expectations." One important additional concept to introduce here is that of covariance. Covariance captures interrelations between the variations in individual channels.

For example, our background might be an ocean scene in which we expect very little variation in the red channel, but quite a bit in the green and blue channels. If we follow our intuition that the ocean is just one color really, and that the variation we see is primarily a result of lighting effects, we might conclude that if there were a gain or loss of intensity in the green channel, there should be a *corresponding* gain or loss of intensity in the blue channel. The corollary of this is that if there were a substantial gain in the blue channel without an accompanying gain in the green channel, we might not want to consider this to be part of the background. This intuition is captured by the concept of covariance.

In Figure 15-3, we visualize what might be the blue and green channel data for a particular pixel in our ocean background example. On the left, only the variance has been computed. On the right, the covariance between the two channels has also been computed, and the resulting model fits the data much more tightly.

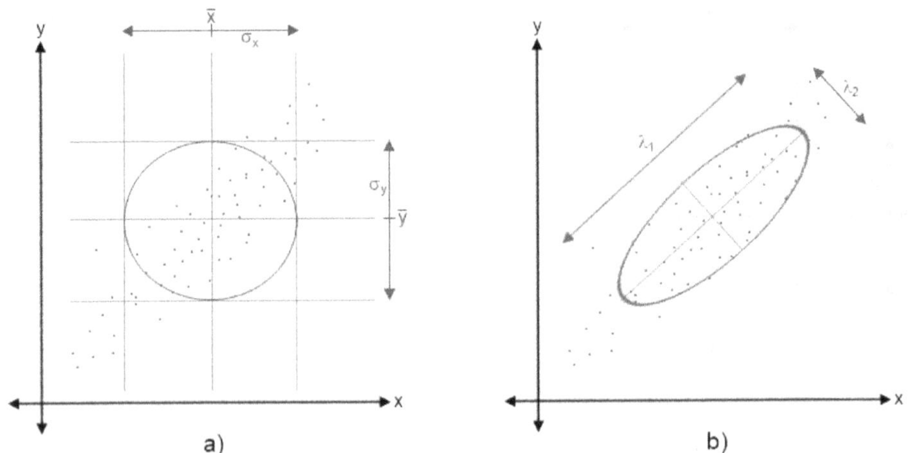


*Figure 15-3. The same data set is visualized on the left and the right. On the left (a), the (square root of the) variance of the data in the x and y dimensions is shown, and the resulting model for the data is visualized. On the right (b), the covariance of the data is captured in the visualized model. The model has become an ellipsoid that is narrower in one dimension and wider in the other than the more simplistic model on the left*

Mathematically, the covariance between any two different observables is given by the formula:

$$Cov(x, y) = \left(\frac{1}{N}\sum_{i=0}^{N-1}(x_i \cdot y_i)\right) - \left(\frac{1}{N}\sum_{i=0}^{N-1}x_i\right)\left(\frac{1}{N}\sum_{i=0}^{N-1}y_i\right)$$

As you can see then, the covariance between any observable $x$ and itself—$Cov(x, x)$—is the same as the variance of that same observable $\sigma_x^2$. In a $d$-dimensional space (such as the RGB values for a pixel, for which $d = 3$), it is convenient to talk about the *covariance matrix* $\Sigma_{x,y}$, whose components include all of the covariances between the variables as well as the variances of the variables individually. As you can see from the preceding formula, the covariance matrix is symmetric—that is, $\Sigma_{x,y} = \Sigma_{y,x}$.

In Chapter 5, we encountered a function that we could use when dealing with individual vectors of data (as opposed to whole arrays of individual vectors), `cv::calcCovarMatrix()`. This function will allow us to provide $N$ vectors of dimension $d$ and will spit out the $d \times d$ covariance matrix. Our problem now, however, is that we would like to compute such a matrix for every point in an array (or at least, in the case of a three-dimensional RGB image, we would like to compute the six unique entries in that matrix).

In practice, the best way to do this is simply to compute the variances using the code we already developed, and to compute the three new objects (the off-diagonal elements of $\Sigma_{x,y}$) separately. Looking at the formula for the covariance, we see that `cv::accumulateSquare()` will not quite work here, as we need to accumulate the $(x_i \cdot y_i)$ terms (i.e., the product of two different channel values from a particular pixel in each image). The function that does this for us in OpenCV is `cv::accumulateProduct()`:

```
void accumulateProduct(
  cv::InputArray       src1,                  // Input, 1 or 3 channels, 8U or F32
  cv::InputArray       src2,                  // Input, 1 or 3 channels, 8U or F32
  cv::InputOutputArray dst,                   // Result image, F32 or F64
  cv::InputArray       mask = cv::noArray() // Use src pixel if mask pixel != 0
);
```

This function works exactly like `cv::accumulateSquare()`, except that rather than squaring the individual elements of `src`, it multiplies the corresponding elements of `src1` and `src2`. What it does not do (unfortunately) is allow us to pluck individual channels out of those incoming arrays. In the case of multichannel arrays in `src1` and `src2`, the computed result is done on a per-channel basis.

For our current need to compute the off-diagonal elements of a covariance model, this is not really what we want. Instead, we want different channels of the *same* image. To do this, we will have to split our incoming image apart using `cv::split()`, as shown in Example 15-3.

*Example 15-3. Computing the off-diagonal elements of a covariance model*

```
vector<cv::Mat> planes(3);
vector<cv::Mat> sums(3);
vector<cv::Mat> xysums(6);

int image_count = 0;

void accumulateCovariance(
  cv::Mat& I
) {

  int i, j, n;
  if( sums.empty ) {
    for( i=0; i<3; i++ ) {         // the r, g, and b sums
      sums[i]   = cv::Mat::zeros( I.size(), CV::F32C1 );
    }
    for( n=0; n<6; n++ ) {         // the rr, rg, rb, gg, gb, and bb elements
      xysums[n] = cv::Mat::zeros( I.size(), CV::F32C1 ) );
    }
```

```
    }
    cv::split( I, rgb );
    for( i=0; i<3; i++ ) {
      cv::accumulate( rgb[i], sums[i] );
    }
    n = 0;
    for( i=0; i<3; i++ ) {            // "row" of Sigma
      for( j=i; j<3; j++ ) {          // "column" of Sigma
        n++;
        cv::accumulateProduct( rgb[i], rgb[j], xysums[n] );
      }
    }
    image_count++;

  }
```

The corresponding compute function is also just a slight extension of the compute function for the variances we saw earlier.

```
// note that 'variance' is sigma^2
//
void computeVariance(
  cv::Mat& covariance          // a six-channel array, channels are the
                               // rr, rg, rb, gg, gb, and bb elements of Sigma_xy
) {
  double one_by_N = 1.0 / image_count;

  // reuse the xysum arrays as storage for individual entries
  //
  int n = 0;
  for( int i=0; i<3; i++ ) {       // "row" of Sigma
    for( int j=i; j<3; j++ ) {     // "column" of Sigma
      n++;
      xysums[n] = one_by_N  * xysums[n]
        - (one_by_N * one_by_N) * sums[i].mul(sums[j]);
    }
  }

  // reassemble the six individual elements into a six-channel array
  //
  cv::merge( xysums, covariance );
}
```

### A brief note on model testing and cv::Mahalanobis()

In this section, we introduced some slightly more complicated models, but did not discuss how to test whether a particular pixel in a new image is in the predicted domain of variation for the background model. In the case of the variance-only model (Gaussian models on all channels with an implicit assumption of statistical independence between the channels) the problem is complicated by the fact that the variances for the individual dimensions will not necessarily be equal. In this case, however, it is common to compute a *z-score* (the distance from the mean divided by the standard deviation: $(x - \bar{x}) / \sigma_x$) for each dimension separately. The z-score tells us something about the probability of the individual pixel originating from the distribution in question. The z-scores for multiple dimensions are then summarized as the square root of sum of squares; for example:

$$\sqrt{z_{red}^2 + z_{green}^2 + z_{blue}^2}$$

In the case of the full covariance matrix, the analog of the z-score is called the *Mahalanobis distance*. This is essentially the distance from the mean to the point in question measured in constant-probability contours such as that shown in Figure 15-3. Looking back at Figure 15-3, we see that a point up and to the left of the mean in the model in (a) will appear to have a low Mahalanobis distance by that model. The same point would have a much higher Mahalanobis distance by the model in (b). It is worth noting that the z-score formula for the simplified model just given is precisely the Mahalanobis distance under the model in Figure 15-3(a), as one would expect.

OpenCV provides a function for computing Mahalanobis distances:

```
double cv::Mahalanobis(              // Return distance as F64
  cv::InputArray vec1,               // First vector (1-dimensional, length n)
  cv::InputArray vec2,               // Second vector (1-dimensional, length n)
  cv::InputArray icovar              // Inverse covariance matrix, n-by-n
);
```

The `cv::Mahalanobis()` function expects vector objects for `vec1` and `vec2` of dimension $d$, and a $d \times d$ matrix for the *inverse* covariance `icovar`. (The inverse covariance is used because inverting this matrix is costly, and in most cases you have many vectors you would like to compare with the same covariance—so the assumption is that you will invert it once and pass the inverse covariance to `cv::Mahalanobis()` many times for each such inversion.)

In our context of background subtraction, this is not entirely convenient, as `cv::Mahalanobis()` wants to be called on a per-element basis. Unfortunately, there is no array-sized version of this capability in OpenCV. As a result, you will have to loop through each pixel, create the covariance matrix from the individual elements, invert that matrix, and store the inverse somewhere. Then, when you want to make a comparison, you will need to loop through the pixels in your image, retrieve the inverse covariance you need, and call `cv::Mahalanobis()` for each pixel.

# A More Advanced Background Subtraction Method

Many background scenes contain complicated moving objects such as trees waving in the wind, fans turning, curtains fluttering, and so on. Often, such scenes also contain varying lighting, such as clouds passing by or doors and windows letting in different light.

A nice method to deal with this would be to fit a time-series model to each pixel or group of pixels. This kind of model deals with the temporal fluctuations well, but its disadvantage is the need for a great deal of memory [Toyama99]. If we use 2 seconds of previous input at 30 Hz, this means we need 60 samples for each pixel. The resulting model for each pixel would then encode what it had learned in the form of 60 different adapted *weights*. Often we'd need to gather background statistics for much longer than 2 seconds, which means that such methods are typically impractical on present-day hardware.

To get fairly close to the performance of adaptive filtering, we take inspiration from the techniques of video compression and attempt to form a YUV[9] *codebook*[10] to represent significant states in the background.[11] The simplest way to do this would be to compare a new value observed for a pixel with prior observed values. If the value is close to a prior value, then it is modeled as a perturbation on that color. If it is not close, then it can seed a new group of colors to be associated with that pixel. The result could be envisioned as a bunch of blobs floating in RGB space, each blob representing a separate volume considered likely to be background.

In practice, the choice of RGB is not particularly optimal. It is almost always better to use a color space whose axis is aligned with brightness, such as the YUV color space. (YUV is the most common choice, but spaces such as HSV, where *V* is essentially brightness, would work as well.) The reason for this is that, empirically, most of the

9 YUV is a color space developed for early color television that needed to be backward-compatible with black-and-white monochrome TV. The first signal is pixel brightness or "luma" *Y* that black-and-white television used. To save space, only two chrominance signals then needed to be transmitted *U* (blue—luma) and *V* (red—luma) from which RGB color could be reconstructed from a conversion formula.

10 The method OpenCV implements is derived from Kim, Chalidabhongse, Harwood, and Davis [Kim05], but rather than using learning-oriented cylinders in RGB space, for speed, the authors use axis-aligned boxes in YUV space. Fast methods for cleaning up the resulting background image can be found in Martins [Martins99].

11 There is a large literature for background modeling and segmentation. OpenCV's implementation is intended to be fast and robust enough that you can use it to collect foreground objects mainly for the purposes of collecting data sets on which to train classifiers. Additional later work in background subtraction allows arbitrary camera motion [Farin04; Colombari07] and dynamic background models using the mean-shift algorithm [Liu07].

natural variation in the background tends to be along the brightness axis, not the color axis.

The next detail is how to model these "blobs." We have essentially the same choices as before with our simpler model. We could, for example, choose to model the blobs as Gaussian clusters with a mean and a covariance. It turns out that the simplest case, in which the "blobs" are simply boxes with a learned extent in each of the three axes of our color space, works out quite well. It is the simplest in terms of both the memory required and the computational cost of determining whether a newly observed pixel is inside any of the learned boxes.

Let's explain what a codebook is by using a simple example (Figure 15-4). A codebook is made up of boxes that grow to cover the common values seen over time. The upper panel of Figure 15-4 shows a waveform over time; you could think of this as the brightness of an individual pixel. In the lower panel, boxes form to cover a new value and then slowly grow to cover nearby values. If a value is too far away, then a new box forms to cover it and likewise grows slowly toward new values.

In the case of our background model, we will learn a codebook of boxes that cover three dimensions: the three channels that make up our image at each pixel. Figure 15-5 visualizes the (intensity dimension of the) codebooks for six different pixels learned from the data in Figure 15-1.[12] This codebook method can deal with pixels that change levels dramatically (e.g., pixels in a windblown tree, which might alternately be one of many colors of leaves, or the blue sky beyond that tree). With this more precise method of modeling, we can detect a foreground object that has values between the pixel values. Compare this with Figure 15-2, where the averaging method cannot distinguish the hand value (shown as a dotted line) from the pixel fluctuations. Peeking ahead to the next section, we see the better performance of the codebook method versus the averaging method shown later in Figure 15-8.

12 In this case, we have chosen several pixels at random from the scan line to avoid excessive clutter. Of course, there is actually a codebook for every pixel.

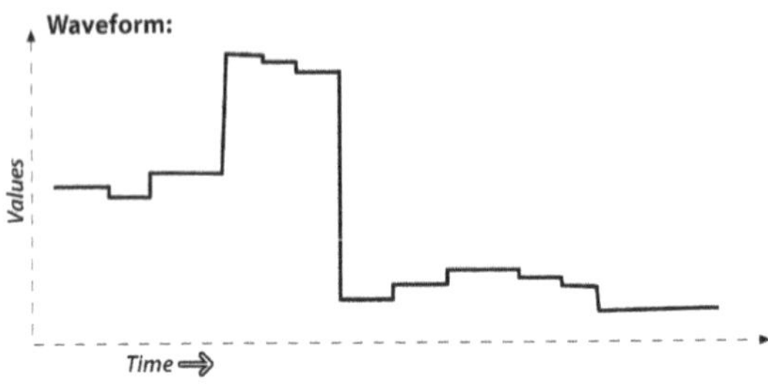


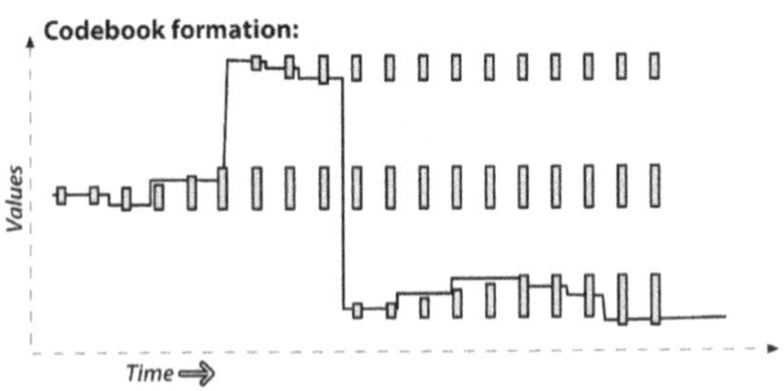


*Figure 15-4. Codebooks are just "boxes" delimiting intensity values: a box is formed to cover a new value and slowly grows to cover nearby values; if values are too far away then a new box is formed*

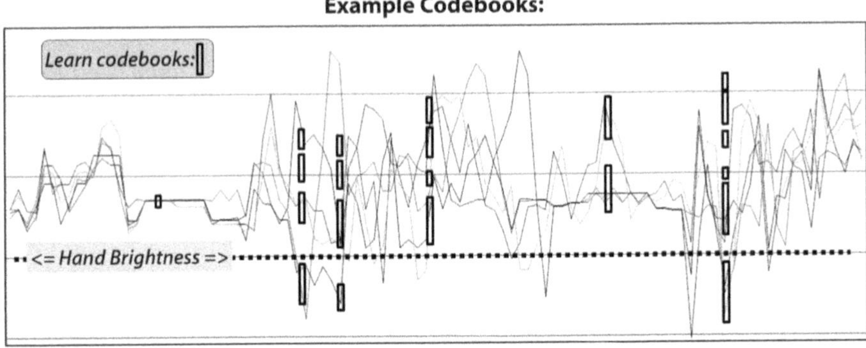


*Figure 15-5. The intensity portion of learned codebook entries for fluctuations of six chosen pixels (shown as vertical boxes): codebook boxes accommodate pixels that take on multiple discrete values and so can better model discontinuous distributions; thus, they can detect a foreground hand (value at dotted line) whose average value is between the values that background pixels can assume. In this case, the codebooks are one-dimensional and represent only variations in intensity*

In the codebook method of learning a background model, each box is defined by two thresholds (`max` and `min`) over each of the three-color axes. These box boundary thresholds will expand (`max` getting larger, `min` getting smaller) if new background samples fall within a learning threshold (`learnHigh` and `learnLow`) above `max` or below `min`, respectively. If new background samples fall outside of the box and its learning thresholds, then a new box will be started. In the *background difference* mode, there are acceptance thresholds `maxMod` and `minMod`; using these threshold values, we say that if a pixel is "close enough" to a `max` or a `min` box boundary, then we count it as if it were inside the box. At runtime, the threshold for inclusion in a "box" can be set to a different value than was used in the construction of the boxes; often this threshold is simply set to `0` in all three dimensions.

A situation we will not cover is a pan-tilt camera surveying a large scene. When working with a large scene, we must stitch together learned models indexed by the pan and tilt angles.

## Structures

It's time to look at all of this in more detail, so let's create an implementation of the codebook algorithm. First, we need our codebook structure, which will simply point to a bunch of boxes in YUV space (see Example 15-4).

*Example 15-4. Codebook algorithm implementation*

```
class CodeBook : public vector<CodeElement> {

public:

  int t;                                            // count every access

  CodeBook() { t=0; }                               // Default is an empty book
  CodeBook( int n ) : vector<CodeElement>(n) { t=0; } // Construct book of size n

};
```

The codebook is derived from an STL vector of `CodeElement` objects as follows.[13] The variable `t` counts the number of points we've accumulated since the start or the last clear operation. Here's how the actual codebook elements are described:

13 Using an STL vector for each pixel (to represent the codebook) is quite inefficient; for a real-world implementation a more efficient representation should be used. For example, limit the number of codebook entries to `MAX_CODES` and use a statically allocated array `CodeElement[MAX_CODES]` instead.

```
#define CHANNELS 3

class CodeElement {

public:

  uchar learnHigh[CHANNELS];  // High side threshold for learning
  uchar learnLow[CHANNELS];   // Low side threshold for learning
  uchar max[CHANNELS];        // High side of box boundary
  uchar min[CHANNELS];        // Low side of box boundary
  int   t_last_update;        // Allow us to kill stale entries
  int   stale;                // max negative run (longest period of inactivity)

  CodeElement() {
    for(i = 0; i < CHANNELS; i++)
      learnHigh[i] = learnLow[i] = max[i] = min[i] = 0;
    t_last_update = stale = 0;
  }

  CodeElement& operator=( const CodeElement& ce ) {
    for( i=0; i<CHANNELS; i++ ) {
      learnHigh[i] = ce.learnHigh[i];
      learnLow[i]  = ce.learnLow[i];
      min[i]       = ce.min[i];
      max[i]       = ce.max[i];
    }
    t_last_update = ce.t_last_update;
    stale         = ce.stale;
    return *this;
  }

  CodeElement( const CodeElement& ce ) { *this = ce; }

};
```

Each codebook entry consumes 4 bytes per channel plus two integers, or (4 * CHANNELS + 4 + 4) bytes (20 bytes when we use three channels). We may set CHANNELS to any positive number equal to or less than the number of color channels in an image, but it is usually set to either 1 (*Y*, or brightness only) or 3 (for YUV or HSV images). In this structure, for each channel, max and min are the boundaries of the codebook box. The parameters learnHigh[] and learnLow[] are the thresholds that trigger generation of a new code element. Specifically, a new code element will be generated if a new pixel is encountered whose values do not lie between min - learnLow and max + learnHigh in each of the channels. The time to last update (t_last_update) and stale are used to enable the deletion of seldom-used codebook entries created during learning. Now we can proceed to investigate the functions that use this structure to learn dynamic backgrounds.

## Learning the Background

We will have one CodeBook of CodeElements for each pixel. We will need an array of such codebooks that is equal in length to the number of pixels in the images we'll be learning. For each pixel, updateCodebook() is called for as many images as are sufficient to capture the relevant changes in the background. Learning may be updated periodically throughout, and we can use clearStaleEntries() to learn the background in the presence of (small numbers of) moving foreground objects. This is possible because the seldom-used "stale" entries induced by a moving foreground will be deleted. The interface to updateCodebook() is as follows:

```
// Updates the codebook entry with a new data point
// Note: cbBounds must be of length equal to numChannels
//
//
int updateCodebook(             // return CodeBook index
  const cv::Vec3b& p,           // incoming YUV pixel
  CodeBook&        c,           // CodeBook for the pixel
  unsigned*        cbBounds,    // Bounds for codebook (usually: {10,10,10})
  int              numChannels  // Number of color channels we're learning
) {
  unsigned int high[3], low[3], n;
  for( n=0; n<numChannels; n++ ) {
    high[n] = p[n] + *(cbBounds+n);    if( high[n] > 255 ) high[n] = 255;
    low[n]  = p[n] - *(cbBounds+n);    if( low[n]  < 0   ) low[n]  = 0;
  }
  // SEE IF THIS FITS AN EXISTING CODEWORD
  //
  int i;
  int matchChannel;
  for( i=0; i<c.size(); i++ ) {

    matchChannel = 0;
    for( n=0; n<numChannels; n++ ) {
      if( // Found an entry for this channel
       ( c[i].learnLow[n] <= p[n] ) && ( p[n] <= c[i].learnHigh[n])
      )
      matchChannel++;
    }

    if( matchChannel == numChannels ) {     // If an entry was found
      c[i].t_last_update = c.t;

      // adjust this codeword for the first channel
      //
      for( n=0; n<numChannels; n++ ) {
        if( c[i].max[n] < p[n] )       c[i].max[n] = p[n];
        else if( c[i].min[n] > p[n] )  c[i].min[n] = p[n];
      }
      break;
    }
```

```
    }
  . . .continued below
```

This function grows or adds a codebook entry when the pixel p falls outside the existing codebook boxes. Boxes grow when the pixel is within cbBounds of an existing box. If a pixel is outside the cbBounds distance from a box, a new codebook box is created. The routine first sets high and low levels to be used later. It then goes through each codebook entry to check whether the pixel value p is inside the learning bounds of the codebook "box." If the pixel is within the learning bounds for all channels, then the appropriate max or min level is adjusted to include this pixel and the time of last update is set to the current timed count, c.t. Next, the updateCodebook() routine keeps statistics on how often each codebook entry is hit:

```
  . . . continued from above

    // OVERHEAD TO TRACK POTENTIAL STALE ENTRIES
    //
    for( int s=0; s<c.size(); s++ ) {

      // Track which codebook entries are going stale:
      //
      int negRun = c.t - c[s].t_last_update;
      if( c[s].stale < negRun ) c[s].stale = negRun;

    }

  . . .continued below
```

Here, the variable stale contains the largest *negative runtime* (i.e., the longest span of time during which that code was not accessed by the data). Tracking stale entries allows us to delete codebooks that were formed from noise or moving foreground objects and hence tend to become stale over time. In the next stage of learning the background, updateCodebook() adds a new codeword if needed:

```
  . . . continued from above

    // ENTER A NEW CODEWORD IF NEEDED
    //
    if( i == c.size() ) {          // if no existing codeword found, make one

      CodeElement ce;
      for( n=0; n<numChannels; n++ ) {
        ce.learnHigh[n] = high[n];
        ce.learnLow[n]  = low[n];
        ce.max[n]       = p[n];
        ce.min[n]       = p[n];
      }
      ce.t_last_update = c.t;
      ce.stale = 0;
```

```
        c.push_back( ce );

    }
  . . .continued below
```

Finally, `updateCodebook()` slowly adjusts (by adding 1) the `learnHigh` and `learnLow` learning boundaries if pixels were found outside of the box thresholds but still within the `high` and `low` bounds:

```
  . . . continued from above

    // SLOWLY ADJUST LEARNING BOUNDS
    //
    for( n=0; n<numChannels; n++ ) {

      if( c[i].learnHigh[n] < high[n]) c[i].learnHigh[n] += 1;
      if( c[i].learnLow[n]  > low[n] ) c[i].learnLow[n]  -= 1;

    }
    return i;
  }
```

The routine concludes by returning the index of the modified codebook. We've now seen how codebooks are learned. In order to learn in the presence of moving foreground objects and to avoid learning codes for spurious noise, we need a way to delete entries that were accessed only rarely during learning.

## Learning with Moving Foreground Objects

The following routine, `clearStaleEntries()`, allows us to learn the background even if there are moving foreground objects:

```
// During learning, after you've learned for some period of time,
// periodically call this to clear out stale codebook entries
//
int clearStaleEntries(      // return number of entries cleared
  CodeBook &c               // Codebook to clean up
){
  int staleThresh = c.t>>1;
  int *keep       = new int[c.size()];
  int keepCnt     = 0;

  // SEE WHICH CODEBOOK ENTRIES ARE TOO STALE
  //
  for( int i=0; i<c.size(); i++ ){
    if(c[i].stale > staleThresh)
      keep[i] = 0;  // Mark for destruction
    else
    {
      keep[i]  = 1; // Mark to keep
      keepCnt += 1;
```

```
    }
  }

  // move the entries we want to keep to the front of the vector and then
  // truncate to the correct length once all of the good stuff is saved.
  //
  int k = 0;
  int numCleared = 0
  for( int ii=0; ii<c.size(); ii++ ) {
    if( keep[ii] ) {
      c[k] = c[ii];
      // We have to refresh these entries for next clearStale
      cc[k]->t_last_update = 0;
      k++;
    } else {
      numCleared++;
    }
  }
  c.resize( keepCnt );
  delete[] keep;

  return numCleared;
}
```

The routine begins by defining the parameter `staleThresh`, which is hardcoded (by a rule of thumb) to be half the total running time count, `c.t`. This means that, during background learning, if codebook entry `i` is not accessed for a period of time equal to half the total learning time, then `i` is marked for deletion (`keep[i] = 0`). The vector `keep[]` is allocated so that we can mark each codebook entry; hence, it is `c.size()` long. The variable `keepCnt` counts how many entries we will keep. After recording which codebook entries to keep, we go through the entries and move the ones we want to the front of the vector in the codebook. Finally, we resize that vector so that everything hanging off the end is chopped off.

## Background Differencing: Finding Foreground Objects

We've seen how to create a background codebook model and how to clear it of seldom-used entries. Next we turn to `backgroundDiff()`, where we use the learned model to segment foreground pixels from the previously learned background:

```
// Given a pixel and a codebook, determine whether the pixel is
// covered by the codebook
//
// NOTES:
// minMod and maxMod must have length numChannels,
// e.g. 3 channels => minMod[3], maxMod[3]. There is one min and
//      one max threshold per channel.
//
uchar backgroundDiff(        // return 0 => background, 255 => foreground
  const cv::Vec3b& p,        // Pixel (YUV)
```

```
    CodeBook&  c,              // Codebook
    int        numChannels,    // Number of channels we are testing
    int*       minMod,         // Add this (possibly negative) number onto max level
                               //   when determining whether new pixel is foreground
    int*       maxMod          // Subtract this (possibly negative) number from min
                               //   level when determining whether new pixel is
                               //   foreground
  ) {
    int matchChannel;

    // SEE IF THIS FITS AN EXISTING CODEWORD
    //
    for( int i=0; i<c.size(); i++ ) {
      matchChannel = 0;
      for( int n=0; n<numChannels; n++ ) {
        if(
          (c[i].min[n] - minMod[n] <= p[n] ) && (p[n] <= c[i].max[n] + maxMod[n])
        ) {
          matchChannel++; // Found an entry for this channel
        } else {
          break;
        }
      }
      if(matchChannel == numChannels) {
        break; // Found an entry that matched all channels
      }
    }

    if( i >= c.size() ) return 0;
    return 255;
  }
```

The background differencing function has an inner loop similar to the learning routine `updateCodebook`, except here we look within the learned `max` and `min` bounds plus an offset threshold, `maxMod` and `minMod`, of each codebook box. If the pixel is within the box plus `maxMod` on the high side or minus `minMod` on the low side for each channel, then the `matchChannel` count is incremented. When `matchChannel` equals the number of channels, we've searched each dimension and know that we have a match. If the pixel is not within a learned box, `255` is returned (a positive detection of foreground); otherwise, `0` is returned (the pixel is background).

The three functions `updateCodebook()`, `clearStaleEntries()`, and `backgroundDiff()` constitute a codebook method of segmenting foreground from learned background.

## Using the Codebook Background Model

To use the codebook background segmentation technique, typically we take the following steps:

1. Learn a basic model of the background over a few seconds or minutes using `updateCodebook()`.
2. Clean out stale entries with `clearStaleEntries()`.
3. Adjust the thresholds `minMod` and `maxMod` to best segment the known foreground.
4. Maintain a higher-level scene model (as discussed previously).
5. Use the learned model to segment the foreground from the background via `backgroundDiff()`.
6. Periodically update the learned background pixels.
7. At a much slower frequency, periodically clean out stale codebook entries with `clearStaleEntries()`.

## A Few More Thoughts on Codebook Models

In general, the codebook method works quite well across a wide number of conditions, and it is relatively quick to train and to run. It doesn't deal well with varying patterns of light—such as morning, noon, and evening sunshine—or with someone turning lights on or off indoors. We can account for this type of global variability by using several different codebook models, one for each condition, and then allowing the condition to control which model is active.

# Connected Components for Foreground Cleanup

Before comparing the averaging method to the codebook method, we should pause to discuss ways to clean up the raw segmented image using connected component analysis. This form of analysis is useful for noisy input mask images, and such noise is more the rule than the exception.

The basic idea behind the method is to use the morphological operation `open` to shrink areas of small noise to `0` followed by the morphological operation `close` to rebuild the area of surviving components that was lost in opening. Thereafter, we find the "large enough" contours of the surviving segments and can take statistics of all such segments. Finally, we retrieve either the largest contour or all contours of size above some threshold. In the routine that follows, we implement most of the functions that you would want for this connected component analysis:

- Whether to approximate the surviving component contours by polygons or by convex hulls
- Setting how large a component contour must be in order for it not to be deleted
- Returning the bounding boxes of the surviving component contours
- Returning the centers of the surviving component contours

The connected components header that implements these operations is shown in Example 15-5.

*Example 15-5. Cleanup using connected components*

```
// This cleans up the foreground segmentation mask derived from calls
// to backgroundDiff
//
void findConnectedComponents(
  cv::Mat&   mask,               // Is a grayscale (8-bit depth) "raw" mask image
                                 // that will be cleaned up
  int        poly1_hull0 = 1,    // If set, approximate connected component by
                                 //  (DEFAULT) polygon, or else convex hull (0)
  float      perimScale  = 4,    // Len = (width+height)/perimScale. If contour
                                 //  len < this, delete that contour (DEFAULT: 4)
  vector<cv::Rect>&  bbs         // Ref to bounding box rectangle return vector
  vector<cv::Point>& centers     // Ref to contour centers return vector
);
```

The function body is listed next. First, we do morphological opening and closing in order to clear out small pixel noise, after which we rebuild the eroded areas that survive the erosion of the opening operation. The routine takes two additional parameters, which here are hardcoded via `#define`. The defined values work well, and you are unlikely to want to change them. These additional parameters control how simple the boundary of a foreground region should be (higher numbers simpler) and how many iterations the morphological operators should perform; the higher the number of iterations, the more erosion takes place in opening before dilation in closing.[14] More erosion eliminates larger regions of blotchy noise at the cost of eroding the boundaries of larger regions. Again, the parameters used in this sample code work well, but there's no harm in experimenting with them if you like:

```
// polygons will be simplified using DP algorithm with 'epsilon' a fixed
// fraction of the polygon's length. This number is that divisor.
//
#define DP_EPSILON_DENOMINATOR 20.0
```

14 Observe that the value `CVCLOSE_ITR` is actually dependent on the resolution. For images of extremely high resolution, leaving this value set to `1` is not likely to yield satisfactory results.

```
// How many iterations of erosion and/or dilation there should be
//
#define CVCLOSE_ITR 1
```

We now discuss the connected component algorithm itself. The first part of the routine performs the morphological open and closing operations:

```
void findConnectedComponents(
  cv::Mat&    mask,
  int         poly1_hull0,
  float       perimScale,
  vector<cv::Rect>&  bbs,
  vector<cv::Point>& centers
) {

  // CLEAN UP RAW MASK
  //
  cv::morphologyEx(
    mask, mask, cv::MOP_OPEN,  cv::Mat(), cv::Point(-1,-1), CVCLOSE_ITR
  );
  cv::morphologyEx(
    mask, mask, cv::MOP_CLOSE, cv::Mat(), cv::Point(-1,-1), CVCLOSE_ITR
  );
```

Now that the noise has been removed from the mask, we find all contours:

```
// FIND CONTOURS AROUND ONLY BIGGER REGIONS
//
vector< vector<cv::Point> > contours_all; // all contours found
vector< vector<cv::Point> > contours;     // just the ones we want to keep
cv::findContours(
  mask,
  contours_all,
  CV_RETR_EXTERNAL,
  CV_CHAIN_APPROX_SIMPLE
);
```

Next, we toss out contours that are too small and approximate the rest with polygons or convex hulls:

```
for(
  vector< vector<cv::Point> >::iterator c = contours_all.begin();
  c != contours.end();
  ++c
) {

  // length of this contour
  //
  int len = cv::arcLength( *c, true );

  // length threshold a fraction of image perimeter
  //
  double q = (mask.rows + mask.cols) / DP_EPSILON_DENOMINATOR;
```

```
        if( len >= q ) {          // If the contour is long enough to keep...

          vector<cv::Point> c_new;
          if( poly1_hull0 ) {   // If the caller wants results as reduced polygons...
            cv::approxPolyDP( *c, c_new, len/20.0, true );
          } else {              // Convex Hull of the segmentation
            cv::convexHull( *c, c_new );
          }
          contours.push_back(c_new );

        }

      }
```

In the preceding code, we use the Douglas-Peucker approximation algorithm to reduce polygons (if the user has not asked us to return just convex hulls). All this processing yields a new list of contours. Before drawing the contours back into the mask, we define some simple colors to draw:

```
    // Just some convenience variables
    const cv::Scalar CVX_WHITE   = cv::RGB(0xff,0xff,0xff);
    const cv::Scalar CVX_BLACK   = cv::RGB(0x00,0x00,0x00);
```

We use these definitions in the following code, where we first analyze each contour separately, then zero out the mask and draw the whole set of clean contours back into the mask:

```
      // CALC CENTER OF MASS AND/OR BOUNDING RECTANGLES
      //
      int idx = 0;
      cv::Moments moments;
      cv::Mat scratch = mask.clone();
      for(
        vector< vector<cv::Point> >::iterator c = contours.begin();
        c != contours.end;
        c++, idx++
      ) {

        cv::drawContours( scratch, contours, idx, CVX_WHITE, CV_FILLED );

        // Find the center of each contour
        //
        moments = cv::moments( scratch, true );
        cv::Point p;
        p.x = (int)( moments.m10 / moments.m00 );
        p.y = (int)( moments.m01 / moments.m00 );
        centers.push_back(p);

        bbs.push_back( cv::boundingRect(c) );

        Scratch.setTo( 0 );
```

```
    }

    // PAINT THE FOUND REGIONS BACK INTO THE IMAGE
    //
    mask.setTo( 0 );
    cv::drawContours( mask, contours, -1, CVX_WHITE );

}
```

That concludes a useful routine for creating clean masks out of noisy raw masks. Note that the new function `cv::connectedComponentsWithStats()` from OpenCV 3 can be used before `cv::findContours()` to mark and delete small connected components.

## A Quick Test

We start this section with an example to see how this really works in an actual video. Let's stick with our video of the tree outside the window. Recall from Figure 15-1 that at some point a hand passes through the scene. You might expect that we could find this hand relatively easily with a technique such as frame differencing (discussed previously). The basic idea of frame differencing is to subtract the current frame from a "lagged" frame and then threshold the difference.

Sequential frames in a video tend to be quite similar, so you might expect that, if we take a simple difference of the original frame and the lagged frame, we won't see too much unless there is some foreground object moving through the scene.[15] But what does "won't see too much" mean in this context? Really, it means "just noise." Thus, in practice the problem is sorting out that noise from the signal when a foreground object really does come along.

To understand this noise a little better, first consider a pair of frames from the video in which there is no foreground object—just the background and the resulting noise. Figure 15-6 shows a typical frame from such a video (upper left) and the previous frame (upper right). The figure also shows the results of frame differencing with a threshold value of 15 (lower left). You can see substantial noise from the moving leaves of the tree. Nevertheless, the method of connected components is able to clean up this scattered noise quite well[16] (lower right). This is not surprising, because there

15 In the context of frame differencing, an object is identified as "foreground" mainly by its velocity. This is reasonable in scenes that are generally static or in which foreground objects are expected to be much closer to the camera than background objects (and thus appear to move faster by virtue of the projective geometry of cameras).

16 The size threshold for the connected components has been tuned to give zero response in these empty frames. The real question, then, is whether the foreground object of interest (the hand) survives pruning at this size threshold. We will see (in Figure 15-8) that it does so nicely.

is no reason to expect much spatial correlation in this noise and so its signal is characterized by a large number of very small regions.

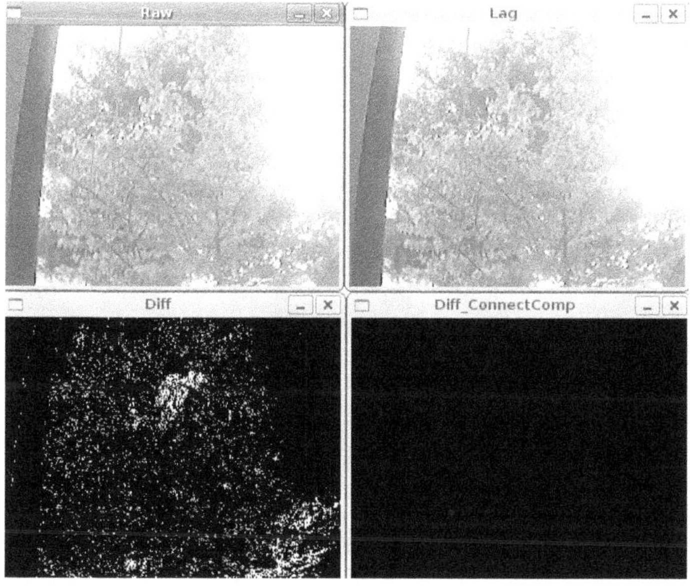


*Figure 15-6. Frame differencing: a tree is waving in the background in the current (upper left) and previous (upper right) frame images; the difference image (lower left) is completely cleaned up (lower right) by the connected component method*

Now consider the situation in which a foreground object (our ubiquitous hand) passes through the frame. Figure 15-7 shows two frames that are similar to those in Figure 15-6 except that now there is a hand moving across from left to right. As before, the current frame (upper left) and the previous frame (upper right) are shown along with the response to frame differencing (lower left) and the fairly good results of the connected component cleanup (lower right).

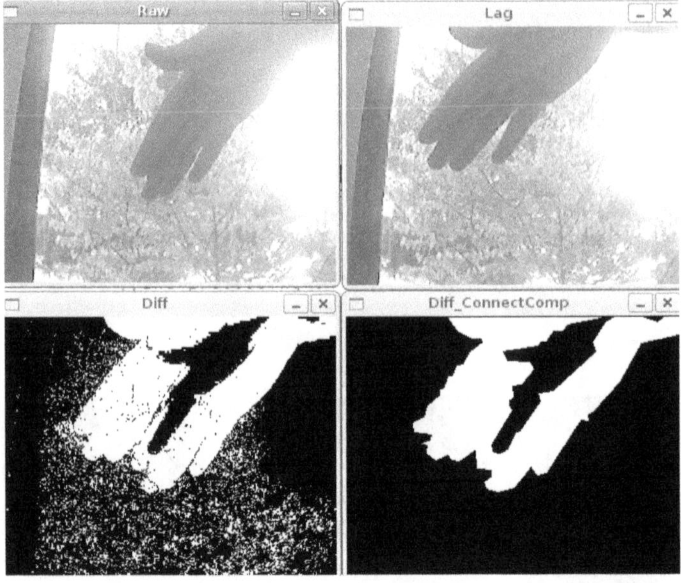


*Figure 15-7. Frame difference method of detecting a hand, which is moving left to right as the foreground object (upper two panels); the difference image (lower left) shows the "hole" (where the hand used to be) toward the left and its leading edge toward the right, and the connected component image (lower right) shows the cleaned-up difference*

We can also clearly see one of the deficiencies of frame differencing: it cannot distinguish between the region from where the object moved (the "hole") and where the object is now. Furthermore, in the overlap region, there is often a gap because "flesh minus flesh" is `0` (or at least below threshold).

Thus, we see that using connected components for cleanup is a powerful technique for rejecting noise in background subtraction. As a bonus, we were also able to glimpse some of the strengths and weaknesses of frame differencing.

# Comparing Two Background Methods

We have discussed two classes of background modeling techniques so far in this chapter: the average distance method (and its variants) and the codebook method. You might be wondering which method is better, or, at least, when you can get away with using the easy one. In these situations, it's always best to just do a straight bake-off[17] between the available methods.

17 For the uninitiated, *bake-off* is actually a bona fide term used to describe any challenge or comparison of multiple algorithms on a predetermined data set.

We will continue with the same tree video that we've been using throughout the chapter. In addition to the moving tree, this film has a lot of glare coming off a building to the right and off portions of the inside wall on the left. It is a fairly challenging background to model.

In Figure 15-8, we compare the average difference method at the top against the codebook method at bottom; on the left are the raw foreground images and on the right are the cleaned-up connected components. You can see that the average difference method leaves behind a sloppier mask and breaks the hand into two components. This is not too surprising; in Figure 15-2, we saw that using the average difference from the mean as a background model often included pixel values associated with the hand value (shown as a dotted line in that figure). Compare this with Figure 15-5, where codebooks can more accurately model the fluctuations of the leaves and branches and so more precisely identify foreground hand pixels (dotted line) from background pixels. Figure 15-8 confirms not only that the background model yields less noise but also that connected components can generate a fairly accurate object outline.

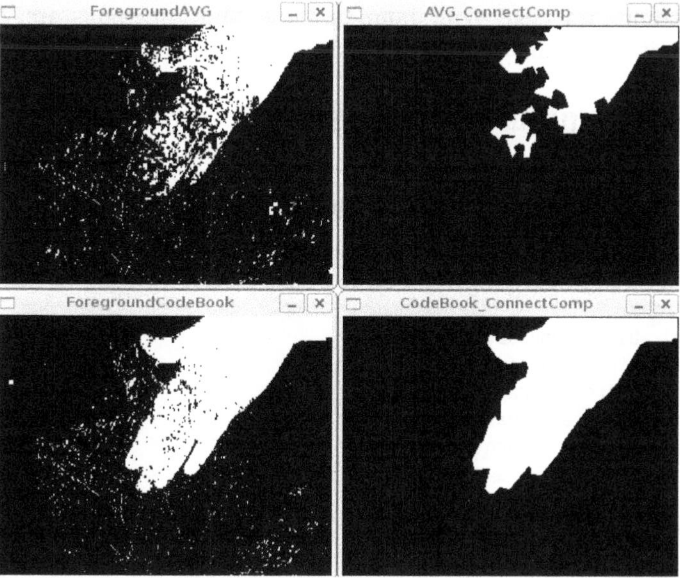


*Figure 15-8. With the averaging method (top row), the connected component cleanup knocks out the fingers (upper right); the codebook method (bottom row) does much better at segmentation and creates a clean connected component mask (lower right)*

# OpenCV Background Subtraction Encapsulation

Thus far, we have looked in detail at how you might implement your own basic background subtraction algorithms. The advantage of that approach is that it is much clearer what is going on and how everything is working. The disadvantage is that as time progresses, newer and better methods are developed that, though rooted in the same fundamental ideas, become sufficiently complicated that you would prefer to regard them as "black boxes" and just use them without getting into the gory details.

To this end, OpenCV now provides a genericized class-based interface to background subtraction. At this time, there are two implementations that use this interface, but more are expected in the future. In this section we will first look at the interface in its generic form, then investigate the two implementations that are available. Both implementations are based on a *mixture of gaussians* (MOG) approach, which essentially takes the statistical modeling concept we introduced for our simplest background modeling scheme (see the section "Accumulating Means, Variances, and Covariances" on page 461) and marries it with the multimodal capability of the codebook scheme (the one developed in the section "A More Advanced Background Subtraction Method" on page 470). Both of these MOG methods are 21st-century algorithms suitable for many practical day-to-day situations.

## The cv::BackgroundSubtractor Base Class

The `cv::BackgroundSubtractor` (abstract) base class specifies only the minimal number of necessary methods. It has the following definition:

```
class cv::BackgroundSubtractor {

public:
  virtual ~BackgroundSubtractor();
  virtual void apply()(
    cv::InputArray  image,
    cv::OutputArray fgmask,
    double          learningRate = -1
  );

  virtual void getBackgroundImage(
    cv::OutputArray backgroundImage
  ) const;

};
```

As you can see, after the constructor, there are only two methods defined. The first is the function operator, which in this context is used to both ingest a new image and to produce the calculated foreground mask for that image. The second function produces an image representation of the background. This image is primarily for visualization and debugging; recall that there is much more information associated with any single pixel in the background than just a color. As a result the image produced by `getBackgroundImage()` can be only a partial presentation of the information that exists in the background model.

One thing that might seem to be a glaring omission is the absence of a method that accumulates background images for training. The reason for this is that there came to be (relative) consensus in the academic literature that any background subtraction algorithm that was not essentially continuously training was undesirable. The reasons for this are many, with the most obvious being the effect of gradual illumination change on a scene (e.g., as the sun rises and sets outside the window). The subtler issues arise from the fact that in many practical scenes there is no opportunity to expose the algorithm to a prolonged period in which no foreground objects are present. Similarly, in many cases, things that seem to be background for an extended period (such as a parked car) might finally move, leaving a permanent foreground "hole" at the location of their absence. For these reasons, essentially all modern background subtraction algorithms do not distinguish between training and running modes; rather, they continuously train and build models in which those things that are seen rarely (and can thus be understood to be foreground) are removed and those things that are seen a majority of the time (which are understood to be the background) are retained.

## KaewTraKuPong and Bowden Method

The first of the available algorithms, KaewTraKuPong and Bowden (KB), brings us several new capabilities that address real-world challenges in background subtraction. These are: a multimodal model, continuous online training, two separate (automatic) training modes that improve initialization performance, and explicit detection and rejection of shadows [KaewTraKulPong2001]. All of this is largely invisible to the user. Not unexpectedly, however, this algorithm does have some parameters that you may want to tune to your particular application. They are the history, the number of Gaussian mixtures, the background ratio, and the noise strength.[18]

The first of these, the *history*, is the point at which the algorithm will switch out of initialization mode and into its nominal run mode. The default value for this parameter is `200` frames. The *number of Gaussian mixtures* is the number of Gaussian components to the overall mixture model that is used to approximate the background in any given pixel. The default value for this parameter is `5`.

Given some number of Gaussian components to the model, each will have a weight indicating the portion of the observed values of that pixel that are explained by that

18 If you find yourself looking up the citation given for this algorithm, the first three parameters—history, number of Gaussian mixtures, and background ratio—are referred to in the paper as $L$, $K$, and $T$, respectively. The last, noise strength, can be thought of as the initialization value of $\Theta_k$ for a newly created component.

particular component of the model. They are not all necessarily "background"; some are likely to be foreground objects that have passed by at one point or another. With the components ordered by weight, the ones that are included as true background are the first *b* of them, where *b* is the minimum number required to "explain" some fixed percentage of the total model. This percentage is called the *background ratio*, and its default value is `0.7` (or `70%`). Thus, by way of example, if there are five components, with weights `0.40`, `0.25`, `0.20`, `0.10`, and `0.05`, then *b* would be `3`, because it required the first three `0.40` + `0.25` + `0.20` to exceed the required background ratio of `0.70`.

The last parameter, *noise strength*, sets the uncertainty assigned to a new Gaussian component when it is created. New components are created whenever new unexplained pixels appear, either because not all components have been assigned yet, or because a new pixel value has been observed that is not explained by any existing component (in which case the least valuable existing component is recycled to make room for this new information). In practice, the effect of increasing the noise strength is to allow the given number of Gaussian components to "explain" more. Of course, the trade-off is that they will tend to explain perhaps even more than has been observed. The default value for the noise strength is `15` (measured in units of 0–255 pixel intensities).

### cv::BackgroundSubtractorMOG

The primary factor that distinguishes the object implementation of the KB background segmentation algorithm from the generic interface is the constructor. You might have noticed that the constructor was not defined in the base class; this is because every implementation will have its own parameters to set, and so no truly generic prototype is possible. The prototype for the constructor for the `cv::bgsegm::BackgroundSubtractorMOG` class from the `bgsegm` module from the *opencv_contrib* repository looks like this:

```
cv::Ptr<cv::bgsegm::BackgroundSubtractorMOG>
    cv::bgsegm::createBackgroundSubtractorMOG(
        int history = 200,
        int nmixtures = 5,
        double backgroundRatio = 0.7,
        double noiseSigma = 0
    );
```

Here, the second constructor allows you to set all four of the parameters that the algorithm needs to operate to whatever values you like. Alternatively, there is the default constructor, which will set all four values to the default values given earlier (i.e., `200`, `5`, `0.70`, and `15`).

## Zivkovic Method

This second background subtraction method, the Zivkovic method, is similar to the KB algorithm in that it also uses a *Gaussian mixture model* to model the distribution of colors observed in any particular pixel. One particularly notable distinction between the two algorithms, however, is that the Zivkovic method does not use a fixed number of Gaussian components; rather, it adapts the number dynamically to give the best overall explanation of the observed distribution [Zivkovic04, Zivkovic06]. This has the downside that the more components there are, the more compute resources are consumed updating and comparing with the model. On the other hand, it has the upside that the model is capable of potentially much higher fidelity.

This algorithm has some parameters in common with the KB method, but introduces many new parameters as well. Fortunately, only two of the parameters are especially important, while the others we can mostly leave at their default values. The two particularly critical parameters are the history (also called the *decay parameter*) and the variance threshold.

The first of these, the *history*, sets the amount of time over which some "experience" of a pixel color will last. Essentially, it is the time it takes for the influence of that pixel to decay away to nothing. The default value for this period is `500` frames. That value is approximately the time before a measurement is "forgotten." Internally to the algorithm, however, it is slightly more accurate to think of this as an exponential *decay parameter* whose value is $\alpha = 1/500 = 0.002$ (i.e., the influence of a measurement decays like $(1-\alpha)^t$.

The second parameter, the *variance threshold*, sets the confidence level that a new pixel measurement must be within, relative to an existing Gaussian mixture component, to be considered part of that component. The units of the variance threshold are in squared-Mahalanobis distance. This means essentially that if you wish to include a pixel that is three sigma from the center of a component into that component, then you would set the variance threshold to `3 * 3 = 9`.[19] The default value for this parameter is actually `4 * 4 = 16`.

---

19 Recall that the Mahalanobis distance is essentially a z-score (i.e., a measurement of how far you are from the center of a Gaussian distribution—measured in units of that distribution's uncertainty) that takes into account the complexities of a distribution in an arbitrary number of dimensions with arbitrary covariance matrix σ.

$r_M^2 = (\vec{x} - \vec{\mu})\Sigma^{-1}(\vec{x} - \vec{\mu})$

You can also see why computing the squared Mahalanobis distance is more natural, which is why you provide the threshold as $r_M^2$ rather than $r_M$.

### cv::BackgroundSubtractorMOG2

The constructor for `cv::BackgroundSubtractorMOG2` allows us to set these two most important parameters. The others may be set from outside once the background subtractor object has been constructed (meaning, they are public members):

```
Ptr<BackgroundSubtractorMOG2> createBackgroundSubtractorMOG2(
  int    history       = 500,
  double varThreshold  = 16,
  bool   detectShadows = true
);
```

The history and variance threshold parameters `history` and `varThreshold` are just as described earlier. The new parameter, `bShadowDetection`, allows optional shadow detection and removal to be turned on. When operational, pixels that are not obviously part of the background are reconsidered to determine if they are just darker versions of the background. If this is found to be the case, these pixels are marked with a special value (typically distinct from the value for background or foreground pixels).

In addition to the two most important parameters, there are the following parameters, which you can tinker with if you are feeling particularly brave:

```
class cv::BackgroundSubtractorMOG2 {

  ...

public:

  ...
  int    getNMixtures() const;           // Minimum number of mixture components
  void   setNMixtures( int nmixtures );

  double getBackgroundRatio() const;     // If component is significant enough
  void   setBackgroundRatio( double backgroundRatio );

  double getVarInit() const;             // Initial variance for new components
  void   setVarInit( double varInit ) const;

  double getVarMin() const;              // Smallest allowed variance
  void   setVarMin(double varMin);

  double getVarMax() const;              // Largest allowed variance
  void   setVarMax(double varMax);

  double getComplexityReductionThreshold() const; // Samples needed to prove that
                                                   //   the component exists
  void   setComplexityReductionThreshold(double CT);

  bool   getDetectShadows() const;       // true try to detect shadows
  void   setDetectShadows(bool detectShadows);
```

```
    int    getShadowValue() const;          // value of shadow pixels in
    void   setShadowValue(int value);       //   output mask

    double getShadowThreshold() const;      // Shadow threshold
    void   setShadowThreshold(double shadowThreshold);
    ...

};
```

The meaning of these parameters is as follows: `nmixtures` is the maximum number of Gaussian components any pixel model can have (the default is `5`). Increasing this improves model fidelity at the cost of runtime. `backgroundRatio` has the same meaning as in the KB algorithm (the default for this algorithm is `0.90`). `varInit`, `varMin`, and `varMax` are the initialization minimum and maximum values for the variance ($\sigma^2$) of any particular Gaussian component (their default values are `15`, `4`, and `75`, respectively). `varInit` is analogous to `noiseSigma` in the KB algorithm. `CT` is what Zivkovic et al. call the *complexity reduction prior*. It is related to the number of samples needed to accept that a component actually exists. The default value for this parameter is `0.05`. Probably the most important thing to know about this value is that if you set it to `0.00`, then the entire algorithm simplifies substantially[20] (both in terms of speed and result quality). `shadowValue` is the value to which shadow pixels will be set in the foreground image (assuming that the constructor argument `detectShadows` was set to `true`). The default for this value is 127. Finally, `shadowThreshold` is a parameter that is used to determine whether a pixel is a shadow. The interpretation of `shadowThreshold` is that it is the relative brightness threshold for a pixel to be considered a shadow relative to something already in the model (e.g., if `fTau` is `0.60`, then any pixel that has the same color as an existing component and is between `0.60` and `1.0` times as bright is considered a shadow). The default value for this parameter is `0.50`.

# Summary

In this chapter, we looked at the specific problem of background subtraction. This problem plays a major role in a vast array of practical computer vision applications, ranging from industrial automation to security to robotics. Starting with the basic theory of background subtraction, we developed two models of how such subtraction could be accomplished based on simple statistical methods. From there we showed

20 For a more technical definition of "simplifies substantially," what really happens is that Zivkovic's algorithm simplifies into something very similar to the algorithm of Stauffer and Grimson. We do not discuss that algorithm here in detail, but it is cited in Zivkovic's paper and was a relatively standard benchmark upon which Zivkovic's algorithm improved.

how connected component analysis could be used to increase the utility of background subtraction results and compared the two basic methods we had developed.

We concluded the chapter by looking at the more advanced background subtraction methods supplied by the OpenCV library as complete implementations. These methods are similar in spirit to the simpler methods we developed in detail at the beginning of the chapter, but contain improvements that make them suitable for more challenging real-world applications.

## Exercises

1. Using `cv::accumulateWeighted()`, reimplement the averaging method of background subtraction. In order to do so, learn the running average of the pixel values in the scene to find the mean and the running average of the absolute difference, `cv::absdiff()`, as a proxy for the standard deviation of the image.

2. Shadows are often a problem in background subtraction because they can show up as a foreground object. Use the averaging or codebook method of background subtraction to learn the background. Then have a person walk in the foreground. Shadows will "emanate" from the bottom of the foreground object.

   a. Outdoors, shadows are darker and bluer than their surroundings; use this fact to eliminate them.

   b. Indoors, shadows are darker than their surroundings; use this fact to eliminate them.

3. The simple background models presented in this chapter are often quite sensitive to their threshold parameters. In Chapter 17, we'll see how to track motion, and this can be used as a reality check on the background model and its thresholds. You can also use it when a known person is doing a "calibration walk" in front of the camera: find the moving object and adjust the parameters until the foreground object corresponds to the motion boundaries. We can also use distinct patterns on a calibration object itself (or on the background) for a reality check and tuning guide when we know that a portion of the background has been occluded.

   a. Modify the code to include an autocalibration mode. Learn a background model and then put a brightly colored object in the scene. Use color to find the colored object and then use that object to automatically set the thresholds in the background routine so that it segments the object. Note that you can leave this object in the scene for continuous tuning.

   b. Use your revised code to address the shadow-removal problem of Exercise 2.

4. Use background segmentation to segment a person with arms held out. Investigate the effects of the different parameters and defaults in the `cv::findConnectedComponents()` routine. Show your results for different settings of:
   a. `poly1_hull0`
   b. `DP_EPSILON_DENOMINATOR`
   c. `perimScale`
   d. `CVCLOSE_ITR`
5. From the directory where you installed OpenCV, using the *.../samples/data/tree.avi* video file, compare and contrast how `cv::BackgroundSubtractorMOG2` and `cv::BackgroundSubtractorMOG2` work on segmenting the moving hand. Use the first part of the video to learn the background, and then segment the moving hand in the later part of the video.
6. Although it might be a little slow, try running background segmentation when the video input is first prefiltered by using `cv::bilateralFilter()`. That is, the input stream is first smoothed and then passed for background learning—and later testing for foreground—by the codebook background segmentation routine.
   a. Show the results compared to not running the bilateral filtering.
   b. Try systematically varying the `spatialSigma` and `colorSigma` of the bilateral filter (e.g., add sliders allowing you to change them interactively). Compare those results.

CHAPTER 16

# Keypoints and Descriptors

## Keypoints and the Basics of Tracking

This chapter is all about informative feature points in images. We will begin by describing what are called *corners* and exploring their definition in the subpixel domain. We will then learn how to track such corners with optical flow. Historically, the tracking of corners evolved into the theory of *keypoints*, to which we will devote the remainder of this chapter, including extensive discussion of keypoint feature detectors and *descriptors* implemented in the OpenCV library for you to use.[1]

The concept of corners, as well as that of keypoints, is based on the intuition that it would be useful in many applications to be able to represent an image or object in an invariant form that will be the same, or at least very similar, in other similar images of the same scene or object. Corner and keypoint representations are powerful methods for doing this. A corner is a small patch of an image that is rich in local information and therefore likely to be recognized in another image. A keypoint is an extension of this concept that encodes information from a small local patch of an image such that the keypoint is highly recognizable and, at least in principle, largely unique. The descriptive information about a keypoint is summarized in the form of its descriptor, which is typically much lower-dimensional than the pixel patch that formed the keypoint. The descriptor represents that patch so as to make it much easier to recognize that patch when it appears in another, different image.

1 In this chapter we restrict our attention primarily to the features in the main library that have the main library's licensing terms. There are more "nonfree" features and feature detectors and descriptors in `xfeatures2D` module (see Appendix B), as well as new experimental features. Of the nonfree features, we will cover only SIFT and SURF in this chapter due to their great historical, practical, and pedagogical importance.

From an intuitive point of view, you can think of a keypoint like a piece from a jigsaw puzzle. When you begin the puzzle, some pieces are easily recognized: the handle of a door, a face, the steeple of a church. When you go to assemble the puzzle, you can immediately relate these keypoints to the image on the puzzle box and know immediately where to place them. In addition, if you and a friend had two unassembled puzzles and you wanted to know if they were, for example, both different images of the beautiful Neuschwanstein Castle, you could assemble both puzzles and compare, but you could also just pick out the most salient pieces from each puzzle and compare them. In this latter case, it would take only a few matches before you were convinced that they were either literally the same puzzle, or two puzzles made from two different images of the same castle.

In this chapter, we will start by building up the basics of the theory of keypoints by discussing the earliest ancestor of the modern keypoint: the Harris corner. From there we will discuss the concept of *optical flow* for such corners, which captures the basic idea of tracking such features from one frame to another in a video sequence. After that we will move on to more modern keypoints and their descriptors and discuss how OpenCV helps us find them as well as how the library will help us match them between frames. Finally, we will look at a convenient method that allows us to visualize keypoints overlaid on top of the images in which they were detected.

## Corner Finding

There are many kinds of local features that you can track. It is worth taking a moment to consider what exactly constitutes such a feature. Obviously, if we pick a point on a large blank wall, then it won't be easy to find that same point in the next frame of a video.

If all points on the wall are identical or even very similar, then we won't have much luck tracking that point in subsequent frames. On the other hand, if we choose a point that is unique, then we have a pretty good chance of finding that point again. In practice, the point or feature we select should be unique, or nearly unique, and should be parameterizable such that it can be compared to other points in another image (see Figure 16-1).

Returning to our intuition from the large blank wall, we might be tempted to look for points that have some significant change in them—for example, a strong derivative. It turns out that this is not quite enough, but it's a start. A point to which a strong derivative is associated may be on an edge of some kind, but it still may look like all of the other points along that same edge (see the aperture problem diagrammed in Figure 16-8 and discussed in the section "Introduction to Optical Flow" on page 502).

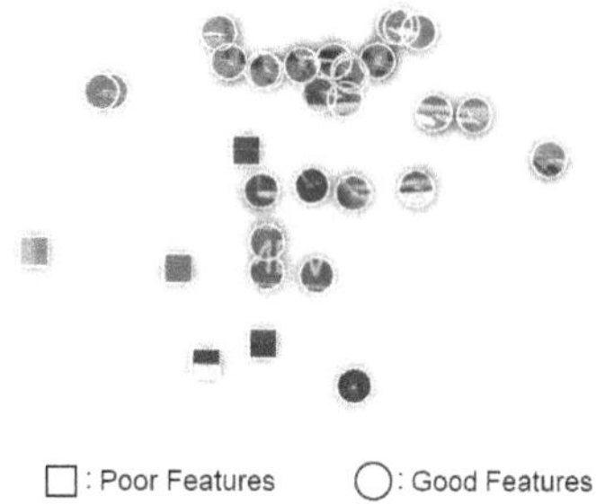


*Figure 16-1. The points marked with circles here are good points to track, whereas those marked with boxes—even the ones that are sharply defined edges—are poor choices*

However, if strong derivatives are observed nearby in two different directions, then we can hope that this point is more likely to be unique. For this reason, many trackable features are called *corners*. Intuitively, corners—not edges—are the points that contain enough information to be picked out from one frame to the next.

The most commonly used definition of a corner was provided by Harris [Harris88]. This definition captures the intuition of the previous paragraph in a mathematically specific form. We will look at the details of this method shortly, but for the moment, what is important to know is that you can ask OpenCV to simply find the points in the image that are good candidates for being tracked, and it will use Harris's method to identify them for you.

#### Finding corners using cv::goodFeaturesToTrack()

The `cv::goodFeaturesToTrack()` routine implements Harris's method and a slight improvement credited to Shi and Tomasi [Shi94]. This function conveniently computes the necessary derivative operators, analyzes them, and returns a list of the points that meet our definition of being good for tracking:

```
void cv::goodFeaturesToTrack(
  cv::InputArray  image,                          // Input, CV_8UC1 or CV_32FC1
  cv::OutputArray corners,                        // Output vector of corners
  int             maxCorners,                     // Keep this many corners
  double          qualityLevel,                   // (fraction) rel to best
  double          minDistance,                    // Discard corner this close
  cv::InputArray  mask              = noArray(),  // Ignore corners where mask=0
  int             blockSize         = 3,          // Neighborhood used
  bool            useHarrisDetector = false,      // false='Shi Tomasi metric'
  double          k                 = 0.04        // Used for Harris metric
);
```

The input `image` can be any 8-bit or 32-bit (i.e., `8U` or `32F`), single-channel image. The output `corners` will be a vector or array (depending on what you provide) containing all of the corners that were found. If it is a `vector<>`, it should be a vector of `cv::Point2f` objects. If it is a `cv::Mat`, it will have one row for every corner and two columns for the *x* and *y* locations of the points. You can limit the number of corners that will be found with `maxCorners`, the quality of the returned points with `qualityLevel` (typically between `0.10` and `0.01`, and never greater than `1.0`), and the minimum separation between adjacent corners with `minDistance`.

If the argument `mask` is supplied, it must be the same dimension as `image`, and corners will not be generated anywhere `mask` is `0`. The `blockSize` argument indicates how large an area is considered when a corner is computed; a typical value is `3` but, for high-resolution images, you may want to make this slightly larger. The `useHarrisDetector` argument, if set to `true`, will cause `cv:: goodFeaturesToTrack()` to use an exact corner strength formula of Harris's original algorithm; if set to `false`, Shi and Tomasi's method will be used. The parameter `k` is used only by Harris's algorithm, and is best left at its default value.[2]

### Subpixel corners

If you are processing images for the purpose of extracting geometric measurements, as opposed to extracting features for recognition, then you will normally need more resolution than the simple pixel values supplied by `cv::goodFeaturesToTrack()`. Another way of saying this is that such pixels come with integer coordinates whereas we sometimes require real-valued coordinates—for example, a pixel location of (8.25, 117.16).

If you imagine looking for a particular small object in a camera image, such as a distant star, you would invariably be frustrated by the fact that the point's location will almost never be in the exact center of a camera pixel element. Of course, in this circumstance, some of the light from the object will appear in the neighboring pixels as well. To overcome this, you might try to fit a curve to the image values and then use a little math to find where the peak occurred between the pixels. Subpixel corner detection techniques all rely on approaches of this kind (for a review and newer techniques, see Lucchese [Lucchese02] and Chen [Chen05]). Common uses of such measurements include tracking for three-dimensional reconstruction, calibrating a camera, warping partially overlapping views of a scene to stitch them together in the most natural way, and finding an external signal such as the precise location of a building in a satellite image.

2 Later in this chapter, we will look at many ways to compute keypoints, which are essentially a generalization of "corners." In that section, we will discuss many keypoint finding algorithms in detail, including Harris's algorithm. We will get into more detailed descriptions of the algorithms and these parameters there.

One of the most common tricks for subpixel refinement is based on the mathematical observation that the dot product between a vector and an orthogonal vector is `0`; this situation occurs at corner locations, as shown in Figure 16-2.

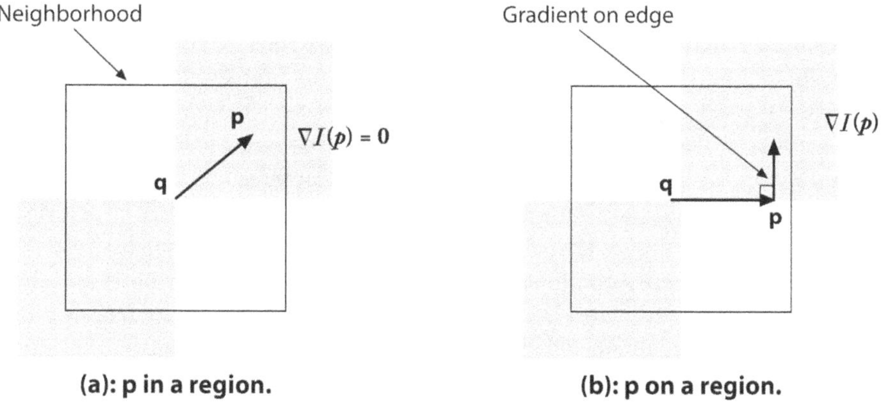


*Figure 16-2. Finding corners to subpixel accuracy: (a) the image area around the point p is uniform and so its gradient is 0; (b) the gradient at the edge is orthogonal to the vector q-p along the edge; in either case, the dot product between the gradient at p and the vector q-p is 0 (see text)*

In Figure 16-2, we assume a starting corner location $q$ that is near the actual subpixel corner location. We examine vectors starting at point $q$ and ending at $p$. When $p$ is in a nearby uniform or "flat" region, the gradient there is `0`. On the other hand, if the vector $q$-$p$ aligns with an edge, then the gradient at $p$ on that edge is orthogonal to the vector $q$-$p$. In either case, the dot product between the gradient at $p$ and the vector $q$-$p$ is `0`. We can assemble many such pairs of the gradient at a nearby point $p$ and the associated vector $q$-$p$, set their dot product to `0`, and solve this assemblage as a system of equations; the solution will yield a more accurate subpixel location for $q$, the exact location of the corner.

The function that does subpixel corner finding is `cv::cornerSubPix()`:

```
void cv::cornerSubPix(
  cv::InputArray        image,                  // Input image
  cv::InputOutputArray  corners,                // Guesses in, and results out
  cv::Size              winSize,                // Area is NXN; N=(winSize*2+1)
  cv::Size              zeroZone,               // Size(-1,-1) to ignore
  cv::TermCriteria      criteria                // When to stop refinement
);
```

The input `image` is the original image from which your corners were computed. The `corners` array contains the integer pixel locations, such as those obtained from routines like `cv::goodFeaturesToTrack()`, which are taken as the initial guesses for the corner locations.

As described earlier, the actual computation of the subpixel location uses a system of dot-product expressions that express the combinations that should sum to zero (see Figure 16-2). Each of these equations arises from considering a single pixel in the region around $p$. The parameter `winSize` specifies the size of window from which these equations will be generated. This window is centered on the original integer corner location and extends outward in each direction by the number of pixels specified in `winSize` (e.g., if `winSize.width = 4`, then the search area is actually 4 + 1 + 4 = 9 pixels wide). These equations form a linear system that can be solved by the inversion of a single autocorrelation matrix.[3] In practice, this matrix is not always invertible owing to small eigenvalues arising from the pixels very close to $p$. To protect against this, it is common to simply reject from consideration those pixels in the immediate neighborhood of $p$. The parameter `zeroZone` defines a window (analogously to `winSize`, but always with a smaller extent) that will *not* be considered in the system of constraining equations and thus the autocorrelation matrix. If no such zero zone is desired, then this parameter should be set to `cv::Size(-1,-1)`.

Once a new location is found for $q$, the algorithm will iterate using that value as a starting point and will continue until the user-specified termination criterion is reached. Recall that this criterion can be of type `cv::TermCriteria::MAX_ITER` or of type `cv::TermCriteria::EPS` (or both) and is usually constructed with the `cv::TermCriteria()` function. Using `cv::TermCriteria::EPS` will effectively indicate the accuracy you require of the subpixel values. Thus, if you specify `0.10`, then you are asking for subpixel accuracy down to one-tenth of a pixel.

## Introduction to Optical Flow

The *optical flow* problem involves attempting to figure out where many (and possibly all) points in one image have moved to in a second image—typically this is done in sequences of video, for which it is reasonable to assume that most points in the first frame can be found somewhere in the second. Optical flow can be used for motion estimation of an object in the scene, or even for ego-motion of the camera relative to the scene as a whole. In many applications, such as video security, it is motion itself that indicates that a portion of the scene is of specific interest, or that something interesting is going on. Optical flow is illustrated in Figure 16-3.

3 Later, we will encounter another autocorrelation matrix in the context of the inner workings of Harris corners. The two are unrelated, however.

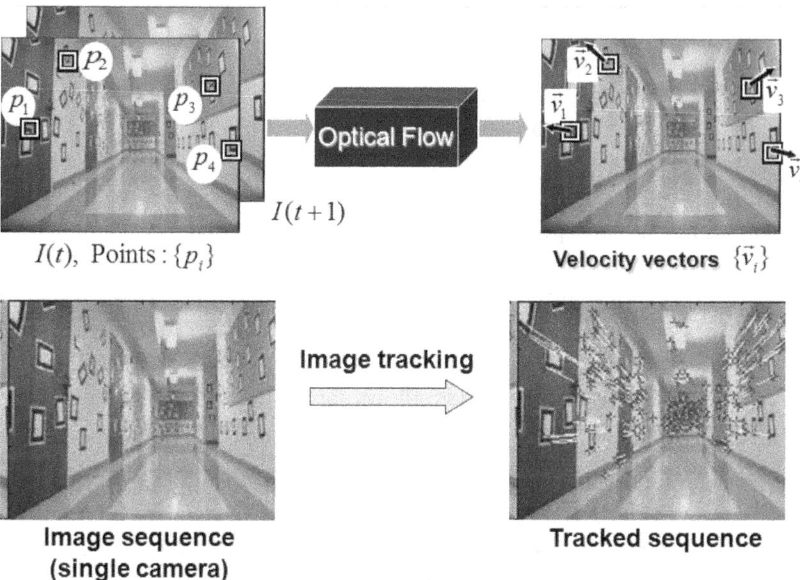


*Figure 16-3. Optical flow: target features (left) are tracked over time and their movement is converted into velocity vectors (right); original images courtesy of Jean-Yves Bouguet*

The ideal output of an optical flow algorithm would be the association of some estimate of velocity for each and every pixel in a frame pair or, equivalently, a displacement vector for every pixel in one image that indicates the relative location of that pixel in the other image. Such a construction, when it applies to every pixel in the image, is usually referred to as *dense optical flow*. There is an alternative class of algorithms, called *sparse optical flow* algorithms, that track only some subset of the points in the image. These algorithms are often fast and reliable because they restrict their attention to specific points in the image that will be easier to track. OpenCV has many ways of helping us identify points that are well suited for tracking, with the corners introduced earlier being only one among a long list. For many practical applications, the computational cost of sparse tracking is so much less than dense tracking that the latter is relegated to only academic interest.[4] In this section, we will look at one sparse optical flow technique. Later, we will look at more powerful tools for sparse optical flow, and then finally move on to dense optical flow.

4 Black and Anandan have created dense optical flow techniques [Black93; Black96] that are often used in movie production where, for the sake of visual quality, the movie studio is willing to spend the time necessary to obtain detailed flow information. These techniques are slated for inclusion in later versions of OpenCV (see Chapter 23).

## Lucas-Kanade Method for Sparse Optical Flow

The Lucas-Kanade (LK) algorithm [Lucas81], as originally proposed in 1981, was an attempt to produce *dense* optical flow (i.e., flow for every pixel). Yet, because the method is easily applied to a subset of the points in the input image, it has become an important technique for *sparse* optical flow. The algorithm can be applied in a sparse context because it relies only on local information that is derived from some small window surrounding each point of interest. The disadvantage of using small local windows in Lucas-Kanade is that large motions can move points outside of the local window and thus become impossible for the algorithm to find. This problem led to development of the "pyramidal" LK algorithm, which tracks starting from highest level of an image pyramid (lowest detail) and working down to lower levels (finer detail). Tracking over image pyramids allows large motions to be caught by local windows.[5]

Because this is an important and effective technique, we will go into some mathematical detail; readers who prefer to forgo such details can skip to the function description and code. However, it is recommended that you at least scan the intervening text and figures, which describe the assumptions behind Lucas-Kanade optical flow, so that you'll have some intuition about what to do if tracking isn't working well.

#### How Lucas-Kanade works

The basic idea of the LK algorithm rests on three assumptions:

*Brightness constancy*
: A pixel from the image of an object in the scene does not change in appearance as it (possibly) moves from frame to frame. For grayscale images (LK can also be done in color), this means we assume that the brightness of a pixel does not change as it is tracked from frame to frame.

*Temporal persistence, or "small movements"*
: The image motion of a surface patch changes slowly in time. In practice, this means the temporal increments are fast enough relative to the scale of motion in the image that the object does not move much from frame to frame.

*Spatial coherence*
: Neighboring points in a scene belong to the same surface, have similar motion, and project to nearby points on the image plane.

5 The definitive description of Lucas-Kanade optical flow in a pyramid framework implemented in OpenCV is an unpublished paper by Bouguet [Bouguet04].

We now look at how these assumptions, illustrated in Figure 16-4, lead us to an effective tracking algorithm. The first requirement, brightness constancy, is just the requirement that pixels in one tracked patch look the same over time, defining:

$$f(x, t) \equiv I(x(t), t) = I(x(t + dt), t + dt)$$

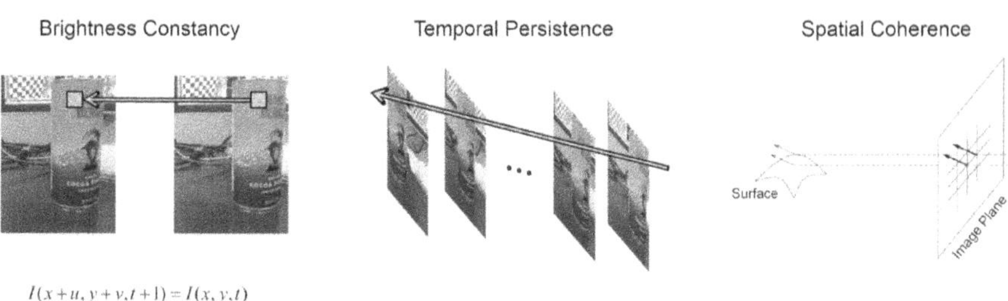


*Figure 16-4. Assumptions behind Lucas-Kanade optical flow: for a patch being tracked on an object in a scene, the patch's brightness doesn't change (left); motion is slow relative to the frame rate (center); and neighboring points stay neighbors (right); component images courtesy of Michael Black [Black92]*

The requirement that our tracked pixel intensity exhibits no change over time can simply be expressed as:

$$\frac{\partial f(x)}{\partial t} = 0$$

The second assumption, temporal persistence, essentially means that motions are small from frame to frame. In other words, we can view this change as approximating a derivative of the intensity with respect to time (i.e., we assert that the change between one frame and the next in a sequence is *differentially small*). To understand the implications of this assumption, first consider the case of a single spatial dimension.

In this case, we can start with our brightness consistency equation, substitute the definition of the brightness *f*(*x*, *t*) while taking into account the implicit dependence of *x* on *t*, *I*(*x*(*t*)*t*), and then apply the chain rule for partial differentiation. This yields:

$$I_x \cdot v + I_t = \left.\frac{\partial I}{\partial x}\right|_t \left(\frac{\partial x}{\partial t}\right) + \left.\frac{\partial I}{\partial t}\right|_{x(t)} = 0$$

where $I_x$ is the spatial derivative across the first image, $I_t$ is the derivative between images over time, and $v$ is the velocity we are looking for. We thus arrive at the simple equation for optical flow velocity in the simple one-dimensional case:

$$v = -\frac{I_t}{I_x}$$

Let's now try to develop some intuition for this one-dimensional tracking problem. Consider Figure 16-5, which shows an "edge"—consisting of a high value on the left and a low value on the right—that is moving to the right along the x-axis. Our goal is to identify the velocity $v$ at which the edge is moving, as plotted in the upper part of Figure 16-5. In the lower part of the figure, we can see that our measurement of this velocity is just "rise over run," where the rise is over time and the run is the slope (spatial derivative). The negative sign corrects for the slope of $x$.

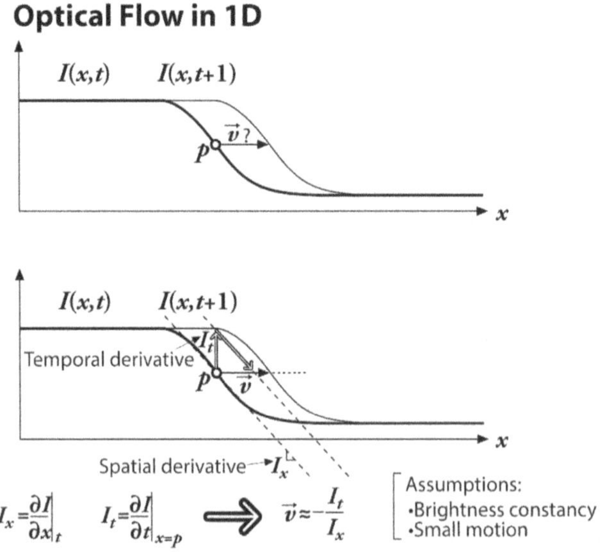


*Figure 16-5. Lucas-Kanade optical flow in one dimension: we can estimate the velocity of the moving edge (upper panel) by measuring the ratio of the derivative of the intensity over time divided by the derivative of the intensity over space*

Figure 16-5 reveals another aspect to our optical flow formulation: our assumptions are probably not quite true. That is, image brightness is not really stable; and our time steps (which are set by the camera) are often not as fast relative to the motion as we'd like. Thus, our solution for the velocity is not exact. However, if we are "close enough," then we can iterate to a solution. Iteration is shown in Figure 16-6 where we use our first (inaccurate) estimate of velocity as the starting point for our next iteration and then repeat. Note that we can keep the same spatial derivative in $x$ as computed on the first frame because of the brightness constancy assumption—pixels moving in $x$ do not change. This reuse of the spatial derivative already calculated yields significant computational savings. The time derivative must still be recomputed each iteration and each frame, but if we are close enough to start with, then these iterations will converge to near exactitude within about five iterations. This is known as *Newton's method*. If our first estimate was not close enough, then Newton's method will actually diverge.

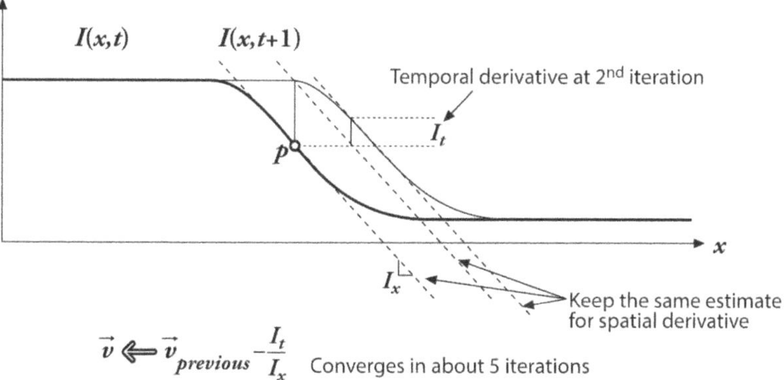


*Figure 16-6. Iterating to refine the optical flow solution (Newton's method): using the same two images and the same spatial derivative (slope) we solve again for the time derivative; convergence to a stable solution usually occurs within a few iterations*

Now that we've seen the one-dimensional solution, let's generalize it to images in two dimensions. At first glance, this seems simple: just add in the y-coordinate. Slightly changing notation, we'll call the y-component of velocity $v$ and the x-component of velocity $u$; then we have:

$$I_x u + I_y v + I_t = 0$$

This is often written as a single vector equation:

$$\vec{\nabla} I \cdot \vec{u} = -I_t$$

where:

$$\vec{\nabla} I = \begin{bmatrix} I_x \\ I_y \end{bmatrix}, and\ \vec{u} = \begin{bmatrix} u \\ v \end{bmatrix}$$

Unfortunately, for this single equation there are two unknowns for any given pixel. This means that measurements at the single-pixel level are underconstrained and cannot be used to obtain a unique solution for the two-dimensional motion at that point. Instead, we can solve only for the motion component that is perpendicular or "normal" to the line described by our flow equation. Figure 16-7 illustrates the geometry.

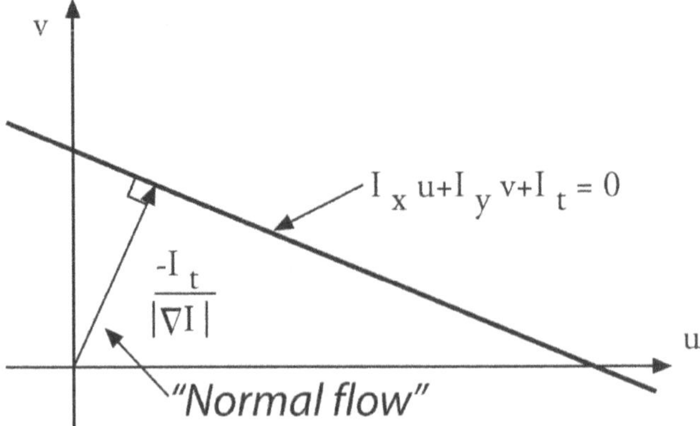


*Figure 16-7. Two-dimensional optical flow at a single pixel: optical flow at one pixel is underdetermined and so can yield at most motion, which is perpendicular ("normal") to the line described by the flow equation (figure courtesy of Michael Black)*

Normal optical flow results from the *aperture problem*, which arises when you have a small aperture or window in which to measure motion. When motion is detected with a small aperture, you often see only an edge, not a corner. But an edge alone is insufficient to determine exactly how (i.e., in what direction) the entire object is moving; see Figure 16-8.

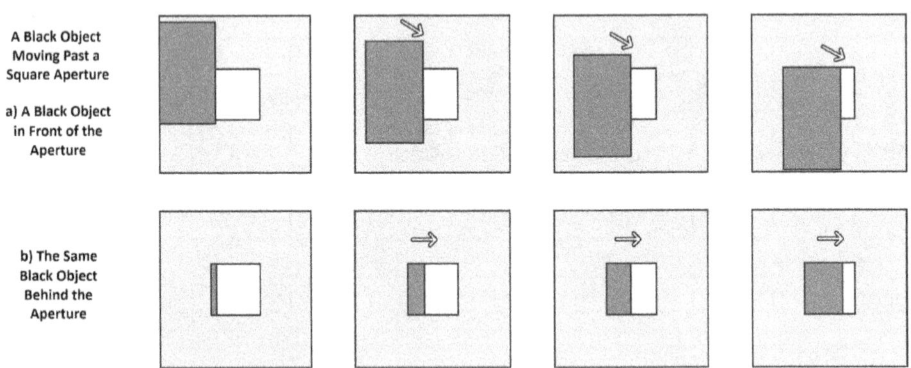


*Figure 16-8. Aperture problem: a) An object is moving to the right and down. (b) Through a small aperture, we see an edge moving to the right but cannot detect the downward part of the motion*

So then how do we get around this problem that, at one pixel, we cannot resolve the full motion? We turn to the last optical flow assumption for help. If a local patch of pixels moves coherently, then we can easily solve for the motion of the central pixel by using the surrounding pixels to set up a system of equations. For example, if we use a 5 × 5[6] window of brightness values (you can simply triple this for color-based optical flow) around the current pixel to compute its motion, we can then set up 25 equations as follows:

$$A_{25\times 2}\cdot d_{2\times 1} = \begin{bmatrix} I_x(p_1) & I_y(p_1) \\ I_x(p_2) & I_y(p_2) \\ \vdots & \vdots \\ I_x(p_{25}) & I_y(p_{25}) \end{bmatrix} \begin{bmatrix} u \\ v \end{bmatrix} = -\begin{bmatrix} I_t(p_1) \\ I_t(p_2) \\ \vdots \\ I_t(p_{25}) \end{bmatrix} = b_{25\times 1}$$

We now have an overconstrained system for which we can solve provided it contains more than just an edge in that 5 × 5 window. To solve for this system, we set up a least-squares minimization of the equation, whereby $\min\| Ad - b\|^2$ is solved in standard form as:

$$(A^T A)_{2\times2} \cdot d_{2\times1} = (A^T b)_{2\times2}$$

From this relation we obtain our $u$ and $v$ motion components. Writing this out in more detail yields:

$$\left(A^T A\right)\cdot d = \begin{bmatrix} \sum I_x I_x & \sum I_x I_y \\ \sum I_y I_x & \sum I_y I_y \end{bmatrix} \begin{bmatrix} u \\ v \end{bmatrix} = -\begin{bmatrix} \sum I_x I_t \\ \sum I_y I_t \end{bmatrix} = \left(A^T b\right)$$

The solution to this equation is then:

$$\begin{bmatrix} u \\ v \end{bmatrix} = \left(A^t A\right)^{-1} A^T b$$

When can this be solved? When $(A^TA)$ is invertible. And $(A^TA)$ is invertible when it has full rank (2), which occurs when it has two large eigenvectors. This will happen in image regions that include texture running in at least two directions. In this case, $(A^TA)$ will have the best properties when the tracking window is centered over a corner region in an image. This ties us back to our earlier discussion of the Harris corner detector. In fact, those corners were "good features to track" (see our previous

6 Of course, the window could be 3 × 3, 7 × 7, or anything you choose. If the window is too large, then you will end up violating the coherent motion assumption and will not be able to track well. If the window is too small, you will encounter the aperture problem again.

remarks concerning `cv::goodFeaturesToTrack()`) for precisely the reason that $(A^TA)$ had two large eigenvectors there! We'll see shortly how all this computation is done for us by the `cv::calcOpticalFlowPyrLK()` function.

The reader who understands the implications of our assuming small and coherent motions will now be bothered by the fact that, for most video cameras running at 30 Hz, large and noncoherent motions are commonplace. In fact, Lucas-Kanade optical flow by itself does not work very well for exactly this reason: we want a large window to catch large motions, but a large window too often breaks the coherent motion assumption! To circumvent this problem, we can track first over larger spatial scales using an image pyramid and then refine the initial motion velocity assumptions by working our way down the levels of the image pyramid until we arrive at the raw image pixels.

Hence, the recommended technique is first to solve for optical flow at the top layer and then to use the resulting motion estimates as the starting point for the next layer down. We continue going down the pyramid in this manner until we reach the lowest level. Thus we minimize the violations of our motion assumptions and so can track faster and longer motions. This more elaborate function is known as *pyramid Lucas-Kanade optical flow* and is illustrated in Figure 16-9. The OpenCV function that implements Pyramid Lucas-Kanade optical flow is `cv::calcOpticalFlowPyrLK()`, which we examine next.

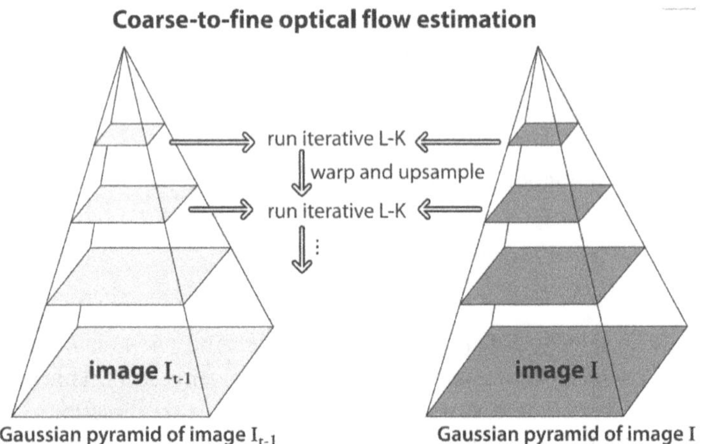


*Figure 16-9. Pyramid Lucas-Kanade optical flow: running optical flow at the top of the pyramid first mitigates the problems caused by violating our assumptions of small and coherent motion; the motion estimate from the preceding level is taken as the starting point for estimating motion at the next layer down*

### Pyramid Lucas-Kanade code: cv::calcOpticalFlowPyrLK()

We come now to OpenCV's algorithm that computes Lucas-Kanade optical flow in a pyramid, `cv::calcOpticalFlowPyrLK()`. As we will see, this optical flow function makes use of "good features to track" and also returns indications of how well the tracking of each point is proceeding.

```
void cv::calcOpticalFlowPyrLK(
  cv::InputArray        prevImg,              // Prior image (t-1), CV_8UC1
  cv::InputArray        nextImg,              // Next image (t), CV_8UC1
  cv::InputArray        prevPts,              // Vector of 2d start points (CV_32F)
  cv::InputOutputArray nextPts,              // Results: 2d end points (CV_32F)
  cv::OutputArray       status,               // For each point, found=1, else=0
  cv::OutputArray       err,                  // Error measure for found points
  cv::Size              winSize        = Size(15,15),  // size of search window
  int                   maxLevel       = 3,            // Pyramid layers to add
  cv::TermCriteria      criteria       = TermCriteria(  // How to end search
                          cv::TermCriteria::COUNT | cv::TermCriteria::EPS,
                          30,
                          0.01
                        ),
  int                   flags          = 0,      // use guesses, and/or eigenvalues
  double                minEigThreshold = 1e-4  // for spatial gradient matrix
);
```

This function has a lot of inputs, so let's take a moment to figure out what they all do. Once we have a handle on this routine, we can move on to the problem of which points to track and how to compute them. The basic plan is simple, however: you supply the images, list the points you want to track in `prevPts`, and call the routine. When the routine returns, you check the `status` array to see which points were successfully tracked and then check `nextPts` to find the new locations of those points. Now let's proceed into the details.

The first two arguments of `cv::calcOpticalFlowPyrLK()`, `prevImg` and `nextImg`, are the initial and final images. Both should be the same size and have the same number of channels.[7] The next two arguments, `prevPts` and `nextPts`, are the input list of features from the first image, and the output list to which matched points in the second image will be written. These can be either N × 2 arrays or vectors of points. The arrays `status` and `err` will be filled with information to tell you how successful the matching was. In particular, each entry in `status` will tell whether the corresponding feature in `prevPts` was found at all (`status[i]` will be nonzero if and only if `prevPts[i]` was found in `nextImg`). Similarly, `err[i]` will indicate an error measure

7 In older versions of the library, only single- or three-channel images could be used. The new implementation (as of v2.4) can handle images with arbitrary numbers of channels. This enables the use of textural descriptors or other dense descriptors to pixel tracking (so long as the descriptors can be compared using Euclidean norm).

for any point prevPts[i] that was found in nextImg (if point i was not found, then err[i] is not defined).

The window used for computing the local coherent motion is given by winSize. Because we are constructing an image pyramid, the argument maxLevel is used to set the depth of the stack of images. If maxLevel is set to 0, then the pyramids are not used. The argument criteria is used to tell the algorithm when to quit searching for matches; recall that cv::TermCriteria is the structure used by many OpenCV algorithms that iterate to a solution:

```
struct cv::TermCriteria(

public:
  enum {
    COUNT    = 1,
    MAX_ITER = COUNT,
    EPS      = 2
  };

  TermCriteria();
  TermCriteria( int _type, int_maxCount, double _epsilon );

  int    type,      // one of the enum types above
  int    max_iter,
  double epsilon
);
```

The default values will be satisfactory for most situations. As is often the case, however, if your image is unusually large, you may want to slightly increase the maximum allowed number of iterations.

The argument flags can have one or both of the following values:

cv::OPTFLOW_LK_GET_MIN_EIGENVALS
: Set this flag for a somewhat more detailed error measure. The default error measure for the error output is the average per-pixel change in intensity between the window around the previous corner and the window around the new corner. With this flag set to true, that error is replaced with the minimum eigenvalue of the Harris matrix associated with the corner.[8]

cv::OPTFLOW_USE_INITIAL_FLOW
: Use when the array nextPts already contains an initial guess for the feature's coordinates when the routine is called. (If this flag is not set, then the initial guesses will just be the point locations in prevPts.)

8 When we get to the details of the Harris algorithm later in this chapter, the meaning of this flag will become more clear.

The final argument, `minEigThreshold`, is used as a filter for removing points that are, in fact, not such good choices to track after all. In effect, it is somewhat analogous to the `qualityLevel` argument to `cv::goodFeaturesToTrack`, except its exact method of computation is different. The default value of $10^{-4}$ is a good choice; it can be increased in order to throw away more points.

### A worked example

Putting this all together, we now know how to find features that are good ones to track, and we know how to track them. We can obtain good results by using the combination of `cv::goodFeaturesToTrack()` and `cv::calcOpticalFlowPyrLK()`. Of course, you can also use your own criteria to determine which points to track.

Let's now look at a simple example (Example 16-1) that uses both `cv::goodFeaturesToTrack()` and `cv::calcOpticalFlowPyrLK()`; see also Figure 16-10.

*Example 16-1. Pyramid Lucas-Kanade optical flow code*

```
// Pyramid L-K optical flow example
//
#include <opencv2/opencv.hpp>
#include <iostream>

using namespace std;

static const int MAX_CORNERS = 1000;

void help( argv ) {
    cout << "Call: " <<argv[0] <<" [image1] [image2]" << endl;
    cout << "Demonstrates Pyramid Lucas-Kanade optical flow." << endl;
}

int main(int argc, char** argv) {

  if( argc != 3 ) { help( char** argv ); exit( -1 ); }

  // Initialize, load two images from the file system, and
  // allocate the images and other structures we will need for
  // results.
  //
  cv::Mat imgA = cv::imread(argv[1], cv::LOAD_IMAGE_GRAYSCALE);
  cv::Mat imgB = cv::imread(argv[2], cv::LOAD_IMAGE_GRAYSCALE);
  cv::Size img_sz = imgA.size();
  int win_size = 10;
  cv::Mat imgC = cv::imread(argv[2], cv::LOAD_IMAGE_UNCHANGED);

  // The first thing we need to do is get the features
  // we want to track.
  //
```

```
vector< cv::Point2f > cornersA, cornersB;

cv::goodFeaturesToTrack(
  imgA,                           // Image to track
  cornersA,                       // Vector of detected corners (output)
  MAX_CORNERS,                    // Keep up to this many corners
  0.01,                           // Quality level (percent of maximum)
  5,                              // Min distance between corners
  cv::noArray(),                  // Mask
  3,                              // Block size
  false,                          // true: Harris, false: Shi-Tomasi
  0.04                            // method specific parameter
);
cv::cornerSubPix(
  imgA,                           // Input image
  cornersA,                       // Vector of corners (input and output)
  cv::Size(win_size, win_size),   // Half side length of search window
  cv::Size(-1,-1),                // Half side length of dead zone (-1=none)
  cv::TermCriteria(
    cv::TermCriteria::MAX_ITER | cv::TermCriteria::EPS,
    20,                           // Maximum number of iterations
    0.03                          // Minimum change per iteration
  )
);

// Call the Lucas Kanade algorithm
//
vector<uchar> features_found;
cv::calcOpticalFlowPyrLK(
  imgA,                           // Previous image
  imgB,                           // Next image
  cornersA,                       // Previous set of corners (from imgA)
  cornersB,                       // Next set of corners (from imgB)
  features_found,                 // Output vector, elements are 1 for tracked
  cv::noArray(),                  // Output vector, lists errors (optional)
  cv::Size( win_size*2+1, win_size*2+1 ), // Search window size
  5,                              // Maximum pyramid level to construct
  cv::TermCriteria(
    cv::TermCriteria::MAX_ITER | cv::TermCriteria::EPS,
    20,                           // Maximum number of iterations
    0.3                           // Minimum change per iteration
  )
);

// Now make some image of what we are looking at:
// Note that if you want to track cornersB further, i.e.
// pass them as input to the next calcOpticalFlowPyrLK,
// you would need to "compress" the vector, i.e., exclude points for which
// features_found[i] == false.
for( int i = 0; i < (int)cornersA.size(); i++ ) {
  if( !features_found[i] )
    continue;
```

```
    line(imgC, cornersA[i], cornersB[i], Scalar(0,255,0), 2, cv::LINE_AA);
  }
  cv::imshow( "ImageA", imgA );
  cv::imshow( "ImageB", imgB );
  cv::imshow( "LK Optical Flow Example", imgC );
  cv::waitKey(0);

  return 0;
}
```

## Generalized Keypoints and Descriptors

Two of the essential concepts that we will need to understand tracking, object detection, and a number of related topics are *keypoints* and *descriptors*. Our first task is to understand what these two things are and how they differ from one another.

At the highest level of abstraction, a *keypoint* is a small portion of an image that, for one reason or another, is unusually distinctive, and which we believe we might be able to locate in another related image. A *descriptor* is some mathematical construction, typically (but not always) a vector of floating-point values, which somehow describes an individual keypoint, and which can be used to determine whether—in some context—two keypoints are "the same."[9]

Historically, one of the first important keypoint types was the Harris corner, which we encountered at the beginning of this chapter. Recall that the basic concept behind the Harris corner was that any point in an image that seemed to have a strong change in intensity along two different axes was a good candidate for matching another related image (for example, an image in a subsequent frame of a video stream).

Take a look at the images in Figure 16-10; the image is of text from a book.[10] You can see that the Harris corners are found at the locations where lines that make up the individual text characters begin or end, or where there are intersections of lines (such as the middle of an *h* or *b*. You will notice that "corners" do not appear along the edges of long lines in the characters, only at the end. This is because a feature found on such an edge looks very much like any other feature found anywhere on that edge.

---

9 In the next chapter, we will talk extensively about tracking (following things from one frame to another) and about object recognition (finding a thing in an image about which you have some information in a database of prior experience). A third very important case, which we will not talk about much in these chapters, is *sparse stereo*. In sparse stereo, one is interested in locating keypoints in two or more images taken at the same time from different points of view. Until we get to discussing stereo vision in Chapter 19, the only important fact you will need to keep in mind is that from the point of view of the library, sparse stereo will use the same methods and interfaces as tracking; that is, you will have two images in hand and want to find correspondences (matches) between the keypoints in each image.

10 In fact, this is the text from the first edition of the very book you are reading.

It stands to reason that if something is not unique in the current image, it will not be unique in another image, so such pixels do not qualify as good features.

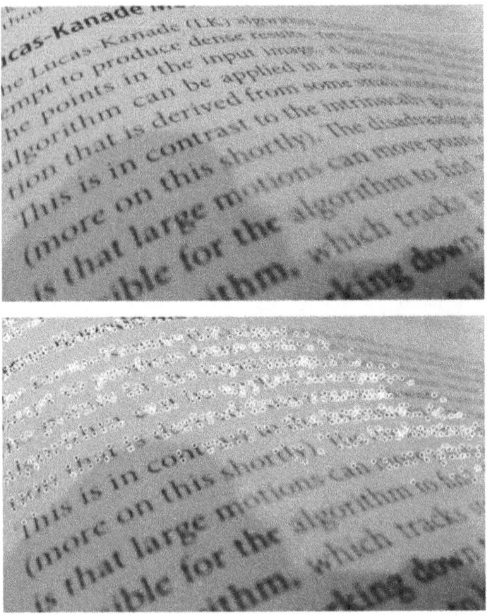

*Figure 16-10. Two images containing text, with the Harris corners shown on the right as white circles; note that in the portion of the image that is slightly out of focus, no corners are detected at all*

To our human eyes, each feature will look somewhat different than any other. The question of *feature descriptors* addresses the problem of how to have the computer make such associations. We mentioned that the three-way intersections that appear in an *h* or a *b* make good "features," but how would the computer tell the difference between them? This is what feature descriptors do.

We could construct a feature descriptor any way we would like—for example, we might make a vector from the intensity values of the 3 × 3 area around the keypoint. A problem with such a prescription is that the descriptor can have different values if the keypoint is seen from even a slightly different angle. In general, rotational invariance[11] is a desirable property for a feature descriptor. Of course, whether or not you

11 The exact meaning of *rotational invariance* is actually that the feature is *invariant under rotation*. Specifically, this means that when a rotation operation is applied to the underlying image, the feature descriptor is unchanged. Such invariances (also called *symmetries*) are of central importance when you are selecting or designing feature descriptors. Ideally, this includes not only rotations in the plane of the image, but also (at least small) three-dimensional rotations of the object that, from the point of view of the imager, will appear as affine transformations in the image plane.

need a rotationally invariant descriptor depends on your application. When detecting and tracking people, gravity plays a strong role in creating an asymmetric world in which people's heads are usually at the top and their feet are usually at the bottom. In such applications, a descriptor lacking rotational symmetry may not be a problem. In contrast, aerial imagery of the ground "rotates" when the aircraft travels in a different direction, and so the imagery may appear in what seems to be a random orientation.

## Optical Flow, Tracking, and Recognition

In the previous section, we discussed optical flow in the context of the Lucas Kanade algorithm. In that case, OpenCV provided us with a single high-level tool that would take a list of keypoints and try to locate the best matches for those points in a new frame. These points did not have a lot of structure or identity, just enough to make them locatable from one frame to the next in two very similar frames. Generalized keypoints can have associated descriptors that are very powerful and will allow those points to be matched not only in sequential frames of video, but even in completely different images. This can allow us to locate known objects in new environments, or to track an object through complex changing scenery, as well as many other applications.

The three major categories of tasks for which keypoints (and their descriptors) are useful are tracking, object recognition, and stereoscopy.

*Tracking* is the task of following the motion of something as the scene evolves in a sequential image stream. Tracking comes up in two major subcategories: the first is tracking objects in a stationary scene, and the other is tracking the scene itself for the purpose of estimating the motion of the camera. Usually, the term *tracking* alone refers to the former, and the latter is referred to as *visual odometry*. Of course, it is very common to want to do both of these things at once.

The second task category is *object recognition*. In this case, one is looking at a scene and attempting to recognize the presence of one or more known objects. The idea here is to associate certain keypoint descriptors with each object, with the reasoning that if you were to see enough keypoints associated with a particular object, you could reasonably conclude that this object was present in the scene.

Finally, there is *stereoscopy*. In this case, we are interested in locating corresponding points in two or more camera views of the same scene or object. Combining these locations with information about the camera locations and the optical properties of the cameras themselves, we can compute the location, in three dimensions, of the individual points we were able to match.

OpenCV provides methods for handling many types of keypoints and many kinds of keypoint descriptors. It also provides methods for matching them, either between pairs of frames (in the manner of sparse optical flow, tracking and visual odometry, or stereoscopy) or between a frame and a database of images (for object recognition).

## How OpenCV Handles Keypoints and Descriptors, the General Case

When you are doing tracking, as well as many other kinds of analysis for which keypoints and their descriptors are useful, there are typically three things you would like to do. The first is to search an image and find all of the keypoints that are in that image, according to some keypoint definition. The second is to create a descriptor for every keypoint you found. The third is to compare those keypoints you found, by means of their descriptors, to some existing set of descriptors, and see whether you can find any matches. In tracking applications, this last step involves looking for features in one frame of a sequence and trying to match those with features in the previous frame. In object-detection applications, one is often searching for features in some (potentially vast) database of "known" features that are "known" to be associated with either individual objects or object classes.

For each of these layers, OpenCV provides a generalized mechanism that follows the "classes that do stuff" (functor) model. Thus, for each of these stages, there is an abstract base class that defines a common interface for a family of objects that are derived from it; each derived class implements a particular algorithm.

### The cv::KeyPoint object

Of course, if we are going to find keypoints, we will need some way to represent them. Recall that the keypoint is not the feature descriptor, so most of what we need to know when we have a keypoint is where it is located. After that, there are some secondary features that some keypoints have and some do not. Here is the actual definition of the `cv::KeyPoint` class:

```
class cv::KeyPoint {

public:

  cv::Point2f pt;       // coordinates of the keypoint
  float       size;     // diameter of the meaningful keypoint neighborhood
  float       angle;    // computed orientation of the keypoint (-1 if none)
  float       response; // response for which the keypoints was selected
  int         octave;   // octave (pyramid layer) keypoint was extracted from
  int         class_id; // object id, can be used to cluster keypoints by object

  cv::KeyPoint(
    cv::Point2f _pt,
    float       _size,
    float       _angle=-1,
    float       _response=0,
    int         _octave=0,
    int         _class_id=-1
```

```
    );
    cv::KeyPoint(
      float       x,
      float       y,
      float       _size,
      float       _angle=-1,
      float       _response=0,
      int         _octave=0,
      int         _class_id=-1
    );
  ...
};
```

As you can see, every keypoint has a `cv::Point2f` member that just tells us where it is located.[12] The concept of `size` tells us something about the region around the keypoint that either was somehow included in the determination that the keypoint existed in the first place, or is going to play a role in that keypoint's descriptor. The `angle` of a keypoint is meaningful only for some keypoints. Many keypoints achieve rotational symmetry not by actually being invariant in the strictest sense, but by having some kind of natural orientation that you can take into account when comparing two descriptors. (This is not a complicated idea; if you were looking at two images of pencils, rotation obviously matters, but if you wanted to compare them, you could easily visualize them both in the same orientation before making the comparison.)

The `response` is used for detectors that can respond "more strongly" to one keypoint than another. In some cases, this can be interpreted as a probability that the feature is in fact present. The `octave` is used when the keypoint was found in an image pyramid. In these cases, it is important to know which scale a keypoint was found at because, in most cases, we would expect to find matches at the same or similar scale in new images. Finally, there is the class ID. You use `class_id` when constructing keypoint databases to distinguish the keypoints that are associated with one object from those that are associated with another (we will return to this point when we discuss the keypoint matching interface in the section "The (abstract) keypoint matching class: cv::DescriptorMatcher" on page 524).

The `cv::KeyPoint` object has two constructors, which are essentially the same; their only difference is that you can set the location of the keypoint either with two floating-point numbers or with a single `cv::Point2f` object. In fact, though, unless you are writing your own keypoint finder, you will not tend to use these functions. If you are using the detection, descriptor construction, and comparison functions avail-

12 For experts, this is a little subtler of a point than it might at first appear. Most, if not all, keypoints are actually extended objects with some number of pixels that contribute to them. For each keypoint type, however, there is a "center" that is defined in terms of that neighborhood (or, perhaps more accurately, the neighborhood is defined in terms of that center). In this sense the center of a keypoint is similar to the anchor point of a filter or morphological operator.

able to you from the library, you will typically never even look inside the keypoint objects.

### The (abstract) class that finds keypoints and/or computes descriptors for them: cv::Feature2D

For finding keypoints or computing descriptors (or performing both these tasks simultaneously), OpenCV provides the `cv::Feature2D` class. There are classes called `cv::FeatureDetector` and `cv::DescriptorExtractor` that used to be separate classes for pure feature detection or descriptor extraction algorithms, but since OpenCV 3.x they are all synonyms of `cv::Feature2D`. This abstract class has just a few methods, described next. Derived classes add some more methods to set and retrieve various properties, as well as static methods that construct the algorithm instance and return a smart pointer to it. There are two methods for actually asking the `cv::Feature2D` class to detect keypoints, functions that allow you to save and restore from disk, and a handy (static) function that allows you to create a feature detector–derived class by providing the name of the detector type (as a string). The two provided detection methods (as well as the two provided "compute" methods) differ only in that one operates on a single image, while the other operates on a set of images. There is some amount of efficiency to be gained by operating on many images at once (this is more true for some detectors than for others). Here is the relevant excerpt from the class description of `cv::Feature2D`:

```
class cv::Feature2D : public cv::Algorithm {

public:

  virtual void detect(
    cv::InputArray                  image,      // Image on which to detect
    vector< cv::KeyPoint >&         keypoints,  // Array of found keypoints
    cv::InputArray                  mask      = cv::noArray()
  ) const;

  virtual void detect(
    cv::InputArrayOfArrays          images,     // Images on which to detect
    vector<vector< cv::KeyPoint > >& keypoints, // keypoints for each image
    cv::InputArrayOfArrays          masks     = cv::noArray ()
  ) const;

  virtual void compute(
    cv::InputArray image,                  // Image where keypoints are located
    std::vector<cv::KeyPoint>& keypoints,  // input/output vector of keypoints
    cv::OutputArray descriptors );         // computed descriptors, M x N matrix,
                                           // where M is the number of keypoints
                                           // and N is the descriptor size
  virtual void compute(
    cv::InputArrayOfArrays image,            // Images where keypoints are located
    std::vector<std::vector<cv::KeyPoint> >& keypoints, //I/O vec of keypnts
    cv::OutputArrayOfArrays descriptors ); // computed descriptors,
```

```
                                    // vector of (Mi x N) matrices, where
                                    // Mi is the number of keypoints in
                                    // the i-th image and N is the
                                    // descriptor size
  virtual void detectAndCompute(
    cv::InputArray image,             // Image on which to detect
    cv::InputArray mask,              // Optional region of interest mask
    std::vector<cv::KeyPoint>& keypoints, // found or provided keypoints
    cv::OutputArray descriptors,      // computed descriptors
    bool useProvidedKeypoints=false ); // if true,
                                      // the provided keypoints are used,
                                      // otherwise they are detected

  virtual int descriptorSize() const;  // size of each descriptor in elements
  virtual int descriptorType() const;  // type of descriptor elements
  virtual int defaultNorm() const;     // the recommended norm to be used
                                       // for comparing descriptors.
                                       // Usually, it's NORM_HAMMING for
                                       // binary descriptors and NORM_L2
                                       // for all others.

  virtual void read( const cv::FileNode& );
  virtual void write( cv::FileStorage& ) const;

  ...
};
```

The actual implementation may implement just `cv::Feature2D::detect()`, if it's a pure keypoint detection algorithm (like FAST); just `cv::Feature2D::compute()`, if it's a pure feature description algorithm (like FREAK); or `cv::Feature2D::detectAndCompute()` in the case of "complete solution" algorithm, like SIFT, SURF, ORB, BRISK, and so on, in which case `detect()` and `compute()` will call it implicitly:

- `detect(image, keypoints, mask) ~ detectAndCompute(image, mask, keypoints, noArray(), false)`
- `compute(image, keypoints, descriptors) ~ detectAndCompute(image, noArray(), keypoints, descriptors, true)`

`cv::Feature2D::detect()` methods are the ones that do the basic work of finding keypoints for you—directly or through a call to `detectAndCompute()`. The first takes an image, a vector of keypoints, and an optional mask. It then searches the image (or the portion corresponding to the mask, if you provided one) for keypoints, and places whatever it finds in the vector you provided. The second variant does exactly the same thing, except that it expects a vector of images, a vector of masks (or none at all), and a vector of vectors of keypoints. The number of images, the number of masks (if not zero), and the number of keypoint vectors must all be equal. Then every image will be searched, and the corresponding keypoint vector in `keypoints` will be filled with the found keypoints.

The actual method used to find the keypoints is, of course, different for each of the many available derived classes from `cv::Feature2D`. We will get to the details of these shortly. In the meantime, it is important to keep in mind that the actual keypoints detected may be different (if and how some of the internal parameters are used) from one detector to another. This will mean that keypoints detected by one method may not be universally usable by every kind of feature descriptor.

Once you have found your keypoints, the next thing to do is to compute descriptors for them. As was mentioned earlier, these descriptors will allow you to compare keypoints to one another (based on their appearance, as opposed to by their location). This capability can then serve as the basis for tracking or object recognition.

The descriptors are computed by the `compute()` methods (that may also use `detectAndCompute()`). The first `compute()` method requires an image, a list of keypoints (presumably produced by the `detect()` method of the same class, but possibly a different one), and the output `cv::Mat` (passed as `cv::OutputArray`) that is used by `compute()` as a place to put the computed features. All descriptors generated by objects derived from `cv::Feature2D` are representable as vectors of fixed length. As a result, it's possible to arrange them all into an array. The convention used is that each row of `descriptors` is a separate descriptor, and the number of such rows is equal to the number of elements in `keypoints`.

The second `compute()` method is used when you have several images you would like to process at once. In this case the `images` argument expects an STL vector containing all the images you would like to process. The `keypoints` argument expects a vector containing vectors of keypoints, and the `descriptors` argument expects a vector containing arrays that can be used to store all of the resulting descriptors. This method is a companion to the multiple-image `detect()` method and expects the images as you would have given them to that `detect()` method, and the keypoints as you would have received them from that method.

Depending on when the algorithm was implemented, many of the keypoint feature detectors and extractors are bundled together into a single object. In this case the method `cv::Feature2D::detectAndCompute()` is implemented. Whenever a certain algorithm provides `detectAndCompute()`, it's strongly recommended to use it instead of subsequent calls to `detect()` and then to `compute()`. The obvious reason for that is a better performance: normally such algorithms require special image representation (called a *scale-space representation*), and it may be quite expensive to compute. When you find keypoints and compute descriptors in two separate steps, the scale-space representation is basically computed twice.

In addition to the main compute methods, there are the methods `descriptorSize()`, `descriptorType()`, and `defaultNorm()`. The first of these, `descriptorSize()`, tells us the length of the vector representation of the descriptor,[13] The `descriptorType()` method returns information about the specific type of the elements of the vector descriptor (e.g., `CV_32FC1` for 32-bit floating-point numbers, `CV_8UC1` for 8-bit descriptors or binary descriptors, etc.).[14] The `defaultNorm()` basically tells you how to compare the descriptors. In the case of binary descriptors, it's `NORM_HAMMING`, and you would rather use it. In the case of descriptors like SIFT or SURF, it's `NORM_L2` (e.g., Euclidean distance), but you can obtain equally good or even better results by using `NORM_L1`.

### The cv::DMatch object

Before we get deep into the topic of how matchers work, we need to know how matches are expressed. In general, a matcher will be an object that tries to match keypoints in one image with either a single other image or a collection of other images called a *dictionary*. When matches are found, OpenCV describes them by generating lists (STL vectors) of `cv::DMatch` objects. Here is the class definition for the `cv::DMatch` object:

```
class cv::DMatch {

public:

  DMatch();                  // sets this->distance
                             // to std::numeric_limits<float>::max()

  DMatch( int _queryIdx, int _trainIdx, float _distance );
  DMatch( int _queryIdx, int _trainIdx, int _imgIdx, float _distance );

  int   queryIdx;            // query descriptor index
  int   trainIdx;            // train descriptor index
  int   imgIdx;              // train image index
  float distance;

  bool operator<( const DMatch &m ) const; // Comparison operator
                                           // based on 'distance'
```

13 In the case of binary descriptor (we will encounter these later in the chapter), this method returns the number of bytes in the descriptor—not the number of bits. Generally, `descriptorSize()` returns the number of columns in the descriptor matrix that `cv::DescriptorExtractor::compute()` would return.

14 In all currently implemented cases, the descriptor type is single channel. The convention is to always "flatten" any channel-like elements of the descriptors into the descriptor's overall length. As a result a descriptor might be, for example, 10-dimensional for a grayscale image or 30-dimensional for a color image, but would never be a 10-dimensional array of three channel objects.

```
}
```

The data members of `cv::DMatch` are `queryIdx`, `trainIdx`, `imgIdx`, and `distance`. The first two identify the keypoints that were matched relative to the keypoint lists in each image. The convention used by the library is to always refer to these two images as the *query image* (the "new" image) and the *training image* (the "old" image), respectively.[15] `imgIdx` is used to identify the particular image from which the training image came in such cases that a match was sought between an image and a dictionary. The last member, `distance`, is used to indicate the quality of the match. In many cases, this is something like a Euclidean distance between the two keypoints in the many-dimensional vector space in which they live. Though this is not always the metric, it is guaranteed that if two different matches have different `distance` values, the one with the lower `distance` is the better match. To facilitate such comparisons (particularly for the purpose of sorting), the operator `cv::DMatch::operator<()` is defined, which allows two `cv::DMatch` objects to be compared directly with the meaning that it is their `distance` members that are actually compared.

#### The (abstract) keypoint matching class: cv::DescriptorMatcher

The third stage in the process of detect-describe-match is also implemented through a family of objects that are all derived from a common abstract base class whose interface they all share. This third layer is based on the `cv::DescriptorMatcher` class.

Before we get deep into how the interface works, it is important to understand that there are two basic situations in which you would want to use a matcher: object recognition and tracking, as described earlier in the chapter. In the *object recognition* case, one first compiles keypoints associated with a variety of objects into a kind of database called a *dictionary*. Thereafter, when a new scene is presented, the keypoints in that image are extracted and compared to the dictionary to estimate what objects from the dictionary might be present in the new scene. In the *tracking* case, the goal is to find all the keypoints in some image, typically from a video stream, and then to look for all of those keypoints in another image, typically the prior or next image in that video stream.

Because there are these two different situations, the matching methods in the class have two corresponding variations. For the object recognition case, we need to first *train* the matcher with a dictionary of descriptors, and then be able to present a single descriptor list to the matcher and have it tell us which (if any) keypoints that the

15 This is a somewhat awkward piece of terminology, and is a bit confusing, as the word *train* is also used to refer to the process of digesting all of the images in the object recognition case into a dictionary that can subsequently be queried. We will use this terminology because it is what is used in the library, but where possible we will always say "training image" to make clear that this noun is distinct from the verb *train*.

matcher has stored are matches with the ones in the list we provided. For the tracking case, we would like to provide two lists of descriptors and have the matcher tell us where the matches are between them. The `cv:DescriptorMatcher` class interface provides three functions, `match()`, `knnMatch()`, and `radiusMatch()`; and for each, there are two different prototypes—one for recognition (takes one list of features and uses the trained dictionary) and another for tracking (takes two lists of features).

Here is the part of the class definition relevant to us for the generic descriptor matcher base class:

```
class cv::DescriptorMatcher {

public:

  virtual void add( InputArrayOfArrays descriptors );  // Add train descriptors
  virtual void clear();                                // Clear train descriptors
  virtual bool empty() const;                          // true if no descriptors
  void train();                                        // Train matcher
  virtual bool isMaskSupported() const = 0;            // true if supports masks
  const vector<cv::Mat>& getTrainDescriptors() const;  // Get train descriptors

  // methods to match descriptors from one list vs. "trained" set (recognition)
  //
  void match(
    InputArray                   queryDescriptors,
    vector<cv::DMatch>&          matches,
    InputArrayOfArrays           masks          = noArray ()
  );
  void knnMatch(
    InputArray                   queryDescriptors,
    vector< vector<cv::DMatch> >& matches,
    int                          k,
    InputArrayOfArrays           masks          = noArray (),
    bool                         compactResult  = false
  );
  void radiusMatch(
    InputArray                   queryDescriptors,
    vector< vector<cv::DMatch> >& matches,
    float                        maxDistance,
    InputArrayOfArrays           masks          = noArray (),
    bool                         compactResult  = false
  );

  // methods to match descriptors from two lists (tracking)
  //
  // Find one best match for each query descriptor
  void match(
    InputArray                   queryDescriptors,
    InputArray                   trainDescriptors,
    vector<cv::DMatch>&          matches,
    InputArray                   mask           = noArray ()
```

```
) const;
// Find k best matches for each query descriptor (in increasing
// order of distances)
void knnMatch(
  InputArray                    queryDescriptors,
  InputArray                    trainDescriptors,
  vector< vector<cv::DMatch> >& matches,
  int                           k,
  InputArray                    mask            = noArray(),
  bool                          compactResult   = false
) const;
// Find best matches for each query descriptor with distance less
// than maxDistance
void radiusMatch(
  InputArray                    queryDescriptors,
  InputArray                    trainDescriptors,
  vector< vector<cv::DMatch> >& matches,
  float                         maxDistance,
  InputArray                    mask            = noArray (),
  bool                          compactResult   = false
) const;

virtual void read( const FileNode& );      // Reads matcher from a file node
virtual void write( FileStorage& ) const; // Writes matcher to a file storage

virtual cv::Ptr<cv::DescriptorMatcher> clone(
       bool emptyTrainData=false
) const = 0;
static cv::Ptr<cv::DescriptorMatcher> create(
       const string& descriptorMatcherType
);
...
};
```

The first set of methods is used to match an image against prestored set of descriptors, one array per image. The purpose is to build up a *keypoint dictionary* that can be referenced when novel keypoints are provided. The first method is the `add()` method, which expects an STL vector of sets of descriptors, each of which is in the form of a `cv::Mat` object. Each `cv::Mat` object should have *N* rows and *D* columns, where *N* is the number of descriptors in the set, and *D* is the dimensionality of each descriptor (i.e., each "row" is a separate descriptor of dimension *D*). The reason that `add()` accepts an array of arrays (which is usually represented as `std::vector<cv::Mat>`) is that in practice, one often computes a set of descriptors from each image in a set of images.[16] That set of images was probably presented to a `cv::Feature2D`-based class as a vector of images, and returned as a vector of sets of keypoint descriptors.

Once you have added some number of keypoint descriptor sets, if you would like to access them, you may do so with the (constant) methods `getTrainDescriptors()`, which will return the descriptors to you in the same way you first provided them (i.e., as a vector of descriptor sets, each of type `cv::Mat`, with each row of each `cv::Mat` being a single descriptor). If you would like to clear the added descriptors, you may do so with the `clear()` method, and if you would like to test whether a matcher has descriptors stored in it, you may do so with the `empty()` method.

Once you have loaded all of the keypoint descriptors you would like to load, you may need to call `train()`. Only some implementations require the `train()` operation (i.e., some classes derived from `cv::DescriptorMatcher`). This method's purpose is to tell the matcher that you are done loading images, and that it can proceed to precompute any internal information that it will need in order to perform matches on the provided keypoints. By way of example, if a matcher performs matches using only the Euclidean distance between a provided new keypoint and those in the existing dictionary, it would be prudent to construct a quad-tree or similar data structure to greatly accelerate the task of finding the closest dictionary keypoint to the provided keypoint. Such data structures can require substantial effort to compute, and are computed only once after all of the dictionary keypoints are loaded. The `train()` method tells the matcher to take the time to compute these sorts of adjunct internal data structures. Typically, if a `train()` method is provided, you must call it before calling any matching method that uses the internal dictionary.

The next set of methods is the set of matching methods used in object recognition. They each take a list of descriptors, called a *query list*, which they compare with the descriptors in the trained dictionary. Within this set, there are three methods: `match()`, `knnMatch()`, and `radiusMatch()`. Each of these methods computes matches in a slightly different way.

The `match()` method expects a single list of keypoint descriptors, `queryDescriptors`, in the usual `cv::Mat` form. In this case, recall that each row represents a single descriptor, and each column is one dimension of that descriptor's vector representation. `match()` also expects an STL vector of `cv::DMatch` objects that it can fill with the

16 Typically, these images contain individual objects that one hopes to be able to find by means of the appearance of their keypoints in subsequent images. For this reason, the keypoints provided as part of the `keypoints` argument should have their `class_id` fields set such that, for each object to be recognized, the `class_id` for the associated keypoints is distinct.

individual detected matches. In the case of the `match()` method, each keypoint on the query list will be matched to the "best match" from the train list.

The `match()` method also supports an optional `mask` argument. Unlike most `mask` arguments in OpenCV, this mask does not operate in the space of pixels, but rather in the space of descriptors. The type of mask, however, should still be `CV_8U`. The `mask` argument is an STL-vector of `cv::Mat` objects. Each entire matrix in that vector corresponds to one of the training images in the dictionary.[17] Each row in a particular mask corresponds to a row in `queryDescriptors` (i.e., one descriptor). Each column in the mask corresponds to one descriptor associated with the dictionary image. Thus, `masks[k].at<uchar>(i,j)` should be nonzero if descriptor `j` from image (object) `k` should be compared with descriptor `i` from the query image.

The next method is the `knnMatch()` function, which expects the same list descriptors as `match()`. In this case, however, for each descriptor in the query list, it will find a specific number of best matches from the dictionary. That number is given by the `k` integer argument. (The "knn" in the function name stands for *k-nearest neighbors.*) The vector of `cv::DMatch` objects from `match()` is replaced by a vector of vectors of `cv::DMatch` objects called `matches` in the `knnMatch()` method. Each element of the top-level vector (e.g., `matches[i]`) is associated with one descriptor from `queryDescriptors`. For each such element, the next-level element (e.g., `matches[i][j]`) is the `j`th best match from the descriptors in `trainDescriptors`.[18] The mask argument to `knnMatch()` has the same meaning as for `match()`. The final argument for `knnMatch()` is the Boolean `compactResult`. If `compactResult` is set to the default value of `false`, the `matches` vector of vectors will contain one vector entry for every entry in `queryDescriptors`—even those entries that have no matches (for which the corresponding vector of `cv::DMatch` objects is empty). If, however, `compactResult` is set to `true`, then such noninformative entries will simply be removed from `matches`.

The third matching method is `radiusMatch()`. Unlike k-nearest neighbor matching, which searches for the `k` best matches, *radius matching* returns all of the matches within a particular distance of the query descriptor.[19] Other than the substitution of the integer `k` for the maximum distance `maxDistance`, the arguments and their meanings for `radiusMatch()` are the same as those for `knnMatch()`.

---

17 Recall that this typically means "one of the objects in your dictionary."

18 The order is always starting with the best and working down to the kth-best (i.e., $j \in [0, k - 1]$).

19 It should be noted, however, that the term *distance* here is defined by the metric used by a particular matcher, and may or may not be the standard Euclidean distance between the vector representations of the descriptors.

Don't forget that in the case of matching, "best" is determined by the individually derived class that implements the `cv::DescriptorMatcher` interface, so the exact meaning of "best" may vary from matcher to matcher. Also, keep in mind that there is typically no "optimal assignment" being done, so one descriptor on the query list could match several on the train list, or vice versa.

The next three methods—the alternate forms of `match()`, `knnMatch()`, and `radiusMatch()`—support two lists of descriptors. These are typically used for tracking. Each of these has the same inputs as their aforementioned counterparts, with the addition of the `trainDescriptors` argument. These methods ignore any descriptors in the internal dictionary, and instead compare the descriptors in the `queryDescriptors` list only with the provided `trainDescriptors`.[20]

After these six matching methods, there are some methods that you will need for general handling of matcher objects. The `read()` and `write()` methods require a `cv::FileNode` and `cv::FileStorage` object, respectively, and allow you to read and write a matcher from or to disk. This is particularly important when you are dealing with recognition problems in which you have "trained" the matcher by loading information in from what might be a very large database of files. This saves you from needing to keep the actual images around and reconstruct the keypoints and their descriptors from every image every time you run your code.

Finally, the `clone()` and `create()` methods allow you to make a copy of a matcher or create a new one by name, respectively. The first method, `clone()`, takes a single Boolean, `emptyTrainData`, which, if `true`, will create a copy with the same parameter values (for any parameters accepted by the particular matcher implementation) but without copying the internal dictionary. Setting `emptyTrainData` to `false` is essentially a deep copy, which copies the dictionary in addition to the parameters. The `create()` method is a static method, which will accept a single string from which a particular derived class can be constructed. The currently available values for the `descriptorMatcherType` argument to `create()` are given in Table 16-1. (The meaning of the individual cases will be described in the next section.)

20 Note that even though this second list is called the "train" list, it is not the list "trained" into the matcher with the `add()` method; rather, it is being used *in place of* the internal "trained" list.

*Table 16-1. Available options for descriptorMatcherType argument to cv::DescriptorMatcher::create() method*

| descriptorMatcherType string | Matcher type |
|---|---|
| "FlannBased" | FLANN (Fast Library for Approximate Nearest Neighbors) method; L2 norm will be used by default |
| "BruteForce" | Element-wise direct comparison using L2 norm |
| "BruteForce-SL2" | Element-wise direct comparison using squared L2 norm |
| "BruteForce-L1" | Element-wise direct comparison using L1 norm |
| "BruteForce-Hamming" | Element-wise direct comparison using Hamming distance[a] |
| "BruteForce-Hamming(2)" | Element-wise direct comparison using Multilevel Hamming distance (two levels) |

[a] All the Hamming distance methods can be applied only to binary descriptors that are encoded with the CV_8UC1 type (i.e., eight components of the descriptor per each descriptor byte).

## Core Keypoint Detection Methods

In the last 10 years, there has been tremendous progress in tracking and image recognition. Within this space, one very important theme has been the development of keypoints that, as you now know, are small fragments of an image that contain the highest density of information about the image and its contents. One of the most important features of the keypoint concept is that it allows an image to be "digested" into a finite number of essential elements, even as the resolution of the image becomes very high. In this sense, keypoints offer a way to get from an image in a potentially very high dimensionality pixel representation into a more compact representation whose quality increases with image size, but whose actual size does not. It is thought that the human visual cortex "chunks" individual retinal responses (essentially pixels) up into higher-level blocks of information, at least some of which are analogous to the kind of information contained in a keypoint.

Early work that focused on concepts like corner detection (which we saw earlier) gave way to increasingly sophisticated keypoints with increasingly expressive descriptors (the latter we will visit in the next section), which exhibit a variety of desirable characteristics—such as rotational or scale invariance, or invariance to small affine transformations—that were not present in the earlier keypoint detectors.

The current state of the art, however, is that there is a large number of keypoint-detection algorithms (and keypoint descriptors), none of which is "clearly better" than the others. As a result, the approach of the OpenCV library has been to provide a common interface to all of the detectors, with the hope of encouraging and facilitating experimentation and exploration of their relative merits within your individual context. Some are fast, and some are comparatively quite slow. Some find features for which very rich descriptors can be extracted, and some do not. Some exhibit one or more useful invariance properties, of which some might be quite necessary in your application, and some might actually work against you.

In this section, we will look at each keypoint detector in turn, discussing its relative merits and delving into the actual science of each detector, at least deeply enough that you will get a feel for what each one is for and what it offers that may be different than the others. As we have learned, for each descriptor type, there will be a detector that locates the keypoints and a descriptor extractor. We will cover each of these as we discuss each detection algorithm.

In general, it is not absolutely necessary to use the feature extractor that is historically associated with a particular keypoint detector. In most cases, it is meaningful to find keypoints using any detector and then proceed to characterize those keypoints with any feature extractor. In practice, however, these two layers are usually developed and published together, and so the OpenCV library uses this pattern as well.

#### The Harris-Shi-Tomasi feature detector and cv::GoodFeaturesToTrackDetector

The most commonly used definition of a *corner* was provided by Harris [Harris88], though others proposed similar definitions even earlier. Such corners, known as *Harris corners*, can be thought of as the prototypical keypoint.[21] Figure 16-11 shows the Harris corners on a pair of images that we will continue to use for other keypoints in this section (for convenient visual comparison). Their definition relies on a notion of autocorrelation between the pixels in a small neighborhood. In plain terms, this means "if the image is shifted a small amount ($\Delta x$, $\Delta y$), how similar is it to its original self?"

21 These features were once associated with the name "Good Features To Track" (i.e., `cvGoodFeaturesToTrack()`) in older versions of the library. This is why the associated detector is now called `cv::GFTTDetector`, rather than something potentially more intuitive like `cv::HarrisCornerDetector`.

*Figure 16-11. Two images of the same vehicle; in each image are the 1,000 strongest Harris-Shi-Tomasi corners. Notice that, in the image on the right, the corners in the background are stronger than those on the car, and so most of the corners in that image originate from there*

Harris began with the following autocorrelation function, for image intensity pixels $I(x, y)$:

$$c(x, y, \Delta x, \Delta y) = \sum_{(i,j)\in W(x,y)} w_{i,j}(I(i, j) - I(i + \Delta x, j + \Delta y))^2$$

This is just a weighted sum over a small window around a point $(x,y)$ of the squared difference between the image at some point $(i,j)$ in the window and some other point displaced by $(\Delta x, \Delta y)$. (The weighting factor $w_{i,j}$ is a Gaussian weighting that makes the differences near the center of the window contribute more strongly than those farther away from the center.)

What follows in Harris's derivation is a small amount of algebra, and the approximation that because $\Delta x$ and $\Delta y$ are assumed to be small, the term $I(i + \Delta x, j + \Delta y)$ can be estimated by $I(i, j) + I_x(i, j)\Delta x + I_y(i, j)\Delta y)$ (here $I_x$ and $I_y$ are the first-order partial derivatives of $I(x,y)$ in $x$ and $y$, respectively).[22] The result is the re-expression of the autocorrelation in the form:

$$c(x, y, \Delta x, \Delta y) = [\Delta x \;\; \Delta y] M(x, y) \begin{bmatrix} \Delta x \\ \Delta y \end{bmatrix}$$

22 That is, a first-order Taylor approximation.

Where $M(x, y)$ is the symmetric *autocorrelation matrix* defined by:

$$M(x, y) =$$

$$\begin{bmatrix} \sum_{-K \le i,j \le K} w_{i,j} I_x^2(x+i, y+j) & \sum_{-K \le i,j \le K} w_{i,j} I_x(x+i, y+j) I_y(x+i, y+j) \\ \sum_{-K \le i,j \le K} w_{i,j} I_x(x+i, y+j) I_y(x+i, y+j) & \sum_{-K \le i,j \le K} w_{i,j} I_y^2(x+i, y+j) \end{bmatrix}$$

Corners, by Harris's definition, are places in the image where the autocorrelation matrix has two large eigenvalues. In essence, this means that moving a small distance in any direction will change the image.[23] This way of looking at things has the advantage that, when we consider only the eigenvalues of the autocorrelation matrix, we are considering quantities that are invariant also to rotation, which is important because objects that we are tracking might rotate as well as translate.

In this case these two eigenvalues of the Harris corner do more than determine whether a point is a good feature to track (i.e., a keypoint); they also provide an identifying signature for the point (i.e., a keypoint descriptor). It is a common, though by no means universal, feature of keypoints that they are intimately tied to their descriptors in this way. In many cases, the keypoint is, in essence, any point for which the associated descriptor (in this case the two eigenvalues of $M(x, y)$) meets some threshold criteria. At the same time, it is also noteworthy that Harris's original threshold criterion was not the same as that later proposed by Shi and Tomasi; the latter turns out to be superior for most tracking applications.

Harris's original definition involved taking the determinant of $M(x, y)$ and subtracting its squared trace (with some weighting coefficient):

$$H = \det(M) - \kappa \operatorname{trace}^2(M) = \lambda_1 \lambda_2 - \kappa(\lambda_1 + \lambda_2)^2$$

One then found the "corners" (what we now call keypoints) by searching for local maxima of this function (and often also comparing this function to a predetermined threshold). This function $H$, known as the *Harris measure*, effectively compares the

23 If there had been only one large eigenvalue, then this point would be something like an edge, in that motion along the edge would not seem to change the image, while motion perpendicular to the edge would change the image. If there were no large eigenvalues at all, this would have meant that you could displace the little window in any direction and nothing would happen at all; in other words, the image intensity is constant here.

eigenvalues of $M$ (which we refer to as $\lambda_1$ and $\lambda_2$ in our definition of $H$) without requiring their explicit computation. This comparison implicitly contains the parameter $\kappa$, termed the *sensitivity*, which can be set meaningfully to any value between `0` and `0.24`, but is typically set to about `0.04`.[24] Figure 16-12 shows an image in which the regions around some individual keypoint candidates are shown enlarged.

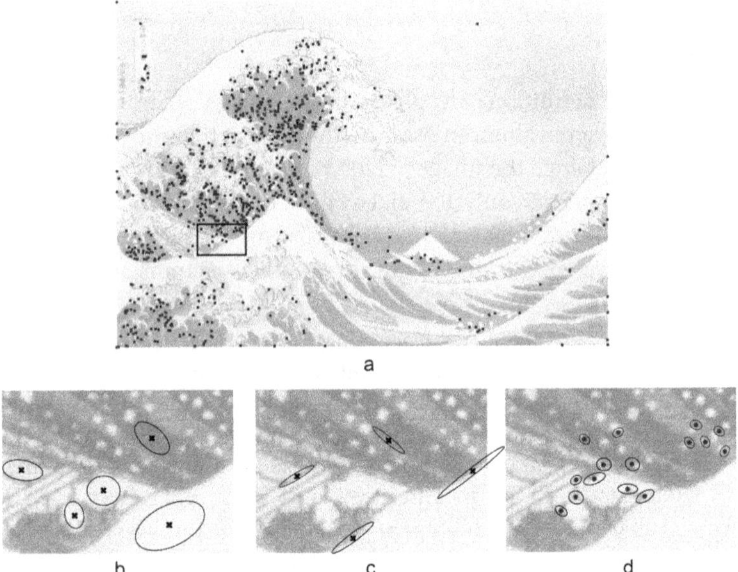


*Figure 16-12. In a classic image (a), keypoints found by the Shi-Tomasi method are shown as black dots. Below that are three images that are enlargements of a small subsection of the original. On the left (b) are shown (as Xs) points that are not keypoints. These points have small eigenvalues in both dimensions. In the center (c) are shown (as Xs) points that are also not keypoints; these are edges, and have one small eigenvalue and one large eigenvalue associated with them. On the right (d) are actual found keypoints; for these points, both eigenvalues are large. The ovals visualize the inverse of these eigenvalues*

It was later found by Shi and Tomasi [Shi94] that good corners resulted as long as the smaller of the two eigenvalues was greater than a minimum threshold. Shi and Tomasi's method was not only sufficient, but in many cases gave more satisfactory results than Harris's method. The OpenCV implementation of `cv::GFTTDetector`, as a default, uses Shi and Tomasi's measure, but other keypoint detectors we will discuss later often use either Harris's original measure or a variation of it.

---

24 Making this number smaller increases the sensitivity of the algorithm, so the value of `0.04` is on the more sensitive side.

**Keypoint finder.** The Harris-Shi-Tomasi corner detector is also the simplest implementation of the `cv::Feature2D` (the detector part) interface:

```
class cv::GFTTDetector : public cv::Feature2D {
public:
  static Ptr<GFTTDetector> create(
    int    maxCorners        = 1000,  // Keep this many corners
    double qualityLevel      = 0.01,  // fraction of largest eigenvalue
    double minDistance       = 1,     // Discard corners if this close
    int    blockSize         = 3,     // Neighborhood used
    bool   useHarrisDetector = false, // If false, use Shi Tomasi
    double k                 = 0.04   // Used for Harris metric
  );
  ...
};
```

The constructor for `cv::GFTTDetector` takes arguments that set all of the basic runtime parameters for the algorithm. The `maxCorners` parameter indicates the maximum number of points that you would like returned.[25] The parameter `qualityLevel` indicates the minimal acceptable lower eigenvalue for a point to be included as a corner. The actual minimal eigenvalue used for the cutoff is the product of the `qualityLevel` and the largest lower eigenvalue observed in the image. Hence, the `qualityLevel` should not exceed `1` (a typical value might be `0.10` or `0.01`). Once these candidates are selected, a further culling is applied so that multiple points within a small region need not be included in the response. In particular, the `minDistance` guarantees that no two returned points are within the indicated number of pixels.

The `blockSize` is the region around a given pixel that is considered when you are computing the autocorrelation matrix of derivatives. It turns out that in almost all cases you will get superior results if you sum these derivatives over a small window than if you simply compute their value at only a single point (i.e., at a `blockSize` of `1`).

If `useHarris` is `true`, then the Harris corner definition is used rather than the Shi-Tomasi definition, and the value `k` is the weighting coefficient used to set the relative weight given to the trace of the autocorrelation matrix Hessian compared to the determinant of the same matrix.

25 Because the method of returning keypoints by the `detect()` method is to fill an STL vector of `cv::KeyPoint` objects, there is no real upper bound on how many keypoints you could ask for. In practice, however, it is often useful to limit the number of keypoints for the purpose of computational efficiency, or to bound computation time in downstream processing (particularly in real-time applications). In any case, the returned corners will be "best" corners found, in terms of the magnitude of the smaller eigenvalue of the autocorrelation matrix $M(x, y)$.

Of course, when you want to actually compute keypoints, you do that with the `detect()` method, which `cv::GFTTDetector` inherits from the `cv::Feature2D` base class.

**Additional functions.** `cv::GFTTDetector` also supports setting and retrieving various properties using set/get methods; for example, you can turn on the Harris detector instead of the default minimum eigenvalue based (Shi-Tomasi) GFTT algorithm by calling the `gfttdetector->setHarrisDetector(true)` method.

### A brief look under the hood

Internally, `cv::goodFeaturesToTrack()` and `cv::GFTTDetector` have a few specific phases: the computation of the autocorrelation matrix $M(x, y)$, the analysis of this matrix, and some kind of threshold applied. The critical steps are accomplished with the functions `cv::cornerHarris()` and `cv::cornerMinEigenVal()`:

```
void cv::cornerHarris(
  cv::InputArray  src,                                // Input array CV_8UC1
  cv::OutputArray dst,                                // Result array CV_32FC1
  int             blockSize,                          // Autocorrelation block sz
  int             ksize,                              // Sobel operator size
  double          k,                                  // Harris's trace weight
  int             borderType = cv::BORDER_DEFAULT     // handle border pix
);
void cv::cornerMinEigenVal(
  cv::InputArray  src,                                // Input array CV_8UC1
  cv::OutputArray dst,                                // Result array CV_32FC1
  int             blockSize,                          // Autocorrelation block sz
  int             ksize,       = 3                    // Sobel operator size
  int             borderType = cv::BORDER_DEFAULT     // handle border pix
);
```

The arguments to these two functions are exactly analogous to the function `cv::goodFeaturesToTrack()`, with the first filling `dst` with the characteristic values used by Harris:

$$dst(x, y) = \det M^{(x,y)} - k \cdot (trM^{(x,y)})^2$$

and the second filling `dst` with the characteristic values used by Shi and Tomasi—that is, the minimal eigenvalue of the autocorrelation matrix $M(x, y)$.

If you would like to implement your own variation of the GFTT algorithm, a final function is provided for you that computes and gives to you the eigenvalues and eigenvectors of the autocorrelation matrix for every point on your image. This function is called `cv::cornerEigenValsAndVecs()`:

```
void cornerEigenValsAndVecs(
  cv::InputArray  src,                          // Input array CV_8UC1
  cv::OutputArray dst,                          // Result array CV_32FC1
  int             blockSize,                    // Autocorrelation block sz
  int             ksize,                        // Sobel operator size
  int             borderType = cv::BORDER_DEFAULT // handle border pix
);
```

The only significant difference between this function and `cv::cornerMinEigenVal()` is the nature of the output. In this case, the results array `dst` will be of type `CV_32FC6`. The six channels will contain the two eigenvalues, the two components of the eivenvector for the first eigenvalue, and the two components in the eigenvector for the second eigenvalue (in that order).

#### The simple blob detector and cv::SimpleBlobDetector

The corner detection concept embodied by the work of Harris, and later Shi and Tomasi, represents what will turn out to be one major approach to the keypoint concept. From this point of view, keypoints are highly localized structures that exist at points in an image where a larger-than-normal amount of information is present. An alternative point of view is the concept of a *blob* (see Figure 16-13). Blobs are, by nature, not so clearly localized, but represent regions of interest that might be expected to have some stability over time (see Figure 16-14).

*Figure 16-13. The simple blob detector run on two similar images of the same vehicle. There is little consistency between the blobs found in the two images. Blob detection works best in simple environments where there are expected to be a few very well-defined objects to locate*

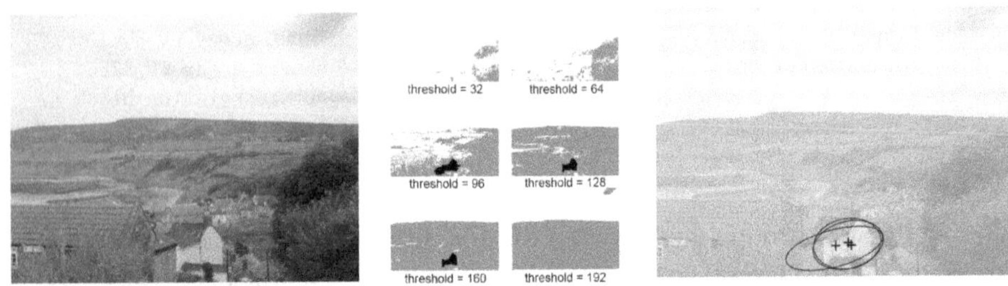


*Figure 16-14. Starting with an image of countryside scene (left), six thresholded images are generated (center). One set of overlapping blob candidates, corresponding to the building in the bottom center of the original image, are shown (right). These candidates will be combined to produce a final estimation of the associated blob (not shown). The contours contributing to these blob candidates are highlighted (in black) in the center thresholded images*

There are many algorithms for blob detection. The `cv::SimpleBlobDetector` class implements just one of them.[26] The simple blob detector works by first converting the input image to grayscale, and then computing a series of thresholded (binary) images from that grayscale image. The number of binary images is determined by parameters to the algorithm: `minimum threshold`, `maximum threshold`, and a `threshold step`. Once converted to binary, connected components are extracted—for example, by `cv::findContours()`—and the centers of each such contour are computed; they are the candidate blob centers. Next, candidate blob centers near one another in space (controlled by a minimum distance parameter, `minDistBetweenBlobs`) and from images with adjacent thresholds (differing by one step in the list of applied thresholds) are grouped together. Once these groups have been determined, the groups are assigned a radius and a center, which is computed from all of the contours that form the group. The resulting objects are the keypoints.

Once the blobs have been located, some built-in filtering can be turned on to reduce the number of blobs. Blobs may be filtered by color (which really means intensity, since this is a grayscale image), by size (area), by circularity (ratio of the area of the actual blob to a circle of the blob's computed effective radius), by what is called the *inertia ratio* (the ratio of the eigenvalues of the second moment matrix), or by the convexity (the ratio of the blob's area to the area of its convex hull).

26 As we will see as we investigate other more complex feature detectors in this section, there are many other possible approaches to blob detection. Many of them will appear as components of more complex algorithms, and we will look at how they work as we get to them. Difference of Gaussian (DoG), Laplacian of Gaussian (LoG), and Determinant of Hessian (DoH) are all examples of blob detection mechanisms.

**Keypoint finder.** Start by taking a look at (a somewhat simplified version of) the blob detector's declaration:

```
class SimpleBlobDetector : public Feature2D {

public:
  struct Params {
      Params();
      float  minThreshold;              // First threshold to use
      float  maxThreshold;              // Highest threshold to use
      float  thresholdStep;             // Step between thresholds

      size_t minRepeatability;          // Blob must appear
                                        // in this many images
      float  minDistBetweenBlobs;       // Blob must be this far
                                        // from others

      bool   filterByColor;             // True to use color filter
      uchar  blobColor;                 // always 0 or 255

      bool   filterByArea;              // True to use area filter
      float  minArea, maxArea;          // min and max area to accept

      // True to filter on "circularity", and min/max
      // ratio to circle area
      bool   filterByCircularity;
      float  minCircularity, maxCircularity;

      // True to filter on "inertia", and min/max eigenvalue ratio
      bool   filterByInertia;
      float  minInertiaRatio, maxInertiaRatio;

      // True to filter on convexity, and min/max ratio to hull area
      bool   filterByConvexity;
      float  minConvexity, maxConvexity;

      void read( const FileNode& fn );
      void write( FileStorage& fs ) const;
  };

  static Ptr<SimpleBlobDetector> create(
    const SimpleBlobDetector::Params &parameters
      = SimpleBlobDetector::Params()
  );

  virtual void read( const FileNode& fn );
  virtual void write( FileStorage& fs ) const;

  ...
};
```

As you can see from scanning over this declaration, it is clear that there is not actually much going on here. There is a definition for `cv::SimpleBlobDetector::Params`, a structure that can hold all of the information needed to actually run a simple blob detector, a constructor (which takes a `Params` argument), and `read()` and `write()` functions that will allow us to store the state of our detector. Of course, there are also the all-important `detect()` routines inherited from the `cv::Feature2D` interface.

We already know how the `detect()` member works, in the sense that what it does is entirely generic to all feature detectors; what really matters to us here is how to set up the parameters in the `Params` argument to the constructor. The first group of five parameters controls the basic functioning of the algorithm. We use `thresholdStep`, `minThreshold`, and `maxThreshold` to configure the set of thresholded images to generate. We do so by starting at `minThreshold` and stepping up by `thresholdStep` each time up until, but not including, `maxThreshold`. It is typical to start with a value around `50` to `64` and step in small increments (e.g., `10`) up to about `220` to `235`, thus avoiding the often-less-informative ends of the intensity distribution. `minRepeatability` determines how many (consecutive) threshold images must contain overlapping blob candidates in order for the candidates to be combined into a blob. This number is typically a small integer, but rarely less than two. The actual meaning of "overlapping" is controlled by `minDistBetweenBlobs`. If two blob candidates have their centers within this distance, they are considered to be related to the same blob. Keep in mind that this one is in pixel units, and so should scale with your image. The default constructor for `cv::SimpleBlobDetector::Params` sets this to `10`, which is probably only suitable for images of about 640 × 480.

The remaining parameters affect the different filtering options, and are arranged into small groups, each of which contains a Boolean that turns on or off the particular filtering feature, and a parameter or two that control the filtering (if it is turned on). The first is `filterByColor`, which has only one associated parameter. That parameter, `blobColor`, is an intensity value required for a blob candidate to be kept. Because the blob candidates are generated on binary thresholded images, only the values `0` and `255` are meaningful. (Use the former to extract only dark blobs and the latter to extract only light blobs; turn off the feature all together to get both kinds of blobs.)

The `filterByArea` parameter, if `true`, will cause only blobs whose area is greater than or equal to `minArea`, yet strictly less than `maxArea`, to be kept. Similarly the `filterByCircularity` parameter, if `true`, will cause only blobs whose circularity is greater than or equal to `minCircularity`, yet strictly less than `maxCircularity`, to be kept.

The same goes for `filterByInertial`, `minInertiaRatio`, and `maxInertiaRatio`, as well as for `filterByConvexity`, `minConvexity`, and `maxConvexity`.[27]

### The FAST feature detector and cv::FastFeatureDetector

The *FAST* (Features from Accelerated Segments Test) feature-detection algorithm, originally proposed by Rosten and Drummond [Rosten06], is based on the idea of a direct comparison between a point *P* and a set of points on a small circle around it (see Figure 16-15). The basic idea is that if only a few of the points nearby are similar to *P*, then *P* is going to be a good keypoint. An earlier implementation of this idea, the *SUSAN algorithm*, compared all of the points in a disk around *P*. FAST, which could be thought of as a successor to SUSAN, improves on this idea in two ways.

*Figure 16-15. Two images of the same vehicle; in each image are the 1,000 strongest FAST features. Notice that in the image on the right, as with the Harris-Shi-Tomasi features, the corners in the background are stronger than those on the car, and so again most of the corners in the right image originate from the background*

The first difference is that FAST only uses the points on a ring around *P*. The second is that individual points on the ring are classified as either darker than *P*, lighter than *P*, or similar to *P*. This classification is done with a threshold *t*, such that the darker pixels are ones that are less bright than $I_P - t$, the lighter pixels are ones that are more bright than $I_P + t$, and the similar pixels are those that are in between $I_P - t$ and $I_P + t$. Once this classification has been done, the FAST detector requires some number of contiguous points on the ring to be either all brighter or all darker than *P*. If the number of points on the ring is *N*, then the arc that contains only lighter or darker pixels must contain at least $N/2 + 1$ pixels (i.e., more than half the total number on the ring).

This algorithm is already very fast, but a moment's thought will also reveal that this test permits a convenient optimization in which only four equidistant points are

27 Recall that the exact definitions of circularity, inertia (or inertia ratio), and convexity were covered previously in the algorithm description for the blob finder.

tested. In this case, if there is not at least a pair of consecutive points that are brighter or darker than *P*, then the point *P* cannot be a FAST feature. In practice this optimization greatly reduces the time required to search an entire image.

One difficulty with the algorithm as described so far is that it will tend to return multiple adjacent pixels all as corners. In Figure 16-16, for example, the pixel directly above *P*, among others, is also a FAST keypoint. In general this is not desirable.

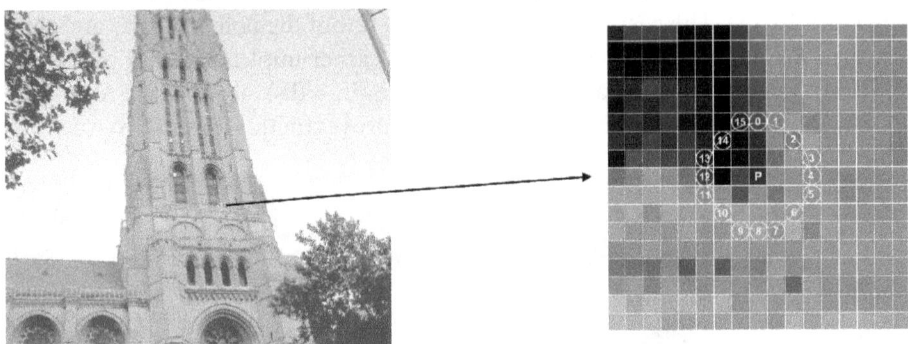


*Figure 16-16. The point P is a keypoint candidate for the FAST algorithm. The ring of points that contribute to the classification of P are identified by a circle around p. In this case there are 16 pixels on that circle, numbered 0–15 here*

To avoid this problem, the FAST algorithm defines a score for each corner, and can remove all keypoints that are adjacent to keypoints of higher score. We construct the score by first computing the sum of absolute differences between the "lighter" pixels and the center pixel, then doing the same for the darker pixels, and finally taking the greater of these two.

$$score = \max\left( \sum_{x \in \{brighter\}} \left| I_x - I_P \right| - t, \sum_{x \in \{darker\}} \left| I_x - I_P \right| - t \right)$$

It is worth noting, mainly because we will come back to this point later when we discuss the ORB feature, that the FAST feature, as defined here, does not have any kind of intrinsic orientation.

**Keypoint finder.** The FAST feature detector is very simple, and looks very much like the Harris corner detector `cv::GoodFeaturesToTrackDetector`:

```
class cv::FastFeatureDetector : public cv::Feature2D {
public:
  enum {
    TYPE_5_8  = 0,                    //  8 points, requires 5 in a row
    TYPE_7_12 = 1,                    // 12 points, requires 7 in a row
    TYPE_9_16 = 2                     // 16 points, requires 9 in a row
```

```
    };

    static Ptr<FastFeatureDetector> create(
      int     threshold         = 10,        // center to periphery diff
      bool    nonmaxSupression = true,      // suppress non-max corners?
      int     type              = TYPE_9_16 // Size of circle (see enum)
    );
  ...
};
```

The constructor for the `cv::FastFeatureDetector` has three arguments: the threshold, a Boolean flag, and the operator *type*. The value of `threshold` is measured in pixel intensity units and is therefore an integer. The Boolean value, `nonMaxSupression`, turns on or off the suppression of neighboring points with inferior scores. The last argument sets the type of operator, with this type determining the circumference of the circle of sampled points. There are three available types, defined in the `cv::FastFeatureDetector` class as enumerations. Each type specifies both the circumference of the circle and the number of contiguous points required for the center of that circle to be considered a keypoint. For example, the value `cv::FastFeatureDetector::TYPE_9_16` conveys that 9 out of 16 points should be either all brighter or all darker than the center point.

In most cases, you will want to set the threshold to a somewhat large number, like `30`. If the threshold is too low, you will get a lot of spurious points in areas of very minor real intensity variation.

### The SIFT feature detector and cv::xfeatures2d::SIFT

The *SIFT* feature (Scale Invariant Feature Transform),[28] originally proposed by David Lowe in 2004 [Lowe04], is widely used and the basis for many subsequently developed features (see Figure 16-17). SIFT features are computationally expensive compared to many other feature types, but they are highly expressive, and thus are well suited to both tracking and recognition tasks.

28 Unlike the open and free algorithms using `cv::features2d`, `cv::xfeatures2d` is suspected of having patent issues and so is relegated to a special *opencv_contrib* directory.

*Figure 16-17. Two images of the same vehicle from different angles. On the left 237 SIFT features are found, while on the right 490 are found. In this case, you can see that the features on the vehicle are relatively stable and can find many correspondences by eye. The density of features on the car is approximately the same in both images, despite the many more features found on the background in the righthand image*

The *scale invariant* property that gives SIFT features their name results from an initial phase of the SIFT algorithm in which a set of convolutions are computed between the input image and Gaussian kernels of increasing size. These convolutions are then combined, each with its *successor*, which is the one convolved with a slightly larger Gaussian. The result of this process is a new set of images that approximate the *difference of Gaussian* (DoG) operator. Given this set of images, which you might visualize as a stack, each pixel in each image in the stack is compared with not only its neighbors in its own image (of which there are eight), but also with itself and its neighbors in the images above and below in the stack (of which there are nine more in the image above and nine more in the image below). If a pixel has a higher value in the difference of Gaussian convolution than all 26 of these neighbors, it is considered a *scale space extremum* of the difference of Gaussian operator (see Figure 16-18).

The intuition here is straightforward. Consider an image that is black except for a white disk in the center. The difference of Gaussians kernel (Figure 16-19) will give the strongest response when the zero crossings are exactly at the edges of the white disk. In this configuration, the positive part of the kernel is multiplied by the positive image values of the white disk, while the negative part of the kernel is entirely multiplied by the zero values of the black background. Neighboring locations or different sizes at the same location will give weaker responses. In this sense, the "feature" of the disk is found, both in position and in scale.

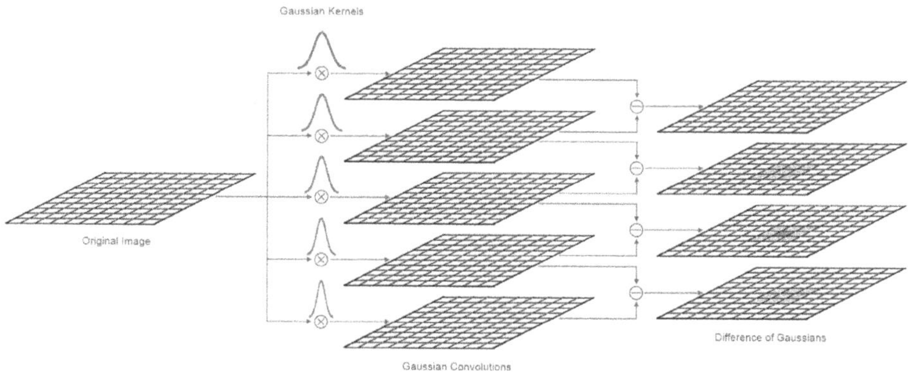


*Figure 16-18. We locate scale space extrema by first convolving the original image with Gaussian kernels of various sizes and then computing difference images between convolutions of neighboring sizes. In the difference images, each pixel (shown as a solid square) is compared to all of its neighbors (shown as Xs) in both the same layer and the adjacent layers. If the difference of Gaussians signal is stronger than all neighbors on all three layers, that pixel is considered a scale space extremum*

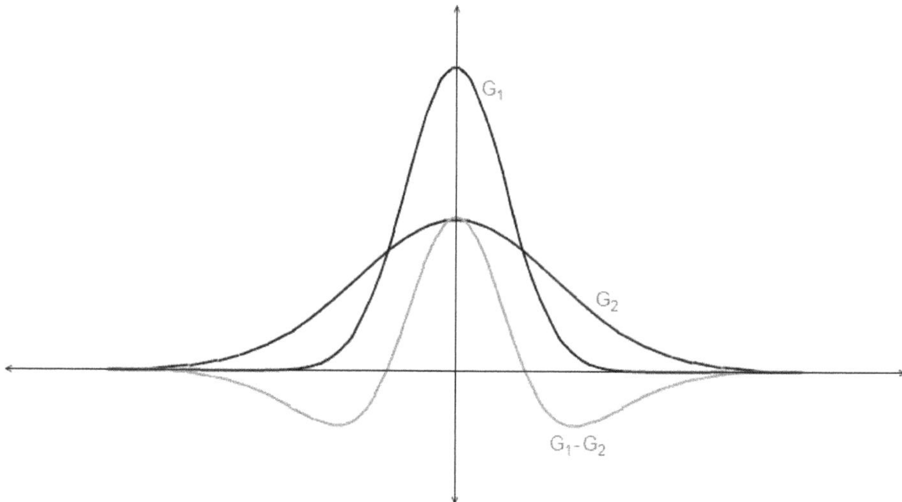


*Figure 16-19. The Gaussian kernels $G_1$ and $G_2$, along with the difference $G_1$–$G_2$*

Once a set of features is found, the algorithm tests each feature both to determine its quality as a feature and to refine the estimate of its location. It does so by fitting a paraboloid to the 3 × 3 × 3 volume around the extremum (the three dimensions being `x`, `y`, and `scale`). From this quadratic form, we can extract two important pieces of information. The first is an offset to be added to the location of the keypoint; this offset is a subpixel correction for spatial location, and interpolates in scale as well as in between the discrete scales available from the original set of Gaussian convolutions. The second is an estimate of the local curvature at this extremal point, in the form of a Hessian matrix whose determinant can be used as a thresholdable value by which to reject keypoints of low discriminating power. Similarly, a large ratio between the eigenvalues of the purely spatial part of this matrix indicates that the feature is primarily an edge (rather than a "corner"), and is also a means by which candidate features can be rejected. These considerations can be compared to the figure of merit used in the Harris corner or Shi-Tomasi feature we saw earlier in this chapter.

Once all such scale space extrema have been found, the SIFT algorithm proceeds to construct a descriptor for each such object as shown in Figure 16-20. The first step here is to assign an orientation to the keypoint. The orientation is based on essentially comparing the directional derivatives over the points around the keypoint and picking the orientation that corresponds to the largest derivatives.[29] Once such an orientation is found, all subsequent properties of the descriptor can be assigned relative to this primary orientation. In this way, SIFT features are not only scale invariant, but also orientation invariant—the latter in the sense that given a new rotated image, the same keypoint would be found; its orientation would be found to be different, but the rest of the feature descriptors, because they are measured relative to the orientation, would be found to match.

Finally, with the scale and orientation computed, the *local image descriptor* can be computed. The local image descriptor is also formed from local image gradients, but this time after the local region has been rotated to a fixed orientation relative to the descriptor orientation. Next, they are clumped into regions (typically 16 in a 4 × 4 pattern around the keypoint, or more), and for each region an angle histogram is created from all of the points in the associated region. Typically this histogram will have 8 entries. Combining the 8 entries per region with the 16 regions gives a vector of 128 components. This 128-component vector is the SIFT keypoint descriptor. This large number of components is essential to the highly descriptive nature of the SIFT keypoints.

29 For the curious, the way the algorithm does this is to first rescale the pixels around the keypoint using the scale already determined. Then, in this scale-normalized image, Sobel derivatives are used in the x- and y-directions, and then converted to a polar form (magnitude and orientation). These derivatives, one for each point in the area of the feature, are then put into a histogram of orientations, each weighted by its magnitude. Finally, the maximum of the histogram is found, a parabola is fit to that maximum and its immediate neighbors, and the maximum of that parabola serves as an interpolated angle for the overall orientation of the feature.

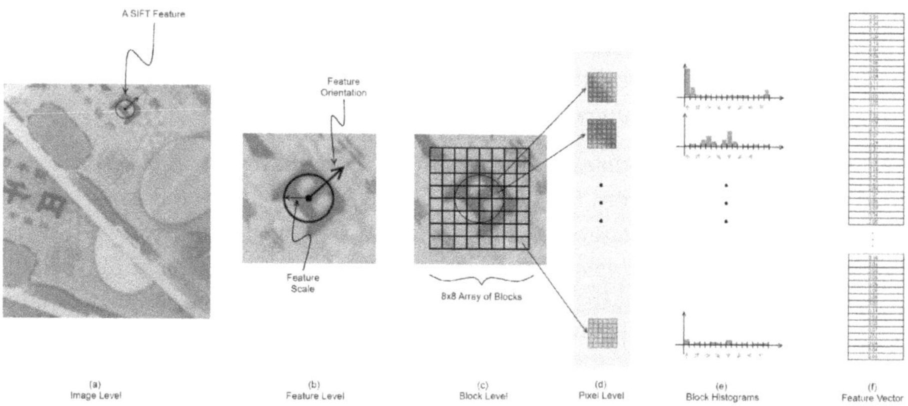


*Figure 16-20. A SIFT feature is extracted from an image (a). That feature has a size and an orientation, shown in (b). The area around the feature is divided up into blocks (c), and for each block, a directional derivative is computed for every pixel in the cell (d). These directional derivatives are aggregated into a histogram for each block (e), and the magnitudes in each bin in all of the histograms for all of the blocks are concatenated into a vector descriptor for the feature (f)*

**Keypoint finder and feature extractor.** The SIFT implementation in OpenCV implements both the feature detection and descriptor extraction parts of `cv::Feature2D` interface, and, as usual, we recommend using the `detectAndCompute()` method, which combines the keypoint detection and feature extraction into one step. Note that because SIFT algorithm is patented, it's placed in the *opencv_contrib* repository, in the module `xfeatures2d`, starting from OpenCV 3.0. Here is (a slightly abbreviated version of) the class definition for the `cv::xfeatures2d::SIFT` object:

```
class SIFT : public Feature2D {

public:

  static Ptr<SIFT> create (
    int    nfeatures         = 0,    // Number of features to use
    int    nOctaveLayers     = 3,    // Layers in each octave
    double contrastThreshold = 0.04, // to filter out weak features
    double edgeThreshold     = 10,   // to filter out "edge" features
    double sigma             = 1.6   // variance of level-0 Gaussian
  );

  int descriptorSize() const;        // descriptor size, always 128
  int descriptorType() const;        // descriptor type, always CV_32F
   ...
};
```

All of the parameters required by the `cv::xfeatures2d::SIFT` constructor are used in the construction of the scale-space image representation and also at the keypoint-detection portion of the algorithm. In the current implementation, the actual feature descriptors are always computed with a fixed set of parameters.[30]

The first argument, `nfeatures`, indicates the maximum number of features you would like be computed from the image. If it is left set to the default value of `0`, the algorithm will return all features that can be found. The next argument, `nOctaveLayers`, determines how many *layers* (different scales of Gaussian convolutions) should be computed for each *octave* (images in the image pyramid). The number of layers actually computed is the value of the `nOctaveLayers` argument plus three. Thus, the illustration in Figure 16-18 shows the case of five layers for each image in the pyramid, which corresponds to a value for `nOctaveLayers` of two.

The next two parameters are threshold values, which are used to determine whether a found keypoint candidate should be retained. Recall that once a keypoint candidate is generated by scale space search, that keypoint is then subjected to two tests. The first test is whether or not the local extremum of the DoG operator is sufficiently distinct from the surrounding region. This test is against the value of `contrastThreshold`; typical values of this argument are of the order of `0.04` (the default value). The second test has to do with the ratio of spatial eigenvalues, and serves the purpose of rejecting edges. This test is against the value of `edgeThreshold`; typical values of this argument are of the order `10.0` (the default value).

The final parameter, `sigma`, is used to create a presmoothing of the image, which incidentally effectively also sets the scale of the very first layer in the scale space. Typical values are of the order of a pixel (the default is 1.6 pixels), but it is often useful to make this value slightly larger for images that contain a bit of noise or other artifacts.

You could just prefilter an image by convolving with your own Gaussian filter, instead of using the sigma parameter, but if you do this, the algorithm will not be aware that there is no information below the scale you used in the filtering. As a result, it will waste time computing layers whose purpose is to search for features that cannot exist. Therefore, it is much more efficient to use the sigma parameter provided for presmoothing. In effect, you are telling the algorithm "I don't care about anything smaller than this size."

30 In general, it has been found that SIFT works best with 128 element descriptors. Similarly, some other values used by the descriptor computation (such as the *magnification*, if you are a SIFT expert) have been found to have values that are essentially always left unchanged. In future implementations, some of these may be exposed if there is found to be value in doing so.

Once you have created your `cv::xfeatures2d::SIFT` object, you can use the `descriptorSize()` and `descriptorType()` functions to query the size of the features it will compute and the type of elements the feature vector will contain. These two functions always return `128` and `CV_32F`, respectively, but are often handy when you are using many kinds of feature objects and are handling them by a base class pointer, but need to query an individual one to find out about the feature vectors it will return.

The main function you will be using is the overloaded `detectAndCompute()` method. Depending on the arguments given, this operator will either just compute keypoints, or compute keypoints and their associated descriptors. The keypoints-only case requires just three arguments: `img`, `mask`, and `keypoints`. The first argument is the image you want keypoints extracted from (which can be color or grayscale, but in the former case it will be changed to a grayscale representation internally before the algorithm begins). The `image` argument should always be of type `CV_8U`. The `mask` argument is used in order to restrict the keypoints generated to only those within some specified area. The `mask` array should be a single channel of type `CV_8U`, but can be set to `cv::noArray()` if no filtering of this kind is required. The next argument to `cv::xfeatures2d::SIFT::detectAndCompute()` is `keypoints`, which must be a reference to an STL-style vector of `cv::KeyPoint` objects. This is where `SIFT` will put the keypoints it finds.

The next argument, `descriptors`, is an output array, analogous to other feature-point descriptors we have seen already. If `descriptors` is an array, then each row of the array will be a separate descriptor, and the number of such rows is equal to the number of keypoints.

The final argument is the Boolean `useProvidedKeypoints`. If this argument is set to `true`, then a keypoint search will not be undertaken, and the `keypoints` argument will instead be treated as an input. In this case, descriptors will be generated for every keypoint indicated in the `keypoints` vector.

#### The SURF feature detector and cv::xfeatures2d::SURF

The *SURF* feature (Speeded-Up Robust Features)[31] was originally proposed in 2006 by Bay et al. [Bay06, Bay08], and is in many ways an evolution of the SIFT feature we just discussed (see Figure 16-21). The creators of SURF were interested in ways in which the different components of the SIFT features could be replaced with more computationally efficient techniques that might give similar or better performance in (primarily) recognition tasks. The resulting features are not only much faster to

31 As was the case with SIFT, because SURF is a patented algorithm, it's placed in the `xfeatures2d` module of the *opencv_contrib* repository as of the OpenCV 3.0 release.

compute; in many cases, their slightly simpler nature results in greater robustness to changes in orientation or lighting than is observed with SIFT features.

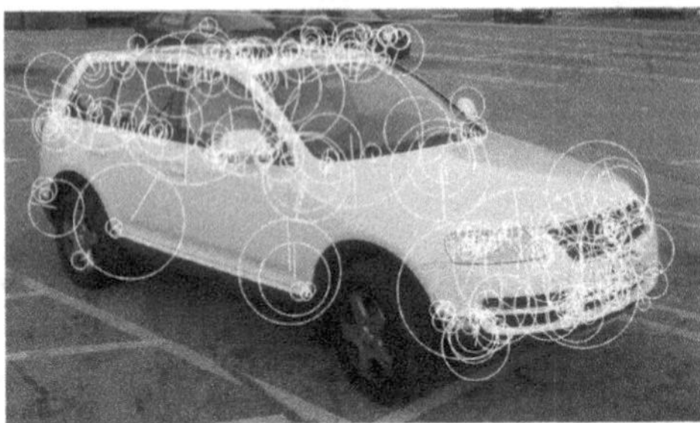

*Figure 16-21. SURF features are computed for the same vehicle from two different orientations. On the left, 224 features are found, while on the right 591 features are found, with many new features being associated with the visible background in the righthand image. SURF features are orientable like SIFT features. This image was generated with the hessianThreshold set to 1500*

Relevant to several phases of the SURF feature detector's operation is the concept of the integral image. Recall that we encountered integral images in Chapter 12; where we saw that they enable us to make a single transform of a whole image and, thereafter, that transformed image allows us to compute sums over any rectangular area with only a few simple operations. The SURF feature relies heavily on computations that can be greatly accelerated by the use of the integral image technique.

Like many other detectors, SURF defines a keypoint in terms of the local Hessian[32] at a given point. Previously, we saw that in order to introduce the concept of scale, the SIFT detector computed the local Hessian using a difference of Gaussian convolutions with slightly differing widths (Figure 16-19). The SURF detector, however, computes the local Hessian by convolving with a *box filter* that approximates the difference of the two Gaussian kernels (Figure 16-22).[33] The primary advantage of these box filters is that we can evaluate them quickly using the integral image technique.

32 The Hessian is normally understood to be the matrix of second-order derivatives. In this case, however, it is really the matrix of so-called "Gaussian second-order derivatives," which are defined by $\frac{\partial^2}{\partial x_i \partial x_j} G(\vec{x}, \sigma)$, where $G(\vec{x}, \sigma)$ is a normalized Gaussian of size $\sigma$ by which the image is convolved before the derivative is approximated.

33 Note that in the SIFT case the actual technique was to convolve with the two different Gaussian kernels and then subtract the results. In the SURF case the kernels are first subtracted and a single convolution is done with the differenced kernel. The two operations are equivalent, but the second operation is more natural and more efficient in the case of the box filter approximation used by SURF.

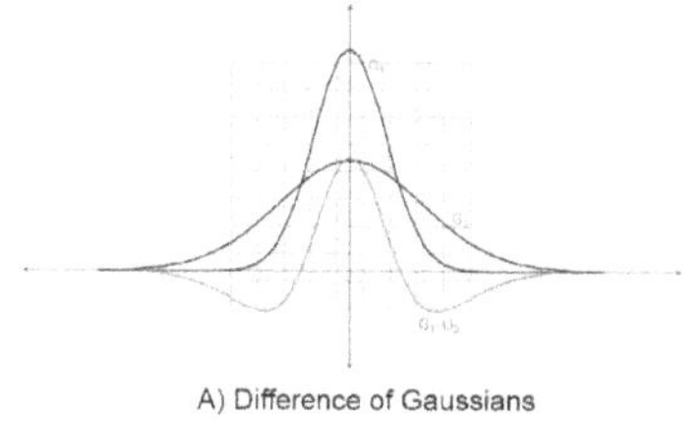


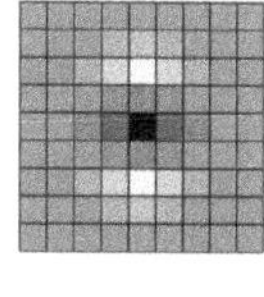


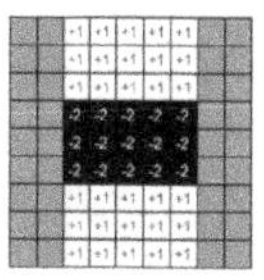


*Figure 16-22. The difference of two continuous Gaussian kernels is shown on the left (a). A discrete 9 × 9 filter kernel is shown in the center (b) that approximates the second derivative in the vertical direction. A box filter approximation of the DoG filter kernel is shown on the right (c)*

Because the cost of the box filters does not change with the size of the filter (because of the integral image), it is not necessary to generate a scale pyramid of the images as was done with SIFT. Instead, a variety of ever larger box filters can be used to evaluate the Hessian at different scales. From the response to these box filters, keypoint features are defined in SURF to be the local extrema of the determinant of this Hessian that also exceed some threshold.

Like SIFT, SURF includes the notion of orientation for a feature, which we compute by using the integral image again to estimate a local bulk gradient of the region around the feature. We do so using a pair of simple Haar wavelets (Figure 16-23c) to approximate local gradients, and by applying these wavelets to different areas of the region around where the scale space extremum was found. If the scale of the feature was found to be $s$, then we calculate the gradients using wavelets of size $4s$, spaced at intervals a distance $s$ apart in a region of radius $6s$ centered on the feature (Figure 16-23b). We then aggregate these gradient estimations by considering a sliding window in angle of size $\frac{\pi}{3}$. By summing all of the gradients in this orientation window (with a weighting factor determined by the area's distance from the feature center), we choose the maximum orientation found as the orientation of the feature (Figure 16-23d).[34] Once an orientation has been computed, the feature vector can then be generated in a manner relative to that orientation and, as was the case with SIFT, the feature becomes effectively invariant to orientation.

34 This procedure might seem quite convoluted, but note that the actual number of evaluations required to compute this orientation is actually very small. With only nine cells, and each cell requiring only six additions, the 81 points are computed in fewer than 500 operations.

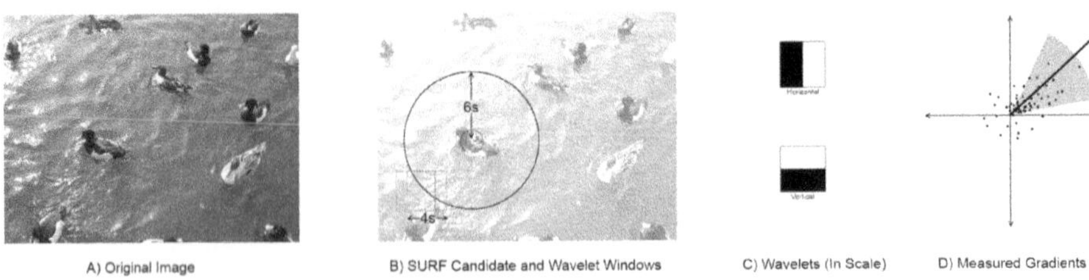


*Figure 16-23. We determine the SURF orientation of an image (a) by analyzing the region around where a scale space extremum was found (b). Two simple wavelets (c) are used to approximate a local gradient and are convolved with the image in many positions near that extremum (dashed boxes in b regularly sampled from the region shown by the solid circle in b). The final orientation is extracted from an analysis of all of the gradients measured in this way (d)*

The features themselves are also computed in a manner analogous to the SIFT feature design. For a feature of scale $s$, the $20s \times 20s$ region centered on the feature is first divided into 16 cells in a $4 \times 4$ grid. This grid is rotated relative to the feature by an angle given by the orientation just computed. For every such cell, 25 pairs of Haar wavelets (identical to those shown in Figure 16-23c, other than being much smaller) are used to approximate the x- and y-gradients of the image in each of a $5 \times 5$ array (within each cell of the $4 \times 4$ grid).[35] For each cell of this kind, the 25 x-direction wavelet convolutions are summed, as are the 25 y-direction wavelet convolutions. For each direction, both the sum as well as the sum of absolute values are computed. This gives 4 numbers for each cell in the $4 \times 4$ grid, or a total of 64 numbers. These 64 values form the entries in a 64-dimensional feature vector for the individual SURF feature (see Figure 16-24).

There is a variant SURF feature, called an "extended" SURF feature, that sums separately the sums of the convolutions into eight sums rather than four. It does so by summing those values of the x-direction wavelet convolutions for which the corresponding y-direction wavelet convolution was positive from those for which the corresponding y-direction wavelet convolution was negative (and similarly for the y-direction wavelet convolutions sums depending on the sign of the corresponding x-direction convolution). The resulting descriptors are larger, which means that matching them will be slower, but in some cases the greater descriptive power of the extended features has been found to improve recognition performance.

35 Note that x and y as used here are in the rotated coordinate system of the feature descriptor, not the coordinate system of the image.

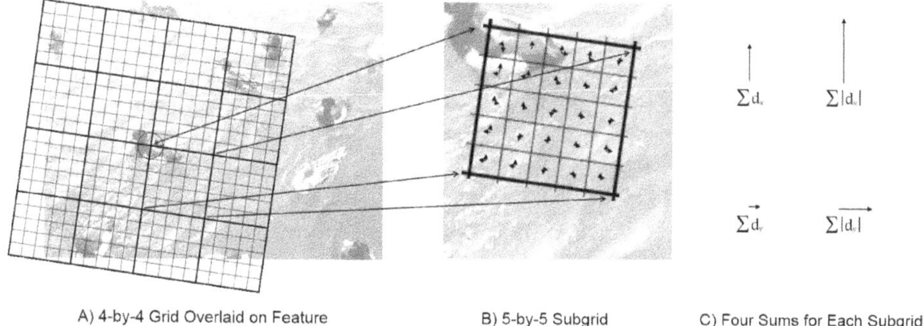


*Figure 16-24. We compute the SURF feature from estimating a gradient in each of 400 subcells. The area around the feature is first divided into a 4 × 4 grid of cells (a). Each cell is then divided into 25 subcells, and directional derivatives are estimated for each subcell (b). The directional derivatives for the subcells are then summed to compute four values for each cell in the large grid (c)*

**Keypoint finder and feature extractor.** As with SIFT, the best and most recent implementation of SURF in OpenCV uses the `cv::Feature2D` interface. The construction process is done by a single constructor for the `cv::xfeatures2d::SURF` object, which then provides an interface for keypoint finding as well as feature descriptor extraction—as well as a few other useful functions. Here is the (slightly abbreviated) definition of the SURF object:

```
class cv::xfeatures2d::SURF : public cv::Feature2D {

public:
  static Ptr<SURF> create (
    double hessianThreshold = 100,    // Keep features above this
    int    nOctaves         = 4,      // Num of pyramid octaves
    int    nOctaveLayers    = 3,      // Num of images in each octave
    bool   extended         = false,  // false: 64-element,
                                      //  true: 128-element descriptors
    bool   upright          = false,  // true: don't compute orientation
                                      //  (w/out is much faster)
  );

  int descriptorSize() const;         // descriptor size, 64 or 128
  int descriptorType() const;         // descriptor type, always CV_32F

  ...
};
typedef SURF SurfFeatureDetector;
typedef SURF SurfDescriptorExtractor;
```

The constructor method `create()` for the `cv::xfeatures2d::SURF` object has five arguments that are used to configure the algorithm. The first, `hessianThreshold`, sets the threshold value for the determinant of the Hessian that is required in order for a particular local extremum to be considered a keypoint. The value assigned by the default constructor is `100`, but this is a very low value that could be interpreted as meaning "all of them." A typical value for reasonable selectivity would be something like `1500`.[36]

The `extended` parameter tells the feature extractor to use the extended (128-dimensional) feature set (described in the previous section). The parameter `upright` indicates that orientations should not be computed for features, and they should all be treated as "vertical"; this is also known as "upright SURF" or just "U-SURF."

In uses such as automotive or mobile robot applications, it is often safe to assume that the orientation of the camera is fixed relative to the orientation of the objects one wants to detect. Consider, for example, the case of an automobile detecting road signs. In this case, the use of the `upright` argument will improve speed and most likely improve matching performance as well.

The final arguments, `nOctaves` and `nOctaveLayers`, are closely analogous to the corresponding arguments for `cv::xfeatures2d::SIFT()`. The `nOctaves` argument determines how many "doublings" of scale will be searched for keypoints. In the case of SURF, the minimal size feature that can be found is calculated by convolution with a 9 × 9 pixel filter. The default number of octaves is four, which is normally sufficient for most applications. When using very high resolution imagers, however, you might wish to increase this number. Reducing it to three produces only a very small improvement in speed, however, as the scale search for the higher octaves is very inexpensive compared to the lower octaves.[37]

For each octave, several different kernels will be evaluated. Unlike SIFT however, the kernels are not distributed to evenly subdivide the octaves. In fact, if more than two octave layers are used, there will be overlap between the size of the kernels used in successive octaves. (This does not mean that a larger value is not useful, only that the effect of larger numbers of octave layers is not entirely intuitive.) The default value

36 For the car images in Figure 16-21, the Hessian threshold was empirically adjusted to give a similar number of features to the number found by SIFT in Figure 16-11. For comparison, the OpenCV default value of `100` would have generated 2,017 and 2,475 features, respectively, for the left and right images in Figure 16-21.

37 It is worth noting here that unlike SIFT, which actually reduces the size of the image with each octave, SURF is instead increasing the size of the kernels with which it is convolving. In addition, the step between "adjacent" evaluations of the kernels is also being increased as the kernels themselves are enlarged. As a result, because the kernels have fixed cost regardless of size (remember the integral image trick?), the cost of evaluating higher octaves decreases rapidly.

for `nOctaveLayers` is 3, but some studies have found increasing it to as high as four useful (though at increasing computational cost).

The `descriptorSize()` and `descriptorType()` methods return the number of elements in the descriptor vector (64 normally, and 128 for extended SURF features) and the type of the descriptor vector. (Currently the latter is always `CV_32F`.)

There are overloaded methods `SURF::detect()`, `SURF::compute()`, and `SURF::detectAndCompute()`. As usual, when you need both the keypoints and their descriptors, it's recommended to use the latter method. The parameters are absolutely identical to those of `SIFT::detectAndCompute()`.

**Additional functions provided by cv::xfeatures2d::SURF.** `cv::xfeatures2d::SURF` also provides a bunch of methods to set and retrieve the algorithm's parameters on the fly. Make sure that you do not alter parameters while processing a set of images. Find the optimal parameters and keep using them; otherwise, you may get incomparable descriptors.

### The Star/CenSurE feature detector and cv::xfeatures2d::StarDetector

The *Star* features (see Figure 16-25) were developed originally for visual odometry, measuring the self-motion of a video camera from the image data alone [Agarwal08]. In this context, features such as Harris corner or FAST are desirable because they are highly localized. In contrast, features like SIFT, which rely on image pyramids, can become poorly localized in the original image space as you move higher up the pyramid. Unfortunately, features like the Harris corner or FAST are not scale invariant like SIFT is, precisely because of the lack of a scale space search. The Star feature, also known as the *Center Surround Extremum* (or *CenSurE*) feature, attempts to solve the problem of providing the level of localization of Harris corners or FAST features while also providing scale invariance. There is no associated unique descriptor attached to the Star/CenSurE feature; in the paper in which it was first introduced, the authors used a variant of the "Upright SURF" or U-SURF feature descriptors.

Conceptually, the approach of Star is to compute all variants of some feature at all scales and select the extrema across scale and location. At the same time, the goal was to have the features be very fast to compute (recall that visual odometry has applications in robotic and many other real-time environments). The CenSurE feature addresses these competing goals with a two-stage process. The first stage is a very fast approximation to something like the difference of Gaussians (DoG) operation used by SIFT (and others), and the extraction of the local extrema of this operation. The second stage attempts to cull things that look too much like edges (as opposed to corners) using a scale-adapted version of the Harris measure.

*Figure 16-25. Star features computed for the same vehicle from two slightly different orientations. At default parameters, 137 are found on the left image and 336 are found on the right. In both cases, the number on the automobile is about the same, with the additional features on the background accounting for the greater number in the second image. The found features on the vehicle are few compared to other methods, but you can easily see correspondences, suggesting the features are very stable*

To understand how the fast DoG approximation is done, it is useful to think back to how SURF computed the box approximation to similar objects (Figure 16-22). In this case however, the DoG is approximating a feature that looks like the difference of two similarly sized Gaussians that are both rotationally symmetric in the image plane. As a result the feature is particularly simple.

The approximation used is square (Figure 16-26) and so it can be constructed at any size. It has only two regions, which can both be computed via integral images. Thus, the entire evaluation of the feature amounts to three operations for the outer square, three for the inner square, two more to scale them, and one final operation to add the two terms: nine operations. This simplicity is why every point can be tested at many scales. In practice, the smallest feature that can be constructed this way has a side length of 4 and, in principle, every size above that can be computed. In practice, of course, it is natural to distribute the sizes of the features actually computed in an exponential, rather than linear, way. The result of this stage of the process is that for every point in the image, there is a particular size DoG kernel for which the response was the largest and the particular magnitude for that response.

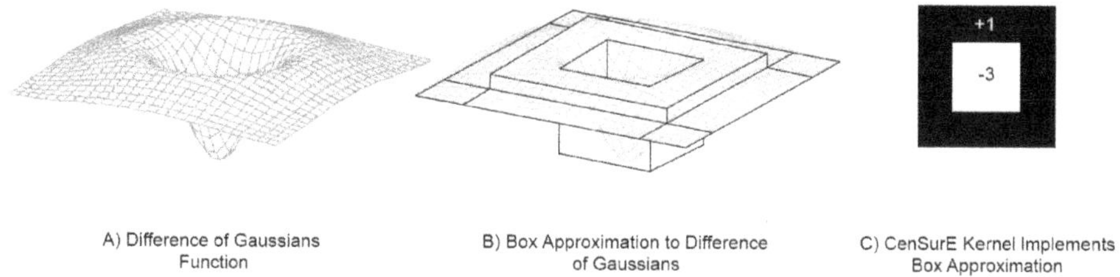


*Figure 16-26. Box representation of DoG kernel used by CenSureE keypoint finder; for a particular size S, the center portion of the feature is of size S/2*

Once the DoG kernel has been approximated everywhere, the next step is to threshold this value and then reject those points that are not local extrema. We do this by comparing to the usual 3 × 3 × 3 cube of neighbors in (*x*, *y*, scale)–space and keeping only those with the highest (or lowest) value in that 27-element set.

Finally, because features of this kind can still respond rather strongly to edges, the Star algorithm computes a *scale-adapted Harris measure*. The scale-adapted Harris measure is computed from a matrix very similar to the one we encountered in the Harris-Shi-Tomasi corners discussion earlier, with two important exceptions. The first is that the window over which the summations are done for the individual elements of the autocorrelation matrix is sized proportional to the scale of the feature. The second is that the autocorrelation matrix is constructed from the maximal responses to the censure features rather than the image intensity. The actual test then performed is the test used by Harris, which compares the determinant of this matrix to the squared trace multiplied by the *sensitivity* constant.

In the OpenCV implementation, there is also a second test, which is similar to the scale-adapted Harris measure described except that it constructs an autocorrelation matrix from the size values associated with the response at each point in the window. This measure is called the *binarized scale-adapted Harris measure*. It is called "binarized" because for each point in the window, a value of `1`, `0`, or `-1` is assigned based on the rate of change of the size of the maximal response at that point relative to its neighbors. Recall that the original Harris measure used a rate of change of image intensity, and the scale-adapted Harris measure used a rate of change of response to the DoG operator; the binarized measure uses the rate of change of the size of the maximal DoG operator. This binarized test is a way of quantifying the extent to which a particular point is a scale space extremum.

**Keypoint finder.** As was mentioned earlier, there is no specific feature descriptor extractor associated with the Star algorithm. The detector `cv::StarDetector` is derived directly from the `cv::Feature2D` base class. Here is the slightly abbreviated `cv::StarDetector` object definition:

```
// Constructor for the Star detector object:
//
class cv::xfeatures2d::StarDetector : public cv::Feature2D {

public:

  static Ptr<StarDetector> create(
    int maxSize                = 45,  // Largest feature considered
    int responseThreshold      = 30,  // Minimum wavelet response
    int lineThresholdProjected = 10,  // Threshold on Harris measure
    int lineThresholdBinarized = 8,   // Threshold on binarized Harris
    int suppressNonmaxSize     = 5    // Keep only best features
                                      //  in this size space
  );

  ...
};
```

The Star detector constructor takes five arguments. The first is the largest size of feature that will be searched for. The `maxSize` argument can be set to only one of a finite list of values: `4`, `6`, `8`, `11`, `12`, `16`, `22`, `23`, `32`, `45`, `46`, `64`, `90`, or `128`. For any value you might choose, all of the lower values in this list will also be checked.

The `responseThreshold` argument indicates the threshold applied to the convolution with the CenSurE kernel (Figure 16-26c) in order to find keypoint candidates. For all scales above the smallest, the kernel is normalized such that the threshold at all larger scales is equivalent to the given value on the smallest kernel (i.e., of size 4).

The next two arguments, `lineThresholdProjected` and `lineThresholdBinarized`, are thresholds associated with the scale-adapted Harris measure described earlier. The projected line threshold is effectively the inverse of the sensitivity constant in the Harris test on the response values; raising `lineThresholdProjected` will reject more features as lines. The `lineThresholdBinarized` value does something very similar, except that it is the sensitivity constant for the binarized scale-adapted Harris measure. This second parameter enforces the requirement that the CenSurE feature be a scale space extrema. Both comparisons are made, and a keypoint candidate must succeed relative to both criteria in order to be accepted.

The final argument, `supressNonmaxSize`, sets the region over which Star features will be rejected if they are not the strongest feature within that distance.

### The BRIEF descriptor extractor and cv::BriefDescriptorExtractor

BRIEF, which stands for *Binary Robust Independent Elementary Features*, is a relatively new algorithm for assigning a novel kind of feature to a keypoint (see Figure 16-27). BRIEF features were introduced by Calonder et al. and for this reason are also often known as *Calonder features* [Calonder10]. BRIEF does not locate keypoints; rather, it is used to generate descriptors for keypoints that can be located through any of the other available feature-detector algorithms.[38]

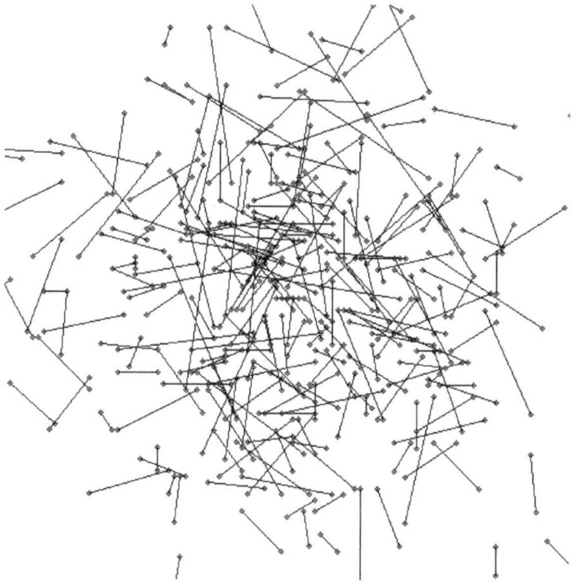

*Figure 16-27. A visualization of pixel-to-pixel tests that collectively comprise a single BRIEF descriptor; each line connects a pair of pixels that are compared by the test*

The basic concept behind the BRIEF descriptor is that a feature is described as a series of tests, each of which simply compares a single pixel in the area of the feature to some other single pixel, yielding a simple binary result (i.e., `0` or `1`) based on which portion was brighter (Figure 16-27). The BRIEF descriptor is simply the result of $n$ such tests arranged into a bit string. In order to keep the descriptor from being overly sensitive to noise, the BRIEF descriptor first smooths the image by convolution with a Gaussian kernel. Because the descriptors are binary strings, they not only can be computed quickly and stored efficiently, but they can also be compared to one another extremely efficiently.[39]

There are many ways to generate the actual pairs that will be matched to form the BRIEF descriptor. One of the best ways is simply to randomly generate all of the pairs by first drawing a point from a Gaussian distribution around the center of the

38 In the paper in which they were introduced, BRIEF descriptors were used with features found through U-SURF.

39 Many modern processors contain single-cycle instructions that will perform an XOR on a 256-bit word (e.g., the Intel SSE4™ instruction set).

feature, and then computing the second point by drawing from a Gaussian distribution around the first (with one half the standard deviation). The area within which the points are drawn (the total footprint of the feature) is called the *patch size*, while the standard deviation of the distribution from which the points are drawn is called the *kernel size*. In the case of Figure 16-27, the ratio of the kernel size to the patch size is approximately 1:5. The current OpenCV implementation fixes these sizes, but in principle they are tunable parameters of the algorithm. The number of such tests generated overall is typically 128, 256, or 512, but following the style of the original creators of the feature, it is traditional to refer to this size in terms of the number of *bytes* in the descriptor (i.e., 16, 32, or 64 bytes, respectively).

**Feature extractor.** As was mentioned earlier, the BRIEF algorithm is specifically for extracting feature descriptors, and so the associated method is derived directly from the `cv::Feature2D` base class, implementing only the descriptor extraction part. The relevant parts of the class definition are the following:

```
class cv::xfeatures2d::BriefDescriptorExtractor : public cv::Feature2D {

public:

  static Ptr<BriefDescriptorExtractor> create(
    int bytes             = 32,        // can be equal 16, 32 or 64 bytes
    bool use_orientation = false       // true if point pairs are "rotated"
                                       // according to keypoint orientation
  );


  virtual int descriptorSize() const;  // number of bytes for features
  virtual int descriptorType() const;  // Always returns CV_8UC1
};
```

At this time, the only two user-configurable parameters of the BRIEF descriptor extractor are the number of bytes of information comprising the feature (equal to the total number of tests divided by eight) and the `use_orientation` flag which is analogous to the role of the upright parameter of SURF algorithm. The same considerations are applicable here also—when the features are unlikely to rotate much—for example, you recognize road signs or stitch images—`use_orientation` should likely be set to `false`. Otherwise, you may wish to set it to `true`.

In order to actually compute descriptors from an image and a set of keypoint locations, `cv::xfeatures2d::BriefDescriptorExtractor` uses the `compute()` interfaces defined in the `cv::Feature2D` base class.

### The BRISK algorithm

Not long after the introduction of the BRIEF descriptor, several new techniques appeared that use a similar notion of point-wise comparison as a means to produce a compact descriptor that could be compared quickly. The *BRISK*[40] descriptor (see Figure 16-28), introduced by Leutenegger et al., attempts to improve on BRIEF in two distinct ways [Leutenegger11]. Firstly, BRISK introduces a feature detector of its own (recall that BRIEF is only a method for computing descriptors). Second, the BRISK feature itself, though similar to BRIEF in principle, attempts to organize the binary comparisons in a manner that improves robustness of the feature as a whole.

*Figure 16-28. Here the BRISK feature detector is used on our two reference images. The left image contains 232 features, while the right contains 734. The more complex visible background in the right image contributes most of the new features, however, and the features on the car are relatively stable in both number and location*

The feature-detector portion of BRISK is essentially based on the FAST-alike *AGAST*[41] detector, with the improvement that it attempts to identify a scale for the feature as well as an orientation. BRISK identifies the scale by first creating a scale space pyramid with a fixed number of scales (factors of two in size), and then computing a fixed number of *intra-octaves* per scale.[42] The first step of the BRISK feature detector is to apply FAST (or actually AGAST) to find features at all of these scales. Once this is done, nonmaxima suppression is applied; that is, features whose score (called $\rho_0$ in our discussion of FAST) is not the largest of all of its neighbors are removed; this leaves only the "maximal" features.

Once a list of features is found in this way, BRISK then goes on to compute the AGAST score at the corresponding locations in the image immediately larger and immediately smaller (Figure 16-29). At this point, a simple quadratic is fit to the

40 "BRISK" does not appear to stand for anything; it is not an acronym. Rather, the name is essentially a play on words similar to BRIEF.

41 The feature detector called "AGAST" (Adaptive and Generic Corner Detection Based on the Accelerated Segment Test) [Mair10] is an improvement on FAST and a precursor to BRISK. It is mentioned here for completeness, but is not implemented separately in the OpenCV library.

42 In the original implementation there was just one intra-octave per scale, thus creating just some number $N$ of scales, and $N - 1$ intra-octaves for a total of $2N - 1$ images.

AGAST scores (as a function of scale), and the maximum point of that quadratic is taken to be the true scale of the BRISK feature. In this way, a continuous value is extracted and the BRISK feature is not forced to be associated with one of the discrete images computed in the pyramid. A similar interpolation method is applied in pixel coordinates to assign a subpixel location to the feature.

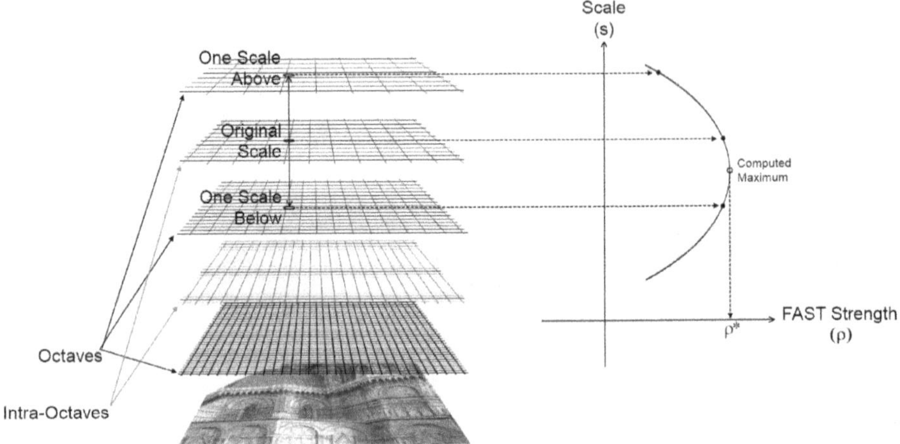


*Figure 16-29. BRISK constructs the scale space using several octaves with "intra-octaves" in between. When a FAST feature is found at one scale, its FAST strength is computed at that scale and at the scale immediately above and below (left). These strengths are used to fit a quadratic curve and the scale at which the maximum score is expected is extrapolated from that curve (right)*

In addition to a scale, BRISK features also have an orientation. To see how, we need to first understand how the sampling pattern used by BRISK differs from the random sampling pattern used by BRIEF. The BRISK descriptor is constructed by a series of rings around the center point. Each ring has some number $K_i$ of sample points allocated to it, and each sample point is assigned a circular area of diameter equal to the circumference of the particular circle $C_i$ divided by $K_i$ (Figure 16-30). This area corresponds to the image being convolved by a Gaussian of that particular radius ($\sigma_i = C_i/2K_i$) and sampled at the indicated point.

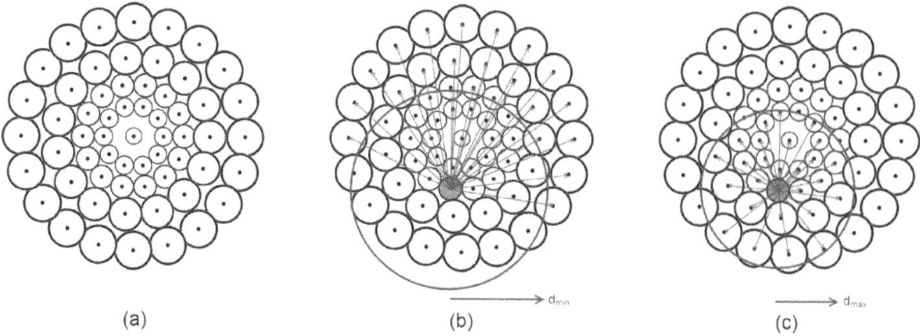


*Figure 16-30. The test points in the BRISK descriptor are identified in the figure by the small solid dots. The regions contributing to each test point are shown as circles around each point. Note that as the test points move out from the center of the descriptor the size of the associated regions is increased. The left image shows only the points and their associated regions (a). The center image shows all long-range pairings associated with one particular point (b). The right image shows all of the short-range pairings associated with that same point (c)*

The brightness comparisons that comprise the bitwise descriptor (analogous to BRIEF) are computed between pairings in all of the circles. In particular, these pairings are divided into two subsets, called short- and long-range pairings. The short-range pairings are all the pairings between points that are less than some specific distance $d_{max}$ apart, while the long-range pairings are all of the pairings between points that are more than a specific distance $d_{min}$ apart. The short-range pairings form the descriptor, while the long-range pairings are used to compute a dominant orientation.

To see how this dominant orientation is constructed, first note that the BRISK descriptor, like the BRIEF descriptor, computes differences in intensity between point pairs. The BRISK descriptor, however, goes on to normalize those differences by the distance between the points, thus creating what is in effect a local gradient. By summing these local gradients over all of the long-range pairs, we compute an orientation that can be used to orient the descriptor. The short-range features are then computed relative to this orientation such that the descriptor, made by thresholding these short-range gradients, is effectively orientation independent. By tuning the number of points per circle and the value of $d_{max}$, we can give the descriptor any length. By convention, these values are chosen so as to make the descriptor 512 bits long, the same as the (typical) BRIEF descriptor.

**Keypoint finder and feature extractor.** The `cv::BRISK` object inherits from the `cv::Feature2D` interface, and thus provides both a feature-finding capability and a descriptor-extraction capability. Here is the (abbreviated as usual) definition of the `cv::BRISK` object:

```
class cv::BRISK : public cv::Feature2D {

public:
  static Ptr<BRISK> create(
    int                   thresh       = 30,   // Threshold passed to FAST
    int                   octaves      = 3,    // N doublings in pyramid
    float                 patternScale = 1.0f  // Rescale default pattern
  );

  int descriptorSize() const;                  // descriptor size
  int descriptorType() const;                  // descriptor type


   static Ptr<BRISK> create(                   // Compute BRISK features
    const vector<float>& radiusList,           // Radii of sample circles
    const vector<int>&   numberList,           // Sample points per circle
    float                dMax        = 5.85f, // Max distance for short pairs
    float                dMin        = 8.2f,  // Min distance for long pairs
    const vector<int>&   indexChange = std::vector<int>() // Unused
  );

};
```

The `cv::BRISK::create()` constructor method accepts three arguments: the AGAST threshold, the number of octaves, and an overall scale factor for the pattern. If you use this constructor, the locations of the sample points will be taken from a fixed lookup table inside of the library. The threshold argument, `thresh`, sets the threshold that will be used by the AGAST feature detector.[43] The number of octaves, `octaves`, sets the number of full octaves. Remember that if you set this to some number $N$, then the total number of levels computed will be $2N - 1$ (including the intra-octaves). The final `patternScale` argument applies an overall scale factor to the built-in pattern.

The overloaded `cv::BRISK::detectAndCompute()` method implements the usual feature detection and descriptor extraction, which are inherited from the `cv::Feature2D` interface.

43 Deep inside of the BRISK code, BRISK calls `cv::FAST` and passes precisely this threshold value to the FAST algorithm.

**Additional functions provided by cv::BRISK.** In addition to the preceding methods, which you would expect from any function that inherits from `cv::Feature2D`, `cv::BRISK` has an extra extended constructor: `cv::BRISK::create()`. This function is used if you do not want to rely on the built-in pattern for the sample points, which is provided by the library. If you would rather build your own pattern, then you must provide a list of radii for the circles in the form of an STL-style vector of numbers for the `radiusList` argument. Similarly, you must provide an STL-style list of integers (of the same length as `radiusList`) to `numberList` that indicates the number of sample points to be used at each radius. You can then (optionally) specify `dMax` and `dMin`, the maximum distance for short-range pairings and the minimum distance for long-range pairings ($d_{max}$ and $d_{min}$ from the previous section). The final argument, `indexChange`, currently has no effect, and should best be omitted.

### The ORB feature detector and cv::ORB

For many applications, feature detector speed is not only helpful, but essential. This is particularly true for tasks that are expected to run in real time on video data, such as augmented reality or robotics applications. For this reason, the SURF feature was developed with the goal of providing similar capability to SIFT, but at a much higher speed. Similarly, the ORB feature [Rublee11] was created with the goal of providing a higher-speed alternative to either SIFT or SURF (see Figure 16-31). The ORB feature uses a keypoint detector that is very closely based on FAST (which we saw earlier in this chapter), but uses a substantially different descriptor, based primarily on BRIEF. The BRIEF descriptor is augmented in ORB by an orientation computation that essentially gives ORB features the same sort of rotational invariance enjoyed by SIFT and SURF.[44]

This first stage of the ORB algorithm is to use FAST to locate a candidate set of features. FAST features are inexpensive to locate, but have several shortcomings. One is that they tend to respond to edges as well as corners. In order to overcome this, the ORB algorithm computes the Harris corner measure for the located FAST points. This measure, you may recall, is a constraint on the eigenvalues of an autocorrelation matrix formed from pixels near the location of the feature.[45] Using this process, we then construct an image pyramid so that a scale space search can be done. Because the Harris corner measure provides not only a test for the quality of a FAST feature,

---

44 The feature used in ORB is also referred to (by its creators as well as others) as an rBRIEF (or *rotation-aware BRIEF*) feature, and is closely related to the BRIEF feature we just encountered. In fact this is the origin of the name ORB, which is derived from "*O*riented FAST and *R*otation Aware *B*RIEF."

45 You will recall from our earlier discussion that Harris originally proposed a slightly different constraint than the one later proposed by Shi and Tomasi. The `cv::GoodFeaturesToTrackDetector()` algorithm (by default) uses the latter, while ORB uses the former.

but also a better metric for feature quality, it can also be used to select for the "best" features in an image. When some particular number of features is desired (which is commonly the case in practical applications), the features are ordered by the Harris corner measure; those that are the best are retained until the desired number is found.

*Figure 16-31. Two images of the same vehicle each produce 500 ORB features. An interesting characteristic of ORB is visible here, namely that if a corner is large in the image, many ORB features of differing sizes will be found on that same corner*

An important contribution of the ORB algorithm relative to FAST (or the Harris corners) is the introduction of an orientation to the located keypoints. The orientation is assigned in a two-step process. First the first moments (in $x$ and $y$) are computed for the distribution of intensities inside of a box around the feature. This box is of side length equal to twice the scale at which the feature was found (an approximation to a disk of radius given by the scale). Figure 16-32 shows the normalized x- and y-gradients (they are divided by their mean) that gives the orientation of the gradient direction relative to the center of the feature. For this reason, this ORB feature is also known as an *oFAST feature*, or *oriented-FAST feature*.

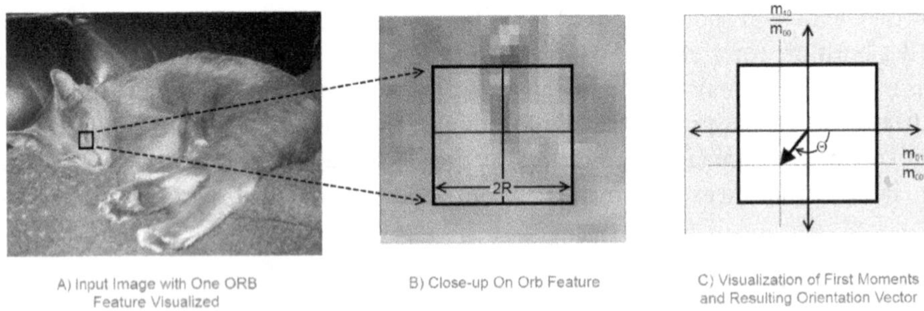


*Figure 16-32. We compute the orientation of the ORB feature by analyzing the first moments (average intensity) of the image in a box whose size is given by the scale at which the FAST feature was found; the orientation of the feature is given by the direction vector from the center of the feature to the point indicated by those moments*

Once the feature has been located and an orientation has been assigned, it is possible to compute a feature vector relative to that orientation. The resulting feature can then be used in a rotationally invariant way.[46] The feature descriptor used by ORB, as mentioned earlier, is based on the descriptor used in the BRIEF algorithm, but the introduction of orientation information is an important differentiator of the ORB feature relative to its predecessor BRIEF.

The second significant difference between the ORB and BRIEF features is that BRIEF's authors actually produced the rotation-aware BRIEF descriptor by analyzing a large data set of images and looking for test pairs that had particular properties: a high variance, an average result close to `0.5`, and a minimal correlation with the other tests pairs.[47] In order to do this analysis, however, they converted each descriptor to a representation that located the test points relative to the orientation of the feature. This analysis was done once in the construction of the ORB descriptor by its authors, and is hereafter "built into" the descriptor. The data set they used was a well-known image data set containing many kinds of images.[48]

**Keypoint finder and feature extractor.** As with SIFT and SURF, the ORB algorithm is implemented in OpenCV via the `cv::Feature2D` interface. Here is the (slightly abbreviated) definition of the `cv::ORB` object that implements the ORB algorithm:

```
class ORB : public Feature2D {
public:
  // the size of the signature in bytes
  enum { kBytes = 32, HARRIS_SCORE = 0, FAST_SCORE = 1 };

  static Ptr<ORB> create(
    int   nfeatures     = 500,    // Maximum features to compute
    float scaleFactor   = 1.2f,   // Pyramid ratio (greater than 1.0)
    int   nlevels       = 8,      // Number of pyramid levels to use
    int   edgeThreshold = 31,     // Size of no-search border
    int   firstLevel    = 0,      // Always '0'
    int   WTA_K         = 2,      // Pts in each comparison: 2, 3, or 4
    int   scoreType     = 0,      // Either HARRIS_SCORE or FAST_SCORE
    int   patchSize     = 31,     // Size of patch for each descriptor
    int   fastThreshold = 20      // Threshold for FAST detector
```

46 Recall from our earlier discussion of BRIEF that the BRIEF descriptor used a random array of "tests." This array, however, had no ability to be "aligned" with the feature (e.g., in the manner of SIFT).

47 The astute reader will recognize that the first two properties are actually the same property for a distribution of binary variables.

48 This data set is known as the PASCAL-2006 data set, and is publicly available on the Internet. It is a well-known benchmark data set widely used in computer vision research and widely cited in computer vision papers. It is an open question, however, whether performance on any specialized type of data set might be affected by the choice to train the ORB feature set on PASCAL-2006 (rather generic) data rather than data from the specialized type.

```
    );

    int descriptorSize() const;       // descriptor size (bytes), always 32
    int descriptorType() const;       // descriptor type, always CV_8U
  };
```

The ORB constructor supports a somewhat intimidating list of arguments. Most of these can safely be left at their default values and you will get satisfactory results. The first argument, `nfeatures`, is probably the one you are most likely to change; it simply determines the number of keypoints you would like `cv::ORB` to find at a time.

Because the ORB detector uses an image pyramid, you must tell the detector both what the scale factor is between each layer in the pyramid and how many levels the pyramid is to have. You do not want to use a coarse factor-of-two type pyramid of the kind that would be created by `cv::buildPyramid()` because too many features would get lost in between. The default value for the scale factor is only `1.2`. The variables that control the scale factor and the number of levels in the pyramid are `scaleFactor` and `nlevels`, respectively.

Because the keypoints are of a specific size in pixels, it is necessary to avoid the boundaries of the image. This distance is set by `edgeThreshold`. The size of the patch used for individual features can also be set with the `patchSize` argument. You will notice that they have the same default value of `31`. If you change `patchSize`, you should make sure that `edgeThreshold` remains equal to or greater than `patchSize`.

The `firstLevel` argument allows you to set the scale pyramid such that the level whose scale is unity is not necessarily the first level. In effect, setting `firstLevel` to a value other than zero means that some number of images in the pyramid will actually be larger than the input image. This is meaningful with ORB features because their descriptors rely intrinsically on a smoothed version of the image anyhow. In most cases, however, making the `firstLevel` very high will produce features at the resulting smaller scales that are mainly driven by noise.

The argument `WTA_K` controls the *tuple size*, which in turn controls precisely how the descriptor is constructed from the binary tests. The case of `WTA_K=2` is the scenario we described before in which each bit of each descriptor byte is a separate comparison between pairs of *test points*. These test points are drawn from a pregenerated list. In the case of `WTA_K=3`, the bits of the descriptor are set two at a time through a three-way comparison between sets of three test points from that list. Similarly, for `WTA_K=4`, the bits of the descriptor are set two at a time through a four-way comparison between sets of four test points.

It is important to understand, however, that the pregenerated list of test points described in the beginning of this section is only really meaningful for a tuple size of 2, and for a feature size of `31`. If you use features of any other size, the test points will be generated randomly (instead of the precomputed list). If you use a tuple size other than `2`, the test points will also be arranged randomly into tuples of the correct length. (So if you use a feature size of `31`, but a tuple size other than `2`, you will be using the precomputed list of test points, but in a random manner, which is arguably no better than just using random test points altogether.)

The last argument to the `cv::ORB` constructor is the score type. The `scoreType` argument can be set to one of two values: `cv::ORB::HARRIS_SCORE` or `cv::ORB::FAST_SCORE`. The former case was described in the beginning of this section, in which a large number of features are found, all have their scores recomputed using the Harris metric, and only the best ones by that metric are kept. Unfortunately, this incurs compute cost in two ways. One is that the Harris metric takes time to compute, and the other is the need to compute more feature metrics in the first place.[49] The alternative is to use the metric natively associated with FAST. The features are not as good, but it improves run speed slightly.

**Additional functions provided by cv::ORB.** `cv::ORB` also includes a bunch of get*/set* methods that can be used to retrieve or modify various algorithm parameters after the class instance is constructed.

#### The FREAK descriptor extractor and cv::xfeatures2d::FREAK

As with the BRIEF descriptor, the FREAK algorithm (see Figure 16-33) computes a descriptor only, and does not have a naturally associated keypoint detector. Originally introduced as an advancement on BRIEF, BRISK, and ORB, the FREAK descriptor is a biologically inspired descriptor that functions much like BRIEF, differing primarily in the manner in which it computes the areas for binary comparison [Alahi12]. The second, more subtle, distinction is that rather than making point comparisons of pixels around a uniformly smoothed image, FREAK uses points for comparison that each correspond to a different sized region of integration, with points farther from the center of the descriptor being assigned larger regions. In this way FREAK captures an essential feature of the human visual system, and thus derives its name *Fast Retinal Keypoint.*

49 As implemented, when you request some specific number of features, exactly twice that many will be retained using the FAST metric, and then for those, the Harris metric will be computed and the best half by that metric will be kept.

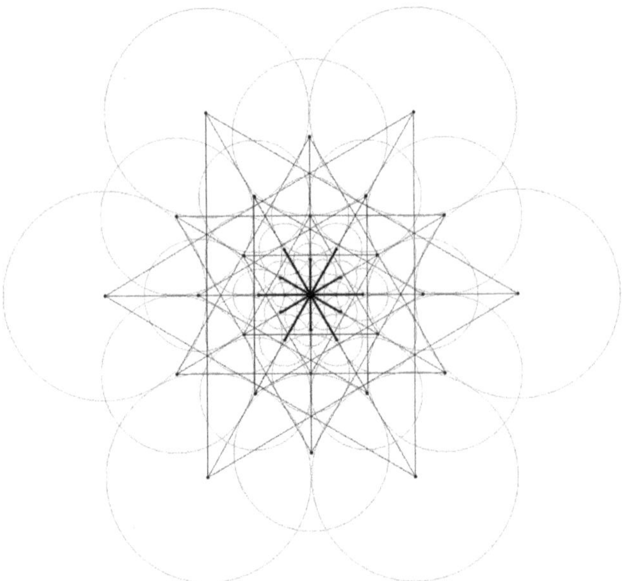

*Figure 16-33. A diagram representing the operation of the FREAK descriptor. Straight lines represent possible comparisons between vertices. The circles represent the "receptive field" associated with each vertex. Notice that the sizes of the receptive fields grow as the vertices get farther from the center of the descriptor;. Figure taken with permission from [Alahi12]*

To understand the FREAK feature, it is useful to first revisit the BRIEF feature, but from a slightly different perspective than when it was introduced earlier. Recall that the BRIEF feature involves bitwise comparisons between a large number of pairs of individual pixel intensities in the immediate proximity of the feature. To improve robustness to noise, the BRIEF algorithm first applies a Gaussian filter to the image before computing these pixel intensities.

Alternatively, we can think of each of the pixels used in the comparison as representing Gaussian weighted sums over the pixels in the input image corresponding to the immediate vicinity of the output pixel. Mathematically these two are exactly equivalent, but intuitively, this new picture naturally introduces the idea of the *receptive field* of a comparison pixel, and calls to mind the relationship between the ganglion cells of the retina and the set of individual photoreceptors to which that ganglion cell responds.

One substantial difference, however, between the receptive fields of the BRIEF feature and those of the human eye is that while the receptive fields in BRIEF are of uniform size (recall the Gaussian convolution), those of the human retina are increasingly large as one moves farther from the center of the retina. Inspired by this biological observation, the FREAK descriptor employs a series of receptive fields that increase in size with distance from the center of the feature (the circles in Figure 16-33). In the standard construction there are 43 such visual fields.

As was the case with the ORB descriptor, the creators of the FREAK descriptor then went on to use a learning technique to organize the possible comparisons between these receptive fields in order of their decreasing utility. In this way the comparisons with the greatest discriminative power (across a large training set of features) could be given priority over those with comparatively less discriminative power. Once this ordering was complete, pairs were retained only if they showed strong decorrelation with pairs of higher individual utility. In application, given a few dozen fields of varying sizes and thousands of possible comparisons, only the most useful 512 were found to be worth keeping.

Of these 512 comparisons, the FREAK descriptor organizes them into four sets of 128. Empirically, it was observed that each group appears to successively employ more of the small-scale receptive fields near the center, but that the most initially discriminative comparisons are between the large field areas. As a result, it is possible to first make comparisons only with these large field values and then, if enough similarity is found, to proceed to the finer areas to refine the match. Each set of 128 comparisons corresponds only to an XOR operation (and bit summation) across a single 16-byte value. This number is significant, as many modern processors can make such a comparison in a single cycle. Because it is possible to reject the overwhelming majority of possible matches with just this single comparison, the FREAK descriptor has been found to be extremely efficient.

**Feature extractor.** The FREAK feature is implemented in OpenCV by the class of the same name: `cv::xfeatures2d::FREAK`, which inherits from the `cv::Features2D` interface and implements the descriptor extraction part.

```
class FREAK : public Features2D {

public:

  static Ptr<FREAK> create(
    bool orientationNormalized = true,  // enable orientation normalization
    bool scaleNormalized       = true,  // enable scale normalization
    float patternScale         = 22.0f, // scaling of the description pattern
    int nOctaves               = 4,     // octaves covered by detected keypoints
    const vector<int>& selectedPairs = vector<int>() // user selected pairs
  );

  virtual int descriptorSize() const;   // returns the descriptor length in bytes
  virtual int descriptorType() const;   // returns the descriptor type

  ...
};
```

The FREAK constructor has a number of arguments that, as is often the case, can be left safely at their default arguments most of the time. The first two represent slight extensions to the published algorithm. As originally proposed, the FREAK features have a fixed orientation and scale. By setting `orientationNormalized` to `true`, you can ask the OpenCV FREAK object to build descriptors that are orientation invariant. This option reduces the discriminatory power of the keypoints, but in exchange you get orientation invariance.[50]

To understand the roles of the argument `scaleNormalized`, you must recall first that (most) keypoints have a size, which is set by the keypoint detector that located them in the first place. If `scaleNormalized` is `true`, then the image patch around the feature will first be rescaled by this keypoint size before the feature vector is computed.

The `patternScale` argument is used to uniformly rescale the FREAK receptor field pattern. It is unlikely you will want to change this. Closely related to the `patternScale` argument is the `nOctaves` argument. When the FREAK descriptor object is created, a lookup table is generated containing all of the necessary information to compute the FREAK descriptor at a range of scales. The combination of the `patternScale` and `nOctaves` arguments allows you to control the span of this lookup table (in terms of the size of the patterns). The exact scales generated are given by the following formula:

$$scale_i = patternScale * 2^{i*\left(\frac{nOctaves}{nbScales}\right)}, \quad for\ i \in \{0,\ 1,\ \ldots,\ nbScales\}$$

Here `nbScales` is the total number of scales, equal to 64 in the current implementation. Note that in every case the number of scales generated is the same, but the spacing between the scales is increased or reduced when `nOctaves` is increased or reduced.[51]

The final argument to the constructor is `selectedPairs`. This argument is for real experts, and allows you to override the lists of pairs used for comparison in the descriptor's construction. If present, `selectedPairs` must be a vector of exactly 512

50 Recall that a similar trade-off was possible between the intrinsically oriented SURF features and the "upright" (U-SURF) features.

51 If you are wondering why the terminology for this is "octave," remember that the `cv::KeyPoint` object has an element `octave`, which indicates the scale at which that keypoint was found. The `nOctaves` argument to `cv::FREAK::FREAK()` corresponds to this same octave value. `nOctaves` should be set to (at least) the maximum octave of the keypoint detector used, or to the max octave in the set of keypoints for which we want to compute a description.

integers. These integers index into an internal table of all possible pairings of fields.[52] In this way, you can give your own pairs as integers. It is unlikely that most users would ever use this feature; it is exposed only for serious power users—those who, after reading the original paper on FREAK, would conclude that it was necessary to repeat the learning process used by the creators of FREAK with some hope of maximizing the efficacy of the descriptor on their own unique data set.

### Dense feature grids and the cv::DenseFeatureDetector class

There is one other feature detector, which is really a sort of "almost" feature detector in that it does not really detect features, it just generates them. The purpose of the `cv::DenseFeatureDetector` class[53] is just to generate a regular array of features in a grid across your image (see Figure 16-34). In this case, nothing is done yet with these features (i.e., no descriptors are computed). Once these keypoints are generated, however, you can then compute any descriptor for them that you like. It turns out that, in many applications, it is not only sufficient, but especially desirable, to compute a descriptor everywhere, in the sense that "everywhere" is represented by a uniform grid of some density you choose.

*Figure 16-34. Here dense features are generated on the two views of an automobile; in this case, there are three levels, with a starting spatial step of 50, a scale step of 2.0, and feature step being scaled as well as feature size.*[54]

52 The pairs are indexed from 0 to 902, starting at (1,0), (2,0), (2,1), and so on up to (42,41). There is no natural explanation for this particular ordering, other than that the authors of this module opted to enclose a loop running from index `i = 1` to `i < 43` around a loop running from `j = 0` to `j < i`, and to construct all of the pairs as `pair[k] = (i,j)`. What is most important, however, is that you will be unlikely to interact with these indices directly. If you are using these indices, chances are that OpenCV constructed them for you with the `cv::FREAK::selectPairs()` function.

53 `DenseFeatureDetector` is not (yet?) available in OpenCV 3.0.

54 This density of features is unrealistically low, but was chosen for clarity of illustration.

It is also often useful to be able to compute not only a uniform grid of features in image space, but also to sample that space at a uniform number of scales. `cv::DenseFeatureDetector` will also do this for you if you so desire.

**Keypoint finder.** The keypoint finder for the dense feature class is just derived from the `cv::FeatureDetector` interface, as it has no descriptor extraction functionality. Thus, the main thing we need to understand about it is how to use its constructor in order to configure it:

```
class cv::DenseFeatureDetector : public cv::FeatureDetector {

public:
  explicit DenseFeatureDetector(
    float initFeatureScale      = 1.f,  // Size of first layer
    int   featureScaleLevels    = 1,    // Number of layers
    float featureScaleMul       = 0.1f, // Scale factor for layers
    int   initXyStep            = 6,    // Spacing between features
    int   initImgBound          = 0,    // No-generate boundary
    bool  varyXyStepWithScale   = true, // if true, scale 'initXyStep'
    bool  varyImgBoundWithScale = false // if true, scale 'initImgBound'
  );

  cv::AlgorithmInfo* info() const;

  ...
};
```

The constructor for the dense feature detector has a lot of arguments because there are a lot of different ways you might want to generate feature grids. The first argument, `initFeatureScale`, sets the size of the first layer of features. The default value of `1.0` is almost certainly not what you want, but remember that the right size for features depends not only on the nature of your image, but also on the descriptor type you are going to use.

By default, the dense feature detector will generate a single grid of keypoints. You can generate a pyramid of such keypoints, however, by setting the `featureScaleLevels` to any value larger than `1`. Each grid generated after the first will assign the scale of the generated features to be the `initFeatureScale` multiplied by one factor of `featureScaleMul` for each level above the first that scale was generated on.[55] Figure 16-35 shows a representation of the use of this detector.

55 The default value for the scale multiplier is `0.1`. It can be a number larger than `1` as well. It is essentially a matter of personal style as to whether you like to think of the grids as getting "finer" or "coarser" with each step.

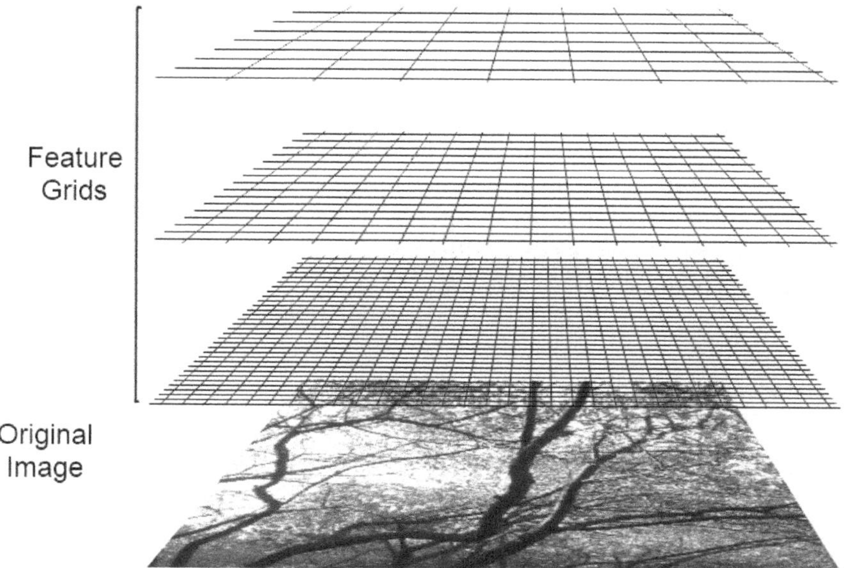


*Figure 16-35. cv::DenseFeatureDetector generates one or more grids of keypoints of varying scale*

The spacing between the features is set by `initXyStep`. If `featureScaleLevels` is larger than one, then the locations of the features will, by default, also be scaled by one power of `featureScaleMul` for each level after the first. If `varyXyStepWithScale` is set to `false`, then only the scale value is assigned to the features will change (and not their locations).

Features will not be generated on the boundary of the image. You can increase the width of the area in which features will not be generated by setting `initImgBound`. In most cases, it makes sense to set `initImgBound` to be equal to `initFeatureScale` (i.e., such that generated features do not spill over the edges of the image). In order to have this boundary scale up in the case of multiple levels, however, you must set `varyImgBoundWithScale` to `true`; otherwise, it will keep a constant size.

## Keypoint Filtering

It is a relatively common situation that you find yourself with a list of keypoints that needs to be pruned in some way. Sometimes this happens because the list is just too long, and you want to throw out some of the lower-quality ones until you get a manageable number. Other times, you need to remove duplicates or all of the keypoints that are outside of some region. OpenCV handles all of these sorts of tasks with an object called a *keypoint filter*. The keypoint filter provides a variety of methods you can use to cull lists of keypoints. These functions are heavily used internally by the implementations of various keypoint finders, but you will probably find them very useful as standalone tools in many of your applications as well.

### The cv::KeyPointsFilter class

All keypoint filtering is done by methods of a single `cv::KeyPointsFilter` class. The `cv::KeypointFilter` class is used internally by many of the keypoint detectors we have already encountered in order to filter by location (apply a mask) or reduce the number of keypoints to some number that are the "best" from a list. The salient part of the definition of this class is as follows:

```
class cv::KeyPointsFilter {
public:
  static void runByImageBorder(
    vector< cv::KeyPoint >& keypoints,  // in/out list of keypoints
    cv::Size                imageSize,  // Size of original image
    int                     borderSize  // Size of border in pixels
  );
  static void runByKeypointSize(
    vector< cv::KeyPoint >& keypoints,  // in/out list of keypoints
    float                   minSize,    // Smallest keypoint to keep
    float                   maxSize  = FLT_MAX // Largest one to keep
  );
  static void runByPixelsMask(
    vector< cv::KeyPoint >& keypoints,  // in/out list of keypoints
    const cv::Mat&          mask        // Keep where mask is nonzero
  );
  static void removeDuplicated(
    vector< cv::KeyPoint >& keypoints   // in/out list of keypoints
  );
  static void retainBest(
    vector< cv::KeyPoint >& keypoints,  // in/out list of keypoints
    int                     npoints     // Keep this many
  );
}
```

The first thing you probably noticed was that all of these methods were `static`, so really `cv::KeyPointsFilter` is more of a namespace than an object. Each of the five filter methods takes a reference to an STL-style vector of `cv::KeyPoint` objects called `keypoints`. This is both the input and the output of the function (i.e., you will give the function your keypoints, and when you check, you will find some of them are now missing).

The `cv::KeyPointsFilter::runByImageBorder()` function removes all of the `key points` that are within `borderSize` of the edge of the image. You must also tell `cv::KeyPointsFilter::runByImageBorder()` how big the image was in the first place with the `imageSize` argument.

The `cv::KeyPointsFilter::runByKeypointSize()` function removes all of the `key points` that are either smaller than `minSize` or larger than `maxSize`.

The `cv::KeyPointsFilter::runByPixelsMask()` function removes all of the `key points` that are associated with zero-valued pixels in `mask`.

The `cv::KeyPointsFilter::removeDuplicated()` function removes all duplicate keypoints.

The `cv::KeyPointsFilter::retainBest()` function removes `keypoints` until the target number given by `npoints` is reached. Keypoints are removed in ascending order of their quality, as indicated by `response`.

## Matching Methods

Once you have your keypoints, you will want to use them to do something useful. As we discussed early on, the two most common applications for keypoint methods are object recognition and tracking. In both cases, we saw that it was objects derived from the `cv::DescriptorMatcher` base class that would provide this functionality for us. In this section, we will look at the options available to us for doing this kind of matching.

At this time, there are essentially two different matching methods you can use. The first is *brute force matching*, which is the most basic and obvious choice—just compare everything in set A to everything in set B. The second is called *FLANN*, and is really an interface to a collection of methods for locating nearest neighbors.

### Brute force matching with cv::BFMatcher

The (important part of the) declaration of the `cv::BFMatcher` class is shown here:

```
class cv::BFMatcher : public cv::DescriptorMatcher {

public:

  BFMatcher( int normType, bool crossCheck=false );

  virtual ~BFMatcher() {}

  virtual bool isMaskSupported() const { return true; }
  virtual Ptr<DescriptorMatcher> clone(
    bool emptyTrainData=false
  ) const;
  ...

};
```

The brute force matcher essentially implements the most straightforward solution to the matching problem. It takes each descriptor in the query set and attempts to match it with each descriptor in the training set (either the internal dictionary, or a set provided with the query set). The only thing you need to decide when you create a brute force matcher is what distance metric it will use in order to compute distances for comparison. The available options are shown in Table 16-2.

*Table 16-2. Available metrics for the brute force matcher, with their associated formulas; the summations are over the dimensions of the feature vector*

| Metric | Function |
|---|---|
| NORM_L2 | $dist(\vec{a}, \vec{b}) = \left[\sum_i (a_i - b_i)^2\right]^{1/2}$ |
| NORM_L2SQR | $dist(\vec{a}, \vec{b}) = \sum_i (a_i - b_i)^2$ |
| NORM_L1 | $dist(\vec{a}, \vec{b}) = \sum_i abs(a_i - b_i)$ |
| NORM_HAMMING | $dist(\vec{a}, \vec{b}) = \sum_i (a_i = = b_i)?1:0$ |
| NORM_HAMMING2 | $dist(\vec{a}, \vec{b}) = \sum_{i(even)} [(a_i = = b_i)\&\&(a_{i+1} = = b_{i+1}]?1:0$ |

It is worth noting that the member function `isMaskSupported()` always returns `true`. We will see that this is not the case for the FLANN matcher in the next section. Only the brute force matcher (at this time) supports the mask construct described when we discussed the `cv::DescriptorMatcher` base class.[56]

The final feature of the brute force matching object is what is called *cross-checking*. Cross-checking is turned on with the argument `crosscheck` to the `cv::BFMatcher` constructor. When cross-checking is enabled, a match between object `i` of the query set and object `j` of the training set will be reported only if both `train[j]` is `query[i]`'s closest neighbor in the training set and `query[i]` is `train[j]`'s closest neighbor in the query set. This is very useful for eliminating false matches, but costs additional time to calculate.

56 Recall that these masks were the ones that did not mask the image, but rather indicated which features should and should not be compared during matching.

### Fast approximate nearest neighbors and cv::FlannBasedMatcher

The term *FLANN* refers to the *Fast Library for Approximate Nearest Neighbor* computation. OpenCV provides an interface to FLANN, which itself provides a variety of algorithms for finding (or at least approximately finding) the nearest neighbors of points in high-dimensional spaces. Conveniently, this is precisely what we need for descriptor matching. The common interface to FLANN is through the `cv::FlannBasedMatcher` object, which of course is derived from the `cv::DescriptorMatcher` base class:

```
class cv::FlannBasedMatcher : public cv::DescriptorMatcher {

public:

  FlannBasedMatcher(
    const cv::Ptr< cv::flann::IndexParams>&  indexParams
      = new cv::flann::KDTreeIndexParams(),
    const cv::Ptr< cv::flann::SearchParams>& searchParams
      = new cv::flann::SearchParams()
  );

  virtual void add( const vector<Mat>& descriptors );
  virtual void clear();
  virtual void train();
  virtual bool isMaskSupported() const;

  virtual void read( const FileNode& );      // Read from file node
  virtual void write( FileStorage& ) const; // Write to file storage

  virtual cv::Ptr<DescriptorMatcher> clone(
    bool emptyTrainData = false
  ) const;
  ...
};
```

Pretty much everything in the preceding declaration is what you should have expected by now. The one important feature is the constructor for `cv::FlannBasedMatcher`, which takes some special structures that are used to actually configure the matching. These arguments determine both what method the FLANN matcher will use, as well as how to parameterize the selected method. For example, the default value for the `indexParams` argument is `new cv::flann::KDTreeIndexParams()`. This tells the FLANN matcher that the index method it should use is *kd-trees* as well as how many kd-trees to use (the number of trees is an argument to the `cv::flann::KDTreeIndexParams()` constructor, whose default happens to be 4). The `searchParams` argument is somewhat more generic, and plays a role somewhat analogous to `cv::TermCriteria`, but with a little more general function. We will look at each of the indexing methods, and then return to the `cv::flann::SearchParams` object.

**Linear indexing with cv::flann::LinearIndexParams.** You can ask the FLANN library to do essentially the same thing that `cv::BFMatcher` would have done. This is usually useful for benchmark comparisons, but also can be used as a comparison to verify that other, more approximating methods are giving satisfactory results. The constructor for `cv::LinearIndexParams` takes no arguments. By way of example, the following code generates a matcher that is essentially equivalent to `cv::BFMatcher`:

```
cv::FlannBasedMatcher matcher(
  new cv::flann::LinearIndexParams(), // Default index parameters
  new cv::flann::SearchParams()       // Default search parameters
);
```

**KD-tree indexing with cv::flann::KDTreeIndexParams.** The `cv::flann::KDTreeIndexParams` index parameters tell the FLANN matcher that you would like to use randomized kd-trees for matching. FLANN's normal behavior is to assume that you want many such randomized trees to be generated as the indexing method, and then it will search them all when you attempt to match.[57] The constructor for the `cv::flann::KDTreeIndexParams` accepts one argument, which specifies the number of such trees to construct. The default value is 4, but it is common to make this number as large as 16 or so. The declaration for `cv::flann::KDTreeIndexParams` is the following:

```
struct cv::flann::KDTreeIndexParams : public cv::flann::IndexParams {
  KDTreeIndexParams( int trees = 4 ); // kd-tree needs number of trees
};
```

An example declaration of a matcher using `cv::flann::KDTreeIndexParams` is shown here:

```
cv::FlannBasedMatcher matcher(
  new cv::flann::KDTreeIndexParams( 16 ), // Index using 16 kd-trees
  new cv::flann::SearchParams()           // Default search parameters
);
```

**Hierarchical k-means tree indexing with cv::flann::KMeansIndexParams.** Another option for constructing the index is to use hierarchical k-means clustering.[58] The advantage of k-means clustering is that it makes some intelligent use of the density of points in the database. Hierarchical k-means clustering is a recursive scheme by which the data points are first grouped into some number of clusters, and then each cluster is grou-

57 This is kind of a subtle point, but what happens is that there is one master list of nearest neighbors, and as each tree is descended, the nearest neighbor candidates are compared not only with what has been found so far on *that* tree, but with what has been found on *all* trees thus far.

58 We will get to the k-means algorithm when we get to the machine learning library in Chapter 20. For now, it is enough to know that it attempts to organize some large number of points into *k* distinct clusters.

ped into some number of subclusters, and so on. Obviously this is going to be more helpful when you have reason to believe that such structure exists in the data in the first place. The declaration for `cv::flann::KMeansIndexParams` is the following:

```
struct cv::flann::KMeansIndexParams : public cv::flann::IndexParams {

  KMeansIndexParams(
    int             branching    = 32,  // Branching factor for tree
    int             iterations   = 11,  // Max for k-means stage
    float           cb_index     = 0.2, // Probably don't mess with
    cv::flann::flann_centers_init_t centers_init
                                 = cv::flann::CENTERS_RANDOM
  );

};
```

The default parameters for the `cv::flann::KMeansIndexParams` structure are all perfectly reasonable values, so in many cases, you will just leave these. The first argument, `branching`, is the branching factor that is used in the hierarchical k-means tree. It determines how many clusters will be formed at each level of the tree. The next argument, `iterations`, determines how many iterations will be allowed to the k-means algorithm for the formation of each cluster.[59] It can be set to `-1` to force the clustering algorithm to run to completion at every node in the tree. The third argument controls how the cluster centers are initialized. Traditionally, random initialization was common (`cv::flann::CENTERS_RANDOM`), but in recent years it has been shown that in most cases, a prudent choice of starting centers can give substantially better results. The two additional options `cv::flann::CENTERS_GONZALES` (Gonzales's algorithm [Tou77]) and `cv::flann::CENTERS_KMEANSPP` (the so called "k-means++" algorithm [Arthur07]) are available, with the latter being an increasingly standard choice. The final argument, `cb_index` (cluster boundary index), is really there for true experts in the FLANN library; it is used when the tree is being searched, and controls how the tree will be explored. It is best left at its default value, or set to `0` (which indicates that searches that exhaust one domain should move directly to the closest unexplored domain[60]).

**Combining KD-trees and k-means with cv::flann::CompositeIndexParams.** This method simply combines the random kd-tree and k-means methods described earlier and attempts to find the best matches as found by either method. Because these are all

59 Recall that k-means is an NP-hard problem, so most algorithms used to compute k-means clustering are approximate, and use an iterative scheme to find a locally optimal solution given some particular starting set of cluster center candidates.

60 If this parameter is nonzero, it indicates that the algorithm should take into account the overall size of the domain as part of its consideration of where to go next.

approximate methods, there is always potential benefit to searching another way. (You could think of this as an extension of the logic behind having multiple random trees in the kd-tree method.) The constructor for the `cv::flann::CompositeIndexParams` object combines the arguments for the kd-tree and k-means methods:

```
struct cv::flann::CompositeIndexParams : public cv::flann::IndexParams {

  CompositeIndexParams(
    int              trees          = 4,    // Number of trees
    int              branching      = 32,   // Branching factor for tree
    int              iterations     = 11,   // Max for k-means stage
    float            cb_index       = 0.2,  // Usually leave as-is
    cv::flann::flann_centers_init_t centers_init
                                    = cv::flann::CENTERS_RANDOM

  );

};
```

**Locality-sensitive hash (LSH) indexing with cv::flann::LshIndexParams.** Another, very different method of indexing the space of known objects is to attempt to use hash functions to map similar objects into the same buckets. To the extent that this can be done, these hash functions can be used to generate a list of candidate objects very quickly, which can then be evaluated and compared to one another. This technique is called *locality-sensitive hashing* (LSH). The variation of LSH implemented in OpenCV as part of the FLANN library was first proposed by Lv et al. [Lv07]:

```
struct cv::flann::LshIndexParams : public cv::flann::IndexParams {

  LshIndexParams(
    unsigned int table_number,       // Number of hash tables to use
    unsigned int key_size,           // key bits, usually '10' to '20'
    unsigned int multi_probe_level   // Best to just set this to '2'
  );

};
```

The first argument, `table_number`, is the number of actual hash tables to use. This number is typically tens of tables, with `10` to `30` being reasonable numbers. The second argument, `key_size`, is the size of the hash key (in bits). This number is typically also more than `10` and usually less than `20`. The last argument, `multi_probe_level`, controls how neighboring buckets are searched; it is part of what distinguishes the multiprobe algorithm (cited earlier) from previous LSH implementations. The recommended value for `multi_probe_level` is 2. If it is set to `0`, the algorithm will degenerate into non-multiprobe LSH.

LSH indexing in FLANN works only for binary features (using Hamming distances); it should not be applied to other distance metrics.

**Automatic index selection with cv::flann::AutotunedIndexParams.** You can also ask OpenCV/FLANN to attempt to identify for you what the best indexing scheme is. Needless to say, this will take a while. The basic idea behind this approach is that you set a *target precision*, which is the percentage of nearest neighbor searches you would like to return the correct exact solution. Of course, the higher you make this, the more difficult it will be for the algorithm to find an indexing scheme that can deliver, and the longer it will take to actually generate a full index for all of your data of that type:

```
struct cv::flann::AutotunedIndexParams : public cv::flann::IndexParams {

  AutotunedIndexParams(
    float target_precision = 0.9,     // Percentage of searches required
                                      //  to return an exact result
    float build_weight     = 0.01,    // Priority for building fast
    float memory_weight    = 0.0,     // Priority for saving memory
    float sample_fraction  = 0.1      // Fraction of training data to use
  );

};
```

The target precision is set by the argument `targetPrecision`. When you create a FLANN-based matcher using `cv::flann::AutotunedIndexParams`, you will also have to indicate how important it is to you that the index builds quickly; this is controlled by the `build_weight` argument. If you do not care much how long it takes to build the index, so long as your returns are quick, you can set this to a very small value (such as the default of `0.01`). If you need to build your index often, you will want to set this number higher. Similarly, the `memory_weight` controls the priority you want to put on minimizing the amount of memory consumed by the indexes. The default value of this parameter is `0`, which means you don't care.

Finally, there is the question of how much of the training data to actually use in this search. This is controlled by the `sample_fraction` argument. Clearly, if you make this fraction too big, it will take an enormous amount of time to find a satisfactory solution. On the other hand, if you make it too small, the observed performance on your full data set may be much worse than you saw when you generated the index. For large data sets, the default value of `0.1` is generally found to be a good choice.

**FLANN search parameters and cv::flann::SearchParams.** In addition to the previous arguments used for the `indexParams` argument, the `cv::FlannBasedMatcher` constructor requires an object of type `cv::flann::SearchParams`. This is a straightforward structure that controls some of the matcher's general behavior. It has the following simple definition:

```
struct cv::flann::SearchParams : public cv::flann::IndexParams {

  SearchParams(
    int   checks = 32,    // Limit on NN candidates to check
    float eps    = 0,     // (Not used right now)
    bool  sorted = true   // Sort multiple returns if 'true'
  );

};
```

The parameter `checks` is used by the kd-tree and k-means algorithms differently, but in each case it essentially limits the number of nearest neighbor candidates that are evaluated in the attempt to find the truly nearest neighbor(s). The `eps` parameter is currently not used.[61] The `sorted` parameter indicates that, in the case of searches that can return multiple hits (e.g., radius search), the hits returned should be in ascending order of distance from the query point.[62]

## Displaying Results

Now that you can compute all of these feature types and compute matches between them from one image to another, the next logical thing to want to do is to actually display the keypoints and matches on the screen. OpenCV provides one function for each of these tasks.

### Displaying keypoints with cv::drawKeypoints

```
void cv::drawKeypoints(
  const cv::Mat&                   image,       // Image to draw keypoints
  const vector< cv::KeyPoint >&    keypoints,   // List of keypoints to draw
  cv::Mat&                         outImg,      // image and keypoints drawn
  const Scalar&                    color     = cv::Scalar::all(-1),
  int                              flags     = cv::DrawMatchesFlags::DEFAULT
);
```

Given an image and a set of keypoints, `cv::drawKeypoints` will annotate all of the keypoints onto the image and put the result in `outImg`. The color of the annotations

61 If you are an expert in the FLANN library, you will know that the eps parameter is used by the variant of kd-tree called `KDTreeSingleIndex` (which is not currently exposed in the OpenCV interface). In that context, it determines when a search down a particular branch can be terminated because a found point is considered close enough that a yet closer point is unlikely to be found.

62 The `sorted` argument has no effect on kNN search, because in that case the returned hits will already always be in ascending order.

can be set with color, which can be set to the special value cv::Scalar::all(-1), indicating that they should be all different colors. The flags argument can be set to cv::DrawMatchesFlags::DEFAULT or to cv::DrawMatchesFlags:: DRAW_RICH_KEY POINTS. In the former case, keypoints will be visualized as small circles. In the latter, they will be visualized as circles with radius equal to their size member (if available), and an orientation line given by their angle member (if available).[63]

### Displaying keypoint matches with cv::drawMatches

Given a pair of images, the associated keypoints, and a list of cv::DMatch objects generated by one of the matchers, cv::drawMatches() will compose an image for you containing the two input images, visualizations of all of the keypoints (in the style of cv::drawKeypoints()), and indicate for you which keypoints in the first image matched which keypoints in the second image (Figure 16-36). There are two variations of cv::drawMatches(); they are the same except for two arguments.

*Figure 16-36. Here SIFT keypoints and their descriptors are extracted for two views of the same automobile. Matches were generated using a FLANN-based matcher and visualized with cv::drawMatches(). Matches that were found are marked in white with a line connecting the corresponding features. Keypoints in either image that were not found to have a match are indicated in black*

```
void cv::drawMatches(
  const cv::Mat&                   img1,         // "Left" image
  const vector< cv::KeyPoint >& keypoints1,   // Keypoints (lt. img)
  const cv::Mat&                   img2,         // "Right" image
  const vector< cv::KeyPoint >& keypoints2,   // Keypoints (rt. img)
  const vector< cv::DMatch >&   matches1to2,  // List of matches
```

63 Many of the figures appearing earlier in this chapter were made with cv::drawKeypoints. In these figures, DrawMatchesFlags::DRAW_RICH_KEYPOINTS was always used, but where scale or angle information was not present for the feature type, it was not displayed by cv::drawKeypoints. Figures 16-11, 16-13, and 16-17 are examples where only location data was available; only location and scale were available; and location, scale, and orientation date were available, respectively.

```
    cv::Mat&                      outImg,       // Result images
    const cv::Scalar&             matchColor        = cv::Scalar::all(-1),
    const cv::Scalar&             singlePointColor = cv::Scalar::all(-1),
    const vector<char>&           matchesMask      = vector<char>(),
    int                           flags
                                       = cv::DrawMatchesFlags::DEFAULT
  )

  void cv::drawMatches(
    const cv::Mat&                img1,         // "Left" image
    const vector< cv::KeyPoint >& keypoints1,   // Keypoints (lt. img)
    const cv::Mat&                img2,         // "Right" image
    const vector< cv::KeyPoint >& keypoints2,   // Keypoints (rt. img)
    const vector< vector<cv::DMatch> >& matches1to2, // List of lists
                                                      // of matches
    cv::Mat&                      outImg,       // Result images
    const cv::Scalar&             matchColor    // and connecting line
                                        = cv::Scalar::all(-1),
    const cv::Scalar&             singlePointColor   // unmatched ones
                                        = cv::Scalar::all(-1),
    const vector< vector<char> >& matchesMask   // only draw for nonzero
                                        = vector< vector<char> >(),
    int                           flags  = cv::DrawMatchesFlags::DEFAULT
  );
```

In both cases the two images are supplied by the `img1` and `img2` arguments, while the corresponding keypoints are supplied by the `keypoints1` and `keypoints2` arguments. An argument that differs in the two cases is `matches1to2`. It has the same meaning in both versions of `cv::drawMatches()`, but in one case it is an STL vector of `cv::DMatch` objects, while in the other it is a vector of such vectors. The second form is just a convenience, which is useful when you want to visualize the response to many different match computations at once.

The results are placed in the image `outImg`. When the output is drawn, those features that have matches will be drawn in the color `matchColor` (along with a line connecting them), while those features that are not matched will be drawn in `singlePointColor`. The vector `matchesMask` indicates which matches should be visualized; only those matches for which `matchesMask[i]` is nonzero will be drawn. The variation of `cv::drawMatches()` that expects a vector of vectors for the matches also expects a vector of vectors for the `matchesMask` argument.

The final argument to `cv::drawMatches()` is `flags`. The `flags` argument can have any of four values, which can be combined (where relevant) with the OR operator.

If flags is set to cv::DrawMatchesFlags::DEFAULT, then the output image will be created for you in outImg, and the keypoints will be visualized as small circles (without additional size or orientation information).[64]

If flags contains cv::DrawMatchesFlags::DRAW_OVER_OUTIMG, then the output image will not be reallocated, but the annotations will be drawn onto it. This is useful when you have several sets of matches you would like to visualize in different colors; in this case, you can make multiple calls to cv::drawMatches() and use cv::DrawMatchesFlags::DRAW_OVER_OUTIMG for all of the calls after the first.

By default, keypoints that are not part of any match will be drawn in the color indicated by singlePointColor. If you would prefer to not have them drawn at all, you can set the flag cv::DrawMatchesFlags::NOT_DRAW_SINGLE_POINTS.

Finally, the flag cv::DrawMatchesFlags::DRAW_RICH_KEYPOINTS has the same function as in cv::drawKeypoints; it causes the keypoints to be visualized with scale and orientation information (as in Figure 16-36).

# Summary

We started this chapter by reviewing the basic concepts of subpixel corner location and of sparse optical flow, and then we discussed the essential role played by keypoints. From there we saw how OpenCV handles keypoints as a general concept, and looked at a wide array of keypoint-detection schemes implemented by the library. We also saw that the process of identifying keypoints was distinct from the process of characterizing them. This characterization was accomplished by descriptors. As with keypoint identification methods, there were many descriptor types, and we saw how OpenCV handles these descriptors as a general class.

Thereafter, we considered how keypoints and their descriptors could be matched in an efficient manner for object recognition or object tracking. We concluded this chapter by looking at a useful function in the library that allows us to easily visualize the keypoints in the context of the image in which they were found. Note that there are more feature detectors and descriptors in xfeatures2d, mentioned in Appendix B.

64 Actually, the value cv::DrawMatchesFlags::DEFAULT is numerically equal to zero, so combining it with the others, while legal, will be meaningless.

## Exercises

There are sample code routines included with OpenCV in its *.../samples/cpp directory* that demonstrate many of the algorithms discussed in this chapter. Use these examples in the following exercises:

- *matchmethod_orb_akaze_brisk.cpp* (feature matching in *samples/cpp*)
- *videostab.cpp* (feature tracking to stabilize video in *samples/cpp*)
- *video_homography.cpp* (planar tracking in *opencv_contrib/modules/xfeatures2d/samples*)
- *lkdemo.cpp* (optical flow in *samples/cpp*)

1. The covariance Hessian matrix used in `cv::goodFeaturesToTrack()` is computed over some square region in the image set by `block_size` in that function.
   a. Conceptually, what happens when block size increases? Do we get more or fewer "good features"? Why?
   b. Dig into the *lkdemo.cpp* code, search for `cv::goodFeaturesToTrack()`, and try playing with the `block_size` to see the difference.
2. Refer to Figure 16-2 and consider the function that implements subpixel corner finding, `cv::findCornerSubPix()`.
   a. What would happen if, in Figure 16-2, the corner was twisted so that the straight dark-light lines formed curves that met in a point? Would subpixel corner finding still work? Explain.
   b. If you expand the window size around the twisted checkerboard's corner point (after expanding the `win` and `zero_zone` parameters), does subpixel corner finding become more accurate or does it rather begin to diverge? Explain your answer.
3. Modify *matchmethod_orb_akaze_brisk.cpp* to train on a planar object (a magazine or book cover, for instance) and track it with a video camera. Study and report how well it finds the correct keypoints using `AKAZE`, `ORB`, and `BRISK` features in the modified code.
4. Run *video_homography.cpp* on the same planar pattern as in Exercise 3. Learn the pattern using the key control in the program and then track it, noting how the homography is used to generate a stable output. How many features must be found in order to compute the homography matrix $H$?
5. Using what you learned in Exercise 3, test out *videostab.cpp* and describe how it works to stabilize video.

6. Features and their descriptors can also be used for recognition. Take one top-down photo of three different book covers against a blank background, then take 10 different off-center photos of each book cover against varied backgrounds, and finally take 10 photos with no books in varied backgrounds. Modify *matchmethod_orb_akaze_brisk.cpp* to separately store the descriptors for each book. Then modify *matchmethod_orb_akaze_brisk.cpp* again to detect (as best you can) the correct book in the off-center photos and to indicate no book in the photos with no books. Report the results.

7. Optical flow:

   a. Describe an object that would be better tracked by block matching than by Lucas-Kanade optical flow.

   b. Describe an object that would be better tracked by Lucas-Kanade optical flow than by block matching.

8. Compile *lkdemo.cpp*. Set up a web camera to view a well-textured object (or use a previously captured sequence of a textured moving object). In running the program, note that `r` autoinitializes tracking, `c` clears tracking, and a mouse click will enter a new point or turn off an old point. Run *lkdemo.cpp* and initialize the point tracking by typing **`r`**. Observe the effects.

   a. Now go into the code and remove the subpixel point placement function `cv::findCornerSubPix()`. Does this hurt the results? In what way?

   b. Go into the code again and, in place of `cv::goodFeaturesToTrack()`, just put down a grid of points in an ROI around the object. Describe what happens to the points and why.

   Hint: part of what happens is a consequence of the aperture problem—given a fixed window size and a line, we can't tell how the line is moving.

9. Modify the *lkdemo.cpp* program to create a program that performs simple image stabilization for moderately moving cameras. Display the stabilized results in the center of a much larger window than the one output by your camera (so that the frame may wander while the first points remain stable).

# CHAPTER 17
# Tracking

## Concepts in Tracking

When we are dealing with a video source, as opposed to individual still images, we often have a particular object or objects that we would like to follow through the visual field. In previous chapters, we saw various ways we might use to isolate a particular shape, such as a person or an automobile, on a frame-by-frame basis. We also saw how such objects could be represented as collections of keypoints, and how those keypoints could be related between different images or different frames in a video stream.

In practice, the general problem of tracking in computer vision appears in one of two forms. Either we are tracking objects that we have already identified, or we are tracking unknown objects and, in many cases, identifying them based on their motion. Though it is often possible to identify an object in a frame using techniques from previous chapters, such as moments or color histograms, on many occasions we will need to analyze the motion itself in order to infer the existence or the nature of the objects in which we are interested.

In the previous chapter, we studied keypoints and descriptors, which form the basis of *sparse* optical flow. In this chapter we will introduce several techniques that can be applied to *dense* optical flow. An optical flow result is said to be dense if it applies to every pixel in a given region.

In addition to tracking, there is the problem of *modeling*. Modeling helps us address the fact that tracking techniques, even at their best, give us noisy measurements of an object's actual position from frame to frame. Many powerful mathematical techniques have been developed for estimating the trajectory of an object measured in such a noisy manner. These methods are applicable to two- or three-dimensional models of objects and their locations. In the latter part of this chapter, we will explore tools that OpenCV provides to help you with this problem and discuss some of the theory behind them.

We will begin this chapter with a detailed discussion of dense optical flow, including several different algorithms available in the OpenCV library. Each algorithm has a slightly different definition of dense optical flow and, as a result, gives slightly different results and works best in slightly different circumstances. From there we will move on to tracking. The first tracking solutions we will look at are used to track regions in what is essentially a dense manner for those regions. These methods include the mean-shift and camshift tracking algorithms as well as motion templates. We will conclude the chapter with a discussion of the Kalman filter, a method for building a model of a tracked object's motion that will help us integrate our many, typically noisy, observations with any prior knowledge we might have about the object's behavior to generate a best estimate of what the object is actually doing in the real world. Note that further optical flow and tracking algorithms are referenced in `optflow` and `tracking` functions in the separate *opencv_contrib* directory described in Appendix B.

# Dense Optical Flow

So far, we have looked at techniques that would allow us to locate individual features from one image in another image. When applied to the problem of optical flow, this necessarily gives us a sparse representation of the overall motion of objects in the scene. In this context, if a car is moving in a video, we will learn where certain parts of the car went, and perhaps make reasonable conclusions about the bulk motion of the car, but we do not necessarily get a sense of the overall activity in the scene. In particular, it is not always so easy to determine which car features go with which car in a complex scene containing many cars. The conceptual alternative to sparse optical flow is a dense construction in which a motion vector is assigned to each and every pixel in our images. The result is a velocity vector field that allows us many new ways to analyze that data.

In practice, calculating dense optical flow is not easy. Consider the motion of a white sheet of paper. Many of the white pixels in the previous frame will simply remain white in the next. Only the edges may change, and even then only those perpendicular to the direction of motion. The result is that dense methods must have some method of interpolating between points that are more easily tracked so as to solve for those points that are more ambiguous. These difficulties manifest themselves most clearly in the high computational costs of dense optical flow.

The early *Horn-Schunck* algorithm [Horn81] attempted to compute just such a velocity field and address these interpolation issues. This method was an improvement over the straightforward but problematic strategy called *block matching*, in which one simply attempted to match windows around each pixel from one frame to the next.

Both of these algorithms were implemented in early versions of the OpenCV library, but neither was either particularly fast or reliable. Much later work in this field, however, has produced algorithms that, while still slow compared to sparse keypoint methods, are fast enough to be used and accurate enough to be useful. The current version of the OpenCV library supports two of these newer algorithms, called *Polynomial Expansion*, and *The Dual TV-$L^1$ algorithm*.

Both the Horn-Schunck and the block matching algorithms are still supported in the legacy portion of the library (i.e., they have C interfaces), but are officially deprecated. We will investigate only the Polynomial Expansion algorithm here, and some more modern algorithms that, though they ultimately owe their genesis to the work of Horn and Schunck, have evolved significantly and show much better performance than the original algorithm.

## The Farnebäck Polynomial Expansion Algorithm

The Polynomial Expansion algorithm, developed by G. Farnebäck [Farnebäck03], attempts to compute optical flow based on an analytical technique that starts with an approximation of the image as a continuous surface. Of course, real images are discrete, so there are also some additional complexities of the Farnebäck method that allow us to apply the basic method to real images. The basic idea of the Farnebäck algorithm is to approximate the image, as a function, by locally fitting a polynomial to the image at every point.

The first phase of the algorithm, from which it gets its name, is the transformation of the image into a representation that associates a quadratic polynomial with each point. This polynomial is approximated based on a window around a pixel in which a weighting is applied to make the fitting more sensitive to the points closer to the center of the window. As a result, the scale of the window determines the scale of the features to which the algorithm is sensitive.

In the idealized case, in which the image can be treated as a smooth continuous function, a small displacement of a portion of the image results in an analytically calculable change in the coefficients of the polynomial expansion at that same point. From this change it is possible to work backward and compute the magnitude of that displacement (Figure 17-1). Of course, this makes sense only for small displacements. However, there is a nice trick for handling larger displacements.

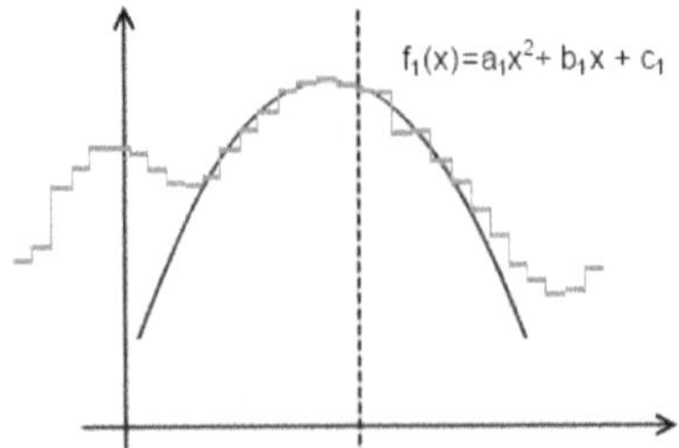

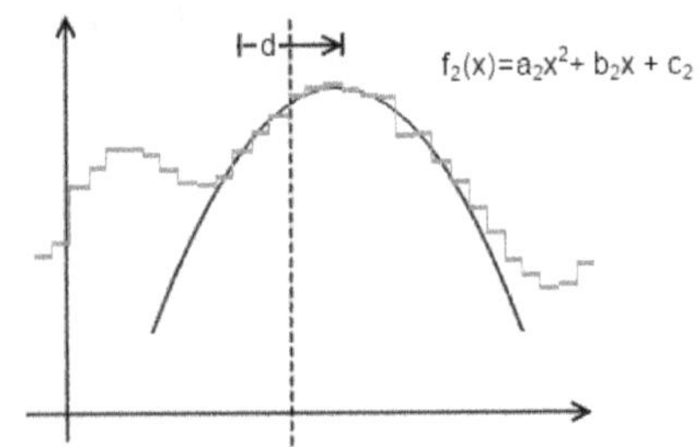


*Figure 17-1. In the one-dimensional case the image is represented by the gray histogram I(x), both before (left) and after (right) a small displacement d. At any given point (dashed line) a parabolic curve can be fitted to the intensity values nearby. In the approximations of smooth functions I(x) and small displacement, the resulting second-order polynomials, $f_1$ and $f_2$, are related by the following analytical formulae: $a_2 = a_1$, $b_2 = b_1 - 2a_1 d$, and $c_2 = a_1 d^2 - b_1 d + c_1$. Given the coefficients of the fit before and after, and making use of the second of these relations, d can be solved for analytically:* $d = -\frac{1}{2}a_1^{-1}(b_2 - b_1)$

The trick is to first notice that if you knew something about a displacement, you could compare the two images not at the same point, but at points related by your estimate of the displacement. In this case, the analytical technique would then compute only a (hopefully small) correction to your original estimate of the displacement. In fact, this mechanism can be used to simply iterate the algorithm and get successive improvements on the motion estimation at each iteration.

This insight can also be used to help find larger motions. Consider the case of first estimating these displacements on a pair of lower-resolution images from image pyramids. In this case, motions appear smaller and the necessary "small displacement" approximation may hold. Then, moving down the pyramid, the result of each prior computation can be used as an initial guess for the next. In the end, it is possible to get results on the scale of the original image by accumulating these successive corrections.

#### Computing dense optical flow with cv::calcOpticalFlowFarneback

OpenCV provides a complete implementation of the dense Farnebäck method; see Figure 17-2 for an example output. This functionality is entirely contained in the `cv::calcOpticalFlowFarneback()` method, which has the following prototype:

```
void cv::calcOpticalFlowFarneback(
  cv::InputArray       prevImg,    // An input image
  cv::InputArray       nextImg,    // Image immediately subsequent to 'prevImg'
  cv::InputOutputArray flow,       // Flow vectors will be recorded here
  double               pyrScale,   // Scale between pyramid levels (< '1.0')
  int                  levels,     // Number of pyramid levels
  int                  winsize,    // Size of window for pre-smoothing pass
```

```
    int                   iterations, // Iterations for each pyramid level
    int                   polyN,      // Area over which polynomial will be fit
    double                polySigma,  // Width of fit polygon, usually '1.2*polyN'
    int                   flags       // Option flags, combine with OR operator
  );
```

*Figure 17-2. Motion from the left and center frames are compared, and represented with a field of vectors on the right. The original images were 640x480, winsize=13, numIters=10, polyN=5, polySigma=1.1, and a box kernel was used for pre-smoothing. Motion vectors are shown only on a lower density subgrid, but the actual results are valid for every pixel*

The first two arguments are the pair of previous and next images you want to compute a displacement between. Both should be 8-bit, single-channel images and be the same size. The next argument, `flow`, is the result image; it will be the same size as `prevImg` and `nextImg` but be two channels and of the 32-bit floating-point type (`CV_32FC2`). `pyrScale` and `levels` affect how the image pyramid is constructed. `pyrScale` must be less than `1`, and indicates the size of each level of the pyramid relative to its predecessor (e.g., if `pyrScale` is `0.5`, then the pyramid will be a "typical" factor-of-two scaling pyramid). `levels` determines how many levels the pyramid will have.

The `winsize` argument controls a presmoothing pass done before the fitting. If you make this an odd number greater than `5`, then image noise will not cause as much trouble for the fitting and it will be possible to detect larger (faster) motions. On the other hand, it will also tend to blur the resulting motion field and make it difficult to detect motion on small objects. This smoothing pass can be either a Gaussian blurring or a simple sliding-average window (controlled by the `flags` argument, described shortly).

The `iterations` argument controls how many iterations are used at each level in the pyramid. In general, increasing the number of iterations will increase accuracy of the final result. Though in some cases, three or even one iteration is sufficient, the algorithm's author found six to be a good number for some scenes.

The `polyN` argument determines the size of the area considered when fitting the polynomial around a point. This is different from `winsize`, which is used only for presmoothing. `polyN` could be thought of as analogous to the window size associated with Sobel derivatives. If this number is large, high-frequency fluctuations will not contribute to the polynomial fitting. Closely related to `polyN` is `polySigma`, which is the source of the intrinsic scale for the motion field. The derivatives computed as part of the fit use a Gaussian kernel (not the one associated with the smoothing) with variance `polySigma` and whose total extent is `polyN`. The value of `polySigma` should be a bit more than 20% of `polyN`. (The pairings `polyN=5`, `polySigma=1.1`, and `polyN=7`, `polySigma=1.5` have been found to work well and are recommended in source code.)

The final argument to `cv::calcOpticalFlowFarneback()` is `flags`, which (as usual) supports several options that may be combined with the logical OR operator. The first option is `cv::OPTFLOW_USE_INITIAL_FLOW`, which indicates to the algorithm that the array `flow` should also be treated as an input, and should be used as an initial estimation for the motion in the scene. This is commonly used when one is analyzing sequential frames in video, on the theory that one frame is likely to contain similar motion to the next. The second option is `cv::OPTFLOW_FARNEBACK_GAUSSIAN`, which tells the algorithm to use a Gaussian kernel in the presmoothing (the one controlled by `winsize`). In general, one gets superior results with the Gaussian presmoothing kernel, but at a cost of somewhat greater compute time. The higher cost comes not just from the multiplication of the kernel weight values in the smoothing sum, but also because one tends to need a larger value for `winsize` with the Gaussian kernel.[1]

## The Dual TV-L$^1$ Algorithm

The Dual TV-L$^1$ algorithm is an evolution on the algorithm of Horn and Schunck (HS), which we encountered briefly earlier in this chapter. The implementation in OpenCV is based on the original paper by Christopher Zach, Thomas Pock, and Horst Bischof [Zach07] as well as improvements proposed later by Javier Sánchez, Enric Meinhardt-Llopis, and Gabriele Facciolo [Sánchez13]. Whereas the original HS algorithm was based on a formulation of the optical flow problem that made solving the problem amenable to straightforward (though not necessarily fast) numerical methods, the Dual TV- L$^1$ algorithm relies on a slightly different formulation of the problem that turns out to be solvable in a much more efficient manner. Because the HS algorithm plays a central role in the evolution of the Dual TV- L$^1$ algorithm, we will review it very briefly, and then explain the differences that define the Dual TV- L$^1$ algorithm.

The HS algorithm operated by defining a flow vector field and by defining an energy cost that was a function of the intensities in the prior frame, the intensities in the sub-

1 The actual smoothing kernel applied has sigma equal to 0.3 · (*winsize* / 2). In practice, this means that the area in which the bulk of the mass of the Gaussian is contained is only about 60–70% of the kernel area. As a result, one expects that the kernel will need to be about 30% larger in order to see the benefit over a box kernel.

sequent frame, and that flow vector field. This energy was defined by the following functional:

$$E(\vec{x}, \vec{u}) = \sum_{\vec{x} \in image} I_{t+1}(\vec{x} + \vec{u}) - I_t(\vec{x})^2 + \alpha^2(\| \nabla u_x \|^2 + \| \nabla u_y \|^2)$$

In this expression, $I_t(\vec{x})$ is the intensity of the image at time $t$ and at location $\vec{x} = (x, y)$. The values $u_x = u_x(\vec{x})$ and $u_y = u_y(\vec{x})$ are the x- and y-components of the flow field at location $\vec{x}$. The vector $\vec{u}$ is simply a convenient shorthand for $(u_x, u_y)$. The value $\alpha^2$ is a weight parameter that affects the relative influence of the first (fidelity) constraint with respect to the second (smoothness) constraint.[2] The HS algorithm attempts to minimize this energy functional across all possible flow fields $\vec{u}$. The method of actually achieving this minimization (proposed by the original authors) was to convert the energy functional to the associated Euler-Lagrange equations and solve those iteratively. The primary problem with this method is that because the Euler-Lagrange equations are entirely local, each iteration only solves the problem relative to the nearest neighbors of a pixel, and thus the practical number of these computationally expensive iterations can be prohibitively large, but hierarchical approaches might help.

This brings us to the Dual TV-$L^1$ algorithm, which differs from the HS algorithm both in terms of the formulation of the energy functional and the method used to solve that problem. The name "TV-$L^1$" is meant to convey that the fidelity constraint is replaced with a *total variation* (TV) while the smoothness constraint uses an $L^1$-norm. Both of these are in contrast to the HS algorithm. The term *total variation* means that the differences are simply summed, rather than squared and summed, while the $L^1$-norm is applied to the gradients in the smoothness constraint instead of the $L^2$-norm used by the HS algorithm. Thus the energy functional used by the TV-$L^1$ algorithm is:

$$E(\vec{x}, \vec{u}) = \sum_{\vec{x} \in image} \lambda \mid I_{t+1}(\vec{x} + \vec{u}) - I_t(\vec{x}) \mid + ( \mid \nabla u_x \mid + \mid \nabla u_y \mid )$$

The primary advantage of the shift to the $L^1$-norm is that local gradients are not so severely penalized and so the algorithm performs much better at discontinuities.[3] The importance of the shift to the total variation rather than the sum of squared differences is in the effect that this has on the solution method for this alternative energy

2 In general, the fidelity constraint (the first squared difference in the energy functional) represents one constraint per pixel. Because the flow field has two dimensions, such a problem is always underspecified. Fundamentally, this is why some kind of continuity constraint is always required.

3 The change from $\alpha^2$ to $\lambda$ is simply to maintain consistency with the variables used by the original authors of both methods. The two are otherwise exactly equivalent.

functional. Whereas Horn and Schunck relied on an iterative Euler-Lagrange method, the Dual TV-$L^1$ algorithm relies on a clever trick that separates the energy minimization into two separate problems, one of which has a known solution (which happens to be the origin of the "dual" in the Dual TV- $L^1$ algorithm's name), while the other has the very desirable property of being entirely local to each pixel, and so can be solved pointwise. Let's see how this works.

First, assume for the moment that we have a flow field $\vec{u}^0$ that is very close to the final flow field $\vec{u}$.[4] Using this assumption, we can approximate the difference in the energy equation using the first-order Taylor expansion, as follows:

$$\rho(u) \equiv \nabla I_{t+1}(\vec{x} + \vec{u}^0) \cdot (\vec{u} - \vec{u}^0) + I_{t+1}(\vec{x} + \vec{x}^0) - I_t(x, y)$$

Thus, under this approximation, we can write the energy as:

$$E(\vec{x}, \vec{u}) = \sum_{\vec{x} \varepsilon image} \lambda \mid \rho(\vec{u}) \mid + \mid \nabla u_x \mid + \mid \nabla u_y \mid$$

What comes next is the heart of the method, which is to introduce a new field $\vec{v}$, such that the energy becomes:

$$E(\vec{x}, \vec{u}, \vec{v}) = \sum_{\vec{x} \varepsilon image} \mid \nabla u_x \mid + \mid \nabla u_y \mid + \frac{1}{2\theta} \mid \vec{u} - \vec{v} \mid + \lambda \mid \rho(\vec{v}) \mid$$

This decouples the fidelity and smoothness terms in what will turn out to be a very useful way. Of course, we now have another field to solve for, but in the limit, with a very small $\theta$, $\vec{u}$ and $\vec{v}$ are effectively forced to become equal. However, the biggest benefit of this change is that we can now solve the fields $\vec{u}$ and $\vec{v}$ first by fixing one and solving for the other, then fixing the other and solving for the first (and iterating in this manner). When $\vec{v}$ is fixed, we find the value of $\vec{u}$, which minimizes:

$$\sum_{\vec{x} \varepsilon image} \mid \nabla u_x \mid + \mid \nabla u_y \mid + \frac{1}{2\theta} \mid \vec{u} - \vec{v} \mid$$

When $\vec{u}$ is fixed, we find the value of $\vec{v}$, which minimizes:

$$\sum_{\vec{x} \varepsilon image} \frac{1}{2\theta} \mid \vec{u} - \vec{v} \mid + \lambda \mid \rho(\vec{v}) \mid$$

4 In practice, the algorithm is implemented on a pyramid of images, such that the coarsest scales can be solved for first, and the results can be propagated to ever-finer scales. In this way, the estimate $\vec{u}^0$ is always available.

The first of these is the problem that has a known solution,[5] while the second is the one that is entirely local and can be solved on a per-pixel basis. This first procedure, however, introduces two new parameters: the *time step* and the *stopping criterion*. These parameters control how and when convergence is reached for the computation of $\vec{u}$ when $\vec{v}$ is fixed.

#### Computing dense optical flow with cv::createOptFlow_DualTVL1

The OpenCV implementation of the Dual TV- $L^1$ algorithm uses a slightly different interface style than the other optical flow algorithms in this section. There is a separate factory-like function in the library called `cv::createOptFlow_DualTVL1()`, which constructs an object of type `cv::OpticalFlow_DualTVL1` (derived from the base class `cv::DenseOpticalFlow`) and returns a pointer to it:

```
cv::Ptr<cv::DenseOpticalFlow> createOptFlow_DualTVL1();
```

The resulting object has member variables that you will need to override directly if you want to change them from their default values. Here is the relevant portion of the definition of `cv::OpticalFlow_DualTVL1`:

```
// Function to get a dense optical flow object
//
cv::Ptr<cv::DenseOpticalFlow> createOptFlow_DualTVL1();

class OpticalFlowDual_TVL1 : public DenseOpticalFlow {

public:

  OpticalFlowDual_TVL1();

  void calc( InputArray I0, InputArray I1, InputOutputArray flow );
  void collectGarbage();

  double tau;             // timestep for solver          (default = 0.25)
  double lambda;          // weight of smoothness term    (default = 0.15)
  double theta;           // tightness parameter          (default = 0.3)
  int    nscales;         // scales in pyramid            (default = 5)
  int    warps;           // warpings per scale           (default = 5)
  double epsilon;         // stopping criterion           (default = 0.01)
  int    iterations;      // max iterations               (default = 300)
  bool   useInitialFlow; // use 'flow' as starting guess (default = false)

};
```

The variables you can set to configure the algorithm are the arguments to the create function. `tau` is the timestep used by the numerical solver. It can be set to any value

5 This is the so-called *total variation denoising model*, and can be solved by Chambolle's duality-based algorithm [Chambolle04].

less than `0.125` and convergence is guaranteed, but empirically it can be set as high as `0.25` for faster convergence (in fact, this is the default value). `lambda` is the most important parameter; it sets the weight of the smoothness term in the energy. The ideal value of `lambda` will vary depending on the image sequence, with smaller values corresponding to smoother solutions. The default value for `lambda` is `0.15`. The parameter `theta` is called (by the authors of the original paper) the "tightness parameter." This is the parameter that couples the two stages of the overall solver. In principle it should be very small, but the algorithm is stable for a wide range of values. The default tightness parameter is `0.30`.

The number of scales in the image pyramid is set by `nscales`. For each scale, the number of `warps` is the number of times $\nabla I_{t+1}(\vec{x} + \vec{u}^0)$ and $I_{t+1}(\vec{x} + \vec{u}^0)$ are computed per scale. This parameter allows a trade-off between speed (fewer warps) and accuracy (more warps). By default there are five scales and five warps per scale.

`epsilon` is the stopping criterion used by the numerical solver. Additionally there is an `iterations` criterion, which sets the maximum number of iterations allowed. The default for `epsilon` is `0.01`, while the default for `iterations` is `300`. Making `epsilon` smaller will give more accurate solutions, but at the cost of more computational time.

The final parameter that you might want to set is `useInitialFlow`. If this parameter is set to `true`, then when you call the `calc()` method for the `cv::OpticalFlow_DualTVL1` object, the `flow` parameter you pass will be used as a starting point for the computation. In many cases it makes sense to use the previous result when you are running on sequential video.

The method `cv::OpticalFlow_DualTVL1::calc()` is what you use when you want to actually compute optical flow. The `calc()` method expects two 8-bit, single-channel images as input and will compute the `flow` output for you from those. As just described, `flow` is an input-output parameter, and if you set the `useInitialFlow` member variable of the `cv::OpticalFlow_DualTVL1` object to `true`, then any data currently in `flow` will be used as a starting point for the next calculation. In any case, `flow` will be the same size as the input images and will be of type `CV_32FC2`.

The final method of `cv::OpticalFlow_DualTVL1` is `collectGarbage()`. This method takes no arguments and simply releases any internally allocated memory inside of the `cv::OpticalFlow_DualTVL1` object.

## The Simple Flow Algorithm

Another, recent algorithm for computing optical flow is the *Simple Flow* algorithm. Originally proposed by Michael Tao et al. [Tao12], this algorithm has the important feature of requiring time that is sublinear[6] in the number of pixels in the image. This is achieved by a pyramid-based scheme that, as it moves from coarse to fine layers in the pyramid, ascertains whether optical flow calculation is needed at the pixels in the new layer. Where it is found that no new information will be gained at the finer level, no optical flow calculation is done. Instead, the flow is simply propagated to and interpolated on the new level.

The Simple Flow algorithm attempts to establish a local flow vector for each point that best explains the motion of the neighborhood around that point. It does this by computing the (integer) flow vector that optimizes an energy function. This energy function is essentially a sum over terms for each pixel in the neighborhood in which the energy grows quadratically with the difference between the intensities of the pixel in the neighborhood at time $t$ and the corresponding pixel (i.e., displaced by the flow vector) at time $t + 1$. Defining the pixelwise energy function $e(x, y, u, v)$ (where $u$ and $v$ are the components of the flow vector):

$$e(x, y, u, v) = \| \ I_t(x, y) - I_{t+1}(x + u, y + v)\| \ ^2$$

we can then express the actual energy, which is minimized as:

$$E(x, y, u, v) = \sum_{(i,j)\varepsilon N} w_d w_c e(x + i, y + j, u, v)$$

Here the parameters $w_d$ and $w_c$ have the following definitions:

$$w_d = \exp\left( - \frac{\| \ (x, y) - (x + i, y + j)\| \ ^2}{2\sigma_d} \right)$$

and:

$$w_c = \exp\left( - \frac{\| \ I_t(x, y) - I_t(x + i, y + j)\| \ ^2}{2\sigma_c} \right)$$

The result of these two terms is to create the effect of a bilateral filter.[7] Notice how for small values of $\sigma_d$, the term $w_d$ becomes very small for pixels far from the centerpoint

6 Technically speaking, the Simple Flow algorithm is not sublinear, as it has a component that must operate on every pixel. However, the expensive operations in the algorithm need not be applied to every pixel, and so in practice the cost is effectively sublinear. In their paper [Tao12], the authors demonstrate sublinear time behavior for images as large as 4K images (i.e., "quad hd," 4,096 × 2,160).

7 For a brief review of what a bilateral filter is, return to the discussion in Chapter 10 of the function `cv::bilateralFilter()`.

$(x, y)$. Similarly, for small values of $\sigma_c$, the term $w_c$ becomes very small for pixels whose intensity is substantially different from that of the center point $(x, y)$.

The energy $E(x, y, u, v)$ is first minimized with respect to a range of possible integer values for $(u, v)$. Once this is complete, a parabola is fit to a 3 × 3 set of cells around the integer $(x, y)$ found to minimize $E(x, y, u, v)$. In this way, the optimal noninteger value can be interpolated. The resulting flow field (i.e., the set of all $(u, v)$ pairs found in this way) is then passed through another bilateral filter.[8] Notice, however, that this filter is now operating on the $(u, v)$ vectors rather than the energy density $e(x, y, u, v)$. Thus this is a separate filter with separate parameters $\sigma_d^{fix}$ and $\sigma_c^{fix}$ that are not only independent of $\sigma_d$ and $\sigma_c$, but don't even have the same units. (The former is a velocity variance, while the latter is an intensity variance.)

This solves part of the problem, but in order to compute motions that are more than a few pixels in a frame, it would be necessary to search large windows in velocity space in order to find the optimal values of $E(x, y, u, v)$. Simple Flow solves this problem, as is often the case in computer vision, with an image pyramid. In this way, gross motions can be found at higher levels in the pyramid and refined at finer levels of the pyramid. Starting at the coarsest image in the pyramid, we begin computing each subsequent level by upsampling the prior level and interpolating all of the new pixels. This upsampling is done through a method called *joint bilateral upsampling* [Kopf07]. This technique makes use of the fact that though we are upsampling the solution for the velocity field, we have access to an image at the higher resolution already. The joint bilateral filter introduces two more parameters that characterize the spatial and color extent of the filter: $\sigma_d^{up}$ and $\sigma_c^{up}$.

An important contribution of the Simple Flow algorithm is what the authors call a *flow irregularity map*. The essential idea behind this map is to compute for every point the amount by which the neighboring pixel flows differ from the flow at that pixel:

$$\mathrm{H}(x, y) = \max_{(i,j)\in N} \| (u(x+i, y+j), v(x+i, y+j)) - (u(x, y), v(x, y)) \|$$

Where the flow irregularity is found to be small, relative to some cutoff parameter $\tau$, the flow is computed at the corners of the patch $N$ and interpolated within the patch. Where the irregularity exceeds $\tau$, the flow computation just described is repeated at this finer level of hierarchy.

8 For real experts, there is a subtle point about this second filter, which is that it is not applied to regions in which there was found to be occlusion. For details on how this is done and what it means exactly, visit the original paper [Tao12].

### Computing Simple Flow with cv::optflow::calcOpticalFlowSF()

We are now ready to look at the OpenCV function that implements the Simple Flow algorithm.[9] It is called `cv::optflow::calcOpticalFlowSF()`.

```
void cv::optflow::calcOpticalFlowSF(
  InputArray  from,                       // Initial image (input)
  InputArray  to,                         // Subsequent image (input)
  OutputArray flow,                       // Output flow field, CV_32FC2
  int         layers,                     // Number of layers in pyramid
  int         averaging_block_size,       // Size of neighborhoods (odd)
  int         max_flow                    // Velocities search region sz (odd)
);

void cv::calcOpticalFlowSF(
  InputArray  from,                       // Initial image (input)
  InputArray  to,                         // Subsequent image (input)
  OutputArray flow,                       // Output flow field, CV_32FC2
  int         layers,                     // Number of layers in pyramid
  int         averaging_block_size,       // Size of neighborhoods (odd)
  int         max_flow                    // Velocities search region sz (odd)
  double      sigma_dist,                 // sigma_d
  double      sigma_color,                // sigma_c
  int         postprocess_window,         // Velocity filter window sz (odd)
  double      sigma_dist_fix,             // Sigma_d^fix
  double      sigma_color_fix,            // Sigma_c^fix
  double      occ_thr,                    // Threshold used in occlusion detection
  int         upscale_averaging_radius,   // Window for joint bilateral upsampling
  double      upscale_sigma_dist,         // Sigma_d^up
  double      upscale_sigma_color,        // Sigma_c^up
  double      speed_up_thr                // Tao
);
```

The first form of the `cv::calcOpticalFlowSF()` function does not require you to have a particularly deep understanding of the Simple Flow algorithm. It requires the previous and current images, `from` and `to`, and returns the velocity field to you in the array `flow`. The input images should be 8-bit, three-channel images (`CV_8UC3`); the result array, `flow`, will be a 32-bit, two-channel image (`CV_32FC2`). The only parameters you need to specify for this version of the function are the number of layers, the neighborhood size, and the largest flow velocity that you would like the algorithm to consider when running the initial velocity solver (at each level). These three parameters—`layers`, `averaging_block_size`, and `max_flow` (respectively)—can reasonably be set to `5`, `11`, and `20`.

The second form of `cv::calcOpticalFlowSF()` allows you to really get in there and tune every parameter of the algorithm. In addition to the arguments used by the

9 Starting from OpenCV 3.0, `calcOpticalFlowSF()` is moved to *opencv_contrib*, in the `optflow` module.

short form, the long form allows you to set the bilateral filter parameters used in the energy minimization (`sigma_dist` and `sigma_color`, the same as $\sigma_d$ and $\sigma_c$ in the previous discussion), the window size for the velocity field cross-bilateral filters (`postprocess_window`), the parameters for the velocity field cross-bilateral filter (`sigma_dist_fix` and `sigma_color_fix`, the same as $\sigma_d^{fix}$ and $\sigma_c^{fix}$ in the previous discussion), the threshold used for occlusion detection, the window size for the upsampling joint bilateral filter (`upscale_averaging_radius`), the parameters for the upsampling joint bilateral filter (`upscale_sigma_dist` and `upscale_sigma_color`, the same as $\sigma_d^{up}$ and $\sigma_c^{up}$ in the previous discussion), and the threshold used to determine when the irregularity map dictates that flows must be recalculated at a finer pyramid level (`speed_up_thr`, the same as $\tau$ in the previous discussion).

Detailed tuning of these parameters is clearly a topic for experts, and you should read the original paper if you want to understand them deeply. For our needs here, however, it is helpful to know the nominal values—used by default by the short argument version of `cv::calcOpticalFlowSF()`—listed in Table 17-1.

*Table 17-1. Nominal values for the detail arguments of cv::calcOpticalFlowSF()*

| Argument | Nominal value |
|---|---|
| sigma_dist | 4.1 |
| sigma_color | 25.5 |
| postprocess_window | 18 |
| sigma_dist_fix | 55.0 |
| sigma_color_fix | 25.5 |
| occ_thr | 0.35 |
| upscale_averaging_radius | 18 |
| upscale_sigma_dist | 55.0 |
| upscale_sigma_color | 25.5 |
| speed_up_thr | 10 |

# Mean-Shift and Camshift Tracking

In this section, we will look at two techniques, *mean-shift* and *Camshift* (where *Camshift* stands for "continuously adaptive mean-shift"). The former is a general technique for data analysis in many applications (discussed in Chapter 12 in the context of segmentation), of which computer vision is only one. After introducing the general theory of mean-shift, we'll describe how OpenCV allows you to apply it to tracking in images. The latter technique, Camshift, builds on mean-shift to allow for the tracking of objects whose size may change during a video sequence.

## Mean-Shift

The mean-shift algorithm[10] is a robust method of finding local extrema in the density distribution of a data set. This is an easy process for continuous distributions; in that context, it is essentially just *hill climbing* applied to a density histogram of the data.[11] For discrete data sets, however, this is a somewhat less trivial problem.

The term *robust* is used here in its statistical sense; that is, mean-shift ignores outliers, data points that are far away from peaks in the data. It does so by processing only points within a local window of the data and then moving that window.

The mean-shift algorithm runs as follows:

1. Choose a search window:
   - its initial location;
   - its type (uniform, polynomial, exponential, or Gaussian);
   - its shape (symmetric or skewed, possibly rotated, rounded, or rectangular);
   - its size (extent at which it rolls off or is cut off).
2. Compute the window's (possibly weighted) center of mass.
3. Center the window at the center of mass.
4. Return to Step 2 until the window stops moving (it always will).[12]

To give a more formal sense of what the mean-shift algorithm is: it is related to the discipline of *kernel density estimation*, where by *kernel* we refer to a function that has mostly local focus (e.g., a Gaussian distribution). With enough appropriately weighted and sized kernels located at enough points, one can express a distribution of data entirely in terms of those kernels. Mean-shift diverges from kernel density estimation in that it seeks only to estimate the gradient (direction of change) of the data distribution. When this change is `0`, we are at a stable (though perhaps local) peak of the distribution. There might be other peaks nearby or at other scales.

---

10 Because mean-shift is a fairly deep topic, our discussion here is aimed mainly at developing intuition for the user. For the original formal derivation, see Fukunaga [Fukunaga90] and Comaniciu and Meer [Comaniciu99].

11 The word *essentially* is used because there is also a scale-dependent aspect of mean-shift. To be exact: mean-shift is equivalent in a continuous distribution to first convolving with the mean-shift kernel and then applying a hill-climbing algorithm.

12 Iterations are typically restricted to some maximum number or to some epsilon change in center shift between iterations; however, they are guaranteed to converge eventually.

Figure 17-3 shows the equations involved in the mean-shift algorithm. We can simplify these equations by considering a *rectangular* kernel,[13] which reduces the mean-shift vector equation to calculating the center of mass of the image pixel distribution:

$$x_c = \frac{M_{10}}{M_{00}},\ y_c = \frac{M_{01}}{M_{00}}$$

Here the zeroth moment is calculated as:

$$M_{00} = \sum_x \sum_y I(x, y)$$

and the first moments are:

$$M_{01} = \sum_x \sum_y x \cdot I(x, y) \quad and \quad M_{10} = \sum_x \sum_y y \cdot I(x, y)$$

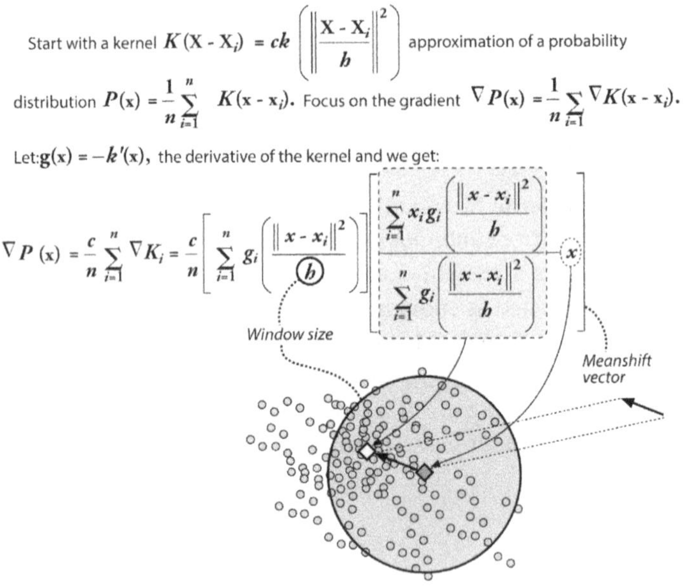

Start with a kernel $K(\mathbf{X} - \mathbf{X}_i) = ck\left(\left\|\frac{\mathbf{X} - \mathbf{X}_i}{h}\right\|^2\right)$ approximation of a probability distribution $P(\mathbf{x}) = \frac{1}{n}\sum_{i=1}^{n} K(\mathbf{x} - \mathbf{x}_i)$. Focus on the gradient $\nabla P(\mathbf{x}) = \frac{1}{n}\sum_{i=1}^{n} \nabla K(\mathbf{x} - \mathbf{x}_i)$.

Let: $\mathbf{g}(\mathbf{x}) = -\mathbf{k}'(\mathbf{x})$, the derivative of the kernel and we get:

$$\nabla P(\mathbf{x}) = \frac{c}{n}\sum_{i=1}^{n} \nabla K_i = \frac{c}{n}\left[\sum_{i=1}^{n} g_i\left(\left\|\frac{x - x_i}{h}\right\|^2\right)\right]\left[\frac{\sum_{i=1}^{n} x_i g_i\left(\left\|\frac{x - x_i}{h}\right\|^2\right)}{\sum_{i=1}^{n} g_i\left(\left\|\frac{x - x_i}{h}\right\|^2\right)} - x\right]$$

Window size

Meanshift vector

*Figure 17-3. Mean-shift equations and their meaning*

13 A *rectangular kernel* is a kernel with no falloff with distance from the center, until a single sharp transition to `0` value. This is in contrast to the exponential falloff of a Gaussian kernel and the falloff with the square of distance from the center in the commonly used Epanechnikov kernel.

The mean-shift vector in this case tells us to recenter the mean-shift window over the calculated center of mass within that window. This movement will, of course, change what is "under" the window and so we iterate this recentering process. Such recentering will always converge to a mean-shift vector of `0` (i.e., where no more centering movement is possible). The location of convergence is at a local maximum (peak) of the distribution under the window. Different window sizes will find different peaks because "peak" is fundamentally a scale-sensitive construct.

In Figure 17-4, we see an example of a two-dimensional distribution of data and an initial (in this case, rectangular) window. The arrows indicate the process of convergence on a local mode (peak) in the distribution. Observe that, as promised, this peak finder is statistically robust in the sense that points outside the mean-shift window do not affect convergence—the algorithm is not "distracted" by faraway points.

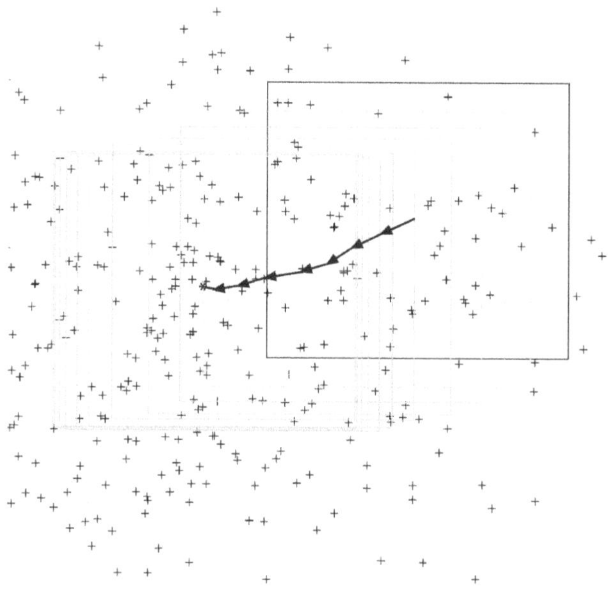

*Figure 17-4. Mean-shift algorithm in action: an initial window is placed over a two-dimensional array of data points and is successively recentered over the mode (or local peak) of its data distribution until convergence*

In 1998, it was realized that this mode-finding algorithm could be used to track moving objects in video [Bradski98a; Bradski98b], and the algorithm has since been greatly extended [Comaniciu03]. The OpenCV function that performs mean-shift is implemented in the context of image analysis. This means in particular that, rather than taking some arbitrary set of data points (possibly in some arbitrary number of dimensions), the OpenCV implementation of mean-shift expects as input an image representing the density distribution being analyzed. You could think of this image as a two-dimensional histogram measuring the density of points in some two-dimensional space. It turns out that, for vision, this is precisely what you want to do most of the time: it's how you can track the motion of a cluster of interesting features.

```
int cv::meanShift(                    // Return number of iterations to converge
    cv::InputArray     probImage,  // density of locations (CV_8U or CV_32F)
    cv::Rect&          window,     // initial location (and size) of kernel window
    cv::TermCriteria   criteria    // limits location update iterations
);
```

In `cv::meanShift()`, the `probImage`, which represents the density of probable locations, may be only one channel but of either type (byte or float). The `window` is set at the initial desired location and size of the kernel window. On completion, it will contain the final location of the kernel window. (Its size will remain unchanged.) The termination `criteria` sets the maximum limit on the number of mean-shift location update iterations and a minimal movement for which we consider the window locations to have converged.[14] The return value will be the number of iterations before convergence.

The function `cv::meanShift()` is one expression of the mean-shift algorithm for rectangular windows, but it may also be used for tracking. In this case, you first choose the feature distribution to represent an object (e.g., color + texture), then start the mean-shift window over the feature distribution generated by the object, and finally compute the chosen feature distribution over the next video frame. Starting from the current window location, the mean-shift algorithm will find the new peak or mode of the feature distribution, which (presumably) is centered over the object that produced the color and texture in the first place. In this way, the mean-shift window tracks the movement of the object frame by frame.

## Camshift

A related algorithm is the *Camshift*[15] tracker. It differs primarily from the mean-shift algorithm in that the search window adjusts itself in size. If you have well-segmented distributions (say, face features that stay compact), then this algorithm will, for example, automatically adjust itself for the size of face as the person moves closer to and farther from the camera. The function that implements the Camshift algorithm is:

---

14 Again, mean-shift will always converge, but convergence may be very slow near the local peak of a distribution if that distribution is fairly "flat" there.

15 For the curious, the original algorithm was named CAMSHIFT (all capitals), but is not an acronym. Because the mean-shift algorithm (on which is it based) is hyphenated, it is common to see "cam-shift," even though this is not correct. The name in the OpenCV library, `cv::CamShift`, is actually derived from the incorrect hyphenation. Because the algorithm name is not an acronym, it is also common to see it written as Camshift (including by its authors), and so we use that representation here.

```
RotatedRect cv::CamShift(           // Return final window size and rotation
    cv::InputArray    probImage, // Density of locations (CV_8U or CV_32F)
    cv::Rect&         window,    // Initial location (and size) of kernel window
    cv::TermCriteria  criteria   // Limits location update iterations
);
```

The three parameters used by the `cv::CamShift()` function have the same interpretations as for the `cv::meanShift()` algorithm. The return value will contain the newly resized box, which also includes the orientation of the object as computed via second-order moments. For tracking applications, we would use the resulting resized box in the previous frame as the `window` in the next frame.

Many people think of mean-shift and Camshift as tracking using color features, but this is not entirely correct. Both of these algorithms track the distribution of any kind of feature that is expressed in the `probImage`; hence, they make for very lightweight, robust, and efficient trackers.

# Motion Templates

Motion templates were invented in the MIT Media Lab by Bobick and Davis [Bobick96; Davis97] and were further developed jointly with one of the authors of this book [Davis99; Bradski00]. This more recent work forms the basis for the implementation in OpenCV.

Motion templates are an effective way to track general movement and are especially applicable to gesture recognition. Using motion templates requires a silhouette (or part of a silhouette) of an object. Object silhouettes can be obtained in a number of ways:

- The simplest method of obtaining object silhouettes is to use a reasonably stationary camera and then employ frame-to-frame differencing (as discussed in Chapter 15). This will give you the moving edges of objects, which is enough to make motion templates work.
- You can use chroma keying. For example, if you have a known background color such as bright green, you can simply take as foreground anything that is not bright green.
- As also discussed in Chapter 15, you can learn a background model from which you can isolate new foreground objects/people as silhouettes.
- You can use active silhouetting techniques—for example, creating a wall of near-infrared light and having a near-infrared-sensitive camera look at the wall. Any intervening object will show up as a silhouette.

- You can use thermal imagers; any hot object (such as a face) can be taken as foreground.
- Finally, you can generate silhouettes by using the segmentation techniques (e.g., pyramid segmentation or mean-shift segmentation) described in Chapter 15.

For now, assume that we have a good, segmented object silhouette as represented by the white rectangle of Figure 17-5(A). Here we use white to indicate that all the pixels are set to the floating-point value of the most recent system timestamp. As the rectangle moves, new silhouettes are captured and overlaid with the (new) current timestamp; the new silhouette is the white rectangle of Figures 17-5(B) and 17-5(C). Older motions are shown in Figure 17-5(C) as successively darker rectangles. These sequentially fading silhouettes record the history of previous movement and thus are referred to as the *motion history image*.

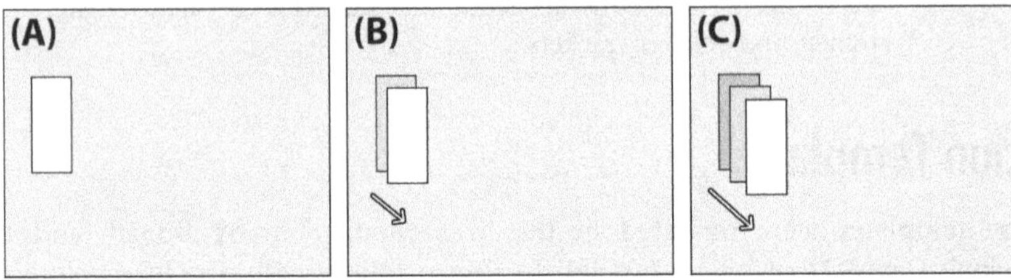


*Figure 17-5. Motion template diagram: (a) a segmented object at the current timestamp (white); (b) at the next time step, the object moves and is marked with the (new) current timestamp, leaving the older segmentation boundary behind; (c) at the next time step, the object moves farther, leaving older segmentations as successively darker rectangles whose sequence of encoded motion yields the motion history image*

Silhouettes whose timestamp is more than a specified `duration` older than the current system timestamp are set to `0`, as shown in Figure 17-6. The OpenCV function that accomplishes this motion template construction is `cv::motempl::updateMotionHistory()`:[16]

```
void cv::motempl::updateMotionHistory(
   cv::InputArray        silhouette, // Nonzero pixels where motion occurs
   cv::InputeOutputArray mhi,        // Motion history image
   double                timestamp,  // Current time (usually milliseconds)
   double                duration    // Max track duration ('timestamp' units)
);
```

16 Starting from OpenCV 3.0 motion templates functions, described in this section, are moved to the `optflow` module of *opencv_contrib* repository, under the `motempl` namespace.

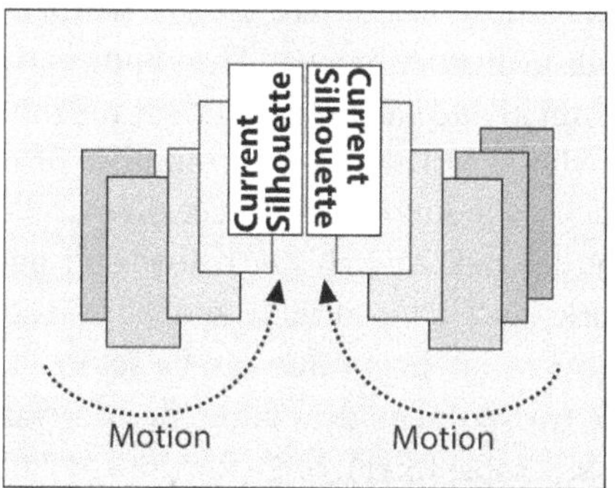


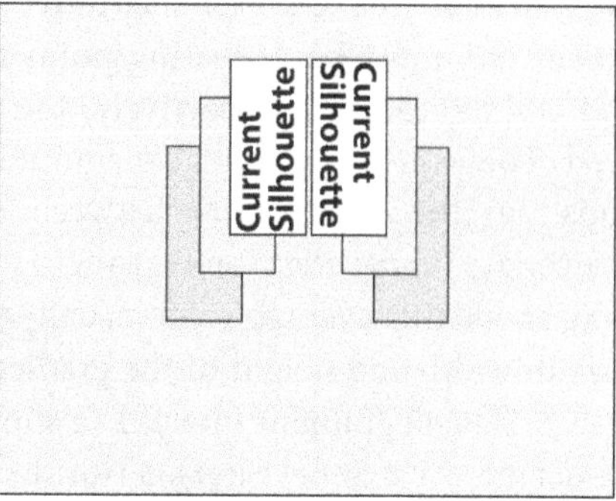


*Figure 17-6. Motion template silhouettes for two moving objects (left); silhouettes older than a specified duration are set to 0 (right)*

In `cv::motempl::updateMotionHistory()`, all image arrays consist of single-channel images. The `silhouette` image is a byte image in which nonzero pixels represent the most recent segmentation silhouette of the foreground object. The argument `mhi` is a floating-point image that represents the motion template (a.k.a. motion history image). Here `timestamp` is the current system time (typically a millisecond count) and `duration`, as just described, sets how long motion history pixels are allowed to remain in the `mhi`. In other words, any `mhi` pixels that are older (less) than `timestamp` minus `duration` are set to `0`.

Once the motion template has a collection of object silhouettes overlaid in time, we can derive an indication of overall motion by taking the gradient of the `mhi`. When we take these gradients (e.g., by using the Scharr or Sobel gradient functions discussed in Chapter 12), some gradients will be large and invalid. Gradients are invalid when older or inactive parts of the `mhi` are set to `0`, which produces artificially large gradients around the outer edges of the silhouettes; see Figure 17-6. Because we know the time-step duration with which we've been introducing new silhouettes into the `mhi` via `cv::motempl::updateMotionHistory()`, we know how large our gradients (which are just *dx* and *dy* step derivatives) should be. We can therefore use the gradient magnitude to eliminate gradients that are too large, as in Figure 17-6. Finally, we can collect a measure of global motion; see Figure 17-6. The function that does all of this for us is `cv::motempl::calcMotionGradient()`:

```
void cv::motempl::calcMotionGradient(
   cv::InputArray  mhi,               // Motion history image
   cv::OutputArray mask,              // Nonzero where valid gradients were found
   cv::OutputArray orientation,       // Orientation of found gradients
   double          delta1,            // Minimal gradient allowed
   double          delta2,            // Maximal gradient allowed
   int             apertureSize = 3 // Size of gradient operator ('-1'=SCHARR)
);
```

In cv::motempl::calcMotionGradient(), all image arrays are single-channel. The argument mhi must be a floating-point motion history image. The input variables delta1 and delta2 are (respectively) the minimal and maximal gradient magnitudes allowed. The expected gradient magnitude will be just the average number of milliseconds in the timestamp between each silhouette in successive calls to cv::motempl::updateMotionHistory(). Setting delta1 halfway below and delta2 halfway above this average value usually works well. The variable apertureSize sets the size in width and height of the gradient operator. This value can be set to -1 (the $3 \times 3$ cv::SCHARR gradient filter), 1 (a simple two-point central difference derivative), 3 (the default $3 \times 3$ Sobel filter), 5 (for the $5 \times 5$ Sobel filter), or 7 (for the $7 \times 7$ filter). The function results will be placed in mask and orientation. The former will be a single-channel, 8-bit image in which nonzero entries indicate where valid gradients were found, and the latter will be a floating-point image that gives the gradient direction's angle at each point. Entries in orientation will be in degrees and confined to the range from 0 to 360.

The function cv::motempl::calcGlobalOrientation() finds the overall direction of motion as the vector sum of the valid gradient directions:

```
double cv::motempl::calcGlobalOrientation(
   cv::InputArray orientation, // Orientation image from calcMotionGradient()
   cv::InputArray mask,        // Nonzero where direction is to be calculated
   cv::InputArray mhi,         // Motion history img from updateMotionHistory()
   double timestamp,           // Current time (usually milliseconds)
   double duration             // Maximum duration of track ('timestamp' units)
);
```

When using cv::motempl::calcGlobalOrientation(), we pass in the orientation and mask images computed in cv::motempl::calcMotionGradient() along with the timestamp, duration, and resulting mhi from cv::motempl::updateMotionHistory(). The vector-sum global orientation is returned, as in Figure 17-7. This orientation will be in degrees and confined to the range from 0 to 360.

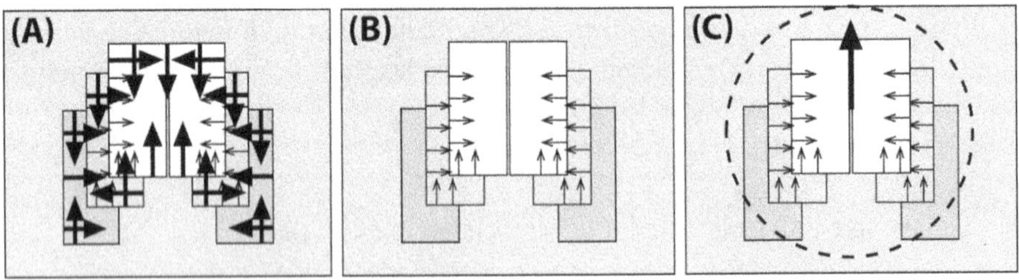


*Figure 17-7. Motion gradients of the MHI image: (a) gradient magnitudes and directions; (b) large gradients are eliminated; (c) overall direction of motion is found*

The `timestamp`, together with `duration`, tells the function how much motion to consider from the `mhi` and motion `orientation` images. You could compute the global motion from the center of mass of each of the `mhi` silhouettes, but summing up the precomputed motion vectors is much faster.

We can also isolate regions of the motion template `mhi` and determine the local motion within that region, as shown in Figure 17-8. In the figure, the `mhi` is scanned for current silhouette regions. When a region marked with the most current timestamp is found, the region's perimeter is searched for sufficiently recent motion (recent silhouettes) just outside its perimeter. When such motion is found, a downward-stepping flood fill is performed to isolate the local region of motion that "spilled off" the current location of the object of interest. Once found, we can calculate local motion gradient direction in the spill-off region, then remove that region, and repeat the process until all regions are found (as diagrammed in Figure 17-8).

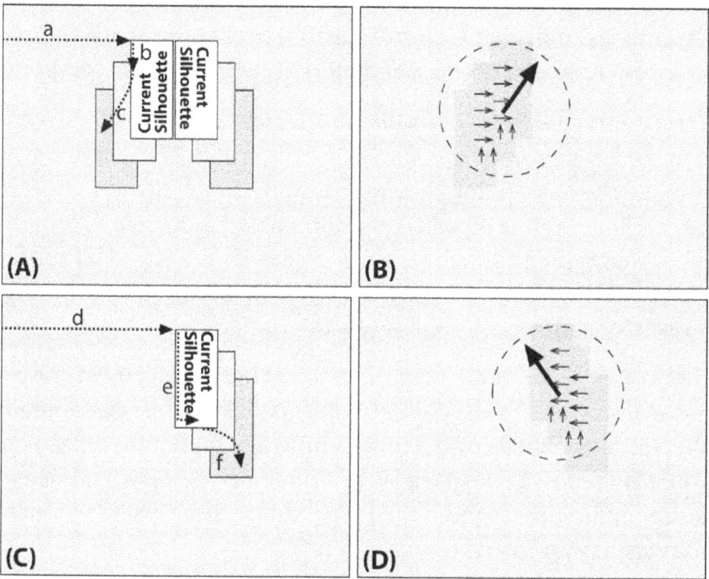


*Figure 17-8. Segmenting local regions of motion in the MHI: In (A) (a) scan the MHI for current silhouettes and, when they're found, go around the perimeter looking for other recent silhouettes; when a recent silhouette (b) is found, perform downward-stepping flood fills (c) to isolate local motion. In (B) use the gradients found within the isolated local motion region to compute local motion. In (C) remove the previously found region and (d) search for the next current silhouette region (e), scan along it and (f) perform downward-stepping flood fill on it. (D) Compute motion within the newly isolated region and continue the process until no current silhouette remains*

The function that isolates and computes local motion is cv::motempl::segmentMotion():

```
void cv::motempl::segmentMotion(
    cv::InputArray    mhi,            // Motion history image
    cv::OutputArray   segMask,        // Output image, found segments (CV_32FC1)
    vector<cv::Rect>& boundingRects, // ROI's for motion connected components
    double            timestamp,      // Current time (usually milliseconds)
    double            segThresh       // >= interval between motion history steps
);
```

In cv::motempl::segmentMotion(), the mhi must be a single-channel, floating-point input. The argument segMask is used for output; when returned it will be a single-channel, 32-bit floating-point image. The individual segments will be "marked" onto this image, with each segment being given a distinct nonzero identifier (e.g., 1, 2, etc.; zero (0) is reserved to mean "no motion"). Similarly, the vector boundingRects will be filled with regions of interest (ROIs) for the motion-connected components. (This allows you to use cv::motempl::calcGlobalOrientation separately on each such connected component to determine the motion of that particular component.)

The timestamp input should be the value of the most current silhouettes in the mhi (the ones from which you want to segment local motions). The last argument is segThresh, which is the maximum downward step (from current time to previous motion) that you'll accept as attached motion. This parameter is provided because there might be overlapping silhouettes from recent and much older motion that you don't want to connect together. It's generally best to set segThresh to something like 1.5 times the average difference in silhouette timestamps.

Given the discussion so far, you should now be able to understand the *motempl.cpp* example that ships with OpenCV in the *opencv_contrib/modules/optflow/samples/* directory. We will now extract and explain some key points from the update_mhi() function in *motempl.cpp*. The update_mhi() function extracts templates by thresholding frame differences and then passing the resulting silhouette to cv::motempl::updateMotionHistory():

```
...
cv::absdiff( buf[idx1], buf[idx2], silh );

cv::threshold( silh, silh, diff_threshold, 1, cv::THRESH_BINARY );

cv::updateMotionHistory( silh, mhi, timestamp, MHI_DURATION );
...
```

The gradients of the resulting mhi are then taken, and a mask of valid gradients is produced via `cv::motempl::calcMotionGradient()`. Then resulting local motions are segmented into `cv::Rect` structures:

```
...
cv::motempl::calcMotionGradient(
  mhi,
  mask,
  orient,
  MAX_TIME_DELTA,
  MIN_TIME_DELTA,
  3
);

vector<cv::Rect> brects;

cv::motempl::segmentMotion(
  mhi,
  segmask,
  brects,
  timestamp,
  MAX_TIME_DELTA
);
...
```

A `for` loop then iterates through the bounding rectangles for each motion. The iteration starts at `-1`, which has been designated as a special case for finding the global motion of the whole image. For the local motion segments, small segmentation areas are first rejected and then the orientation is calculated via `cv::motempl::calcGlobalOrientation()`. Instead of using exact masks, this routine restricts motion calculations to ROIs that bound the local motions; it then calculates where valid motion within the local ROIs was actually found. Any such motion area that is too small is rejected. Finally, the routine draws the motion. Examples of the output for a person flapping their arms is shown in Figure 17-9, where the output is drawn above the raw image for eight sequential frames (arranged in two rows). (For the full code, see *opencv_contrib* (as described in Appendix B). If you download *opencv_contrib*, the code is in *.../opencv_contrib/modules/optflow/samples/motempl.cpp*.) In the same sequence, "Y" postures were recognized by the shape descriptors (Hu moments) discussed in Chapter 14, although the shape recognition is not included in the *samples* code.

```
...
  for( i = -1; i < (int)brects.size(); i++ ) {

    cv::Rect roi; Scalar color; double magnitude;
    cv::Mat  maski = mask;
    if( i < 0 ) {

      // case of the whole image
```

```
      //
      roi       = Rect( 0, 0, img.cols, img.rows );
      color     = Scalar::all( 255 );
      magnitude = 100;

    } else {

      // i-th motion component
      //
      roi       = brects[i];
      if( roi.area() < 3000 ) continue;    // reject very small components
      color     = Scalar( 0, 0, 255 );
      magnitude = 30;
      maski     = mask(roi);

    }

    double angle = cv::motempl::calcGlobalOrientation(
      orient(roi),
      maski,
      mhi(roi),
      timestamp,
      MHI_DURATION
    );

    // ...[find regions of valid motion]...

    // ...[reset ROI regions]...

    // ...[skip small valid motion regions]...

    // ...[draw the motions]...
  }
...
```

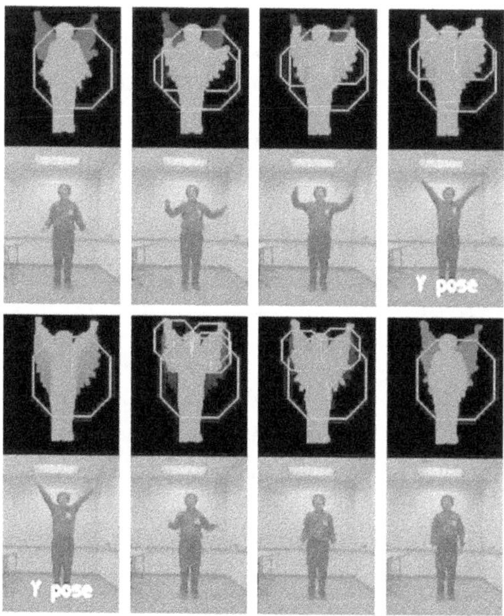


*Figure 17-9. Results of motion template routine: going across and top to bottom, a person moving and the resulting global motions indicated in large octagons and local motions indicated in small octagons; also, the "Y" pose can be recognized via shape descriptors (Hu moments)*

# Estimators

Suppose we are tracking a person who is walking across the view of a video camera. At each frame, we make a determination of the person's location. We could do this any number of ways, as we have seen, but for now all that is important is that we determine an estimate of the person's position at each frame. This estimation is not likely to be extremely accurate. The reasons for this are many—for example, inaccuracies in the sensor, approximations in earlier processing stages, issues arising from occlusion or shadows, or the apparent shape change due to the person's legs and arms swinging as they walk. Whatever the source, we expect that these measurements will vary, perhaps somewhat randomly, about the "actual" values that might be received from some more ideal sensor or sensing pathway. We can think of all these inaccuracies, taken together, as simply contributing *noise* to our measurement process.

We'd like to be able to estimate this person's motion in a way that makes maximal use of the measurements we've made. Thus, the cumulative effect of our many measurements could allow us to detect the part of the person's observed trajectory that does not arise from noise. The trivial example would be if we knew the person was not moving. In this case, our intuition tells us that we could probably average all of the measurements we had made to get the best notion of the person's actual location. The problem of *motion estimation* addresses both why our intuition gives this prescription for a stationary person, and more importantly, how we would generalize that result to moving objects.

The key additional ingredient we will need is a *model* for the person's motion. For example, we might model the person's motion with the following statement: "A person enters the frame from one side and walks across the frame at constant velocity." Given this model, we can ask not only where the person is but also what parameters of the model (in this case, the person's velocity) are best supported by our observations.

This task is divided into two phases (see Figure 17-10). In the first phase, typically called the *prediction phase*, we use information learned in the past to further refine our model for what the next location of the person (or object) will be. In the second phase, the *correction phase*, we make a measurement and then reconcile that measurement with the predictions based on our previous measurements (i.e., our model).

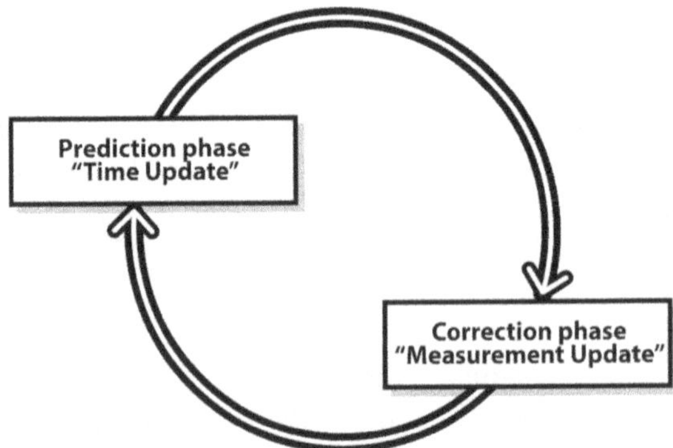


*Figure 17-10. Two-phase estimator cycle: prediction based on prior data followed by reconciliation of the newest measurement*

The machinery for accomplishing the two-phase estimation task falls generally under the heading of *estimators*, with the *Kalman filter* [Kalman60] being the most widely used technique. In addition to the Kalman filter, another important method is the *condensation algorithm*, which is a computer-vision implementation of a broader class of methods known as *particle filters*. The primary difference between the Kalman filter and the condensation algorithm is how the state probability density is described. In the following sections, we will explore the Kalman filter in detail and then touch on some related techniques. We will not dwell on the condensation algorithm, as there is no implementation of it in the OpenCV library. We will, however, touch on what it is and how it differs from the Kalman filter–related techniques as we progress.

## The Kalman Filter

First introduced in 1960, the Kalman filter has risen to great prominence in a wide variety of signal-processing contexts. The basic idea behind the Kalman filter is that, under a strong but reasonable set of assumptions,[17] it is possible—given a history of measurements of a system—to build a model for the current state of the system that maximizes the a posteriori[18] probability given those previous measurements. In addition, it turns out that we can do this estimation without keeping a long history of the previous measurements themselves. Instead, we iteratively update our model of a system's state as we make successive measurements and keep only the most recent model for use in the next iteration. This greatly simplifies the computational implications of this method. We will provide a basic introduction to the Kalman filter here; for a more detailed (but still very accessible) introduction, see Welsh and Bishop [Welsh95].

### What goes in and what comes out

The Kalman filter is an estimator. This means that it helps us integrate information we have about the state of a system, information we have about the dynamics of the system, and new information we might learn by observation while the system is operating. In practice, there are important limitations to how we can express each of these things. The first important limitation is that the Kalman filter will help us only with systems whose state can be expressed in terms of a vector, representing what we think the current value is for each of its degrees of freedom, and a matrix, expressing our uncertainty about those same degrees of freedom. (It is a matrix because such an uncertainty includes not only the variance of individual degrees of freedom, but also possible covariance between them.)

The *state vector* can contain any variables we think are relevant to the system we are working with. In the example of a person walking across an image, the components of this state vector might be their current position and their velocity: $\vec{x}_i^* = (x^*, v_x^*)$. In some cases, it is appropriate to have two dimensions each for position and velocity: $\vec{x}_i^* = (x^*, y^*, v_x^*, v_y^*)$. In complex systems there may be many more components to this vector. State vectors are normally accompanied by a subscript indicating the time step; the little "star" indicates that we are talking about our best estimate at that

17 By "reasonable," we mean something like "sufficiently unrestrictive that the method is useful for a variety of actual problems arising in the real world." "Reasonable" just seemed like less of a mouthful.

18 The modifier *a posteriori* is academic jargon for "with hindsight." Thus, when we say that such and such a distribution "maximizes the *a posteriori* probability," what we mean is that that distribution, which is essentially a possible explanation of "what really happened," is actually the most likely one given the data we have observed—that is, looking back on it all in retrospect.

particular time.[19] The state vector is accompanied by a *covariance matrix*, normally written as $\Sigma_i^*$. This matrix will be an $n \times n$ matrix if $\vec{x}_i^*$ has $n$ elements.

The Kalman filter will allow us to start with a state $(\vec{x}_0^*, \Sigma_0^*)$, and to estimate subsequent states $(\vec{x}_i^*, \Sigma_i^*)$ at later times. Two very important observations are in order about this statement. The first is that every one of these states has the same form; that is, it is expressed as a mean, which represents what we think the most likely state of the system is, and the covariance, which expresses our uncertainty about that mean. This means that only certain kinds of states can be handled by a Kalman filter. Consider, for example, a car that is driving down the road. It is perfectly reasonable to model such a system with a Kalman filter. Consider, however, the case of the car coming to a fork in the road. It could proceed to the right, or to the left, but not straight ahead. If we do not yet know which way the car has turned, the Kalman filter will not work for us. This is because the states represented by the filter are Gaussian distributions—basically, blobs with a peak in the middle. In the fork-in-the-road case, the car is likely to have gone right, and likely to have gone left, but unlikely to have gone straight and crashed. Unfortunately, any Gaussian distribution that gives equal weight to the right and left sides must give yet greater weight to the middle. This is because the Gaussian distribution is *unimodal*.[20]

The second important assumption is that the system must have started out in a state we can describe by such a Gaussian distribution. This state, written $(\vec{x}_0^*, \Sigma_0^*)$, is called a *prior distribution* (note the zero subscript). We will see that it is always possible to start a system with what is called an *uninformative prior*, in which the mean is arbitrary and the covariance is taken to have huge values in it—which will basically express the idea that we "don't know anything." But what is important is that the prior must be Gaussian and that we must always provide one.

### Assumptions required by the Kalman filter

Before we go into the details of how to work with the Kalman filter, let's take a moment to look at the "strong but reasonable" set of assumptions we mentioned earlier. There are three important assumptions required in the theoretical construction of the Kalman filter: (1) the system being modeled is linear, (2) the noise that measurements are subject to is "white," and (3) this noise is also Gaussian in nature. The first assumption means (in effect) that the state of the system at time $k$ can be expressed as a vector, and that the state of the system a moment later, at time $k + 1$ (also a vector), can be expressed as the state at time $k$ multiplied by some matrix $F$.

19 Many authors will use a "hat" instead of the "star." We opt for the latter as it allows us to make vectors explicit without the visual clutter of putting hats on top of arrows.

20 The particle filter or condensation algorithm alluded to earlier is most primarily an alternative way of formulating the state such that the state need not be unimodal, or even Gaussian.

The additional assumptions that the noise is both white and Gaussian means that any noise in the system is not correlated in time and that its amplitude can be accurately modeled with only an average and a covariance (i.e., the noise is completely described by its first and second moments). Although these assumptions may seem restrictive, they actually apply to a surprisingly general set of circumstances.

What does it mean to "maximize the a posteriori probability given the previous measurements"? It means that the new model we construct after making a measurement—taking into account both our previous model with its uncertainty and the new measurement with *its* uncertainty—is the model that has the highest probability of being correct. For our purposes, this means that the Kalman filter is, given the three assumptions, the best way to combine data from different sources or from the same source at different times. We start with what we know, we obtain new information, and then we decide to change what we know based on how certain we are about the old and new information using a weighted combination of the old and the new.

Let's work all this out with a little math for the case of one-dimensional motion. You can skip the next section if you want, but linear systems and Gaussians are so friendly that Dr. Kalman might be upset if you didn't at least give it a try.

## Information fusion

So what's the gist of the Kalman filter? *Information fusion.* Suppose you want to know where some point is on a line—a simple one-dimensional scenario.[21] As a result of noise, suppose you have two unreliable (in a Gaussian sense) reports about where the object is: locations $x_1$ and $x_2$. Because there is Gaussian uncertainty in these measurements, they have means of $\bar{x}_1$ and $\bar{x}_2$ together with standard deviations $\sigma_1$ and $\sigma_2$. The standard deviations are, in fact, expressions of our uncertainty regarding how good our measurements are. For each measurement, the implied probability distribution as a function of location is the *Gaussian distribution:*[22]

$$p_i(x) = N(x; \bar{x}_i, \sigma_i^2) \equiv \frac{1}{\sigma_i\sqrt{2\pi}}\exp\left(-\frac{(x - \bar{x}_i)^2}{2\sigma_i^2}\right), \ (i = 1, 2)$$

Given two such measurements, each with a Gaussian probability distribution, we would expect that the probability density for some value of $x$ given both measurements would be proportional to $p_{12}(x) = p_1(x) \cdot p_2(x)$. It turns out that this product is

21 For a more detailed explanation that follows a similar trajectory, refer to J. D. Schutter, J. De Geeter, T. Lefebvre, and H. Bruyninckx, "Kalman Filters: A Tutorial" (*http://citeseer.ist.psu.edu/443226.html*).

22 The notation $N(\bar{x}, \sigma^2)$ is common shorthand for the *Gaussian* or *normal* distribution, and means "A normal distribution with mean $\bar{x}$ and variance $\sigma^2$." We will use the notation $N(x; \bar{x}, \sigma^2)$ when it is helpful to explicitly describe a function of $x$ that happens to be a normal distribution.

another Gaussian distribution, and we can compute the mean and standard deviation of this new distribution as follows. Given that:

$$p_{12}(x) \propto \exp\left(-\frac{(x-x_1)^2}{2\sigma_1^2}\right)\exp\left(-\frac{(x-x_2)^2}{2\sigma_2^2}\right) = \exp\left(-\frac{(x-x_1)^2}{2\sigma_1^2} - \frac{(x-x_2)^2}{2\sigma_2^2}\right)$$

Given also that a Gaussian distribution is maximal at the average value, we can find that average value simply by computing the derivative of $p(x)$ with respect to $x$. Where a function is maximal its derivative is 0, so:

$$\left.\frac{dp_{12}}{dx}\right|_{x_{12}} = -\left[\frac{x_{12}-x_1}{\sigma_1^2} + \frac{x_{12}-x_2}{\sigma_2^2}\right] \cdot p_{12}(\bar{x}_{12}) = 0$$

Since the probability distribution function $p(x)$ is never 0, it follows that the term in brackets must be 0. Solving that equation for $x$ gives us this very important relation:

$$\bar{x}_{12} = \left(\frac{\sigma_2^2}{\sigma_1^2+\sigma_2^2}\right)\bar{x}_1 + \left(\frac{\sigma_1^2}{\sigma_1^2+\sigma_2^2}\right)\bar{x}_2$$

Thus, the new mean value $\bar{x}_{12}$ is just a weighted combination of the two measured means. What is critically important, however, is that the weighting is determined by (and only by) the relative uncertainties of the two measurements. Observe, for example, that if the uncertainty $\sigma_2$ of the second measurement is particularly large, then the new mean will be essentially the same as the mean $x_1$, the more certain previous measurement.

With the new mean $\bar{x}_{12}$ in hand, we can substitute this value into our expression for $p_{12}(x)$ and, after substantial rearranging,[23] identify the uncertainty $\sigma_{12}^2$ as:

$$\sigma_{12}^2 = \frac{\sigma_1^2\sigma_2^2}{\sigma_1^2+\sigma_2^2}.$$

At this point, you are probably wondering what this tells us. Actually, it tells us a lot. It says that when we make a new measurement with a new mean and uncertainty, we can combine that measurement with the mean and uncertainty we already have to obtain a new state characterized by a still newer mean and uncertainty. (We also now have numerical expressions for these things, which will come in handy momentarily.)

23 The rearranging is a bit messy. If you want to verify all this, it is much easier to (1) start with the equation for the Gaussian distribution $p12(x)$ in terms of $\bar{x}_{12}$ and $\sigma_{12}$, (2) substitute in the equations that relate $\bar{x}_{12}$ to $\bar{x}_1$ and $\bar{x}_2$ and those that relate $\sigma_{12}$ to $\sigma_1$ and $\sigma_2$, and (3) verify that the result can be separated into the product of the Gaussians with which we started.

This property that two Gaussian measurements, when combined, are equivalent to a single Gaussian measurement (with a computable mean and uncertainty) is the most important feature for us. It means that when we have $M$ measurements, we can combine the first two, then the third with the combination of the first two, then the fourth with the combination of the first three, and so on. This is what happens with tracking in computer vision: we obtain one measurement followed by another followed by another.

Thinking of our measurements $(x_i, \sigma_i)$ as time steps, we can compute the current state of our estimation $(x_i^*, \sigma_i^*)$ as follows. First, assume we have some initial notion of where we think our object is; this is the prior probability. This initial notion will be characterized by a mean and an uncertainty $x_0^*$ and $\sigma_0^*$. (Recall that we introduced the little stars or asterisks to denote that this is a current estimate, as opposed to a measurement.)

Next, we get our first measurement $(x_1, \sigma_1)$ at time step 1. All we have to go on is this new measurement and the prior, but we can substitute these into our optimal estimation equation. Our optimal estimate will be the result of combining our previous state estimate—in this case the prior $(x_0^*, \sigma_0^*)$—with our new measurement $(x_1, \sigma_1)$.

$$x_1^* = \left(\frac{\sigma_1^2}{\sigma_0^{*2} + \sigma_1^2}\right)x_0^* + \left(\frac{\sigma_0^{*2}}{\sigma_0^{*2} + \sigma_1^2}\right)x_1$$

Rearranging this equation gives us the following useful iterable form:

$$x_1^* = x_0^* + \left(\frac{\sigma_0^{*2}}{\sigma_0^{*2} + \sigma_1^2}\right)(x_1 - x_0^*)$$

Before we worry about just what this is useful for, we should also compute the analogous equation for $\sigma_1^*$.

$$\sigma_1^{*2} = \frac{\sigma_0^{*2}\sigma_1^2}{\sigma_0^{*2} + \sigma_1^2}$$

A rearrangement similar to what we did for $x_1^*$ yields an iterable equation for estimating variance given a new measurement:

$$\sigma_1^{*2} = \left(1 - \frac{\sigma_0^{*2}}{\sigma_0^{*2} + \sigma_1^2}\right)\sigma_0^{*2}$$

In their current form, these equations allow us to separate clearly the "old" information (what we knew before a new measurement was made—the part with the stars) from the "new" information (what our latest measurement told us—no stars). The new information $(x_1 - x_0^*)$, seen at time step 1, is called the *innovation*. We can also see that our optimal iterative update factor is now:

$$K = \left(\frac{\sigma_0^{*2}}{\sigma_0^{*2} + \sigma_1^2}\right)$$

This factor is known as the *update gain*. Using this definition for $K$, we obtain a convenient recursion form. This is because there is nothing special about the derivation up to this point or about the prior distribution or the new measurement. In effect, after our first measurement, we have a new state estimate $(x_1^*, \sigma_1^*)$, which is going to function relative to the new measurement $(x_2, \sigma_2)$ in exactly the same way as $(x_0^*, \sigma_0^*)$ did relative to $(x_1, \sigma_1)$. In effect, $(x_1^*, \sigma_1^*)$ is a new prior (i.e., what we "previously believed" before the next measurement arrived). Putting this realization into equations:

$$x_k^* = x_{k-1}^* + K(x_k - x_{k-1}^*)$$

$$\sigma_k^{*2} = (1 - K)\sigma_{k-1}^{*2}$$

and

$$K = \left(\frac{\sigma_{k-1}^{*2}}{\sigma_{k-1}^{*2} + \sigma_k^2}\right)$$

In the Kalman filter literature, if the discussion is about a general series of measurements, then it is traditional to refer to the "current" time step as $k$, and the previous time step is thus $k - 1$ (as opposed to $k + 1$ and $k$, respectively). Figure 17-11 shows this update process sharpening and converging on the true distribution.

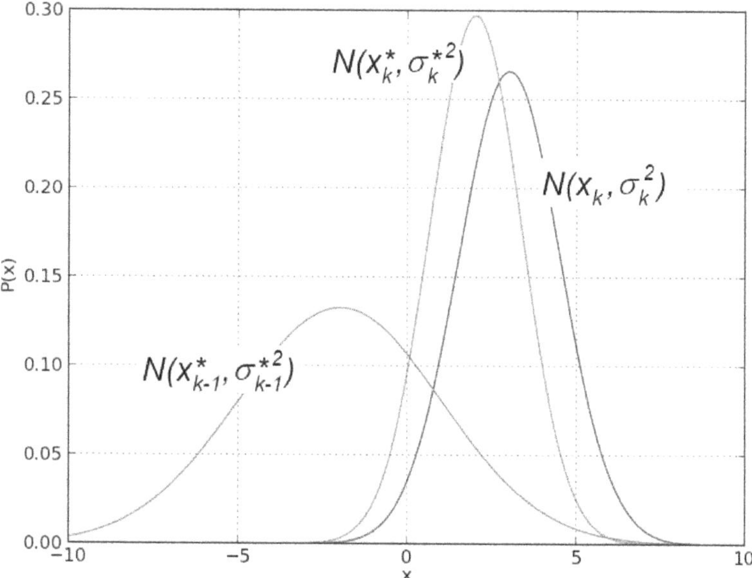


*Figure 17-11. Combining our prior belief* $N(x_{k-1}^*, \sigma_{k-1}^{*2})$ *[left] with our measurement observation* $N(x_k, \sigma_k^2)$ *[right], the result is our new estimate:* $N(x_k^*, \sigma_k^{*2})$ *[center]*

#### Systems with dynamics

In our simple one-dimensional example, we considered the case of an object being located at some point $x$, and a series of successive measurements of that point. In that case we did not specifically consider the case in which the object might actually be moving in between measurements. In this new case we will have what is called the *prediction phase*. During the prediction phase, we use what we know to figure out where we expect the system to be before we attempt to integrate a new measurement.

In practice, the prediction phase is done immediately after a new measurement is made, but before the new measurement is incorporated into our estimation of the state of the system. An example of this might be when we measure the position of a car at time $t$, then again at time $t + dt$. If the car has some velocity $v$, then we do not just incorporate the second measurement directly. We first *fast-forward* our model based on what we knew at time $t$ so that we have a model not only of the system at time $t$ but also of the system at time $t + dt$, the instant before the new information is incorporated. In this way, the new information, acquired at time $t + dt$, is fused not with the old model of the system, but with the old model of the system projected forward to time $t + dt$. This is the meaning of the cycle depicted in Figure 17-10. In the context of Kalman filters, there are three kinds of motion that we would like to consider.

The first is *dynamical motion*. This is motion that we expect as a direct result of the state of the system when last we measured it. If we measured the system to be at position $x$ with some velocity $v$ at time $t$, then at time $t + dt$ we would expect the system to be located at position $x + v \cdot dt$, possibly still with velocity $v$.

The second form of motion is called *control motion*. Control motion is motion that we expect because of some external influence applied to the system of which, for whatever reason, we happen to be aware. As the name implies, the most common example of control motion is when we are estimating the state of a system that we ourselves have some control over, and we know what we did to bring about the motion. This situation arises often in robotic systems where the control is the system telling the robot, for example, to accelerate or to go forward. Clearly, in this case, if the robot was at $x$ and moving with velocity $v$ at time $t$, then at time $t + dt$ we expect it to have moved not only to $x + v \cdot dt$ (as it would have done without the control), but also a little farther, since we know we told it to accelerate.

The final important class of motion is *random motion*. Even in our simple one-dimensional example, if whatever we were looking at had a possibility of moving on its own for whatever reason, we would want to include random motion in our prediction step. The effect of such random motion will be to simply increase the uncertainty of our state estimate with the passage of time. Random motion includes any motions that are not known or under our control. As with everything else in the Kalman filter framework, however, there is an assumption that this random motion is either Gaussian (i.e., a kind of random walk) or that it can at least be effectively modeled as Gaussian.

These dynamical elements are input into our simulation model by the *update step* before a new measurement can be fused. This update step first applies any knowledge we have about the motion of the object according to its prior state, then any additional information resulting from actions that we ourselves have taken or that we know have been taken on the system from another outside agent, and then finally our estimation of random events that might have changed the state of the system along the way. Once those factors have been applied, we can incorporate our next new measurement.

In practice, the dynamical motion is particularly important when the state of the system is more complex than our simulation model. Often when an object is moving, there are multiple components to the state vector, such as the position as well as the velocity. In this case, of course, the state is evolved according to the velocity that we believe it to have. Handling systems with multiple components to the state is the topic of the next section. In order to handle these new aspects of the situation, however, we will need to develop our mathematical machinery a little further as well.

### Kalman equations

We are now ready to put everything together and understand how to handle a more general model. Such a model will include a multidimensional state vector, dynamical motion, control motion, random motion, measurement, measurement uncertainty, and the techniques necessary to fuse together our estimates of where the system has gone with what we subsequently measure.

The first thing we need to do is to generalize our discussion to states that contain many state variables. The simplest example of this might be in the context of video tracking, where objects can move in two or three dimensions. In general, the state might contain additional elements, such as the velocity of an object being tracked. We will generalize the description of the state at time step $k$ to be the following function of the state at time step $k - 1$:

$$\vec{x}_k = F\vec{x}_{k-1} + B\vec{u}_k + \vec{w}_k$$

Here $\vec{x}_k$ is now the $n$-dimensional vector of state components and $F$ is an $n \times n$ matrix, sometimes called the *transfer matrix* or *transition matrix*, which multiplies $\vec{x}_{k-1}$. The vector $\vec{u}_k$ is new. It's there to allow external controls on the system, and it consists of a $c$-dimensional vector referred to as the *control inputs*; $B$ is an $n \times c$ matrix that relates these control inputs to the state change.[24] The variable $\vec{w}_k$ is a random variable (usually called the *process noise*) associated with random events or forces that directly affect the actual state of the system. We assume that the components of $\vec{w}_k$ have a zero mean Gaussian distribution, $N(\vec{0}, Q_k)$ for some $n \times n$ covariance matrix $Q_k$ (which could, in principle, vary with time, but typically it does not).

In general, we make measurements $\vec{Z}_k$ that may or may not be direct measurements of the state variable $\vec{X}_k$. (For example, if you want to know how fast a car is moving then you could either measure its speed with a radar gun or measure the sound coming from its tailpipe; in the former case, $\vec{Z}_k$ will be $\vec{X}_k$ with some added measurement noise, but in the latter case, the relationship is not direct in this way.) We can summarize this situation by saying that we measure the $m$-dimensional vector of measurements $\vec{Z}_k$ given by:

$$\vec{z}_k = H\vec{x}_k + \vec{v}_k$$

24 The astute reader, or one who already knows something about Kalman filters, will notice another important assumption we slipped in—namely, that there is a linear relationship (via matrix multiplication) between the controls $u_k$ and the change in state. In practical applications, this is often the first assumption to break down.

Here $H$ is an $m \times n$ matrix and $\vec{v}_k$ is the measurement error, which is also assumed to have Gaussian distributions $N(0, R_k)$ for some $m \times m$ covariance matrix $R_k$.[25]

Before we get totally lost, let's consider a particular realistic situation of taking measurements on a car driving in a parking lot. We might imagine that the state of the car could be summarized by two position variables, $x$ and $y$, and two velocities, $V_x$ and $V_y$. These four variables would be the elements of the state vector $\vec{X}_k$. This suggests that the correct form for $F$ is:

$$\vec{x}_k = \begin{bmatrix} x \\ y \\ v_x \\ v_y \end{bmatrix}, F = \begin{bmatrix} 1 & 0 & dt & 0 \\ 0 & 1 & 0 & dt \\ 0 & 0 & 1 & 0 \\ 0 & 0 & 0 & 1 \end{bmatrix}$$

However, when using a camera to make measurements of the car's state, we probably measure only the position variables:

$$\vec{z}_k = \begin{bmatrix} z_x \\ x_y \end{bmatrix}$$

This implies that the structure of $H$ is something like:

$$H = \begin{bmatrix} 1 & 0 & 0 & 0 \\ 0 & 1 & 0 & 0 \end{bmatrix}$$

In this case, we might not really believe that the velocity of the car is constant and so would assign a value of $Q_k$ to reflect this. The actions of a driver not under our control are, in a sense, random from our point of view. We would choose $R_k$ based on our estimate of how accurately we have measured the car's position using (for example) our image-analysis techniques on a video stream.

All that remains now is to plug these expressions into the generalized forms of the update equations. The basic idea is the same, however. First we compute the a priori estimate $x_k^-$ of the state. It is relatively common (though not universal) in the literature to use the superscript minus sign to mean "at the time immediately prior to the new measurement"; we'll adopt that convention here as well. Thus the a priori estimate is given by:

$$\vec{x}_k^- = F\vec{x}_{k-1} + B\vec{u}_{k-1} + \vec{w}_k$$

25 The $k$ in these terms allows for them to vary with time but does not require this. In actual practice, it's common for $H$ and $R$ not to vary with time.

Using $\Sigma_k^-$ to denote the error covariance, the a priori estimate for this covariance at time $k$ is obtained from the value at time $k - 1$ by:

$$\Sigma_k^- = F\Sigma_{k-1}F^T + Q_{k-1}$$

This equation forms the basis of the predictive part of the estimator, and it tells us "what we expect" based on what we've already seen. From here we'll state (without derivation) what is often called the *Kalman gain* or the *blending factor*, which tells us how to weight new information against what we think we already know:

$$K_k = \Sigma_k^- H_k^T (H_k \Sigma_k^- H_k^T + R_k)^{-1}$$

Though this equation looks intimidating, it's really not so bad. We can understand it more easily by considering various simple cases. For our one-dimensional example in which we measured one position variable directly, $H_k$ is just a $1 \times 1$ matrix containing only a `1`! Thus, if our measurement error is $\sigma_{k+1}^2$, then $R_k$ is also a $1 \times 1$ matrix containing that value. Similarly, $\Sigma_k$ is just the variance $\sigma_k^2$. So that big equation boils down to just this:

$$K_k = \frac{\sigma_k^2}{\sigma_k^2 + \sigma_{k+1}^2}$$

Note that this is exactly what we thought it would be. The gain, which we first saw in the previous section, allows us to optimally compute the updated values for $\vec{x}_k$ and $\sigma_k$ when a new measurement is available:

$$\vec{x}_k = \vec{x}_k^- + K_k(\vec{z}_k^- - H_k\vec{x}_k^-)$$

and:

$$\Sigma_k = (I - K_k H_k)\Sigma_k^-$$

Once again, these equations look intimidating at first, but in the context of our simple one-dimensional discussion, it's really not as bad as it looks. The optimal weights and gains are obtained by the same methodology as for the one-dimensional case, except this time we minimize the uncertainty of our position state $x$ by setting to `0` the partial derivatives with respect to $x$ before solving. We can show the relationship with the simpler one-dimensional case by first setting $F = I$ (where $I$ is the identity matrix), $B = 0$, and $Q = 0$. The similarity to our one-dimensional filter derivation is then revealed as we make the following substitutions in our more general equations:

$$\vec{x}_k \leftarrow \hat{x}_2$$

$$\vec{x}_k^- \leftarrow \hat{x}_2^-$$

$$K_k \leftarrow K$$

$$\vec{z}_k \leftarrow x_2$$

$$H_k \leftarrow I$$

$$\Sigma_k \leftarrow \hat{\sigma}_2^2$$

$$\Sigma_k^- \leftarrow \hat{\sigma}_1^2$$

$$R_k \leftarrow \sigma_2^2$$

$$I \leftarrow 1$$

### Tracking in OpenCV with cv::KalmanFilter

With all of this at our disposal, you probably feel either that you don't need OpenCV to do anything for you, or that you desperately need OpenCV to do all of this for you. Fortunately, the library and its authors are amenable to either interpretation. The Kalman filter is (not surprisingly) represented in OpenCV by an object called `cv::KalmanFilter`. The declaration for the `cv::KalmanFilter` object has the following form:

```
class cv::KalmanFilter {

public:
  cv::KalmanFilter();

  cv::KalmanFilter(
    int dynamParams,              // Dimensionality of state vector
    int measureParams,            // Dimensionality of measurement vector
    int controlParams = 0,        // Dimensionality of control vector
    int type          = CV_32F    // Type for matrices (CV_32F or F64)
  );

  //! re-initializes Kalman filter. The previous content is destroyed.
  void init(
    int dynamParams,              // Dimensionality of state vector
    int measureParams,            // Dimensionality of measurement vector
    int controlParams = 0,        // Dimensionality of control vector
    int type          = CV_32F    // Type for matrices (CV_32F or F64)
  );
```

```
    //! computes predicted state
    const cv::Mat& predict(
      const cv::Mat& control = cv::Mat() // External applied control vector (u_k)
    );

    //! updates the predicted state from the measurement
    const cv::Mat & correct(
      const cv::Mat& measurement  // Measurement vector (z_k)
    );

    cv::Mat transitionMatrix;      // state transition matrix (F)
    cv::Mat controlMatrix;         // control matrix (B) (w/o control)
    cv::Mat measurementMatrix;     // measurement matrix (H)
    cv::Mat processNoiseCov;       // process noise covariance matrix (Q)
    cv::Mat measurementNoiseCov;   // measurement noise covariance matrix (R)

    cv::Mat statePre;          // predicted state (x'_k):
                               //   x_k        = F * x_(k-1) + B * u_k

    cv::Mat statePost;         // corrected state (x_k):
                               //   x_k        = x'_k + K_k * ( z_k - H * x'_k )

    cv::Mat errorCovPre;       // a-priori error estimate covariance matrix (P'_k):
                               //   \Sigma'_k = F * \Sigma_(k-1) * F^t + Q

    cv::Mat gain;              // Kalman gain matrix (K_k):
                               //   K_k = \Sigma'_k * H^t * inv(H*\Sigma'_k*H^t+R)

    cv::Mat errorCovPost;      // a posteriori error covariance matrix (\Sigma_k):
                               //   \Sigma_k  = ( I - K_k * H ) * \Sigma'_k

    ...
};
```

You can either create your filter object with the default constructor and then configure it with the `cv::KalmanFilter::init()` method, or just call the constructor that shares the same argument list with `cv::KalmanFilter::init()`. In either case, there are four arguments you will need.

The first is `dynamParams`; this is the number of dimensions of the state vector $\vec{x}_k$. It does not matter what dynamical parameters you will have—only the number of them matters. Remember that their interpretation will be set by the various other components of the filter (notably the state transition matrix $F$). The next parameter is `measureParams`; this is the number of dimensions that are present in a measurement, the dimension of $z_k$. As with $x_k$, it is the other components of the filter that ultimately give $\vec{z}_k$ its meaning, so all we care about right now is the total dimension of the vector (in this case the meaning of $\vec{z}_k$ is primarily coming from the way we define the measurement matrix $H$, and its relationship to $\vec{x}_k$). If there are to be external controls to the system, then you must also specify the dimension of the control vector $\vec{u}_k$. By default, all of the internal components of the filter will be created as 32-bit floating-point type numbers. If you would like the filter to run in higher precision, you can set the final argument, `type`, to `cv::F64`.

The next two functions implement the Kalman filter process (Figure 17-10). Once the data is in the structure (we will talk about this in a moment), we can compute the prediction for the next time step by calling `cv::KalmanFilter::predict()` and then integrate our new measurements by calling `cv::KalmanFilter::correct()`.[26] The prediction method (optionally) accepts a control vector $\vec{u}_k$, while the correction method requires the measurement vector $\vec{z}_k$. After running each of these routines, we can read the state of the system being tracked. The result of `cv::KalmanFilter::correct()` is placed in `statePost`, while the result of `cv::KalmanFilter::predict()` is placed in `statePre`. You can read these values out of the member variables of the filter, or you can just use the return values of the two methods.

You will notice that, unlike many objects in OpenCV, all of the member variables are public. There are not "get/set" routines for them, so you just access them directly when you need to do so. For example, the state-transition matrix *F*, which is called `cv::KalmanFilter::transitionMatrix` in this context, is something you will just set up yourself for your system. This will all make a little more sense with an example.

#### Kalman filter example code

Clearly it is time for a good example. Let's take a relatively simple one and implement it explicitly. Imagine that we have a point moving around in a circle, like a car on a racetrack. The car moves with mostly a constant velocity around the track, but there is some variation (i.e., process noise). We measure the location of the car using a method such as tracking it via our vision algorithms. This generates some (unrelated and probably different) noise as well (i.e., measurement noise).

So our model is quite simple: the car has a position and an angular velocity at any moment in time. Together these factors form a two-dimensional state vector $\vec{x}_k$. However, our measurements are only of the car's position and so form a one-dimensional "vector" $\vec{z}_k$.

We'll write a program (Example 17-1) whose output will show the car circling around (in red) as well as the measurements we make (in yellow) and the location predicted by the Kalman filter (in white).

26 In recent OpenCV versions, the correction step may be omitted when there is no measurement available (e.g., the tracked object is occluded). In that case, `statePost` will set to the value of `statePre`.

We begin with the usual calls to include the library header files. We also define a macro that will prove useful when we want to transform the car's location from angular to Cartesian coordinates so we can draw on the screen.

*Example 17-1. Kalman filter example code*

```
#include "opencv2/opencv.hpp"
#include <iostream>

using namespace std;

#define phi2xy( mat )                                                  \
  cv::Point( cv::cvRound( img.cols/2 + img.cols/3*cos(mat.at<float>(0)) ),   \
             cv::cvRound( img.rows/2 - img.cols/3*sin(mat.at<float>(0)) ))

int main(int argc, char** argv) {

...
```

Next, we will create a random-number generator, an image to draw to, and the Kalman filter structure. Notice that we need to tell the Kalman filter how many dimensions the state variables are (2) and how many dimensions the measurement variables are (1).

```
    ...

        // Initialize, create Kalman filter object, window, random number
        // generator etc.
        //
        cv::Mat img( 500, 500, CV_8UC3 );
        cv::KalmanFilter kalman(2, 1, 0);

    ...
```

Once we have these building blocks in place, we create a matrix (really a vector, but in OpenCV we call everything a matrix) for the state `x_k` ($\vec{x}_k$), the process noise `w_k` ($\vec{w}_k$), the measurements `z_k` ($\vec{z}_k$), and the all-important transition matrix `transitionMatrix` ($F$). The state needs to be initialized to something, so we fill it with some reasonable random numbers that are narrowly distributed around zero.

The transition matrix is crucial because it relates the state of the system at time $k$ to the state at time $k + 1$. In this case, the transition matrix will be $2 \times 2$ (since the state vector is two-dimensional). It is, in fact, the transition matrix that gives meaning to the components of the state vector. We view `x_k` as representing the angular position of the car ($\varphi$) and the car's angular velocity ($\omega$). In this case, the transition matrix has the components [[1, `dt`], [0, 1]]. Hence, after we multiply by $F$, the state $(\phi, \omega)$ becomes $(\phi + \omega\, dt, \omega)$—that is, the angular velocity is unchanged but the angular position increases by an amount equal to the angular velocity multiplied by the time step. In our example we choose `dt = 1.0` for convenience, but in practice we'd need to use something like the time between sequential video frames.

```
...

    // state is (phi, delta_phi) - angle and angular velocity
    // Initialize with random guess.
    //
    cv::Mat x_k( 2, 1, CV_32F );
    randn( x_k, 0., 0.1 );

    // process noise
    //
    cv::Mat w_k( 2, 1, CV_32F );

    // measurements, only one parameter for angle
    //
    cv::Mat z_k = cv::Mat::zeros( 1, 1, CV_32F );

    // Transition matrix 'F' describes relationship between
    // model parameters at step k and at step k+1 (this is
    // the "dynamics" in our model.
    //
    float F[] = { 1, 1, 0, 1 };
    kalman.transitionMatrix = Mat( 2, 2, CV_32F, F ).clone();

...
```

The Kalman filter has other internal parameters that must be initialized. In particular, the $1 \times 2$ measurement matrix $H$ is initialized to $[1, 0]$ by a somewhat unintuitive use of the identity function. The covariances of the process noise and the measurement noise are set to reasonable but interesting values (you can play with these yourself), and we initialize the posterior error covariance to the identity as well (this is required to guarantee the meaningfulness of the first iteration; it will subsequently be overwritten).

Similarly, we initialize the posterior state (of the hypothetical step previous to the first one!) to a random value since we have no information at this time.

```
...

    // Initialize other Kalman filter parameters.
    //
    cv::setIdentity( kalman.measurementMatrix,   cv::Scalar(1)    );
    cv::setIdentity( kalman.processNoiseCov,     cv::Scalar(1e-5) );
    cv::setIdentity( kalman.measurementNoiseCov, cv::Scalar(1e-1) );
    cv::setIdentity( kalman.errorCovPost,        cv::Scalar(1)    );

    // choose random initial state
    //
    randn(kalman.statePost, 0., 0.1);
```

```
  for(;;) {
...
```

Finally we are ready to start up on the actual dynamics. First we ask the Kalman filter to predict what it thinks this step will yield (i.e., before giving it any new information); we call this y_k ($\vec{x}_k^-$). Then we proceed to generate the new value of z_k ($\vec{z}_k$, the measurement) for this iteration. By definition, this value is the "real" value x_k ($\vec{x}_k$) multiplied by the measurement matrix $H$ with the random measurement noise added. We must remark here that, in anything but a toy application such as this, you would not generate z_k from x_k; instead, a generating function would arise from the state of the world or your sensors. In this simulated case, we generate the measurements from an underlying "real" data model by adding random noise ourselves; this way, we can see the effect of the Kalman filter.

```
...

    // predict point position
    //
    cv::Mat y_k = kalman.predict();

    // generate measurement (z_k)
    //
    cv::randn(z_k, 0., sqrt((double)kalman.measurementNoiseCov.at<float>(0,0)));
    z_k = kalman.measurementMatrix*x_k + z_k;

...
```

Draw the three points corresponding to the observation we synthesized previously, the location predicted by the Kalman filter, and the underlying state (which we happen to know in this simulated case).

```
...

    // plot points (e.g., convert to planar co-ordinates and draw)
    //
    img = Scalar::all(0);
    cv::circle( img, phi2xy(z_k), 4, cv::Scalar(128,255,255) );     // observed
    cv::circle( img, phi2xy(y_k), 4, cv::Scalar(255,255,255), 2 ); // predicted
    cv::circle( img, phi2xy(x_k), 4, cv::Scalar(0,0,255) );         // actual

    cv::imshow( "Kalman", img );

...
```

At this point we are ready to begin working toward the next iteration. The first thing to do is again call the Kalman filter and inform it of our newest measurement. Next we will generate the process noise. We then use the transition matrix $F$ to time-step x_k forward one iteration, and then add the process noise we generated; now we are ready for another trip around.

```
...

    // adjust Kalman filter state
    //
    kalman.correct( z_k );

    // Apply the transition matrix 'F' (e.g., step time forward)
    // and also apply the "process" noise w_k
    //
    cv::randn(w_k, 0., sqrt((double)kalman.processNoiseCov.at<float>(0,0)));
    x_k = kalman.transitionMatrix*x_k + w_k;

    // exit if user hits 'Esc'
    if( (cv::waitKey( 100 ) & 255) == 27 ) break;
  }

  return 0;
}
```

As you can see, the Kalman filter part was not that complicated; half of the required code was just generating some information to push into it. In any case, we should summarize everything we've done, just to be sure it all makes sense.

We started out by creating matrices to represent the state of the system and the measurements we would make. We defined both the transition and measurement matrices, and then initialized the noise covariances and other parameters of the filter.

After initializing the state vector to a random value, we called the Kalman filter and asked it to make its first prediction. Once we read out that prediction (which was not very meaningful this first time through), we drew to the screen what was predicted. We also synthesized a new observation and drew that on the screen for comparison with the filter's prediction. Next we passed the filter new information in the form of that new measurement, which it integrated into its internal model. Finally, we synthesized a new "real" state for the model so that we could iterate through the loop again.

When we run the code, the little red ball orbits around and around. The little yellow ball appears and disappears about the red ball, representing the noise that the Kalman filter is trying to "see through." The white ball rapidly converges down to moving in a small space around the red ball, showing that the Kalman filter has given a reasonable estimate of the motion of the particle (the car) within the framework of our model.

One topic that we did not address in our example is the use of control inputs. For example, if this were a radio-controlled car and we had some knowledge of what the person with the controller was doing, we could include that information into our model. In that case it might be that the velocity is being set by the controller. We'd then need to supply the matrix $B$ (`kalman.controlMatrix`) and also to provide a second argument for `cv::KalmanFilter::predict()` to accommodate the control vector $\vec{u}_k$.

## A Brief Note on the Extended Kalman Filter

You might have noticed that requiring the dynamics of the system to be linear in the underlying parameters is quite restrictive. It turns out that the Kalman filter is still useful to us when the dynamics are nonlinear, and the OpenCV Kalman filter routines remain useful as well.

Recall that "linear" meant (in effect) that the various steps in the definition of the Kalman filter could be represented with matrices. When might this not be the case? There are actually many possibilities. For example, suppose our control measure is the amount by which our car's gas pedal is depressed; the relationship between the car's velocity and the gas pedal's depression is not a linear one. Another common problem is a force on the car that is more naturally expressed in Cartesian coordinates, while the motion of the car (as in our example) is more naturally expressed in polar coordinates. This might arise if our car were instead a boat moving along a circular path but in uniform water current moving to some particular direction.

In all these cases, the Kalman filter is not, by itself, sufficient; the state at time $t + 1$ will not be a linear function of the state at time $t$. One way to address these nonlinearities is to *locally linearize* the relevant processes (e.g., the update $F$ or the control input response $B$). Thus, we'd need to compute new values for $F$ and $B$ at every time step based on the state $x$. These values would only approximate the real update and control functions in the vicinity of the particular value of $x$, but in practice this is often sufficient. This extension to the Kalman filter is known simply enough as the *extended Kalman filter* or simply *EKF* [Schmidt66].

OpenCV does not provide any specific routines to implement the EKF, but none are actually needed. All we have to do is recompute and reset the values of `kalman.transitionMatrix` and `kalman.controlMatrix` before each update.

In general, the matrix update equations we encountered earlier when we introduced the Kalman filter are now understood as a special case of the more general form:

$$\vec{x}_k = \vec{f}(\vec{x}_{k-1}, \vec{u}_k) + \vec{w}_k$$

$$\vec{z}_k = \vec{h}(\vec{x}_k) + \vec{v}_k$$

Here $f(x_{k-1}, u_k)$ and $h(x_k)$ are arbitrary nonlinear functions of their arguments. In order to use the linear formulation of the Kalman filter provided, we recomputed the matrices $F$ and $H$ at each time step according to the matrices $F_k$ and $H_k$ defined by:

$$[F_k]_{i,j} = \left.\frac{\partial f_i}{\partial x_j}\right|_{x^*_{k-1}, u_k}$$

$$[H_k]_{i,j} = \left.\frac{\partial h_i}{\partial x_j}\right|_{x^*_{k-1}}$$

Note that the evaluations of the partial derivatives that form the new matrices are performed at the estimated position from the previous time step $x^*_{k-1}$ (note the star). We can then substitute the values of $F_k$ and $H_k$ into our usual update equations.[27]

$$\vec{x}_k = F_k \vec{x}_{k-1} + \vec{w}_k$$

$$\vec{z}_k = H_k \vec{x}_k + \vec{v}_k$$

The Kalman filter has since been more elegantly extended to nonlinear systems in a formulation called the *unscented particle filter* [Merwe00]. A very good overview of the entire field of Kalman filtering, including the recent advances, as well as particle filters (which we touched on only briefly along the way), is given in [Thrun05].

# Summary

In this chapter we continued our investigation of tracking, adding dense optical flow techniques, mean-shift, Camshift, and motion templates to the sparse tracking methods that we learned in the previous chapter. We concluded the chapter with a discussion of recursive estimators such as the Kalman filter. After an overview of the mathematical theory of the Kalman filter, we saw how OpenCV makes this construct available to us. We concluded with a brief note on more advanced filters that can be used instead of the Kalman filter to handle more complex situations that do not conform perfectly to the assumptions required by the Kalman filter. Further optical flow and tracking algorithms are referenced in the `optflow` and `tracking` functions in the *opencv_contrib* directory described in Appendix B.

# Exercises

There are sample code routines included with the library that demonstrate many of the algorithms discussed in this chapter. Use these examples in the following exercises:

27 You will have noticed that the matrix $B$ disappeared. This is because in the general nonlinear case, it is not uncommon for the state $\vec{x}_k$ and the control $\vec{u}_k$ to be coupled in the update. For this reason, the generalization of $F\vec{x}_k + B\vec{u}_k$ is expressed as $\vec{f}(\vec{x}_k, \vec{u}_k)$—rather than, for example, $\vec{f}(x_{k-1}) + \vec{b}(\vec{u}_k)$—in the EKF case, and the effect of $\vec{u}_k$ is thus absorbed into $F_k$.

- *lkdemo.cpp* (sparse optical flow in *opencv/samples/cpp*)
- *fback.cpp* (Farnebäck dense optical flow in *samples/cpp*)
- *tvl1_optical_flow.cpp* (dense optical flow in *samples/cpp*)
- *camshiftdemo.cpp* (mean-shift tracking of colored regions in *samples/cpp*)
- *motempl.cpp* (motion templates, in *opencv_contrib/modules/optflow/samples*)
- *kalman.cpp* (Kalman filter in *samples/cpp*)
- *768x576.avi* (movie data files in *samples/data*)

1. Use a motion model that posits that the current state depends on the previous state's location and velocity. Combine the *lkdemo.cpp* (using only a few click points) with the Kalman filter to track Lucas-Kanade points better. Display the uncertainty around each point. Where does this tracking fail?

   Hint: use Lucas-Kanade as the observation model for the Kalman filter, and adjust noise so that it tracks. Keep motions reasonable.

2. Compile and run *fback.cpp* demo with Farnebäck's dense optical flow algorithm on *768x576.avi*. Modify it to print the pure algorithm execution time. Install OpenCL drivers on your machine (if you're using a Mac, you are probably already good to go), and see if it performs faster with OpenCL (make sure you are using OpenCV 3.0 or later version). Try to substitute Farnebäck's with TV-$L^1$ (see *tvl1_optical_flow.cpp*) or Simple Flow calls and measure time again.

3. Compile and run *camshiftdemo.cpp* using a web camera or color video of a moving colored object. Use the mouse to draw a (tight) box around the moving object; the routine will track it.

   a. In *camshiftdemo.cpp*, replace the `cv::camShift()` routine with `cv::meanShift()`. Describe situations where one tracker will work better than another.

   b. Write a function that will put down a grid of points in the initial `cv::meanShift()` box. Run both trackers at once.

   c. How can these two trackers be used together to make tracking more robust? Explain and/or experiment.

4. Compile and run the motion template code *motempl.cpp* with a web camera or using a previously stored movie file.

   a. Modify *motempl.cpp* so that it can do simple gesture recognition.

   b. If the camera was moving, explain how you would use motion stabilization code to enable motion templates to work also for moderately moving cameras.

5. A Kalman filter depends on linear dynamics and on Markov independence (i.e., it assumes the current state depends only on the immediate past state, not on all past states). Suppose you want to track an object whose movement is related to its previous location and its previous velocity but that you mistakenly include a dynamics term only for state dependence on the previous location—in other words, forgetting the previous velocity term.

   a. Do the Kalman assumptions still hold? If so, explain why; if not, explain how the assumptions were violated.

   b. How can a Kalman filter be made to still track when you forget some terms of the dynamics?

   Hint: think of the noise model.

6. Describe how you can track circular (nonlinear) motion using a linear state model (not extended) Kalman filter.

   Hint: how could you preprocess this to get back to linear dynamics?

# CHAPTER 18
# Camera Models and Calibration

Vision begins with the detection of light from the world. That light begins as rays emanating from some source (e.g., a light bulb or the sun), which travel through space until striking some object. When that light strikes the object, much of the light is absorbed, and what is not absorbed we perceive as the color of the object. Reflected light that makes its way to our eye (or our camera) is collected on our retina (or our imager). The geometry of this arrangement—particularly of the rays' travel from the object, through the lens in our eye or camera, and to the retina or imager—is of particular importance to practical computer vision.

A simple but useful model of how this happens is the pinhole camera model.[1] A *pinhole* is an imaginary wall with a tiny hole in the center that blocks all rays except those passing through the tiny aperture in the center. In this chapter, we will start with a pinhole camera model to get a handle on the basic geometry of projecting rays. Unfortunately, a real pinhole is not a very good way to make images because it does not gather enough light for rapid exposure. This is why our eyes and cameras use lenses to gather more light than what would be available at a single point. The downside, however, is that gathering more light with a lens not only forces us to move beyond the simple geometry of the pinhole model but also introduces distortions from the lens itself.

1 Knowledge of lenses goes back at least to Roman times. The pinhole camera model goes back almost 1,000 years to al-Hytham (1021) and is the classic way of introducing the geometric aspects of vision. Mathematical and physical advances followed in the 1600s and 1700s with Descartes, Kepler, Galileo, Newton, Hooke, Euler, Fermat, and Snell (see O'Connor [O'Connor02]). Some key modern texts for geometric vision include those by Trucco [Trucco98], Jaehne (also sometimes spelled Jähne) [Jaehne95; Jaehne97], Hartley and Zisserman [Hartley06], Forsyth and Ponce [Forsyth03], Shapiro and Stockman [Shapiro02], and Xu and Zhang [Xu96].

In this chapter, we will learn how, using *camera calibration*, to correct (mathematically) for the main deviations from the simple pinhole model that the use of lenses imposes on us. Camera calibration is also important for relating camera measurements to measurements in the real, three-dimensional world. This is important because scenes are not only three-dimensional; they are also physical spaces with physical units. Hence, the relation between the camera's natural units (pixels) and the units of the physical world (e.g., meters) is a critical component of any attempt to reconstruct a three-dimensional scene.

The process of camera calibration gives us both a model of the camera's geometry and a *distortion* model of the lens. These two informational models define the *intrinsic parameters* of the camera. In this chapter, we use these models to correct for lens distortions; in Chapter 19, we will use them to interpret the entire geometry of the physical scene.

We will begin by looking at camera models and the causes of lens distortion. From there, we will explore the *homography transform*, the mathematical instrument that allows us to capture the effects of the camera's basic behavior and of its various distortions and corrections. We will take some time to discuss exactly how the transformation that characterizes a particular camera can be calculated mathematically. Once we have all this in hand, we'll move on to the OpenCV functions that handle most of this work for us.

Just about all of this chapter is devoted to building enough theory that you will truly understand what is going into (and coming out of) the OpenCV function `cv::calibrateCamera()` as well as what that function is doing "under the hood." This is important stuff if you want to use the function responsibly. Having said that, if you are already an expert and simply want to know how to use OpenCV to do what you already understand, jump right ahead to "Calibration function" on page 675 and get to it. Note that Appendix B references other calibration patterns and techniques in the `ccalib` function group.

# Camera Model

We begin by looking at the simplest model of a camera, the pinhole. In this simple model, light is envisioned as entering from the scene or a distant object, but only a single ray enters the pinhole from any particular point in that scene. In a physical pinhole camera, this point is then "projected" onto an imaging surface. As a result, the image on this *image plane* (also called the *projective plane*) is always in focus, and the size of the image relative to the distant object is given by a single parameter of the camera: its *focal length*. For our idealized pinhole camera, the distance from the pinhole aperture to the screen is precisely the focal length.[2] This is shown in Figure 18-1, where $f$ is the focal length of the camera, $Z$ is the distance from the camera to the object, $X$ is the length of the object, and $x$ is the object's image on the imaging plane. In the figure, we can see by similar triangles that $-x/f = X/Z$, or:

$$-x = f \cdot \frac{X}{Z}$$

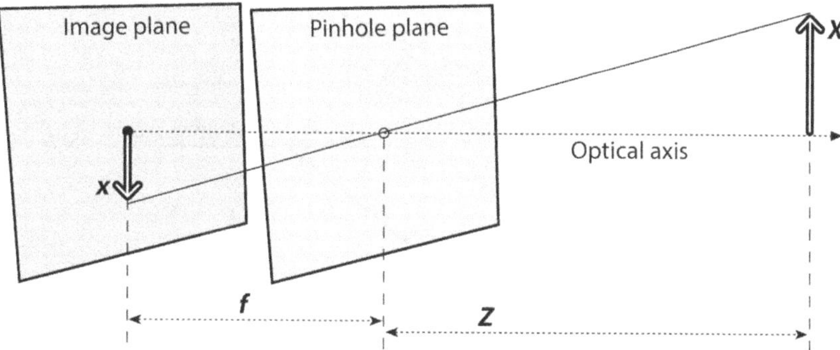


*Figure 18-1. Pinhole camera model: a pinhole (the pinhole aperture) lets through only those light rays that intersect a particular point in space; these rays then form an image by "projecting" onto an image plane*

We will now rearrange our pinhole camera model to a form that is equivalent but in which the math comes out easier. In Figure 18-2, we swap the pinhole and the image plane.[3] The main difference is that the object now appears right side up. The point in the pinhole is reinterpreted as the *center of projection*. In this way of looking at things, every ray leaves a point on the distant object and heads for the center of projection. The point at the intersection of the image plane and the optical axis is referred to as the *principal point*. On this *new* frontal image plane (see Figure 18-2), which is the equivalent of the old projective or image plane, the image of the distant object is exactly the same size as it was on the image plane in Figure 18-1. The image is generated by intersecting these rays with the image plane, which happens to be

2 You are probably used to thinking of the focal length in the context of lenses, in which case the focal length is a property of a particular lens, not the projection geometry. This is the result of a common abuse of terminology. It might be better to say that the "projection distance" is the property of the geometry and the "focal length" is the property of the lens. The $f$ in the previous equations is really the projection distance. For a lens, the image is in focus only if the focus length of the configuration matches the focal length of the lens, so people tend to use the terms interchangeably.

3 Typical of such mathematical abstractions, this new arrangement is not one that can be built physically; the image plane is simply a way of thinking of a "slice" through all of those rays that happen to strike the center of projection. This arrangement is, however, much easier to draw and do math with.

exactly a distance $f$ from the center of projection. This makes the similar triangles relationship $x/f = X/Z$ more directly evident than before. The negative sign is gone because the object image is no longer upside down.

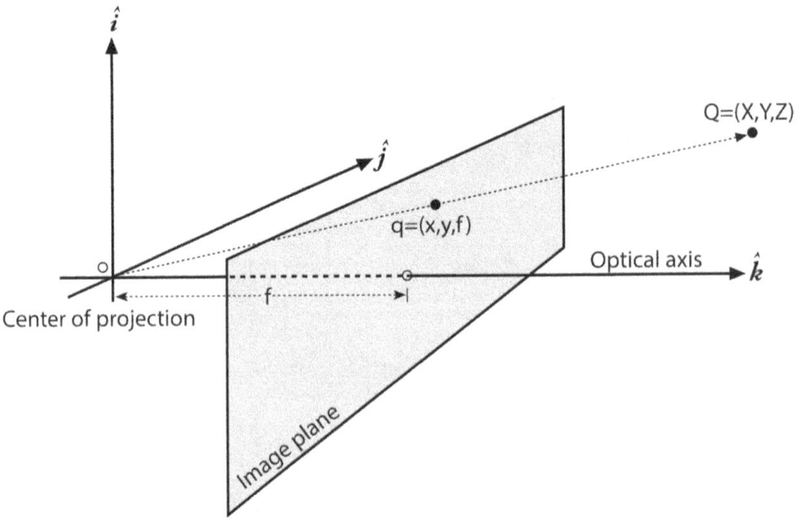


*Figure 18-2. A point $\vec{Q} = (X, Y, Z)$ is projected onto the image plane by the ray passing through the center of projection, and the resulting point on the image is $\vec{q} = (x, y, f)$; the image plane is really just the projection screen "pushed" in front of the pinhole (the math is equivalent but simpler this way)*

You might think that the principal point is equivalent to the center of the imager, but this would imply that some guy with tweezers and a tube of glue was able to attach the imager in your camera to micron accuracy. In fact, the center of the imager chip is usually not on the optical axis. We thus introduce two new parameters, $c_x$ and $c_y$, to model a possible displacement (away from the optic axis) of the center of coordinates on the projection screen. The result is that a relatively simple model in which a point $\vec{Q}$ in the physical world, whose coordinates are $(X, Y, Z)$, is projected onto the imager at some pixel location given by $(x_{screen}, y_{screen})$ in accordance with the following equations:[4]

$$x_{screen} = f_x \cdot \frac{X}{Z} + c_x \text{, and } y_{screen} = f_y \cdot \frac{Y}{Z} + c_y$$

4 Here the subscript *screen* is intended to remind you that the coordinates being computed are in the coordinate system of the screen (i.e., the imager). The difference between $(x_{screen}, y_{screen})$ in the equation and $(x,y)$ in Figure 18-2 is precisely the point of $c_x$ and $c_y$. Having said that, we will subsequently drop the "screen" subscript and simply use lowercase letters to describe coordinates on the imager.

Note that we have introduced two different focal lengths; the reason for this is that the individual pixels on a typical low-cost imager are rectangular rather than square. The focal length $f_x$, for example, is actually the product of the physical focal length of the lens and the size $s_x$ of the individual imager elements (this should make sense because $s_x$ has units of pixels per millimeter,[5] while $f$ has units of millimeters, which means that $f_x$ is in the required units of pixels). Of course, similar statements hold for $f_y$ and $s_y$. It is important to keep in mind, though, that $s_x$ and $s_y$ cannot be measured directly via any camera calibration process, and neither is the physical focal length $f$ directly measurable. We can derive only the combinations $f_x = F \cdot s_x$ and $f_y = F \cdot s_y$ without actually dismantling the camera and measuring its components directly.

## The Basics of Projective Geometry

The relation that maps a set of points $\vec{Q}_i$ in the physical world with coordinates ($X_i$, $Y_i$, $Z_i$) to the points on the projection screen with coordinates ($x_i$, $y_i$) is called a *projective transform*. When you are working with such transforms, it is convenient to use what are known as *homogeneous coordinates*. The homogeneous coordinates associated with a point in a projective space of dimension $n$ are typically expressed as an $(n + 1)$-dimensional vector (e.g., $x, y, z$ becomes $x, y, z, w$), with the additional restriction that any two points whose values are proportional are, in fact, equivalent points. In our case, the image plane is the projective space and it has two dimensions, so we will represent points on that plane as three-dimensional vectors $\vec{q} = (q_1, q_2, q_3)$. Recalling that all points having proportional values in the projective space are equivalent, we can recover the actual pixel coordinates by dividing through by $q_3$. This allows us to arrange the parameters that define our camera (i.e., $f_x$, $f_y$, $c_x$ and $c_y$) into a single $3 \times 3$ matrix, which we will call the *camera intrinsics matrix*.[6] The projection of the points in the physical world into the camera is now summarized by the following simple form:

$$\vec{q} = M \cdot \vec{Q}$$

where:

$$\vec{q} = \begin{bmatrix} x \\ y \\ w \end{bmatrix}, M = \begin{bmatrix} f_x & 0 & c_x \\ 0 & f_y & c_y \\ 0 & 0 & 1 \end{bmatrix}, \text{and} \quad \vec{Q} = \begin{bmatrix} X \\ Y \\ Z \end{bmatrix}$$

5 Of course, *millimeter* is just a stand-in for any physical unit you like. It could just as easily be *meter*, *micron*, or *furlong*. The point is that $s_x$ converts physical units to pixel units.

6 The approach OpenCV takes to camera intrinsics is derived from Heikkila and Silven; see [Heikkila97].

Multiplying this out, you will find that $w = Z$ and so, since the point $\vec{q}$ is in homogeneous coordinates, we can divide through by $w$ (or $Z$) in order to recover our earlier definitions. The minus sign is gone because we are now looking at the noninverted image on the projective plane in front of the pinhole rather than the inverted image on the projection screen behind the pinhole.

While we are on the topic of homogeneous coordinates, there are a few functions in the OpenCV library that are appropriate to introduce here. The functions `cv::convertPointsToHomogeneous()` and `cv::convertPointsFromHomogeneous()` allow us to convert to and from homogeneous coordinates.[7] They have the following prototypes:

```
void cv::convertPointsToHomogeneous(
  cv::InputArray  src,           // Input vector of N-dimensional points
  cv::OutputArray dst            // Result vector of (N+1)-dimensional points
);

void cv::convertPointsFromHomogeneous(
  cv::InputArray  src,           // Input vector of N-dimensional points
  cv::OutputArray dst            // Result vector of (N-1)-dimensional points
);
```

The first function expects a vector of $N$-dimensional points (in any of the usual representations) and constructs a vector of $(N + 1)$-dimensional points from that vector. All of the entries of the newly constructed vector associated with the added dimension are set to `1`. The result is that:

$$\overrightarrow{dst_i} = \begin{pmatrix} src_{i,0} & src_{i,1} & \dots & src_{i,N-1} & 1 \end{pmatrix}$$

The second function does the conversion back from homogeneous coordinates. Given an input vector of points of dimension $N$, it constructs a vector of $(N - 1)$-dimensional points by first dividing all of the components of each point by the value of the last components of the point's representation and then throwing away that component. The result is that:

$$\overrightarrow{dst_i} = \left( \frac{src_{i,0}}{src_{i,N-1}} \quad \frac{src_{i,1}}{src_{i,N-1}} \quad \dots \quad \frac{src_{i,N-2}}{src_{i,N-1}} \right)$$

7 There is also a third function, `cv::convertPointsHomogeneous()`, which is just a convenient way to call either of the former two. It will look at the dimensionality of the points provided in `dst` and determine automatically whether you want it to convert to or from homogeneous coordinates. This function is now considered obsolete. It is primarily present for backward compatibility. It should not be used in new code, as it substantially reduces code clarity.

With the ideal pinhole, we have a useful model for some of the three-dimensional geometry of vision. Remember, however, that very little light goes through a pinhole; thus, in practice, such an arrangement would make for very slow imaging while we wait for enough light to accumulate on whatever imager we are using. For a camera to form images at a faster rate, we must gather a lot of light over a wider area and bend (i.e., focus) that light to converge at the point of projection. To accomplish this, we use a lens. A lens can focus a large amount of light on a point to give us fast imaging, but it comes at the cost of introducing distortions.

## Rodrigues Transform

When dealing with three-dimensional spaces, one most often represents rotations in that space by 3 × 3 matrices. This representation is usually the most convenient because multiplying a vector by this matrix is equivalent to rotating the vector in some way. The downside is that it can be difficult to intuit just what 3 × 3 matrix goes with what rotation. Briefly, we are going to introduce an alternative representation for such rotations that is used by some of the OpenCV functions in this chapter, as well as a useful function for converting to and from this alternative representation.

This alternate, and somewhat easier-to-visualize,[8] representation for a rotation is essentially a vector about which the rotation operates together with a single angle. In this case it is standard practice to use only a single vector whose direction encodes the direction of the axis to be rotated around, and to use the length of the vector to encode the amount of rotation in a counterclockwise direction. This is easily done because the direction can be equally well represented by a vector of any magnitude; hence, we can choose the magnitude of our vector to be equal to the magnitude of the rotation. The relationship between these two representations, the matrix and the vector, is captured by the Rodrigues transform.[9]

Let $\vec{r}$ be the three-dimensional vector $\vec{r} = [r_x \quad r_y \quad r_z]$; this vector implicitly defines $\theta$, the magnitude of the rotation by the length (or magnitude) of $\vec{r}$. We can then convert from this axis-magnitude representation to a rotation matrix $R$ as follows:

$$R = \cos\theta \cdot I_3 + (1 - \cos\theta) \cdot \vec{r} \cdot \vec{r}^T + \sin\theta \cdot \begin{bmatrix} 0 & -r_z & r_y \\ r_z & 0 & -r_x \\ r_y & r_x & 0 \end{bmatrix}$$

8 This "easier" representation is not just for humans. Rotation in three-dimensional space has only three components. For numerical optimization procedures, it is more efficient to deal with the three components of the Rodrigues representation than with the nine components of a 3 × 3 rotation matrix.

9 Rodrigues was a 19th-century French mathematician.

We can also go from a rotation matrix back to the axis-magnitude representation by using:

$$\sin\theta \cdot \begin{bmatrix} 0 & -r_z & r_y \\ r_z & 0 & -r_x \\ r_y & r_x & 0 \end{bmatrix} = \frac{R - R^T}{2}$$

Thus we find ourselves in the situation of having one representation (the matrix representation) that is most convenient for computation and another representation (the Rodrigues representation) that is a little easier on the brain. OpenCV provides us with a function for converting from either representation to the other:

```
void cv::Rodrigues(
  cv::InputArray  src,                        // Input rotation vector or matrix
  cv::OutputArray dst,                        // Output rotation matrix or vector
  cv::OutputArray jacobian = cv::noArray() // Optional Jacobian (3x9 or 9x3)
);
```

Suppose we have the vector $\vec{r}$ and need the corresponding rotation matrix representation $R$; we set `src` to be the $3 \times 1$ vector $\vec{r}$ and `dst` to be the $3 \times 3$ rotation matrix $R$. Conversely, we can set `src` to be a $3 \times 3$ rotation matrix $R$ and `dst` to be a $3 \times 1$ vector $\vec{r}$. In either case, `cv::Rodrigues()` will do the right thing. The final argument is optional. If `jacobian` is something other than `cv::noArray()`, then it should be a pointer to a $3 \times 9$ or a $9 \times 3$ matrix that will be filled with the partial derivatives of the output array components with respect to the input array components. The `jacobian` outputs are mainly used for the internal optimization of the `cv::solvePnP()` and `cv::calibrateCamera()` functions; your use of the `cv::Rodrigues()` function will mostly be limited to converting the outputs of `cv::solvePnP()` and `cv::calibrateCamera()` from the Rodrigues format of $1 \times 3$ or $3 \times 1$ axis-angle vectors to rotation matrices. For this, you can leave `jacobian` set to `cv::noArray()`.

## Lens Distortions

In theory, it is possible to define a lens that will introduce no distortions. In practice, however, no lens is perfect. This is mainly for reasons of manufacturing; it is much easier to make a "spherical" lens than to make a more mathematically ideal "parabolic" lens. It is also difficult to mechanically align the lens and imager exactly. Here we describe the two main lens distortions and how to model them.[10] *Radial distortions* arise as a result of the shape of lens, whereas *tangential distortions* arise from the assembly process of the camera as a whole.

10 The approach to modeling lens distortion taken here derives mostly from Brown [Brown71] and earlier Fryer and Brown [Fryer86].

We start with radial distortion. The lenses of real cameras often noticeably distort the location of pixels near the edges of the imager. This bulging phenomenon is the source of the "barrel" or "fisheye" effect (see the room-divider lines at the top of Figure 18-17 for a good example). Figure 18-3 gives some intuition as to why this radial distortion occurs. With some lenses, rays farther from the center of the lens are bent more than those closer in. A typical inexpensive lens is, in effect, stronger than it ought to be as you get farther from the center. Barrel distortion is particularly noticeable in cheap web cameras but less apparent in high-end cameras, where a lot of effort is put into fancy lens systems that minimize radial distortion.

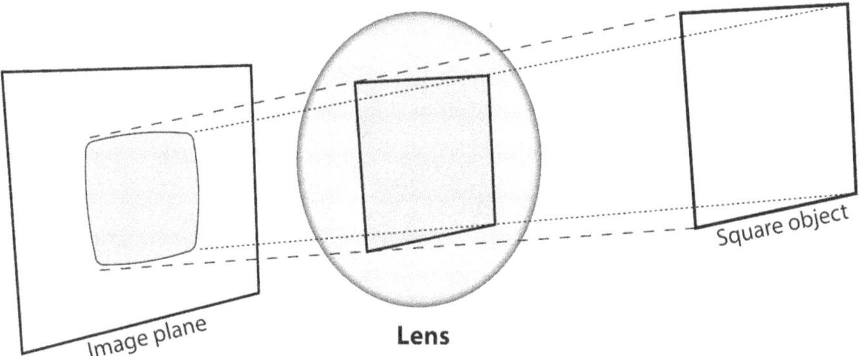


*Figure 18-3. Radial distortion: rays farther from the center of a simple lens are bent too much compared to rays that pass closer to the center; thus, the sides of a square appear to bow out on the image plane (this is also known as barrel distortion)*

For radial distortions, the distortion is `0` at the (optical) center of the imager and increases as we move toward the periphery. In practice, this distortion is small and can be characterized by the first few terms of a Taylor series expansion around $r = 0$.[11] For cheap web cameras, we generally use the first two such terms; the first of which is conventionally called $k_1$ and the second $k_2$. For highly distorted cameras such as fish-

11 If you don't know what a Taylor series is, don't worry too much. The Taylor series is a mathematical technique for expressing a (potentially) complicated function in the form of a polynomial of similar value to the approximated function in at least a small neighborhood of some particular point (the more terms we include in the polynomial series, the more accurate the approximation). In our case we want to expand the distortion function as a polynomial in the neighborhood of $r = 0$. This polynomial takes the general form $f(r) = a_0 + a_1 r + a_2 r^2 + \ldots$, but in our case the fact that $f(r) = 0$ at $r = 0$ implies $a_0 = 0$. Similarly, because the function must be symmetric in $r$, only the coefficients of even powers of $r$ will be nonzero. For these reasons, the only parameters that are necessary for characterizing these radial distortions are the coefficients of $r^2$, $r^4$, and (sometimes) higher even powers of $r$.

eye lenses, we can use a third radial distortion term, $k_3$. In general, the radial location of a point on the imager will be rescaled according to the following equations:[12]

$$x_{corrected} = x \cdot (1 + k_1 r^2 + k_2 r^4 + k_3 r^6)$$

and:

$$y_{corrected} = y \cdot (1 + k_1 r^2 + k_2 r^4 + k_3 r^6)$$

Here, $(x, y)$ is the original location (on the imager) of the distorted point and $(x_{corrected}, y_{corrected})$ is the new location as a result of the correction. Figure 18-4 shows displacements of a rectangular grid that are due to radial distortion. External points on a front-facing rectangular grid are increasingly displaced inward as the radial distance from the optical center increases.

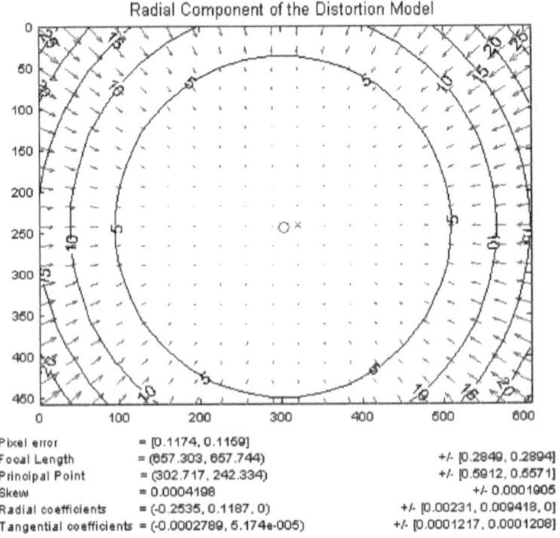


*Figure 18-4. Radial distortion plot for a particular camera lens: the arrows show where points on an external rectangular grid are displaced in a radially distorted image (courtesy of Jean-Yves Bouguet)*[13]

---

12 Now it's possible to use much more complex "rational" and "thin prism" models in OpenCV. There is also support for omni (180-degree) cameras, omni stereo, and multicamera calibration. See samples of use in the *opencv_contrib/modules/ccalib/samples* directory.

13 Some old cameras had nonrectangular sensors. These sensors were actually parallelograms, due to some imperfect manufacturing technology (and some other possible reasons). The camera-intrinsic matrices for such cameras had the form $[[f_x \; skew \; c_x] \; [0 \; f_y \; c_y] \; [0, 0, 1]]$. This is the origin of the "skew" in Figure 18-4. Almost all modern cameras have no skew, so the OpenCV calibration functions assume it is zero and do not compute it.

The second-largest common distortion is *tangential distortion*. This distortion is due to manufacturing defects resulting from the lens not being exactly parallel to the imaging plane; see Figure 18-5.

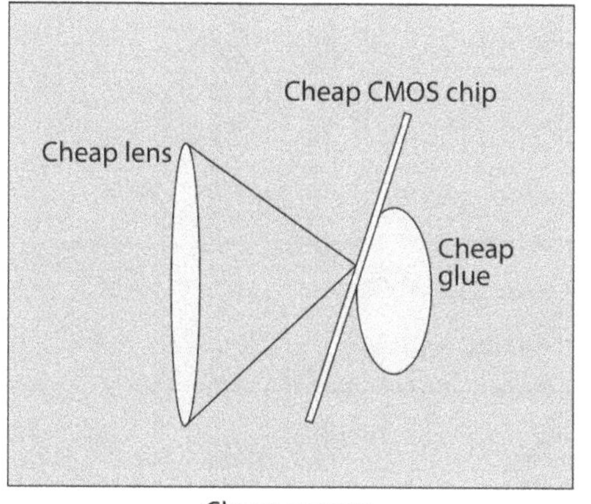


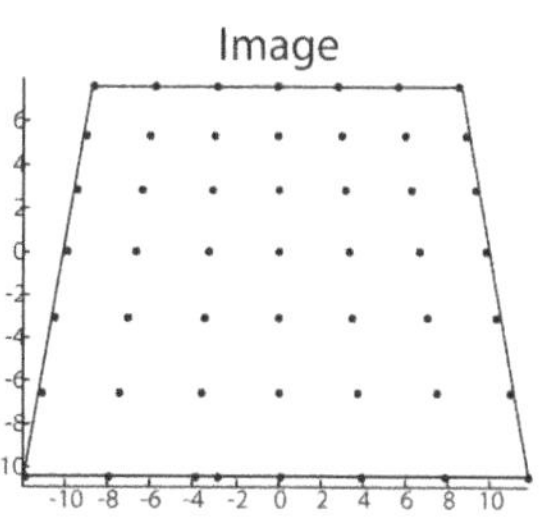


*Figure 18-5. Tangential distortion results when the lens is not fully parallel to the image plane; in cheap cameras, this can happen when the imager is glued to the back of the camera (image courtesy of Sebastian Thrun)*

Tangential distortion is minimally characterized by two additional parameters: $p_1$ and $p_2$, such that:[14]

$$x_{corrected} = x + [2p_1xy + p_2(r^2 + 2x^2)]$$

and:

$$y_{corrected} = y + [p_1(r^2 + 2y^2) + 2p_2xy]$$

Thus in total there are five distortion coefficients that we require. Because all five are necessary in most of the OpenCV routines that use them, they are typically bundled into one *distortion vector*; this is just a 5 × 1 matrix containing $k_1$, $k_2$, $p_1$, $p_2$, and $k_3$ (in that order). Figure 18-6 shows the effects of tangential distortion on a front-facing external rectangular grid of points. The points are displaced elliptically as a function of location and radius.

14 The derivation of these equations is beyond the scope of this book, but the interested reader is referred to the "plumb bob" model; see D. C. Brown, "Decentering Distortion of Lenses," *Photometric Engineering* 32, no. 3 (1966): 444–462.

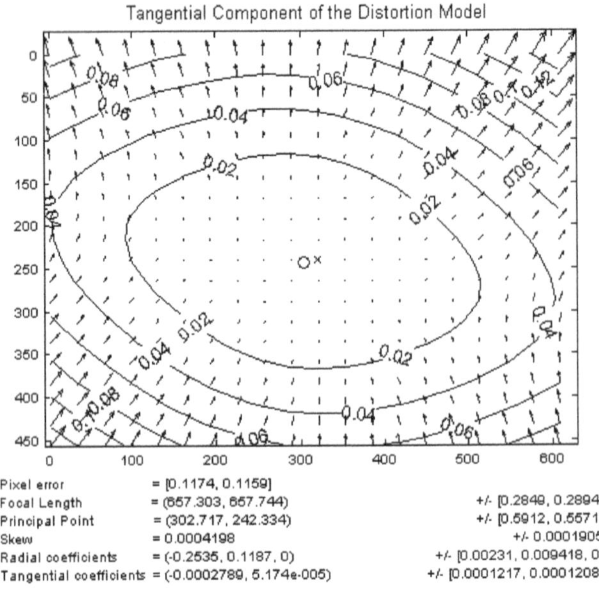


*Figure 18-6. Tangential distortion plot for a particular camera lens: the arrows show where points on an external rectangular grid are displaced in a tangentially distorted image (courtesy of Jean-Yves Bouguet)*

There are many other kinds of distortions that occur in imaging systems, but they typically have lesser effects than radial and tangential distortions. Hence, neither we nor OpenCV will deal with them further.

# Calibration

Now that we have some idea of how we'd describe the intrinsic and distortion properties of a camera mathematically, the next question that naturally arises is how we can use OpenCV to compute the intrinsics matrix and the distortion vector.[15]

OpenCV provides several algorithms to help us compute these intrinsic parameters. The actual calibration is done via `cv::calibrateCamera()`. In this routine, the method of calibration is to target the camera on a known structure that has many individual and identifiable points. By viewing this structure from a variety of angles, we can then compute the (relative) location and orientation of the camera at the time of each image as well as the intrinsic parameters of the camera (see Figure 18-10 in the section "Finding chessboard corners with cv::findChessboardCorners()" on page

15 For a great online tutorial of camera calibration, see Jean-Yves Bouguet's calibration website (*http://www.vision.caltech.edu/bouguetj/calib_doc*).

659). To provide multiple views, we rotate and translate the object, so let's pause to learn a little more about rotation and translation.

OpenCV continues to improve its calibration techniques; there are now many different types of calibration board patterns, as described in the section "Calibration Boards" on page 656. There are also specialized calibration techniques for "out of the ordinary" cameras. For fisheye lenses, you want to use the fisheye methods in the `cv::fisheye` class in the user documentation.

There are also techniques for omnidirectional (180-degree) camera and multicamera calibration; see *opencv_contrib/modules/ccalib/samples*, *opencv_contrib/modules/ccalib/tutorial/omnidir_tutorial.markdown*, *opencv_contrib/modules/ccalib/tutorial/multi_camera_tutorial.markdown*, and/or search on "omnidir" and "multiCameraCalibration," respectively, in the user documentation; see Figure 18-7.[16]

*Figure 18-7. An omnidirectional camera*[17]

16 As these are highly specialized cases, their details are beyond the scope of this book. However, you should find them easily understandable once you are comfortable with the calibration of more common cameras as we describe here.

17 This image was originally contributed to the OpenCV library by Baisheng Lai.

## Rotation Matrix and Translation Vector

For each image the camera takes of a particular object, we can describe the *pose* of the object relative to the camera coordinate system in terms of a rotation and a translation; see Figure 18-8.

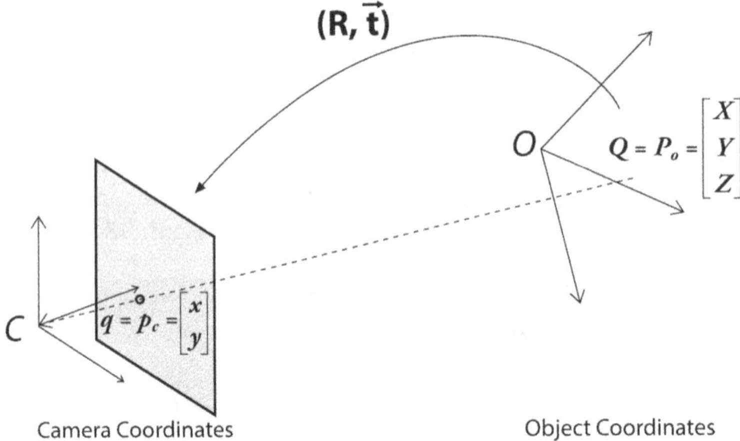


*Figure 18-8. Converting from object to camera coordinate systems: the point P on the object is seen as point p on the image plane; we relate the point p to point P by applying a rotation matrix R and a translation vector t to P*

In general, a rotation in any number of dimensions can be described in terms of multiplication of a coordinate vector by a square matrix of the appropriate size. Ultimately, a rotation is equivalent to introducing a new description of a point's location in a different coordinate system. Rotating the coordinate system by an angle $\theta$ is equivalent to counter-rotating our target point around the origin of that coordinate system by the same angle $\theta$. The representation of a two-dimensional rotation as matrix multiplication is shown in Figure 18-9. Rotation in three dimensions can be decomposed into a two-dimensional rotation around each axis in which the pivot axis measurements remain constant. If we rotate around the x-, y-, and z-axes in sequence[18] with respective rotation angles $\psi$, $\varphi$, and $\theta$, the result is a total rotation matrix $R$ that is given by the product of the three matrices $R_x(\psi)$, $R_y(\varphi)$, and $R_z(\theta)$, where:

18 Just to be clear: the rotation we are describing here is first around the z-axis, then around the *new* position of the y-axis, and finally around the new position of the x-axis.

$$R_x(\psi) = \begin{bmatrix} 1 & 0 & 0 \\ 0 & \cos\psi & \sin\psi \\ 0 & -\sin\psi & \cos\psi \end{bmatrix}$$

$$R_y(\psi) = \begin{bmatrix} \cos\varphi & 0 & -\sin\varphi 0 \\ 0 & 1 & 0 \\ \sin\varphi & 0 & \cos\varphi \end{bmatrix}$$

$$R_z(\theta) = \begin{bmatrix} \cos\theta & \sin\theta & 0 \\ -\sin\theta & \cos\theta & 0 \\ 0 & 0 & 1 \end{bmatrix}$$

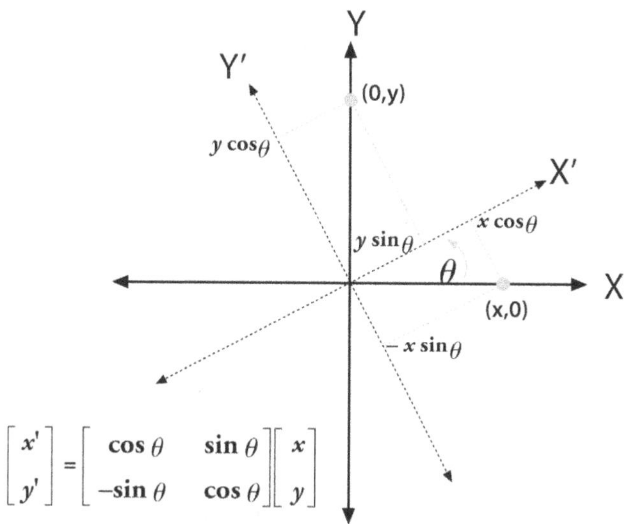


*Figure 18-9. Rotating points by θ (in this case, around the z-axis) is the same as counter-rotating the coordinate axis by θ; by simple trigonometry, we can see how rotation changes the coordinates of a point*

Thus $R = R_x(\psi) \cdot R_y(\varphi) \cdot R_z(\theta)$. The rotation matrix $R$ has the property that its inverse is its transpose (we just rotate back); hence, we have $R^T \cdot R = R \cdot R^T = I_3$, where $I_3$ is the $3 \times 3$ identity matrix consisting of 1s along the diagonal and 0s everywhere else.

The *translation vector* is how we represent a shift from one coordinate system to another system whose origin is displaced to another location; in other words, the translation vector is just the offset from the origin of the first coordinate system to the origin of the second coordinate system. Thus, to shift from a coordinate system centered on an object to one centered at the camera, the appropriate translation vector is simply $\vec{T} = origin_{object} - origin_{camera}$. We then know (with reference to Figure 18-8) that a point in the object (or world) coordinate frame $\vec{P}_o$ has coordinates $\vec{P}_c$ in the camera coordinate frame:

$$\vec{P}_c = R \cdot (\vec{P}_o - \vec{T})$$

Combining this equation for $\vec{P}_c$ with the camera intrinsic-corrections will form the basic system of equations that we will be asking OpenCV to solve. The solution to these equations will contain the camera calibration parameters we seek.

We have just seen that a three-dimensional rotation can be specified with three angles and that a three-dimensional translation can be specified with the three parameters ($x$, $y$, $z$); thus we have six parameters so far. The OpenCV intrinsics matrix for a camera has four parameters ($f_x$, $f_y$, $c_x$, and $c_y$), yielding a grand total of 10 parameters that must be solved for each view (but note that the camera-intrinsic parameters stay the same between views). Using a planar object, we'll soon see that each view fixes eight parameters. Because the six parameters of rotation and translation change between views, for each view we have constraints on two additional parameters that we then use to resolve the camera-intrinsic matrix. Thus, we need (at least) two views to solve for all the geometric parameters.

We'll provide more details on the parameters and their constraints later in the chapter, but first we'll discuss the *calibration object*. The calibration objects used in OpenCV are flat patterns of several types described next. The first OpenCV calibration object was a "chessboard" as shown in Figure 18-10. The following discussion will mostly refer to this type of pattern, but other patterns are available, as we will see.

## Calibration Boards

In principle, any appropriately characterized object could be used as a calibration object. One practical choice is a regular pattern on a flat surface, such as a *chessboard*[19] (see Figure 18-10), *circle-grid* (see Figure 18-15), *randpattern*[20] (see Figure 18-11), *ArUco* (see Figure 18-12), or *ChArUco patterns*[21] (see Figure 18-13). Appendix C has usable examples of all the calibration patterns available in OpenCV.

19 The specific use of this calibration object—and much of the calibration approach itself—comes from Zhang [Zhang99; Zhang00] and Sturm [Sturm99].

20 There are also random-pattern calibration pattern generators in the *opencv_contrib/modules/ccalib/samples* directory. These patterns can be used for multicamera calibration, as seen in the tutorial in that directory.

21 These boards came from Augmented Reality 2D barcodes called *ArUco* [Garrido-Jurado]. Their advantage is that the whole board does not have to be in view to get labeled corners on which to calibrate. The authors recommend you use the ChArUco pattern. The OpenCV documentation describes how to make and use these patterns (search on "ChArUco"). The code and tutorials are in the *opencv_contrib/modules/aruco* directory.

Some calibration methods in the literature rely on three-dimensional objects (e.g., a box covered with markers), but flat chessboard patterns are much easier to deal with; among other things, it is rather difficult to make (and to store and distribute) precise three-dimensional calibration objects. OpenCV thus opts for using multiple views of a planar object rather than one view of a specially constructed three-dimensional object. For the moment, we will focus on the chessboard pattern. The use of a pattern of alternating black and white squares (see Figure 18-10) ensures that there is no bias toward one side or the other in measurement. Also, the resulting grid corners lend themselves naturally to the subpixel localization function discussed in Chapter 16. We will also discuss another alternative calibration board, called a circle-grid (see Figure 18-15), which has some desirable properties, and which in some cases may give superior results to the chessboard. For the other patterns, please see the documentation referenced in the figures in this chapter. The authors have had particularly good success using the ChArUco pattern.

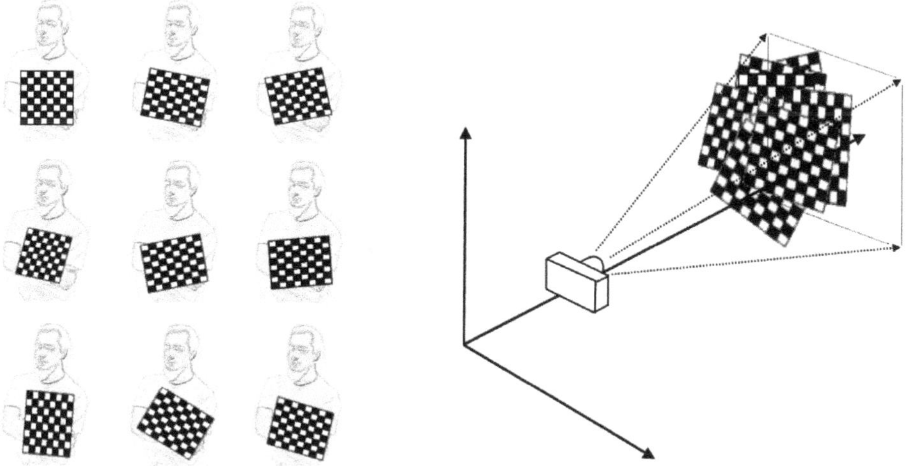

*Figure 18-10. Images of a chessboard being held at various orientations (left) provide enough information to completely solve for the locations of those images in global coordinates (relative to the camera) and the camera intrinsics*

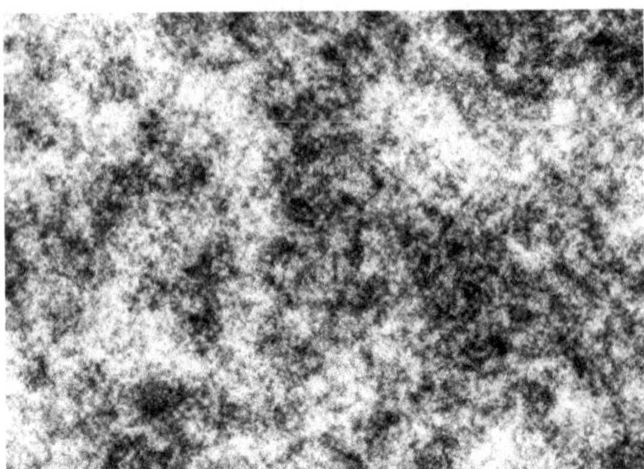

*Figure 18-11. A calibration pattern made up of a highly textured random pattern; see the multicamera calibration tutorial in the opencv_contrib/modules/ccalib/tutorial directory*[22]

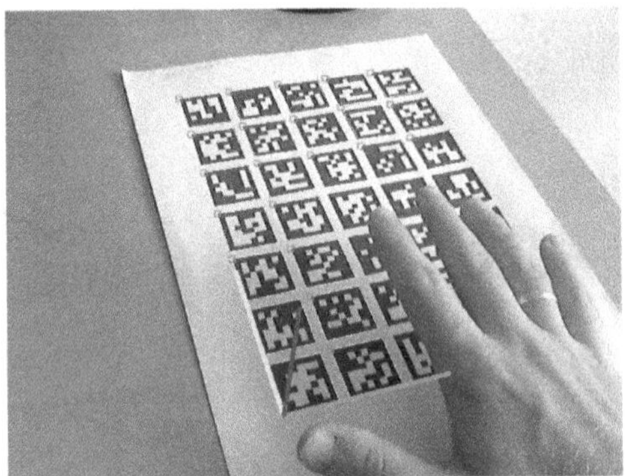

*Figure 18-12. A calibration pattern made up of a grid of ArUco (2D barcode) squares. Note that because each square is identified by its ArUco pattern, much of the board can be occluded and yet still have enough spatially labeled points to be used in calibration. See "ArUco marker detection (aruco module)" in the OpenCV documentation*[23]

22 This image was originally contributed to the OpenCV library by Baisheng Lai.

23 This image was originally contributed to the OpenCV library by Sergio Garrido.

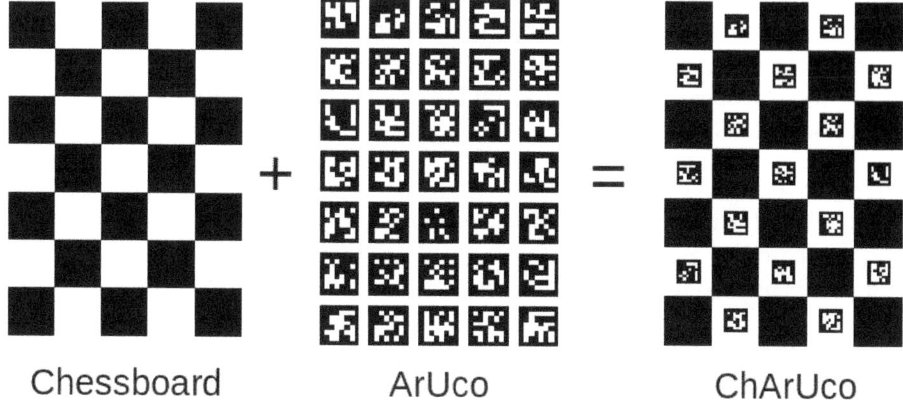


*Figure 18-13. Checkerboard with embedded ArUco (ChArUco). A checkerboard calibration pattern where each corner is labeled with an ArUco (2D barcode) pattern. This allows much of the checkerboard to be occluded while allowing for the higher positional accuracy of corner intersections. See "ArUco marker detection (aruco module)" in the OpenCV documentation*[24]

#### Finding chessboard corners with cv::findChessboardCorners()

Given an image of a chessboard (or a person holding a chessboard, or any other scene with a chessboard and a reasonably uncluttered background), you can use the OpenCV function `cv::findChessboardCorners()` to locate the corners of the chessboard:

```
bool cv::findChessboardCorners(     // Return true if corners were found
  cv::InputArray  image,            // Input chessboard image, 8UC1 or 8UC3
  cv::Size        patternSize,      // corners per row, and per column
  cv::OutputArray corners,          // Output array of detected corners
  int             flags        = cv::CALIB_CB_ADAPTIVE_THRESH
                               | cv::CALIB_CB_NORMALIZE_IMAGE
);
```

This function takes as arguments a single image containing a chessboard. This image must be an 8-bit image. The second argument, `patternSize`, indicates how many corners are in each row and column of the board (e.g., `cv::Size(cols,rows)`). This count is the number of *interior* corners; thus, for a standard chess game board, the correct value would be `cv::Size(7,7)`.[25] The next argument, `corners`, is the output

24 This image was originally contributed to the OpenCV library by Sergio Garrido.

25 In practice, it is often more convenient to use a chessboard grid that is asymmetric and of even and odd dimensions—for example, (5, 6). Using such even-odd asymmetry yields a chessboard that has only one symmetry axis, so the board orientation can always be defined uniquely.

array where the corner locations will be recorded. The individual values will be set to the locations of the located corners in pixel coordinates. The final `flags` argument can be used to implement one or more additional filtration steps to help find the corners on the chessboard. You may combine any or all of the following arguments using a Boolean OR:

`cv::CALIB_CB_ADAPTIVE_THRESH`
: The default behavior of `cv::findChessboardCorners()` is first to threshold the image based on average brightness, but if this flag is set, then an adaptive threshold will be used instead.

`cv::CALIB_CB_NORMALIZE_IMAGE`
: If set, this flag causes the image to be normalized via `cv::equalizeHist()` before the thresholding is applied.

`cv::CALIB_CB_FILTER_QUADS`
: Once the image is thresholded, the algorithm attempts to locate the quadrangles resulting from the perspective view of the black squares on the chessboard. This is an approximation because the lines of each edge of a quadrangle are assumed to be straight, which isn't quite true when there is radial distortion in the image. If this flag is set, then a variety of additional constraints are applied to those quadrangles in order to reject false quadrangles.

`cv::CALIB_CV_FAST_CHECK`
: When this option is present, a fast scan will be done on the image to make sure that there actually are any corners in the image. If there are not, then the image is skipped entirely. This is not necessary if you are absolutely certain that your input data is "clean" and has no images without the chessboard in them. On the other hand, you will save a great deal of time using this option if there actually turn out to be images without chessboards in them in your input.

The return value of `cv::findChessboardCorners()` will be set to `true` if all of the corners in the pattern could be found and ordered;[26] otherwise, it will be `false`.

### Subpixel corners on chessboards and cv::cornerSubPix()

The internal algorithm used by `cv::findChessboardCorners()` gives only the approximate location of the corners. Therefore, `cv::cornerSubPix()` is automatically called by `cv::findChessboardCorners()` in order to give more accurate results. What this means in practice is that the locations are going to be relatively accurate.

26 In this context, *ordered* means that a model of the found points could be constructed that was consistent with the proposition that they are in fact sets of collinear points on a plane. Obviously not all images containing 49 points, for example, are generated by a regular 7 × 7 grid on a plane.

However, if you would like them located to very high precision, you will want to call `cv::cornerSubPix()` yourself (effectively calling it again) on the output, but with tighter termination criteria.

### Drawing chessboard corners with cv::drawChessboardCorners()

Particularly when one is debugging, it is often desirable to draw the found chessboard corners onto an image (usually the image that we used to compute the corners in the first place); this way, we can see whether the projected corners match up with the observed corners. Toward this end, OpenCV provides a convenient routine to handle this common task. The function `cv::drawChessboardCorners()` draws the corners found by `cv::findChessboardCorners()` onto an image that you provide. If not all of the corners were found, then the available corners will be represented as small red circles. If the entire pattern was found, then the corners will be painted into different colors (each row will have its own color) and connected by lines representing the identified corner order.

```
void cv::drawChessboardCorners(
  cv::InputOutputArray image,          // Input/output chessboard image, 8UC3
  cv::Size             patternSize,    // Corners per row, and per column
  cv::InputArray       corners,        // corners from findChessboardCorners()
  bool                 patternWasFound // Returned from findChessboardCorners()
);
```

The first argument to `cv::drawChessboardCorners()` is the image to which the drawing will be done. Because the corners will be represented as colored circles, this must be an 8-bit color image. In most cases, this will be a copy of the image you gave to `cv::findChessboardCorners()` (but you must convert it to a three-channel image yourself, if it wasn't already). The next two arguments, `patternSize` and `corners`, are the same as the corresponding arguments for `cv::findChessboardCorners()`. Finally, the argument `patternWasFound` indicates whether the entire chessboard pattern was successfully found; this can be set to the return value from `cv::findChessboardCorners()`. Figure 18-14 shows the result of applying `cv::drawChessboardCorners()` to a chessboard image.

We now turn to what a planar object such as the calibration board can do for us. Points on a plane undergo a *perspective transformation* when viewed through a pinhole or lens. The parameters for this transform are contained in a 3 × 3 *homography* matrix, which we will describe shortly, after a brief discussion of an alternative pattern to square grids.

*Figure 18-14. Result of cv::drawChessboardCorners(); once you find the corners using cv::findChessboardCorners(), you can project where these corners were found (small circles on corners) and in what order they belong (as indicated by the lines between circles)*

### Circle-grids and cv::findCirclesGrid()

An alternative to the chessboard is the *circle-grid*. Conceptually, the circle-grid is similar to the chessboard, except that rather than an array of alternating black and white squares, the board contains an array of black circles on a white background.

Calibration with a circle-grid proceeds exactly the same as with `cv::findChessboardCorners()` and the chessboard, except that a different function is called and a different calibration image is used. That different function is `cv::findCirclesGrid()`, and it has the following prototype:

```
bool cv::findCirclesGrid(// Return true if corners were found
  cv::InputArray  image,            // Input chessboard image, 8UC1 or 8UC3
  cv::Size        patternSize,      // corners per row, and per column
  cv::OutputArray centers,          // Output array of detected circle centers
  int             flags      = cv::CALIB_CB_SYMMETRIC_GRID,
  const cv::Ptr<cv::FeatureDetector>& blobDetector
                             = new SimpleBlobDetector()
);
```

Like `cv::findChessboardCorners()`, it takes an image and a `cv::Size` object defining the number (and arrangement) of the circle pattern. It outputs the location of the centers, which are equivalent to the corners in the chessboard.

The `flags` argument tells the function what sort of array the circles are arranged into. By default, `cv::findCirclesGrid()` expects a *symmetric grid* of circles. A "symmetric" grid is a grid in which the circles are arranged neatly into rows and columns in the same way as the chessboard corners. The alternative is an *asymmetric grid*. We use the asymmetric grid by setting the `flags` argument to `cv::CALIB_CB_ASYMMETRIC_GRID`. In an "asymmetric" grid, the circles in each row are staggered transverse to the row. (The grid shown in Figure 18-15 is an example of an asymmetric grid.)

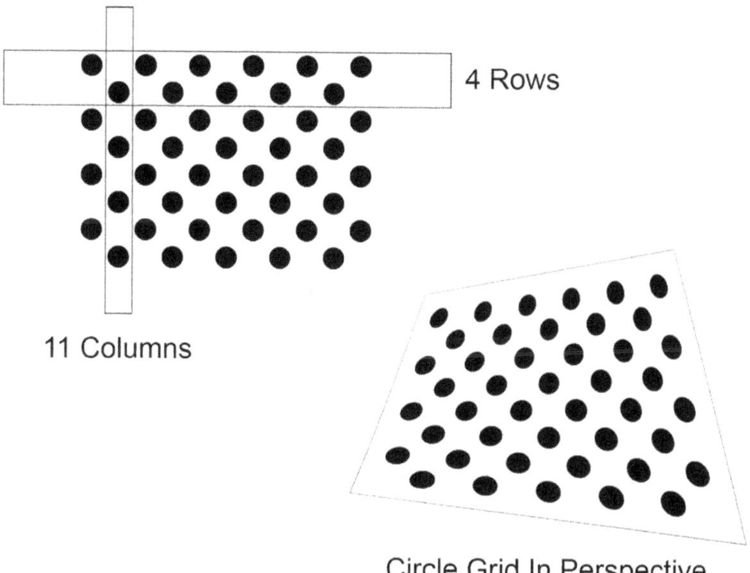


*Figure 18-15. With the regular array of circles (upper left), the centers of the circles function analogously to the corners of the chessboard for calibration. When seen in perspective (lower right), the deformation of the circles is regular and predictable*

When you are using an asymmetric grid, it is important to remember how rows and columns are counted. By way of example, in the case shown in Figure 18-15, because it is the rows that are "staggered," then the array shown has only 4 rows, and 11 columns. The final option for flags is `cv::CALIB_CB_CLUSTERING`. It can be set along with `cv::CALIB_CB_SYMMETRIC_GRID` or `cv::CALIB_CB_ASYMMETRIC_GRID` with the logical OR operator. If this option is selected, then `cv::findCirclesGrid()` will use a slightly different algorithm for finding the circles. This alternate algorithm is more robust to perspective distortions, but (as a result) is also a lot more sensitive to background clutter. This is a good choice when you are trying to calibrate a camera with an unusually wide field of view.

In general, one often finds the asymmetric circle-grid to be superior to the chessboard, both in terms of the quality of final results, as well as the stability of those results between multiple runs. For these reasons, asymmetric circle-grids increasingly became part of the standard toolkit for camera calibration. In very modern times, patterns such as *ChArUco* (see the `contrib` experimental code section of the library) are gaining significant traction as well.

## Homography

In computer vision, we define *planar homography* as a projective mapping from one plane to another.[27] Thus, the mapping of points on a two-dimensional planar surface to the imager of our camera is an example of planar homography. It is possible to express this mapping in terms of matrix multiplication if we use homogeneous coordinates to express both the viewed point $\vec{Q}$ and the point $\vec{q}$ on the imager to which $\vec{Q}$ is mapped. If we define:

$$\vec{Q} = \begin{bmatrix} X \\ Y \\ Z \\ 1 \end{bmatrix}, \text{ and } \vec{q} = \begin{bmatrix} x \\ y \\ 1 \end{bmatrix}$$

then we can express the action of the homography simply as:

$$\vec{q} = s \cdot H \cdot \vec{Q}$$

Here we have introduced the parameter $s$, which is an arbitrary scale factor (intended to make explicit that the homography is defined only up to that factor). It is conventionally factored out of $H$, and we'll stick with that convention here.

With a little geometry and some matrix algebra, we can solve for this transformation matrix. The most important observation is that $H$ has two parts: the physical transformation, which essentially locates the object plane we are viewing, and the projection, which introduces the camera intrinsics matrix. See Figure 18-16.

27 The term *homography* has different meanings in different sciences; for example, it has a somewhat more general meaning in mathematics. The homographies of greatest interest in computer vision are a subset of the other more general meanings of the term.

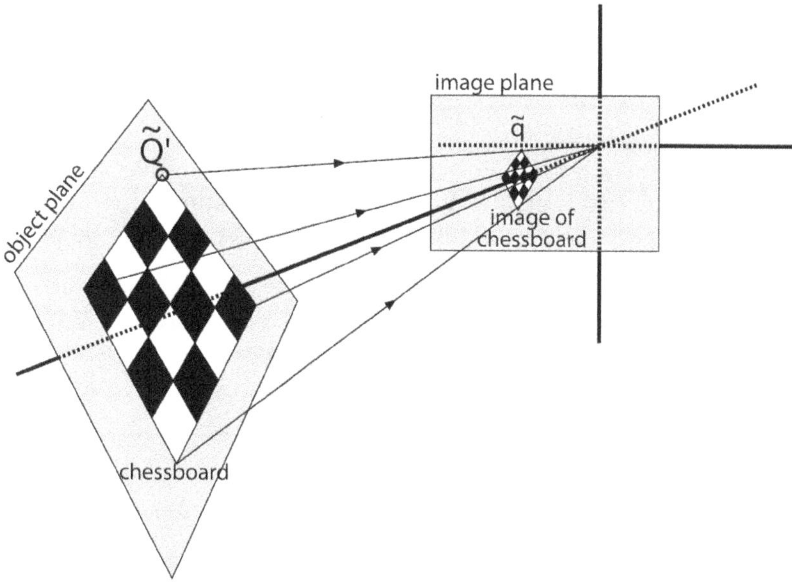


*Figure 18-16. View of a planar object as described by homography: a mapping—from the object plane to the image plane—that simultaneously comprehends the relative locations of those two planes as well as the camera projection matrix*

The physical transformation part is the sum of the effects of some rotation $R$ and some translation $\vec{t}$ that relate the plane we are viewing to the image plane. Because we are working in homogeneous coordinates, we can combine these within a single matrix as follows:[28]

$$W = \begin{bmatrix} R & \vec{t} \end{bmatrix}$$

Then, the action of the camera matrix $M$, which we already know how to express in projective coordinates, is multiplied by $\vec{Q}$, which yields:

$$\vec{q} = s \cdot M \cdot W \cdot \vec{Q}$$

where:

28 Here $W = \begin{bmatrix} R & \vec{t} \end{bmatrix}$ is a 3 × 4 matrix whose first three columns comprise the nine entries of $R$ and whose last column consists of the three-component vector $\vec{t}$.

$$M = \begin{bmatrix} f_x & 0 & c_x \\ 0 & f_x & c_y \\ 0 & 0 & 1 \end{bmatrix}$$

It would seem that we are done. However, it turns out that in practice our interest is not the coordinate $\vec{Q}$ that is defined for all of space, but rather a coordinate $\vec{Q}'$ that is defined only on the plane we are looking at. This allows for a slight simplification.

Without loss of generality, we can choose to define the object plane so that $Z = 0$. We do this because, if we also break up the rotation matrix into three $3 \times 1$ columns (i.e., $R = [\vec{r}_1 \ \vec{r}_2 \ \vec{r}_3]$), then one of those columns is no longer needed. In particular:

$$\begin{bmatrix} x \\ y \\ 1 \end{bmatrix} = s \cdot M \cdot \begin{bmatrix} \vec{r}_1 & \vec{r}_2 & \vec{r}_3 & \vec{t} \end{bmatrix} \cdot \begin{bmatrix} X \\ Y \\ 0 \\ 1 \end{bmatrix} = s \cdot M \cdot \begin{bmatrix} \vec{r}_1 & \vec{r}_2 & \vec{t} \end{bmatrix} \cdot \begin{bmatrix} X \\ Y \\ 1 \end{bmatrix}$$

The homography matrix $H$ that maps a planar object's points onto the imager is then described completely by $H = s \cdot M \cdot [\vec{r}_1 \ \vec{r}_2 \ \vec{t}]$, where:

$$\vec{q} = s \cdot H \cdot \vec{Q}'$$

Observe that $H$ is now a $3 \times 3$ matrix.[29]

OpenCV uses the preceding equations to compute the homography matrix. It uses multiple images of the same object to compute both the individual translations and rotations for each view as well as the intrinsics (which are the same for all views). As we have discussed, rotation is described by three angles and translation is defined by three offsets; hence there are six unknowns for each view. This is OK, because a known planar object (such as our chessboard) gives us eight equations—that is, the mapping of a square into a quadrilateral can be described by four $(x, y)$ points. Each new frame gives us eight equations at the cost of six new extrinsic unknowns, so given enough images we should be able to compute any number of intrinsic unknowns (more on this shortly).

The homography matrix $H$ relates the positions of the points on a source image plane to the points on the destination image plane (usually the imager plane) by the following simple equations:

29 The astute reader will have noticed that we have been a little cavalier with the factor $s$. The $s$ that appears in this final expression is not identical to those that appear in the previous equations, but rather the product of them. In any case, the product of two arbitrary scale factors is itself an arbitrary scale factor, so no harm is done.

$$\overrightarrow{p_{dst}} \equiv \begin{bmatrix} x_{dst} \\ y_{dst} \\ 1 \end{bmatrix} = H \cdot \overrightarrow{p_{src}}, \quad \overrightarrow{p_{src}} \equiv \begin{bmatrix} x_{src} \\ y_{src} \\ 1 \end{bmatrix} = H^{-1} \cdot \overrightarrow{p_{dst}}$$

Notice that we can compute $H$ without knowing anything about the camera intrinsics. In fact, computing multiple homographies from multiple views is the method OpenCV uses to solve for the camera intrinsics, as we'll see.

OpenCV provides us with a handy function, `cv::findHomography()`, that takes a list of correspondences and returns the homography matrix that best describes those correspondences. We need a minimum of four points to solve for $H$, but we can supply many more if we have them[30] (as we will with any chessboard bigger than 3 × 3). Using more points is beneficial, because invariably there will be noise and other inconsistencies whose effect we would like to minimize:

```
cv::Mat cv::findHomography(
  cv::InputArray  srcPoints,                  // Input array source points (2-d)
  cv::InputArray  dstPoints,                  // Input array result points (2-d)
  cv::int         method                 = 0, // 0, cv::RANSAC, cv::LMEDS, etc.
  double          ransacReprojThreshold = 3, // Max reprojection error
  cv::OutputArray mask                   = cv::noArray() // use only non-zero pts
);
```

The input arrays `srcPoints` and `dstPoints` contain the points in the original plane and the target plane, respectively. These are all two-dimensional points, so they must be N × 2 arrays, N × 1 arrays of `CV_32FC2` elements, or STL vectors of `cv::Point2f` objects (or any combination of these).

The input called `method` determines the algorithm that will be used to compute the homography. If left as the default value of `0`, all of the points will be considered and the computed result will be the one that minimizes the *reprojection error*. In this case, the reprojection error is the sum of squared Euclidean distances between $H$ times the "original" points and the target points.

Conveniently, fast algorithms exist to solve such problems in the case of an error metric of this kind. Unfortunately, however, defining the error in this way leads to a system in which outliers—individual data points that seem to imply a radically different solution than the majority—tend to have a drastic effect on the solution. In practical cases such as camera calibration, it is common that measurement errors will produce outliers, and the resulting solution will often be very far from the correct answer because of these outliers. OpenCV provides three *robust fitting methods* that

30 Of course, an exact solution is guaranteed only when there are four correspondences. If more are provided, then what's computed is a solution that is optimal in the sense of least-squares error (possibly also with some points rejected; see the upcoming RANSAC discussion).

can be used as an alternative, and tend to give much better behavior in the presence of noise.

The first such option, which we select by setting `method` to `cv::RANSAC`, is the *RANSAC method* (also known as the "random sampling with consensus" method). In the RANSAC method, subsets of the provided points are selected at random, and a homography matrix is computed for just that subset. It is then refined by all of the remaining data points that are roughly consistent with that initial estimation. The "inliers" are those that are consistent, while the "outliers" are those that are not. The RANSAC algorithm computes many such random samplings, and keeps the one that has the largest portion of inliers. This method is extremely efficient in practice for rejecting noisy outlier data and finding the correct answer.

The second alternative is the *LMeDS algorithm* (also known as the "least median of squares" algorithm). As the name suggests, the idea behind LMeDS is to minimize the median error, as opposed to what is essentially the mean squared error minimized by the default method.[31]

The advantage of LMeDS is that it does not need any further information or parameters to run. The disadvantage is that it will perform well only if the inliers constitute at least a majority of the data points. In contrast, RANSAC can function correctly and give a satisfactory answer given almost any signal-to-noise ratio. The cost of this, however, is that you will have to tell RANSAC what constitutes "roughly consistent" —the maximum distance reprojected points can be from its source and still be considered worth including in the refined model. If you are using the `cv::RANSAC` method, then the input `ransacReprojThreshold` controls this distance. If you are using any other method, this parameter can be ignored.

The value of `ransacReprojThreshold` is measured in pixels. For most practical cases, it is sufficient to set it to a small integer value (i.e., less than `10`) but, as is often the case, this number must be increased for very high-resolution images.

Finally, there is RHO algorithm, introduced in [Bazargani15] and available in OpenCV 3, which is based on a "weighted" RANSAC modification called PROSAC and runs faster in the case of many outliers.

31 For more information on robust methods, consult the original papers: Fischler and Bolles [Fischler81] for RANSAC; Rousseeuw [Rousseeuw84] for least median squares; and Inui, Kaneko, and Igarashi [Inui03] for line fitting using LMedS.

The final argument, `mask`, is used only with the robust methods, and it is an output. If an array is provided, `cv::findHomography()` will fill that array indicating which points were actually used in the best computation of *H*.

The return value will be a 3 × 3 matrix. Because there are only eight free parameters in the homography matrix, we chose a normalization where $H_{33} = 1$ (which is usually possible except for the quite rare singular case $H_{33} = 0$). Scaling the homography could be applied to the ninth homography parameter, but usually prefer to instead scale by multiplying the entire homography matrix by a scale factor, as described earlier in this chapter.

## Camera Calibration

We finally arrive at camera calibration for camera intrinsics and distortion parameters. In this section, we'll explain how to compute these values using `cv::calibrateCamera()` and also how to use these models to correct distortions in the images that the calibrated camera would have otherwise produced. First we will say a little more about just how many views of a chessboard are necessary in order to solve for the intrinsics and distortion. Then we'll offer a high-level overview of how OpenCV actually solves this system before moving on to the code that makes it all easy to do.

### How many chess corners for how many parameters?

To begin, it will prove instructive to review our unknowns; that is, how many parameters are we attempting to solve for through calibration? In the OpenCV case, we have four parameters associated with the *camera intrinsic matrix* ($f_x$, $f_y$, $c_x$, $c_y$) and five (or more) *distortion* parameters—the latter consisting of three (or more) radial parameters ($k_1$, $k_2$, $k_3$ [, $k_4$, $k_5$, $k_6$]) and the two tangential ($p_1$, $p_2$).[32] The intrinsic parameters control the linear projective transform that relates a physical object to the produced image. As a result, they are entangled with the *extrinsic* parameters, which tell us where that object is actually located.

The distortion parameters are tied to the two-dimensional geometry of how a pattern of points gets distorted in the final image. In principle, then, it would seem that just three corner points in a known pattern, yielding six pieces of information, might be all that is needed to solve for our five distortion parameters. Thus a single view of our calibration chessboard could be enough.

32 It is commonplace to refer to the total set of the camera intrinsic matrix parameters and the distortion parameters as simply *the intrinsic parameters* or *the intrinsics*. In some cases, the matrix parameters will also be referred to as the *linear intrinsic parameters* (because they collectively define a linear transformation), while the distortion parameters are referred to as the *nonlinear intrinsic parameters*.

However, because of the coupling between the intrinsic parameters and the extrinsic parameters, it turns out that one will not be enough. To understand this, first note that the extrinsic parameters include three rotation parameters ($\psi$, $\phi$, $\theta$) and three translation parameters ($T_x$, $T_y$, $T_z$) for a total of six per view of the chessboard. Together, the four parameters of the camera intrinsic matrix and six extrinsic parameters make 10 altogether that we must solve for, in the case of a single view, and 6 more for each additional view.

Let's say we have $N$ corners and $K$ images of the chessboard (in different positions). How many views and corners must we see so that there will be enough constraints to solve for all these parameters?

- $K$ images of the chessboard provide $2 \cdot N \cdot K$ constraints (the factor of 2 arises because each point on the image has both an x- and a y-coordinate).
- Ignoring the distortion parameters for the moment, we have 4 intrinsic parameters and $6 \cdot K$ extrinsic parameters (since we need to find the 6 parameters of the chessboard location in each of the $K$ views).
- Solving then requires that we have: $2 \cdot N \cdot K \geq 6 \cdot K + 4$ (or, equivalently, $(N - 3) \cdot K \geq 2$).

So it would seem that if $N = 5$, then we need only $K = 1$ image, but watch out! For us, $K$ (the number of images) must be more than 1. The reason for requiring $K > 1$ is that we are using chessboards for calibration to fit a homography matrix for each of the $K$ views. As discussed previously, a homography can yield at most eight parameters from four $(x, y)$ pairs. This is because only four points are needed to express everything that a planar perspective view can do: it can stretch a square in four different directions at once, turning it into any quadrilateral (see the perspective images in Chapter 11). So, no matter how many corners we detect on a plane, we only get four corners' worth of information. Per chessboard view, then, the equation can give us only four corners of information or $(4 - 3) \cdot K > 1$, which means $K > 1$. This implies that two views of a $3 \times 3$ chessboard (counting only internal corners) are the minimum that could solve our calibration problem. Consideration for noise and numerical stability is typically what requires the collection of more images of a larger chessboard. In practice, for high-quality results, you'll need at least 10 images of a $7 \times 8$ or larger chessboard (and that's only if you move the chessboard enough between images to obtain a "rich" set of views).

This disparity between the theoretically minimal 2 images and the practically required 10 or more views is a result of the very high degree of sensitivity that the intrinsic parameters have on even very small noise.

### What's under the hood?

This subsection is for those who want to go deeper; it can be safely skipped if you just want to call the calibration functions.

If you're still with us, the question remains: how does the actual mathematics work for calibration? Although there are many ways to solve for the camera parameters, OpenCV chose one that works well on planar objects. The algorithm OpenCV uses to solve for the focal lengths and offsets is based on Zhang's method [Zhang00], but OpenCV uses a different method based on Brown [Brown71] to solve for the distortion parameters.

To get started, we pretend that there is no distortion in the camera while solving for the other calibration parameters. For each view of the chessboard, we collect a homography $H$, as described previously (i.e., a map from the physical object to the imager). We'll write $H$ out as column vectors, $H = [\vec{h}_1, \vec{h}_2, \vec{h}_3]$, where each $h$ is a $3 \times 1$ vector. Then, in view of the preceding homography discussion, we can set $H$ equal to the camera intrinsics matrix $M$ multiplied by a combination of the first two rotation matrix columns, $\vec{r}_1$ and $\vec{r}_2$, and the translation vector $\vec{t}$. After we include the scale factor $s$, this yields:

$$H = [\vec{h}_1, \vec{h}_2, \vec{h}_3] = s \cdot M \cdot [\vec{r}_1, \vec{r}_2, \vec{t}\,]$$

Reading off these equations, we have:

$$\vec{h}_1 = s \cdot M \cdot \vec{r}_1 \text{ or } \quad \vec{r}_1 = \lambda \cdot M^{-1} \cdot \vec{h}_1$$

$$\vec{h}_2 = s \cdot M \cdot \vec{r}_2 \text{ or } \vec{r}_2 = \lambda \cdot M^{-1} \cdot \vec{h}_2$$

$$\vec{h}_3 = s \cdot M \cdot \vec{t} \text{ or } \vec{t} = \lambda \cdot M^{-1} \cdot \vec{h}_3$$

with:

$$\lambda = \frac{1}{s}$$

The rotation vectors are orthogonal to each other by construction, and since the scale is extracted we can take $\vec{r}_1$ and $\vec{r}_2$ to be orthonormal. Orthonormal implies two things: the rotation vector's dot product is 0, and the vectors' magnitudes are equal. Starting with the dot product, we have:

$$\vec{r}_1^{\,T} \cdot \vec{r}_2 = 0$$

For any vectors $\vec{a}$ and $\vec{b}$ we have $(\vec{a} \cdot \vec{b})^T = \vec{b}^T \cdot \vec{a}^T$, so we can substitute for $\vec{r}_1$ and $\vec{r}_2$ to derive our first constraint:

$$\vec{h}_1^T \cdot M^{-T} \cdot M^{-1} \cdot \vec{h}_2 = 0$$

where $M^{-T}$ is shorthand for $(M^{-1})^T$. We also know that the magnitudes of the rotation vectors are equal:

$$\| \vec{r}_1 \| = \| \vec{r}_2 \| \quad or \quad \vec{r}_1^T \cdot \vec{r}_1 = \vec{r}_2^T \cdot \vec{r}_2$$

Substituting for $\vec{r}_1$ and $\vec{r}_2$ yields our second constraint:

$$\vec{h}_1^T \cdot M^{-T} \cdot M^{-1} \cdot \vec{h}_1 = \vec{h}_2^T \cdot M^{-T} \cdot M^{-1} \cdot \vec{h}_2$$

To make things easier to manage, we define $B = M^{-T} \cdot M^{-1}$. Writing this out, we have:

$$B = M^{-T} \cdot M^{-1} \equiv \begin{bmatrix} B_{11} & B_{12} & B_{13} \\ B_{21} & B_{22} & B_{23} \\ B_{13} & B_{23} & B_{33} \end{bmatrix}$$

It so happens that this matrix $B$ has a general closed-form solution:

$$B = \begin{bmatrix} \frac{1}{f_x^2} & 0 & -\frac{c_x}{f_x^2} \\ 0 & \frac{1}{f_y^2} & -\frac{c_y}{f_y^2} \\ -\frac{c_x}{f_x^2} & -\frac{c_y}{f_y^2} & \left(\frac{c_x}{f_x^2} + \frac{c_y}{f_y^2} + 1\right) \end{bmatrix}$$

Using the $B$-matrix, both constraints have the general form $\vec{h}_i^T \cdot B \cdot \vec{h}_j$ in them. Let's multiply this out to see what the components are. Because $B$ is symmetric, it can be written as one six-dimensional vector dot product. Arranging the necessary elements of $B$ into the new vector $\vec{b}$, we have:

$$\vec{h}_i^T \cdot B \cdot \vec{h}_j = \vec{v}_{i,j}^T \cdot \vec{b} =$$

$$\begin{bmatrix} h_{i,1}h_{j,1} & (h_{i,1}h_{j,2} + h_{i,2}h_{j,1}) & h_{i,2}h_{j,2} & (h_{i,3}h_{j,1} + h_{i,1}h_{j,3}) & (h_{i,3}h_{j,2} + h_{i,2}h_{j,3}) & h_{i,3}h_{j,3} \end{bmatrix}$$

$$\cdot \begin{bmatrix} B_{11} \\ B_{12} \\ B_{22} \\ B_{13} \\ B_{23} \\ B_{33} \end{bmatrix}$$

Using this definition for $\vec{v}_{i,j}^T$, our two constraints may now be written as:

$$\begin{bmatrix} v_{12}^T \\ (v_{11} - v_{22})^T \end{bmatrix} \cdot \vec{b} = 0$$

If we collect *K* images of chessboards together, then we can stack *K* of these equations together:

$$V \cdot \vec{b} = 0,$$

where *V* is a $2 \cdot K \times 6$ matrix. As before, if $K \geq 2$ then this equation can be solved for our vector $\vec{b} = [B_{11} \;\; B_{12} \;\; B_{22} \;\; B_{13} \;\; B_{23} \;\; B_{33}]^T$. The camera intrinsics are then pulled directly out of our closed-form solution for the *B*-matrix:

$$f_x = \sqrt{\lambda / B_{11}}$$

$$f_y = \sqrt{\frac{\lambda B_{11}}{B_{11}B_{22} - B_{12}^2}}$$

$$c_x = \frac{B_{13} f_x^2}{\lambda}$$

and:

$$c_y = \frac{B_{12}B_{13} - B_{11}B_{23}}{B_{11}B_{22} - B_{12}^2}$$

with:

$$\lambda = B_{33} - \frac{\left(B_{13}^2 + c_y(B_{12}B_{13} - B_{11} - B_{23})\right)}{B_{11}}$$

The extrinsics (rotation and translation) are then computed from the equations we read off of the homography condition:

$$\vec{r}_1 = \lambda \cdot M^{-1} \cdot \vec{h}_1$$

$$\vec{r}_2 = \lambda \cdot M^{-1} \cdot \vec{h}_2$$

$$\vec{r}_3 = \vec{r}_1 \times \vec{r}_2$$

and:

$$\vec{t} = \lambda \cdot M^{-1} \cdot \vec{h}_3$$

Here the scaling parameter is determined from the orthonormality condition $\lambda - 1/\| M^{-1} \cdot \vec{h} \|$ . Some care is required because, when we solve using real data and put the *r*-vectors together ($R = [\vec{r}_1 \;\; \vec{r}_2 \;\; \vec{r}_3]$), we will not likely end up with an exact rotation matrix for which $R^T R = R R^T = I_3$ holds.[33]

To get around this problem, the usual trick is to take the singular value decomposition (SVD) of *R*. As discussed in Chapter 5, SVD is a method of factoring a matrix into two orthonormal matrices, *U* and *V*, and a middle matrix *D* of scale values on its diagonal. This allows us to turn *R* into $R = U \cdot D \cdot V^T$. Because *R* is itself orthonormal, the matrix *D* must be the identity matrix $I_3$ such that $R = U \cdot I_3 \cdot V^T$. We can thus "coerce" our computed *R* into being a rotation matrix by taking *R*'s singular value decomposition, setting its *D* matrix to the identity matrix, and multiplying by the SVD again to yield our new, conforming rotation matrix *R*.

Despite all this work, we have not yet dealt with lens distortions. We use the camera intrinsics found previously—together with the distortion parameters set to `0`—for our initial guess to start solving a larger system of equations.

The points we "perceive" on the image are really in the wrong place owing to distortion. Let $(x_p, y_p)$ be the point's location if the pinhole camera were perfect and let $(x_d, y_d)$ be its distorted location; then:

33 This is mainly due to precision errors.

$$\begin{bmatrix} x_p \\ y_p \end{bmatrix} = \begin{bmatrix} f_x \dfrac{X_W}{Z_W} + c_x \\ f_y \dfrac{Y_W}{Z_W} + c_y \end{bmatrix}$$

We use the results of the calibration without distortion using the following substitution:[34]

$$\begin{bmatrix} x_p \\ y_p \end{bmatrix} = \left(1 + k_1 r^2 + k_2 r^4 + k_3 r^6\right) \begin{bmatrix} x_d \\ y_d \end{bmatrix} + \begin{bmatrix} 2p_1 x_d y_d + p_2\left(r^2 + 2x_d^2\right) \\ p_1\left(r^2 + 2y_d^2\right) + 2p_2 x_d y_d \end{bmatrix}$$

A large list of these equations are collected and solved to find the distortion parameters, after which the intrinsics and extrinsics are re-estimated. That's the heavy lifting that the single function `cv::calibrateCamera()`[35] does for you!

### Calibration function

Once we have the corners for several images, we can call `cv::calibrateCamera()`. This routine will do the number crunching and give us the information we want. In particular, the results we receive are the *camera intrinsics matrix*, the *distortion coefficients*, the *rotation vectors*, and the *translation vectors*. The first two of these constitute the intrinsic parameters of the camera, and the latter two are the extrinsic measurements that tell us where the objects (i.e., the chessboards) were found and what their orientations were. The distortion coefficients ($k_1$, $k_2$, $p_1$, $p_2$, and any higher orders of $k_j$) are the coefficients from the radial and tangential distortion equations we encountered earlier; they help us when we want to correct that distortion away. The camera intrinsic matrix is perhaps the most interesting final result, because it is what allows us to transform from three-dimensional coordinates to the image's two-dimensional coordinates. We can also use the camera matrix to do the reverse operation, but in this case we can only compute a line in the three-dimensional world to which a given image point must correspond. We will return to this shortly.

Let's now examine the camera calibration routine itself:

```
double cv::calibrateCamera(
  cv::InputArrayOfArrays  objectPoints,     // K vecs (N pts each, object frame)
  cv::InputArrayOfArrays  imagePoints,      // K vecs (N pts each, image frame)
```

34 Camera calibration functionality is now very expanded to include fisheye and omni cameras; see the *opencv_contrib/modules/ccalib/src* and the *opencv_contrib/modules/ccalib/samples* directories.

35 The `cv::calibrateCamera()` function is used internally in the stereo calibration functions we will see in Chapter 19. For stereo calibration, we'll be calibrating two cameras at the same time and will be looking to relate them together through a rotation matrix and a translation vector.

```
    cv::Size                   imageSize,       // Size of input images (pixels)
    cv::InputOutputArray       cameraMatrix,    // Resulting 3-by-3 camera matrix
    cv::InputOutputArray       distCoeffs,      // Vector of 4, 5, or 8 coefficients
    cv::OutputArrayOfArrays rvecs,              // Vector of K rotation vectors
    cv::OutputArrayOfArrays tvecs,              // Vector of K translation vectors
    int                        flags       = 0, // Flags control calibration options
    cv::TermCriteria           criteria    = cv::TermCriteria(
      cv::TermCriteria::COUNT | cv::TermCriteria::EPS,
      30,                                       // ...after this many iterations
      DBL_EPSILON                               // ...at this total reprojection error
    )
  );
```

When calling `cv::calibrateCamera()`, you have many arguments to keep straight. The good news is that we've covered (almost) all of them already, so hopefully they'll all make sense.

The first argument is `objectPoints`. It is a vector of vectors, each of which contains the coordinates of the points on the calibration pattern for a particular image. Those coordinates are in the coordinate system of the object, so it is acceptable to make them simply integers in the x- and y-dimensions and zero in the z-dimension.[36]

The `imagePoints` argument follows. It is also a vector of vectors, and contains the location of each point as it was found in each image. If you are using the chessboard, each vector will be the `corners` output array from the corresponding image.

When you are defining the `objectPoints` input, you are implicitly altering the scale for some of the outputs of `cv::calibrateCamera`. Specifically, you are affecting the `tvecs` output. If you say that one corner on the chessboard is at (0, 0, 0), the next is at (0, 1, 0), and the next is at (0, 2, 0), and so on, then you are implicitly saying that you would like the distances measured in "chessboard squares." If you want physical units for the outputs, you must measure the chessboard in physical units. For example, if you want distances in meters, then you will have to measure your chessboard and use the correct square size in meters. If the squares turn out to be 25mm across, then you should set the same corners to (0, 0, 0), (0, 0.025, 0), (0, 0.050, 0), and so on. In contrast, the camera intrinsic matrix parameters are always reported in *pixels*.

The `imageSize` argument just tells `cv::calibrateCamera()` how large the images were (in pixels) from which the points in `imagePoints` were extracted.

36 In principle, you could use a different object for each calibration image. In practice, the outer vector will typically just contain *K* copies of the same list of point locations where *K* is the number of views.

The camera intrinsics are returned in the `cameraMatrix` and `distCoeffs` arrays. The former will contain the linear intrinsics, and should be a 3 × 3 matrix. The latter may be 4, 5, or 8 elements. If `distCoeffs` is of length `4`, then the returned array will contain the coefficients ($k_1$, $k_2$, $p_1$, and $p_2$). If the length is `5` or `8`, then the elements will be either ($k_1$, $k_2$, $p_1$, $p_2$, and $k_3$) or ($k_1$, $k_2$, $p_1$, $p_2$, $k_3$, $k_4$, $k_5$, and $k_6$), respectively. The five-element form is primarily for use fisheye lenses, and is generally useful only for them. The eight-element form is run only if you set the `cv::CALIB_RATIONAL_MODEL`, and is for very high-precision calibration of exotic lenses. It is important to remember, however, that the number of images you require will grow dramatically with the number of parameters you wish to solve for.

The `rvecs` and `tvecs` arrays are vectors of vectors, like the input points arrays. They contain a representation of the rotation matrix (in Rodrigues form—that is, as a three-component vector) and the translation matrix for each of the chessboards shown.

Because precision is very important in calibration, the `cameraMatrix` and `distCoeffs` arrays (as well as the `rvecs` and `tvecs` arrays) will always be computed and returned in double precision, even if you do not initially allocate these input arrays in this form.

Finding parameters through optimization can be something of an art. Sometimes trying to solve for all parameters at once can produce inaccurate or divergent results, especially if your initial starting position in parameter space is far from the actual solution. Thus, it is often better to "sneak up" on the solution by getting close to a good parameter starting position in stages. For this reason, we often hold some parameters fixed, solve for other parameters, then hold the other parameters fixed and solve for the original, and so on. Finally, when we think all of our parameters are close to the actual solution, we use our close parameter setting as the starting point and solve for everything at once. OpenCV allows you this control through the `flags` setting.

The `flags` argument allows for some finer control of exactly how the calibration will be performed. The following values may be combined together with a Boolean OR operation as needed:

`cv::CALIB_USE_INTRINSIC_GUESS`
: Normally the intrinsic matrix is computed by `cv::calibrateCamera()` with no additional information. In particular, the initial values of the parameters $c_x$ and $c_y$ (the image center) are taken directly from the `imageSize` argument. If this argument is set, then `cameraMatrix` is assumed to contain valid values that will be used as an initial guess to be further optimized by `cv::calibrateCamera()`.

In many practical applications, we know the focal length of a camera because we can read it off the side of the lens.[37] In such cases, it is usually a good idea to leverage this information by putting it into the camera matrix and using `cv::CALIB_USE_INTRINSIC_GUESS`. In most such cases, it is also safe (and a good idea) to use `cv::CALIB_FIX_ASPECT_RATIO`, discussed shortly.

`cv::CALIB_FIX_PRINCIPAL_POINT`
: This flag can be used with or without `cv::CALIB_USE_INTRINSIC_GUESS`. If used without, then the principal point is fixed at the center of the image; if used with, then the principal point is fixed at the supplied initial value in the `cameraMatrix`.

`cv::CALIB_FIX_ASPECT_RATIO`
: If this flag is set, then the optimization procedure will vary only $f_x$ and $f_y$ together and will keep their ratio fixed to whatever value is set in the `cameraMatrix` when the calibration routine is called. (If the `cv::CALIB_USE_INTRINSIC_GUESS` flag is not also set, then the values of $f_x$ and $f_y$ in `cameraMatrix` can be any arbitrary values and only their ratio will be considered relevant.)

`cv::CALIB_FIX_FOCAL_LENGTH`
: This flag causes the optimization routine to just use the $f_x$ and $f_y$ that were passed in the `cameraMatrix`.

`cv::CALIB_FIX_K1`, `cv::CALIB_FIX_K2`, ... `cv::CALIB_FIX_K6`
: Fix the radial distortion parameters $k_1$, $k_2$, up through $k_6$. You may set the radial parameters in any combination by adding these flags together.

`cv::CALIB_ZERO_TANGENT_DIST`
: This flag is important for calibrating high-end cameras that, as a result of precision manufacturing, have very little tangential distortion. Trying to fit parameters that are near `0` can lead to noisy, spurious values and to problems of numerical stability. Setting this flag turns off fitting the tangential distortion parameters $p_1$ and $p_2$, which are thereby both set to `0`.

`cv::CALIB_RATIONAL_MODEL`
: This flag tells OpenCV to compute the $k_4$, $k_5$, and $k_6$ distortion coefficients. This is here because of a backward compatibility issue; if you do not add this flag, only

37 This is a slight oversimplification. Recall that the focal lengths that appear in the intrinsics matrix are measured in units of pixels. So if you have a 2,048 × 1,536 pixel imager with a 1/1.8" sensor format, that gives you 3.45μm pixels. If your camera has a 25.0mm focal length lens, then the initial guesses for $f$ are not 25 (mm) but 7,246.38 (pixels) or, if you are realistic about significant digits, just 7,250.

the first three $k_j$ parameters will be computed (even if you gave an eight-element array for `distCoeffs`).

The final argument to `cv::calibrateCamera()` is the termination criteria. As usual, the termination criteria can be a number of iterations, an "epsilon" value, or both. In the case of the epsilon value, what is being computed is called the *reprojection error*. The reprojection error, as with the case of `cv::findHomography()`, is the sum of the squares of the distances between the computed (projected) locations of the three-dimensional points onto the image plane and the actual location of the corresponding points on the original image.

It is increasingly common to use asymmetric circle-grids for camera calibration. In this case, it is important to remember that the `objectPoints` argument must be set accordingly. For example, a possible set of coordinates for the object points in Figure 18-15 would be (0, 0, 0), (1, 1, 0), (2, 1, 0), (3, 1, 0), and so on, with the ones on the next row being (0, 2, 0), (1, 3, 0), (2, 2, 0), (3, 3, 0), and so on through all of the rows.

### Computing extrinsics only with cv::solvePnP()

In some cases, you will already have the intrinsic parameters of the camera and therefore need only to compute the location of the object(s) being viewed. This scenario clearly differs from the usual camera calibration, but it is nonetheless a useful task to be able to perform.[38] In general, this task is called the *Perspective N-Point* or PnP problem:

```
bool cv::solvePnP(
  cv::InputArray  objectPoints,                // Object points (object frame)
  cv::InputArray  imagePoints,                 // Found pt locations (img frame)
  cv::InputArray  cameraMatrix,                // 3-by-3 camera matrix
  cv::InputArray  distCoeffs,                  // Vector of 4, 5, or 8 coeffs
  cv::OutputArray rvec,                        // Result rotation vector
  cv::OutputArray tvec,                        // Result translation vector
  bool            useExtrinsicGuess = false, // true='use vals in rvec and tvec'
  int             flags             = cv::ITERATIVE
);
```

The arguments to `cv::solvePnP()` are similar to the corresponding arguments for `cv::calibrateCamera()` with two important exceptions. First, the `objectPoints` and `imagePoints` arguments are those from just a single view of the object (i.e., they are of type `cv::InputArray`, not `cv::InputArrayOfArrays`). Second, the intrinsic matrix

38 In fact, this task is a component of the overall task of camera calibration, and this function is called internally by `cv::calibrateCamera()`.

and the distortion coefficients are supplied rather than computed (i.e., they are inputs instead of outputs). The resulting rotation output is again in the Rodrigues form: three-component rotation vector that represents the three-dimensional axis around which the chessboard or points were rotated, with the vector magnitude or length representing the counterclockwise angle of rotation. This rotation vector can be converted into the 3 × 3 rotation matrix we've discussed before via the `cv::Rodrigues()` function. The translation vector is the offset in camera coordinates to where the chessboard origin is located.

The `useExtrinsicGuess` argument can be set to `true` to indicate that the current values in the `rvec` and `tvec` arguments should be considered as initial guesses for the solver. The default is `false`.

The final argument, `flags`, can be set to one of three values—`cv::ITERATIVE`, `cv::P3P`, or `cv::EPNP`—to indicate which method should be used for solving the overall system. In the case of `cv::ITERATIVE`, a Levenberg-Marquardt optimization is used to minimize reprojection error between the input `imagePoints` and the projected values of `objectPoints`. In the case of `cv::P3P`, the method used is based on [Gao03]. In this case, exactly four object and four image points should be provided. The return value for `cv::SolvePNP` will be true only if the method succeeds. Finally, in the case of `cv::EPNP`, the method described in [Moreno-Noguer07] will be used. Note that neither of the latter two methods is iterative and, as a result, should be much faster than `cv::ITERATIVE`.

Though we introduced `cv::solvePnP()` as a way to compute the pose of an object (e.g., the chessboard) in each of many frames relative to which the camera is imagined to be stationary, the same function can be used to solve what is effectively the inverse problem. In the case of, for example, a mobile robot, we are more interested in the case of a stationary object (maybe a fixed object, maybe just the entire scene) and a moving camera. In this case, you can still use `cv::solvePnP()`; the only difference is in how you interpret the resulting `rvec` and `tvec` vectors.

### Computing extrinsics only with cv::solvePnPRansac()

One shortcoming with `cv::solvePnP` is that it is not robust to outliers. In camera calibration, this is not as much of a problem, mainly because the chessboard itself gives us a reliable way to find the individual features we care about and to verify that we are looking at what we think we are looking at through their relative geometry. However, in cases in which we are trying to localize the camera relative to points not on a chessboard, but in the real world (e.g., using sparse keypoint features), mismatches are likely and will cause severe problems. Recall from our discussion in "Homography" on page 664 that the RANSAC method can be an effective way to handle outliers of this kind:

```
bool cv::solvePnPRansac(
  cv::InputArray  objectPoints,             // Object points (object frame)
  cv::InputArray  imagePoints,              // Found pt locations (img frame)
  cv::InputArray  cameraMatrix,             // 3-by-3 camera matrix
  cv::InputArray  distCoeffs,               // Vector of 4, 5, or 8 coeffs
  cv::OutputArray rvec,                     // Result rotation vector
  cv::OutputArray tvec,                     // Result translation vector
  bool            useExtrinsicGuess = false, // read vals in rvec and tvec ?
  int             iterationsCount   = 100,  // RANSAC iterations
  float           reprojectionError = 8.0,  // Max error for inclusion
  int             minInliersCount   = 100,  // terminate if this many found
  cv::OutputArray inliers           = cv::noArray(), // Contains inlier indices
  int flags                         = cv::ITERATIVE  // same as solvePnP()
)
```

All of the arguments for `cv::solvePnP()` that are shared by `cv::solvePnPRansac()` have the same interpretation. The new arguments control the RANSAC portion of the algorithm. In particular, the `iterationsCount` argument sets the number of RANSAC iterations and the `reprojectionError` argument indicates the maximum reprojection error that will still cause a configuration to be considered an inlier.[39] The argument `minInliersCount` is somewhat misleadingly named; if at any point in the RANSAC process the number of inliers exceeds `minInliersCount`, the process is terminated, and this group is taken to be the inlier group. This can improve performance substantially, but can also cause a lot of problems if set too low. Finally, the `inliers` argument is an output that, if provided, will be filled with the indices of the points (from `objectPoints` and `imagePoints`) selected as inliers.

# Undistortion

As we have alluded to already, there are two things that one often wants to do with a calibrated camera: correct for distortion effects and construct three-dimensional representations of the images it receives. Let's take a moment to look at the first of these before diving into the more complicated second task in the next chapter.

OpenCV provides us with a ready-to-use undistortion algorithm that takes a raw image and the distortion coefficients from `cv::calibrateCamera()` and produces a corrected image (Figure 18-17). We can access this algorithm either through the function `cv::undistort()`, which does everything we need in one shot, or through

39 For the PnP problem, what is effectively sought is a perspective transform, which is fully determined by four points. Thus, for a RANSAC iteration, the initial number of selected points is four. The default value of `reprojectionError` corresponds to an average distance of $\sqrt{2}$ between each of these points and its corresponding reprojection.

the pair of routines `cv::initUndistortRectifyMap()` and `cv::remap()`, which allow us to handle things a little more efficiently for video or other situations where we have many images from the same camera.[40]

*Figure 18-17. Camera image before undistortion (left) and after (right)*

## Undistortion Maps

When performing undistortion on an image, we must specify where every pixel in the input image is to be moved in the output image. Such a specification is called an *undistortion map* (or sometimes just a *distortion map*). There are several representations available for such maps.

The first and most straightforward representation is the *two-channel float* representation. In this representation, a remapping for an $N \times M$ image is represented by an $N \times M$ array of two-channel floating-point numbers as shown in Figure 18-18. For any given entry $(i, j)$ in the image, the value of that entry will be a pair of numbers $(i^*, j^*)$ indicating the location to which pixel $(i, j)$ of the input image should be relocated. Of course, because $(i^*, j^*)$ are floating-point numbers, interpolation in the target image is implied.[41]

40 We should take a moment to clearly make a distinction here between *undistortion*, which mathematically removes lens distortion, and *rectification*, which mathematically aligns the two (or more) images with respect to each other. The latter will be important in the next chapter.

41 As is typically the case, the interpolation is actually performed in the opposite direction. This means that, given the map, we compute a pixel in the final image by determining which pixels in the original image map into its vicinity, and then interpolating between those pixel values appropriately.

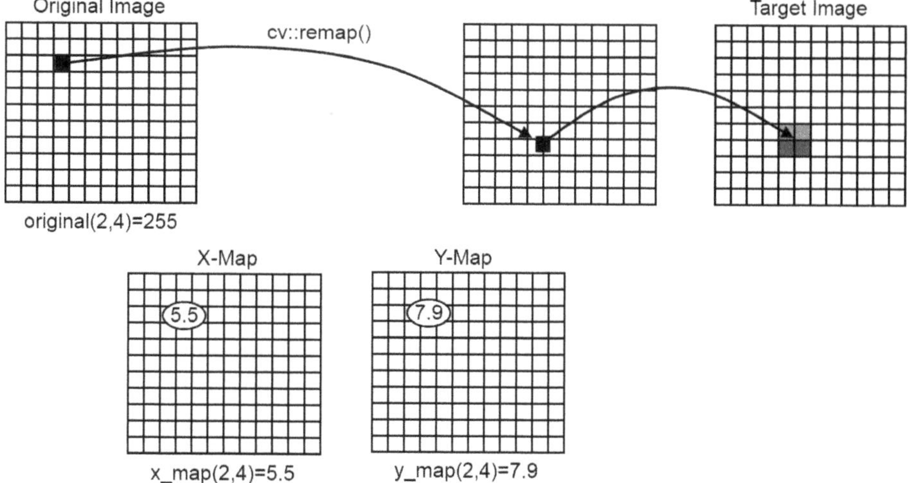


*Figure 18-18. In the float-float representation, two different floating-point arrays, the X-map and the Y-map, encode the destination of a pixel at (i, j) in the original image. A pixel at (i, j) is mapped to (x_map(i, j), y_map(i, j)) in the destination image. Because that destination is not necessarily integer, the interpolation is used to compute the final image pixel intensities*

The next representation is the *two-array float* representation. In this representation, the remapping is described by a pair of $N \times M$ arrays, each of which is a single-channel floating-point array. The first of these arrays contains at location $(i, j)$ the value $i^*$, the x-coordinate of the remapped location of pixel $(i, j)$ from the original array. Similarly, the second array contains at the same location $j^*$, the y-coordinate of the remapped location of pixel $(i, j)$.

The final representation is the *fixed-point* representation. In this representation, a mapping is specified by a two-channel signed integer array (i.e., of type `CV_16SC2`). The interpretation of this array is the same as the *two-channel float* representation, but operations with this format are much faster. In the case in which higher precision is required, the necessary information required for interpolation during the remapping is encoded in a second single-channel unsigned integer array (i.e., `CV_16UC1`). The entries in this array refer to an internal lookup table, which is used for the interpolation (and thus there are no "user-serviceable parts" in this array).

## Converting Undistortion Maps Between Representations with cv::convertMaps()

Because there are multiple representations available for undistortion maps, it is natural that one might want to convert between them. We do this with the `cv::convert`

Maps() function. This function allows you to provide a map in any of the four formats we just discussed, and convert it into any of the others. The prototype for cv::convertMaps() is the following:

```
void cv::convertMaps(
  cv::InputArray  map1,          // First in map: CV_16SC2/CV_32FC1/CV_32FC2
  cv::InputArray  map2,          // Second in map: CV_16UC1/CV_32FC1 or none
  cv::OutputArray dstmap1,       // First out map
  cv::OutputArray dstmap2,       // Second out map
  int             dstmap1type,   // dstmap1 type: CV_16SC2/CV_32FC1/CV_32FC2
  bool            nninterpolation = false  // For conversion to fixed point types
);
```

The inputs map1 and map2 are your existing maps, and the outputs dstmap1 and dstmap2 are the outputs where the converted maps will be stored. The argument dstmap1type tells cv::convertMaps() what sort of maps to create for you; from the type specified here, the remapping you want is then inferred. You use the final argument, nninterpolation, when converting to the fixed point type to indicate whether you would like the interpolation tables to be computed as well. The possible conversions are shown in Table 18-1, along with the correct settings for dstmap1type and nninterpolation for each conversion.

*Table 18-1. Possible map type conversions for cv::convertMaps()*

| map1 | map2 | dstmap1type | nninterpolation | dstmap1 | dstmap2 |
|---|---|---|---|---|---|
| CV_32FC1 | CV_32FC1 | CV_16SC2 | true | CV_16SC2 | CV_16UC1 |
| CV_32FC1 | CV_32FC1 | CV_16SC2 | false | CV_16SC2 | cv::noArray() |
| CV_32FC2 | cv::noArray() | CV_16SC2 | true | CV_16SC2 | CV_16UC1 |
| CV_32FC2 | cv::noArray() | CV_16SC2 | false | CV_16SC2 | cv::noArray() |
| CV_16SC2 | CV_16UC1 | CV_32FC1 | true | CV_32FC1 | CV_32FC1 |
| CV_16SC2 | cv::noArray() | CV_32FC1 | false | CV_32FC1 | CV_32FC1 |
| CV_16SC2 | CV_16UC1 | CV_32FC2 | true | CV_32FC2 | cv::noArray() |
| CV_16SC2 | cv::noArray() | CV_32FC2 | false | CV_32FC2 | cv::noArray() |

Note that, because the conversion from floating point to fixed point naturally loses precision (even if nninterpolation is used), if you convert one way and then convert back again, you should not expect to necessarily receive back exactly what you started with.

## Computing Undistortion Maps with cv::initUndistortRectifyMap()

Now that we understand undistortion maps, the next step is to see how to compute them from camera parameters. In fact, distortion maps are much more general than just this one specific situation, but for our current purpose, the interesting thing is to know how we can use the results of our camera calibrations to create nicely rectified images. So far, we have been discussing the situation of monocular imaging. In fact, one of the more important applications of rectification is in preparing a pair of images to be used for stereoscopic computation of depth images. We will get back to the topic of stereo in the next chapter, but it is worth keeping this application in mind, partially because some of the arguments to the functions that do undistortion are primarily for use in that stereo vision context.

The basic process is to first compute the undistortion maps, and then to apply them to the image. The reason for this separation is that, in most practical applications, you will compute the undistortion maps for your camera only once, and then you will use those maps again and again as new images come streaming in from your camera. The function `cv::initUndistortRectifyMap()` computes the distortion map from camera-calibration information:

```
void cv::initUndistortRectifyMap(
  cv::InputArray  cameraMatrix,        // 3-by-3 camera matrix
  cv::InputArray  distCoeffs,          // Vector of 4, 5, or 8 coefficients
  cv::InputArray  R,                   // Rectification transformation
  cv::InputArray  newCameraMatrix,     // New camera matrix (3-by-3)
  cv::Size        size,                // Undistorted image size
  int             m1type,              // 'map1' type: 16SC2, 32FC1, or 32FC2
  cv::OutputArray map1,                // First output map
  cv::OutputArray map2,                // Second output map
);
```

The function `cv::initUndistortRectifyMap()` computes the undistortion map (or maps). The first two arguments are the camera intrinsic matrix and the distortion coefficients, both in the form you received them from `cv::calibrateCamera()`.

The next argument, `R`, may be used or, alternatively, set to `cv::noArray()`. If used, it must be a 3 × 3 rotation matrix, which will be preapplied before rectification. The function of this matrix is to compensate for a rotation of the camera relative to some global coordinate system in which that camera is embedded.

Similar to the rotation matrix, `newCameraMatrix` can be used to affect how the images are undistorted. If used, it will "correct" the image before undistortion to how it would have looked if taken from a different camera with different intrinsic parameters. In practice, the aspect that one changes in this way is the camera center, not the focal length. You will typically not use this when dealing with monocular imaging, but it is important in the case of stereo image analysis. For monocular images, you will typically just set this argument to `cv::noArray()`.

It is unlikely that you will use the rotation or newCameraMatrix arguments unless you are dealing with stereo images. In that case, the correct arrays to give to cv::initUndistortRectifyMap() will be computed for you by cv::stereoRectify(). After calling cv::stereoRectify(), you will rectify the image from the first camera with the R1 and P1 arrays, and the image from the second camera with the R2 and P2 arrays. We will revisit this topic in the next chapter when we discuss stereo imaging.[42]

The argument size simply sets the size of the output maps; this should correspond to the size of images you will be undistorting.

The final three arguments—m1type, map1, and map2—specify the final map type and provide a place for that map to be written, respectively. The possible values of m1type are CV_32FC1 or CV_16SC2, and correspond to the type that will be used to represent map1. In the case of CV_32FC1, map2 will also be of type CV_32FC1. This corresponds to the *two-array float* representation of the map we encountered in the previous section. In the case of CV_16SC2, map1 will be in the fixed-point representation (recall that this single-channel array contains the interpolation table coefficients).

## Undistorting an Image with cv::remap()

Once you have computed the undistortion maps, you can apply them to incoming images using cv::remap(). We encountered cv::remap() earlier, when discussing general image transforms; this is one specific, but very important, application for that function.

As we saw previously, the cv::remap() function has two map arguments that correspond to the undistortion maps, such as those computed by cv::initUndistortRectifyMap(). cv::remap() will accept any of the distortion map formats we have discussed: the two-channel float, the two-array float, or the fixed-point format (with or without the accompanying array of interpolation table indexes).

If you use the cv::remap() with the two different two-channel float representations, or the fixed-point representation without the interpolation table array, you should just pass cv::noArray() for the map2 argument.

42 The exceptionally astute reader may have noticed that the camera-matrix is a 3 × 3 matrix, while the return values from cv::stereoRectify() called P1 and P2 are in fact 3 × 4 projection matrices. Not to worry. In fact, the first three columns of P1 and P2 contain the same information as the camera matrix, and this is the part that is actually used internally by cv::initUndistortRectifyMap().

## Undistortion with cv::undistort()

In some cases, you will either have only one image to rectify, or need to recompute the undistortion maps for every image. In such cases, you can use the somewhat more compact `cv::undistort()`, which effectively computes the maps and applies them in a single go.

```
void cv::undistort(
  cv::InputArray  src,                       // Input distorted image
  cv::OutputArray dst,                       // Result corrected image
  cv::InputArray  cameraMatrix,              // 3-by-3 camera matrix
  cv::InputArray  distCoeffs,                // Vector of 4, 5, or 8 coeffs
  cv::InputArray  newCameraMatrix = noArray() // Optional new camera matrix
);
```

The arguments to `cv::undistort()` are identical to the corresponding arguments to `cv::initUndistortRectifyMap()`.

## Sparse Undistortion with cv::undistortPoints()

Another situation that arises occasionally is that, rather than rectifying an entire image, you have a set of points you have collected from an image, and you care only about the location of those points. In this case, you can use `cv::undistortPoints()` to compute the "correct" locations for your specific list of points:

```
void cv::undistortPoints(
  cv::InputArray  src,                         // Input array, N pts. (2-d)
  cv::OutputArray dst,                         // Result array, N pts. (2-d)
  cv::InputArray  cameraMatrix,                // 3-by-3 camera matrix
  cv::InputArray  distCoeffs,                  // Vector of 4, 5, or 8 coeffs
  cv::InputArray  R           = cv::noArray(), // 3-by-3 rectification mtx.
  cv::InputArray  P           = cv::noArray()  // 3-by-3 or 3-by-4 new camera
                                               //   or new projection matrix
);
```

As with `cv::undistort()`, the arguments to `cv::undistortPoints()` are analogous to the corresponding arguments to `cv::initUndistortRectifyMap()`. The primary difference is that the `src` and `dst` arguments are vectors of two-dimensional points, rather than two-dimensional arrays. (As always, these vectors can be of any of the usual forms: $N \times 1$ array of `cv::Vec2i` objects, $N \times 1$ array of float objects, an STL-style vector of `cv::Vec2f` objects, etc.)

The argument `P` of `cv::undistortPoints()` corresponds to `newCameraMatrix` of `cv::undistortPoints()`. As before, these two extra parameters relate primarily to the function's use in stereo rectification, which we will discuss in Chapter 19. The rectified camera matrix `P` can have dimensions of $3 \times 3$ or $3 \times 4$ deriving from the first three or four columns of `cv::stereoRectify()`'s return value for camera matrices P1 or P2 (for the left or right camera; see Chapter 19). These parameters are by default cv::noArray(), which the function interprets as identity matrices.

## Putting Calibration All Together

Now it's time to put all of this together in an example. Example 18-1 presents a program that performs the following tasks: it looks for chessboards of the dimensions that the user specified, grabs as many full images (i.e., those in which it can find all the chessboard corners) as the user requested, and computes the camera intrinsics and distortion parameters. Finally, the program enters a display mode whereby an undistorted version of the camera image can be viewed.

When using this program, you'll want to substantially change the chessboard views between successful captures. Otherwise, the matrices of points used to solve for calibration parameters may form an ill-conditioned (rank-deficient) matrix and you will end up with either a bad solution or no solution at all.

*Example 18-1. Reading a chessboard's width and height, reading and collecting the requested number of views, and calibrating the camera*

```
#include <opencv2/opencv.hpp>
#include <iostream>

using namespace std;

void help( char *argv[] ) {
  ...
}

int main(int argc, char* argv[]) {

  int   n_boards = 0; // Will be set by input list
  float image_sf = 0.5f;
  float delay    = 1.f;
  int   board_w  = 0;
  int   board_h  = 0;

  if(argc < 4 || argc > 6) {
    cout << "\nERROR: Wrong number of input parameters";
    help( argv );
    return -1;
  }
  board_w  = atoi( argv[1] );
  board_h  = atoi( argv[2] );
  n_boards = atoi( argv[3] );
  if( argc > 4 ) delay    = atof( argv[4] );
```

```
if( argc > 5 ) image_sf = atof( argv[5] );

int      board_n  = board_w * board_h;
cv::Size board_sz = cv::Size( board_w, board_h );

cv::VideoCapture capture(0);
if( !capture.isOpened() ) {
  cout << "\nCouldn't open the camera\n";
  help( argv );
  return -1;
}

// ALLOCATE STORAGE
//
vector< vector<cv::Point2f> > image_points;
vector< vector<cv::Point3f> > object_points;

// Capture corner views: loop until we've got n_boards successful
// captures (all corners on the board are found).
//
double   last_captured_timestamp = 0;
cv::Size image_size;

while( image_points.size() < (size_t)n_boards ) {

  cv::Mat image0, image;
  capture >> image0;
  image_size = image0.size();
  cv::resize(image0, image, cv::Size(), image_sf, image_sf, cv::INTER_LINEAR );

  // Find the board
  //
  vector<cv::Point2f> corners;
  bool found = cv::findChessboardCorners( image, board_sz, corners );

  // Draw it
  //
  drawChessboardCorners( image, board_sz, corners, found );

  // If we got a good board, add it to our data
  //
  double timestamp = (double)clock()/CLOCKS_PER_SEC;

  if( found && timestamp - last_captured_timestamp > 1 ) {

    last_captured_timestamp = timestamp;
    image ^= cv::Scalar::all(255);

    cv::Mat mcorners(corners);                  // do not copy the data
    mcorners *= (1./image_sf);                  // scale the corner coordinates
    image_points.push_back(corners);
    object_points.push_back(vector<Point3f>());
```

```
    vector<cv::Point3f>& opts = object_points.back();
    opts.resize(board_n);
    for( int j=0; j<board_n; j++ ) {
      opts[j] = cv::Point3f((float)(j/board_w), (float)(j%board_w), 0.f);
    }
    cout << "Collected our " <<  (int)image_points.size() <<
      " of " << n_boards << " needed chessboard images\n" << endl;
  }
  cv::imshow( "Calibration", image );  //show in color if we did collect the image

  if((cv::waitKey(30) & 255) == 27)
    return -1;
}
// END COLLECTION WHILE LOOP.


cv::destroyWindow( "Calibration" );
cout << "\n\n*** CALIBRATING THE CAMERA...\n" << endl;

// CALIBRATE THE CAMERA!
//
cv::Mat intrinsic_matrix, distortion_coeffs;
double err = cv::calibrateCamera(
  object_points,
  image_points,
  image_size,
  intrinsic_matrix,
  distortion_coeffs,
  cv::noArray(),
  cv::noArray(),
  cv::CALIB_ZERO_TANGENT_DIST | cv::CALIB_FIX_PRINCIPAL_POINT
);

// SAVE THE INTRINSICS AND DISTORTIONS
cout << " *** DONE!\n\nReprojection error is " << err <<
  "\nStoring Intrinsics.xml and Distortions.xml files\n\n";
cv::FileStorage fs( "intrinsics.xml", FileStorage::WRITE );

fs << "image_width" << image_size.width << "image_height" << image_size.height
  <<"camera_matrix" << intrinsic_matrix << "distortion_coefficients"
  << distortion_coeffs;
fs.release();

// EXAMPLE OF LOADING THESE MATRICES BACK IN:
fs.open( "intrinsics.xml", cv::FileStorage::READ );
cout << "\nimage width: " << (int)fs["image_width"];
cout << "\nimage height: " << (int)fs["image_height"];

cv::Mat intrinsic_matrix_loaded, distortion_coeffs_loaded;
fs["camera_matrix"] >> intrinsic_matrix_loaded;
fs["distortion_coefficients"] >> distortion_coeffs_loaded;
cout << "\nintrinsic matrix:" << intrinsic_matrix_loaded;
```

```
  cout << "\ndistortion coefficients: " << distortion_coeffs_loaded << endl;

  // Build the undistort map which we will use for all
  // subsequent frames.
  //
  cv::Mat map1, map2;
  cv::initUndistortRectifyMap(
    intrinsic_matrix_loaded,
    distortion_coeffs_loaded,
    cv::Mat(),
    intrinsic_matrix_loaded,
    image_size,
    CV_16SC2,
    map1,
    map2
  );

  // Just run the camera to the screen, now showing the raw and
  // the undistorted image.
  //
  for(;;) {
    cv::Mat image, image0;
    capture >> image0;
    if( image0.empty() ) break;
    cv::remap(
      image0,
      image,
      map1,
      map2,
      cv::INTER_LINEAR,
      cv::BORDER_CONSTANT,
      cv::Scalar()
    );
    cv::imshow("Undistorted", image);
    if((cv::waitKey(30) & 255) == 27) break;
  }

  return 0;
}
```

# Summary

We began the chapter with a brief review of the pinhole camera model and an overview of the basics of projective geometry. After introducing the Rodrigues transform as an alternate representation of rotations, we introduced the concept of lens distortions and learned how they are modeled in OpenCV and that this model is summarized by the camera intrinsics matrix.

With this model in hand, we proceeded to learn how to calibrate a camera using chessboards or circle-grids (and that there are yet other calibration patterns and techniques, referenced in Appendix B, in the `ccalib` function group). We saw how we could have OpenCV compute these intrinsic and extrinsic parameters for us using the results of intersection or circle finding from many calibration images. The calibration function led us to the topic of image homography. We learned the difference between intrinsic and extrinsic parameters in calibration, and related those extrinsics to the general "PnP" pose estimation problem. We saw finally how to undistort images by making use of the computed intrinsic parameters of a camera to correct away common distortions that arise from real-world lenses.

We finished off with a complete example in which images were captured, calibration data was extracted from those images, camera intrinsics were computed from that data, and incoming video was corrected using that camera information.

# Exercises

1. If you had only a ruler, how could you determine the focal length of a camera? Assume that the camera has negligible distortion and that the principal point is the center of the image.
2. Use Figure 18-2 to derive the equations $x = f_x \cdot (X/Z) + c_x$ and $y - f_y \cdot (Y/Z) + c_y$ using similar triangles with a center-position offset.
3. Will errors in estimating the true center location $(c_x, c_y)$ affect the estimation of other parameters such as focus?

   Hint: See the $q = MQ$ equation.
4. Draw an image of a square:
   a. under radial distortion;
   b. under tangential distortion; and
   c. under both distortions.
5. Refer to Figure 18-19. For perspective views, explain the following.
   a. Where does the "line at infinity" come from?
   b. Why do parallel lines on the object plane converge to a point on the image plane?
   c. Assume that the object and image planes are perpendicular to one another. On the object plane, starting at a point $p_1$, move 10 units directly away from the image plane to $p_2$. What is the corresponding movement distance on the image plane?
6. Figure 18-3 shows the outward-bulging "barrel distortion" effect of radial distortion, which is especially evident in the left panel of Figure 18-12. Could some lenses generate an inward-bending effect? How would this be possible?

7. Using a cheap web camera or cell phone, take pictures that show examples of radial and tangential distortion using images of concentric squares or chessboards.

   a. Calibrate the camera using `cv::calibrateCamera()` and at least 15 images of chessboards. Then use `cv::projectPoints()` to project an arrow orthogonal to the chessboards (the surface normal) into each of the chessboard images using the rotation and translation vectors from the camera calibration.

   b. Display the distorted pictures before and after undistortion.

8. What would happen to your calibration for focal length and principal point if you subsampled the image array by a factor of 2 (skip every other pixel in the x- and y-direction)?

9. Experiment with numerical stability and noise by collecting many images of chessboards and doing a "good" calibration on all of them. Then see how the calibration parameters change as you reduce the number of chessboard images. Graph your results: camera parameters as a function of number of chessboard images.

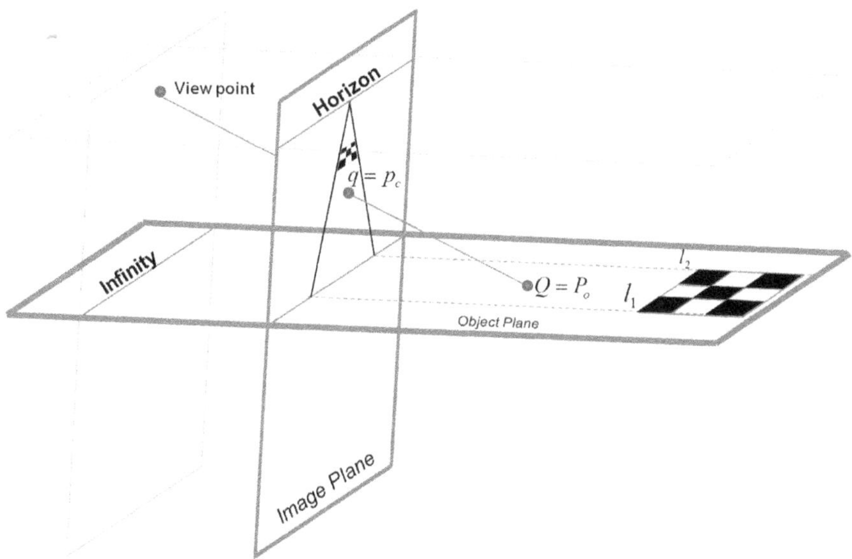


*Figure 18-19. Homography diagram showing intersection of the object plane with the image plane and a viewpoint representing the center of projection*

10. High-end cameras typically have systems of lenses that correct physically for distortions in the image. What might happen if you nevertheless use a multiterm distortion model for such a camera?

    Hint: This condition is known as *overfitting*.

11. *Three-dimensional joystick trick.* Calibrate a camera. Using video, wave a chessboard around and use `cv::solvePnP()` as a three-dimensional joystick. Remember that `cv::solvePnP()` outputs rotation as a 3 × 1 or 1 × 3 vector axis of rotation, where the magnitude of the vector represents the counterclockwise angle of rotation along with a three-dimensional translation vector.

    a. Output the chessboard's axis and angle of the rotation along with where it is (i.e., the translation) in real time as you move the chessboard around. Handle cases where the chessboard is not in view.

    b. Use `cv::Rodrigues()` to translate the output of `cv::solvePnP()` into a 3 × 3 rotation matrix and a translation vector. Use this to animate a simple three-dimensional stick figure of an airplane rendered back into the image in real time as you move the chessboard in view of the video camera.

# CHAPTER 19
# Projection and Three-Dimensional Vision

In this chapter, we'll move into three-dimensional vision. First, we will investigate three- to two-dimensional projections and the inverse operation (as much as this operation can be inverted), and then move on to multicamera stereo depth perception. To do this, we'll have to carry along some of the concepts from Chapter 18. We'll need the *camera intrinsics* matrix $M$, the *distortion coefficients*, the rotation matrix $R$, the translation vector $\vec{T}$, and especially the *homography matrix* $H$.[1]

We'll start by discussing projection into the three-dimensional world using a calibrated camera and reviewing affine and projective transforms (which we first encountered in Chapter 11); then we'll move on to an example of how to get a bird's-eye view of a ground plane.[2] We'll also discuss, in a little more detail, `cv::solvePnP()`, which we first saw in Chapter 18. In this context, we will see how this algorithm can be used to find the three-dimensional pose (position and rotation) of a known three-dimensional object in an image.

With those concepts in hand, we will then move into the three-dimensional geometry and multiple imagers. In general, there is no reliable way to do calibration or to extract three-dimensional information without multiple images. The most obvious case in which we use multiple images to reconstruct a three-dimensional scene is *stereo vision*. In stereo vision, features in two (or more) images taken at the same time from separate cameras are matched with the corresponding features in the other

1 Also that, as alluded to in Chapter 18, and as described in Appendix B, *opencv_contrib* contains further calibration algorithms for omni cameras and multicameras (in `ccalib`), different types of calibration patterns (in `aruco` and `ccalib`), and finally color balance and denoising algorithms (`xphoto`). As this is experimental code, we will not cover it in detail here, but many of these tools can be very useful in improving your work in stereo calibration and vision.

2 This is a recurrent problem in robotics as well as many other vision applications.

images, and the differences are analyzed to yield depth information. Another case is *structure from motion*. In this case, we may have only a single camera, but we have multiple images taken at different times and from different places. In the former case, we are primarily interested in *disparity effects* (triangulation) as a means of computing distance. In the latter, we compute the *fundamental matrix* (which relates two different views together) as the source of our scene understanding.

Because all of these problems rely, in one way or another, on the ability to project from the three-dimensional world into the two-dimensional image or the camera, we'll start with what we learned in Chapter 18 and build our way up from there.

# Projections

Once we have calibrated the camera (see Chapter 18), it is possible to unambiguously project points in the physical world to points in the image. This means that, given a location in the three-dimensional physical coordinate frame attached to the camera, we can compute where on the imager, in pixel coordinates, an external three-dimensional point should appear. This transformation is accomplished by the OpenCV routine `cv::projectPoints()`:

```
void cv::projectPoints(
  cv::InputArray  objectPoints, // 3xN/Nx3 Nc=1, 1xN/Nx1 Nc=3,
                                //  or vector<Point3f>
  cv::InputArray  rvec,         // Rotation *vector*
                                //  (see cv::Rodrigues())
  cv::InputArray  tvec,         // Translation vector
  cv::InputArray  cameraMatrix, // 3x3 Camera intrinsics matrix
  cv::InputArray  distCoeffs,   // 4, 5, or 8 elements vector,
                                //  or cv::noArray()
  cv::OutputArray imagePoints,  // 2xN/Nx2 Nc=1, 1xN/Nx1 Nc=2,
                                //  or vector<Point2f>
  cv::OutputArray jacobian    = cv::noArray(), // Optional,
                                               //  2N x (10+nDistCoeff)
  double          aspectRatio = 0              // If nonzero, fix
                                               //  fx/fy at this value
);
```

At first glance, the number of arguments might be a little intimidating, but in fact this is a simple function to use. The `cv::projectPoints()` routine was designed to accommodate the (very common) circumstance where the points you want to project are located on some rigid body. In this case, it is natural to represent the points not just as a list of locations in the camera coordinate system but rather as a list of locations in the object's own body-centered coordinate system; then we can add a rotation and a translation to specify the relationship between the object coordinates and the camera's coordinate system. In fact, `cv::projectPoints()` is used internally in `cv::calibrateCamera()`, and of course this is the way `cv::calibrateCamera()` organizes its own internal operation. All of the optional arguments are primarily there for use by `cv::calibrateCamera()`, but sophisticated users might find them handy for their own purposes as well.

The first argument, `objectPoints`, is the list of points you want projected; it can be in any of the usual forms: $N \times 3$ array, $3 \times N$ array, $N \times 1$ or $1 \times N$ array of `cv::Vec3f` objects, or just a plain old STL-style vector of `cv::Vec3f` objects containing the point locations. You supply these in the object's own local coordinate system and then provide the vectors `rotationVector`[3] and `translationVector` that relate the object coordinates to the camera coordinates. If, in your particular context, it is easier to work directly in the camera coordinates, then you can just give `objectPoints` in that system and set both `rotationVector` and `translationVector` to contain `0`s.[4]

The `cameraMatrix` and `distortionCoeffs` are just the camera intrinsic information and the distortion coefficients that come from `cv::calibrateCamera()` discussed in Chapter 18. The `imagePoints` argument is an array of two-dimensional points (in any of the usual forms) to which the results of the computation will be written.

Next is the optional argument `jacobian` which, if provided, will be filled with values corresponding to the partial derivatives of each point location with respect to the components of the rotation and translation vectors, the elements of the camera matrix, and the distortion coefficients. As a result, `jacobian` will be an $(2 \cdot N_p) \times (10 + N_d)$ array—where $N_p$ is the number of points and $N_d$ is the number of distortion coefficients. The exact entries in `jacobian` are shown in Figure 19-1. If you do not need `jacobian` computed (and in most cases, you probably will not) you can just leave it set to `cv::noArray()` and these values will not be computed.

The last parameter, `aspectRatio`, is also optional; it is used for derivatives only when the aspect ratio is fixed in `cv::calibrateCamera()` or `cv::stereoCalibrate()`. If this parameter is not `0.0`, then the derivatives appearing in `jacobian` are adjusted.

---

3 The "rotation vector" is in the usual Rodrigues representation that we learned about in the beginning of Chapter 18.

4 Remember that this rotation vector is an axis-angle representation of the rotation, so being set to all `0`s means it has zero magnitude and thus "no rotation."

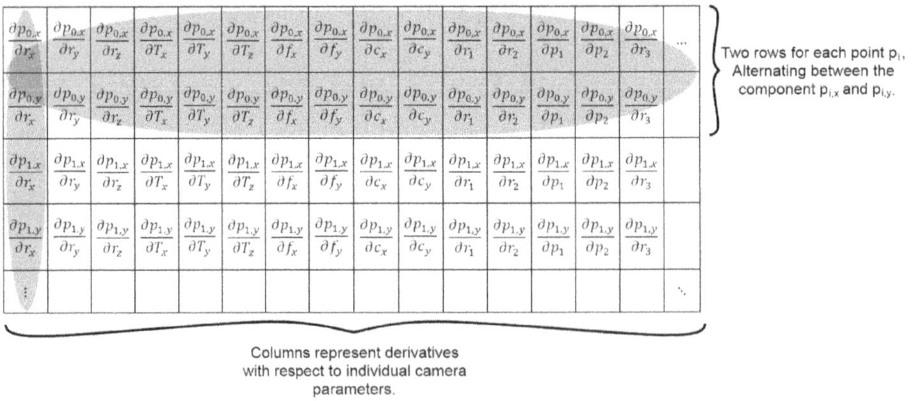

| $\frac{\partial p_{0,x}}{\partial r_x}$ | $\frac{\partial p_{0,x}}{\partial r_y}$ | $\frac{\partial p_{0,x}}{\partial r_z}$ | $\frac{\partial p_{0,x}}{\partial T_x}$ | $\frac{\partial p_{0,x}}{\partial T_y}$ | $\frac{\partial p_{0,x}}{\partial T_z}$ | $\frac{\partial p_{0,x}}{\partial f_x}$ | $\frac{\partial p_{0,x}}{\partial f_y}$ | $\frac{\partial p_{0,x}}{\partial c_x}$ | $\frac{\partial p_{0,x}}{\partial c_y}$ | $\frac{\partial p_{0,x}}{\partial r_1}$ | $\frac{\partial p_{0,x}}{\partial r_2}$ | $\frac{\partial p_{0,x}}{\partial p_1}$ | $\frac{\partial p_{0,x}}{\partial p_2}$ | $\frac{\partial p_{0,x}}{\partial r_3}$ | ... |
|---|---|---|---|---|---|---|---|---|---|---|---|---|---|---|---|
| $\frac{\partial p_{0,y}}{\partial r_x}$ | $\frac{\partial p_{0,y}}{\partial r_y}$ | $\frac{\partial p_{0,y}}{\partial r_z}$ | $\frac{\partial p_{0,y}}{\partial T_x}$ | $\frac{\partial p_{0,y}}{\partial T_y}$ | $\frac{\partial p_{0,y}}{\partial T_z}$ | $\frac{\partial p_{0,y}}{\partial f_x}$ | $\frac{\partial p_{0,y}}{\partial f_y}$ | $\frac{\partial p_{0,y}}{\partial c_x}$ | $\frac{\partial p_{0,y}}{\partial c_y}$ | $\frac{\partial p_{0,y}}{\partial r_1}$ | $\frac{\partial p_{0,y}}{\partial r_2}$ | $\frac{\partial p_{0,y}}{\partial p_1}$ | $\frac{\partial p_{0,y}}{\partial p_2}$ | $\frac{\partial p_{0,y}}{\partial r_3}$ | |
| $\frac{\partial p_{1,x}}{\partial r_x}$ | $\frac{\partial p_{1,x}}{\partial r_y}$ | $\frac{\partial p_{1,x}}{\partial r_z}$ | $\frac{\partial p_{1,x}}{\partial T_x}$ | $\frac{\partial p_{1,x}}{\partial T_y}$ | $\frac{\partial p_{1,x}}{\partial T_z}$ | $\frac{\partial p_{1,x}}{\partial f_x}$ | $\frac{\partial p_{1,x}}{\partial f_y}$ | $\frac{\partial p_{1,x}}{\partial c_x}$ | $\frac{\partial p_{1,x}}{\partial c_y}$ | $\frac{\partial p_{1,x}}{\partial r_1}$ | $\frac{\partial p_{1,x}}{\partial r_2}$ | $\frac{\partial p_{1,x}}{\partial p_1}$ | $\frac{\partial p_{1,x}}{\partial p_2}$ | $\frac{\partial p_{1,x}}{\partial r_3}$ | |
| $\frac{\partial p_{1,y}}{\partial r_x}$ | $\frac{\partial p_{1,y}}{\partial r_y}$ | $\frac{\partial p_{1,y}}{\partial r_z}$ | $\frac{\partial p_{1,y}}{\partial T_x}$ | $\frac{\partial p_{1,y}}{\partial T_y}$ | $\frac{\partial p_{1,y}}{\partial T_z}$ | $\frac{\partial p_{1,y}}{\partial f_x}$ | $\frac{\partial p_{1,y}}{\partial f_y}$ | $\frac{\partial p_{1,y}}{\partial c_x}$ | $\frac{\partial p_{1,y}}{\partial c_y}$ | $\frac{\partial p_{1,y}}{\partial r_1}$ | $\frac{\partial p_{1,y}}{\partial r_2}$ | $\frac{\partial p_{1,y}}{\partial p_1}$ | $\frac{\partial p_{1,y}}{\partial p_2}$ | $\frac{\partial p_{1,y}}{\partial r_3}$ | |
| ⋮ | | | | | | | | | | | | | | | ⋱ |

Two rows for each point $p_i$, Alternating between the component $p_{i,x}$ and $p_{i,y}$.

Columns represent derivatives with respect to individual camera parameters.

*Figure 19-1. Entries in the jacobian array*

# Affine and Perspective Transformations

Two transformations we have discussed that come up often in the OpenCV routines —as well as in other applications you might write yourself—are the affine and perspective transformations. We first encountered these in Chapter 11. Recall that, as implemented in OpenCV, these routines affect either lists of points or entire images, and they map points on one location in the image to a different location, often performing subpixel interpolation along the way. You may also recall that an affine transform can produce any parallelogram from a rectangle; the perspective transform is more general and can produce any trapezoid from a rectangle.

The *perspective transformation* is closely related to the *perspective projection*. Recall that the perspective projection maps points in the three-dimensional physical world onto points on the two-dimensional image plane along a set of projection lines that all meet at a single point called the *center of projection*. The perspective transformation, which is a specific kind of *homography*,[5] relates two different images that are alternative projections of the same three-dimensional object onto two different *projective planes*. Importantly, for nondegenerate configurations (such as the plane physically intersecting the three-dimensional object) we could equally say that the homography relates two different centers of projection.

Though these projective transformation-related functions were discussed in detail in Chapter 11, for convenience, we summarize them here in Table 19-1.

5 Recall from Chapter 18 that this special kind of homography is known as a *planar homography*.

*Table 19-1. Affine and perspective transform functions*

| Function | Use |
| --- | --- |
| `cv::transform()` | Affine transform a list of points |
| `cv::warpAffine()` | Affine transform a whole image |
| `cv::getAffineTransform()` | Calculate affine matrix from points |
| `cv::getRotationMatrix2D()` | Calculate affine matrix to achieve rotation |
| `cv::perspectiveTransform()` | Perspective transform a list of points |
| `cv::warpPerspective()` | Perspective transform a whole image |
| `cv::getPerspectiveTransform()` | Fill in perspective transform matrix parameters |

## Bird's-Eye-View Transform Example

A common task in robotic navigation, typically used for planning purposes, is to convert the robot's camera view of the scene into a top-down "bird's-eye" view. In Figure 19-2, a robot's view of a scene is turned into a bird's-eye view so that it can be subsequently used for planning and navigation in that plane, possibly also overlaid with alternative representations of the world created from sonar, scanning laser range finders, or similar sensors natively operating in the plane of motion. Using what we've learned so far, we'll look in detail at some code to use our calibrated camera to compute such a view.

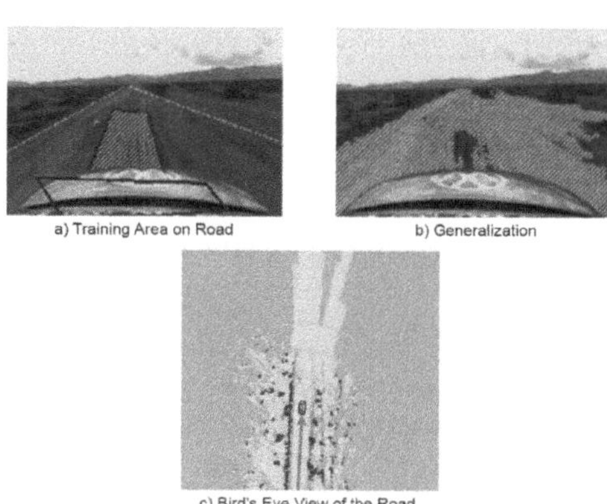


*Figure 19-2. Bird's-eye view: A camera on a robot car looks out at a road scene where laser range finders have identified a region of "road" in front of the car and marked it with a box (a); vision algorithms have segmented the flat, road-like areas (b); the segmented road areas are converted to a bird's-eye view and merged with the bird's-eye-view laser map (c)*

To get a bird's-eye view,[6] we'll need our camera matrix and distortion parameters from the calibration routine. Just for the sake of variety, we'll read these from data files on disk. For our example, we will put a chessboard on the floor and use that to obtain a ground plane image for a miniature robot car; we then remap that image into a bird's-eye view. The algorithm runs as follows:

1. Read the intrinsics and distortion models for the camera.
2. Find a known object on the ground plane (in this case, a chessboard). Get at least four points at subpixel accuracy.
3. Enter the found points into `cv::getPerspectiveTransform()` (see Chapter 11) to compute the homography matrix *H* for the ground plane view.
4. Use `cv::warpPerspective()` (Chapter 11) with the flags `cv::WARP_INVERSE_MAP | cv::INTER_LINEAR` to obtain a frontal parallel (bird's-eye) view of the ground plane.

Example 19-1 shows the full working code for bird's-eye view.

*Example 19-1. Bird's-eye view*

```
#include <opencv2/opencv.hpp>
#include <iostream>
using namespace std;

void help( char *argv[] ){
  ...
}

// args: [board_w] [board_h] [intrinsics.xml] [checker_image]
//
int main( int argc, char* argv[] ) {

  if( argc != 5 ) {
    cout << "\nERROR: too few parameters\n";
    help( argv );
    return -1;
  }

  // Input Parameters:
  //
  int               board_w = atoi(argv[1]);
  int               board_h = atoi(argv[2]);
  int               board_n = board_w * board_h;
```

6 The bird's-eye view technique also works for transforming perspective views of any plane (e.g., a wall or ceiling) into frontal parallel views.

```
cv::Size        board_sz( board_w, board_h );
cv::FileStorage fs( argv[3], cv::FileStorage::READ );
cv::Mat         intrinsic, distortion;

fs["camera_matrix"] >> intrinsic;
fs["distortion_coefficients"] >> distortion;
if( !fs.isOpened() || intrinsic.empty() || distortion.empty() )
{
  cout << "Error: Couldn't load intrinsic parameters from "
       << argv[3] << endl;
  return -1;
}
fs.release();

cv::Mat gray_image, image, image0 = cv::imread(argv[4], 1);
if( image0.empty() )
{
  cout << "Error: Couldn't load image " << argv[4] << endl;
  return -1;
}

// UNDISTORT OUR IMAGE
//
cv::undistort( image0, image, intrinsic, distortion, intrinsic );
cv::cvtColor( image, gray_image, cv::BGR2GRAY );

// GET THE CHECKERBOARD ON THE PLANE
//
vector<cv::Point2f> corners;
bool found = cv::findChessboardCorners(  // True if found
  image,                                 // Input image
  board_sz,                              // Pattern size
  corners,                               // Results
  cv::CALIB_CB_ADAPTIVE_THRESH | cv::CALIB_CB_FILTER_QUADS
);
if( !found ) {
  cout << "Couldn't acquire checkerboard on " << argv[4]
    <<", only found " << corners.size() << " of " << board_n
    << " corners\n";
  return -1;
}

// Get Subpixel accuracy on those corners
//
cv::cornerSubPix(
  gray_image,          // Input image
  corners,             // Initial guesses, also output
  cv::Size(11,11),     // Search window size
  cv::Size(-1,-1),     // Zero zone (in this case, don't use)
  cv::TermCriteria(
    cv::TermCriteria::EPS | cv::TermCriteria::COUNT,
    30, 0.1
```

```
  )
);

// GET THE IMAGE AND OBJECT POINTS:
// Object points are at (r,c):
//  (0,0), (board_w-1,0), (0,board_h-1), (board_w-1,board_h-1)
// That means corners are at: corners[r*board_w + c]
//
cv::Point2f objPts[4], imgPts[4];
objPts[0].x = 0;         objPts[0].y = 0;
objPts[1].x = board_w-1; objPts[1].y = 0;
objPts[2].x = 0;         objPts[2].y = board_h-1;
objPts[3].x = board_w-1; objPts[3].y = board_h-1;
imgPts[0]   = corners[0];
imgPts[1]   = corners[board_w-1];
imgPts[2]   = corners[(board_h-1)*board_w];
imgPts[3]   = corners[(board_h-1)*board_w + board_w-1];

// DRAW THE POINTS in order: B,G,R,YELLOW
//
cv::circle( image, imgPts[0], 9, cv::Scalar( 255,   0,   0), 3);
cv::circle( image, imgPts[1], 9, cv::Scalar(   0, 255,   0), 3);
cv::circle( image, imgPts[2], 9, cv::Scalar(   0,   0, 255), 3);
cv::circle( image, imgPts[3], 9, cv::Scalar(   0, 255, 255), 3);

// DRAW THE FOUND CHECKERBOARD
//
cv::drawChessboardCorners( image, board_sz, corners, found );
cv::imshow( "Checkers", image );

// FIND THE HOMOGRAPHY
//
cv::Mat H = cv::getPerspectiveTransform( objPts, imgPts );

// LET THE USER ADJUST THE Z HEIGHT OF THE VIEW
//
double Z = 25;
cv::Mat birds_image;
for(;;) {                                     // escape key stops
  H.at<double>(2, 2) = Z;
  // USE HOMOGRAPHY TO REMAP THE VIEW
  //
  cv::warpPerspective(
    image,                                    // Source image
    birds_image,                              // Output image
    H,                                        // Transformation matrix
    image.size(),                             // Size for output image
    cv::WARP_INVERSE_MAP | cv::INTER_LINEAR,
    cv::BORDER_CONSTANT,
    cv::Scalar::all(0)                        // Fill border with black
  );
  cv::imshow("Birds_Eye", birds_image);
```

```
    int key = cv::waitKey() & 255;
    if(key == 'u') Z += 0.5;
    if(key == 'd') Z -= 0.5;
    if(key ==  27) break;
  }

  // SHOW ROTATION AND TRANSLATION VECTORS
  //
  vector<cv::Point2f> image_points;
  vector<cv::Point3f> object_points;
  for( int i=0; i<4; ++i ){
    image_points.push_back( imgPts[i] );
    object_points.push_back(
      cv::Point3f( objPts[i].x, objPts[i].y, 0)
    );
  }

  cv::Mat rvec, tvec, rmat;
  cv::solvePnP(
    object_points,        // 3-d points in object coordinate
    image_points,         // 2-d points in image coordinates
    intrinsic,            // Our camera matrix
    cv::Mat(),            // Since we corrected distortion in the
                          //   beginning,now we have zero distortion
                          //   coefficients
    rvec,                 // Output rotation *vector*.
    tvec                  // Output translation vector.
  );
  cv::Rodrigues( rvec, rmat );

  // PRINT AND EXIT
  cout << "rotation matrix: "             << rmat    << endl;
  cout << "translation vector: "          << tvec    << endl;
  cout << "homography matrix: "           << H       << endl;
  cout << "inverted homography matrix: " << H.inv() << endl;

  return 1;
}
```

Once we have the homography matrix and the height parameter set as we wish, we could then remove the chessboard and drive the miniature car around, making a bird's-eye-view video of the path, but we'll leave that as an exercise for the reader. Figure 19-3 shows the input at left and output at right for the bird's-eye-view code.

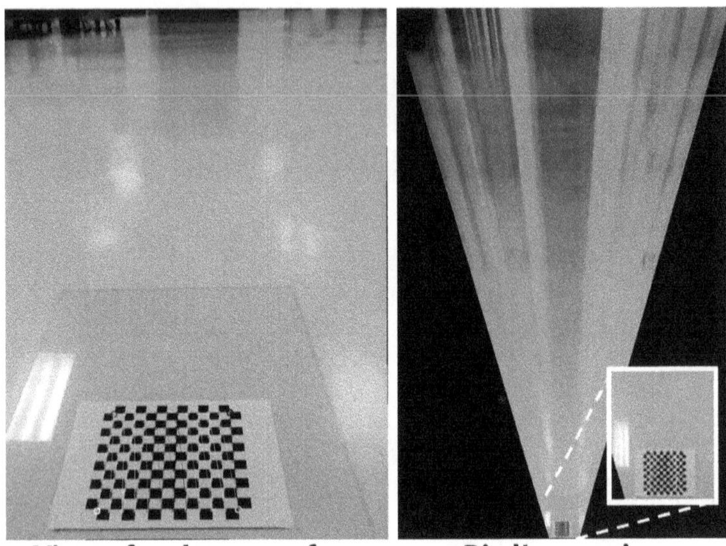


*Figure 19-3. Bird's-eye-view example*

# Three-Dimensional Pose Estimation

The problem of estimating the pose of three-dimensional objects can be tackled with a single camera as well as with multiple cameras. In the multicamera case, we use correspondences between what is seen from each of the separate cameras to draw conclusions about where the object is (i.e., by triangulation). The advantage of such a technique is that it will work with even unknown objects or entire unknown scenes. The disadvantage is that it requires multiple cameras. It is possible, however, to compute the pose of a known object with only one camera. We will consider that case first. In addition to being a useful technique in its own right, understanding the single-camera pose estimation problem will give us important insights into the multiple camera problem, which we will consider next.

## Pose Estimation from a Single Camera

To understand how this problem is solved, consider Figure 19-4. An object is "known" to the extent that we have identified some number of keypoints on the object (Chapter 16), whose location we know in the coordinate system of the object (Figure 19-4a). Now if we are presented with the same object in a novel pose, we can look for those same keypoints (Figure 19-4b). If we now want to figure out the relationship between the pose of the object and the camera, the essential observation is that for each point that we find, that point must lie on a particular ray emanating from a pixel location on the camera's imager out through the aperture of the camera.

Of course, individually we cannot know the distance from the camera to a particular point, but given many such constraints, a rigid object will only be able to meet all of those constraints one way (Figure 19-4c).[7]

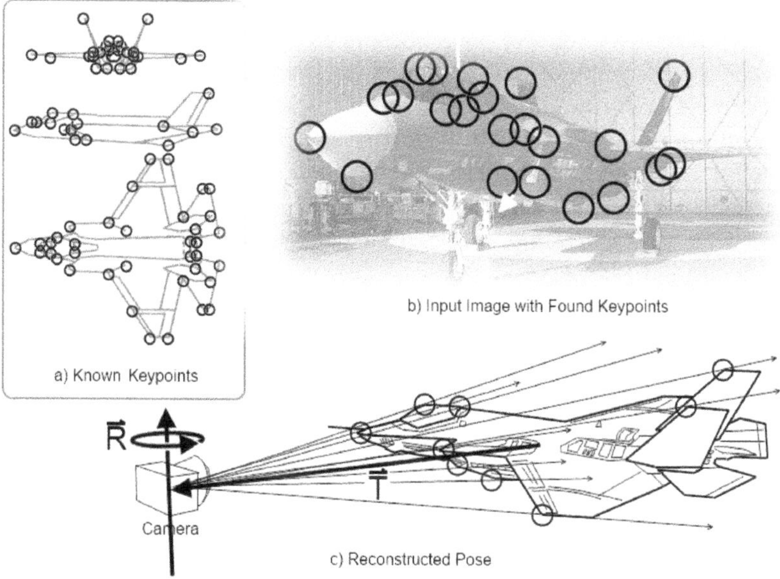


*Figure 19-4. Given a known set of keypoints (a), which can be found in an image of the same object (b), it is possible to reconstruct the pose of the object relative to the camera (c)*

### Computing the pose of a known object with cv::solvePnP()

The function in OpenCV that solves this problem is one we have already encountered; it is `cv::solvePnP()`, or the closely related `cv::solvePnPRansac()`.[8] In Chapter 18, we introduced this function primarily as a way to solve the problem of the pose of a chessboard or other calibration device. In fact, however, this function can be used to solve the general *Perspective N-Point (PNP) problem*. To understand the generality of the problem, consider Figure 19-4. The inset (Figure 19-4a) shows a schematic representation of an airplane; overlaid are circles representing features on the aircraft that we both know the exact location of (in the body coordinates of the

7 Actually, it is a little more complicated than this for objects that have some kind of intrinsic symmetry. If this is the case, there may be multiple solutions that meet the constraints. Depending on the symmetry of the object, these solutions may be discrete or form a continuous family.

8 `cv::solvePnP()` is a more general way to solve the rigid-object pose problem than the older POSIT routine [DeMenthon92] in the C version of the library. POSIT is still in OpenCV; you can find a tutorial for it by searching on "POSIT tutorial" in the OpenCV online documentation.

aircraft) and believe we could successfully recognize if the plane were seen from some arbitrary viewpoint.

With this information, we can extract features from some image (e.g., Figure 19-4b) and try to compute the pose of the object that would map the individual points we know to the locations where we observe them in the image. Because each of those points must lie on a ray that passes through the camera aperture and lands on the imager at some specific point, this problem will typically have a unique solution.[9] To get some intuition for this, consider Figure 19-4c; the found features are each constrained to be on a particular ray. Even though we don't know where on that ray the point is actually located, there is only one way in which the object can be placed such that all of these ray constraints are simultaneously satisfied. Example 19-1 shows `cv::solvePnP()` in use.

It is not necessary that we be able to see all—or even a majority—of the features on an object in order to recognize it and compute its pose. It is also not necessary that we associate only one feature with a particular physical location. In practice, keypoint detectors will recognize a feature only from a small window of angles, so it is often helpful to have multiple descriptors at a location to capture the way that feature is perceived from different locations.

Note that the PNP problem does not always have a unique solution. There are two important cases in which PNP cannot provide reliable results. The first case is when you just don't have enough points. In theory, the problem can be solved with as little as three matches. In practice, with so few points, and the natural noise in their location resulting from the accuracy of whatever method was used to find them (e.g., keypoint matching), the pose can be off by a significant amount. As a good rule of thumb, it is better to have a dozen or more matches. The second case is when the object is very far away. In this situation, the rays that constrain the locations of the features become effectively parallel. It is the divergence of the rays that guarantees a unique scale for the object—or equivalently a unique solution for the distance to the object.[10]

This monocular method of estimating the pose (and this distance) of an object is very similar to how your own eyes work when looking at distant objects. This is why it is

9 If the object is symmetric in some way, then there may be multiple solutions corresponding to the possible symmetric arrangements of the object.

10 As a practical tip, in many cases you can estimate the orientation of a distant object very roughly simply by considering what keypoints can be seen at all. This does not use `cv::solvePnP()`, but is a useful technique in real systems that can often derive great benefit from a less accurate fallback when a full solution using `cv::solvePnP()` is not possible.

impossible to gauge the distance to an object whose actual size (and thus the location of its features in its own coordinate frame) is not known without other context. It is also the basis of "forced perspective" illusions, such as why some buildings have increasingly small windows on higher floors, in order to make the building appear much taller from the ground. In the next section, we will discuss stereo imaging, which uses two or more cameras as a means of removing this final ambiguity, and as a result also allows us to simultaneously deduce both the structure and pose of a novel[11] object.

# Stereo Imaging

Now we are in a position to address *stereo imaging*.[12] We are all familiar with the stereo imaging capability that our eyes give us. To what degree can we emulate this capability in computational systems? Computers accomplish this task by finding correspondences between points that are seen by both imagers. With such correspondences and a known baseline separation between cameras, we can compute the three-dimensional location of the points. Although the search for corresponding points can be computationally expensive, we can use our knowledge of the geometry of the system to narrow down the search space as much as possible. In practice, stereo imaging involves four steps when you are using two cameras:

1. Mathematically remove radial and tangential lens distortion; this is called *undistortion* and is detailed in Chapter 18. The outputs of this step are undistorted images.
2. Adjust for the angles and distances between cameras, a process called *rectification*. The outputs of this step are images that are rectified and row-aligned (the latter meaning that the two image planes are coplanar and that corresponding image rows on the two imagers are in fact collinear relative to each other).
3. Find the same features in the left and right[13] camera views, a process known as *correspondence*. The output of this step is a *disparity* map, where the disparities are the differences in x-coordinates on the image planes of the same feature viewed in the left and right cameras: $x_l - x_r$.

11 In this context, and often in machine learning and computer vision, the word *novel* means a situation that has never been encountered by the system before and about which the system has no prior knowledge.

12 Here we give just a high-level understanding. For details, we recommend the following texts: Trucco and Verri [Trucco98], Hartley and Zisserman [Hartley06], Forsyth and Ponce [Forsyth03], and Shapiro and Stockman [Shapiro02]. The stereo rectification sections of these books will give you the background to tackle the original papers cited in this chapter.

13 Every time we refer to left and right cameras, you can also use vertically oriented up and down cameras, where disparities are in the y-direction rather than the x-direction.

4. If we know the geometric arrangement of the cameras, then we can turn the disparity map into distances by *triangulation*. This step is called *reprojection*, and the output is a depth map.

We start with the last step to motivate the first three.

## Triangulation

Consider first an "ideal" stereo rig as shown in Figure 19-5. In this case, we imagine that we have a perfectly undistorted, aligned, and measured system: two cameras whose image planes are exactly coplanar with each other, with exactly parallel optical axes (the optical axis is the ray from the center of projection $O$ through the principal point $c$ and is also known as the *principal ray*) that are a known distance apart, and with equal focal lengths: $f_l = f_r$. Also, assume that the *principal points* $c_x^{left}$ and $c_x^{right}$ have been calibrated to have the same pixel coordinates in their respective left and right images.[14]

Further, let's also assume for the moment that the imagers are perfectly row-aligned, such that every pixel row of one camera aligns exactly with the corresponding row in the other camera[15] (we will call such a camera arrangement *frontal parallel*). We will also assume that we can find a point $\vec{P}$ in the physical world in the left and the right image views at $\vec{p}_l$ and $\vec{p}_r$, which we will assign respective horizontal coordinates $\vec{x}_l$ and $\vec{x}_r$.

14 Still, don't confuse these principal points with the centers of the images. A principal point is defined as the place where the principal ray intersects the imaging plane. This intersection depends on the optical axis of the lens and, as we saw in Chapter 18, the image plane is essentially never aligned exactly with the lens, so the center of the imager is not going to be exactly aligned with the principal point, other than perhaps in an "ideal" stereo rig.

15 This makes for quite a few assumptions, but we are just looking at the basics right now. Remember that the process of rectification (to which we will return shortly) is how we get things done mathematically when these assumptions are not physically true. Similarly, in the next sentence we will temporarily "assume away" the correspondence problem.

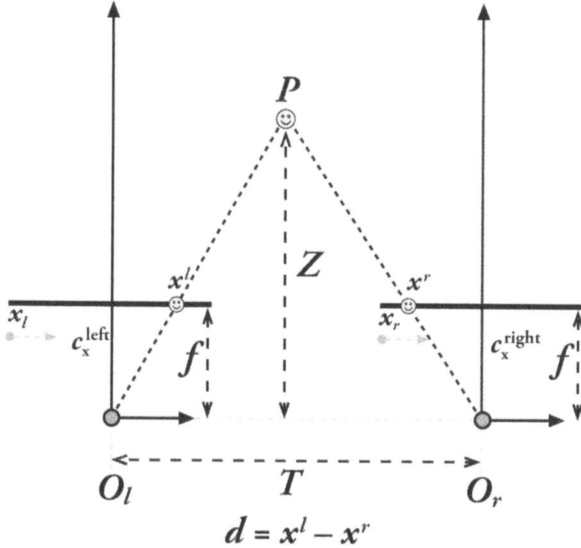


*Figure 19-5. With a perfectly undistorted, aligned stereo rig and known correspondence, the depth Z can be found by similar triangles; the principal rays of the imagers begin at the centers of projection $O_l$ and $O_r$ and extend through the principal points of the two image planes at $c_l$ and $c_r$*

In this simplified case, we can see that the depth is inversely proportional to the disparity between these views, where the disparity is defined simply by $d = \vec{x}_l - \vec{x}_r$. This situation is shown in Figure 19-5, where we can easily derive the depth $Z$ by using similar triangles. Referring to the figure, we have:[16]

$$\frac{T - (x_l - x_r)}{Z - f} = \frac{T}{Z} \Rightarrow Z = \frac{f \cdot T}{x_l - x_r}$$

Since depth is inversely proportional to disparity, there is obviously a nonlinear relationship between these two terms. When disparity is near zero, small disparity differences make for large depth differences. When disparity is large, small disparity differences do not change the depth by much. The consequence is that stereo vision systems have high depth resolution only for objects relatively near the camera, as Figure 19-6 makes clear.

16 This formula is predicated on the principal rays intersecting at infinity. However, as you will see in the section "Stereo Rectification" on page 730, we derive stereo rectification relative to the principal points $c_x^{left}$ and $c_x^{right}$. In our derivation, if the principal rays intersect at infinity, then the principal points have the same coordinates and so the formula for depth holds as is. However, if the principal rays intersect at a finite distance, then the principal points will not be equal and so the equation for depth becomes $Z = \frac{fT_x}{d - (c_x^{left} - c_x^{right})}$.

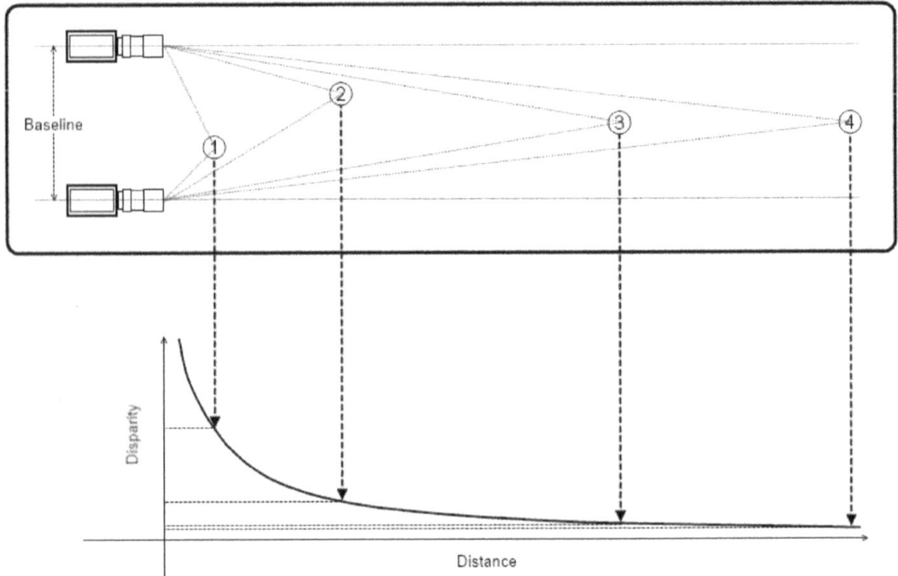


*Figure 19-6. Depth and disparity are inversely related, so fine depth measurements are restricted to nearby objects*

Figure 19-7 shows the two- and three-dimensional coordinate systems used in OpenCV for stereo vision. Note that it is a right-handed coordinate system: if you point your right index finger in the direction of the x-axis and bend your right middle finger in the direction of the y-axis, then your thumb will point in the direction of the principal ray. The left and right imager pixels have image origins at upper-left in the image, and pixels are denoted by coordinates $(x_l, y_l)$ and $(x_r, y_r)$, respectively. The centers of projection are at $\vec{O}_l$ and $\vec{O}_r$ with principal rays intersecting the image plane at the principal point (not the center) $(c_x, c_y)$. After mathematical rectification, the cameras are row-aligned (coplanar and horizontally aligned), displaced from one another by $\vec{T}$, and of the same focal length $f$. With this arrangement, it is relatively easy to solve for distance.

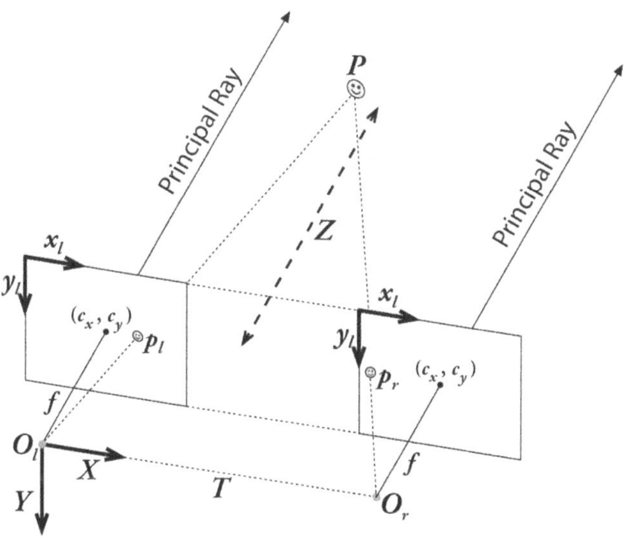


*Figure 19-7. Stereo coordinate system used by OpenCV for undistorted rectified cameras: the pixel coordinates are relative to the upper-left corner of the image, and the two planes are row-aligned; the camera coordinates are relative to the left camera's center of projection*

Now, with this simplified model in mind, we can get down to the more serious business of understanding how we can map a real-world camera setup into a geometry that resembles this ideal arrangement. In the real world, cameras will not be exactly aligned in the frontal parallel configuration depicted in Figure 19-5. Instead, we will mathematically find image projections and distortion maps that will rectify the left and right images into a frontal parallel arrangement. When designing your stereo rig, it is usually best to arrange the cameras approximately frontal parallel and as close to horizontally aligned as practical (though there is utility in some contexts to deliberately creating a converging geometry). Such a frontal parallel physical alignment will make the mathematical transformations more tractable. If you don't align the cameras at least approximately, then the resulting mathematical alignment can produce extreme image distortions and so reduce or eliminate the stereo overlap area of the resulting images.[17] For good results, you'll also need synchronized cameras. If they

17 The exception to this advice is applications where we want more resolution at close range; in this case, we tilt the cameras slightly in toward each other so that their principal rays intersect at a finite distance. After mathematical alignment, the effect of such inward-verging cameras is to introduce an x-offset that is subtracted from the disparity. This may result in negative disparities, but we can thus gain finer depth resolution at the nearby depths of interest.

don't capture their images at the same time,[18] then you will have problems if anything is moving in the scene (including the cameras themselves) and you will be limited to using stationary cameras viewing static scenes.

Figure 19-8 depicts the real situation between two cameras and the mathematical alignment we want to achieve. To perform this mathematical alignment, we need to learn more about the geometry of two cameras viewing a scene. Once we have that geometry defined and some terminology and notation to describe it, we can return to the problem of alignment.

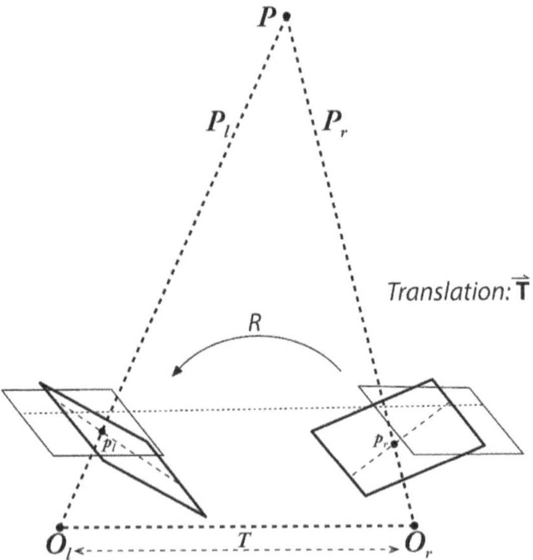


*Figure 19-8. Our goal will be to mathematically (rather than physically) align the two cameras into one viewing plane so that pixel rows between the cameras are exactly aligned with each other*

## Epipolar Geometry

The basic geometry of a stereo imaging system is referred to as *epipolar geometry*. In essence, this geometry combines two pinhole models (one for each camera[19]) and some interesting new points called the *epipoles* (see Figure 19-9). Before explaining

18 Of course "same" is really a matter of context. What is critical here is that no moving object in the scene, nor any motion of the cameras themselves, be sufficiently fast that the two imagers capture the scene at sufficiently different times that objects will appear to have moved in between captures.

19 Since we are actually dealing with real lenses and not pinhole cameras, it is important that the two images be undistorted; see Chapter 18.

what epipoles are good for, we will start by taking a moment to define them clearly and to add some new related terminology. When we are done, we will have a concise understanding of this overall geometry and will also find that we can narrow down considerably the possible locations of corresponding points on the two stereo cameras. This added discovery will have important implications for practical stereo implementations.

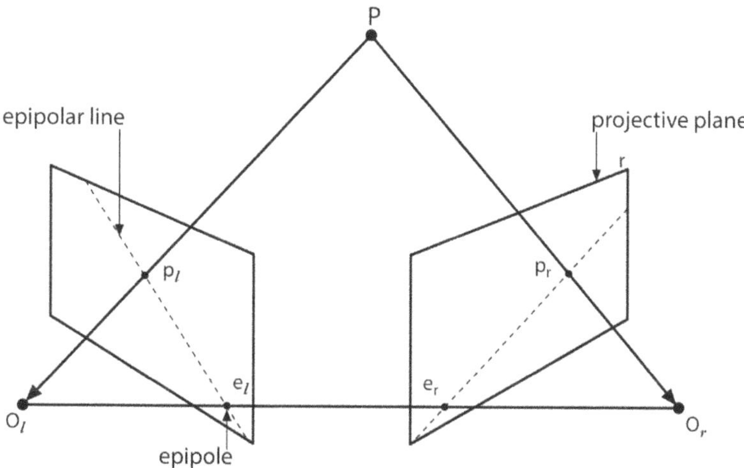


*Figure 19-9. The epipolar plane is defined by the observed point P and the two centers of projection, $O_l$ and $O_r$; the epipoles are located at the point of intersection of the line joining the centers of projection and the two projective planes*

For each camera there is now a separate center of projection, $\vec{O}_l$ and $\vec{O}_r$, and a pair of corresponding projective planes, $\Pi_l$ and $\Pi_r$. The point $\vec{R}$ in the physical world has a projection onto each of the projective planes that we can label $\vec{p}_l$ and $\vec{p}_r$. The new points of interest are the epipoles. An epipole $\vec{e}_l$ (or $\vec{e}_r$) on image plane $\Pi_l$ (or $\Pi_r$) is defined as the image of the center of projection of the other camera $\vec{O}_l$ (or $\vec{O}_r$) on that plane. The plane in space formed by the actual viewed point $\vec{P}$ and the two epipoles $\vec{e}_l$ and $\vec{e}_r$ (or, equivalently, through the two centers of projection $\vec{O}_l$ and $\vec{O}_r$) is called the *epipolar plane*, and the lines $\overline{p_l e_l}$ and $\overline{p_r e_r}$ (from the points of projection to the corresponding epipoles) are called the *epipolar lines.*[20]

To understand the utility of the epipoles first recall that, when we see a point in the physical world projected onto our right (or left) image plane, that point could actually be located anywhere along an entire line of points formed by the ray going

20 You can see why the epipoles did not come up before: as the planes approach being perfectly parallel, the epipoles head out toward infinity!

from $\vec{O}_r$ out through $\vec{p}_r$ (or $\vec{O}_l$ through $\vec{p}_l$) because, with just that single camera, we do not know the distance to the point we are viewing. More specifically, take for example the point $\vec{p}$ as seen by the camera on the right. Because that camera sees only $\vec{p}_r$ (the projection of $\vec{P}$ onto $\Pi_r$), the actual point $\vec{P}$ could be located anywhere on the line defined by $\vec{p}_r$ and $\vec{O}_r$. This line obviously contains $\vec{P}$, but it contains a lot of other points, too. What is interesting, though, is to ask what that line looks like projected onto the left image plane $\Pi_l$; in fact, it is the epipolar line defined by $\vec{p}_l$ and $\vec{e}_l$. To put that into English, the image of all of the possible locations of a *point* seen in one imager is the *line* that goes through the corresponding point and the epipole on the other imager.

We'll now summarize some facts about stereo camera epipolar geometry (and why we care):

- Every three-dimensional point in view of the cameras is contained in an epipolar plane that intersects each image. The resulting line of intersection is an epipolar line.
- Given a feature in one image, its matching view in the other image must lie along the corresponding epipolar line. This is known as the *epipolar constraint*.
- The epipolar constraint means that the possible two-dimensional search for matching features across two imagers becomes a one-dimensional search along the epipolar lines once we know the epipolar geometry of the stereo rig. This is not only a vast computational savings; it also allows us to reject a lot of points that could otherwise lead to spurious correspondences.
- Order is preserved. If points $\vec{A}$ and $\vec{B}$ are visible in both images and occur horizontally in that order in one imager, then they occur horizontally in that order in the other imager.[21]

## The Essential and Fundamental Matrices

You might think that the next step would be to introduce some OpenCV function that computes these epipolar lines for us, but we actually need two more ingredients before we can arrive at that point. These ingredients are the essential matrix $E$ and the fundamental matrix $F$.[22] The matrix $E$ contains information about the translation

21 Because of occlusions and areas of overlapping view, it is certainly possible that both cameras do not see the same points. Nevertheless, order is maintained. If points $\vec{A}$, $\vec{B}$, and $\vec{C}$ are arranged left to right on the left imager and if $\vec{B}$ is not seen on the right imager owing to occlusion, then the right imager will still see points $\vec{A}$ and $\vec{C}$ left to right.

22 The next subsections are a bit mathy. If you do not like math, then just skim over them; at least you'll have confidence that somewhere, someone understands all of this stuff. For simple applications, you can just use the machinery that OpenCV provides without the need for all of the details in these next few pages.

and rotation that relate the two cameras in physical space (see Figure 19-10), and $F$ contains the same information as $E$ in addition to information about the intrinsics of both cameras.[23] Because $F$ embeds information about the intrinsic parameters, it relates the two cameras in pixel coordinates.

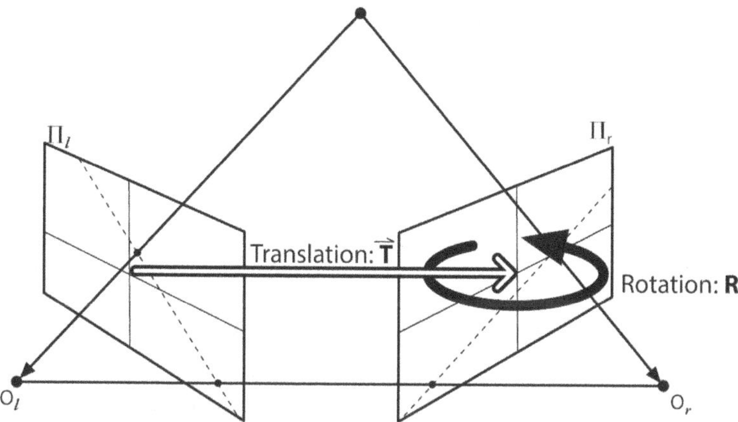


*Figure 19-10. The essential geometry of stereo imaging is captured by the essential matrix E, which contains all of the information about the translation T and the rotation R, which describe the location of the second camera relative to the first in global coordinates*

To reinforce the differences between $E$ and $F$: the essential matrix $E$ is purely geometrical and knows nothing about imagers. It relates the location, in physical coordinates, of the point $P$ as seen by the left camera to the location of the same point as seen by the right camera (i.e., it relates $\vec{p}_l$ and $\vec{p}_r$). The fundamental matrix $F$ relates the points on the image plane of one camera in image coordinates (pixels) to the points on the image plane of the other camera in image coordinates (for which we will use the notation $\vec{q}_l$ and $\vec{q}_r$).

23 The astute reader will recognize that $E$ was described in almost the exact same way as the homography matrix $H$ in the previous section. Although both are constructed from similar information, they are not the same matrix and should not be confused. An essential part of the definition of $H$ is that we were considering a plane viewed by a camera and thus could relate one point in that plane to the point on the camera plane. The matrix $E$ makes no such assumption and so will only be able to relate a point in one image to a line in the other.

## Essential matrix math

We will now dive into some math so we can better understand the OpenCV function calls that do the hard work for our stereo geometry problems.

Given a point $\vec{P}$, we would like to derive a relation that connects the observed locations $\vec{p}_l$ and $\vec{p}_r$ of $\vec{P}$ on the two imagers. This relationship will turn out to serve as the definition of the essential matrix. We begin by considering the relationship between $\vec{p}_l$ and $\vec{p}_r$, the physical locations of the point we are viewing in the coordinates of the two cameras. We can relate these by using epipolar geometry, as we have already seen.[24]

Let's pick one set of coordinate systems, left or right, to work in and do our calculations there. Either one is just as good, but we'll choose the coordinates centered on $\vec{O}_l$ of the left camera. In these coordinates, the location of the observed point is $\vec{p}_l$ and the origin of the other camera is located at $\vec{O}$. The point $\vec{P}$ as seen by the right camera is $\vec{p}_r$ in that camera's coordinates, where $\vec{p}_r = R \cdot (\vec{P}_l - \vec{T})$. The key step is the introduction of the epipolar plane, which we already know relates all of these things. We could, of course, represent a plane any number of ways, but for our purpose it is most helpful to recall that all points $\vec{x}$ on a plane with normal vector $\vec{n}$ and passing through point $\vec{a}$ obey the following constraint:

$$(\vec{x} - \vec{a}) \cdot \vec{n} = 0$$

Recall now that the epipolar plane contains the vectors $\vec{P}_l$ and $\vec{T}$; thus, if we had a vector (e.g., $\vec{P}_l \times \vec{T}$) perpendicular to both,[25] then we could use that for $\vec{n}$ in our plane equation. This gives us an equation for all possible points $\vec{P}_l$ through the point $\vec{T}$ and containing both vectors:[26]

$$(\vec{P}_l - \vec{T})^T (\vec{P}_l \times \vec{T}) = 0$$

Remember that our goal was to relate $\vec{q}_l$ and $\vec{q}_r$ by first relating $\vec{P}_l$ and $\vec{P}_r$. We bring $\vec{P}_r$ into the picture via our equality $\vec{P}_r = R \cdot (\vec{P}_l - \vec{T})$, which we can conveniently rewrite as: $(\vec{P}_l - \vec{T}) = R^{-1} \cdot \vec{P}_r$. Making this substitution and using that $R^{-1} = R^T$ yields:

24 Please do not confuse $\vec{p}_l$ and $\vec{p}_r$, which are points on the projective image planes, with $\vec{P}_l$ and $\vec{P}_r$, which are the locations of the point $\vec{P}$ in the coordinate frames of the two cameras.

25 The cross product of vectors produces a third vector orthogonal to the first two. Its direction is defined by the "right hand rule": if you point in the direction $\hat{a}$ and bend your middle finger in the direction $\hat{b}$, then the cross product $\hat{a} \times \hat{b}$ points perpendicular to $\hat{a}$ and $\hat{b}$ in the direction of your thumb.

26 Here we have replaced the dot product with matrix multiplication by the transpose of the normal vector.

$$(R^T \cdot \vec{P}_r)^T(\vec{P}_l \times \vec{T}) = 0$$

It is always possible to rewrite a cross-product as a (somewhat bulky) matrix multiplication. We use this fact to define the matrix $S$ such that:

$$\vec{T} \times \vec{P}_l = S \cdot \vec{P}_l \Rightarrow S = \begin{bmatrix} 0 & -T_x & T_y \\ T_z & 0 & -T_x \\ -T_y & T_x & 0 \end{bmatrix}$$

This leads to our first result. Making this substitution for the cross-product gives:

$$\vec{P}_r^{\,T} \cdot R \cdot S \cdot \vec{P}_l = 0$$

This product $R \cdot S$ is what we define to be the essential matrix $E$, which leads to the compact equation:

$$\vec{P}_r^{\,T} \cdot E \cdot \vec{P}_l = 0$$

Of course, what we really wanted was a relation between the points as we observe them on the imagers, but this is just a step away. We can simply substitute using the projection equations $\vec{p}_l = (f_l / z_l) \cdot \vec{P}_l$ and $\vec{p}_r = (f_r / z_r) \cdot \vec{P}_r$ and then divide the whole thing by $Z_l\, Z_r / f_l\, f_r$ to obtain our final result:

$$\vec{p}_r^{\,T} \cdot E \cdot \vec{p}_l$$

This might look at first like it completely specifies one of the $p$-terms if the other is given, but $E$ turns out to be a rank-deficient matrix[27] (the $3 \times 3$ essential matrix has rank 2) and so this actually ends up being an equation for a line. There are five parameters in the essential matrix—three for rotation and two for the direction of translation (scale is not set)—along with two other constraints. The two additional constraints on the essential matrix are: (1) the determinant is `0` because it is rank-deficient; and (2) its two nonzero singular values are equal because the matrix $S$ is skew-symmetric and $R$ is a rotation matrix. This yields a total of seven constraints. Note again that $E$ contains nothing intrinsic to the cameras; thus, it relates points to each other in physical or camera coordinates, not pixel coordinates.

27 For a square $n \times n$ matrix like $E$, *rank deficient* essentially means that there are fewer than $n$ nonzero eigenvalues. As a result, a system of linear equations specified by a rank-deficient matrix does not have a unique solution. If the rank (number of nonzero eigenvalues) is $n - 1$, then there will be a line formed by a set of points, all of which satisfy the system of equations. A system specified by a matrix of rank $n - 2$ will form a plane, and so forth.

### Fundamental matrix math

The matrix $E$ contains all of the information about the geometry of the two cameras relative to each other but no information about the cameras themselves. In practice, we are usually interested in pixel coordinates. In order to find a relationship between a pixel in one image and the corresponding epipolar line in the other image, we will have to introduce intrinsic information about the two cameras. To do this, for $\vec{p}$ (the pixel coordinate) we substitute $\vec{q}$ and the camera intrinsics matrix that relates them. Recall that $\vec{q} = M \cdot \vec{p}$ (where $M$ is the camera intrinsics matrix) or equivalently: $\vec{p} = M^{-1} \cdot \vec{q}$. Hence our equation for $E$ becomes:

$$\vec{q_r}^T \cdot (M_r^{-1})^T \cdot E \cdot M_l^{-1} \cdot \vec{q_l} = 0$$

Though this looks like a bit of a mess, we clean it up by defining the fundamental matrix $F$ as:

$$F = (M_r^{-1})^T \cdot E \cdot M_l^{-1}$$

so that:

$$\vec{q_r}^T \cdot F \cdot \vec{q_l} = 0$$

In a nutshell: the fundamental matrix $F$ is just like the essential matrix $E$, except that $F$ operates in image pixel coordinates whereas $E$ operates in physical coordinates.[28] Just like $E$, the fundamental matrix $F$ is of rank 2. The fundamental matrix has seven parameters, two for each epipole and three for the homography that relates the two image planes (the scale aspect is missing from the usual four parameters).

### How OpenCV handles all of this

We can compute $F$ in a manner analogous to computing the image homography in the previous section, by providing a number of known correspondences. In this case, we don't even have to calibrate the cameras separately because we can solve directly for $F$, which contains implicitly the fundamental matrices for both cameras. The routine that does all of this for us is called `cv::findFundamentalMat()`:

```
cv::Mat cv::findFundamentalMat(             // computed fundamental mtx
  cv::InputArray  points1,                  // Points image 1 (floats)
  cv::InputArray  points2,                  // Points image 2 (floats)
```

28 Note the equation that relates the fundamental matrix to the essential matrix. If we have rectified images and we normalize the points by dividing by the focal lengths, then the intrinsic matrix $M$ becomes the identity matrix and $F = E$.

```
    int               method = cv::FM_RANSAC, // Method for computing mtx
    double            param1 = 3.0,           // RANSAC max distance
    double            param2 = 0.99,          // RANSAC/LmedS confidence
    cv::OutputArray mask   = cv::noArray()  // (Optional) array
                                              // indicates used elements
  );
```

The first two arguments are arrays of two- or three-dimensional points, arranged in any of the usual ways.[29]

The third argument determines the method to be used in computing the fundamental matrix from the corresponding points, and it can take one of four values. For each value, there are particular restrictions on the number of points required (or allowed) in `points1` and `points2`, as shown in Table 19-2.

*Table 19-2. Restrictions on argument for method in cvFindFundamentalMat()*

| Value of method | Number of points | Algorithm |
|---|---|---|
| `cv::FM_7POINT` | $N = 7$ | Seven-point algorithm |
| `cv::FM_8POINT` | $N \geq 8$ | Eight-point algorithm |
| `cv::FM_RANSAC` | $N \geq 8$ | RANSAC algorithm |
| `cv::FM_LMEDS` | $N \geq 8$ | LMedS algorithm |

The seven-point algorithm uses exactly seven points, and it uses the fact that the matrix $F$ must be of rank 2 to fully constrain the matrix. The advantage of this constraint is that $F$ is then always exactly of rank 2 and so cannot have one very small eigenvalue that is not quite zero. The disadvantage is that this constraint is not absolutely unique and so three different matrices might be returned (this is the case where the returned array may be a 9 × 3 matrix, in which all three returns are placed). The eight-point algorithm just solves $F$ as a linear system of equations. If more than eight points are provided, then a linear least-squares error is minimized across all points. The problem with both the seven-point and eight-point algorithms is that they are extremely sensitive to outliers (even if you have many more than eight points in the eight-point algorithm). This is addressed by the *RANSAC* and *LMedS* algorithms which, as we saw in Chapter 18, are considered *robust methods* because they have

29 You might be wondering what the $N \times 3$ or three-channel matrix is for. The algorithm will deal just fine with actual three-dimensional points $(x, y, z)$ measured on the calibration object. Three-dimensional points will end up being scaled to $(x/z, y/z)$ or you could enter two-dimensional points in homogeneous coordinates, $(x, y, 1)$, which will be treated in the same way. If you enter $(x, y, 0)$, then the algorithm will just ignore the 0. Using actual three-dimensional points would be rare because usually you have only the two-dimensional points detected on the calibration object.

some capacity to recognize and remove outliers.[30] For both these methods, it is desirable to have many more than the minimal eight points.

The next two arguments are parameters used only by RANSAC and/or LMedS. The first, param1, used by RANSAC, is the maximum distance from a point to the epipolar line (in pixels) beyond which the point is considered an outlier. The second parameter, param2, used by RANSAC and LMeDS, is the desired confidence (between 0 and 1), which essentially tells the algorithm how many times to iterate.

The result is returned as a cv::Mat array, which will typically be a 3 × 3 matrix of the same precision as the points. (Recall that in the special case of the seven-point algorithm, it is possible also for the returned array to be a 9 × 3). The calculated fundamental matrix will typically next be passed to cv::computeCorrespondEpilines() to find the epipolar lines corresponding to the specified points or to cv::stereoRectifyUncalibrated() to compute the rectification transformation. Example 19-2 gives example code for finding the fundamental matrix.

*Example 19-2. Computing the fundamental matrix using RANSAC*

```
#include <opencv2/opencv.hpp>
#include <iostream>

using namespace std;

void help( char* argv[] ) {
  ...
}

// args: [board_w] [board_h] [number_of_boards] [delay]? [scale]?
//
int main( int argc, char* argv[] ) {

  int   n_boards = 0;      // Will be set by input list
  float image_sf = 0.5f;
  float delay    = 1.f;
  int   board_w  = 0;
  int   board_h  = 0;

  if( argc < 4 || argc > 6 ) {
    cout << "\nERROR: Wrong number of input parameters";
    help( argv );
    return -1;
  }
  board_w  = atoi( argv[1] );
```

30 For more information, consult the original papers: Fischler and Bolles [Fischler81] for RANSAC; Rousseeuw [Rousseeuw84] for least median squares; and Inui, Kaneko, and Igarashi [Inui03] for line fitting using LMedS.

```
board_h  = atoi( argv[2] );
n_boards = atoi( argv[3] );
if( argc > 4 ) delay    = atof( argv[4] );
if( argc > 5 ) image_sf = atof( argv[5] );

int board_n        = board_w * board_h;
cv::Size board_sz = cv::Size( board_w, board_h );
cv::VideoCapture capture(0);
if( !capture.isOpened() ) {
  cout << "\nCouldn't open the camera\n"; help();
  return -1;
}

// Allocate Storage
//
vector< vector< cv::Point2f> > image_points;
vector< vector< cv::Point3f> > object_points;

// Capture corner views; loop until we've got n_boards number of
// successful captures (meaning: all corners on each
// board are found).
//
double   last_captured_timestamp = 0;
cv::Size image_size;
while( image_points.size() < (size_t) n_boards ) {

  cv::Mat image0, image;
  capture >> image0;
  image_size = image0.size();
  resize(
    image0,   image,    cv::Size(),
    image_sf, image_sf, cv::INTER_LINEAR
  );

  // Find the board
  //
  vector< cv::Point2f > corners;
  bool found = cv::findChessboardCorners(
   image, board_sz, corners
  );

  // Draw it
  //
  cv::drawChessboardCorners( image, board_sz, corners, found );

  // If we got a good board, add it to our data
  //
  double timestamp = (double) clock() / CLOCKS_PER_SEC;
  if( found && timestamp - last_captured_timestamp > 1 ) {

    last_captured_timestamp = timestamp;
    image ^= cv::Scalar::all(255);
```

```
      cv::Mat mcorners( corners );        // do not copy the data
      mcorners *= ( 1./ image_sf );       // scale corner coordinates
      image_points.push_back( corners );
      object_points.push_back( vector< cv::Point3f >() );
      vector< cv::Point3f >& opts = object_points.back();
      opts.resize( board_n );
      for( int j=0; j<board_n; j++ ) {
        opts[j] = cv::Point3f(
          (float)(j/board_w), (float)(j%board_w), 0.f
        );
      }
      cout << "Collected our " <<(int) image_points.size()
        <<" of " << n_boards <<" needed chessboard images\n" << endl;
    }

    // in color if we did collect the image
    //
    cv::imshow( "Calibration", image );
    if( (cv::waitKey(30) & 255) == 27 )
      return -1;

  }             // end collection while() loop.

  cv::destroyWindow("Calibration");
  cout <<"\n\n*** CALIBRATING THE CAMERA...\n" << endl;

  // Calibrate the camera!
  //
  cv::Mat intrinsic_matrix, distortion_coeffs;
  double err = cv::calibrateCamera(
    object_points,       // Vector of vectors of points
                         //  from the calibration pattern
    image_points,        // Vector of vectors of projected
                         //  locations (on images)
    image_size,          // Size of images used
    intrinsic_matrix,    // Output camera matrix
    distortion_coeffs,   // Output distortion coefficients
    cv::noArray(),       // We'll pass on the rotation vectors...
    cv::noArray(),       // ...and the translation vectors
    cv::CALIB_ZERO_TANGENT_DIST | cv::CALIB_FIX_PRINCIPAL_POINT
  );

  // Save the intrinsics and distortions
  cout << " *** DONE!\n\nReprojection error is " << err
    <<"\nStoring Intrinsics.xml and Distortions.xml files\n\n";
  cv::FileStorage fs( "intrinsics.xml", cv::FileStorage::WRITE );

  fs << "image_width"     << image_size.width << "image_height"
     << image_size.height << "camera_matrix"  << intrinsic_matrix
     << "distortion_coefficients" << distortion_coeffs;
  fs.release();
```

```
// Example of loading these matrices back in:
//
fs.open( "intrinsics.xml", cv::FileStorage::READ );
cout << "\nimage width: "  << (int)fs["image_width"];
cout << "\nimage height: " << (int)fs["image_height"];

cv::Mat intrinsic_matrix_loaded, distortion_coeffs_loaded;
fs["camera_matrix"]           >> intrinsic_matrix_loaded;
fs["distortion_coefficients"] >> distortion_coeffs_loaded;
cout << "\nintrinsic matrix:"         << intrinsic_matrix_loaded;
cout << "\ndistortion coefficients: " << distortion_coeffs_loaded
     << endl;

// Compute Fundamental Matrix Between the first
// and the second frames:
//
cv::undistortPoints(
  image_points[0],     // Observed point coordinates (from frame 0)
  image_points[0],     // undistorted coordinates (in this case,
                       //  the same array as above)
  intrinsic_matrix,    // Intrinsics, from cv::calibrateCamera()
  distortion_coeffs,   // Distortion coefficients, also
                       //  from cv::calibrateCamera()
  cv::Mat(),           // Rectification transformation (but
                       //  here, we don't need this)
  intrinsic_matrix     // New camera matrix
);
cv::undistortPoints(
  image_points[1],     // Observed point coordinates (from frame 1)
  image_points[1],     // undistorted coordinates (in this case,
                       //  the same array as above)
  intrinsic_matrix,    // Intrinsics, from cv::calibrateCamera()
  distortion_coeffs,   // Distortion coefficients, also
                       //  from cv::calibrateCamera()
  cv::Mat(),           // Rectification transformation (but
                       //  here, we don't need this)
  intrinsic_matrix     // New camera matrix
);

// Since all the found chessboard corners are inliers, i.e., they
// must satisfy epipolar constraints, here we are using the
// fastest, and the most accurate (in this case) 8-point algorithm.
//
cv::Mat F = cv::findFundamentalMat(  // Return computed matrix
  image_points[0],                   // Points from frame 0
  image_points[1],                   // Points from frame 1
  cv::FM_8POINT                      // Use the 8-point algorithm
);
cout << "Fundamental matrix: " << F << endl;
```

```
  // Build the undistort map which we will use for all
  // subsequent frames.
  //
  cv::Mat map1, map2;
  cv::initUndistortRectifyMap(
    intrinsic_matrix_loaded,  // Our camera matrix
    distortion_coeffs_loaded, // Our distortion coefficients
    cv::Mat(),                // (Optional) Rectification, don't
                              //  need.
    intrinsic_matrix_loaded,  // "New" matrix, here it's the same
                              //  as the first argument.
    image_size,               // Size of undistorted image we want
    CV_16SC2,                 // Specifies the format of map to use
    map1,                     // Integerized coordinates
    map2                      // Fixed-point offsets for
                              //  elements of map1
  );
  // Just run the camera to the screen, now showing the raw and
  // the undistorted image.
  //
  for(;;) {
    cv::Mat image, image0;
    capture >> image0;
    if( image0.empty() ) break;
    cv::remap(
      image0,                 // Input image
      image,                  // Output image
      map1,                   // Integer part of map
      map2,                   // Fixed point part of map
      cv::INTER_LINEAR,
      cv::BORDER_CONSTANT,
      cv::Scalar()            // Set border values to black
    );
    cv::imshow( "Undistorted", image );
    if( ( cv::waitKey(30) & 255 ) == 27 ) break;
  }
  return 1;
}
```

One word of warning—related to the possibility of returning `cv::noArray()`—is that these algorithms can fail if the points supplied form *degenerate configurations.* These degenerate configurations arise when the points supplied provide less than the required amount of information, such as when one point appears more than once or when multiple points are collinear or coplanar with too many other points. It is important to always check the return value of `cv::findFundamentalMat()`.

## Computing Epipolar Lines

Now that we have the fundamental matrix, we want to be able to compute epipolar lines. The OpenCV function `cv::computeCorrespondEpilines()` computes, for a list of points in one image, the epipolar lines in the other image. Recall that, for any given point in one image, there is a different corresponding epipolar line in the other image. Each computed line is encoded in the form of a vector of three points $(a, b, c)$ such that the epipolar line is defined by the equation:

$$a \cdot x + b \cdot y + c = 0$$

To compute these epipolar lines, the function requires the fundamental matrix that we computed with `cv::findFundamentalMat()`:

```
void cv::computeCorrespondEpilines(
  cv::InputArray  points,      // Input points, Nx1 or 1xN (Nc=2)
                               //  or vector<Point2f>
  int             whichImage,  // Index of image which contains
                               //  points ('1' or '2')
  cv::InputArray  F,           // Fundamental matrix
  cv::OutputArray lines        // Output vector of lines, encoded as
                               //  tuples (a,b,c)
);
```

Here the first argument, `points`, is the input array of two- or three-dimensional points—it can be in any of the usual forms, but the points should be of floating-point type. The argument `whichImage` must be either `1` or `2`, and indicates which image the points are defined on, relative to the `points1` and `points2` arrays in `cv::findFundamentalMat()`. `F` is the $3 \times 3$ matrix returned by `cv::findFundamentalMat()`. Finally, `lines` will be a floating-point array in which the result lines will be written. Each line is encoded as a three-component vector $\vec{L} \equiv (a, b, c)$, containing the coefficients for the line equation $a \cdot x + b \cdot y + c = 0$. Because the line equation is independent of the overall normalization of the parameters $a$, $b$, and $c$, they are normalized by default such that $a^2 + b^2 = 1$.

## Stereo Calibration

We've built up a lot of theory and machinery behind cameras and three-dimensional points that we can now put to use. This section will cover stereo calibration, and the next section will cover stereo rectification. *Stereo calibration* is the process of computing the geometrical relationship between the two cameras in space. In contrast, *stereo rectification* is the process of "correcting" the individual images so that they appear as if they had been taken by two cameras with row-aligned image planes (review Figures 19-5 and 19-8). With such a rectification, the optical axes (or principal rays) of the two cameras are parallel (and so we say that they intersect at infinity). We could, of course, calibrate the two camera images to be in many other configurations, but here (and in OpenCV generally) we focus on the more common and simpler case of setting the principal rays to intersect at infinity.

Stereo calibration depends on finding the rotation matrix $R$ and translation vector $\vec{T}$ between the two cameras, as depicted back in Figure 19-10. Both $R$ and $\vec{T}$ are calculated by the function `cv::stereoCalibrate()`, which is similar to the function `cv::calibrateCamera()` that we saw in Chapter 18 except that we now have two cameras and our new function can compute (or make use of any prior computation of) the camera, distortion, essential, or fundamental matrices. The other main difference between stereo and single-camera calibration is that, in `cv::calibrateCamera()`, we ended up with a list of rotation and translation vectors between the camera and the chessboard views. In `cv::stereoCalibrate()`, we seek a single rotation matrix and translation vector that relate the right camera to the left camera.

We've already shown how to compute the essential and fundamental matrices. The next problem is how to compute $R$ and $\vec{T}$ between the left and right cameras. We begin with the observation that for any given three-dimensional point $\vec{P}$ in object coordinates, we can separately use single-camera calibration for the two cameras to put $\vec{P}$ into the camera coordinates for either camera: $\vec{P}_l = R_l \cdot \vec{P} + \vec{T}_l$ and $\vec{P}_r = R_r \cdot \vec{P} + \vec{T}_r$ (for the left and right cameras, respectively). It should be evident from Figure 19-10 that the two views of $\vec{P}$ (from the two cameras) are related by $\vec{P}_l = R^T \cdot (\vec{P}_r - \vec{T})$,[31] where $R$ and $\vec{T}$ are, respectively, the rotation matrix and translation vector between the cameras. Taking these three equations and solving for the rotation and translation separately yields the following simple relations:[32]

$$R = R_r \cdot R_l^T$$

and:

$$\vec{T} = \vec{T}_r - R \cdot \vec{T}_l$$

Given many joint views of chessboard corners or a similar calibration object, `cv::stereoCalibrate()` uses `cv::calibrateCamera()` to solve for rotation and translation parameters of the views for each camera separately (see Chapter 18's discussion in "What's under the hood?" on page 671 to recall how this is done). The routine then plugs these left and right rotation and translation solutions into the preceding equations to solve for the rotation and translation parameters between the two

---

31 Let's be careful about what these terms mean: $\vec{P}_l$ and $\vec{P}_r$ denote the locations of the three-dimensional point $\vec{P}$ from the coordinate system of the left and right cameras, respectively; $R_l$ and $\vec{T}_l$ (respectively, $R_r$ and $\vec{T}_r$) denote the rotation and translation vectors from the camera to the three-dimensional point for the left (resp. right) camera; and $R$ and $\vec{T}$ are the rotation and translation that bring the right-camera coordinate system into the left.

32 You can reverse the left and right cameras in these equations either by reversing the subscripts in both equations or by reversing the subscripts and dropping the transpose of $R$ in the translation equation only.

cameras. Because of image noise and rounding errors, each chessboard pair results in slightly different values for $R$ and $\vec{T}$. The `cv::stereoCalibrate()` routine then takes the median values for the $R$ and $\vec{T}$ parameters as the initial approximation of the true solution and runs a robust Levenberg-Marquardt iterative algorithm to find the (local) minimum of the reprojection error of the calibration points for both camera views, and the final solution for $R$ and $\vec{T}$ is returned. To be clear on what stereo calibration gives you: the rotation matrix will put the right camera in the same plane as the left camera; this renders the two image planes parallel but not row-aligned (we'll see how row-alignment is accomplished in the section "Stereo Rectification" on page 730).

The function `cv::stereoCalibrate()` has a lot of parameters, but they are all fairly straightforward and many are the same as for `cv::calibrateCamera()` from Chapter 18:

```
double cv::stereoCalibrate(                   // Return reprojection error
  cv::InputArrayOfArrays objectPoints,   // Vector of vectors of
                                         //  calib. pattern points
  cv::InputArrayOfArrays imagePoints1,   // Vector of vectors of
                                         //  image points (cam 1)
  cv::InputArrayOfArrays imagePoints2,   // Vector of vectors of
                                         //  image points (cam 2)
  cv::InputOutputArray   cameraMatrix1,  // Intrinsics for cam 1
                                         //  (input/output)
  cv::InputOutputArray   distCoeffs1,    // Distortion coeffs for
                                         //  cam 1 (input/output)
  cv::InputOutputArray   cameraMatrix2,  // Intrinsics for cam 2 (
                                         //  input/output)
  cv::InputOutputArray   distCoeffs2,    // Distortion coeffs for
                                         //  cam 2 (input/output)
  cv::Size               imageSize,      // Size of images (assume both
                                         //  cams are same)
  cv::OutputArray        R,              // Computed relative
                                         //  rotation *matrix*
  cv::OutputArray        T,              // Computed relative
                                         //  translation vector
  cv::OutputArray        E,              // Computed essential matrix
  cv::OutputArray        F,              // Computed fundamental matrix
  cv::TermCriteria       criteria     = cv::TermCriteria(
                                           cv::TermCriteria::COUNT
                                             | cv::TermCriteria::EPS,
                                           30,
                                           1e-6
                                         ),
  int                    flags        = cv::CALIB_FIX_INTRINSIC
);
```

The first parameter, `objectPoints`, is an array of arrays of points. Each entry in the top-level arrays is associated with one of the calibration images. Each such entry is itself an array containing the locations of the points on the calibration image (in the coordinate system of the calibration image itself). These should be three-dimensional points, though in most cases the z-coordinate of each point location will be zero. (This is precisely the same as the situation for `cv::calibrateCamera()`; you don't actually have to use a flat calibration object, but usually it is the most convenient thing to do.)

`imagePoints1` and `imagePoints2` are also arrays of arrays, and again the top level contains entries corresponding to the input images. Each such entry contains the observed locations of the calibration points; `imagePoints1` contains the points as seen by the first (typically the left) camera, while `imagePoints2` contains the points as seen by the second (typically the right) camera.[33] Unlike the points in `objectPoints`, the points in `imagePoints1` and `imagePoints2` are two-dimensional points, as they are pixel locations on the images.

If you performed calibration for the two cameras using a chessboard or circle grid, then `imagePoints1` and `imagePoints2` will be just the returned values for the corresponding calls to the `cv::findChessboardCorners()` function, or to one of the other corner (or circle) grid-finding functions, for the left and right camera views, respectively.

The parameters `cameraMatrix1` and `cameraMatrix2` are the $3 \times 3$ camera matrices, and `distCoeffs1` and `distCoeffs2` are the four-entry (or five- or seven-entry) vectors of distortion coefficients for cameras 1 and 2, respectively. Remember that, in these matrices, the first two radial parameters come first; these are followed by the two tangential parameters and finally the subsequent radial parameters (see the discussion in Chapter 18 on distortion coefficients).[34]

The way in which the camera intrinsics are used by `cv::stereoCalibrate()` is controlled by the `flags` parameter. If `flags` contains `cv::CALIB_FIX_INTRINSIC`, then the values in these matrices are used as they were found by the calibration process (i.e., they are not computed by this algorithm). If `flags` is set to `cv::CALIB_USE_INTRINSIC_GUESS`, then these matrices are used as a starting point to optimize further the intrinsic and distortion parameters for each camera and will be set to the refined values on return from `cv::stereoCalibrate()`. In the case in

33 For simplicity, we normally think of "1" as denoting the left camera and "2" as denoting the right camera. In fact, you can interchange these as long as you consistently treat the resulting rotation and translation solutions in the opposite fashion to the text discussion. The most important thing is to physically align the cameras so that their scan lines approximately match in order to achieve good calibration results.

34 The third and higher radial distortion parameters are last because they were added later in OpenCV's development.

which neither of these flags is used, the camera intrinsics will be computed from scratch by `cvStereoCalibrate()`. Thus, if you like, you can compute the intrinsic, extrinsic, and stereo parameters in a single pass using `cvStereoCalibrate()`.[35]

You may also add any of the `flags` values provided to `cv::calibrateCamera()` (as discussed in Chapter 18). In addition to these, one new flag is available: `cv::CALIB_SAME_FOCAL_LENGTH`. This flag is a somewhat less strict alternative to `cv::CALIB_FIX_FOCAL_LENGTH`. While the latter forces the focal lengths to be equal to those found in `cameraMatrix1` and `cameraMatrix2`, the former only enforces the requirement that the two be equal. In this case, `cv::stereoCalibrate()` will solve for the single unknown focal length shared by the two cameras.

The parameter `imageSize` is the image size in pixels. It is used only if you are refining or computing intrinsic parameters, as when `flags` is not equal to `cv::CALIB_FIX_INTRINSIC`.

The terms `R` and `T` are output parameters that are filled on function return with the rotation matrix and translation vector (relating the right camera to the left camera) that we seek. The parameters `E` and `F` are optional. If they are not set to `cv::noArray()`, then `cv::stereoCalibrate()` will calculate and fill these arrays with the 3 × 3 essential and fundamental matrices. Finally, there is `termCrit`, which we have seen many times before. It sets the internal optimization either to terminate after a certain number of iterations or to stop when the computed parameters change by less than the threshold indicated in the `termCrit` structure. A typical argument for this function is `cv::TermCriteria( cv::TermCriteria::COUNT | cv::TermCriteria::EPS, 30, 1e-6 )`. In the case of `cv::TERMCRIT_EPS`, the value of the associated termination criterion is the value of the total reprojection error given by the current estimate of the parameters. Just as with `cv::calibrateCamera()`, the algorithm is minimizing the total reprojection error for all the points in all the available views, but now for both cameras. The return value of `cv::stereoCalibrate()` is the final value of the reprojection error.

35 Be careful. Trying to solve for too many parameters at once will sometimes cause the solution to diverge to nonsense values. Solving systems of equations is something of an art, and you must verify your results. You can see some of these considerations in the calibration and rectification code example, where we check our calibration results by using the epipolar constraint in Example 19-2.

If you've calibrated both cameras and are sure of the result, then you can "hard set" the previous single-camera calibration results by using `cv::CALIB_FIX_INTRINSIC`. If you think the two cameras' initial calibrations were OK but not great, you can ask the algorithm to refine the intrinsic and distortion parameters by setting `flags` to `cv::CALIB_USE_INTRINSIC_GUESS`. If the cameras have not been individually calibrated, then you can use the same settings as we used for the `flags` parameter in `cv::calibrateCamera()` in Chapter 18.

Once we have either the rotation and translation values $(R, \vec{T})$ or the fundamental matrix `F`, we may use these results to rectify the two stereo images so that the epipolar lines are arranged along image rows and the scan lines are the same across both images. Though `R` and `T` don't define a unique stereo rectification, we'll see how to use these terms together with other constraints to rectify our image pairs. We will explore how to do this in the next section.

## Stereo Rectification

It is easiest to compute the stereo disparity when the two image planes align exactly (as shown in Figure 19-5). Unfortunately, as discussed previously, a perfectly aligned configuration is rare with a real stereo system, since the two cameras almost never have exactly coplanar, row-aligned imaging planes. Figure 19-8 shows the goal of stereo rectification: we want to reproject the image planes of our two cameras so that they reside in the exact same plane, with image rows perfectly aligned into a frontal parallel configuration. How we choose the specific plane in which to mathematically align the cameras depends on the algorithm being used. In what follows, we discuss two cases addressed by OpenCV.

Ultimately, after rectification, we want the image rows between the two cameras to be aligned so that stereo correspondence (finding the same point in the two different camera views) will be more reliable and computationally tractable. Specifically, reliability and computational efficiency are both enhanced by having to search only one row for a match with a point in the other image. The result of aligning horizontal rows within a common image plane containing each image is that the epipoles themselves are then located at infinity; that is, the image of the center of projection in one image is parallel to the other image plane. However, because there are an infinite number of possible frontal parallel planes to choose from, we will still need to add a few more constraints; such constraints are based on maximizing view overlap and/or minimizing distortion.

The result of the process of aligning the two image planes will be eight terms, four each for the left and the right cameras. For each camera we'll get a distortion vector `distCoeffs`, a rotation matrix $R_{rect}$ (to apply to the image), and the rectified and unrectified camera matrices ($M_{rect}$ and $M$, respectively). From these terms, we can make a map, using `cv::initUndistortRectifyMap()` (to be discussed shortly), which will tell us where to interpolate pixels from the original image in order to create a new rectified image.[36]

There are many ways to compute our rectification terms, of which OpenCV implements two: (1) Hartley's algorithm [Hartley98], which can yield uncalibrated stereo using just the fundamental matrix; and (2) Bouguet's algorithm,[37] which uses the rotation and translation parameters from two calibrated cameras. Hartley's algorithm can be used to derive structure from motion recorded by a single camera but may (when stereo-rectified) produce more distorted images than Bouguet's calibrated algorithm.

In situations where you can employ calibration patterns—such as on a robot arm or for security camera installations—Bouguet's algorithm is the natural one to use.

#### Uncalibrated stereo rectification: Hartley's algorithm

Hartley's algorithm attempts to find homographies that map the epipoles to infinity while minimizing the computed disparities between the two stereo images; it does this by matching points between two image pairs. With this approach we bypass having to compute the camera intrinsics for the two cameras because such intrinsic information is implicitly contained in the point matches. Thus, we need only compute the fundamental matrix, which can be obtained from any matched set of seven or more points between the two views of the scene via `cv::findFundamentalMat()` as already described. Alternatively, the fundamental matrix can be computed from `cv::stereoCalibrate()`.

The advantage of Hartley's algorithm is that we can perform online stereo calibration simply by observing points in the scene. The disadvantage is that we have no sense of image scale. For example, if we used a chessboard for generating point matches, then we would not be able to tell whether the chessboard were 100 meters on each side and far away or 100 centimeters on each side and nearby. Neither do we explicitly learn the intrinsic camera matrices, without which the cameras might have different focal

36 Stereo rectification of an image in OpenCV is possible only when the epipole is outside of the image rectangle. Hence, this rectification algorithm may not work with stereo configurations that are characterized by either a very wide baseline or when the cameras point toward each other too much.

37 The Bouguet algorithm is a completion and simplification of the method first presented by Tsai [Tsai87] and Zhang [Zhang99; Zhang00]. Jean-Yves Bouguet never published this algorithm beyond its well-known implementation in his Camera Calibration Toolbox Matlab.

lengths, skewed pixels, different centers of projection, and/or different principal points. As a result, we can determine three-dimensional object reconstruction only up to a projective transform. This means that different scales or projections of an object can appear the same to us (i.e., the feature points have the same two-dimensional coordinates even though the three-dimensional objects differ). Both of these issues are illustrated in Figure 19-11.

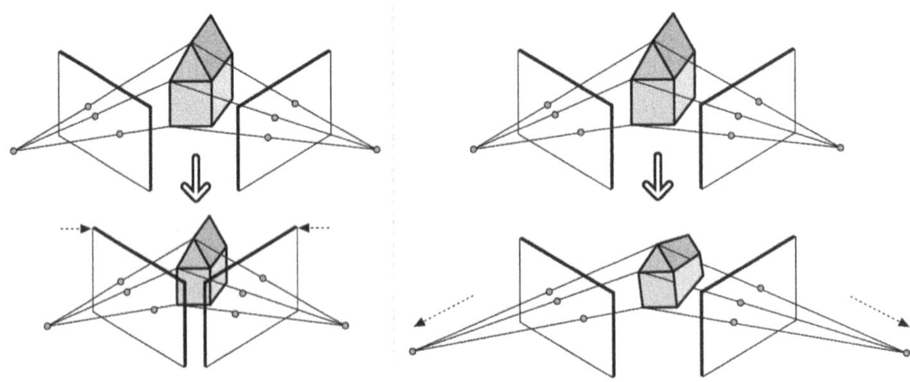

*Figure 19-11. Stereo reconstruction ambiguity: if we do not know object size, then different size objects can appear the same depending on their distance from the camera (left); if we don't know the camera intrinsics, then different projections can appear the same—for example, by having different focal lengths and principal points (right)*

Assuming we have the fundamental matrix $F$ (which required seven or more points to compute), Hartley's algorithm proceeds as follows (see Hartley's original paper [Hartley98] for more details):

1. We use the fundamental matrix to compute the two epipoles via the relations $F \cdot \vec{e}_l = 0$ and $\vec{e}_r^{\,T} \cdot F = 0$ for the left and right epipoles, respectively.
2. We seek a first homography $H_r$, which will map the right epipole to the two-dimensional homogeneous point at infinity $(1, 0, 0)^T$. Since a homography has seven constraints (scale is missing), and we use three to do the mapping to infinity, we have four degrees of freedom left in which to choose our $H_r$. These four degrees of freedom are mostly freedom to make a mess since most choices of $H_r$ will result in highly distorted images. To find a good $H_r$, we choose a point in the image where we want minimal distortion to happen, allowing only rigid rotation and translation (but specifically not shearing) there. A reasonable choice for such a point is the image origin, and we'll further assume that the epipole $\vec{e}_r = (k, 0, 1)^T$ lies on the x-axis (a rotation matrix will accomplish this, as explained later). Given these coordinates, the matrix:

$$G = \begin{pmatrix} 1 & 0 & 0 \\ 0 & 1 & 0 \\ -1/k & 0 & 1 \end{pmatrix}$$

will take such an epipole to infinity.

3. For a selected point of interest in the right image (we chose the origin), we compute the translation $\vec{T}$ that will take that point to the left image origin (0 in our case) and the rotation $R$ that will take the epipole to $\vec{e}_r = (k, 0, 1)^T$. The homography we want will then be given by $H_r = G \cdot R \cdot T$. (Here, $T$ is a $4 \times 4$ transformation matrix constructed from $\vec{T}$).
4. We next search for a matching homography $H_l$ that will send the left epipole to infinity and align the rows of the two images. We easily send the left epipole to infinity by using up three constraints as in Step 2. To align the rows, we use the fact that aligning the rows minimizes the total distance between all matching points between the two images. That is, we find the $H_l$ that minimizes the total disparity in left-right matching points $\sum_i d(H_l \cdot \vec{p}_{i,l}, H_r \cdot \vec{p}_{i,r})$. These two homographies define the stereo rectification.

Although the details of this algorithm are a bit tricky, `cv::stereoRectifyUncalibrated()` does all the hard work for us. The function is a bit misnamed because it does not rectify uncalibrated stereo images; rather, it computes homographies that may be used for rectification. The algorithm prototype is:

```
bool cv::stereoRectifyUncalibrated(
  cv::InputArray  points1,          // Feature points from image 1
  cv::InputArray  points2,          // Corresponding points in image 2
  cv::InputArray  F,                // Fundamental matrix
  cv::Size        imgSize,          // Size of images used
  cv::OutputArray H1,               // Rectification (homography) matrix
                                    //  for image 1
  cv::OutputArray H2,               // Rectification (homography) matrix
                                    //  for image 2
  double          threshold = 5.0 // (Optional) outlier
                                    //  threshold ('0'=ignore)
);
```

In `cv::stereoRectifyUncalibrated()`, the algorithm takes as input two arrays of two-dimensional feature points, which are corresponding points between the left and right images, in the arrays `points1` and `points2`. The fundamental matrix we just calculated is passed as the array `F`($F$). We are familiar with `imageSize`, which just describes the width and height of the images that were used during calibration. Our results, the rectifying homographies, are returned in the function variables `H1` ($H_l$)

and `H2` ($H_r$). Finally, if the distance from points to their corresponding epilines exceeds a set `threshold`, the corresponding point is eliminated by the algorithm.[38]

If our cameras have roughly the same parameters and are set up in an approximately horizontally aligned frontal parallel configuration, then our eventual rectified outputs from Hartley's algorithm will look very much like the calibrated case described next. If we know the size or the three-dimensional geometry of objects in the scene, we can obtain the same results as the calibrated case.

### Calibrated stereo rectification: Bouguet's algorithm

Given the rotation and translation ($R$, $\vec{T}$) that relate the two stereo images, Bouguet's algorithm for stereo rectification attempts to minimize the amount of change reprojection produces for each of the two images (and thereby minimize the resulting reprojection distortions) while maximizing common viewing area.

To minimize image reprojection distortion, the rotation matrix $R$ that rotates the right camera's image plane into the left camera's image plane is split in half between the two cameras; we call the two resulting rotation matrices $r_r$ and $r_l$ for the left and right camera, respectively. Each camera rotates half a rotation, so their principal rays each end up parallel to the vector sum of where their original principal rays had been pointing. As we have noted, such a rotation puts the cameras into coplanar alignment but not into row alignment. To compute the matrix $R_{rect}$ that will take the left camera's epipole to infinity and align the epipolar lines horizontally, we create a rotation matrix by starting with the direction of the epipole $\vec{e}_1$ itself. Taking the principal point ($c_x$, $c_y$) as the left image's origin, the (unit normalized) direction of the epipole is directly along the translation vector between the two cameras' centers of projection:

$$\vec{e}_1 = \frac{\vec{T}}{\|\vec{T}\|}$$

The next vector, $\vec{e}_2$, must be orthogonal to $\vec{e}_1$, but is otherwise unconstrained. For $\vec{e}_2$, selecting a direction orthogonal to the principal ray (which tends to be along the image plane) is a good choice. We accomplish this by using the cross-product of $\vec{e}_1$ with the direction of the principal ray, and then normalizing so that we've got another unit vector:

$$\vec{e}_2 = \frac{1}{\sqrt{T_x^2 + T_y^2}}(-T_y, T_x, 0)^T$$

38 Hartley's algorithm works best for images that have been rectified previously by single-camera calibration. It won't work at all for images with high distortion. It is rather ironic that our "calibration-free" routine works only for undistorted image inputs whose parameters are typically derived from prior calibration. For another uncalibrated three-dimensional approach, see Pollefeys [Pollefeys99a].

We can always find a third vector, $\vec{e}_3$, that is orthogonal to $\vec{e}_1$ and $\vec{e}_2$ by using the cross-product operation:

$$\vec{e}_3 = \vec{e}_1 \times \vec{e}_2.$$

Our matrix that takes the epipole in the left camera to infinity is then:

$$R_{rect} = \begin{bmatrix} \vec{e}_1^{\,T} \\ \vec{e}_2^{\,T} \\ \vec{e}_3^{\,T} \end{bmatrix}$$

This matrix rotates the left camera about the center of projection so that the epipolar lines become horizontal and the epipoles are at infinity. The row alignment of the two cameras is then achieved by setting:

$$R_l = R_{rect} \cdot r_l$$

and:

$$R_r = R_{rect} \cdot r_r$$

We will also compute the rectified left and right camera matrices $M_{rect,l}$ and $M_{rect,r}$ but return them combined with projection matrices $P_l'$ and $P_r'$:

$$P_l = M_{rect,l} \cdot P_l' = \begin{bmatrix} f_{rect,l} & \alpha_l & c_{x,l} \\ 0 & f_{y,l} & c_{y,l} \\ 0 & 0 & 1 \end{bmatrix} \begin{bmatrix} 1 & 0 & 0 & 0 \\ 0 & 1 & 0 & 0 \\ 0 & 0 & 1 & 0 \end{bmatrix}$$

and:

$$P_r = M_{rect,r} \cdot P_r' = \begin{bmatrix} f_{rect,r} & \alpha_r & c_{x,r} \\ 0 & f_{y,r} & c_{y,r} \\ 0 & 0 & 1 \end{bmatrix} \begin{bmatrix} 1 & 0 & 0 & T_x \\ 0 & 1 & 0 & 0 \\ 0 & 0 & 1 & 0 \end{bmatrix}$$

(Here $\alpha_l$ and $\alpha_r$ allow for a pixel skew factor that, as we saw in Chapter 18, is almost always effectively `0` in modern cameras.) The projection matrices take a three-dimensional point in homogeneous coordinates to a two-dimensional point in homogeneous coordinates as follows:

$$P \cdot \begin{bmatrix} X \\ Y \\ Z \\ 1 \end{bmatrix} = \begin{bmatrix} x \\ y \\ w \end{bmatrix}$$

The screen coordinates can be calculated according to $(x, y) = (x/w, y/w)$. Points in two dimensions can also be reprojected into three dimensions given their screen coordinates and the camera intrinsics matrix. The reprojection matrix is:

$$Q = \begin{bmatrix} 1 & 0 & 0 & -c_x \\ 0 & 1 & 0 & -c_y \\ 0 & 0 & 0 & f \\ 0 & 0 & -\frac{1}{T_x} & \frac{c_x - c'_x}{T_x} \end{bmatrix}$$

Here the parameters are from the left image except for $c'_x$, which is the principal point x-coordinate in the right image. If the principal rays intersect at infinity, then $c_x = c'_x$ and the term in the lower-right corner is `0`. Given a two-dimensional homogeneous point and its associated disparity *d*, we can project the point into three dimensions using:

$$Q \cdot \begin{bmatrix} x \\ y \\ d \\ 1 \end{bmatrix} = \begin{bmatrix} X \\ Y \\ Z \\ W \end{bmatrix}$$

The three-dimensional coordinates are then $(X/W, Y/W, Z/W)$.

Applying the Bouguet rectification method just described yields our ideal stereo configuration as per Figure 19-5. New image centers and new image boundaries are then chosen for the rotated images so as to maximize the overlapping viewing area. Mainly this just sets a uniform camera center and a common maximal height and width of the two image areas as the new stereo viewing planes.

In the context of the OpenCV library, the Bouget algorithm is implemented by the function `cv::stereoRectify()`. Given the intrinsics and distortion coefficients for the two cameras, as well as the translation and rotation that related the location of the two cameras, `cv::stereoRectify()` will compute for us the rectification, projection, and disparity maps that we need to extract depth information from stereo images from that camera pair.

```
void cv::stereoRectify(
  cv::InputArray  cameraMatrix1, // Intrinsics (cam 1)
  cv::InputArray  distCoeffs1,   // Distortion coefficients (cam 1)
```

```
    cv::InputArray  cameraMatrix2, // Intrinsics (cam 2)
    cv::InputArray  distCoeffs2,   // Distortion coefficients (cam 2)
    cv::Size        imageSize,     // Size of imgs used for calibration
    cv::InputArray  R,             // Rotation *matrix* between
                                   //  camera coordinates
    cv::InputArray  T,             // Translation vector between
                                   //   camera coordinates
    cv::OutputArray R1,            // 3x3 Rectification xform (cam 1)
    cv::OutputArray R2,            // 3x3 Rectification xform (cam 2)
    cv::OutputArray P1,            // 3x4 (New) projection mtx (cam 1)
    cv::OutputArray P2,            // 3x4 (New) projection mtx (cam 2)
    cv::OutputArray Q,             // 4x4 Disparity to depth
                                   //  mapping matrix
    int             flags        = cv::CALIB_ZERO_DISPARITY,
    double          alpha        = -1,           // [0,1] output crop
    cv::Size        newImageSize = cv::Size(),   // Output mig size (use
                                                 //  '0,0' = "as input")
    cv::Rect*       validPixROI1 = 0,            // (Optional) guaranteed
                                                 //  valid pix (img1)
    cv::Rect*       validPixROI2 = 0             // (Optional) guaranteed
                                                 //  valid pix (img2)
  );
```

For `cv::stereoRectify()`,[39] we first input the familiar camera matrices and distortion vectors returned by `cv::stereoCalibrate()`. These are followed by `imageSize`, the size of the chessboard images used to perform the calibration. We also pass in the rotation matrix `R` ($R$) and translation vector `T`($\vec{T}$) between the right and left cameras that was also returned by `cv::stereoCalibrate()`.

Return parameters are `R1` ($R_l$) and `R2` ($R_r$), the $3 \times 3$ row-aligned rectification rotations for the left and right image planes as derived in the preceding equations. Similarly, we get back the $3 \times 4$ left and right projection equations `P1` ($P_l$) and `P2` ($P_r$). An optional return parameter is `Q` ($Q$), the $4 \times 4$ reprojection matrix described previously.

The `flags` parameter is defaulted to set disparity at infinity, the normal case as per Figure 19-5. Unsetting `flags` means that we want the cameras verging toward each other (i.e., slightly "cross-eyed") so that zero disparity occurs at a finite distance (this might be necessary for greater depth resolution in the proximity of that particular distance).

If the `flags` parameter was not set to `cv::CALIB_ZERO_DISPARITY`, then we must be more careful about how we achieve our rectified system. Recall that we rectified our system relative to the principal points ($c_x$, $c_y$) in the left and right cameras. Thus, our measurements in Figure 19-5 must also be relative to these positions. Basically, we

39 `cv::stereoRectify()` is a bit of a misnomer because the function computes the terms that we can use for rectification but doesn't actually rectify the stereo images.

have to modify the distances so $\tilde{x}_l = x_r - c_x^{right}$ and $\tilde{x}_l = x_r - c_x^{left}$. When disparity has been set to infinity, we have $c_x^{right} = c_x^{left}$ (i.e., when `cv::CALIB_ZERO_DISPARITY` is passed to `cv::stereoRectify()`), and we can pass plain pixel coordinates (or disparity) to the formula for depth. But if `cv::stereoRectify()` is called without `cv::CALIB_ZERO_DISPARITY` then $c_x^{right} \neq c_x^{left}$ in general. Therefore, even though the formula $Z = f \cdot T_x / (x_l - x_r)$ remains the same, one should keep in mind that $x_l$ and $x_r$ are not counted from the image center but rather from the respective principal points $c_x^{left}$ and $c_x^{right}$, which could differ from $x_l$ and $x_r$. Hence, if you computed disparity $d = x_l - x_r$ then you should adjust it before computing $Z = f \cdot T_x / (d - c_x^{left} - c_x^{right})$.

### Rectification map

Once we have our stereo calibration terms, we can precompute left and right rectification lookup maps for the left and right camera views using separate calls to `cv::initUndistortRectifyMap()`. As with any image-to-image mapping function, a forward mapping (in which we just compute where pixels go from the source image to the destination image) will not, owing to floating-point destination locations, hit all the pixel locations in the destination image—and the destination image will thus look like Swiss cheese. So instead we follow the usual procedure of working backward: for each integer pixel location in the destination image, we look up what floating-point coordinate it came from in the source image and then interpolate from its surrounding source pixels a value to use in that integer destination location. This source lookup typically uses bilinear interpolation, which we encountered with `cv::remap()` in Chapter 11.

The process of rectification is illustrated in Figure 19-12. As shown by the equation flow in that figure, the actual rectification process proceeds backward from (c) to (a) in a process known as *reverse mapping*. For each integer pixel in the rectified image (c), we find its coordinates in the undistorted image (b) and use those to look up the actual (floating-point) coordinates in the raw image (a). The floating-point coordinate pixel value is then interpolated from the nearby integer pixel locations in the original source image, and that value is used to fill in the rectified integer pixel location in the destination image (c). After the rectified image is filled in, it is typically cropped to emphasize the overlapping areas between the left and right images.

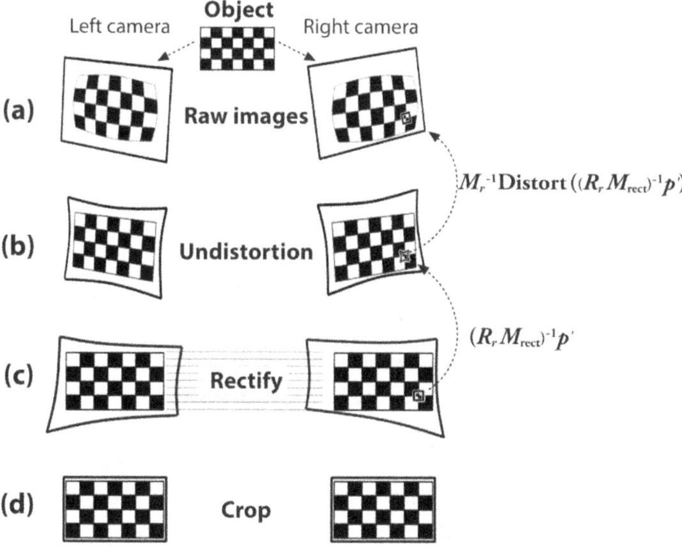


*Figure 19-12. Stereo rectification: for the left and right camera, the raw image (a) is undistorted (b) and rectified (c) and finally cropped (d) to focus on overlapping areas between the two cameras; the rectification computation actually works backward from (c) to (a)*

The function that implements the math depicted in Figure 19-12 is called `cv::initUndistortRectifyMap()`. We call this function twice, once for the left and once for the right image of stereo pair:

```
void cv::initUndistortRectifyMap(
  cv::InputArray  cameraMatrix,     // Camera intrinsics matrix
  cv::InputArray  distCoeffs,       // Camera distortion coefficients
  cv::InputArray  R,                // (Optional) 3x3 rectification
                                    //  transformation
  cv::InputArray  newCameraMatrix,  // New camera matrix usually
                                    //  from cv::stereoRectify()
  cv::Size        size,             // Undistorted image size
  int             m1type,           // Method for encoding result maps
  cv::OutputArray map1,             // First output undistortion map
  cv::OutputArray map2              // Second output undistortion map
);
```

The `cv::initUndistortRectifyMap()` function takes as input the 3 × 3 camera matrix `cameraMatrix`, the 5 × 1 camera distortion parameters in `distCoeffs`, the 3 × 3 rotation matrix `R`, and the rectified 3 × 3 camera matrix `newCameraMatrix`.

If we calibrated our stereo cameras using `cv::stereoRectify()`, then we can read our input to `cv::initUndistortRectifyMap()` straight out of `cv::stereoRectify()`

using first the left parameters to rectify the left camera and then the right parameters to rectify the right camera. For R, use $R_l$ and $R_r$ from cv::stereoRectify(); for cameraMatrix, use cameraMatrix1 or cameraMatrix2. For newCameraMatrix we could use the first three columns of the $3 \times 4$ $P_l$ or $R_r$ from cv::stereoRectify(), but as a convenience the function allows us to pass $P_l$ or $P_r$ directly and it will read newCameraMatrix from them.

If, on the other hand, we used cv::stereoRectifyUncalibrated() to calibrate our stereo cameras, then we must preprocess the homography a bit. Although we could—in principle and in practice—rectify stereo without using the camera intrinsics, OpenCV does not have a function for doing this directly. If we do not have newCameraMatrix from some prior calibration, the proper procedure is to set newCameraMatrix equal to cameraMatrix. Then, for R in cv::initUndistortRectifyMap(), we need to compute $R_{rect,l} = M_{rect,l}^{-1} \cdot H_l \cdot M_l$ (or just $R_{rect,l} = M_l^{-1} \cdot H_l \cdot M_l$ if $M_{rect,l}^{-1}$ is unavailable) and $R_{rect,l} = M_{rect,l}^{-1} \cdot H_l \cdot M_l$ (or just $R_{rect,r} = M_r^{-1} \cdot H_r \cdot M_r$ if $M_{rect,r}^{-1}$ is unavailable) for the left and the right rectification, respectively. Finally, we will also need the distortion coefficients for each camera to fill in the $5 \times 1$ distCoeffs parameters.

The argument size indicates the size of the image to be undistorted (and thus the size of the map arrays to be generated). The m1type argument is used to determine the format of the maps generated. As discussed in Chapter 18, it can be set to either CV_32C1 or CV_16SC2, and the result will be generated maps being in either the floating point or integer form, respectively.

The function cv::initUndistortRectifyMap() returns lookup maps map1 and map2 as output. These maps indicate from where we should interpolate source pixels for each pixel of the destination image; the maps can then be plugged directly into cv::remap(), the function we first saw in Chapter 11. As we mentioned, the function cv::initUndistortRectifyMap() is called separately for the left and the right cameras so that we can obtain their distinct map1 and map2 remappings. We may then call the function cv::remap(), using the left and then the right maps each time we have new left and right stereo images to rectify. Figure 19-13 shows the results of stereo undistortion and rectification of a stereo pair of images. Note how feature points become horizontally aligned in the undistorted rectified images.

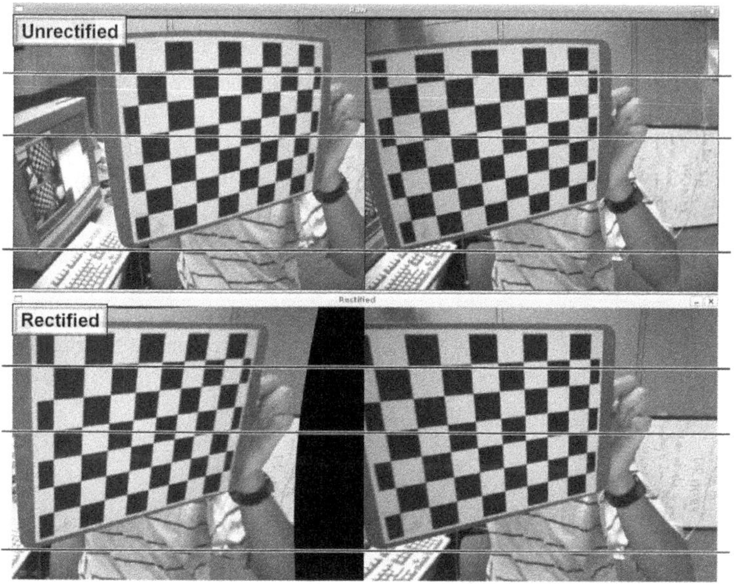


*Figure 19-13. Stereo rectification: original left and right image pair (upper panels) and the stereo rectified left and right image pair (lower panels); note that the barrel distortion (in top of chessboard patterns) has been corrected and the scan lines are aligned in the rectified images*

## Stereo Correspondence

Stereo correspondence—matching a three-dimensional point in the two different camera views—can be computed only over the visual areas in which the views of the two cameras overlap. Once again, this is one reason why you will tend to get better results if you arrange your cameras to be as nearly frontal parallel as possible (at least until you become quite an expert at stereo vision). Then, once we know the physical coordinates of the cameras or the sizes of objects in the scene, we can derive depth measurements from the triangulated disparity measures $d = x^l - x^r$ (or $d = x^l - x^r - (c_x^{left} - c_x^{right})$, if the principal rays intersect at a finite distance) between the corresponding points in the two different camera views. Without such physical information, we can compute depth only up to a scale factor. If we don't have the camera intrinsics, as when using Hartley's algorithm, we can compute point locations only up to a projective transform (recall Figure 19-11).

OpenCV implements two different stereo correspondence algorithms that share a (nearly) common object interface. The first, called the *block matching (BM) algorithm*, is a fast and effective algorithm that is similar to the one developed by Kurt Konolige [Konolige97]. It works by using small "sum of absolute difference" (SAD) windows to find matching points between the left and right stereo-rectified images.[40] This algorithm finds only strongly matching (high-texture) points between the two images. Thus, in a highly textured scene such as might occur outdoors in a forest, every pixel might have computed depth. In a very low-textured scene, such as an indoor hallway, very few points might register depth. The second algorithm is called the *semi-global block matching (SGBM) algorithm*. SGBM, a variation of SGM introduced in [Hirschmuller08], differs from BM primarily in two respects. The first is that matching is done at subpixel level using the Birchfield-Tomasi metric [Birchfield99]. The second difference is that SGBM attempts to enforce a global smoothness constraint on the computed depth information that it approximates by considering many one-dimensional smoothness constraints through the region of interest. These two methods are complementary, in the sense that BM is quite a bit faster, but does not provide the reliability and accuracy of SGBM.

### The stereo matching classes: cv::StereoBM and cv::StereoSGBM

The different stereo matching algorithms both serve the same basic function: to convert two images, one left and one right, into a single depth image. The depth image will associate with each pixel a distance from the cameras to the object this pixel represents.[41] OpenCV provides implementations for both of the different stereo matching algorithms just described: block matching (BM), and semi-global block matching (SGBM). Though these two algorithms are historically related (as their names indicate, they serve an essentially identical purpose), each has its own interface.[42]

### Block matching

The block matching stereo algorithm implemented in OpenCV is a slightly modified version of what has come to be regarded as a (and perhaps *the*) canonical technique for stereo computation. The basic mechanism is to rectify and align the images such that comparisons need be made only in individual rows, and then to have the algorithm search rows in the two images for matching groups of pixels. Of course, there are some extra details that variously make the algorithm either work a little better or run a little faster. The net result is a fast, relatively reliable algorithm that is still heavily used in a wide variety of applications.

---

40 This algorithm is available in an FPGA stereo hardware system from Videre.

41 Typically in OpenCV, the pixels in the depth image will correspond to the pixels in the left camera image, though this convention is far from universal.

42 Because there is such a commonality, you might expect that some virtual base class would be used to define a common interface for objects implementing both methods, as is done elsewhere in the library. In fact, however, because the information required to initialize each algorithm is slightly different, the actual implementation is not to use a common interface (at least not for the time being), but rather two separate but otherwise very similar-looking classes.

There are three stages to the block matching stereo correspondence algorithm, which works on undistorted, rectified stereo image pairs:

1. Prefiltering to normalize image brightness and enhance texture.
2. Correspondence search along horizontal epipolar lines using an SAD window.
3. Postfiltering to eliminate bad correspondence matches.

In the prefiltering step, we normalize the input images to reduce lighting differences and to enhance image texture. We do this by running a window—of size 5 × 5, 7 × 7 (the default), ..., 21 × 21 (the maximum)—over the image. The center pixel $I_c$ under the window is replaced by $min(max(I_c - \bar{I},\ - I_{cap}),\ I_{cap})$, where $\bar{I}$ is the average value in the window and $I_{cap}$ is a positive numeric limit whose default value is `30`. The alternative is to essentially convert the incoming images to their x-Sobel derivatives. This also has the effect of removing many lighting-related artifacts.

Next, we compute correspondence by sliding the SAD window. For each feature in the left image, we search the corresponding row in the right image for a best match. After rectification, each row is an epipolar line, so the matching location in the right image must be along the same row (same y-coordinate) as in the left image; this matching location can be found if the feature has enough texture to be detectable and if it is not occluded in the right camera's view (see Figure 19-17). If the left feature pixel coordinate is at $(x_0, y_0)$, then, for a horizontal frontal parallel camera arrangement, the match (if any) must be found on the same row and at, or to the left of, $x_0$ (see Figure 19-14). For frontal parallel cameras, $x_0$ is at zero disparity and larger disparities are to the left. For cameras that are angled toward each other, the match may occur at negative disparities (to the right of $x_0$). The algorithm needs to be told the minimum disparity it will encounter.

The disparity search is then carried out over a preselected number of disparities, which are counted in pixels (the default is 64 pixels). Disparities have a discrete subpixel resolution that is equal to 4 bits of resolution below the individual pixel level. When the output image is a 32-bit floating point image, noninteger disparities will be returned. When the output image is a 16-bit integer, the disparity will be returned in 4-bit fixed-point form (i.e., multiplied by 16 and rounded to an integer).

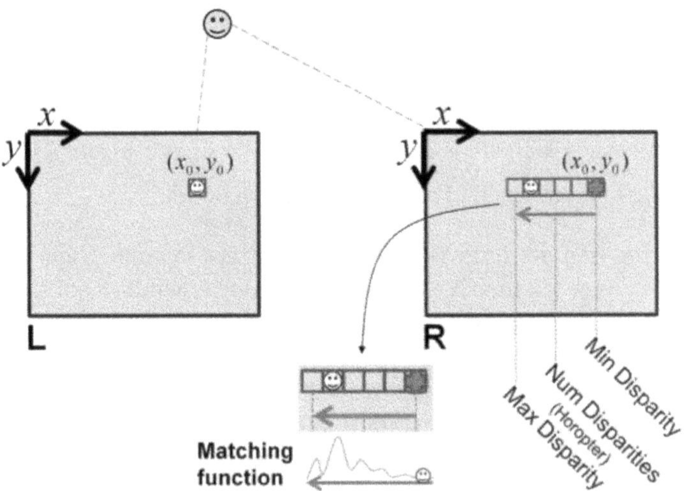


*Figure 19-14. Any right-image match of a left-image feature must occur on the same row and at (or to the left of) the same coordinate point, where the match search starts at the minDisparity point (here, 0) and moves to the left for the set number of disparities; the characteristic matching function of window-based feature matching is shown in the lower part of the figure*

Setting the minimum disparity and the number of disparities to be searched establishes the *horopter*, the three-dimensional volume that is covered by the search range of the stereo algorithm. Figure 19-15 shows disparity search limits of five pixels starting at three different disparity limits: 20, 17, and 16. Each disparity limit defines a plane at a fixed depth from the cameras (see Figure 19-16). As shown in Figure 19-15, each disparity limit—together with the number of disparities—sets a different horopter at which depth can be detected. Outside of this range, depth will not be found and will represent a "hole" in the depth map where depth is not known. You can make horopters larger by decreasing the baseline distance $\| \vec{T} \|$ between the cameras, by making the focal length smaller, by increasing the stereo disparity search range, or by increasing the pixel width.

Correspondence within the horopter has one in-built constraint, called the *order constraint*, which simply states that the order of the features cannot change from the left view to the right. There may be *missing* features—where, owing to occlusion and noise, some features found on the left cannot be found on the right—but the ordering of those features that are found remains the same. Similarly, there may be many features on the right that were not identified on the left (these are called *insertions*), but insertions do not change the order of features although they may spread those features out. The procedure illustrated in Figure 19-17 reflects the ordering constraint when features are being matched on a horizontal scan line.

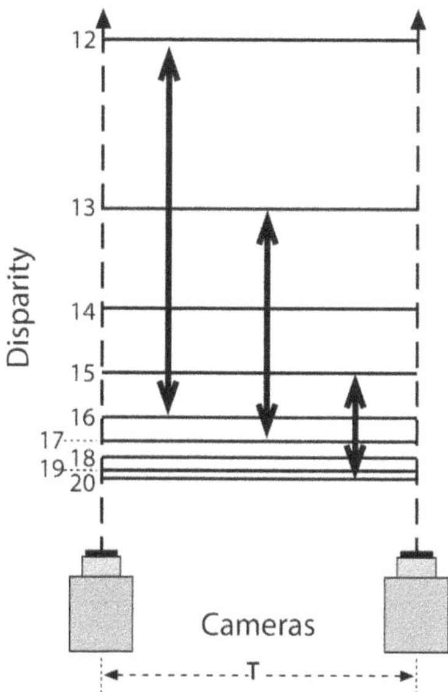


*Figure 19-15. Each line represents a plane of constant disparity in integer pixels from 20 to 12; a disparity search range of five pixels will cover different horopter ranges, as shown by the vertical arrows, and different maximal disparity limits establish different horopters*

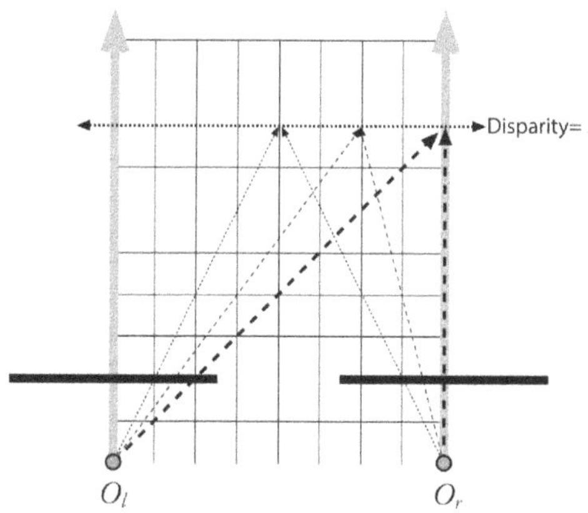


*Figure 19-16. A fixed disparity forms a plane of fixed distance from the cameras*

Given the smallest allowed disparity increment $\Delta_d$, we can determine smallest achievable depth range resolution $\Delta_z$ by using the formula:

$$\Delta_z = \frac{z^2}{f \cdot T_x} \Delta_d$$

It is useful to keep this formula in mind so that you know what kind of depth resolution to expect from your stereo rig.

After correspondence, we turn to postfiltering. The lower part of Figure 19-13 shows a typical matching function response as a feature is "swept" from the minimum disparity out to maximum disparity. Note that matches often have the characteristic of a strong central peak surrounded by side lobes. Once we have candidate feature correspondences between the two views, we use postfiltering to prevent false matches. OpenCV makes use of the matching function pattern via the concept of a *uniqueness ratio*. This ratio essentially enforces the requirement that the match value for the current pixel is more than the minimum match value observed by some margin.

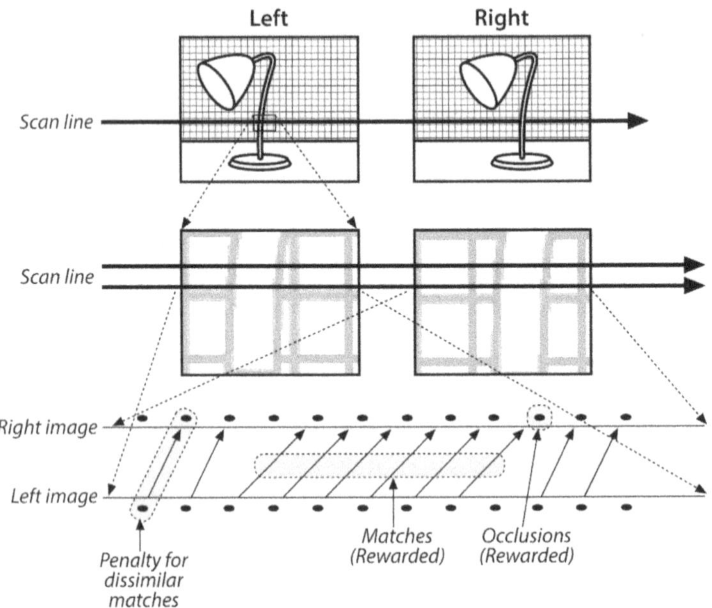


*Figure 19-17. Stereo correspondence starts by assigning point matches between corresponding rows in the left and right images: left and right images of a lamp (upper panel); an enlargement of a single scan line (middle panel); visualization of the correspondences assigned (lower panel)*

To make sure that there is enough texture to overcome random noise during matching, OpenCV also employs a texture threshold. This is just a limit on the SAD window response such that no match is considered whose response is below some minimal value.

Finally, block-based matching can have problems near the boundaries of objects because the matching window catches the foreground on one side and the background on the other side. This results in a local region of large and small disparities that we call *speckle*. To prevent these borderline matches, we can set a speckle detector over a *speckle window*. The way this works is that each pixel is used as a basis for the construction of a connected component defined by a variable range flood fill. The variable range flood fill includes a neighboring pixel only if it is within some range of the current pixel. Once that connected component is computed, if it is smaller than the speckle window, then it is considered speckle. The size of that range is called the *speckle range*. (This range is typically a small number; `1` or `2` works most of the time, though values as large as `4` are not uncommon.)

### Computing stereo depths with cv::StereoBM

In OpenCV, the block matching algorithm is implemented as an object that holds all of the necessary parameters and provides an overloaded `compute()` method that is used to compute disparity images. The algorithm is represented by `cv::StereoBM` class, which is derived from `cv::StereoMatcher` class. Here is how the classes are defined:

```
class StereoMatcher : public Algorithm {

public:
    // the returned depth map has CV_16UC1 type and the elements are
    // actually fixed-point numbers with DISP_SHIFT=4 fractional bits
    enum {
      DISP_SHIFT = 4,
      DISP_SCALE = (1 << DISP_SHIFT)
    };

    // the main method, takes two grayscale 8-bit rectified images and
    // outputs 16-bit fixed-point disparity map
    virtual void compute(
      InputArray  left,
      InputArray  right,
      OutputArray disparity
    ) = 0;

    // the minimum disparity, usually 0 (points are at the infinity
    virtual int  getMinDisparity() const                    = 0;
    virtual void setMinDisparity(int minDisparity)          = 0;

    // the range of disparities, between minimum (inclusive) and
    // maximum (exclusive) disparity
    virtual int  getNumDisparities() const                  = 0;
    virtual void setNumDisparities(int numDisparities)      = 0;

    // size of blocks used by algorithm
    virtual int  getBlockSize() const                       = 0;
    virtual void setBlockSize(int blockSize)                = 0;

    // maximum size of a speckle to be considered as speckle
```

```
    // and marked as such
    virtual int  getSpeckleWindowSize() const              = 0;
    virtual void setSpeckleWindowSize(int speckleWindowSize) = 0;

    // the allowed difference between neighbor pixels;
    // used by  floodfill-based speckle filtering algorithm
    virtual int  getSpeckleRange() const                   = 0;
    virtual void setSpeckleRange(int speckleRange)         = 0;
    ...
};

class cv::StereoBM : public cv::StereoMatcher {

    enum {
      PREFILTER_NORMALIZED_RESPONSE = 0,
      PREFILTER_XSOBEL              = 1
    };

    // choose between PREFILTER_NORMALIZED_RESPONSE and
    // PREFILTER_XSOBEL
    //
    virtual int getPreFilterType() const                    = 0;
    virtual void setPreFilterType(int preFilterType)        = 0;

    // size of block used in PREFILTER_NORMALIZED_RESPONSE mode
    //
    virtual int getPreFilterSize() const                    = 0;
    virtual void setPreFilterSize(int preFilterSize)        = 0;

    // saturation threshold applied after pre-filtering
    //
    virtual int getPreFilterCap() const                     = 0;
    virtual void setPreFilterCap(int preFilterCap)          = 0;

    // textureness threshold, the blocks with textureness
    // characteristics (sum of abs derivatives) less than the
    // threshold are marked as areas without defined disparity
    //
    virtual int getTextureThreshold() const                 = 0;
    virtual void setTextureThreshold(int textureThreshold)  = 0;

    // if there is no clear winner in the cost function across the
    // disparity range, the pixel is having no defined disparity.
    // Uniqueness threshold defines what it means to be a
    // "clear winner," the margin in %% between the best and
    // second-best.
    //
    virtual int getUniquenessRatio() const                  = 0;
    virtual void setUniquenessRatio(int uniquenessRatio)    = 0;

    // the constructor function, the first parameter is made default
    // for techinical reasons, but you should actually specify it
```

```
        // here or later using setNumDisparities()
        //
        static Ptr<StereoBM> create(
          int numDisparities = 0,
          int blockSize      = 21
        );
    };
```

The most important elements of `cv::StereoMatcher` and `cv::StereoBM` are the `compute()` and `create()` methods.

The static `create()` method takes two arguments, `numDisparities` and `blockSize`.

The argument `numDisparities` is the total number of distinct possible disparities to be returned. In effect, this sets the range over which the algorithm will even attempt to find a correspondence.

The second argument is `blockSize`. This sets the size of the region around each pixel around which the "sign of absolute difference" metric will be computed. The larger this value is, the fewer false matches you are likely to find. You should keep in mind, however, that not only does the computational cost of the algorithm scale with the area of the window (i.e., the square of the window size), but there is also a problem arising from the implicit assumption that the disparity is actually the same over the area of the window. Near to discontinuities (the edges of objects) this assumption will not hold, and you are likely to find no match at all. The result will be empty regions where you have no disparity at all near the edges of objects. The thickness of these empty regions will grow with increased window size. Also, with larger `blockSize` you will likely get vaguer depth maps, meaning that the objects' silhouettes in the disparity map will be smoother, capturing the real silhouettes in only an approximate manner.

After you call `create()` and obtain a smart pointer to the `cv::StereoBM` object, you can further configure it by calling various `set*` methods. Of the various members of this structure, the ones you might want to modify are the primarily the prefilter parameters and the postfilter parameters.

The prefilters attempt to remove variations from illumination or other sources that would cause a mismatch between the two images under the SAD metric. The possible values for `preFilterType` are `cv::StereoBM::PREFILTER_NORMALIZED_RESPONSE` and `cv::StereoBM::PREFILTER_XSOBEL`. The former just normalizes the intensities in the window, while the latter actually converts the images into (clipped versions of) their first Sobel derivative (in the x-direction). The value of `preFilterSize` just sets the size of the filter used (for the `NORMALIZED_RESPONSE` case only), and `preFilterCap` is the value of the parameter $I_{cap}$ used to clamp the output of the prefiltering (as described in the previous section).

The post filters attempt to remove outliers and noise in the output image. The `textureThreshold` sets a minimum amount of texture that must be present before disparity will be computed between two regions.

The `uniquenessRatio` applies to SAD windows, and is interpreted as the degree of difference between the best match and the second-best match that is required in order for the disparity to be considered unambiguous. The threshold is defined by the relation $SAD(d) \geq SAD(d^{*}) \cdot (1.0 + \frac{uniquenessRatio}{100.0})$. Here $d$ and $d^{*}$ are the current disparity and the next-best disparity, respectively. Typical values for the uniqueness ratio are between 5 and 15.

The `speckleWindowSize` and `speckleRange` parameters work together. They enable a postfilter that will attempt to remove any small, isolated blobs that are substantially different from their surrounding values. The `speckleWindowSize` sets the size of such blobs, while `speckleRange` sets the largest difference between disparities that will include them in the same blob. This parameter is compared directly to the values of the disparities. This means that if you are using fixed-precision representation for disparity then this value will, in effect, be multiplied by 16. You should take this into account when setting this parameter.

Once you have configured your `cv::StereoBM` object, you can compute disparity images with `compute()`. The overloaded method expects three arguments: the left and right images (`left` and `right`) and the output image (`disparity`). The produced disparity will have fixed-point representation, with 4 bits of fractional precision, so you will want to divide by 16 when using these disparities.

### Semi-global block matching

The alternative to block matching provided by OpenCV is the semi-global block matching algorithm (SGBM), a variation of SGM algorithm [Hirschmuller08]. Developed almost a decade later than block matching, the SGM algorithm applied several new ideas, but at a computational cost far greater than that of BM.[43] The most important new ideas introduced by SGM are the use of *mutual information* (OpenCV's implementation used Birchfield-Tomasi metrics, also used in the original paper as a simpler option) as a superior measure of local correspondence and the enforcement of *consistency constraints* along directions other than the horizontal (epipolar) line. At a high level, the effects of these additions are to provide much greater robustness

43 At current processor speeds, it is typically possible to run BM on relatively high-resolution frames at a frame rate acceptable for real-time processing of video, even in a CPU-only implementation. In contrast, SGBM takes approximately an order of magnitude more compute time, and so is typically considered not yet suitable for real-time video applications. (However, there have been FPGA implementations of SGBM that will run at very high speed on even very large images. Similarly, BM has been implemented on FPGAs, as well as on GPU—with support for the latter being available in OpenCV already.)

to lighting and other variations between the left and right images, and to help eliminate errors by enforcing stronger geometrical constraints across the image.

Like BM, SGBM operates on undistorted, rectified stereo image pairs. SGBM has the following basic steps:

1. Preprocess each image just like in StereoBM with `PREFILTER_XSOBEL` mode. Precompute a $C(x, y, d)$ per-pixel cost map that matches `left_image`$(x, y)$ and `right_image`$(x\text{-}d, y)$ using Birchfield-Tomasi metrics. Initialize the accumulator 3D cost map $S(x, y, d)$ with zeros.
2. For each of the three-, five-, or eight-direction $(r)$ (see Figure 19-18), compute $S^{(r)}(x, y, d)$ using an iterative procedure. Add $S^{(r)}(x, y, d)$ for all $r$'s to $S(x, y, d)$. In order to optimize the memory flow and minimize the memory footprint, the first three or the first five directions (W, E, N[, NW, NE]) are processed together in the forward pass, and in the case of eight-directional mode there is a second pass that processes the remaining three directions (S, SW, SE). In the case of three- or five-directional algorithm, $C(x, y, d)$ and $S(x, y, d)$ are not explicitly stored for all pixels; we only need to store the last three or four rows of the buffers at once.
3. Once $S(x, y, d)$ is complete, we find $d^*(x, y)$ as `argmin` of $S(x, y, d)$. We use the same uniqueness check and the same subpixel interpolation as in the StereoBM algorithm.
4. Do the left-right check to make sure that left-to-right and right-to-left correspondences are consistent. Mark pixels without perfect matches as "invalid disparity."
5. Filter speckles using `cv::filterSpecles`, just like in the StereoBM algorithm.

The key element of the algorithm (as well as practically any other stereo-correspondence algorithm) is how we assign a cost to each pixel for any possible disparity, denoted here as $S(x, y, d)$. Essentially, this is analogous to what we did in block matching, but there are some new twists. The first twist is that we use some more optimistic subpixel Birchfield-Tomasi metrics to compare pixels, instead of simple absolute difference. The second twist is that we engage a very important disparity continuity assumption (neighbor pixels are likely to have the same or similar disparity) and at the same time use much smaller block size (sometimes $3 \times 3$ or $5 \times 5$) instead of using the much bigger, completely independent, windows in StereoBM ($11 \times 11$ and above). Those big windows turn out to be a real problem for BM because to compare two windows, you must first assume that they ought to match when you have the right disparity; but this means that there exists a single disparity that explains the relationship of the two windows. Near a discontinuity (the edge of something) this assumption breaks down, and so you get a lot of problems in BM near such discontinuities.

In SGBM, the alternative to large windows is the use of smaller windows (to compensate for the noise) combined with *paths* (see Figure 19-18). These paths, in principle, extend from the edge of the image to the individual pixel and (again in principle) include every possible path from the edge to the pixel. What SGBM is concerned with is the path of this kind with the lowest cost along the path (analogous to the lowest-cost window in BM). The cost along any path is a cost for each individual pixel (i.e. small block surrounding it) plus a little penalty when a pixel has a slight change in disparity from its neighbor or a big penalty for those that have big changes in disparity from their neighbors.

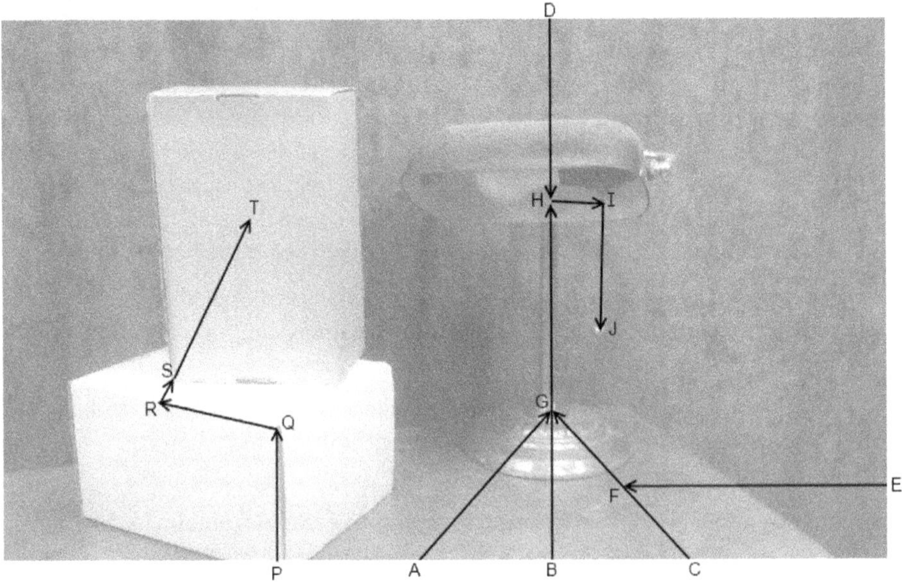


*Figure 19-18. In a simple scene, some possible paths are visualized to several points. The path AGHIJ leads to the end of the lamp chain. Disparities along this chain change only slowly, so it is likely to be preferred over paths like EFGHIJ or DHIJ. To compute the cost at J, all we need to know about I is that BGHI is the lowest cost to I. In practice, the links in these paths are on the pixel scale*

To see how the cost for a particular disparity at some pixel can be computed without an impossible number of summations along a vast number of paths, we need only realize that if we knew the costs for neighboring pixels (at various disparities), the only possible case is that the minimum cost path to this new pixel comes from one of those nearby pixels. Thus we can start at the edges and work our way across the image. First we compute the cost for all possible disparities for those pixels we can compute for, find the best disparity, and then move on to its neighbors.[44]

Ideally, we would consider every possible route into a new pixel. In principle, this is not limited to just the eight nearest neighbors of the pixel, but in fact could include a route into this pixel along a line of any angle. In practice, we are going to have to limit the number of neighbors we consider to compute for each pixel (see Figure 19-19). In theory, it's best to set the number of paths to 8 or 16 (as in the original SGM paper) for high-quality results, with the latter giving substantially better results in many real scenes, but at a substantial cost in computation time. In practice, one can use 5, or even as few as 3, directions with very similar results in most cases. Naively, the computational cost of the algorithm will grow linearly in the number of paths considered but there are significant nonlinear effects that can easily be overlooked. When we move from 5 to 8 directions, however, because of the necessity to store $C(x, y, d)$ and $S(x, y, d)$ for all of the pixels (in order to do two passes over these huge arrays), the algorithm can run much slower than expected.

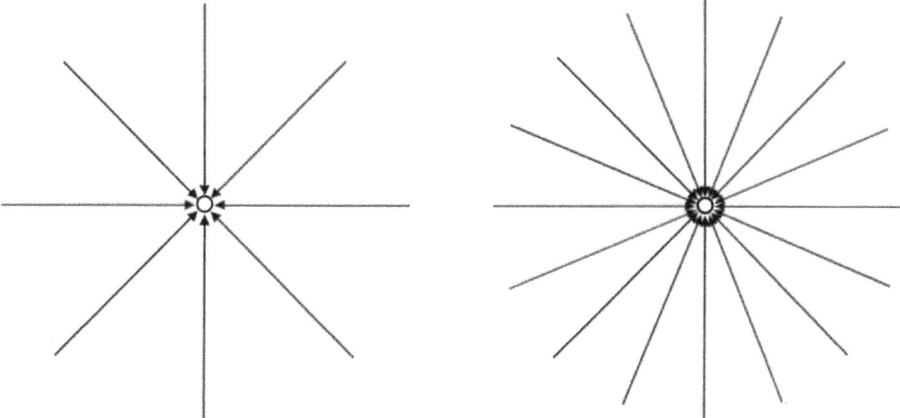

*Figure 19-19. The original semi-global block matching used 8-path or 16-path computations. OpenCV uses 8 paths for quality, but 5 or 3 paths for speed*

44 It should be clear that we are glossing over an important issue here, which is that even an edge pixel could have its lowest-cost path coming from the opposite side of the image. In effect, as we have described it, there is no right place to start. In fact, it turns out to be possible to process the entire image for each direction sequentially. So the actual algorithm does not proceed on a pixel-by-pixel basis. However, the exact details of how this is implemented are well beyond the scope of this book.

### Computing stereo depths with cv::StereoSGBM

As with `cv::StereoBM`, the semi-global block matching algorithm is presented by OpenCV as an object that holds all of the necessary parameters and provides an overloaded `compute()` for actually computing disparities.

```
class StereoSGBM : public StereoMatcher {

public:

  enum {
      MODE_SGBM      = 0,// 5-dir mode
      MODE_HH        = 1,// 8-dir mode (slow, eats lots of memory!)
      MODE_SGBM_3WAY = 2 // 3-dir mode, the fastest one
  };

  // same as in StereoBM
  //
  virtual int  getPreFilterCap() const                   = 0;
  virtual void setPreFilterCap(int preFilterCap)         = 0;

  // same as in StereoBM
  //
  virtual int  getUniquenessRatio() const                = 0;
  virtual void setUniquenessRatio(int uniquenessRatio) = 0;

  // penalty for the difference in disparity between
  //  neighbor pixels = 1.
  //
  virtual int  getP1() const                             = 0;
  virtual void setP1(int P1)                             = 0;

  // penalty for the difference in disparity between
  //  neighbor pixels > 1.
  //
  virtual int  getP2() const                             = 0;
  virtual void setP2(int P2)                             = 0;

  // choose between MODE_SGBM, MODE_HH, MODE_SGBM_3WAY
  //
  virtual int getMode() const                            = 0;
  virtual void setMode(int mode)                         = 0;

  // the constructor function
  //
  static Ptr<StereoSGBM> create(
    int minDisparity,
    int numDisparities,
    int blockSize,
    int P1                = 0,
    int P2                = 0,
    int disp12MaxDiff     = 0,
```

```
        int preFilterCap      = 0,
        int uniquenessRatio   = 0,
        int speckleWindowSize = 0,
        int speckleRange      = 0,
        int mode              = StereoSGBM::MODE_SGBM
      );

  };
```

When we are creating an `cv:: StereoSGBM()` object, there are three required parameters, and then a whole host of optional parameters. The first, `minDisparity`, is the smallest disparity that will be considered (typically `0`). `numDisparities` is the total number of disparities to be considered, so the maximum possible disparity will be equal to `minDisparity` plus `numDisparities`.

The parameter `blockSize` has the same meaning as in StereoBM, but here it's recommended to set it to a much lower value; usually `3` or `5` would be enough, but it can actually be set to `1` to mimic the behavior of the original paper (although OpenCV does not implement the more complex metrics based on the mutual information). This value must always be odd.

`P1`, `P2` parameters correspond to the respective parameters of the SGM algorithm; leaving them as zeros makes the implementation compute some optimal values for them based on the image resolution and `blockSize`. Of the subsequent arguments, `preFilterCap`, `uniquenessRatio`, `speckleWindowSize`, and `speckleRange` all have the same meanings as for `cv::StereoBM`.

The value `disp12MaxDiff` is used in the final comparison between the disparities computed in the left-to-right computation and those computed in the right-to-left computation. If the two are mismatched by more than `disp12MaxDiff`, then the pixel is declared unknown.

The value `mode` defaults to `StereoSGBM::MODE_SGBM` (the five-directional version of the algorithm). If set to `StereoSGBM::MODE_SGBM_3WAY`, it will be even faster; if set to `StereoSGBM::MODE_HH`, the algorithm will proceed as described in the previous section, and it will resolve all eight directions of propagation using a double pass method. The problem with the double pass method is that, as we described earlier, it requires a great deal of memory to store the intermediate results between passes.[45]

As an alternative, leaving `mode` at `StereoSGBM::MODE_SGBM` or `Stereo SGBM::MODE_SGBM_3WAY` will effectively compute the costs per pixel using only three

45 The actual memory required is approximately the area of the image multiplied by the number of possible disparities. For a 640 × 480 image with 1,024 possible disparities, this is 600MB at 16 bits per disparity. For an HD image, this increases to over 4GB.

or five of the eight directions. This reduces accuracy of the final result but does not incur the cost of the huge memory buffer needed for double pass.

## Stereo Calibration, Rectification, and Correspondence Code Example

Let's put this all together with code in an example program that will read in a number of circle grid patterns from a file called *list.txt*. This file contains a list of alternating left and right stereo image pairs, which are used to calibrate the cameras and then rectify the images. Note once again that we're assuming you've arranged the cameras so that their image scan lines are roughly physically aligned and such that each cameras have essentially the same field of view. This will help avoid the problem of the epipole being within the image[46] and will also tend to maximize the area of stereo overlap while minimizing the distortion from reprojection.

In the code (Example 19-3), we first read in the left and right image pairs, find the circles to subpixel accuracy, and then set object and image points for the images where all the circle grids could be found. This process may optionally be displayed. Given this list of found points on the good circle grid images, the code calls `cv::stereoCalibrate()` to calibrate the camera. This calibration gives us the camera matrix `_M` and the distortion vector `_D` for the two cameras; it also yields the rotation matrix `_R`, the translation vector `_T`, the essential matrix `_E`, and the fundamental matrix `_F`.

Next comes a little interlude where we assess the accuracy of calibration by checking how nearly the points in one image lie on the epipolar lines of the other image. To do this, we undistort the original points using `cv::undistortPoints()` (see Chapter 18), compute the epilines using `cv::computeCorrespondEpilines()`, and then compute the dot product of the points with the lines (in the ideal case, these dot products would all be `0`). The accumulated absolute distance forms the error.

The code then optionally moves on to computing the rectification maps using the uncalibrated (Hartley) method `cv::stereoRectifyUncalibrated()` or the calibrated (Bouguet) method `cv::stereoRectify()`. If uncalibrated rectification is used, the code further allows for either computing the needed fundamental matrix from scratch or for just using the fundamental matrix from the stereo calibration. We then compute the rectified images using `cv::remap()`. In our example, lines are drawn across the image pairs to help you see how well the rectified images are aligned. An example result is shown back in Figure 19-13, where we can see that the barrel distortion in the original images is largely corrected from top to bottom and that the images are aligned by horizontal scan lines.

46 OpenCV does not (yet) deal with the case of rectifying stereo images when the epipole is within the image frame. See, for example, Pollefeys, Koch, and Gool [Pollefeys99b] for a discussion of this case.

Finally, if we have rectified the images, we can then compute the disparity maps by using `cv::StereoSGBM`.[47] Our code example allows you to use either horizontally aligned (left-right) or vertically aligned (top-bottom) cameras; note, however, that for the vertically aligned case the function `cv::StereoSGBM()` can compute disparity only for uncalibrated rectification unless you add code to transpose the images yourself. For horizontal camera arrangements, `cv::StereoSGBM()` can find disparity for calibrated or uncalibrated rectified stereo image pairs. (See Figure 19-20 in the next section for example disparity results.)

*Example 19-3. Stereo calibration, rectification, and correspondence*

```
#pragma warning( disable: 4996 )

#include <opencv2/opencv.hpp>
#include <iostream>
#include <string.h>
#include <stdlib.h>
#include <stdio.h>
#include <math.h>

using namespace std;

void help( char* argv[] ) {
  ...
}

static void StereoCalib(
  const char* imageList,
  int         nx,
  int         ny,
  bool        useUncalibrated
) {

  bool              displayCorners   = false;
  bool              showUndistorted  = true;
  bool              isVerticalStereo = false; // horiz or vert cams
  const int         maxScale         = 1;
  const float       squareSize       = 1.f;   // actual square size
  FILE*             f                = fopen( imageList, "rt" );
  int               i, j, lr;
  int               N                = nx*ny;
  vector<string>                 imageNames[2];
  vector< cv::Point3f >          boardModel;
  vector< vector<cv::Point3f> >  objectPoints;
```

47 The code also contains (commented out) the case of using `cv::StereoBM`. In this case `cv::StereoBM::BASIC_PRESET` is used, and then some additional parameters of the `state` member variable in the `cv::StereoBM` object are tuned directly.

```
vector< vector<cv::Point2f> >  points[2];
vector< cv::Point2f >          corners[2];
bool                           found[2]        = {false, false};
cv::Size                       imageSize;

// READ IN THE LIST OF CIRCLE GRIDS:
//
if( !f ) {
  cout << "Cannot open file " << imageList << endl;
  return;
}

for( i = 0; i < ny; i++ )
  for( j = 0; j < nx; j++ )
    boardModel.push_back(
      cv::Point3f((float)(i*squareSize), (float)(j*squareSize), 0.f)
    );

i = 0;
for(;;) {

  char buf[1024];
  lr = i % 2;
  if( lr == 0 ) found[0] = found[1] = false;


  if( !fgets( buf, sizeof(buf)-3, f )) break;
  size_t len = strlen(buf);
  while( len > 0 && isspace(buf[len-1]))  buf[--len] = '\0';
  if( buf[0] == '#') continue;

  cv::Mat img = cv::imread( buf, 0 );
  if( img.empty() ) break;
  imageSize = img.size();
  imageNames[lr].push_back(buf);

  i++;

  // If we did not find board on the left image,
  // it does not make sense to find it on the right.
  //
  if( lr == 1 && !found[0] )
    continue;

  // Find circle grids and centers therein:
  for( int s = 1; s <= maxScale; s++ ) {

    cv::Mat timg = img;
    if( s > 1 )
      resize( img, timg, cv::Size(), s, s, cv::INTER_CUBIC );
    found[lr] = cv::findCirclesGrid(
      timg,
```

```
        cv::Size(nx, ny),
        corners[lr],
        cv::CALIB_CB_ASYMMETRIC_GRID | cv::CALIB_CB_CLUSTERING
      );
      if( found[lr] || s == maxScale ) {
        cv::Mat mcorners( corners[lr] );
        mcorners *= (1./s);
      }
      if( found[lr] ) break;

    }
    if( displayCorners ) {

      cout << buf << endl;
      cv::Mat cimg;
      cv::cvtColor( img, cimg, cv::GRAY2BGR );

      // draw chessboard corners works for circle grids too
      cv::drawChessboardCorners(
        cimg, cv::Size(nx, ny), corners[lr], found[lr]
      );
      cv::imshow( "Corners", cimg );
      if( (cv::waitKey(0)&255) == 27 ) // Allow ESC to quit
        exit(-1);

    }
    else
      cout << '.';

    if( lr == 1 && found[0] && found[1] ) {

      objectPoints.push_back(boardModel);
      points[0].push_back(corners[0]);
      points[1].push_back(corners[1]);

    }
  }
  fclose(f);

  // CALIBRATE THE STEREO CAMERAS
  cv::Mat M1 = cv::Mat::eye( 3, 3, CV_64F );
  cv::Mat M2 = cv::Mat::eye( 3, 3, CV_64F );
  cv::Mat D1, D2, R, T, E, F;
  cout <<"\nRunning stereo calibration ...\n";
  cv::stereoCalibrate(
    objectPoints,
    points[0],
    points[1],
    M1, D1, M2, D2,
    imageSize, R, T, E, F,
    cv::TermCriteria(
      cv::TermCriteria::COUNT | cv::TermCriteria::EPS, 100, 1e-5
```

```
  ),
  cv::CALIB_FIX_ASPECT_RATIO
    | cv::CALIB_ZERO_TANGENT_DIST
    | cv::CALIB_SAME_FOCAL_LENGTH
);
cout <<"Done\n\n";

// CALIBRATION QUALITY CHECK
// because the output fundamental matrix implicitly
// includes all the output information,
// we can check the quality of calibration using the
// epipolar geometry constraint: m2^t*F*m1=0
vector< cv::Point3f > lines[2];

double avgErr = 0;
int nframes = (int)objectPoints.size();

for( i = 0; i < nframes; i++ ) {

  vector< cv::Point2f >& pt0 = points[0][i];
  vector< cv::Point2f >& pt1 = points[1][i];

  cv::undistortPoints( pt0, pt0, M1, D1, cv::Mat(), M1 );
  cv::undistortPoints( pt1, pt1, M2, D2, cv::Mat(), M2 );
  cv::computeCorrespondEpilines( pt0, 1, F, lines[0] );
  cv::computeCorrespondEpilines( pt1, 2, F, lines[1] );

  for( j = 0; j < N; j++ ) {
    double err = fabs(
      pt0[j].x*lines[1][j].x + pt0[j].y*lines[1][j].y + lines[1][j].z
    ) + fabs(
      pt1[j].x*lines[0][j].x + pt1[j].y*lines[0][j].y + lines[0][j].z
    );
    avgErr += err;
  }

}

cout << "avg err = " << avgErr/(nframes*N) << endl;

// COMPUTE AND DISPLAY RECTIFICATION
//
if( showUndistorted ) {

  cv::Mat R1, R2, P1, P2, map11, map12, map21, map22;

  // IF BY CALIBRATED (BOUGUET'S METHOD)
  //
  if( !useUncalibrated ) {
    stereoRectify(
      M1, D1, M2, D2,
      imageSize,
```

```
    R, T, R1, R2, P1, P2,
    cv::noArray(), 0
  );
  isVerticalStereo =
    fabs(P2.at<double>(1, 3)) > fabs(P2.at<double>(0, 3));
  // Precompute maps for cvRemap()
  initUndistortRectifyMap(
    M1, D1, R1, P1, imageSize, CV_16SC2, map11, map12
  );
  initUndistortRectifyMap(
    M2, D2, R2, P2, imageSize, CV_16SC2, map21, map22
  );
}
// OR ELSE HARTLEY'S METHOD
//
else {
  // use intrinsic parameters of each camera, but
  // compute the rectification transformation directly
  // from the fundamental matrix
  vector< cv::Point2f > allpoints[2];
  for( i = 0; i < nframes; i++ ) {
    copy(
      points[0][i].begin(),
      points[0][i].end(),
      back_inserter(allpoints[0])
    );
    copy(
      points[1][i].begin(),
      points[1][i].end(),
      back_inserter(allpoints[1])
    );
  }
  cv::Mat F = findFundamentalMat(
    allpoints[0], allpoints[1], cv::FM_8POINT
  );
  cv::Mat H1, H2;
  cv::stereoRectifyUncalibrated(
    allpoints[0], allpoints[1],
    F,
    imageSize,
    H1, H2,
    3
  );

  R1 = M1.inv()*H1*M1;
  R2 = M2.inv()*H2*M2;
  // Precompute map for cvRemap()
  //
  cv::initUndistortRectifyMap(
    M1, D1, R1, P1,
    imageSize,
    CV_16SC2,
```

```
    map11, map12
  );
  cv::initUndistortRectifyMap(
    M2, D2, R2, P2,
    imageSize,
    CV_16SC2,
    map21, map22
  );
}

// RECTIFY THE IMAGES AND FIND DISPARITY MAPS
//
cv::Mat pair;
if( !isVerticalStereo )
  pair.create( imageSize.height, imageSize.width*2, CV_8UC3 );
else
  pair.create( imageSize.height*2, imageSize.width, CV_8UC3 );

// Setup for finding stereo corrrespondences
//
cv::Ptr<cv::StereoSGBM> stereo = cv::StereoSGBM::create(
  -64, 128, 11, 100, 1000,
  32, 0, 15, 1000, 16,
  StereoSGBM::MODE_HH
);

for( i = 0; i < nframes; i++ ) {

  cv::Mat img1 = cv::imread( imageNames[0][i].c_str(), 0 );
  cv::Mat img2 = cv::imread( imageNames[1][i].c_str(), 0 );
  cv::Mat img1r, img2r, disp, vdisp;

  if( img1.empty() || img2.empty() )
    continue;

  cv::remap( img1, img1r, map11, map12, cv::INTER_LINEAR );
  cv::remap( img2, img2r, map21, map22, cv::INTER_LINEAR );
  if( !isVerticalStereo || !useUncalibrated ) {
    // When the stereo camera is oriented vertically,
    // Hartley method does not transpose the
    // image, so the epipolar lines in the rectified
    // images are vertical. Stereo correspondence
    // function does not support such a case.
    stereo->compute( img1r, img2r, disp);
    cv::normalize( disp, vdisp, 0, 256, cv::NORM_MINMAX, CV_8U );
    cv::imshow( "disparity", vdisp );
  }
  if( !isVerticalStereo )
  {
    cv::Mat part = pair.colRange(0, imageSize.width);
    cvtColor( img1r, part, cv::GRAY2BGR);
    part = pair.colRange( imageSize.width, imageSize.width*2 );
```

```
        cvtColor( img2r, part, cv::GRAY2BGR);

        for( j = 0; j < imageSize.height; j += 16 )
          cv::line(
            pair,
            cv::Point(0,j),
            cv::Point(imageSize.width*2,j),
            cv::Scalar(0,255,0)
          );
      }
      else {
        cv::Mat part = pair.rowRange(0, imageSize.height);
        cv::cvtColor( img1r, part, cv::GRAY2BGR );
        part = pair.rowRange( imageSize.height, imageSize.height*2 );
        cv::cvtColor( img2r, part, cv::GRAY2BGR );

        for( j = 0; j < imageSize.width; j += 16 )
          line(
            pair,
            cv::Point(j,0),
            cv::Point(j,imageSize.height*2),
            cv::Scalar(0,255,0)
          );
      }
      cv::imshow( "rectified", pair );
      if( (cv::waitKey()&255) == 27 )
        break;
    }
  }
}

int main(int argc, char** argv) {

  help( argv );
  int board_w = 9, board_h = 6;
  const char* board_list = "ch12_list.txt";
  if( argc == 4 ) {
    board_list = argv[1];
    board_w    = atoi( argv[2] );
    board_h    = atoi( argv[3] );
  }
  StereoCalib(board_list, board_w, board_h, true);
  return 0;

}
```

## Depth Maps from Three-Dimensional Reprojection

Many algorithms will just use the disparity map directly—for example, to detect whether objects are on (stick out from) a table. But for three-dimensional shape matching, three-dimensional model learning, robot grasping, and so on, we need the actual three-dimensional reconstruction or depth map. Fortunately, all the stereo machinery we've built up so far makes this easy. Recall the 4 × 4 reprojection matrix $Q$ introduced in the section on calibrated stereo rectification. Also recall that, given the disparity $d$ and a two-dimensional point $(x, y)$, we can derive the three-dimensional depth using

$$Q \cdot \begin{bmatrix} x \\ y \\ d \\ 1 \end{bmatrix} = \begin{bmatrix} X \\ Y \\ Z \\ W \end{bmatrix}$$

Here the three-dimensional coordinates are now given by: $(X/W, Y/W, Z/W)$. Remarkably, $Q$ encodes whether or not the cameras' lines of sight were converging (cross-eyed) as well as the camera baseline and the principal points in both images. As a result, we need not explicitly account for converging or frontal parallel cameras and may instead simply extract depth by matrix multiplication. OpenCV has two functions that do this for us. The first, which you are already familiar with, operates on an array of points and their associated disparities. It's called `cv::perspectiveTransform()`:

```
void cv::perspectiveTransform(
  cv::InputArray  src,     // Input 2 or 3 channel array (a list
                           //  of 2d or 3d vectors)
  cv::OutputArray dst,     // Output array, same size as 'src'
  cv::InputArray  Q        // 3x3 or 4x4 floaring point
                           //  transformation matrix
);
```

The second function (which we have not yet encountered) is `cv::reprojectImageTo3D()`, which operates on whole images:

```
void cv::reprojectImageTo3D(
  cv::InputArray  disparity, // Input disparity image, any of:
                             //  8U, S16, S32, or F32
  cv::OutputArray _3dImage,  // image: 3d location of each pixel
  cv::InputArray  Q,         // 4x4 Perspective Transformation
                             //  (from stereoRectify())
  cv::bool        handleMissingValues = false,  // map "unknowns" to
                                                //  large distance
  cv::int         ddepth              = -1      // depth for '_3dimage',
                                                //  can be any of:
                                                //  CV_16S, CV_32S,
                                                //  or CV_32F (default)
);
```

This routine takes the single-channel `disparity` image and transforms each pixel's $(x, y)$ coordinates along with that pixel's disparity (i.e., the vector $(x, y, d)^T$) to the corresponding three-dimensional point $(X/W, Y/W, Z/W)$ by using the 4 × 4 reprojection matrix `Q`. The output will be a three-channel image of the same size as the input. By default, this image will be of 32-bit floating-point type, but this can be controlled with the `ddepth` argument, which can be set to any of `CV_32F`, `CV_32S`, or `CV_16S`. The final argument, `handleMissingValues`, controls what `cv::reprojectImageTo3D` does with pixels in the disparity image for which no disparity could be computed. In the case in which `handleMissingValues` is `false`, these points will simply not appear in the output image; if it is `true`, then points will be generated, but they will be assigned a very large depth value (currently `10000`).

Of course, both functions let you pass an arbitrary perspective transformation (e.g., the canonical one) computed by `cv::stereoRectify` or a superposition of that and the arbitrary three-dimensional rotation, translation, and so on. The results of `cv::reprojectImageTo3D()` on an image of a mug and chair are shown in Figure 19-20.

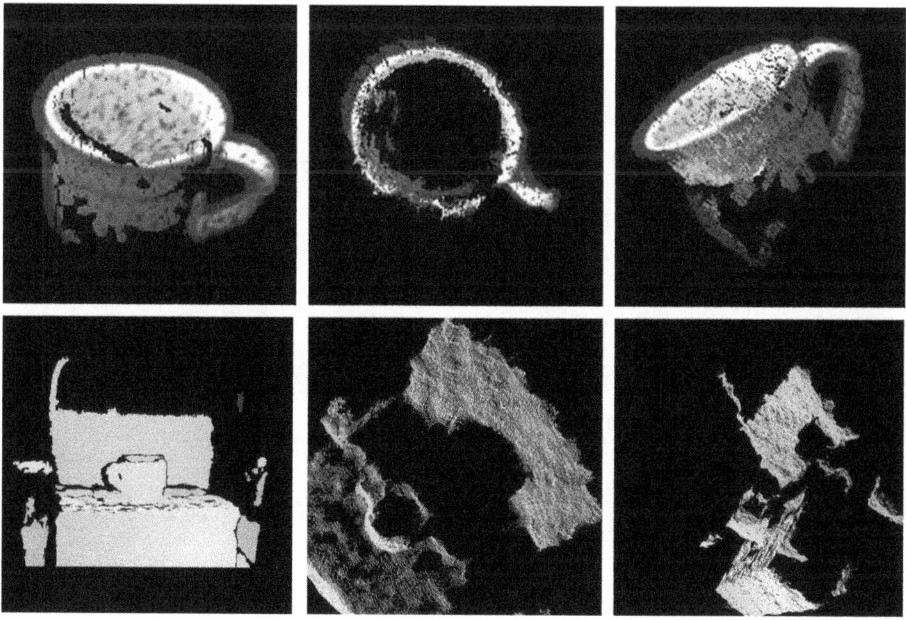

*Figure 19-20. Example output of depth maps (for a mug and a chair) computed using StereoBM and reprojectImageTo3D() (image courtesy of Willow Garage)*

# Structure from Motion

Structure from motion (SfM) is an important topic in mobile robotics as well as in the analysis of more general video imagery such as might come from a handheld video camera. The topic of structure from motion is a broad one, and a great deal of research has been done in this field. However, we can accomplish much by making one simple observation: in a static scene, an image taken by a camera that has moved is no different than an image taken by a second camera. Thus, all of our intuition, as well as our mathematical and algorithmic machinery, is immediately portable to this situation. Of course, the descriptor *static* is crucial, but in many practical situations the scene is either static or sufficiently static that the few moved points can be treated as outliers by robust fitting methods.

Consider the case of a camera moving through a building. If the environment is relatively rich in recognizable features, then we should be able to compute correspondences between consecutive frames. For example, we could find corresponding points with optical flow techniques such as `cv::calcOpticalFlowPyrLK()`. If we could track enough points from frame to frame, we could reconstruct the trajectory of the camera.[48] With this trajectory, we could then proceed to construct the overall three-dimensional structure of the building and the locations of all the aforementioned features in that building. The brand-new (at the time of writing) module in *opencv_contrib/modules/sfm*,[49] contains a ready-to-use SfM pipeline implementation together with tutorials of its use; see Appendix B. This code makes use of the libmv and Ceres libraries with instructions on how to download them.

# Fitting Lines in Two and Three Dimensions

A final topic of interest in this chapter is that of general line fitting. This can arise for many reasons and in many contexts. We have chosen to discuss it here because one especially frequent context in which line fitting arises is analyzing points in three dimensions (although the function described here can also fit lines in two dimensions). Line-fitting algorithms generally use statistically robust techniques [Inui03, Meer91, Rousseeuw87]. The OpenCV line-fitting algorithm `cv::fitLine()` can be used whenever line fitting is needed.

```
void cv::fitLine(
  cv::InputArray  points,    // 2d or 3d, can be Nx2, 2xN,
                             //  vector<Point2d> etc...
  cv::OutputArray line,      // Output lines, array of:
                             //  Vec4f (2d) or Vec6f (3d)
  int             distType,  // Distance type used (see Table 19-3)
  double          param,     // Parameter 'C' used by some distance
                             //  types (see Table 19-3)
  double          reps,      // Sufficient accuracy radius
  double          aeps       // Sufficient accuracy angle
);
```

48 The information we need is encoded in the essential matrix $E$, which can be computed from the fundamental matrix $F$ and the camera intrinsics matrix $M$. We would need to extract this information for every sequential pair of frames in the video stream.

49 This was the result of a Google Summer of Code (GSoC) project in 2015.

The array `points` can be in any of the usual forms, and may contain either two- or three-dimensional points. The argument `distType` indicates the distance metric that is to be minimized across all of the points (see Table 19-3).

*Table 19-3. Metrics used for computing distType values*

| Value of distType | Metric | |
|---|---|---|
| `cv::DIST_L1` | $\rho(r) = r$ | |
| `cv::DIST_L2` | $\rho(r) = \frac{r^2}{2}$ | |
| `cv::DIST_L12` | $\rho(r) = \left[\sqrt{1 + \frac{r^2}{2}} - 1\right]$ | |
| `cv::DIST_FAIR` | $\rho(r) = c^2\left[\frac{r}{c} - \log\left(1 + \frac{r}{c}\right)\right]$ | $c = 1.3998$ |
| `cv::DIST_WELSCH` | $\rho(r) = \frac{c^2}{2}\left[1 - \exp - \left(\frac{r}{c}\right)^2\right]$ | $c = 1.3998$ |
| `cv::DIST_HUBER` | $\rho(r) = \begin{cases} \frac{r^2}{2} & r < c \\ c\left(r - \frac{c}{2}\right) & r \geq c \end{cases}$ | $c = 1.345$ |

The parameter `param` is used to set the parameter $c$ listed in Table 19-3. This can be left set to `0`, in which case the listed value from the table will be selected. We'll get back to `reps` and `aeps` after describing `line`.

The argument `line` is the location where the result is stored. If `points` contains two-dimensional points, then `line` will be an STL-style array of four floating-point numbers (e.g., `cv::Vec4f`). If `points` contains three-dimensional points, then `line` will be an STL-style array of six floating-point numbers (e.g., `cv::Vec6f`). In the former case, the return values will be $(v_x, v_y, x_0, y_0)$, where $(v_x, v_y)$ is a normalized vector parallel to the fitted line and $(x_0, y_0)$ is a point on that line. Similarly, in the latter (three-dimensional) case, the return values will be $(v_x, v_y, v_z, x_0, y_0, z_0)$, where $(v_x, v_y, v_z)$ is a normalized vector parallel to the fitted line and $(x_0, y_0, z_0)$ is a point on that line. Given this line representation, the estimation accuracy parameters `reps` and `aeps` are as follows: `reps` is the requested accuracy of `x0, y0[, z0]` estimates and `aeps` is the requested angular accuracy for `vx, vy[, vz]`. The OpenCV documentation recommends values of `0.01` for both accuracy values.

We will end with a program for line fitting, shown in Example 19-4. In this code, we first synthesize some two-dimensional points noisily around a line, then add some random points that have nothing to do with the line (i.e., *outlier* points), and finally fit a line to the points and display it. The `cv::fitLine()` routine is good at ignoring the outlier points; this is important in real applications, where some measurements might be corrupted by high noise, sensor failure, and so on.

*Example 19-4. Two-dimensional line fitting*

```
#include "opencv2/opencv.hpp"
#include <iostream>
#include <math.h>

using namespace std;

void  help( argv) {
  ...
}

int main( int argc, char** argv ) {

  cv::Mat img(500, 500, CV_8UC3);
  cv::RNG rng(-1);
  help( argv );
  for(;;) {

    char key;
    int   i, count = rng.uniform(0,100) + 3, outliers = count/5;
    float a        = (float) rng.uniform(0., 200.);
    float b        = (float) rng.uniform(0., 40.);
    float angle    = (float) rng.uniform(0., cv::PI);
    float cos_a    = cos(angle), sin_a = sin(angle);
    cv::Point pt1, pt2;
    vector< cv::Point > points( count );
    cv::Vec4f line;
    float d, t;

    b = MIN( a*0.3f, b );

    // generate some points that are close to the line
    for( i = 0; i < count - outliers; i++ ) {
      float x = (float)rng.uniform(-1.,1.)*a;
      float y = (float)rng.uniform(-1.,1.)*b;
      points[i].x = cvRound(x*cos_a - y*sin_a + img.cols/2);
      points[i].y = cvRound(x*sin_a + y*cos_a + img.rows/2);
    }

    // generate outlier points
    for( ; i < count; i++ ) {
      points[i].x = rng.uniform(0, img.cols);
      points[i].y = rng.uniform(0, img.rows);
    }

    // find the optimal line
    cv::fitLine( points, line, cv::DIST_L1, 1, 0.001, 0.001);

    // draw the points
    img = cv::Scalar::all(0);
    for( i = 0; i < count; i++ )
```

```
      cv::circle(
        img,
        points[i],
        2,
        i < count - outliers
          ? cv::Scalar(0, 0, 255)
          : cv::Scalar(0,255,255),
        cv::FILLED,
        cv::AA,
        0
      );

    // ... and the long enough line to cross the whole image
    d = sqrt( (double)line[0]*line[0] + (double)line[1]*line[1] );
    line[0] /= d;
    line[1] /= d;
    t = (float)(img.cols + img.rows);
    pt1.x = cvRound(line[2] - line[0]*t);
    pt1.y = cvRound(line[3] - line[1]*t);
    pt2.x = cvRound(line[2] + line[0]*t);
    pt2.y = cvRound(line[3] + line[1]*t);
    cv::line( img, pt1, pt2, cv::Scalar(0,255,0), 3, cv::AA, 0 );

    cv::imshow( "Fit Line", img );

    key = (char) cv::waitKey(0);
    if( key == 27 || key == 'q' || key == 'Q' ) // 'ESC'
      break;
  }
  return 0;
}
```

# Summary

We began this chapter with a review of the geometry of a camera system, and we learned that the basic mapping that takes points in the three-dimensional world to the two-dimensional world of the imager is a projective transformation. We learned that in some cases—specifically those in which we know that a set of points lies on a plane—this transformation can be inverted.

Even though the mapping from the three-dimensional world to the image plane is not generally invertable, we learned that if we could see the same set of points in many images, we could reconstruct the three-dimensional scene or the pose of a known object. The separately loaded directory, *opencv_contrib*, described in Appendix B, has further calibration algorithms for omni cameras and multicameras (in `ccalib`), for different types of calibration patterns (`aruco` and `ccalib`), and for color balance and denoising routines (`xphoto`).

Finally, we saw how this same geometrical information could be used to construct stereoscopic depth measurements. In order to do this reliably, we had to compute the exact relationship between the stereo imaging cameras—a process called *stereo calibration*. Once the stereo cameras were calibrated, we could use one of two available algorithms provided by OpenCV to compute depth. The block matching algorithm was faster, but provided less complete coverage of the scene, while the semi-global block matching algorithm gave much better results at the cost of substantially more computing time.

Finally, we covered a number of other useful functions for handling points and lines in three dimensions, such as projective transformation, reprojection, and line fitting.

# Exercises

1. Affine and projective (perspective projection) transform (see Figure 11-3):

$$\begin{pmatrix} a_1 & a_2 & b_1 \\ a_3 & a_4 & b_2 \\ c_1 & c_2 & 1 \end{pmatrix}$$

   The *a*s form the rotation, scaling, and skew matrix, and make up the affine transform. The *b*s form an $(x, y)$ translation vector, the *c*s form the perspective projection vector, and together they all make up the full perspective projection matrix.

   a. Imagine a camera facing a chessboard. What camera movements can be modeled equally as both an affine and a perspective projection transform?

   b. How many points on a plane define an affine transform? How many points define a perspective projection transform?

   c. In affine projection: Do lines stay lines? Does their length stay the same? Do parallel lines stay parallel? If two lines intersect in the original image, do they always intersect after affine projection?

   d. In perspective projection: Do lines stay lines? Does their length stay the same? Do parallel lines stay parallel? If two lines intersect in the original image, do they always intersect after perspective projection?

2. Calibrate a camera using `cv::calibrateCamera()` and at least 15 images of chessboards. Then use `cv::projectPoints()` to project an arrow orthogonal to the chessboards (the surface normal) into each of the chessboard images using the rotation and translation vectors from the camera calibration.

3. *Three-dimensional joystick*: Use a simple known object with at least four measured, noncoplanar, trackable feature points as input into the `cv::solvePnP()`

algorithm. Use the object as a three-dimensional joystick to move a little stick figure in the image.

4. In the text's bird's-eye-view example, with a camera above the plane looking out horizontally along the plane, we saw that the homography of the ground plane had a horizon line beyond which the homography wasn't valid. How can an infinite plane have a horizon? Why doesn't it just appear to go on forever?

   Hint: draw lines to an equally spaced series of points on the plane going out away from the camera. How does the angle from the camera to each next point on the plane change from the angle to the point before?

5. Implement a bird's-eye view in a video camera looking at the ground plane. Run it in real time and explore what happens as you move objects around in the normal image versus the bird's-eye-view image.

6. Set up two cameras or a single camera that you move between taking two images.

   a. Compute, store, and examine the fundamental matrix.

   b. Repeat the calculation of the fundamental matrix several times. How stable is the computation?

7. If you had a calibrated stereo camera and were tracking moving points in both cameras, can you think of a way of using the fundamental matrix to find tracking errors?

8. Compute and draw epipolar lines on two cameras set up to do stereo.

9. Set up two video cameras; implement stereo calibration and rectification and experiment with depth accuracy.

   a. What happens when you bring a mirror into the scene?

   b. Vary the amount of texture in the scene and report the results.

   c. Try different disparity methods and report on the results.

10. Set up stereo cameras and wear something that is textured over one of your arms. Fit a line to your arm using all the `distType` methods. Compare the accuracy and reliability of the different methods.

11. Imagine sending a downward-looking camera attached to a helium balloon up on a clear day. The balloon flies upward until at some height the balloon pops. Using just the camera and its frame rate, how could you find out how high the camera was when the balloon popped?

CHAPTER 20

# The Basics of Machine Learning in OpenCV

In this chapter, we'll begin a discussion of the machinery that is used to turn *vision* into perception—in other words, the machinery that turns the visual inputs into meaningful visual semantics.

In the previous chapters we have discussed how to turn 2D or 2D+3D sensor information into features, clusters, or geometric information. In the next three chapters, we'll use the results of these techniques to turn features, segmentations, and their geometry into recognition of scenes or objects; it is this step that turns raw information into a percept: *what* the machine is seeing and *where* it is relative to the camera.

In this chapter we will cover the basics of machine learning, focusing mainly on what it is. We will look at some simple machine learning capabilities of the library that form a good starting point for understanding the basic ideas in machine learning as a whole. In the next chapter, we will get into more detail about how modern machine learning methods are implemented in the library.[1]

1 Note that machine learning, as with so many things, has been extended in the experimental *opencv_contrib* code as described in Appendix B. For more details, see the deep neural network repositories *cnn_3dobj* and *dnn*.

## What Is Machine Learning?

The goal of *machine learning* (ML)[2] is to turn data into information. After learning from a collection of data, we want a machine to be able to answer questions about the data: What other data is most similar to this data? Is there a car in the image? What ad will the user respond to? There is often a cost component, so this question could become: "Of our most profitable products, which one will the user most likely buy if we show them an ad for it?" Machine learning turns data into information by extracting rules or patterns from that data.

### Training and Test Sets

The sort of machine learning we are interested in works on raw numerical data, such as temperature values, stock prices, and color intensities. The data is often preprocessed into *features*.[3] We might, for example, take a database of 10,000 face images, run an edge detector on the faces, and then collect features such as edge direction, edge strength, and offset from face center for each face. We might obtain 500 such values per face or a *feature vector* of 500 entries. We could then use machine learning techniques to construct some kind of model from this collected data. If we want to see only how faces fall into different groups (wide, narrow, etc.), then a *clustering* algorithm would be the appropriate choice. If we want to learn to predict the age of a person from, for example, the pattern of edges detected on his or her face, then a *classifier* algorithm would be appropriate. To meet our goals, machine learning algorithms analyze our collected features and adjust weights, thresholds, and other model parameters to maximize performance according to those goals. This process of parameter adjustment to meet a goal is what we mean by the term *learning*.

It is always important to know how well machine learning methods are working, and this can be a subtle task. Traditionally, one breaks up the available data set into a large training set (perhaps 9,000 faces, in our example) and a smaller test set (the remaining 1,000 faces). We can then run our classifier over the training set to learn our age prediction model given the data feature vectors. When we are done, we can test the age prediction classifier on the remaining images in the test set.

---

2 Machine learning is a vast topic. OpenCV deals mostly with statistical machine learning rather than subjects such as Bayesian networks, Markov random fields, and graphical models. Some good texts in machine learning include those by Hastie, Tibshirani, and Friedman [Hastie01]; Duda and Hart [Duda73]; Duda, Hart, and Stork [Duda00]; and Bishop [Bishop07]. For discussions on how to parallelize machine learning, see Ranger et al. [Ranger07] and Chu et al. [Chu07].

3 As of this writing, OpenCV does not include deep learning since, although promising, such techniques are still too new to know what to include. However, *Convolutional Neural Networks* [Fukushima80; LeCun98a; Ciresan11] are definitely a future candidate.

The test set is not used in training, and we do not let the classifier "see" the test set age labels. Only after training, we run the classifier over each of the 1,000 faces in the test set of data and record how well the ages it predicts (based on the feature vector) match the actual ages. If the classifier does poorly, we might try adding new features to our data or consider a different type of classifier. We'll see in this chapter that there are many kinds of classifiers and many algorithms for training them.

If the classifier does well, we now have a potentially valuable model that we can deploy on data in the real world. Perhaps this system will be used to set the behavior of a video game based on age. As the person prepares to play, his or her face will be processed into 500 features (edge direction, edge strength, offset from face center, etc.). This data will be passed to the classifier; the age it returns will set the game play behavior accordingly. After it has been deployed, the classifier sees faces that it never saw before and makes decisions according to what it learned on the training set.

Finally, when developing a classification system, we often use a validation data set. Sometimes, testing the whole system at the end is too big a step to take. We often want to tweak parameters along the way before submitting our classifier to final testing. We might do this by breaking our example 10,000-face data set into three parts: a training set of 8,000 faces, a validation set of 1,000 faces, and a test set of 1,000 faces. Now, while we're running through the training data set, we can "sneak" pretests on the validation data to see how we are doing. Only when we are satisfied with our performance on the validation set do we run the classifier on the test set for final judgment.

It might strike you that you could do better than to train with 8,000 examples and validate on 1,000 in a 9,000-example training set. If so, you would be correct. The standard practice in such cases is actually to repeat this partitioning multiple times. In this case, nine times would make the most sense. In each case you would set aside a different group of 1,000 points to use for validation and train on the remaining 8,000. This process is called *k-fold cross-validation*, with *k-fold* meaning "done in *k* variations"—in our example, nine.

## Supervised and Unsupervised Learning

Data sometimes has no labels; for example, we might just want to see what kinds of groups faces naturally form based on our edge-detection information. Sometimes the data has labels, such as the age of the person featured in each image. What this means is that machine learning data may be *supervised* (i.e., may utilize a teaching "signal" or "label" that goes with the data feature vectors), or it may be *unsupervised*, in which case the machine learning algorithm has no access to such labels and is expected to figure out the structure of the data on its own (see Figure 20-1).

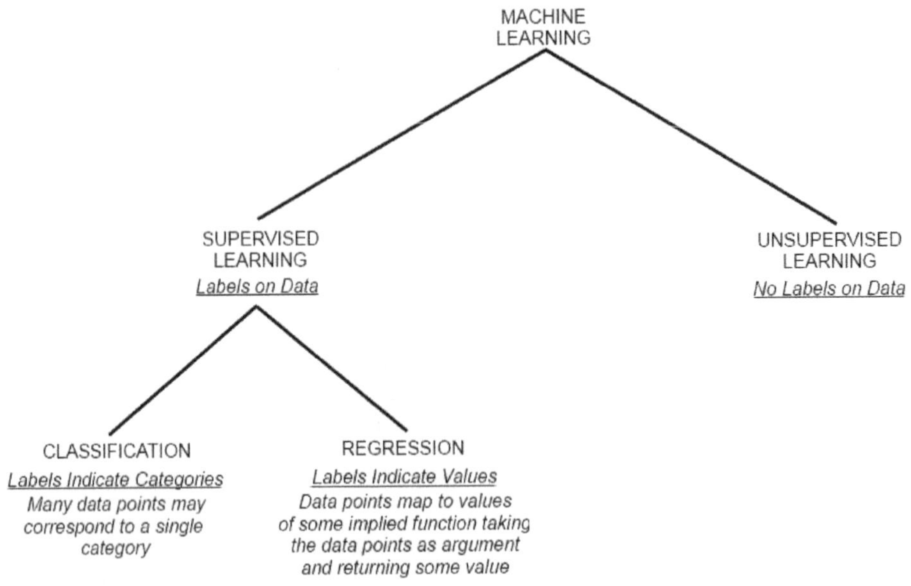


*Figure 20-1. Machine learning includes two broad subcategories: supervised and unsupervised learning. Supervised learning itself has two important subcomponents: classification and regression*

Supervised learning can be *categorical*, such as learning to associate a name with a face, or the data can have *numeric* or *ordered* labels, such as age. When the data has names (categories) as labels, we say we are doing *classification*. When the data is numeric, we say we are doing *regression*: trying to fit a numeric output given some categorical or numeric input data.

Supervised learning also comes in shades of gray: it can involve one-to-one pairing of labels with data vectors or it may consist of *reinforcement learning* (sometimes called *deferred learning*). In reinforcement learning, the data label (also called the *reward* or *punishment*) can come long after the individual data vectors were observed. When a mouse is running down a maze to find food, the mouse may face a series of turns before it finally finds the food, its reward. That reward must somehow cast its influence back on all the sights and actions that the mouse took before finding the food. Reinforcement learning works the same way: the system receives a delayed signal (a reward or a punishment) and tries to infer a behavior (formally called a *policy*) for future runs. In this case, the policy being learned is actually the way of making decisions; e.g., which way to go at each step through the maze. Supervised learning can also have partial labeling, where some labels are missing (this is also called *semisupervised learning*), or it might have noisy labels, where some of the supplied labels are just wrong. Most ML algorithms handle only one or two of the situations just described. For example, the ML algorithms might handle classification but not regression; the algorithm might be able to do semisupervised learning but not reinforcement learning; the algorithm might be able to deal with numeric but not categorical data; and so on.

In contrast, often we don't have labels for our data and are interested in seeing whether the data falls naturally into groups. The algorithms for this kind of unsupervised learning are often called *clustering algorithms*. In this situation, the goal is to group unlabeled data vectors that are "close" (in some predetermined or possibly even learned sense). We might just want to see how faces are distributed: do they form clumps of thin, wide, long, or short faces? If we're looking at cancer data, do some cancers cluster into groups having different chemical signals? Unsupervised clustered data is also often used to form a feature vector for a higher-level supervised classifier. We might first cluster faces into face types (wide, narrow, long, short) and then use that as an input, perhaps with other data, such as average vocal frequency, to predict the gender of a person.

These two common machine learning tasks, classification and clustering, overlap with two of the most common tasks in computer vision, recognition and segmentation. This is sometimes referred to as the "what" and the "where." That is, we often want our computer to name the object in an image (recognition, or "what") and also to say where the object appears (segmentation, or "where"). Because computer vision makes such heavy use of machine learning, OpenCV includes many powerful machine learning algorithms in the ML libraries, located in the *.../opencv/modules/ml* and the *.../opencv/modules/flann* directories.

The OpenCV machine learning code is general; that is, although it is highly useful for vision tasks, the code itself is not specific to vision. One could learn, say, genomic sequences using the appropriate routines. Of course, our concern here is mostly with object recognition given feature vectors derived from images.

## Generative and Discriminative Models

Many algorithms have been devised to perform classification and clustering. OpenCV supports some of the most useful currently available statistical approaches to machine learning. Probabilistic approaches to machine learning, such as Bayesian networks or graphical models, are less well supported in OpenCV, partly because they are newer and still under active development. OpenCV tends to support *discriminative algorithms*, which give us the probability of the label given the data, P(L|D), rather than *generative algorithms*, which give the distribution of the data given the label, P(D|L). Although the distinction is not always clear, discriminative models are good for yielding predictions given the data, while generative models are good for giving you more powerful representations of the data or for conditionally synthesizing new data (think of "imagining" an elephant; you'd be generating data given a condition "elephant").

It is often easier to interpret a generative model because it models (correctly or incorrectly) the cause of the data. Discriminative learning often comes down to making a decision based on some threshold that may seem arbitrary. For example, suppose a patch of road is identified in a scene partly because its color "red" is less than 125. But does this mean that red = 126 is definitely not road? Such issues can be hard to interpret. With generative models you are usually dealing with conditional distributions of data given the categories, so you can develop a feel for what it means to be "close" to the resulting distribution.

# OpenCV ML Algorithms

The machine learning algorithms included in OpenCV are given in Table 20-1. Many of the algorithms listed are in the `ML` module; Mahalanobis and K-means are in the `core` module; face detection and object detection methods are in the `objdetect` module; FLANN takes a dedicated module, `flann`; and other algorithms are placed in *opencv_contrib*.

*Table 20-1. Machine learning algorithms supported in OpenCV; original references to the algorithms are provided after the descriptions*

| Algorithm | Comment |
|---|---|
| Mahalanobis | A distance measure that accounts for the "stretchiness" of the data space by dividing out the covariance of the data. If the covariance is the identity matrix (identical variance), then this measure is identical to the Euclidean distance measure [Mahalanobis36]. |
| K-means | An unsupervised clustering algorithm that represents a distribution of data using *K* centers, where *K* is chosen by the user. The difference between this algorithm and expectation maximization (described shortly) is that here the centers are not Gaussian and the resulting clusters look more like soap bubbles, since centers (in effect) compete to "own" the closest data points. These cluster regions are often used as sparse histogram bins to represent the data. Invented by Steinhaus [Steinhaus56], as used by Lloyd [Lloyd57]. |
| Normal/Naïve Bayes classifier | A generative classifier in which features are assumed to be Gaussian distributed and statistically independent from one another, a strong assumption that is generally not true. For this reason, it's often called a "naïve Bayes" classifier. However, this method often works surprisingly well. Original mention [Maron61; Minsky61]. |
| Decision trees | A discriminative classifier. The tree finds one data feature and a threshold at the current node that best divides the data into separate classes. The data is split and we recursively repeat the procedure down the left and right branches of the tree. Though not usually the top performer, it's often the first thing you should try because it is fast and has high functionality [Breiman84]. |
| Expectation maximization (EM) | A generative unsupervised algorithm that is used for clustering. It will fit *N* multidimensional Gaussians to the data, where *N* is chosen by the user. This can be an effective way to represent a more complex distribution with only a few parameters (means and variances). Often used in segmentation. Compare with K-means listed previously [Dempster77]. |

| Algorithm | Comment |
|---|---|
| Boosting | A discriminative group of classifiers. The overall classification decision is made from the combined weighted classification decisions of the group of classifiers. In training, we learn the group of classifiers one at a time. Each classifier in the group is a "weak" classifier (only just above chance performance). These weak classifiers are typically composed of single-variable decision trees called *stumps*. In training, the decision stump learns its classification decisions from the data and also learns a weight for its "vote" from its accuracy on the data. Between training each classifier one by one, the data points are reweighted so that more attention is paid to data points where errors were made. This process continues until the total error over the data set, arising from the combined weighted vote of the decision trees, falls below a set threshold. This algorithm is often effective when a large amount of training data is available [Freund97]. |
| Random trees | A discriminative forest of many decision trees, each built down to a large or maximal splitting depth. During learning, each node of each tree is allowed to choose splitting variables only from a random subset of the data features. This helps ensure that each tree becomes a statistically independent decision maker. In run mode, each tree gets an unweighted vote. This algorithm is often very effective and can also perform regression by averaging the output numbers from each tree implemented [Ho95; Criminisi13; Breiman01]. |
| K-nearest neighbors | The simplest possible discriminative classifier. Training data are simply stored with labels. Thereafter, a test data point is classified according to the majority vote of its *K* nearest other data points (in a Euclidean sense of nearness). This is probably the simplest thing you can do. It is often effective but it is slow and requires lots of memory [Fix51], but see the FLANN entry. |
| Fast Approximate Nearest Neighbors (FLANN)[a] | OpenCV includes a full fast approximate nearest neighbor library developed by Marius Muja [Muja09]. This allows fast approximations to nearest neighbor and *K* nearest neighbor matching. |
| Support vector machine (SVM) | A discriminative classifier that can also do regression. A distance function between any two data points in a higher-dimensional space is defined. (Projecting data into higher dimensions makes the data more likely to be linearly separable.) The algorithm learns separating hyperplanes that maximally separate the classes in the higher dimension. It tends to be among the best with limited data, losing out to boosting or random trees only when large data sets[b] are available [Vapnik95]. |
| Face detector/cascade classifier | An object detection application based on a clever use of boosting. The OpenCV distribution comes with a trained frontal face detector that works remarkably well. You may train the algorithm on other objects with the software provided. You may also use other features or create your own features for this classifier. It works well for rigid objects and characteristic views. After its inventors, this classifier is also commonly known as a "Viola-Jones Classifier" [Viola04]. |
| Waldboost | A derivative of the cascade method of Viola (see the preceding entry), Waldboost is an object detector that is very fast and often outperforms the traditional cascade classifier for a variety of tasks [Sochman05]. It is in *.../opencv_contrib/modules*. |
| Latent SVM | The Latent SVM method uses a parts-based model to identify composite objects based on recognizing the individual components of the object and learning a model of how those components should expect to be found relative to one another [Felzenszwalb10]. |
| Bag of Words | The Bag of Words method generalizes techniques heavily used in document classification to visual image classification. This method is powerful because it can be used to identify not only individual objects, but often scenes and environments as well. |

[a] Since you are wondering, the *L* in *FLANN* stands for *library* (FLANN = Fast Library for Approximate Nearest Neighbors).

[b] What is "large data"? There is no answer, since it depends on how fast the underlying generating process changes. But here's a very crude rule of thumb: 10 data points per category/object per meaningful dimension (feature). So, two classes, three dimensions needs at least 2*10*10*10 = 2,000 data points to be "large" for that problem.

## Using Machine Learning in Vision

In general, all the algorithms in Table 20-1 take as input a data vector made up of many features, where the number of features might well be in the thousands. Suppose your task is to recognize a certain type of object—for example, a person. The first problem that you will encounter is how to collect and label training data that falls into positive (there is a person in the scene) and negative (no person) cases. You will soon realize that people appear at different scales: their image may consist of just a few pixels, or you may be looking at an ear that fills the whole screen. Even worse, people will often be occluded: a man inside a car; a woman's face; one leg peeking from behind a tree. You need to define what you actually mean by saying a person is in the scene.

Next, you have the problem of collecting data. Do you collect it from a security camera, go to photo-sharing websites and attempt to find "person" labels, or both (and more)? Do you collect movement information? Do you collect other information, such as whether a gate in the scene is open, the time, the season, the temperature? An algorithm that finds people on a beach might fail on a ski slope. You need to capture the variations in the data: different views of people, lightings, weather conditions, shadows, and so on.

After you have collected lots of data, how will you label it? You must first decide what you mean by "label." Do you want to know where the person is in the scene? Are actions (running, walking, crawling, following) important? You might end up with a million images or more. How will you label all that? There are many tricks, such as doing background subtraction in a controlled setting and collecting the segmented foreground humans who come into the scene. You can use data services to help in classification; for example, you can pay people to label your images through Amazon's Mechanical Turk (*http://www.mturk.com/mturk/welcome*). If you arrange tasks to be simple, you can get the cost down to somewhere around a penny per label. Finally, you may use GPUs and/or clusters of computers to render objects/people/faces/hands using computer graphics. Using camera model parameters, you can often generate very realistic images where ground truth is known since you generated the data.

After labeling the data, you must decide which features to extract from the objects. Again, you must know what you are after. If people always appear right side up, there's no reason to use rotation-invariant features and no reason to try to rotate the objects beforehand. In general, you must find features that express some invariance in the objects, such as scale-tolerant histograms of gradients or colors or the popular SIFT features.[4] If you have background scene information, you might want to first

4 See Lowe's SIFT keypoint feature demo (*http://www.cs.ubc.ca/~lowe/keypoints/*).

remove it to make other objects stand out. You then perform your image processing, which may consist of normalizing the image (rescaling, rotation, histogram equalization, etc.) and computing many different feature types. The resulting data vectors are each given the label associated with that object, action, or scene.

Once the data is collected and turned into feature vectors, you often want to break up the data into training, (possibly) validation, and test sets. As we saw earlier, is a "best practice" to do your learning, validation, and testing within a cross-validation framework. Recall that there the data is divided into $K$ subsets and you run many training (possibly validation) and test sessions, where each session consists of different sets of data taking on the roles of training (validation) and test.[5] The test results from these separate sessions are then averaged to get the final performance result. Cross-validation gives a more accurate picture of how the classifier will perform when deployed in operation on novel data. (We'll have more to say about this in what follows.)

Now that the data is prepared, you must choose your classifier. Often this choice is dictated by computational, data, or memory considerations. For some applications, such as online user preference modeling, you must train the classifier rapidly. In this case, nearest neighbors, normal Bayes, or decision trees would be a good choice. If memory is a consideration, decision trees or neural networks are space efficient. If you have time to train your classifier but it must run quickly, neural networks are a good choice, as are naïve Bayes classifiers and support vector machines. If you have time to train and some time to run, but need high accuracy, then boosting and random trees are likely to fit your needs. If you just want an easy, understandable sanity check that your features are chosen well, then decision trees or nearest neighbors are good bets. For best "out of the box" classification performance, try boosting or random trees first.

There is no "best" classifier (see *http://en.wikipedia.org/wiki/No_free_lunch_theorem*). Averaged over all possible types of data distributions, all classifiers perform the same. Thus, we cannot say which algorithm in Table 20-1 is the "best." Over any given data distribution or set of data distributions, however, there is a best classifier. Thus, when faced with real data it's a good idea to try many classifiers. Consider your purpose: Is it just to get the right score, or is it to interpret the data? Do you seek fast computation, small memory requirements, or confidence bounds on the decisions? Different classifiers have different properties along these dimensions.

5 One typically does the train (possibly validation) and test cycle 5 to 10 times.

## Variable Importance

Two of the algorithms in Table 20-1 allow you to assess a variable's importance.[6] Given a vector of features, how do you determine the importance of those features for classification accuracy? Binary decision trees do this directly: you train them by selecting which variable best splits the data at each node. The top node's variable is the most important variable; the next-level variables are the second most important, and so on.[7] Random trees can measure variable importance using a technique developed by Leo Breiman [Breiman02]; this technique can be used with any classifier, but so far it is implemented only for decision and random trees in OpenCV.

One use of variable importance is to reduce the number of features your classifier must consider. Starting with many features, you train the classifier and then find the importance of each feature relative to the other features. You can then discard unimportant features. Eliminating unimportant features improves speed (since it eliminates the processing it took to compute those features) and makes training and testing quicker. Also, if you don't have enough data, which is often the case, then eliminating unimportant variables can increase classification accuracy; this yields faster processing with better results.

Breiman's variable importance algorithm runs as follows.

1. Train a classifier on the training set.
2. Use a validation or test set to determine the accuracy of the classifier.
3. For every data point and a chosen feature, randomly choose a new value for that feature from among the values the feature has in the rest of the data set (called "sampling with replacement"). This ensures that the distribution of that feature will remain the same as in the original data set, but now the actual structure or meaning of that feature is erased (because its value is chosen at random from the rest of the data).[8]
4. Train the classifier on the altered set of training data and then measure the accuracy of classification on the altered test or validation data set. If randomizing a feature hurts accuracy a lot, then that feature is very important. If randomizing a

6 This is commonly known as *variable importance*—meaning the importance of a variable, not importance that varies or fluctuates.

7 This technique by itself would be very sensitive to noise. What binary trees really do is build surrogate splits (other features to split on that result in almost the same decisions) for each node and compute the importance over all splits in all the nodes.

8 The actual implementation for random trees shuffles the feature values instead of generating completely new values.

feature does not hurt accuracy much, then that feature is of little importance and is a candidate for removal.

5. Restore the original test or validation data set and try the next feature until we are done. The result is an ordering of each feature by its importance.

This procedure is built into random trees and decision trees. Thus, you can use random trees or decision trees to decide which variables you will actually use as features; then you can use the slimmed-down feature vectors to train the same (or another) classifier.

## Diagnosing Machine Learning Problems

Getting machine learning to work well can be more of an art than a science. Algorithms often "sort of" work but not quite as well as you need them to. That's where the art comes in; you must figure out what's going wrong in order to fix it. Although we can't go into all the details here, we'll give an overview of some of the more common problems you might encounter.[9]

First, some rules of thumb: more data beats less data, and better features beat better algorithms. If you design your features well—maximizing their independence from one another and minimizing how they vary under different conditions—then almost any algorithm will work well. Beyond that, there are three common problems:

*Bias*
: Your model assumptions are too strong for the data, so the model won't fit well.

*Variance*
: Your algorithm has memorized the data *including* the noise, so it can't generalize.

*Bugs*
: It is not uncommon for machine learning code that contains seemingly severe bugs to "learn its way around" the bugs, often yielding only degraded performance when absolute failure would be expected.

Figure 20-2 shows the basic setup for statistical machine learning. Our job is to model the true function *f* that transforms the underlying inputs to some output. This function may be a regression problem (e.g., predicting a person's age from their face[10]) or

---

9 Professor Andrew Ng at Stanford University gives the details in a web lecture entitled "Advice for Applying Machine Learning" (*http://www.stanford.edu/class/cs229/materials/ML-advice.pdf*).

10 An astute observer might note that since ages are likely to be reported as integers, this could be a classification problem rather than a regression problem. However, it is not the continuousness of the output that makes a problem a regression problem, it is the orderability of the output. Thus, even for integer ages, this is a regression problem.

a category prediction problem (e.g., identifying a person given their facial features). For problems in the real world, noise and unconsidered effects can cause the observed outputs to differ from the theoretical outputs. For example, in face recognition we might learn a model of the measured distance between eyes, mouth, and nose to identify a face. But lighting variations from a nearby flickering bulb might cause noise in the measurements, or a poorly manufactured camera lens might cause a systematic distortion in the measurements that wasn't considered as part of the model. These effects will cause accuracy to suffer.

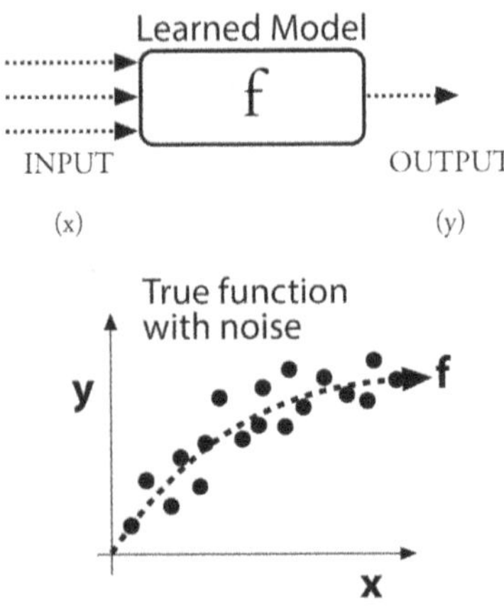


*Figure 20-2. Setup for statistical machine learning: we train a classifier to fit a data set; the true model f is almost always corrupted by noise or unknown influences*

Figure 20-3 shows model bias (resulting in underfitting) and variance (resulting in overfitting) of data in the upper two panels and the consequences in terms of error with training set size in the lower two panels. On the left side of Figure 20-3 we attempt to train a classifier to predict the data in the lower panel of Figure 20-2. If we use a model that's too restrictive—indicated here by the heavy, straight dashed line—then we can never fit the underlying true parabola *f* indicated by the thinner dashed line. Thus, the fit to both the training data and the test data will be poor, even with a lot of data. In this case we have bias because both training and test data are predicted poorly. On the right side of Figure 20-3 we fit the training data exactly, but this produces a nonsense function that fits every bit of noise. Thus, it memorizes the training data as well as the noise in that data. Once again, the resulting fit to the test data is poor. Low training error combined with high test error indicates a variance (overfit) problem.

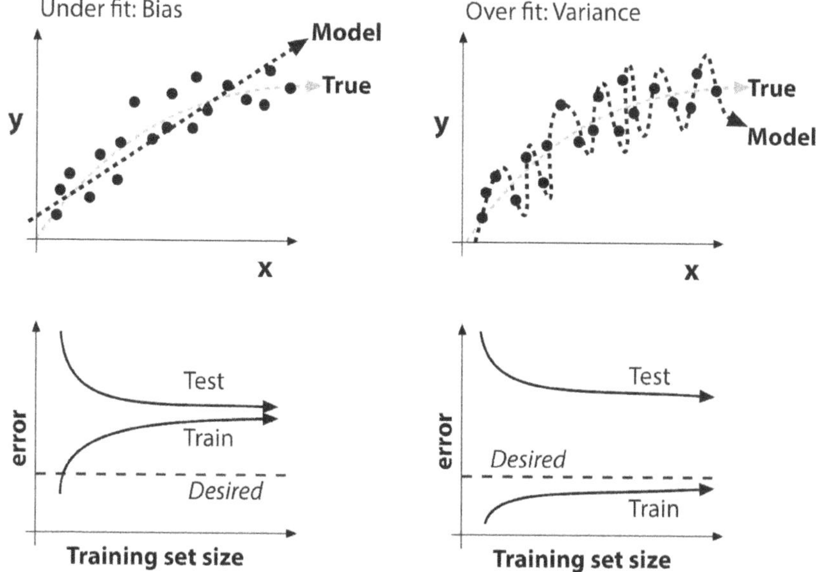


*Figure 20-3. Poor model fitting in machine learning and its effect on training and test prediction performance, where the true function is graphed by the lighter dashed line at top: a biased (underfit) model for the data (upper left) yields high error in predicting the training and the test set (lower left), whereas a variance (overfit) model for the data (upper right) yields low error in the training data but high error in the test data (lower right)*

Sometimes you have to be careful that you are solving the correct problem. If your training and test set error are low but the algorithm does not perform well in the real world, the data set may have been chosen from unrealistic conditions—perhaps because these conditions made collecting or simulating the data easier. If the algorithm just cannot reproduce the test or training set data, then perhaps the algorithm is the wrong one to use, the features that were extracted from the data are ineffective, or the "signal" just isn't in the data you collected. Table 20-2 lays out some possible fixes to the problems we've described here. Of course, this is not a complete list of the possible problems or solutions. It takes careful thought and design of what data to collect and what features to compute in order for machine learning to work well. It can also take some systematic thinking to diagnose machine learning problems.

*Table 20-2. Problems encountered in machine learning and possible solutions to try; coming up with better features will help any problem*

| Problem | Possible solutions |
|---|---|
| Bias | • More features can help make a better fit.<br>• Use a more powerful model/algorithm. |
| Variance | • More training data can help smooth the model.<br>• Fewer features can reduce overfitting.<br>• Use a less powerful model/algorithm. |
| Good test/train, bad real world | • Collect a more realistic set of data. |
| Model can't learn test or train | • Redesign features to better capture invariance in the data.<br>• Collect new, more relevant data.<br>• Use a more powerful model/algorithm. |

### Cross-validation, bootstrapping, ROC curves, and confusion matrices

Finally, there are some basic tools that are used in machine learning to measure results. In supervised learning, one of the most basic problems is simply knowing how well your algorithm has performed: how accurate is it at classifying or fitting the data? You might think: "Easy, I'll just run it on my test or validation data and get the result." But for real problems, we must account for noise, sampling fluctuations, and sampling errors. Simply put, your test or validation set of data might not accurately reflect the actual distribution of data. To get closer to "guessing" the true performance of the classifier, we employ the technique of *cross-validation* and/or the closely related technique of *bootstrapping*.[11]

Recall that, in its most basic form, cross-validation involves dividing the data into $K$ different subsets of data. You train on $K - 1$ of the subsets and test on the final subset of data (the "validation set") that wasn't trained on. You do this $K$ times, where each of the $K$ subsets gets a "turn" at being the validation set, and then average the results.

Bootstrapping is similar to cross-validation, but the validation set is selected at random from the training data. Selected points for that round are used only in test, not training. Then the process starts again from scratch. You do this $N$ times, where each time you randomly select a new set of validation data and average the results in the end. Note that this means some and/or many of the data points are reused in different validation sets, but the results are often superior compared to cross-validation.[12]

11 For more information on these techniques, see "What Are Cross-Validation and Bootstrapping?" (*http://www.faqs.org/faqs/ai-faq/neural-nets/part3/section-12.html*).

12 One of the main goals of bootstrapping is to find the optimal training parameters, because while it's easy to average the training error, it may be nontrivial and/or inefficient to "average" several models into one.

Using either one of these techniques can yield more accurate measures of actual performance. This increased accuracy can in turn be used to tune parameters of the learning system as you repeatedly change, train, and measure.

Two other immensely useful ways of assessing, characterizing, and tuning classifiers are plotting the *receiver operating characteristic* (ROC, often pronounced "rock") curve and the *confusion matrix*; see Figure 20-4. The ROC curve measures the response of the performance parameter of the classifier over the full range of settings of that parameter (each point on a ROC curve may also be computed with cross-validation). Let's say the parameter is a threshold. Just to make this more concrete, suppose we are trying to recognize yellow flowers in an image and that we have a threshold on the color yellow as our detector. Setting the yellow threshold extremely high would mean that the classifier would fail to recognize any yellow flowers, yielding a false positive rate of `0` but at the cost of a true positive rate also at `0` (lower-left part of the curve in Figure 20-4). On the other hand, if the yellow threshold is set to `0`, then any signal at all counts as a recognition. This means that all of the true positives (the yellow flowers) are recognized as well as all the false positives (orange and red flowers); thus, we have a false positive rate of 100% (upper-right part of the curve in Figure 20-4). The best possible ROC curve would be one that follows the y-axis up to 100% and then cuts horizontally over to the upper-right corner. Failing that, the closer the curve comes to the upper-left corner, the better. In practice, one often computes the fraction of area under the ROC curve versus the total area of the ROC plot as a summary statistic of merit: the closer that ratio is to 1, the better is the classifier.[13]

13 It is worth noting that there are endless varieties of "figures of merit" for classifiers in the literature, and equally many reasons to prefer one over another. This particular one is relatively common, however, not because it is necessarily a better representation of the quality of a classifier, but because it is easily understood and well defined for almost any classification problem.

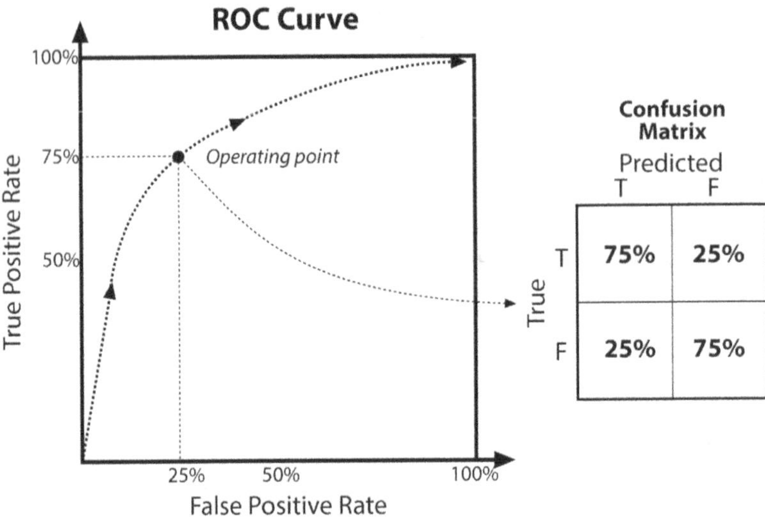


*Figure 20-4. ROC curve and associated confusion matrix: the former shows the response of correct classifications to false positives along the full range of varying a performance parameter of the classifier; the latter shows the false positives (false recognitions) and false negatives (missed recognitions)*

Figure 20-4 also shows a *confusion matrix*—a chart of true and false positives along with true and false negatives. It is another quick way to assess the performance of a classifier: ideally we'd see 100% along the diagonal and 0% elsewhere. If we have a classifier that can learn more than one class (e.g., a multilayer perceptron or random forest classifier, which can learn many different class labels at once), then the confusion matrix is generalized to cover all class labels, and the entries represent the overall percentages of true and false positives and negatives over all class labels. The ROC curve for multiple classes just records the true and false decisions over the test data set.

**Cost of misclassification.** One thing we haven't discussed much here is the cost of misclassification. That is, if our classifier is built to detect poisonous mushrooms (we'll see an example that uses just such a data set in the next chapter), then we are willing to have more false negatives (edible mushrooms mistaken as poisonous) as long as we minimize false positives (poisonous mushrooms mistaken as edible). The ROC curve can help with this; we can set our ROC parameter to choose an operation point lower on the curve—toward the lower left of the graph in Figure 20-4. The other way of doing this is to weight false positive errors more than false negatives when generating the ROC curve. For example, you can set the cost of each false positive error to be equal to 10 false negatives.[14] With some OpenCV machine learning algorithms, such as decision trees and SVM, you can regulate this balance of "hit rate versus false alarm" by specifying prior probabilities of the classes themselves (which classes are expected to be more likely and which less) or by specifying weights of the individual training samples.

**Mismatched feature variance.** Another common problem with training some classifiers arises when the feature vector comprises features of widely different variances. For instance, if one feature is represented by lowercase ASCII characters, then it ranges over only 26 different values. In contrast, a feature that is represented by the count of biological cells on a microscope slide might vary over several billion values. An algorithm such as K-nearest neighbors might then see the first feature as relatively constant (nothing to learn from) compared to the cell-count feature. The way to correct this problem is to preprocess each feature variable by normalizing for its variance. This practice is acceptable provided the features are not correlated with each other; when features are correlated, you can normalize by their average variance or by their covariance. Some algorithms, such as decision trees,[15] are not adversely affected by widely differing variance and so this precaution need not be taken. A rule of thumb is that if the algorithm depends in some way on a distance measure (e.g., weighted values), then you should normalize for variance. One may normalize all features at once and account for their covariance by using the Mahalanobis distance, which is discussed later in this chapter. Readers familiar with machine learning or signal processing might recognize this as the technique called "whitening" the data.

We now turn to discussing some of the machine learning algorithms supported in OpenCV.

# Legacy Routines in the ML Library

Much of the ML library uses a common C++ interface that is object based and implements each algorithm inside of an object derived from this common base class. In this way, the access to the algorithms and their usage is standardized across the library. However, some of the most basic methods in the library do not conform to

14 This is useful if you have some specific a priori notion of the relative cost of the two error types. For example, the cost of misclassifying one product as another in a supermarket checkout would be easy to quantify exactly beforehand.

15 Decision trees are not affected by variance differences in feature variables because each variable is searched only for effective separating thresholds. In other words, it doesn't matter how large the variable's range is as long as a clear separating value can be found.

this standard interface because they were implemented in the early days of the library when the common interface had not yet been designed.[16]

Because it is the most basic methods that do not conform to the object interface, we will look at these first. In the next chapter, we will move on to the object interface and the many algorithms implemented through it. For now, the methods we will cover are K-means clustering, the Mahalanobis distance (and its utility in the context of K-means clustering), and finally a partitioning technique that is normally used as a means of improving both the speed and the accuracy of K-means clustering.

## K-Means

K-means attempts to find the natural clusters in a set of vector-valued data. The user sets the desired number of clusters and then the K-means algorithm rapidly finds a good placement for the centers of those clusters. Here, "good" means that the cluster centers tend to end up located in the middle of the natural clumps of data. The K-means algorithm is one of the most used clustering techniques and has strong similarities to the Expectation Maximization algorithm[17] as well as some similarities to the mean-shift algorithm discussed in Chapter 17 (implemented as `cv::meanShift()` in the CV library). K-means is an iterative algorithm and, as implemented in OpenCV, is also known as Lloyd's algorithm [Lloyd82] or (equivalently) "Voronoi iteration." The algorithm runs as follows.

1. Take as input a data set $D$ and desired number of clusters $K$ (chosen by the user).
2. Randomly assign (sufficiently separated) cluster center locations.
3. Associate each data point with its nearest cluster center.
4. Move cluster centers to the centroid of their data points.
5. Return to Step 3 until convergence (i.e., centroid does not move).

Figure 20-5 shows K-means in action; in this case, it takes just three iterations to converge. In real cases the algorithm often converges rapidly, but there are still times when it will require a large number of iterations.

16 At this time it is an open "to do" item of the library for someone to come along and write new interfaces to these legacy functions that conform to the object-based style used by the rest of MLL.

17 Specifically K-means is similar to the Expectation Maximization algorithm for Gaussian mixture models. This EM algorithm is also implemented in OpenCV (as `cv::EM()` in the ML library). We will encounter this algorithm later in the chapter.

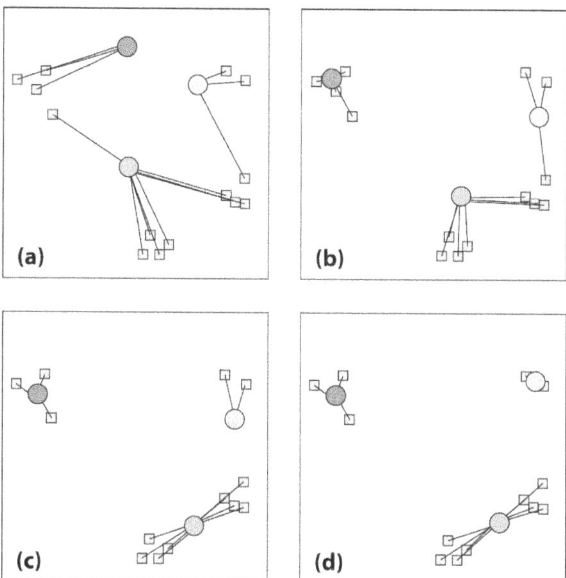


*Figure 20-5. K-means in action for three iterations: (a) cluster centers are placed randomly and each data point is then assigned to its nearest cluster center; (b) cluster centers are moved to the centroid of their points; (c) data points are again assigned to their nearest cluster centers; (d) cluster centers are again moved to the centroid of their points*

### Problems and solutions

K-means is an extremely effective clustering algorithm, but it does have three important shortcomings:

- It isn't guaranteed to find the best possible solution to locating the cluster centers. However, it is guaranteed to converge to some solution (i.e., the iterations won't continue indefinitely).
- It doesn't tell you how many cluster centers you should use. If we had chosen two or four clusters for the example in Figure 20-5, then the results would be different and perhaps less than intuitive.
- It presumes that the covariance in the space either doesn't matter or has already been normalized (we will return to this, and the Mahalanobis distance, shortly).

Each one of these problems has a "solution," or at least an approach that helps. The first two of these solutions depends on "explaining the variance of the data." In

K-means, each cluster center "owns" its data points and we compute the variance of those points.[18]

The best clustering minimizes the variance without causing too much complexity (too many clusters). With that in mind, we can ameliorate the listed problems as follows:

1. Run K-means several times, each with different placement of the cluster centers; then choose the run whose results exhibit the least variance. OpenCV can do this automatically; you need only specify the number of such clustering attempts (see the `attempts` parameter of `cv::kmeans`).
2. Start with one cluster and try an increasing number of clusters (up to some limit), each time employing the method of Step 1 as well. Usually the total variance will shrink quite rapidly, after which an "elbow" will appear in the variance curve; this indicates that a new cluster center does not significantly reduce the total variance. Stop at the elbow and keep that many cluster centers.
3. Multiply the data by the inverse covariance matrix (as described in the section "Mahalanobis Distance" on page 797). For example, if the input data vectors $D$ are organized as rows with one data point per row, then normalize the "stretch" in the space by computing a new data vector $D^*$, where $D^* = D\Sigma^{-1/2}$.

### K-means code

The call for K-means is simple:

```
double cv::kmeans(                                // returns (best) compactness
cv::InputArray       data,                        // Your data, in a float type
int                  K,                           // Number of clusters
cv::InputOutputArray bestLabels,                  // Result cluster indices (int's)
cv::TermCriteria     criteria,                    // iterations and/or min dist
int                  attempts,                    // starts to search for best fit
int                  flags,                       // initialization options
cv::OutputArray      centers    = cv::noArray() // (optional) found centers
);
```

The `data` array is a matrix of multidimensional data points, one per row, where each element of the data is a regular floating-point value (i.e., `CV_32FC1`). Alternatively, `data` may be simply a single column of entries, each of which is a multidimensional

18 In this context the variance of a point is the distance of the point from the cluster center. The variance of the points (plural) is typically the quadrature sum over the points. This sum is also called *compactness*.

point (i.e., of type CV_32FC2 or CV_32FC3 or even CV_32FC(M)).[19] The parameter K is the number of clusters you want to find, and the return vector bestLabels contains the final cluster index for each data point. In the case of criteria, you may specify either the maximum number of iterations you would like the algorithm to run, or the small distance that will be used to determine when a cluster center is effectively stationary (i.e. if it moves less than the given small distance). Of course, you can specify both of these criteria as well.

The parameter attempts tells cv::kmeans() to automatically run some number of times, each time starting with a new set of seed points, and to keep only the best result. The quality of the results is gauged by the *compactness*—that is, the sum of squared distances between every point and the center of the cluster to which that point was associated.

The flags parameter may be any of the following values (with the first being the default): cv::KMEANS_RANDOM_CENTERS, cv::KMEANS_USE_INITIAL_LABELS, or cv::KMEANS_PP_CENTERS. In the case of cv::KMEANS_RANDOM_CENTERS, we assign the starting cluster centers as described earlier, by randomly selecting from the points in the data set. In the case of cv::KMEANS_USE_INITIAL_LABELS, the values stored in the parameter bestLabels at the time the function is called will be used to compute the initial cluster centers. Finally, the cv::KMEANS_PP_CENTERS option instructs cv::kmeans() to use the method of Sergei Vassilvitskii and David Arthur [Arthur07] called K-means++ to assign the cluster centers. The details of this method are not critical to us here, but what is important is that this method more prudently chooses the starting points for the cluster centers and typically gives better results in fewer iterations than the default method. In modern applications, K-means++ is increasingly the standard being used.

19 Recall that this case is in fact exactly equivalent to an $N \times M$ matrix in which the $N$ rows are the data points, the $M$ columns are the individual components of each point's location, and the underlying data type is cv::32FC1. Recall that, owing to the memory layout used for arrays, there is no distinction between these representations.

Even though, in theory, the K-means algorithm can behave rather badly, it is heavily used in practice because most of the time it does quite well. The fact that, in the worst case, the cluster assignment problem is NP-hard means that you are unlikely to ever get a truly optimal answer, but for most applications, a "good" answer is good enough. Still, there are some unsettling problems with the K-means algorithm. One of them is that, in certain cases, it can be tricked into producing what mathematicians call "arbitrarily bad" results. This means that no matter how wrong you might fear the answer might be, some tricky person can come up with a circumstance in which it will be at least that wrong, or worse. As a result there has been some interest in past years (and continues to be) in techniques that provide either some general improvement in performance, or in the best case, some provable bounds that can give the user a little more confidence in the results. One such algorithm is K-means++.

Finally, on completion, the computed centers for the clusters will be placed into the array `centers`. You can omit `centers` if you do not require them (in this case by passing `cv::noArray()`). The function always returns the computed compactness.

It's instructive to see a complete example of K-means in code (Example 20-1). An added benefit of the example is that the data-generation sections can be used to test other machine learning routines as well.

*Example 20-1. Using K-means*

```
#include "opencv2/highgui/highgui.hpp"
#include "opencv2/core/core.hpp"
#include <iostream>

using namespace cv;
using namespace std;

static void help( char* argv[] ) {
  cout << "\nThis program demonstrates kmeans clustering.\n"
   "  It generates an image with random points, then assigns a random number\n"
   "  of cluster centers and uses kmeans to move those cluster centers to their\n"
   "  representative location\n"
   "Usage:\n"
   <<argv[0] <<"\n" << endl;
}

int main( int /*argc*/, char** /*argv*/ ) {

  const int MAX_CLUSTERS = 5;
  cv::Scalar colorTab[] = {
    cv::Scalar(   0,   0, 255 ),
```

```
    cv::Scalar(   0, 255,   0 ),
    cv::Scalar( 255, 100, 100 ),
    cv::Scalar( 255,   0, 255 ),
    cv::Scalar(   0, 255, 255 )
  };

  cv::Mat img( 500, 500, CV_8UC3 );
  cv::RNG rng( 12345 );


  for(;;) {

    int k, clusterCount = rng.uniform(2, MAX_CLUSTERS+1);
    int i, sampleCount = rng.uniform(1, 1001);
    cv::Mat points(sampleCount, 1, CV_32FC2), labels;

    clusterCount = MIN(clusterCount, sampleCount);
    cv::Mat centers(clusterCount, 1, points.type());


    /* generate random sample from multigaussian distribution */
    for( k = 0; k < clusterCount; k++ ) {
      cv::Point center;
      center.x = rng.uniform(0, img.cols);
      center.y = rng.uniform(0, img.rows);
      cv::Mat pointChunk = points.rowRange(
        k*sampleCount/clusterCount,
        k == clusterCount - 1 ? sampleCount : (k+1)*sampleCount/clusterCount
      );
      rng.fill(
        pointChunk,
        RNG::NORMAL,
        cv::Scalar(center.x, center.y),
        cv::Scalar(img.cols*0.05, img.rows*0.05)
      );
    }

    randShuffle(points, 1, &rng);

    kmeans(
      points,
      clusterCount,
      labels,
      cv::TermCriteria(
        cv::TermCriteria::EPS | cv::TermCriteria::COUNT,
        10,
        1.0
      ),
      3,
      KMEANS_PP_CENTERS,
      centers
    );
```

```
    img = Scalar::all(0);

    for( i = 0; i < sampleCount; i++ ) {
      int clusterIdx = labels.at<int>(i);
      cv::Point ipt = points.at<cv::Point2f>(i);
      cv::circle( img, ipt, 2, colorTab[clusterIdx], cv::FILLED, cv::LINE_AA );
    }


    cv::imshow("clusters", img);


    char key = (char)waitKey();
    if( key == 27 || key == 'q' || key == 'Q' ) // 'ESC'
      break;
  }

  return 0;
}
```

In this code we used `highgui` to create a window output interface and include *core.hpp* because it contains `cv::kmeans()`.[20] The basic operation of the program is to first choose a number of clusters to generate, generate centers for those clusters, and then generate a cloud of points around the generated center. We will then come back and see whether `cv::kmeans()` can effectively rediscover this structure we put into the sample data. In `main()`, we first do some minor housekeeping, like setting up the coloring we will later use for displayed clusters. We then set up a main loop, which will let users run over and over again, generating different sets of test data.

This loop begins by determining how many clusters there will be in the underlying data and how many data points will be generated. Then for each cluster the center is generated and points are generated from a Gaussian distribution around that center. The points are then shuffled so that they will not be in cluster order.

At this point we turn the K-means algorithm loose on the data. In this example we don't search the possible cluster counts, we just tell `cv::kmeans()` how many clusters there will be. The resulting labeling is computed and placed in `labels`.

The final `for{}` loop just draws the results. This is followed by deallocating the allocated arrays and displaying the results in the "`clusters`" image. Finally, we wait indefinitely (`cv::waitKey(0)`) to allow the user to do another run or to quit via the Esc key.

20 This is an example of the legacy issue again. `cv::kmeans()` predates the formal creation of the ML library, and so its prototype is in *core.hpp* rather than *ml.hpp* (as you probably imagined it would be).

## Mahalanobis Distance

We encountered the Mahalanobis distance earlier in Chapter 5 as a means of computing the distance between a point and a distribution center that was sensitive to the shape of the distribution. In the context of the K-means algorithm, the concept of the Mahalanobis distance can serve us in two different ways. The first application comes from understanding the Mahalanobis distance from an alternative point of view in which we regard it as measuring Euclidean distance on a deformed space. This allows us to use what we know about the Mahalanobis distance to create a rescaling of data that can substantially improve the performance of the K-means algorithm. The second application of the Mahalanobis distance is as a means of assigning novel data points to the clusters defined by the K-means algorithm.

### Using the Mahalanobis distance to condition input data

In Example 20-1, we mentioned briefly the possibility that the data might be arranged in the space in a highly asymmetrical way. Of course, the entire point of using K-means is to assert that the data is clustered in a nonuniform way and to try to discover something about that clustering. However, there is an important distinction between "asymmetrical" and "nonuniform." If, for example, all of your data is spread out a great distance in some dimensions and relatively little distance in others, then the K-means algorithm will behave poorly. An example of such a situation is shown in Figure 20-6.

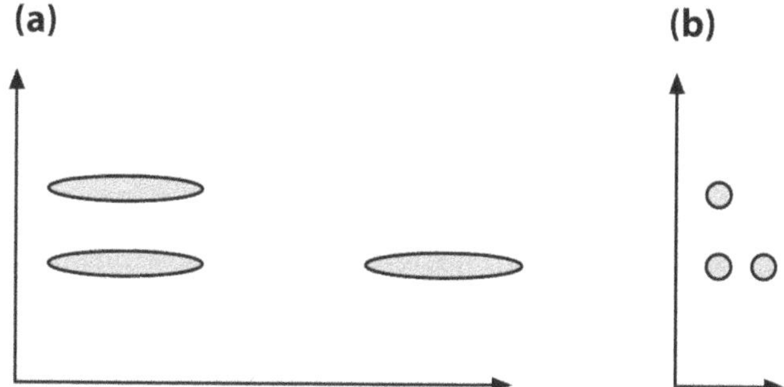


*Figure 20-6. The Mahalanobis computation allows us to reinterpret the data's covariance as a "stretch" of the space: (a) the vertical distance between raw data sets is less than the horizontal distance; (b) after the space is normalized for variance, the horizontal distance between data sets is less than the vertical distance*

Such situations often arise simply because there are different underlying units to different dimensions of a data vector. For example, if persons in a community are represented by their height, age, and the total number of years of schooling they have had, the simple fact that the units of height and age are different will result in a very different dispersion of the data those dimensions. Similarly, age and years of schooling, though they have the same units, have a naturally very different variance among natural populations.

This example, however, suggests a simple technique that can be very helpful. This technique is to look at the data set as a whole, and to compute the covariance matrix for the entire data set. Once this is done, we can rescale the entire data set using that covariance. By using such a technique, we can rescale the data in Figure 20-6(a) into something more like in Figure 20-6(b).

Recall that we encountered the Mahalanobis distance in Chapter 5 when we looked at the function `cv::Mahalonobis()`. The traditional use of the Mahalanobis distance is to measure the distance of a point from a distribution in such a way that the distance is measured in units of the distribution's variance in the particular direction of the point. (For those familiar with the concept of the *Z-score* from statistics, the Mahalanobis distance is the generalization of the Z-score to multidimensional spaces.) We compute the Mahalanobis distance using the inverse of the covariance of the distribution:

$$\Sigma_{i,j} = E[(\vec{X}_i - \vec{\mu})(\vec{X}_i - \vec{\mu})^T] = \frac{1}{N}\sum_i[(\vec{X}_i - \vec{\mu})(\vec{X}_i - \vec{\mu})^T]$$

where $E[\cdot]$ is the "expectation operator." The actual formula for the Mahalanobis distance is then:

$$D_{mahalanobis}(\vec{x}, \vec{y}) = \sqrt{(\vec{x} - \vec{y})^T \Sigma^{-1}(\vec{x} - \vec{y})}$$

At this point, we can look at the situation in one of two ways: we can either say that we would like to use the Mahalanobis distance rather than the Euclidean distance in the K-means algorithm, or we can think of rescaling the data first and then using the Euclidean distance in the rescaled space. The first is probably a more intuitive way to look at things, but the second is much easier computationally—if only because we don't want to actually crack open and modify the nice K-means implementation that is already provided for us. In the end, because the transformation is a linear one, either interpretation is possible.

After a little thought, it should be clear that we can simply rescale the data with the following operation:

$$D^* = D\Sigma^{-1/2}$$

Here $D^*$ is the set of new data vectors we will use, and $D$ is the original data. The factor of $\Sigma^{-1/2}$ is just the square root of the inverse covariance.[21]

In this case, we do not actually make direct use of `cv::Mahalanobis()`. Instead, we compute the set covariance using the `cv::calcCovarMatrix()`, invert it with `cv::invert()` (using `cv::DECOMP_EIG`[22]) and finally compute the square root.[23]

### Using the Mahalanobis distance for classification

Given a set of cluster labels, from K-means clustering or any other method, we can use those labels to attempt to guess what cluster some novel point most likely belongs to. In the case where the clusters themselves are, or are thought to be, Gaussian distributed, it makes sense to apply the concept of the Mahalanobis distance also to this assignment problem.

The first step we need to take in order to make such assignments is to characterize each cluster in terms of its mean and covariance. Once we have done this, we can compute a Mahalanobis distance to each cluster center for any novel point.

From here you might guess that the point with the smallest Mahalanobis distance would be the winner, but it is not quite that simple. This would be true if all of the clusters had the same number of elements (or as a statistician would say, "if the *prior probability* of being a member of each cluster is equal").[24]

This distinction is succinctly captured by Bayes' rule, which states (in words) that, for two propositions $A$ and $B$, the probability that $A$ is true given $B$ is (in general) not equal to the probability that $B$ is true given $A$. In equation form, this looks like:

---

21 If you are wondering why the inverse covariance is on the right in this formula, it is because the convention in the ML library is to represent a data set $D$ as $N$ rows of points and $M$ columns for each point. Thus the data are rows and not columns.

22 `DECOMP_SVD` could also be used in this case, but it is somewhat slower and less accurate than `DECOMP_EIG`. `DECOMP_EIG`, even if it is slower than `DECOMP_LU`, still should be used if the dimensionality of the space is much smaller than the number of data points. In such a case the overall computing time will be dominated by `cv::calcCovarMatrix()` anyway. So it may be wise to spend a little bit more time on computing inverse covariance matrix more accurately (much more accurately, if the set of points is concentrated in a subspace of a smaller dimensionality). Thus, `DECOMP_EIG` is usually the best choice for this task.

23 There is no general function in the library to compute this square root, but because the matrix $\Sigma^{-1}$ is very well behaved (as matrices go), the square root can be computed by first diagonalizing it, then taking the square root of the eigenvalues individually, and then rotating back to the original coordinate frame using the same eigenvectors you used to diagonalize it in the first place. (Not surprisingly this technique is called "The Method of Diagonalization").

24 The problem is that the odds of finding an extremely strange lizard that looks like a dinosaur are still much better than the odds of finding a relatively normal dinosaur. This is because actual dinosaurs are much rarer than lizards.

$$P(A|B) \neq P(B|A)$$

Instead, it says (again in words) that the probability that $A$ is true given $B$ multiplied by the probability that $B$ is true in the first place is equal to the probability that $B$ is true given $A$ multiplied by the probability that $A$ is true in the first place. In equation form, this looks like:

$$P(A|B)P(B) = P(B|A)P(A)$$

If you are trying to figure out how to tie this back to our Mahalanobis distance problem, remember that the Mahalanobis distance is telling us something about the probability that a particular sample came from a particular cluster, but—and here is the key point—that is the probability *provided it came from that cluster at all.* Seen a different way, we want to know the probability that our point is in cluster $C$ given its value of $\vec{x}$. But the Mahalanobis distance is telling us the opposite, namely the probability of getting $\vec{x}$ if we are in cluster $C$. Here it is written out as an equation (Bayes's rule, slightly rearranged):

$$P\left(C|\vec{x}\right) = \frac{P(C)}{P\left(\vec{x}\right)} P\left(\vec{x}|C\right)$$

This means that in order to compare two Mahalanobis distances between two different clusters, we should take into account the sizes of the clusters. Given that the probability is asserted to be Gaussian for each cluster, the figure of merit that we should be comparing, called the *likelihood*, is:[25]

$$P\left(C|\vec{x}\right) \propto \left(\frac{N_c}{N_D}\right)\left(|\Sigma|^{-1/2} e^{-\frac{1}{2}r_M^2}\right)$$

In this equation the ratio $N_c$ divided by $N_D$ is the fraction of data points in cluster $C$ relative to the total number of data points. This ratio is the *prior probability* of cluster $C$. The second term contains the inverse square root of the determinant of the covariance of cluster $C$, and the exponential factor that contains the squared Mahalanobis distance.

25 You will notice that $P(\vec{x})$ has conspicuously disappeared. This is because not only do we have no prior reason to believe that any value of $\vec{x}$ would be more likely than any other (in the absence of a cluster assignment), but also it would not even matter if that were true, because that factor would be common to all of the things we were comparing.

# Summary

In this chapter we started with a basic discussion of what machine learning is and looked at which parts of that large problem space are addressed by the OpenCV library and which are not. We looked at the distinction between training and test data. We learned that *generative* models are those that attempt to find structure in existing data without supervision or labeled training data, while *discriminative* models are those that learn from examples and attempt to generalize from what they have been shown. We then moved on to look at two very fundamental tools that are available in the OpenCV library, K-means clustering and Mahalanobis distance. We saw how these could be used to build simple models and solve interesting problems. We note, in passing, that OpenCV now also supports deep neural networks, but that in their current form they are in the experimental *opencv_contrib*. That code is described in Appendix B (*cnn_3dobj* and *dnn*).

# Exercises

1. If features in each data point vary widely in scale (say the first feature varies from 1 to 100 and the second feature varies from 0.0001 to 0.0002), explain whether and why it would pose a problem for:
   a. SVM
   b. Decision trees
   c. Back-propagation
2. One way of removing the scale differences across features is to normalize the data. Two ways of doing this are to divide by the standard deviation of each feature or to divide by the maximum minus the minimum value. For each method of normalization, describe a set of data this would work well for and one that it would not work well for.
3. Consider the case of rescaling the input data set using the Mahalanobis distance before using the K-means algorithm. Prove that prescaling the data according to the formula $D^* = D\Sigma^{-1/2}$ is exactly equivalent to modifying the algorithm to use the Mahalanobis distance internally.
4. Consider trying to learn the next stock price from several past stock prices. Suppose you have 20 years of daily stock data. Discuss the effects of various ways of turning your data into training and testing data sets. What are the advantages and disadvantages of the following approaches?
   a. Take the even-numbered points as your training set and the odd-numbered points as your test set.
   b. Randomly select points into training and test sets.

c. Divide the data in two, where the first half is for training and the second half for testing.

5. Divide the data into many small windows of several past points and one prediction point. Refer to Figure 20-3. Can you imagine conditions under which the test set error would be lower than the training set error?

6. Figure 20-3 was drawn for a regression problem. Label the first point on the graph *A*, the second point *B*, the third point *A*, the fourth point *B*, and so on. Draw a separation line for these two classes (*A* and *B*) that shows:

   a. bias

   b. variance

7. Refer to Figure 20-4.

   a. Draw the generic best-possible ROC curve.

   b. Draw the generic worst-possible ROC curve.

8. Draw a curve for a classifier that performs randomly on its test data.

9. Consider variable importance.

   a. If two features are exactly the same, will variable importance as described earlier find out whether one or both are important?

   b. If not, what would fix the algorithm to detect that these two identical features are either important or not?

CHAPTER 21

# StatModel: The Standard Model for Learning in OpenCV

In the previous chapter we discussed machine learning broadly and looked at just a few basic algorithms that were implemented in the library long ago. In this chapter we will look at several more modern techniques that will prove to be of very wide application. Before we start on those, however, we will introduce `cv::ml::StatModel`, which forms the basis of the implementation for the interfaces to all of the more advanced algorithms we will see in this chapter. Once armed with an understanding of `cv::ml::StatModel`, we will spend the remainder of the chapter looking at various learning algorithms available in the OpenCV library. The algorithms are presented here in an approximately chronological order relative to their introduction into the computer vision community.[1]

## Common Routines in the ML Library

The contemporary routines in the ML library are implemented within classes that are derived from the common base class `cv::ml::StatModel`. This base class defines the interface methods that are universal to all of the available algorithms. Some of the methods are declared in the base class `cv::Algorithm`, from which `cv::ml::StatModel` itself is derived. Here is the (somewhat abbreviated) `cv::ml::StatModel` base class definition straight from the machine learning (ML) library:

1 Note that OpenCV is currently being extended to support deep neural networks; see Appendix B, repositories *cnn_3dobj* and *dnn*. At the time of this writing, DNNs are emerging as a profoundly important tool for computer vision. However, their implementation in OpenCV is still under development, so they will not be covered here.

```
// Somewhere above...
// namespace cv {
//   namespace ml {

class StatModel : public cv::Algorithm {

public:

    /** Predict options */
    enum Flags {
        UPDATE_MODEL       = 1,
        RAW_OUTPUT         = 1,
        COMPRESSED_INPUT   = 2,
        PREPROCESSED_INPUT = 4
    };

    virtual int  getVarCount() const  = 0; // number training samples
    virtual bool empty() const;            // true if no data loaded

    virtual bool isTrained()    const = 0; // true if the model is trained
    virtual bool isClassifier() const = 0; // true if the model is a classifier

    virtual bool train(
      const cv::Ptr<cv::ml::TrainData>& trainData,     // data to be loaded
      int                               flags    = 0   // (depends on model)
    );

    // Trains the statistical model
    //
    virtual bool train(
      InputArray samples,  // training samples
      int        layout,   // layout See ml::SampleTypes
      InputArray responses // responses associated with the training samples
    );

    // Predicts response(s) for the provided sample(s)
    //
    virtual float predict(
      InputArray  samples,                    // input samples, float matrix
      OutputArray results = cv::noArray(),  // optional output results matrix
      int         flags   = 0               // (model-dependent)
    ) const = 0;

    // Computes error on the training or test dataset
    //
    virtual float calcError(
      const Ptr<TrainData>& data, // training samples
      bool test,                  // true:   compute over test set
                                  //  false: compute over training set
      cv::OutputArray resp        // the optional output responses
    ) const;
```

```
    // In addition, each class must implement static `create()` method with no
    // parameters or with all default parameter values.
    //
    // example:
    // static Ptr<SVM> SVM::create();
};
```

You will notice that `cv::ml::StatModel` inherits from `cv::Algorithm`. Though we will not include everything from that class here, these are a few salient methods that are likely to come up often in the context of `cv::ml::StatModel` and your usage of it:

```
// Somewhere above
// namespace cv:: {
//   namespace ml:: {

class Algorithm {
...
public:

  virtual void save(
   const String& filename
  ) const;

  // Calling example: Ptr<SVM> svm = Algorithm::load<SVM>("my_svm_model.xml");
  //
  template<typename _Tp> static Ptr<_Tp> static load(
   const String& filename,
   const String& objname = String()
  );

  virtual void clear();
...
}
```

The methods of `cv::StatModel` provide mechanisms for reading and writing trained models from and to disk, and a method for clearing the data in the model. These three actions are essentially universal.[2] On the other hand, the routines for training the algorithms and for applying them for prediction vary in interface from algorithm to algorithm. This is natural because the training and prediction aspect of the different algorithms will have different capabilities and, at the very least, will require different parameters to be configured.

2 In the distant past, there were two pairs of functions for reading and writing: `save()`/`load()` and `write()`/`read()`, with the latter pair being lower-level functions that interacted with the now legacy `CvFileStorage` file interface structure. That pair should now be considered deprecated, along with the structure they once accessed, and the only interface that should be used in modern code is the `save()`/`load()` interface.

## Training and the cv::ml::TrainData Structure

The training and prediction methods shown in the `cv::ml::StatModel` prototype will naturally vary from one learning technique to the next. In this section we will look at how those methods are structured and how they are used.

Recall that there were two methods called `train()` in the `cv::ml::StatModel` prototype. The first `train()` method takes the training data in the form of a `cv::ml::TrainData` structure pointer and various algorithm-dependent training flags. The second method is a shortcut variant that constructs that same training data structure using the provided samples and the ground-truth responses directly. As we will see, the `cv::ml::TrainData` interface allows us to prepare the data in some useful ways, so it is generally the more expressive way of training a model.

### Constructing cv::ml::TrainData

The `cv::ml::TrainData` class allows you to package up your data, along with some instructions about how it is to interpreted and used in training. In practice, this additional information is extremely useful. Here is the `create()` method used to generate a new `cv::ml::TrainData` object.

```
// Construct training data from the specified matrix of data points and responses.
// It's possible to use a subset of features (a.k.a. variables) and/or subset of
// samples; it's possible to assign weights to individual samples.
//
static cv::Ptr<cv::ml::TrainData> cv::ml::TrainData::create(
  cv::InputArray samples,                          // Array of samples (CV_32F)
  int            layout,                           // row/col (see ml::SampleTypes)
  cv::InputArray responses,                        // Float array of responses
  cv::Inputarray varIdx        = cv::noArray(), // Specifies training variables
  cv::InputArray sampleIdx     = cv::noArray(), // Specifies training samples
  cv::InputArray sampleWeights = cv::noArray(), // Optional sample wts (CV_32F)
  cv::InputArray varType       = cv::noArray()  // Optional, types for each
                                                   //  input and output
                                                   //  variable (CV_8U)
);
```

This method constructs training data from the preallocated arrays of training samples and the associated responses. The matrix of samples must be of type `CV_32FC1` (32-bit, floating-point, and single-channel). Though the `cv::Mat` class is clearly capable of representing multichannel images, the machine learning algorithms take only a single channel—that is, just a two-dimensional array of numbers. Typically, this array is organized as rows of data points, where each "point" is represented as a vector of features. Hence, the columns contain the individual features for each data point and the data points are stacked to yield the 2D single-channel training matrix. To belabor the topic: the typical data matrix is thus composed of (rows, columns) = (data points, features).

Some of the algorithms can handle transposed matrices directly. The parameter `layout` specifies how the data is stored:

`layout = cv::ml::ROW_SAMPLE`
: Means that the feature vectors are stored as rows (this is the most common layout)

`layout = cv::ml::COL_SAMPLE`
: Means that the feature vectors are stored as columns

You may well ask: What if my training data is not floating-point numbers but instead is letters of the alphabet or integers representing musical notes or names of plants? The answer is: Fine, just turn them into unique 32-bit floating-point numbers when you fill the `cv::Mat`. If you have letters as features or labels, you can cast the ASCII character to floats when filling the data array. The same applies to integers. As long as the conversion is unique, things should work out fine. Remember, however, that some routines are sensitive to widely differing variances among features. It's generally best to normalize the variance of features, as we saw in the previous section. With the exception of the tree-based algorithms (decision trees, random trees, and boosting) that support both categorical and ordered input variables, all other OpenCV ML algorithms work only with ordered inputs. A popular technique for making ordered-input algorithms work with categorical data is to represent them in "1-radix" or "1-hot" notation; for example, if the input variable `color` may have seven different values, then it may be replaced by seven binary variables, where one and only one of the variables may be set to `1`.[3]

The `responses` parameter will contain either categorical labels such as `poisonous` or `nonpoisonous`, in the case of mushroom identification, or are regression values (numbers) such as body temperatures taken with a thermometer. The response values, or "labels," are usually a one-dimensional vector with one value per data point. One important exception is neural networks, which can have a vector of responses for each data point. For categorical responses, the response value type must be an integer (`CV_32SC1`); for regression problems, the response should be of 32-bit floating-point type (`CV_32FC1`). In the special case of neural networks, as alluded to before, it is common to put a little twist on this, and actually perform categorization using a regression framework. In this case, the 1-hot encoding mentioned earlier is used to represent the various categories and floating-point output is used for all of the multiple outputs. In this case the network is, in essence, being trained to regress to something like the probability that the input is in each category.

3 Note that this is different than one input whose value is binary encoded. The value $0100000b = 32$ is very different than a seven-dimensional input vector whose value is [0, 1, 0, 0, 0, 0, 0].

Recall, however, that some algorithms can deal only with classification problems and others only with regression, while still others can handle both. In this last case, the type of output variable is passed either as a separate parameter or through the `varType` vector. This vector can be either a single column or a single row, and must be of type `CV_8UC1` or `cv::S8C1`. The number of entries in `varType` is equal to the number of input variables ($N_f$) plus the number of responses (typically one).[4] The first $N_f$ entries will tell the algorithm the type of the corresponding input feature, while the remainder indicate the types of the output. Each entry in `varType` should be set to one of the following values:

`cv::ml::VAR_CATEGORICAL`
: Means that the output values are discrete class labels

`cv::ml::VAR_ORDERED (= cv::ml::VAR_NUMERICAL)`
: Means that the output values are ordered; that is, different values can be compared as numbers and so this is a regression problem

Algorithms of the regression type can handle only ordered-input variables. Sometimes it is possible to make up an ordering for categorical variables as long as the order is kept consistent, but this can sometimes cause difficulties for regression because the pretend "ordered" values may jump around wildly when they have no philosophical basis for their imposed order.

Many models in the ML library may be trained on a selected feature subset and/or on a selected sample subset of the training set. To make this easier for the user, the `cv::ml::TrainData::create()` method includes the vectors `varIdx` and `sampleIdx` as parameters. The `varIdx` vector can be used to identify specific variables (features) of interest, while `sampleIdx` can identify specific data points of interest. Either of these may simply be omitted or set to `cv::noArray()` (the default value) to indicate that you would like to use "all of the features" or "all of the points." Both vectors are either provided as lists of zero-based indices or as masks of active variables/samples, where a nonzero value signifies active. In the former case, the vector must be of type `CV_32SC1` and may have any length. In the latter case, the array must be of type `CV_8UC1` and must have the same length as the number of features or samples (as appropriate). The parameter `sampleIdx` is particularly helpful when you've read in a chunk of data and want to use some of it for training and some of it for testing without having to first break it into two different vectors.

4 To be clear, this means the number of input features, not the number of input data points.

### Constructing cv::ml::TrainData from stored data

Often, you will have data already saved on disk. If this data is in CSV (comma-separated value) format, or you can put it into this format, you can create a new `cv::ml::TrainData` object from that CSV file using `cv::ml::TrainData::loadFromCSV()`.

```
// Load training data from CSV file; part of each row may be treated as the
// scalar or vector responses; the rest are input values.
//
static cv::Ptr<cv::ml::TrainData> cv::ml::TrainData::loadFromCSV(
   const String& filename,                       // Input file name
   int           headerLineCount,                // Ignore this many lines
   int           responseStartIdx = -1,          // Idx of first out var (-1=last)
   int           responseEndIdx   = -1,          // Idx of last out var plus one
   String&       varTypeSpec      = String(),    // Optional, specifies var types
   char          delimiter        = ',',         // Char used to separate values
   char          missch           = '?'          // Used for missing data in CSV
);
```

The CSV reader skips the first `headerLineCount` lines and then reads the data. The data is read row by row[5] and individual features are separated based on commas. If some other separator is used in the CSV file, the `delimeter` argument may be used to replace the default comma (e.g., by a space or semicolon). Often, the responses will be found in the leftmost or the rightmost column, but the user may specify any column (or a range of columns) as necessary. The responses will be drawn from the interval [`responseStartIdx`, `responseEndIdx`), inclusive of `responseStartIdx` but exclusive of `responseEndIdx`. The variable types, if required, are specified via single compact text string, `vatTypeSpec`. For example:

```
"ord[0-9,11]cat[10]"
```

means that the data has 12 columns, the first 10 columns contain ordered values, then there is column of categorical values, and then there is yet another column of ordered values.

If you do not provide a variable type specification, then the reader tries to "do the right thing" by following a few simple rules. It considers input variables to be ordered (numerical) unless they clearly contain non-numerical values (e.g., `"dog"`, `"cat"`), in which case they are made categorical. If there is only one output variable, then it will follow pretty much the same rule,[6] but if there are multiple, then they are always considered ordered.

5 That is, the `ROW_SAMPLE` layout is assumed.

6 "Pretty much" means that there is a funny exception where, if the output variable is always an integer, then it will still be considered categorical.

It is also possible to specify a special character, using the `missch` argument, to be used for missing measurements. However, it is important to know that some of the algorithms cannot handle such missing values. In such cases missing points should be interpolated or otherwise handled by the user before training or the corrupted records should be rejected in advance.[7]

The problem of missing data comes up in real world problems quite often. For example, when the authors were working with manufacturing data, some measurement features would end up missing during the time that workers took coffee breaks. Sometimes experimental data simply is forgotten, such as forgetting to take a patient's temperature one day during a medical experiment.

#### Secret sauce and cv::ml::TrainDataImpl

Were you to go to the source code and look at the class definition for `cv::ml::TrainData`, you would see that it is full of pure virtual functions. In fact, it is just an interface, from which you can derive and create your own training data containers. This fact immediately leads to two obvious questions: why would you want to do this, and what exactly is going on inside of `cv::ml::TrainData::create()` if `cv::ml::TrainData` is a virtual class type?

As for the why, training data can be very complex in real-life situations and the data itself can be very large. In many cases it is necessary to implement more sophisticated strategies for managing the data and for storing it. For these reasons, you might want to implement your own training data container that, for example, uses a database for the management of the bulk of the available data. Using the `cv::ml::TrainData` interface, you can implement your own data container and the available algorithms will run on that data transparently.

For the second question, the how, the answer is that the `cv::ml::TrainData::create()` method actually creates an object of a different class than it appears to. There is a class called `cv::ml::TrainDataImpl` that is essentially the default implementation of a data container. This object manages the data just the way you would expect—in the form of a few arrays inside that hold the various things you think you have put in there.

7 Various algorithms, such as decision tree and naïve Bayes, handle missing values in different ways. Decision trees use alternative splits (called "surrogate splits" by Breiman [Breiman84]), while the naïve Bayes algorithm infers the values. Unfortunately, the current implementation of ML decision trees/random trees in OpenCV is not yet able to handle such missing measurements.

In fact, the existence of this class will be largely invisible to you unless (until) you start looking at the library source code directly. Of course, if you do find yourself wanting to build your own `cv::ml::TrainData`–derived container class, it will be very useful to look at the implementation of `cv::ml::TrainDataImpl` in *.../opencv/modules/ml/src/data.cpp*.

### Splitting training data

In practice, when you are training a machine learning system, you don't want to use all of the data you have to train the algorithm. You will need to hold some back to test the algorithm when it is done. If you don't do this, you will have no way of estimating how your trained system will behave when presented with novel data. By default, when you construct a new instance of `TrainData`, it's all considered available to be used for training data, and none of it is held back for such testing. Using `cv::ml::TrainData::setTrainTestSplit()`, you can split the data into a training and a test part, and just use the training part for training your model. Using just this training data is the automatic behavior of `cv::ml::StatModel::train()`, assuming you have marked what data you want it to use.

```
// Splits the training data into the training and test parts
//
void cv::ml::TrainData::setTrainTestSplit(
  int    count,
  bool   shuffle = true
);

void cv::ml::TrainData::setTrainTestSplitRatio(
  double ratio,
  bool   shuffle = true
);

void cv::ml::TrainData::shuffleTrainTest();
```

The three members of `cv::ml::TrainData` that will help you out with this are: `setTrainTestSplit()`, `setTrainTestSplitRatio()`, and `shuffleTrainTest()`. The first takes a `count` argument that specifies how many of the vectors in the data set should be labeled as training data (with the remainder being test data). Similarly the second function does the same thing, but allows you to specify the ratio of points (e.g., 0.90 = 90%) that will be labeled as training data. Finally, the third "shuffle" method will randomly assign the train and test vectors (while keeping the number of each fixed). Either of the first two methods supports a `shuffle` argument. If `true`, then the test and train labels will be assigned randomly; otherwise, the train samples will start from the beginning and the test samples will be those vectors thereafter.

Note that internally, the default implementation IMPL has three separate indices that do similar things: the *sample index*, the *train index*, and the *test index*. Each is a list of indices into the overall array of samples in the container that indicates which samples are to be used in a particular context. The sample index is an array listing all samples that will be used. The train index and test index are similar, but list which samples are for training and which are for testing. As implemented, these three indices have a relationship that some might find unintuitive.

If the train index is defined, it should always be the case that the test index is defined. This is the natural result of the use of the functions just described to create these internal indices. If either (both) is defined, then its behavior will always define how `train()` responds; this is regardless of anything that might be in the sample index. Only when these two indices are undefined will the sample index be used. In that case, all data indicated by the sample index will be assumed available for training, and no data will be marked as test data.

### Accessing cv::ml::TrainData

Once the training data is constructed, it's possible to retrieve its parts, with or without preprocessing, using the methods described next. The function `cv::ml::TrainData::getTrainSamples()` retrieves a matrix only of the training data.

```
// Retrieve only the active training data into a cv::Mat array.
//
cv::Mat cv::ml::TrainData::getTrainSamples(
  int  layout          = ROW_SAMPLE,
  bool compressSamples = true,
  bool compressVars    = true
) const;
```

When `compressSamples` or `compressVars` are `true`, the method will retain only rows or columns set by `sampleIdx` and `varIdx`, respectively (typically at construction time). The method also transposes the data if the desired layout is different from the original one. If the sample index or the train index is defined, then only the indicated samples will be returned. Recall, however, that, if both are defined, it will be the train index that determines what is returned.

Similarly, the `cv::ml::TrainData::getTrainResponses()` method extracts only the active response vector elements.

```
// Return the train responses (for the samples selected using sampleIdx).
//
cv::Mat cv::ml::TrainData::getTrainResponses() const;
```

As with `cv::ml::TrainData::getTrainSamples()`, if the sample index or the train index is defined, then only the indicated samples will be returned. Recall, however, that, if both are defined, it will be the train index that determines what is returned.

Similarly, there are two functions—`cv::ml::TrainData::getTestSamples()` and `cv::ml::TrainData::getTestResponses()`—that return the analogous arrays constructed from only the test samples. In this case, however, if the test index is not defined, then an empty array will be returned.

Finally, there are accessors that will simply tell you how many of various kinds of samples are in the data container. We list them here.

```
int getNTrainSamples() const;  // Number of samples indicated by train idx
                               //  or total samples if samples idx is not defined

int getNTestSamples()  const;  // Number of samples indicated by test idx
                               //  or zero test idx is not defined

int getNSamples()      const;  // Number of samples indicated by samples idx
                               //  or total samples if samples idx is not defined

int getNVars()         const;  // Number of features indicated by variable idx
                               //  or total samples if variable idx is not defined

int getNAllVars()      const;  // Number of features total
```

## Prediction

Recall from the prototype that the general form of the predict() method is as follows:

```
float cv::ml::StatModel::predict(
  cv::InputArray  samples,                      // input samples, float matrix
  cv::OutputArray results = cv::noArray(),      // optional output matrix of results
  int             flags   = 0                   // (model-dependent)
) const;
```

This method is used to predict the response for a new input data vector. When you are using a classifier, predict() returns a class label. For the case of regression, this method returns a numerical value. Note that the input sample must have as many components as the train_data that was used for training.[8] In general, samples will be an input floating-point array, with one sample per row, and results will be one result per row. When only a single sample is provided, the predicted result will be returned form the predict() function. Keep in mind, however, that in some cases, these general behaviors will be slightly different for any particular derived classifier. Additional flags are algorithm-specific and allow for such things as missing feature values in tree-based methods. The function suffix const tells us that prediction does not affect the internal state of the model. This method is thread-safe and can be run

8 The var_idx parameter you used with train() is "remembered" and applied to extract only the necessary components from the input sample when you use the predict() method. As a result, the number of columns in the sample should be the same as were in train_data, even if you are using var_idx to ignore some of the columns.

in parallel, which is useful for web servers performing image retrieval for multiple clients and for robots that need to accelerate the scanning of a scene.

In addition to being able to generate a prediction, we can compute the error of the model over the training or test data. When the model is being used for classification, this is the percentage of incorrectly classified samples; when we are using the model for regression, this is the mean squared error. The method that does it is called `cv::ml::StatModel::calcError()`:

```
float cv::ml::StatModel::calcError(
  const cv::Ptr<cv::ml::TrainData>& data, // training samples
  bool                              test, // false: compute over training set
                                          //  true: compute over test set
  cv::OutputArray                   resp  // the optional output responses
) const;
```

In this case, we typically pass in the same `cv::ml::TrainData` data container that we used for training. We then use the `test` argument to determine if we want to know how well the trained algorithm did on either the training data used (`test` set to `false`) or on the test data that we withheld from the training process (`test` set to `true`). Finally, we can use the `resp` array to collect the responses to the individual vectors tested. Though this is optional, the argument is not. If you are not interested in the output responses, you must pass `cv::noArray()` here.

We are now ready to move on to the ML library proper with the normal Bayes classifier, after which we will discuss decision-tree algorithms (decision trees, boosting, random trees, and Haar cascade). For the other algorithms we'll provide short descriptions and usage examples.

# Machine Learning Algorithms Using cv::StatModel

Now that we have a good feel for how the ML library in OpenCV works, we can move on to how to use individual learning methods. This section looks briefly at eight machine learning routines, that latter four of which have recently been added to OpenCV. Each implements a well-known learning technique, by which we mean that a substantial body of literature exists on each of these methods in books, published papers, and on the Internet. In time, it is expected that more new algorithms will appear.

## Naïve/Normal Bayes Classifier

Earlier, we looked at some legacy routines from before the machine learning library was systematized; now we will look at a simple classifier that uses the new `cv::ml::StatModel` interface introduced in this chapter. We'll begin with OpenCV's simplest supervised classifier, `cv::ml::NormalBayesClassifier`, which is alternatively known as a *normal Bayes* classifier or a *naïve Bayes* classifier. It's "naïve" because, in its mathematical implementation, it assumes that all the features we observe are independent variables from one another (even though this is seldom actually the case). For example, finding one eye usually implies that another eye is lurking nearby; these are not uncorrelated observations. However, it is often possible to ignore this correlation in practice and still get good results. Zhang discusses possible reasons for the sometimes surprisingly good performance of this classifier [Zhang04]. Naïve Bayes is not used for regression, but it is an effective classifier that can handle multiple classes, not just two. This classifier is the simplest possible case of what is now the large and growing field of Bayesian networks, or "probabilistic graphical models."[9]

By way of example, consider the case in which we have a collection of images, some of which are images of faces, while others are images of other things (maybe cars and flowers). Figure 21-1 portrays a model in which certain measureable features are *caused to exist* if the object we are looking at is, in fact, a face. In general, Bayesian networks are *causal* models. In the figure, facial features in an image are asserted to be caused by (or not caused by) the existence of an object, which may (or may not) be a face. Loosely translated into words, the graph in the figure says, "An object, which may be of type 'face' or of some other type, would imply either the truth of falsehood of five additional assertions: 'there is a left eye,' 'there is a right eye,' etc., for each of five facial features." In general, such a graph is normally accompanied by additional information that tells us the possible values of each node and the actual probabilities of each value in each bubble as a function of the values of the nodes that have arrows pointing into the bubble. In our case, the node *O* can take values "face," "car," or "flower," and the other five nodes can take the values `present` or `absent`. The probabilities for each feature, given the nature of the object, we will learn from data.

Note that this is precisely where the "uncorrelated" nature of the graph comes in; specifically, the probability that there is a nose depends only on whether the object is a face, and is independent (or at least is asserted to be independent) of whether or not there is a mouth, a hairline, and so on. As a result, there are a lot fewer combinations of cases to learn, because everything we care about essentially factorizes into the question of how each feature is statistically related to the object's presence. This factorization is the precise meaning of *uncorrelated*.

9 For an accessible introduction to the topic see, for example, [Neapolitan04]. For a detailed discussion of the topic of probabilistic graphical models generally, see, for example, [Koller09].

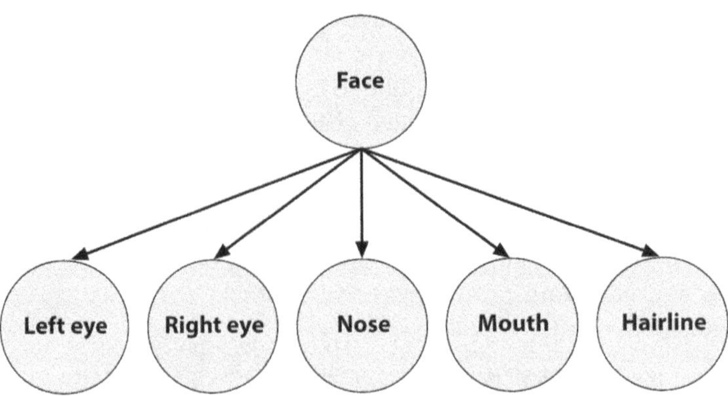


*Figure 21-1. A (naïve) Bayesian network, where the lower-level features are caused by the presence of an object (such as a face)*

In use, the object variable in a naïve Bayes classifier is usually a *hidden variable* and the features—via image processing operations on the input image—constitute the observed evidence that the value of the object variable is of whatever type (i.e., "face"). Models such as this are called *generative* models because the object causally generates (or fails to generate) the face features.[10] Because it is generative, after training, we could instead start by assuming the value of the object node is "face" and then randomly sample what features are probabilistically generated given that we have assumed a face to exist.[11] This top-down generation of data with the same statistics as the learned causal model is a useful capability. For example, one might generate faces for computer graphics display, or a robot might literally "imagine" what it should do next by generating scenes, objects, and interactions. In contrast to Figure 21-1, a discriminative model would have the direction of the arrows reversed.

Bayesian networks, in their generality, are a deep field and initially can be a difficult topic. However, the naïve Bayes algorithm derives from a simple application of Bayes' rule.[12] In this case, the probability (denoted $p$) that an object is a face, given that the features are found (denoted, left to right in Figure 21-1, by $LE$, $RE$, $N$, $M$, and $H$) is:

10 More generically, a model is said to be generative if an entire synthetic data set can be produced from it. In this context the opposite of *generative* is *discriminative*. A discriminative model is any model that can tell you something about any data point provided to it, but cannot be used to synthesize data.

11 Generating a face would be silly with the naïve Bayes algorithm because it assumes independence of features. But a more general Bayesian network can easily build in feature dependence as needed.

12 Recall that we first encountered Bayes' rule in Chapter 20 when we discussed the utility of the Mahalanobis distance in the context of K-means classification.

$$p(O = "face"|LE, RE, N, M, H) = \frac{p(LE, RE, N, M, H|O = "face")p(O = "face")}{p(LE, RE, N, M, H)}$$

In words, the components of this equation are typically read as:

$$posterior \quad probability = \frac{likelihood * prior \quad probability}{evidence}$$

The significance of this equation is that, in practice, we compute some evidence and then decide what object caused it (not the other way around). Since the computed evidence term is the same for any object, we can ignore that term in comparisons. Said another way, if we have many object types then we need only find the one with the maximum numerator. The numerator is exactly the joint probability of the model with the data: *p(O="face", LE, RE, N, M, H)*.

Up to this point, we have not really used the "naïve" part of the naïve Bayes classifier. So far, these equations would be true for any Bayesian classifier. In order to make use of the assumption that the different features are statistically independent of one another (recall that this is the primary informative content of the graph in Figure 21-1), we now use the *chain rule for probability* to derive the joint probability:

$$p(O = "face", LE, RE, N, M, H) =$$

$$p(O = "face") * p(LE|O = "face") * p(RE|O = "face", LE) * p(N|O = "face", LE, RE)$$

$$* p(M|O = "face", LE, RE, N) * p(H|O = "face", LE, RE, N, M)$$

Finally, when we apply our assumption of independence of features, the conditional features drop out. For example, the probability of a nose, given that the object is a face, and that we observe both a left eye and a right eye (i.e., $p(N|O = " face ", LE, RE)$) is, by our assumption, equal to the probability of a nose given just by the fact of a face being present: $p(N|O = " face ")$. Similar logic applies to every term on the righthand side of the preceding equation, with the result that:

$$p(O = "face", LE, RE, N, M, H) = p(O = "face") \prod_{feature}^{\{LE,RE,N,M,H\}} p(feature|O = "face") .$$

So, generalizing face to "object" and our list of features to "all features," we obtain the reduced equation:

$$p(\text{object, all features}) = p(\text{object}) \prod_{i=1}^{all features} p(feature_i \mid object)$$

To use this as an overall classifier, we learn models for the objects that we want. In run mode we compute the features and find particular objects that maximize this equation. Typically, we then test to see whether the probability for that "winning" object is over a given threshold. If it is, then we declare the object to be found; if not, we declare that no object was recognized.

If (as frequently occurs) there is only one object of interest, then you might ask: "The probability I'm computing is the probability relative to what?" In such cases, there is always an implicit second object—namely, the background—which is everything that is *not* the object of interest that we're trying to learn and recognize.

In practice, learning the models is easy. We take many images of the objects; we then compute features over those objects and compute the fraction of how many times a feature occurred over the training set for each object. In general, if you don't have much data, then simple models such as naïve Bayes will tend to outperform more complex models, which will "assume" too much about the data (bias).

### The naïve/normal Bayes classifier and cv::ml::NormalBayesClassifier

The following is the class definition for the normal Bayes classifier. Note that the class name `cv::ml::NormalBayesClassifier` is actually another layer of interface definition, while `cv::ml::NormalBayesClassifierImpl` is the name of the actual class that implements the normal Bayes classifier. For convenience, this definition lists some important inherited methods as comments.

```
// Somewhere above...
// namespace cv {
//   namespace ml {
//
class NormaBayesClassifierImpl : public NormaBayesClassifier {
                                         // cv::ml::NormaBayesClassifier is derived
                                         // from cv::ml::StatModel

public:

  ...

  float predictProb(
    InputArray   inputs,
    OutputArray  outputs,
    OutputArray  outputProbs,
    int          flags      = 0
  );

  ...
```

```
    // From class NormaBayesClassifier
    //
    // Ptr<NormaBayesClassifier> NormaBayesClassifier::create(); // constructor

};
```

The training method for the normal Bayes classifier, inherited from `cv::ml::StatModel`, is:

```
bool cv::ml::NormalBayesClassifier::train(
  const Ptr<cv::ml::TrainData>& trainData,    // your data
  int flags = 0                               // 0=new data or UPDATE_MODEL=add
);
```

The `flags` parameter may be `0` or include the `cv::ml::StatModel::UPDATE_MODEL` flag, which means that the model needs to be updated using the additional training data rather than retrained from scratch.

The `cv::NormalBayesClassifier` implements the inherited `predict()` interface described in `cv::ml::StatModel`, which computes and returns the most probable class for its input vectors. If more than one input data vector (row) is provided in the `samples` matrix, the predictions are returned in corresponding rows of the `results` vector. If there is only a single input in `samples`, then the resulting prediction is also returned as a float value by the `predict()` method and the `results` array may be set to `cv::noArray()`.

```
float cv::ml::NormalBayesClassifier::predict(
  cv::InputArray  samples,                   // input samples, float matrix
  cv::OutputArray results = cv::noArray(),   // optional output results matrix
  int             flags   = 0                // (model-dependent)
) const;
```

Alternatively, the normal Bayes classifier also offers the method `predictProb()`. This method takes the same arguments as `cv::ml::NormalBayesClassifier::predict()`, but also the arrray `resultProbs`. This is a floating-point matrix of `number_of_samples` × `number_of_classes` size, where the computed probabilities (that the corresponding samples belong to the particular classes) will be stored.[13] The format for this prediction method is:

```
float cv::ml::NormalBayesClassifier::predictProb( // prob if single sample
  InputArray  samples,                             // one sample per row
  OutputArray results,                             // predictions, one per row
  OutputArray resultProbs,                         // row=sample, column=class
```

13 Note that you can pass `cv::noArray()` for `resultProbs` if you don't need the probabilities. In fact, `cv::ml::NormalBayesClassifier::predict()` is just a wrapper around `predictProb()` that does exactly this.

```
    int         flags = 0                          // 0 or StatModel::RAW_OUTPUT
  ) const;
```

Though the naïve Bayes classifier is extremely useful for small data sets, it does not generally perform well when the data has a great degree of structure. With this in mind, we move next to a discussion of tree-based classifiers, which can dramatically outperform something as simple as the naïve Bayes classifier, particularly when sufficient data is present.

## Binary Decision Trees

We will go through decision trees in detail, since they are highly useful and use most of the functionality in the machine learning library (and thus serve well as an instructional example more generally). Binary decision trees were invented by Leo Breiman and colleagues,[14] who named them *classification and regression trees* (CART). This is the decision tree algorithm that OpenCV implements. The gist of the algorithm is to define what is called an *impurity metric* relative to the data in every node of a tree of decisions, and to try to minimize the impurity with those decisions. When using CART for regression to fit a function, one often uses the sum of squared differences between the true values and the predicted values; thus, minimizing the impurity means making the predicted function more similar to the data. For categorical labels, one typically defines a measure that is minimal when most values in a node are of the same class. Three common measures to use are *entropy*, *Gini index*, and *misclassification* (all described in this section). Once we have such a metric, a binary decision tree searches through the feature vector to find which feature, combined with which threshold for the value of that feature, most "purifies" the data. By convention, we say that features above the threshold are `true` and that the data thus classified will branch to the left; the other data points branch right.[15] This procedure is then used recursively down each branch of the tree until the data is of sufficient purity at the leaves or until the number of data points in a node reaches a set minimum. Figure 21-2 shows an example.

---

14 Leo Breiman et al., *Classification and Regression Trees* (Belmont, CA: Wadsworth, 1984).

15 Clearly, these two decisions are entirely arbitrary. However, not sticking to them is a good way to do nothing useful while really confusing people who have experience with decision trees.

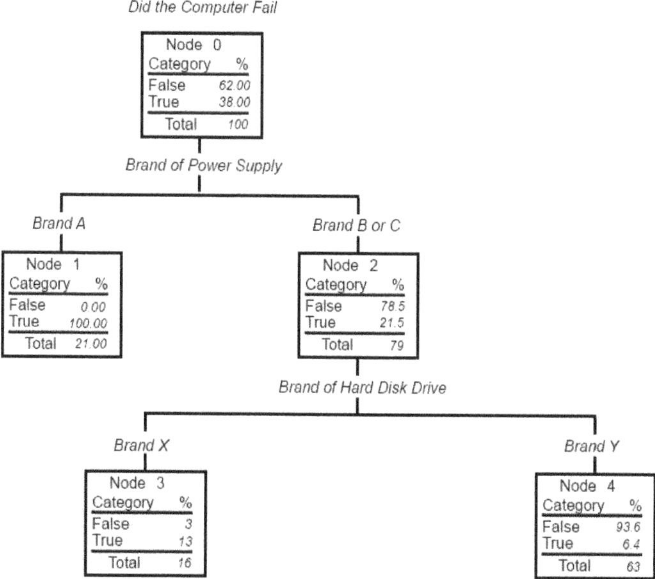


*Figure 21-2. In this example, a hypothetical group of 100 laptop computers is analyzed and the primary factors determining failure rate are used to build the classification tree; all 100 computers are accounted for by the leaf nodes of the tree*

The equations for several available definitions of node impurity $i(N)$ are given next. Different definitions are suited to the distinct problem cases, and for regression versus classification.

### Regression impurity

For regression or function fitting, the equation for node impurity is simply the square of the difference in value between the node value $y$ and the data value $x$. We want to minimize:

$$i(N) = \sum_j (y_j - x_j)^2$$

### Classification impurity

For classification, decision trees often use one of three methods: *entropy impurity, Gini impurity*, or *misclassification impurity*. For these methods, we use the notation $P(\omega_i)$ to denote the fraction of patterns at node $N$ that are in class $\omega_i$. Each of these impurities has slightly different effects on the splitting decision. Gini is the most commonly used, but all the algorithms attempt to minimize the impurity at a node.

Figure 21-3 graphs the impurity measures that we want to minimize. In practice, it is best to just try each impurity to determine on a validation set which one works best.

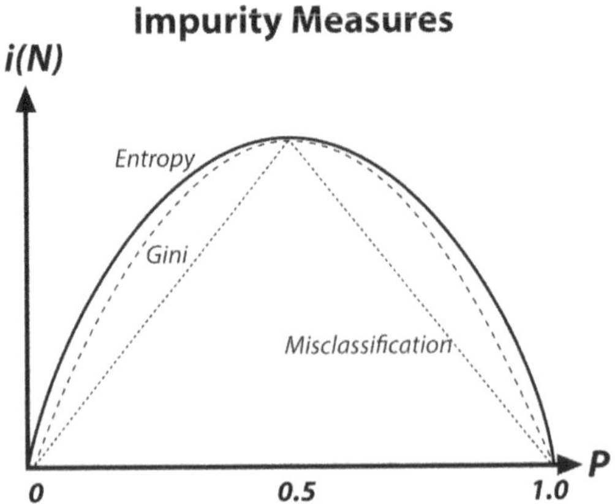


*Figure 21-3. Decision tree impurity measures*

Entropy impurity:

$$i(N) = \sum_j P(\omega_j) \log P(\omega_j)$$

Gini impurity:

$$i(N) = \sum_j 1 - \max P(\omega_j)$$

Misclassification impurity:

$$i(N) = \sum_j P(\omega_i) P(\omega_j)$$

Decision trees are perhaps the most widely used classification technology. This is due to their simplicity of implementation, ease of interpretation of results, flexibility with different data types (categorical, numerical, unnormalized, and mixes thereof), ability to handle missing data through surrogate splits, and natural way of assigning importance to the data features by order of splitting. Decision trees form the basis of other algorithms such as boosting and random trees, which we will discuss shortly.

### OpenCV implementation

The following is the abbreviated declaration of `cv::ml::DTrees`. The `train()` methods are just derived from the base class; most of what we need in the definition is how the parameters to the model are set and loaded.

```
// Somewhere above...
// namespace cv {
//   namespace ml {
//
class DTreesImpl : public Dtrees {      // cv::ml::DTrees is derived
                                        // from cv::ml::StatModel
public:

  // (Inherited from cv::ml::DTrees)
  //
  //enum Flags {
  //  PREDICT_AUTO     = 0,
  //  PREDICT_SUM      = (1<<8),
  //  PREDICT_MAX_VOTE = (2<<8),
  //  PREDICT_MASK     = (3<<8)
  //};

  int     getCVFolds() const;                   // get num cross validation folds
  int     getMaxCategories() const;             // get max number of categories
  int     getMaxDepth() const;                  // get max tree depth
  int     getMinSampleCount() const;            // get min sample count
  Mat     getPriors() const;                    // get priors for categories
  float   getRegressionAccuracy() const;        // get required regression acc.
  bool    getTruncatePrunedTree() const;        // get to truncate pruned trees
  bool    getUse1SERule() const;                // get for use 1SE rule in pruning
  bool    getUseSurrogates() const;             // get to use surrogates

  void    setCVFolds( int val );                // set num cross validation folds
  void    setMaxCategories( int val );          // set max number of categories
  void    setMaxDepth( int val );               // set max tree depth
  void    setMinSampleCount( int val );         // set min sample count
  void    setPriors( const cv::Mat &val );      // set priors for categories
  void    setRegressionAccuracy( float val );   // set required regression acc.
  void    setTruncatePrunedTree( bool val );    // set to truncate pruned trees
  void    setUse1SERule( bool val );            // set for use 1SE rule in pruning
  void    setUseSurrogates( bool val );         // set to use surrogates

  ...

  // Experts can use these, but non-experts can ignore them.
  //
  const std::vector<Node>&  getNodes()   const;
  const std::vector<int>&   getRoots()   const;
  const std::vector<Split>& getSplits()  const;
  const std::vector<int>&   getSubsets() const;
```

```
    ...

    // From class DTrees
    //
    //Ptr<DTrees> DTrees::create(); // The algorithm "constructor",
    //                              //  returns Ptr<DTreesImpl>

};
```

First of all, you might have noticed that the name is plural: `DTrees`. This is because in computer vision it's not standalone decision trees that are primarily used, but rather *ensembles* of decision trees—that is, collections of decision trees that produce some joint decision. Two such popular ensembles, implemented in OpenCV and discussed later in this chapter, are `RTrees` (random trees) and `Boost` (boosting). Both share a lot of internal machinery and so `DTrees` is a sort of base class that, in our current case, can be viewed as an ensemble consisting of a single tree.

The second thing you might have noticed is that again, as we saw with the data container `cv::ml::TrainData`, these prototypes are mostly pure virtual. This is essentially the same thing again. There is actually a hidden class called `cv::ml::DTreesImpl` that is derived from `cv::ml::DTrees` and that contains all of the default implementation. The bottom line is that you can effectively ignore the fact that those functions are pure virtual in the preceding class definition. If, however, at some point you would like to go and look at the default implementation, you will need this information in order to find the member functions you want in *.../modules/ml/src/tree.cpp*.

Once you have constructed a `cv::ml::DTrees` object using `cv::ml::DTrees::create()`, you will need to configure the various runtime parameters. You can do this one of two ways. You will need to construct a structure that contains the needed parameters to configure the tree. This structure is called `cv::ml::TreeParams`. The salient parts of its definition are given next.

You will notice that you can create the structure either with the form of its constructor that has an argument for every component, or you can just create it using the default constructor—in which case everything will be set to default values—and then use individual accessors to set the values you want to customize.

Table 21-1 contains a brief description of the components of `cv::ml::TreeParams()`, their default values, and their meanings.

*Table 21-1. Arguments for cv::ml::TreeParams() constructor and their meanings*

| Params() argument | Value for default constructor | Definition |
|---|---|---|
| `maxDepth` | `INT_MAX` | Tree will not exceed this depth, but may be less deep. |
| `minSampleCount` | `10` | Do not split a node if there are fewer than this number of samples at that node. |
| `regressionAccuracy` | `0.01f` | Stop splitting if difference between estimated value and value in the train samples is less than `regressionAccuracy`. |
| `useSurrogates` | `false` | Allow surrogate splits to handle missing data. [Not yet implemented.] |
| `maxCategories` | `10` | Limits the number of categorical values before which the decision tree will precluster those categories. |
| `CVFolds` | `10` | If (`CVFolds` > 1) then prune the decision tree using K-fold cross-validation where *K* is equal to `CVFolds`. |
| `use1SERule` | `true` | `true` for more aggressive pruning. Resulting tree will be smaller, but less accurate. This can help with overfitting, however. |
| `truncatePrunedTree` | `true` | If `true`, remove pruned branches from the tree. |
| `priors` | `cv::Mat()` | Sets alternative weights for incorrect answers. |

Two of these arguments warrant a little further investigation. `maxCategories` limits the number of categorical values before which the decision tree will precluster those categories so that it will have to test no more than $2^{max\_categories} - 2$ possible value subsets.[16] Those variables that have more categories than `maxCategories` will have their category values clustered down to `maxCategories` possible values. In this way, decision trees will have to test no more than `maxCategories` levels at a time, which results in considering no more than $2^{max\_categories}$ possible decision subsets for each categorical input. This parameter, when set to a low value, reduces computation but at the cost of accuracy.

The last parameter, `priors`, sets the relative weight that you give to misclassification. That is, if we build a two-class classifier and if the weight of the first output class is `1`

16 More detail on categorical versus ordered splits: whereas a split on an ordered variable has the form "if $x < a$ then go left, else go right," a split on a categorical variable has the form "if $x \epsilon \{v_1, v_2, v_3, \dots, v_k\}$ then go left, else go right," where the $v_i$ are some possible values of the variable. Thus, if a categorical variable has $N$ possible values then, in order to find a best split on that variable, one needs to try $2N - 2$ subsets (empty and full subsets are excluded). Thus, an approximate algorithm is used whereby all $N$ values are grouped into $K \leq$ (`max_categories`) clusters (via the K-means algorithm) based on the statistics of the samples in the currently analyzed node. Thereafter, the algorithm tries different combinations of the clusters and chooses the best split, which often gives quite a good result. Note that for the two most common tasks, two-class classification and regression, the optimal categorical split (i.e., the best subset of values) can be found efficiently without any clustering. Hence, the clustering is applied only in $n > 2$-class classification problems for categorical variables with $N >$ (`max_categories`) possible values. Therefore, you should think twice before setting `max_categories` to anything greater than `20`, which would imply more than a million operations for each split!

and the weight of the second output class is 10, then each mistake in predicting the second class is equivalent to making *10* mistakes in predicting the first class. In the example we will look at momentarily, we use edible and poisonous mushrooms. In this context, it makes sense to "punish" mistaking a poisonous mushroom for an edible one 10 times more than mistaking an edible mushroom for a poisonous one. The parameter `priors` is an array of `floats` with the same number of elements as there are classes. The assigned values are in the same order as the classes themselves.

The `train()` method is derived directly from `cv::ml::StatModel`:

```
// Work directly with decision trees:
//
bool cv::ml::DTrees::train(
const cv::Ptr<cv::ml::TrainData>& trainData,    // your data
int                               flags    = 0 // use UPDATE_MODEL to add data
);
```

In the `train()` method, we have the floating-point `trainData` matrix. With decision trees, you can set the `layout` to `cv::COL_SAMPLE` when constructing `trainData` if you want to arrange your data into columns instead of the usual rows, which is the most efficient layout for this algorithm. Example 21-1 details the creation and training of a decision tree.

The function for prediction with a decision tree is the same as for its base class, `cv::ml::StatModel`:

```
float cv::ml::DTrees::predict(
  cv::InputArray  samples,
  cv::OutputArray results = cv::noArray(),
  int             flags   = 0
) const;
```

Here, `samples` is a floating-point matrix, with one row per sample. In the case of a single input, the return value will be enough and `results` can be set to `cv::noArray()`. In the case of multiple vectors to be evaluated, the output `results` will contain a prediction for each input vector. Finally, `flags` specifies various possible options. For example, `cv::ml::StatModel::PREPROCESSED_INPUT` indicates that the values of each categorical variable $j$ are normalized to the range $0..N_j - 1$, where $N_j$ is the number of categories for $j$th variable. For example, if some variable may take just two values, A and B, after normalization A is converted to 0 and B to 1. This is mainly used in ensembles of trees to speed up prediction. Normalizing data to fit within the (0, 1) interval is simply a computational speedup because the algorithm then knows the bounds in which data may fluctuate. Such normalization has no effect on accuracy. This method returns the predicted value, normalized (when `flags` includes `cv::ml::StatModel::RAW_OUTPUT`) or converted to the original label space (when `flags` does not include the `RAW_OUTPUT` flag).

Most users will only train and use the decision trees, but advanced or research users may sometimes wish to examine and/or modify the tree nodes or the splitting criteria. As stated in the beginning of this section, the information for how to do this is in the ML documentation online at *http://docs.opencv.org*. The sections of interest for such advanced analysis are the class structure `cv::ml::DTrees`, the node structure `cv::ml::DTrees::Node`, and its contained split structure `cv::ml::DTrees::Split`.

#### Decision tree usage

We will now explore the details by looking at a specific example. Consider a program whose purpose is to learn to identify poisonous mushrooms. There is a public data set called *agaricus-lepiota.data* that contains information about some 8,000 different kinds of mushrooms. It lists many features that might distinguish a mushroom visually, such as the color of the cap, the size and spacing of the gills, as well as—and this is very important—whether or not that type of mushroom is poisonous.[17] The data file is in CSV format and consists of a label `'p'` or `'e'` (denoting poisonous or edible, respectively) followed by 22 categorical attributes, each represented by a single letter. It should also be noted that the file contains examples in which certain data is missing (i.e., one or more of the attributes is unknown for that particular type of mushroom). In this case, the entry is a `'?'` for that feature.

Let's take the time to look at this program in detail, which will use binary decision trees to learn to recognize poisonous from edible mushrooms based on their various visible attributes (Example 21-1).[18]

*Example 21-1. Creating and training a decision tree*

```
#include <opencv2/opencv.hpp>
#include <stdio.h>
#include <iostream>

using namespace std;
using namespace cv;

int main( int argc, char* argv[] ) {

    // If the caller gave a filename, great. Otherwise, use a default.
```

17 This data set is widely used in machine learning for education and testing of algorithms. It is broadly available on the Web, notably from the UCI Machine Learning Repository (*https://archive.ics.uci.edu/ml/datasets/Mushroom*). The original mushroom records were drawn from G. H. Lincoff (Pres.), *The Audubon Society Field Guide to North American Mushrooms* (New York: Alfred A. Knopf, 1981).

18 Recall the `cv::ml::DTrees` cannot handle missing data at this time. Such support has existed in older implementations, and will likely return in time, but for the moment, you might wish to remove the entries containing the `'?'` marker.

```
//
const char* csv_file_name = argc >= 2
? argv[1]
: "agaricus-lepiota.data";

cout <<"OpenCV Version: " <<CV_VERSION <<endl;

// Read in the CSV file that we were given.
//
cv::Ptr<cv::ml::TrainData> data_set = cv::ml::TrainData::loadFromCSV(
  csv_file_name,     // Input file name
  0,                 // Header lines (ignore this many)
  0,                 // Responses are (start) at thie column
  1,                 // Inputs start at this column
  "cat[0-22]"        // All 23 columns are categorical
);                   // Use defaults for delimeter (',') and missch ('?')

// Verify that we read in what we think.
//
int n_samples = data_set->getNSamples();
if( n_samples == 0 ) {
    cerr <<"Could not read file: " <<csv_file_name <<endl;
    exit( -1 );
} else {
    cout <<"Read " <<n_samples <<" samples from " <<csv_file_name <<endl;
}

// Split the data, so that 90% is train data
//
data_set->setTrainTestSplitRatio( 0.90, false );
int n_train_samples = data_set->getNTrainSamples();
int n_test_samples  = data_set->getNTestSamples();

cout <<"Found " <<n_train_samples <<" Train Samples, and "
<<n_test_samples <<" Test Samples" <<endl;

// Create a DTrees classifier.
//
cv::Ptr<cv::ml::RTrees> dtree = cv::ml::RTrees::create();

// set parameters
//
// These are the parameters from the old mushrooms.cpp code

// Set up priors to penalize "poisonous" 10x as much as "edible"
//
float _priors[] = { 1.0, 10.0 };
cv::Mat priors( 1, 2, CV_32F, _priors );

dtree->setMaxDepth( 8 );
dtree->setMinSampleCount( 10 );
dtree->setRegressionAccuracy( 0.01f );
```

```
dtree->setUseSurrogates( false /* true */ );
dtree->setMaxCategories( 15 );
dtree->setCVFolds( 0 /*10*/ );  // nonzero causes core dump
dtree->setUse1SERule( true );
dtree->setTruncatePrunedTree( true );
//dtree->setPriors( priors );
dtree->setPriors( cv::Mat() ); // ignore priors for now...

// Now train the model
// NB: we are only using the "train" part of the data set
//
dtree->train( data_set );

// Having successfully trained the data, we should be able
// to calculate the error on both the training data, as well
// as the test data that we held out.
//
cv::Mat results;
float train_performance = dtree->calcError(
                                            data_set,
                                            false,      // use train data
                                            results //cv::noArray()
                                            );
std::vector<cv::String> names;
data_set->getNames(names);
Mat flags = data_set->getVarSymbolFlags();

// Compute some statistics on our own:
//
{
    cv::Mat expected_responses = data_set->getResponses();
    int good=0, bad=0, total=0;

    for( int i=0; i<data_set->getNTrainSamples(); ++i ) {
        float received = results.at<float>(i,0);
        float expected = expected_responses.at<float>(i,0);
        cv::String r_str = names[(int)received];
        cv::String e_str = names[(int)expected];

        cout <<"Expected: " <<e_str <<", got: " <<r_str <<endl;

        if( received==expected ) good++; else bad++; total++;
    }
    cout <<"Correct answers:   " <<(float(good)/total) <<"%" <<endl;
    cout <<"Incorrect answers: " <<(float(bad)/total)  <<"%" <<endl;
}

float test_performance  = dtree->calcError(
                                            data_set,
                                            true,       // use test data
                                            results //cv::noArray()
                                            );
```

```
    cout <<"Performance on training data: " <<train_performance <<"%" <<endl;
    cout <<"Performance on test data:     " <<test_performance  <<"%" <<endl;

    return 0;
}
```

We start out by parsing the command line for a single argument, the CSV file to read; if there is no such file, we read the mushroom file by default. We print the OpenCV version number,[19] and then continue on to parse the CSV file. This file has no header, but the results are in the first column, so it is important to specify that. Finally, all of the inputs are categorical, so we state that explicitly. Once we have read the CSV file, we state how many samples were read, which is a good way to verify that the reading went well and that you read the file you thought you were reading.

Next, we set the train-test split. In this case, we do so with `setTrainTestRatio()`, and so we specify what fraction we would like to be training data; in this case it is 90% (`0.90`). Also note that the default behavior of the split is to also shuffle the data. If we do not want to do that shuffling, we need to pass `false` as the second argument. Once this split is done, we print out how many training samples there are and how many test samples.

Once the data is all prepared, we can go on and create the `cv::ml::DTrees` object that we will be training. This object is configured with a series of calls to its various `set*()` methods. Notably, we pass an array of values to `setPriors()`. This allows us to set relative weight of missing a poisonous mushroom as opposed to incorrectly marking an edible mushroom as poisonous. The reason the 1.0 comes first and the 10.0 comes after is because *e* comes before *p* in the alphabet (and thus, once converted to ASCII, then to a floating-point number, it comes first numerically).

In this example, we simply train `DTrees` and then use it to predict results on some test data. In a more realistic application, the decision tree may also be saved to disk via `save()` and loaded via `load()` (see the following). In this way, it is possible to train a classifier and then distribute the trained classifier in your code, without having to distribute the data (or make your users retrain every time!) The following code shows how to save and to load a tree file called *tree.xml*.

```
// To save your trained classifier to disk:
//
dtree->save("tree.xml","MyTree");

// To load a trained classifier from disk:
//
```

19 This can be a good habit, particularly when working with the Machine Learning Library as this portion of OpenCV is under relatively active development.

```
dtree->load("tree.xml","MyTree");

// You can also clear an existing trained classifier.
//
dtree->clear();
```

Using the *.xml* extension stores an XML data file; if we used a *.yml* or *.yaml* extension, it would store a YAML data file. The optional `"MyTree"` is a tag that labels the tree within the *tree.xml* file. As with other statistical models in the machine learning module, you cannot store multiple objects in a single *.xml* or *.yml* file when using `save()`; for multiple storage, you need to use `cv::FileStorage()` and `operator<<()`. However, `load()` is a different story: this function can load an object by its name even if there is some other data stored in the file.

#### Decision tree results

By tinkering with the previous code and experimenting with various parameters, we can learn several things about edible or poisonous mushrooms from the *agaricus-lepiota.data* file. If we just train a decision tree without pruning and without priors, so that it learns the data perfectly, we might get the tree shown in Figure 21-4. Although the full decision tree may learn the training set of data perfectly, remember the lessons from the section "Diagnosing Machine Learning Problems" on page 783 in Chapter 20 about variance/overfitting. What's happened in Figure 21-4 is that the data has been memorized along with its mistakes and noise. Such a tree unlikely to perform well on real data. For this reason, OpenCV decision trees (and CART type trees generally) typically include the additional step of penalizing complex trees and pruning them back until complexity is in balance with performance.[20]

Figure 21-5 shows a pruned tree that will still do quite well (but not perfectly) on the training set but will probably perform better on real data because it has a better balance between bias and variance. However, the classifier shown has a serious shortcoming: although it performs well on the data, it now labels poisonous mushrooms as edible 1.23% of the time.

20 There are other decision tree implementations that grow the tree only until complexity is balanced with performance and so combine the pruning phase with the learning phase. However, during development of the ML library, it was found that trees that are fully grown first and then pruned (as implemented in OpenCV) tended to perform better than those that combine training with pruning in their generation phase.

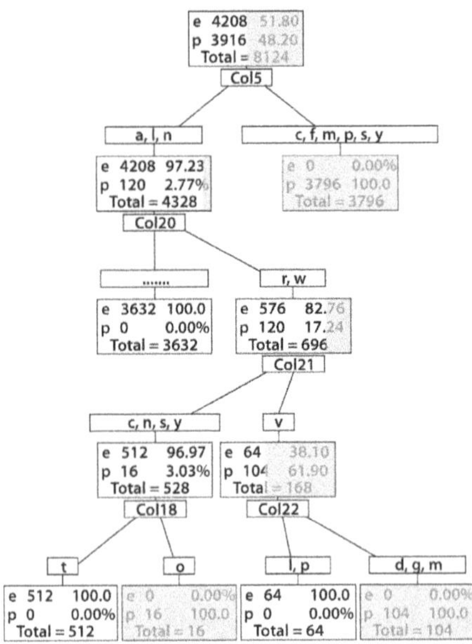


*Figure 21-4. Full decision tree for poisonous (p) or edible (e) mushrooms: this tree was built out to full complexity for 0% error on the training set and so would probably suffer from variance problems on test or real data*

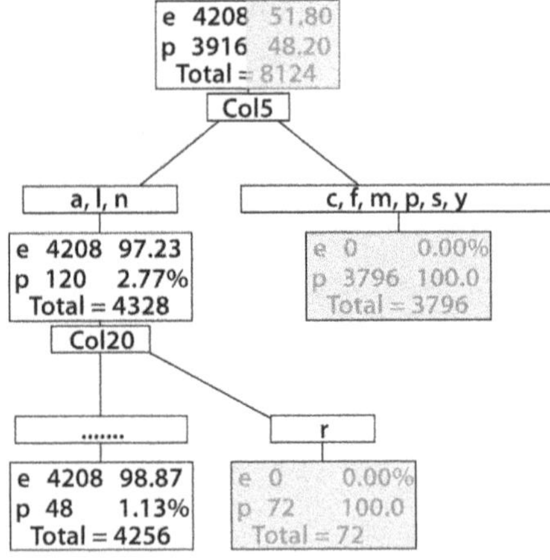


*Figure 21-5. Pruned decision tree for poisonous (p) and edible (e) mushrooms. Despite being pruned, this tree shows low error on the training set and would likely work well on real data*

As you might imagine, there are big advantages to a classifier that may label many edible mushrooms as poisonous but which nevertheless does not invite us to eat a poisonous mushroom! As we saw previously, we can create such a classifier by intentionally biasing (or as *adding a cost* to) the classifier and/or the data. This is what we did in Example 21-1, we added a higher cost for misclassifying poisonous mushrooms than for misclassifying edible mushrooms. By adjusting the `priors` vector, we imposed a cost into the classifier and, as a result, changed the weighting of how much a "bad" data point counts versus a "good" one. Alternatively, if one did not, or could not, modify the classifier code to change the prior, one can equivalently impose additional cost by duplicating (or resampling from) "bad" data. Duplicating "bad" data points implicitly gives a higher weight to the "bad" data, a technique that can work with almost any classifier.

Figure 21-6 shows a tree where a 10× bias was imposed against poisonous mushrooms. This tree makes no mistakes on poisonous mushrooms at a cost of many more mistakes on edible mushrooms, a case of "better safe than sorry." Confusion matrices for the (pruned) unbiased and biased trees are shown in Figure 21-7.

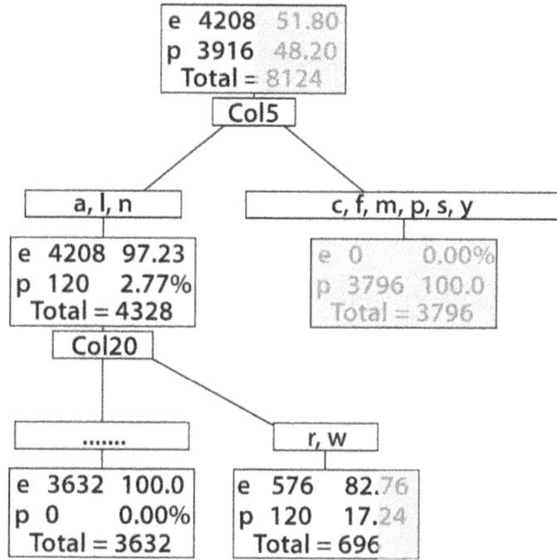


*Figure 21-6. An edible mushroom decision tree with 10× bias against misidentification of poisonous mushrooms as edible; note that the lower-right rectangle, though containing a vast majority of edible mushrooms, does not contain a 10× majority and so would be classified as inedible*

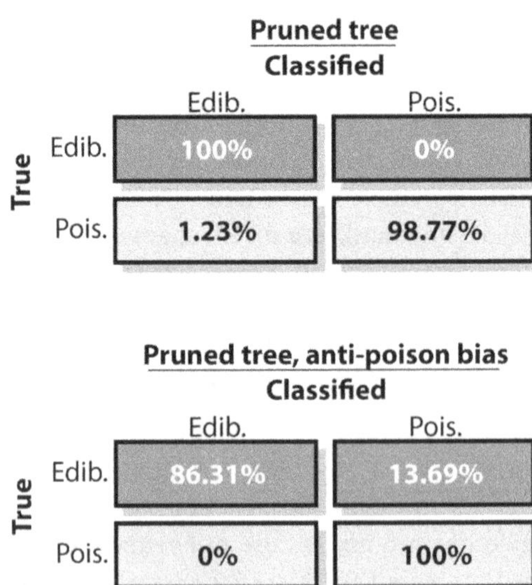


*Figure 21-7. Confusion matrices for (pruned) edible mushroom decision trees: the unbiased tree yields better overall performance (top panel) but sometimes misclassifies poisonous mushrooms as edible; the biased tree does not perform as well overall (lower panel) but never misclassifies poisonous mushrooms*

## Boosting

Decision trees are extremely useful but, used by themselves, are often not the best-performing classifiers. In this and the next section we present two techniques, *boosting* and *random trees*, that use trees in their inner loop and so inherit many of the trees' useful properties (e.g., being able to deal with mixed and unnormalized data types, categorical or ordered). These techniques typically perform at or near the state of the art; thus they are often the best "out of the box" supervised classification techniques available in the library.[21]

In the field of supervised learning there is a *meta-learning* algorithm (first described by Michael Kerns in 1988) called *statistical boosting*. Kerns wondered whether it was possible to learn a *strong classifier* out of many *weak classifiers*. The output of a "weak classifier" is only weakly correlated with the true classifications, whereas that of a "strong classifier" is strongly correlated with true classifications. Thus, weak and strong are defined in a statistical sense.

21 Recall that the "no free lunch" theorem informs us that there is no a priori "best" classifier. But on many data sets of interest in vision, boosting and random trees perform quite well.

The first boosting algorithm, known as *AdaBoost*, was formulated shortly thereafter by Freund and Schapire [Freund97]. Subsequently, other variations of this original boosting algorithm were developed. OpenCV ships with the four types of boosting listed in Table 21-2.

*Table 21-2. Available boosting methods; each is a variant of AdaBoost*

| Boosting method | OpenCV enum value |
|---|---|
| Discrete AdaBoost | `cv::ml::Boost::DISCRETE` |
| Real AdaBoost | `cv::ml::Boost::REAL` |
| LogitBoost | `cv::ml::Boost::LOGIT` |
| Gentle AdaBoost | `cv::ml::Boost::GENTLE` |

Of these, one often finds that the (only slightly different) "real" and "gentle" variants of AdaBoost work best. *Real AdaBoost* is a technique that utilizes confidence-rated predictions and works well with categorical data. *Gentle AdaBoost* puts less weight on outlier data points and for that reason is often good with regression data. *LogitBoost* can also produce good regression fits. Because you need only set a flag, there's no reason not to try all types on a data set and then select the boosting method that works best.[22] Here we'll describe the original AdaBoost. For classification it should be noted that, as implemented in OpenCV, boosting is a two-class (yes or no) classifier,[23] unlike the decision tree or random tree classifiers, which can handle multiple classes at once.

One word of warning: though in theory `LogitBoost` and `GentleBoost` (referenced previously, and in the subsection "Boosting code" on page 836) can be used to perform regression in addition to binary classification, in OpenCV boosting can only be trained for classification as implemented at this time.

### AdaBoost

Boosting algorithms are used to train $N_w$ *weak classifiers* $h_w$, $w \epsilon \{1, \ldots, N_w\}$. These classifiers are generally very simple individually. In most cases these classifiers are decision trees with only one split (called *decision stumps*) or at most a few levels of splits (perhaps up to three). Each of the classifiers is assigned a weighted vote $\alpha_w$ in the final decision-making process for the resulting final *strong classifier*. We use a labeled data

22 This procedure is an example of the machine learning metatechnique known as *voodoo learning* or *voodoo programming*. Although unprincipled, it is often an effective method of achieving the best possible performance. Sometimes, after careful thought, one can figure out why the best-performing method was the best, and this can lead to a deeper understanding of the data. Sometimes not.

23 There is a trick called *unrolling* that can be used to adapt any binary classifier (including boosting) for *N*-class classification problems, but this makes both training and prediction significantly more expensive. See *.../opencv/samples/c/letter_recog.cpp*.

set of input feature vectors $\vec{x}_i$, each with scalar label $y_i$ (where $i \varepsilon \{1, \dots, N_t\}$ indexes the data points). For AdaBoost the label is binary, $y_i \varepsilon \{-1, +1\}$, though it can be any floating-point number in other algorithms. We also initialize a data point weighting distribution $D_w(i)$; this tells the algorithm how much misclassifying a data point will "cost." The key feature of boosting is that, as the algorithm progresses, this cost will evolve so that weak classifiers trained later will focus on the data points that the earlier trained weak classifiers tended to do poorly on. The algorithm is as follows:

1. $D_w(i) = 1 / N_t, \quad i = 1, \dots, N_t$.
2. For $w \varepsilon \{1, \dots, N_w\}$:
   a. Find the classifier $h_w$ that minimizes the $D_w(i)$ weighted error:

      $h = \text{argmin}_{h_j} \varepsilon_j$, where $\varepsilon_j = \Sigma_i^{Nt} D_t(i)$ (for $y_i \neq h_j(\vec{x}_j)$ as long as $\varepsilon < 0.5$; else quit).

   b. Set the $h_w$ "voting weight" $\alpha_i = \frac{1}{2}\log\left[\frac{1-\varepsilon_i}{\varepsilon_i}\right]$, where $\varepsilon_i$ is the argmin error from Step 2a.
   c. Update the data point weights:

$$D_{w+1}(i) = \frac{1}{Z_w} D_w(i) e^{-\alpha_w y_i h_w(\vec{x}_i)}$$

Here, $Z_w$ normalizes the equation over all data points $i \varepsilon \{1, \dots, N_t\}$, so that $\sum_{i=1}^{N_t} D_w(i) = 1$ for all $w$.

Note that, in Step 2a, if we can't find a classifier with less than a 50% error rate then we quit; we probably need better features.

When the training algorithm just described is finished, the final strong classifier takes a new input vector $\vec{x}$ and classifies it using a weighted sum over the learned weak classifiers $h_w$:

$$H(x) = \text{sign}\left( \sum_{w=1}^{T} \alpha_w h_w(\vec{x}) \right)$$

Here, the `sign` function converts anything positive into a `1` and anything negative into a `-1` (zero remains `0`). For performance reasons the leaf values of the just trained $i$th decision tree are scaled by $\alpha_i$ and then $H(x)$ reduces to the sign of the sum of weak classifier responses on $x$.

### Boosting code

The code for boosting is similar to the code for decision trees, and the `cv::ml::Boost` class is derived from `cv::ml::DTrees` with a few extra control parameters. As we have seen elsewhere in the library, when you call `cv::ml::Boost::create()`, you will get back an object of type `cv::Ptr<cv::ml::Boost>`, but this object will actually be a pointer to an object of the (generally invisible) `cv::ml::BoostImpl` class type.

```
// Somewhere above..
// namespace cv {
// namespace ml {
//
class BoostImpl : public Boost {  // cv::ml::Boost is derived from cv::ml::DTrees

public:

  // types of boosting
  // (Inherited from cv::ml::Boost)
  //
  //enum Types {
  //  DISCRETE = 0,
  //  REAL     = 1,
  //  LOGIT    = 2,
  //  GENTLE   = 3
  //};

  // get/set one of boosting types:
  //
  int    getBoostType() const;        // get type: DISCRETE, REAL, LOGIT, GENTLE
  int    getWeakCount() const;        // get the number of weak classifiers
  double getWeightTrimRate() const;   // get the trimming rate, see text

  void setBoostType(int val);         // get type: DISCRETE, REAL, LOGIT, GENTLE
  void setWeakCount(int val);         // get the number of weak classifiers
  void setWeightTrimRate(double val); // get the trimming rate, see text

  ...

  // from class Boost
  //
  //static Ptr<Boost> create(); // The algorithm "constructor",
  //                            //  returns Ptr<BoostImpl>

};
```

The member `setWeakCount()` sets the number of weak classifiers that will be used to form the final strong classifier. The default value for this number is `100`.

The number of weak classifiers is distinct from the maximum complexity that is allowed to each of the individual classifiers. The latter is controlled by `setMaxDepth()`, which sets the maximum number of layers that an individual weak classifier can have. As mentioned earlier, a value of `1` is common, in which case these little trees are just "stumps" and contain only a single decision.

The next parameter, the *weight trim rate*, is used to make the computation more efficient and therefore much faster. As training goes on, many data points become unimportant. That is, the weight $D_t(i)$ for the *i*th data point becomes very small. The `setWeightTrimRate()` function sets a threshold, between `0` and `1` (inclusive), that is implicitly used to throw away some training samples in a given boosting iteration. For example, suppose weight trim rate is set to `0.95` (the default value). This means that the "heaviest" samples (i.e., samples that have the largest weights) with a total weight of at least 95% are accepted into the next iteration of training, while the remaining "lightest" samples with a total weight of at most 5% are temporarily excluded from the next iteration. Note the words "from the next iteration"—the samples are not discarded forever. When the next weak classifier is trained, the weights are computed for all samples and so some previously insignificant samples may be returned to the next training set. Typically, because of the trimming, only about 20% of samples or so take part in each individual round of training and therefore the training accelerates by factor of 5 or so. To turn this functionality off, call `setWeightTrimRate(1.0)`.

For other parameters, note that `cv::ml::BoostImpl` inherits (via `cv::ml::Boost`) from `cv::ml::DTrees`, so we may set the parameters that are related to the decision trees themselves through the inherited interface functions. Overall, training of the boosting model and then running prediction is done in precisely the same way as with `cv::ml::DTrees` or essentially any other `StatModel` derived class from the `ml` module.

The code *.../opencv/samples/cpp/letter_recog.cpp* from the OpenCV package shows an example of the use of boosting. The training code snippet is shown in Example 21-2. This example uses the classifier to try to recognize *a–z* characters, starting with a public data set. That data set has 20,000 entries, each with 16 features and one "result." The features are floating-point numbers and the result is a single character. Because boosting can only be used for two-class discrimination, this program uses the "unrolling" technique that we (briefly) encountered earlier. We will discuss that technique in more detail here.

In unrolling, the data set is essentially expanded from one set of training data to 26 sets, each of which is extended such that what was once the response is now added as a feature. At the same time, the new responses for these extended vectors are now just 1 or 0: `true` or `false`. In this way the classifier is being trained to, in effect, answer the question: is $\vec{x}_i$ equal to $z_i$ by learning the relationships $\{\vec{x}_i, y_i\}$ and $\{\vec{x}_i, not \;\; y_i\} = false$ ?[24] See Example 21-2.

24 Equivalently, one might say that the classifier is being trained to sort propositions of the kind $\vec{x}_i \to y_i$ (i.e., "the vector $\vec{x}_i$ *implies* the result $y_i$") into two classes, those propositions that are true and those propositions that are false.

*Example 21-2. Training snippet for boosted classifiers*

```
...
cv::Mat var_type( 1, var_count + 2, CV_8U );     // var_count is # features (16 here)
var_type.setTo( cv::Scalar::all(VAR_ORDERED) );
var_type.at<uchar>(var_count) = var_type.at<uchar>(var_count+1) = VAR_CATEGORICAL;
...
```

The first thing to do is to create the array `var_type` that indicates how to treat each feature and the results (Example 21-2). Then the training data structure is created. Note that this is wider than you might expect by one. Not only are there `var_count` (in this case, this happens to be 16) features from the original data, there is the one extra column for the response, and there is one more column in between for the extension of the features to include what was once the alphabet-character response (before the unrolling).

```
cv::Ptr<cv::ml::TrainData> tdata = cv::ml::TrainData::create(
  new_data,                  // extended, 26x as many vectors, each contains y_i
  ROW_SAMPLE,                // feature vectors are stored as rows
  new_responses,             // extended, 26x as many vectors, true or false
  cv::noArray(),             // active variable index, here just "all"
  cv::noArray(),             // active sample index, here just "all"
  cv::noArray(),             // sample weights, here just "all the same"
  var_type                   // extended, has 16+2 entries now
);
```

The next thing to do is to construct the classifier. Most of this is pretty usual stuff, but one thing that is unusual is the priors. Note that the price of getting a wrong answer has been inflated to 25 over the price of getting a wrong answer. This is not because some letters are poisonous, but because of the unrolling. What this is saying is that it is 25 times costlier to say that a letter is not something that it is, than to say that it is something that it is not. This needs to be done because there are 25× more vectors effectively enforcing the "negative" rules, so the "positive" rules need correspondingly more weight.[25]

```
vector<double> priors(2);
priors[0] = 1;  // For false (0) answers
priors[1] = 25  // For true (1) answers

model = cv::ml::Boost::create();
model->setBoostType( cv::ml::Boost::GENTLE );
model->setWeakCount( 100 );
model->setWeightTrimRate( 0.95 );
```

25 Recall that we saw earlier that multiplying the number of positive or negative examples was essentially equivalent to weighting the prior. In this case, we multiplied the number of negative vectors for a different reason (the unrolling), so we are inflating the prior for the positive cases for the purpose of compensating the de facto increase of the role of negative examples.

```
model->setMaxDepth( 5 );
model->setUseSurrogates( false );

cout << "Training the classifier (may take a few minutes)...\n";
model->setPriors( cv::Mat(priors) );

model->train( tdata );
```

The prediction function for boosting is also similar to that for decision trees, in this case using `model->predict()`. As described earlier, in the context of boosting, this method computes the weighted sum of weak classifier responses, takes the sign of the sum, and then converts it to the output class label. In some cases, it may be useful to get the actual sum value—for example, to evaluate how confident the decision is. In order to do that, pass the `cv::ml::StatModel::RAW_OUTPUT` flag to the `predict` method. In fact, that is what needs to be done in this case. When dealing with unrolled data, it is not so rare to get two (or more) "true" responses. In this case, one typically chooses the most confident answer.

```
Mat temp_sample( 1, var_count + 1, CV_32F ); // An extended sample "proposition"
float* tptr = temp_sample.ptr<float>();      // Pointer to start of proposition

double correct_train _answers = 0, correct_test _answers = 0;

for( i = 0; i < nsamples_all; i++ ) {

  int          best_class = 0;              // Strongest proposition found so far
  double       max_sum    = -DBL_MAX;       // Strength of current best prop
  const float* ptr        = data.ptr<float>(i); // Points at current sample

  // Copy features from current sample into temp extended sample
  //
  for( k = 0; k < var_count; k++ ) tptr[k] = ptr[k];

  // Add class to sample proposition, then make a prediction for this proposition
  // If this proposition is more true than any previous one, then record this
  // one as the new "best".
  //
  for( j = 0; j < class_count; j++ ) {
    tptr[var_count] = (float)j;
    float s = model->predict(
      temp_sample, noArray(), StatModel::RAW_OUTPUT
    );
    if( max_sum < s ) { max_sum = s; best_class = j + 'A'; }
  }

  // If the strongest (truest) proposition matched the correct response, then
  // score 1, else 0.
  //
  double r = std::abs( best_class - responses.at<int>(i) ) < FLT_EPSILON ? 1 : 0;

  // If we are still in the train samples, record one more correct train result.
```

```
    // Otherwise, record one more correct test result.
    // Hope nobody shuffled the samples!
    //
    if( i < ntrain_samples )
      correct_train _answers += r;
    else
      correct_test  _answers += r;
  }
```

Of course, this isn't the fastest or most convenient method of dealing with many class problems. Random trees may be a preferable solution, and we will consider it next.

## Random Trees

OpenCV contains a *random trees* class, which is implemented following Leo Breiman's theory of *random forests*.[26] Random trees can learn more than one class at a time simply by collecting the class "votes" at the leaves of each of many trees and selecting the class receiving the maximum votes as the winner. We perform regression by averaging the values across the leaves of the "forest." Random trees consist of randomly perturbed decision trees and were among the best-performing classifiers on data sets studied while the ML library was being assembled. Random trees also have the potential for parallel implementation, even on nonshared-memory systems, a feature that lends itself to increased use in the future. The basic subsystem on which random trees are built is once again a decision tree. This decision tree is built all the way down until it's *pure*. Thus (see the upper-right panel of Figure 20-3), each tree is a high-variance classifier that nearly perfectly learns its training data. To counterbalance the high variance, we average together many such trees (hence the name "random trees").

Of course, averaging trees will do us no good if the trees are all very similar to (correlated with) one another. To overcome this, the random trees method attempts to cause each tree to be different (statistically independent) by randomly selecting a different feature subset of the total features from which the tree may learn at each node. For example, an object-recognition tree might have a long list of potential features: colors, textures, gradient magnitudes, gradient directions, variances, ratios of values, and so on. Each node of the tree is allowed to choose from a random subset of these features when determining how best to split the data, and each subsequent node of the tree gets a new, randomly chosen, subset of features on which to split. The size of these random subsets is often chosen as the square root of the number of features. Thus, if we had 100 potential features, then each node would randomly choose 10 of the features and find a best split of the data from among those 10 features. To

26 Most of Breiman's work on random forests is conveniently collected on a single website: *http://www.stat.berkeley.edu/users/breiman/RandomForests/cc_home.htm*.

increase robustness, random trees use an *out of bag* measure to verify splits; that is, at any given node, training occurs on a new subset of the data that is randomly selected *with replacement*,[27] and the rest of the data—those values not randomly selected, or the "out of bag" (OOB) data—are used to estimate the performance of the split. The OOB data is usually set to have about one-third of all the data points.

Like all tree-based methods, random trees inherit many of the good properties of trees: surrogate splits for missing values, handling of categorical and numerical values, no need to normalize values, and easy methods for finding variables that are important for prediction. Also, because random trees use the OOB error results to estimate how well they will do on unseen data, performance prediction can be quite accurate if the training data has a similar distribution to the test data.

Finally, random trees can be used to determine, for any two data points, their *proximity* (which in this context means "how alike" they are, not "how near" they are). The algorithm does this by (1) "dropping" the data points into the trees, (2) counting how many times they end up in the same leaf, and (3) dividing this "same leaf" count by the total number of trees. A proximity result of `1` is exactly similar and `0` means very dissimilar. This proximity measure can be used to identify outliers (those points very unlike any other) and to cluster points (group together points with close proximity).

### Random trees code

We are by now familiar with how the ML module works, and random trees are no exception. We start with the class declaration. Once again, there is a class `cv::ml::RTreesImpl` that inherits from `cv::ml::RTrees`, which itself inherits from the `cv::ml::DTrees` class:

```
// Somewhere above..
// namespace cv {
//   namespace ml {
//
class RTreesImpl: public RTrees {  // cv::ml::RTrees is derived from cv::ml::DTrees

public:

  // whether variable importance should be computed during the training
  // makes the training noticeably slower
  //
  bool getCalculateVarImportance() const;          // Get if importance should be
                                                   //  computed during training
  int  getActiveVarCount() const;                  // Get N Var for each split
  TermCriteria getTermCriteria() const;            // Get max N trees or accuracy

  void setCalculateVarImportance( bool val );      // Get if importance should be
```

27 This means that some data points might be randomly repeated.

```
                                                  //  computed during training
    void setActiveVarCount( int val );            // Set N Var for each split
    void setTermCriteria( const TermCriteria& val ); // Set max N trees or accuracy

    ...

    Mat getVarImportance() const;

    ...

    // from class RTrees
    //
    //static Ptr<RTrees> create(); // The algorithm "constructor",
    //                             //  returns Ptr<RTreesImpl>
  };
```

A new method, `setCalculateVarImportance()`, is a switch to enable calculation of the variable importance of each feature during training (at a slight cost in additional computation time).

Figure 21-8 shows the variable importance computed on a subset of the mushroom data set that we discussed earlier *agaricus-lepiota.data.*

| Variable Name | Random | Boosting | DecisionTree |
|---|---|---|---|
| Col5 | 100.0 | 100.0 | 100.0 |
| Col20 | 35.2 | 58.89 | 57.37 |
| Col21 | 16.47 | 6.11 | 34.51 |
| Col19 | 13.35 | 4.57 | 26.11 |
| Col9 | 13.01 | 43.15 | 45.96 |
| Col13 | 10.02 | 24.47 | 26.85 |
| Col8 | 9.52 | 37.51 | 42.28 |
| Col12 | 9.09 | 27.66 | 28.90 |
| Col22 | 8.29 | 0.28 | 20.00 |
| Col7 | 6.08 | 0.10 | 21.33 |
| Col15 | 4.06 | 1.84 | 21.41 |
| Col11 | 3.52 | 0.44 | 16.29 |
| Col4 | 3.12 | | 14.67 |
| Col14 | 2.98 | 0.25 | 20.81 |
| Col18 | 2.68 | | 0.70 |
| Col3 | 2.56 | 0.11 | 9.15 |
| Col2 | 2.22 | 0.39 | 12.14 |
| Col10 | 1.79 | | 2.67 |
| Col1 | 0.41 | 0.24 | 7.26 |
| Col17 | 0.18 | 0.32 | 0.54 |
| Col0 | | | |
| Col6 | | | |
| Col16 | | | |

*Figure 21-8. Variable importance over the mushroom data set for random trees, boosting, and decision trees: random trees used fewer significant variables and achieved the best prediction (100% correct on a randomly selected test set covering 20% of data). Note that in OpenCV 3.x one can explicitly get the variable importance only for RTrees, and not for decision trees or boosting*

The `setActiveVarCount()` method sets the size of the randomly selected subset of features to be tested at any given node and, if not set by the user explicitly, is typically set to the square root of the total number of features.

Using `setTermCriteria()`, you can set the termination criteria that contains both the maximum number of trees (`cv::TermCriteria::maxIter`) and the OOB error, below which the tree generating need not continue (`cv::TermCriteria::epsilon`). As usual, either or both of the usual two stopping criteria can be applied (usually it's both: `cv::TERMCRIT_ITER | cv::TERMCRIT_EPS`). Otherwise, random trees training has the same form as decision trees training, boosting training, and so on.

The same multiclass learning example that we looked at before in the case of boosting, *.../opencv/samples/cpp/letter_recog.cpp*, also provides for training random forests. The following excerpt shows some salient points. Note that, unlike with boosting, we can train directly on the multiclass data.

```
using namespace cv;
...
Ptr<ml::RTrees> forest = ml::RTrees::create();
forest->setMaxDepth(10);
forest->setMinSampleCount(10);
forest->setMaxCategories(15);
forest->setCalculateVarImportance(true);
forest->setActiveVarCount(4);
forest->setTermCriteria(
  TermCriteria(
    TermCriteria::MAX_ITER+TermCriteria::EPSILON,
    100,
    0.01
  )
);
forest->train(tdata, 0);

...
```

Random trees prediction is no different from all other models from `ml`. Here's an example prediction call from the *letter_recog.cpp* file:

```
...

for( int i = 0; i < nsamples_all; i++ ) {

  cv::Mat sample = mydata.row( i );

  float r = forest->predict( &sample );
r = fabs( (float)r - responses.at<float>[i]) <= FLT_EPSILON ? 1 : 0;

// Accumulate some statistics using 'r'
  if( i < ntrain_samples )
    correct_train _answers += r;
  else
```

```
        correct_test  _answers += r;

    }
```

In this code, the return variable `r` is converted into a count of correct predictions. You can compute the same statistics using the universal method `cv::ml::StatModel::calcError()`.

Finally, there are random tree analysis and utility functions. Assuming that `setCalculateVarImportance( true )` was called before training, we can obtain the relative importance of each variable using the `cv::ml::RTrees` member function:

```
cv::Mat RTrees::getVarImportance() const;
```

In this case, the return will be a vector containing the relative importances of each of the features.

### Using random trees

We've remarked that the random trees algorithm often performs the best (or among the best) on the data sets we tested, but the best policy is still to try many classifiers once you have your training data defined. We ran random trees, boosting, and decision trees on the mushroom data set. From the 8,124 data points we randomly extracted 1,624 test points, leaving the remainder as the training set. After training these three tree-based classifiers with their default parameters, we obtained the results shown in Table 21-3 on the test set. The mushroom data set is fairly easy and so—although random trees did the best—it wasn't such an overwhelming favorite that we can definitively say which of the three classifiers works better on this particular data set.

*Table 21-3. Results of tree-based methods on the OpenCV mushroom data set (1,624 randomly chosen test points with no extra penalties for misclassifying poisonous mushrooms)*

| Classifier | Performance results |
|---|---|
| Random trees | 100% |
| AdaBoost | 99% |
| Decision trees | 98% |

What is more interesting is the variable importance (which we also measured from the classifiers), shown in Figure 21-8. The figure shows that random trees and boosting each used significantly fewer important variables than required by decision trees. Above 15% significance, random trees used only 3 variables and boosting used 6, whereas decision trees needed 13. We could thus shrink the feature set size to save computation and memory and still obtain good results. Of course, for the decision trees algorithm you have just a single tree, while for random trees and AdaBoost you must evaluate multiple trees; thus, which method has the least computational cost depends on the nature of the data being used.

## Expectation Maximization

*Expectation maximization* (EM) is a popular unsupervised clustering technique. The basic concept behind EM is similar to the K-Means algorithm, in that a distribution is modeled by a mixture of Gaussian components.[28] In this case, however, the process of learning those Gaussian components is an iterative one that alternates between two stages called expectation and maximization, respectively (hence the name of the algorithm).

In the formulation of the EM algorithm, each data point in the training set is associated with a latent variable that represents the Gaussian mixture component thought to be responsible for that variable taking its observed value (such variables are typically called *responsibilities*). Ideally, one would like to compute the parameters of the Gaussian components as well as these responsibilities such that they maximize the likelihood of the observed variables. In practice, however, it is not typically possible to maximize all of these variables simultaneously. This is subtly different than K-means in that a responsibility assigned to a data point is not necessarily to the closest cluster center.

The EM algorithm handles these two aspects of the problem by separately addressing the responsibilities (in the expectation or "E-step"), and then the parameters of the Gaussian components (in the maximization or "M-step"), and continuing to alternate between these two steps until convergence is reached.

### Expectation maximization with cv::EM()

In OpenCV, the EM algorithm is implemented in the `cv::ml::EM` class, which has the following declaration:

```
// Somewhere above..
// namespace cv {
//   namespace ml {
//
class EMImpl: public EM {        // cv::ml::EM is derived from cv::ml::StatModel

public:

  // Types of covariation matrices
  // (Inherited from cv::ml::EM)
  //
```

28 Though the expectation maximization algorithm is in fact much more general than just Gaussian mixtures, OpenCV supports only the special case of EM with Gaussian mixtures that we describe here.

```
    //enum Types {
    //  COV_MAT_SPHERICAL = 0,
    //  COV_MAT_DIAGONAL  = 1,
    //  COV_MAT_GENERIC   = 2,
    //  COV_MAT_DEFAULT   = COV_MAT_DIAGONAL
    //};

    // get/set the number of clusters/mixtures (5 by default)
    //
    int  getClustersNumber() const;                  // Get number of clusters
    int  getCovarianceMatrixType() const;            // Get cov. mtx. type
    TermCriteria getTermCriteria() const;            // Get max iter/accuracy

    void setClustersNumber( int val ) ;              // Set number of clusters
    void setCovarianceMatrixType( int val );         // Set cov. mtx. type
    void setTermCriteria( const TermCriteria& val ); // Set max iter/accuracy

    ...

    // the prediction method, see below
    //
    Vec2d predict2(
      InputArray  sample,
      OutputArray probs
    ) const;
    bool trainEM(...); // see below
    bool trainE(...);  // see below
    bool trainM(...);  // see below

    ...

    // from class EM
    //
    //static Ptr<EM> create(); // The algorithm "constructor",
    //                         //  returns Ptr<EMImpl>

};
```

When constructing and configuring a `cv::EM` object, we must tell the algorithm how many clusters there will be. We do this with the `setClusterNumber()` method. As in the case of the K-means algorithm, it is always possible to do tests with different numbers of clusters separately, but the basic `cv::EM` object can handle only one specific number of clusters at a time.

The next member function, `setCovarianceMatrixType()`, specifies the constraints that you wish to have the EM algorithm apply to the covariance matrices associated with the individual components of the Gaussian mixture model. This argument must take one of three possible values:

- `cv::ml::EM::COV_MAT_SPHERICAL`
- `cv::ml::EM::COV_MAT_DIAGONAL`
- `cv::ml::EM::COV_MAT_GENERIC`

In the first case, each mixture component is assumed to be rotationally symmetric. This means that it has only one free parameter that can be maximized in the M-step. Each covariance is just that parameter multiplied by an identity matrix. In the second case, `cv::EM::COV_MAT_DIAGONAL` (which is the default), each matrix is expected to be diagonal, and so the number of parameters is equal to the number of dimensions of the matrix (i.e., the dimensionality of the data). Finally, there is the case of `cv::EM::COV_MAT_GENERIC`, where each covariance matrix is characterized by the $(N_d^2 - N_d)/2$ variables needed to characterize an arbitrary symmetric matrix. In general, the complexity of the model that can be trained is strongly dependent on the amount of data available. In practice, if one has very little data, `cv::EM::COV_MAT_SPHERICAL` is probably a good idea. Conversely, unless one has a vast amount of data, `cv::EM::COV_MAT_GENERIC` is probably not a good idea.

How much is not enough, and how much is a lot? If you are doing EM in $N_d$ dimensions with $N_k$ clusters, then it is easy to compute how many free variables you are solving for. In the case of the spherically symmetric distributions (`COV_MAT_SPHERICAL`), you are trying to compute only one covariance per cluster plus the location of the cluster center, or $N_k(N_d + 1)$ total variables. For the diagonal covariances (`COV_MAT_DIAGONAL`), there are $N_d$ degrees of freedom for the covariance of each cluster or $N_k(N_d + N_d) = 2N_kN_d$ total variables. In the case of the completely general covariance matrices (`COV_MAT_GENERIC`), there are $(N_d^2 - N_d)/2$ degrees of freedom for the covariance of each cluster or $N_k(N_d + (N_d^2 - N_d)/2) = N_kN_d^2 + \frac{1}{2}N_kN_d$ total variables to solve for. Thus, at an absolute minimum there are $O(N_d)$ times more things you are trying to find in the general case than in the most restricted case. Unfortunately, even the most efficient learning algorithms will be much worse than linear in this ratio in terms of the amount of data they require.

The final configuration that EM requires is the termination criteria, set by means of `setTermCriteria()`. The `termCrit` argument required is of the usual `cv::TermCriteria` type, and specifies the maximum number of iterations allowed and (or) the maximum change in likelihood that will be considered "small enough" to terminate the algorithm.

Once you have created an instance of the `cv::EM` object and set the parameters, you can train it with one of three `train*()` methods specific to the EM algorithm—`trainEM()`, `trainE()`, and `trainM()`.

```
bool cv::ml::EM::train(
  cv::InputArray  samples,
  cv::OutputArray logLikelihoods = cv::noArray(),
  cv::OutputArray labels         = cv::noArray(),
  cv::OutputArray probs          = cv::noArray()
);

virtual bool cv::ml::EM::trainE(
  cv::InputArray  samples,
  cv::InputArray  means0,
  cv::InputArray  covs0          = cv::noArray(),
  cv::InputArray  weights0       = cv::noArray(),
  cv::OutputArray logLikelihoods = cv::noArray(),
  cv::OutputArray labels         = cv::noArray(),
cv::OutputArray probs            = cv::noArray()
);

bool cv::ml::EM::trainM(
  cv::InputArray  samples,
  cv::InputArray  probs0,
  cv::OutputArray logLikelihoods = cv::noArray(),
  cv::OutputArray labels         = cv::noArray(),
cv::OutputArray probs            = cv::noArray()
);
```

The `trainEM()` method of `cv::EM` expects the usual input array of samples and returns the responses (`labels`), the likelihoods (`logLikelihoods`), and the assignment probabilities (`probs`). The responses appear in the `labels` array, which will have a single column containing one row for each data point. The value on the *i*th row will be an integer identifying the cluster to which that data point was assigned. This integer is the largest of the membership probabilities computed across the $N_k$ clusters (set by `nclusters` when you called the constructor). The array `probs` will actually contain these individual probabilities, with each row giving the probabilities for a given point, and the column being the probability for that cluster (i.e., `probs.at<int>(i,k)` is the probability of point *i* being a member of cluster *k*). Finally, the likelihoods associated with each point are returned in the `logLikelihoods` array; in essence these likelihoods indicate (in a relative sense) how probable an individual observation was under the final model. As the name suggests, the values returned are actually the natural logarithm of the likelihoods. Any of these three outputs may be replaced with `cv::noArray()` if they are not needed. Once the model is trained, it can be used for prediction.

In addition to the train method just described, there are two additional methods, `trainE()` and `trainM()`. These methods start the algorithm off in the E- or M-step (respectively) and provide, in the case of `trainE()`, an initial model to use, or in the case of `trainM()`, a set of initial cluster assignment probabilities.

In the case of `trainE()`, the model is supplied in the form of the input arrays `means0`, `covs0`, and `weights0`. The form of the means should be an array with $N_k$ rows and $N_d$ columns, where $N_k$ is the number of clusters and $N_d$ is the dimensionality of the sample data. The covariances should be supplied as an STL vector containing $N_k$ separate arrays, each being $N_d \times N_d$ and containing the covariance matrix for Gaussian component *k*. Finally, the array `weights0` should have a single column and $N_k$ rows. The entry in the *k*th row should be the mixture probability associated with Gaussian component *k* (i.e., the marginal probability that any random sample will be drawn from component *k*, as opposed to some other component).

In the case of `trainM()`, you must supply `probs0`, the membership probabilities associated with each point, in the same form that they would be computed by `train()` as just described. This means that `probs0` is an $N_s \times N_k$ array with one row for each sample and the membership probabilities for each Gaussian component in the corresponding columns for that data point.

Once you have trained your model, you can use the `cv::EM::predict()` method to have the trained algorithm attempt to predict what cluster a novel point would most likely be a member of. The return value of `predict()` will be of type `cv::Vec2d`. The first component of the returned vector will give the probability associated with the assignment under the current model, while the second will give the cluster label.

## K-Nearest Neighbors

One of the simplest classification techniques is *K-nearest neighbors* (KNN), which stores all the training data points and labels new points based on proximity to them. When you want to classify a new point, KNN looks up the $N_k$ nearest points that it has stored and then labels the new point according to which class from the training set contains the majority of its $N_k$ neighbors. Alternatively, KNN can be used for regression; in this case the returned result is the average of the values associated with the $N_k$ nearest neighbors.[29] This algorithm is implemented in the `cv::ml::KNearest` class in OpenCV. The KNN classification technique can be very effective, but it requires that you store the entire training set; hence, it can use a lot of memory and become quite slow. People often cluster the training set to reduce its size before using this method. Readers interested in how dynamically adaptive nearest neighbor–type

29 For those of you who are experts in machine learning algorithms, it is noteworthy that the OpenCV implementation of KNN, when doing regression, uses an unweighted average of the $N_k$ neighbor points. (Other implementations often weight these points differently, for example by their inverse distance from the input point.)

techniques might be used in the brain (and in machine learning) can see Grossberg [Grossberg87] or a more recent summary of advances in Carpenter and Grossberg [Carpenter03].

### Using K-nearest neighbors with cv::ml::KNearest()

The KNN algorithm is implemented in OpenCV by the `cv::ml::KNearest` class. The class declaration for `cv::ml::KNearest` has the following form:

```
// Somewhere above..
// namespace cv {
//   namespace ml {
//
class KNearestImpl: public KNearest { // cv::ml::KNearest is derived
                                      // from cv::ml::StatModel
public:

  // (Inherited from cv::ml:: KNearest)
  //
  //enum Types {
  //  BRUTE_FORCE = 1,
  //  KDTREE      = 2
  //};

  // use K-nearest for regression or for classification

  //
  int  getDefaultK() const;         // Get the default number of neighbors
  bool getIsClassifier() const;     // Get if classifier, else it is regression
  int  getEmax() const;             // Get max "close" neighbors (KDTree)
  int  getAlgorithmType() const;    // Get whether brute-force or KD-Tree search
                                    //  (will be either BRUTE_FORCE or KDTREE)

  void setDefaultK( int val );      // Set the default number of neighbors
  void setIsClassifier( bool val ); // Set if classifier, else it is regression
  void setEmax( int val );          // Set max "close" neighbors (KDTree)
  void setAlgorithmType( int val ); // Set whether brute-force or KD-Tree search
                                    //  (must be either BRUTE_FORCE or KDTREE)
  ...

  // find the k nearest neighbors for each sample
  //
  float findNearest(
    InputArray  samples,
    int         k,
    OutputArray results,
    OutputArray neighborResponses = noArray(),
    OutputArray dist              = noArray()
  ) const = 0;

  ...
```

```
    // from class KNearest
    //
    //static Ptr<KNearest> create(); // The algorithm "constructor",
    //                               //  returns Ptr<KNearestImpl>
};
```

The method `setDefaultK()` sets the number of neighbors to be considered if you plan to use `cv::ml::StatModel::predict` instead of `cv::ml::KNearest::findNearest()`, where this parameter is specified explicitly ($k$ corresponds to $N_k$ in our preceding discussion of the KNN algorithm).

The `setIsClassifier()` function is used if you are using KNN for regression (i.e., you plan to approximate some function using the discrete training set). In this case, you should call `setIsClassifier(false)` prior to calling the `train()` method.

You train the KNN model as usual by calling the `cv::ml::StatModel::train()` method. The `trainData` input is the usual array containing $N_s$ rows and $N_d$ columns, with $N_s$ being the number of samples and $N_d$ being the number of dimensions of the data. The `responses` array must contain the usual $N_s$ rows and a single column. How the algorithm then handles this data depends on the `setAlgorithmType()` method. You can call this method with either `BRUTE_FORCE` or `KDTREE`.

If the algorithm is set to `BRUTE_FORCE`, this training data is simply stored internally as an array and then scanned sequentially in order to find the nearest neighbors. In the case of `KDTree`, the BBF (best-bin-first) algorithm (introduced by D. Lowe) will be used, which is much more efficient when $N_d \ll log(N_s)$.

It's important to note that unlike many other models in the `ml` module, the KNN model can be updated with new trained data after it's trained. In order to do this, pass the flag `cv::ml::StatModel::UPDATE_MODEL` to the `train()` method.

Once the data is trained, you can use your `cv::ml::KNearest` object to make predictions about it. The `cv::ml::KNearest` object provides the usual `predict()` method as well as the more model-specific `findNearest()`, which is equivalent to `findNearest( samples, getDefaultK(), results, noArray(), noArray() )`. But the method `findNearest()` method allows you to retrieve some additional information about the neighbors. The `findNearest()` method has the following definition:

```
virtual float cv::ml::KNearest::findNearest(
  cv::InputArray  samples,
  int             k,
  cv::OutputArray results,
  cv::OutputArray neighborResponses = cv::noArray(),
  cv::OutputArray dist              = cv::noArray()
) const = 0;
```

The method takes one or more samples at a time, with the `samples` argument being any number of rows with each row containing a single input. The value of $k$ that is used for the comparison (the argument `k`) can be any value up to the value of $N_k$ given when the object was trained. The predictions will be placed in the array `results`, which will have a single column and one row corresponding to each point (row) in `samples`. If provided, the arrays `neighborResponses` and `dists` will be filled with the responses from and distances to the various neighbors identified for each query point. Each of these will have one row per point (row) in `samples` and one column for each of the *k* neighbors found for that point.[30]

Both methods will return a single floating-point value; this is used when the number of query points in `samples` is just one so only one computed response will be returned. In this case, you do not need to provide a `results` array at all.

## Multilayer Perceptron

The *multilayer perceptron* (MLP; also known as *back propagation*, which actually refers to the weight update rule) is a neural network that ranks among the top-performing classifiers for text recognition as well as a rapidly growing list of other tasks.[31] In fact, MLP plays an important role in the currently evolving topic of deep learning where it, along with other neural network–based methods, has shown significant successes in recent years.

When used to perform predictions, MLP can be very fast; evaluation of a new input requires just a series of dot products followed by a simple nonlinear "squashing" function. On the other hand, it can be rather slow in training because it relies on gradient descent, in a rather difficult environment, to minimize error by adjusting a potentially vast number of weighted connections between the numerical classification nodes within the layers.

The essential idea behind the multilayer perceptron is taken from biology, where studies of mammalian neural networks motivate the idea of layers on neurons, each of which takes in some number of inputs from the prior layer, sums those inputs with learned weights, and outputs some kind of nonlinear transformation on that sum of weighted inputs (Figure 21-9). This simplified model of a biological neuron, often called an *artificial neuron* (or, collectively, an *artificial neural network*) was created with the goal of capturing the basic functionality of the biological neuron with the hope that, when aggregated into networks, those networks would display learning

30 If you wish to use your own weighting function in a regression, you can use this information to do so.

31 Note that this is the core algorithm used in deep neural networks, but this does not implement a deep neural network beyond one or two hidden layers. We will discuss shortly the new support for deep neural networks that is currently being added to the library.

and generalization behavior similar to that observed in biological systems—ideally up to and including human beings.[32]

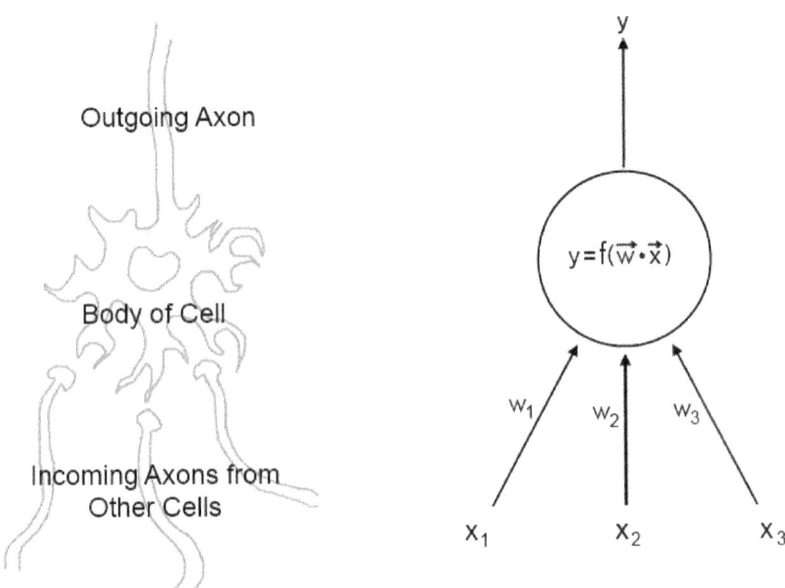


*Figure 21-9. An artificial neuron used in MLP and other neural network applications models the behavior of a physical neuron with a typically nonlinear activation function f, which acts on the weighted sum of inputs. Learning in an MLP is embodied in the tuning of these weights $\vec{w}$*

These artificial neurons, often called simply *nodes*, aggregated into a multilayered network (Figure 21-10), along with an algorithm to train them, form the basis of the MLP algorithm. Such multilayered networks can take many forms, the simplest of which is called a *fully connected multilayer network*.[33] In such a network, all of the inputs to the network can feed any of the nodes in the first layer.[34] Similarly, all of the

32 At best, most current neural networks as well as deep learning architectures represent a possible feed-forward processing stream in the brain, and even that neglects substantial attention and other biases in processing. Most models entirely neglect the 10x greater feedback processes in the brain, which may have to do with simulation, as speculated in the last chapter in this book.

33 For you aficionados, there is a further point here that in this context, we use the term *fully connected network* to also imply that all of the weights are unconstrained, and thus fully independent of one another.

34 In the neural network literature, the term *input layer* is sometimes used to mean the first layer of neurons, and sometimes used to mean the number of inputs before the first layer. To avoid ambiguity, we will simply avoid this terminology altogether. We will always use the term *first layer* to mean the first layer of computational nodes, the number of which is completely independent of the number of inputs to the network.

outputs of the first layer can feed any of the nodes in the second layer, and so on. Such a network can have any number of nodes in any layer, with the last layer having the number of nodes required to express the desired response function.

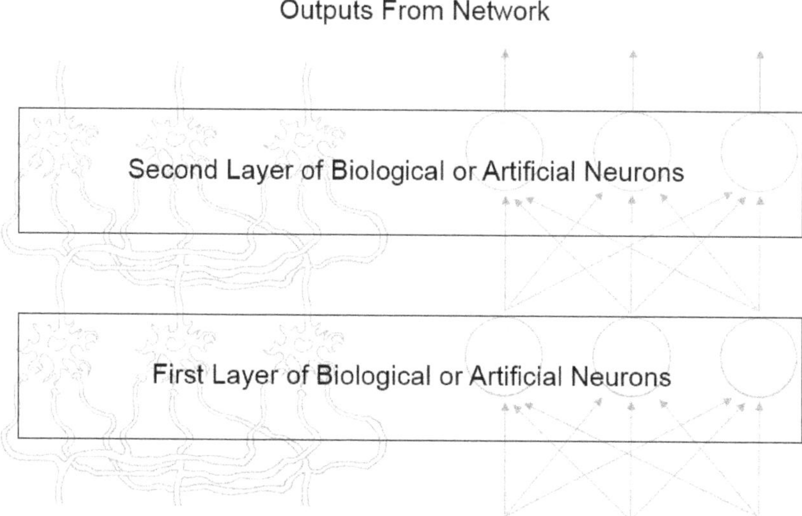


*Figure 21-10. In analogy to biological neurons, artificial neurons can be arranged into layers. The inputs to this network feed all of the neurons in the first layer in a weighted dot product, $\vec{w} \cdot \vec{x}$ which is filtered by a nonlinear function, f(w.x), to produce the output at that node. The outputs of the first-layer neurons feed all of the inputs to the next layer, and so on*

The number of nodes in such a network can be very large, and the number of weights much larger still. Consider a network with $N_x$ inputs, $N_y$ outputs (which defines the number of nodes in the last layer), $N_l$ layers total, and $N_n$ nodes in each layer (except for the last—which has $N_y$ nodes). Such a network would have $N_x N_n$ weights for inputs to the first layer, and $N_n^2$ weights for every layer after the first, up to the last layer, which would require $N_n N_y$ weights. Together, that is $N_n(N_x + N_y + (N_l - 2)N_n)$ different weights, something on the order of $N_l N_n^2$. To train a network of this kind means to find the optimal values of all of these weights such that, for your training data set, the outputs from the network for any given input match—as closely as possible—the results associated with that input.

One way in which you might imagine tackling such a multidimensional optimization problem would be through a simple algorithm such as gradient descent, by means of which you could start with some random set of values and incrementally optimize them in a greedy way until no improvement is possible. This is exactly what was done in the early days of neural networks. However, given the very large number of parameters in even a simple practical network, the most straightforward implementations of such a method proved too slow to be useful in real-world problems (or equivalently, networks that could be trained in practical amounts of time performed too poorly to be useful). It was not until the invention of the back-propagation algorithm [Rumelhart88] that it became possible to train very large networks. Back propagation is a particular variant of gradient descent that makes special use of the structure of the particular problem of training a neural network. It is this algorithm that is embodied in the training step of the OpenCV implementation of MLP.

### Back propagation

Refer to Figure 21-11. We can use a trained back-propagation network to make decisions or to fit functions by feeding an input in at the "bottom" that is weighted and transformed through nonlinearities (usually sigmoid or rectifier functions). This signal is propagated up layer by layer in the network until an output is produced at the "top." In training mode, we use a *loss function* to gauge how well the network is doing. In Figure 21-11, this loss function is the square of target value minus the actual value. The essential issue that back propagation addresses is how to update the weights in the neural network so that the loss function is minimized over the data set.

To minimize the loss function, the chain rule in calculus is used to derive, from top-to-bottom ("back propagation"), how the weights must change in relation to the loss function at the top.[35] The back-propagation algorithm is a *message-passing* algorithm that takes advantage of the topology of the multilayer perceptron. It first computes the error at the output nodes (the difference between what you want and what you get). It then propagates this information backward to those nodes that feed the output nodes. They combine the information they received from the output nodes to compute the necessary derivatives relative to their own weights. This information is then passed on again to the next layer toward the input layer. In this way, we can compute the portion of the gradient that pertains to each weight locally, using only information about that node and information passed to that node from the layer one step closer to the output to which it is connected.

35 Interested readers can find deeper treatments of this algorithm in any good book on machine learning (e.g., [Bishop07]). For more technical details, and for detail on using MLP effectively for text and object, the reader might be directed toward LeCun, Bottou, Bengio, and Haffner [LeCun98a]. Implementation and tuning details are given in LeCun, Bottou, and Muller [LeCun98b]. More recent work on brain-like hierarchical networks that propagate probabilities can be found in Hinton, Osindero, and Teh [Hinton06].

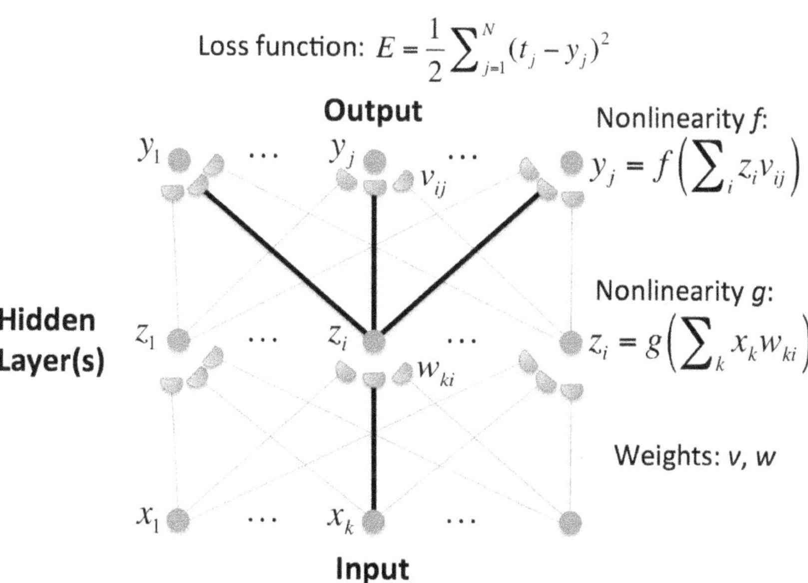


*Figure 21-11. Back-propagation setup. The goal is to update the weights $v_{ij}$, $w_{ki}$ in order to minimize the loss function. Each hidden and output node gets weighted activation $\Sigma_i v_{ij} z_i$, $\Sigma_k w_{ki} x_k$, which then goes through a nonlinearity function g or f. In practice, these nonlinear functions are all the same, often a sigmoid or a rectifier function, but for generality, they can be different*

In the network of Figure 21-11, the output weight change function is:

$$\frac{dE}{dv_{ij}} = -(t_j - y_j)z_i f'\left(\sum_i v_{ij} z_i\right)$$

The hidden unit weight change function is:

$$\frac{dE}{dw_{ki}} = \sum_i -(t_j - y_j)v_{ij} f'\left(\sum_i v_{ij} z_i\right)x_k g'\left(\sum_k w_{ki} x_k\right)$$

where $t_j$ represents the target values, and the input, hidden, and outlook activations are $x_k$, $z_i$,$y_j$. The last two activations are transformed by the nonlinear functions $g$ and $f$ having input weights $w_{ki}$, $v_{ij}$, respectively. This forms the basic core of the back-propagation algorithm.

OpenCV now supports reading and efficiently running the deep neural networks produced by the deep net packages *Caffe*, *TensorFlow*, *CNTK*, *Torch*, and *Theano*; see *...opencv_contrib/modules/dnn* as referenced in Appendix B. Deep networks are often broken up into block convolution layers to save on the number of parameters that must be learned. That—combined with other training tricks (such as batch normalization), dropout and tuning pretrained early layers, increasing compute power, and huge data stores—has allowed the recent remarkable progress toward human-level machine vision performance.

#### The Rprop algorithm

The Rprop (resilient back propagation) was introduced by Martin Riedmiller and Heinrich Braun [Riedmiller93] and provides an often-superior method of doing the update step of the back-propagation algorithm just described. The essential difference between back propagation and Rprop is that, while back propagation explicitly computes the magnitude of each step by which each weight is updated, Rprop uses only the sign (direction) of the computed update. When computing the magnitude of the update, Rprop computes and stores the direction of the update, then makes a change by a standard step (called $\Delta_0$). On the next pass, if the direction changes (meaning that the algorithm essentially overshot the goal), the step is reduced by a constant scale factor (usually called $\eta^-$). If the step does not change sign, then the algorithm increases the step size of the next iteration by multiplying by a different scale factor (usually called $\eta^+$). Clearly $\eta^-$ must be less than `1` and $\eta^+$ must be greater than `1`; canonical values for these parameters are `0.5` and `1.2`, respectively. In the OpenCV implementation, the step is never allowed to be smaller than some absolute minimum, nor larger than some absolute maximum (called $\Delta_{min}$ and $\Delta_{max}$, respectively).

#### Using artificial neural networks and back propagation with cv::ml::ANN_MLP

The OpenCV implementation of multilayer perceptron learning handles the construction and evaluation of the network, and implements the back-propagation algorithm (or Rprop algorithm) just described for training the network. All of this is contained within the `cv::ml::ANN_MLP` class, which implements the usual interface inherited from the statistical learning base class. The `cv::ml::ANN_MLP` class has the usual hidden implementation class, in this case with the following definition:

```
// Somewhere above..
// namespace cv {
//   namespace ml {
//
class ANN_MLPImpl: public ANN_MLP {    // cv::ml::ANN_MLP is derived
                                       // from cv::ml::StatModel
public:
```

```
// (Inherited from cv::ml::ANN_MLP)
//
//enum TrainingMethods {
//  BACKPROP        = 0, // The back-propagation algorithm.
//  RPROP           = 1  // The RPROP algorithm.
//};
//
//enum ActivationFunctions {
//  IDENTITY        = 0,
//  SIGMOID_SYM     = 1,
//  GAUSSIAN        = 2
//};
//
//enum TrainFlags {
//  UPDATE_WEIGHTS  = 1,
//  NO_INPUT_SCALE  = 2,
//  NO_OUTPUT_SCALE = 4
//};


int getTrainMethod() const;          // Get backpropagation or RPROP method
int getActivationFunction() const;   // Get the activation function
                                     // (IDENTITY, SIGMOID_SYM or GAUSSIAN)
Mat getLayerSizes() const;                // Get size of all the layers
TermCriteria getTermCriteria() const;     // Get max it / reproject error
double getBackpropWeightScale() const;    // Set Backprop parameter
double getBackpropMomentumScale() const;  //  "
double getRpropDW0() const;               // Set Rprop parameter
double getRpropDWPlus() const;            //  "
double getRpropDWMinus() const;           //  "
double getRpropDWMin() const;             //  "
double getRpropDWMax() const;             //  "

void setTrainMethod(                 // Set backpropagation or RPROP method
  int    method,                     // either of: BACKPROP or RPROP
  double param1 = 0,
  double param2 = 0
);
void setActivationFunction(          // Set activation function
  int    type,                       // IDENTITY, SIGMOID_SYM, or GAUSSIAN
  double param1 = 0,
  double param2 = 0
);
void setLayerSizes( InputArray _layer_sizes ); // Set size of all the layers
void setTermCriteria( TermCriteria val );      // Set max it / reproject error
void setBackpropWeightScale( double val );     // Get Backprop parameter
void setBackpropMomentumScale( double val );   //  "
void setRpropDW0( double val );                // Set Rprop parameter
void setRpropDWPlus( double val );             //  "
void setRpropDWMinus( double val );            //  "
void setRpropDWMin( double val );              //  "
```

```
    void setRpropDWMax( double val );             //  "

    ...

    Mat getWeights( int layerIdx ) const; // Get the computed weights for
                                          //  the interlayer connections

    ...

    // from class ANN_MLP
    //
    //static Ptr<ANN_MLP> create(); // The algorithm "constructor",
    //                              //  returns Ptr<ANN_MLPImpl>
};
```

To use the multilayer perceptron, you first construct the empty network with the `cv::ml::ANN_MLP::create()` method, and then set the size of all the layers. Next, once you have set all the parameters (activation function, the training method, and its parameters), you can finally call the `cv::ml::StatModel::train()` method, as usual.

The argument of the `setLayerSizes()` method specifies the basic structure of the network.[36] This should be a single-column array in which the rows specify the number of inputs, the number of nodes in the first layer, the number of nodes in the second layer, and so on, until finally, the last row gives the number of nodes in the final layer (and thus the number of outputs from the network).[37]

The method `setActivationFunction()` specifies the function that is applied to the weighted sum of the inputs. The computed output of any node in the network $y_i$ is given by some function $f()$ of the weighted sum of inputs $\vec{x} \cdot \vec{w}$ plus an offset term: $\theta$. The weights and the offset term will be learned in the training phase, but the function itself is a property of the network. At this time, there are three functions you can choose from: a linear function, a sigmoid, and a Gaussian function.[38] The values and corresponding arguments for `setActivationFunction` are shown in Table 21-4.

36 One of the classic "dark arts" in back propagation and in deep learning is how many neurons to set in each layer. You can use fewer neurons in the hidden layer to achieve data compression or more neurons to overrepresent the data, which can improve performance given enough data.

37 By this counting, the network in Figure 21-14 would be described by a column having 3, 3, and 3 in its rows—the first corresponding to the three inputs, the second to the first row, and the third to the final row. Remember that the number of outputs is exactly equal to the number of nodes in the final layer, so you don't need another 3 after the third row.

38 The implementation of the Gaussian function is not entirely complete at this time.

*Table 21-4. Options for the artificial neuron activation function used by cv::ANN_MLP. Each is a function of* $z = \vec{x} \cdot \vec{w} + \theta$

| Value of activateFunc | Activation function |
|---|---|
| `cv::ANN_MLP::IDENTITY` | $f(z) = z$ |
| `cv::ANN_MLP::SIGMOID_SYM` | $f(z) = \beta \frac{(1 - e^{-\alpha z})}{(1 + e^{-\alpha z})}$ |
| `cv::ANN_MLP::GAUSSIAN` | $f(z) = \beta e^{-\alpha z^2}$ |

The final two arguments, `fparam1` and `fparam2`, correspond to the variables $\alpha$ and $\beta$ in Table 21-4. Unless you are an expert however, in almost all cases, you will want to use the default sigmoid activation function, and in the majority of those, you will want to set these parameters at their generic values (of `1.0`).

Once you have created your artificial neural network, you will want to train the network with the data you have. In order to train the network, you will need to construct `cv::ml::TrainData` as usual, and then pass it to the `cv::ml::StatModel::train()` method, overridden in `cv::ml::ANN_MLP`. Unlike all other models from `ml`, neural nets are able to handle vector outputs, not only scalar outputs, so that the response is not necessarily a single-column vector—it can be a matrix, one row per sample. And the vector output can actually be used to implement a multiclass classifier using neural networks.

Artificial neural networks are not natively structured to handle categorical data, either on the input or on the output. The most common solution to this is to use a "one of many" (also known as "one hot") encoding in which a *K*-element class is represented by *K* separate inputs (or outputs) with one being associated with each class. In this scheme, if an input object is a member of class *k*, then network input *k* will be nonzero (typically `1.0`) while all of the others will be `0`. The same system can be used to encode outputs. One interesting feature of such encodings is that in practice, even though inputs are exactly unit vectors, the outputs will typically have an imperfect structure (small values for nonclasses and less than unity value for the selected class). This has the advantage of revealing something about the quality of the categorization.

Also, neural networks can handle the nonempty `sampleWeights` vector, passed to `cv::ml::TrainData::create`, which allows you to assign a relative importance to each data sample (row) in `inputs` and `outputs`. Each row in `sampleWeights` can be set to an arbitrary positive floating-point value, corresponding to the data in the same rows of `inputs` and `outputs`. The weights are automatically normalized, so all that matters is their relative sizes. The `sampleWeights` argument is only respected by the

Rprop algorithm, so if you are using back propagation it will be ignored (see next discussion).

### Parameters for training

The method `setTermCriteria()` specifies when training will terminate, and has the same meaning if you are using back propagation or RProp (see `setTrainMethod()`, next). The number of iterations in the termination criteria is exactly the maximum number of steps the update will take. The epsilon portion of the termination criteria sets the minimum change in the *reprojection error* required for the iteration to continue; this reprojection error is equal to `0.5` times the sum of the squared differences between training set results and computed outputs over the entire training set.[39]

You can set `train_method` to either `cv::ANN_MLP::BACKPROP` or `cv::ANN_MLP::RPROP`. By default, this parameter will be set to `RPROP`, as this method is generally more effective for training the network in most circumstances. In the case of `RPROP`, `param1` and `param2` in the constructor correspond to the initial update step size and the smallest update $\Delta_0$ step size $\Delta_{min}$ (respectively). In the case of `BACKPROP`, `param1` and `param2` correspond to what are called the *weight gradient* and the *momentum*, respectively. Alternatively, you may set these parameters using the `setBackpropWeightScale()` and `setBackpropMomentumScale()` member accessor functions. The weight gradient value multiplies the weight gradient term in the back-propagation update step and essentially controls how fast the updates move in the desired direction. A typical value for this term is `0.1`. The momentum value multiplies an additional term in the update that is proportional to the difference in the value between the prior step and the step prior to that—giving something like a velocity for the weight. The momentum term premultiplies this velocity, which has the effect of smoothing out large fluctuations in the update. If set to `0`, this term is effectively eliminated. However, a small value (typically about `0.1` also) often gives a substantially faster overall convergence. In addition to these two parameters, there are the accessors `setRpropDWPlus()`, `setRpropDWMinus()`, `setRpropDW0()`, `setRpropDWMin()`, and `setRpropDWMax()`, which can be used to set the values $\eta^+$, $\eta^-$, $\Delta_0$, $\Delta_{min}$, and $\Delta_{max}$ respectively (as well as the corresponding `get*()` methods to simply inspect these values).

Once you have trained your network, you can then use it to make predictions in the usual way using the overridden `cv::ml::StatModel::predict()` method. As usual, `samples` must be an array with one row per input data point and the correct number of columns (the same number as the training data). The `results` array will have as many rows as `inputs` did, but the number of columns equal to the number of nodes

39 If you have sample weights, the error terms are also weighted by the sample weights.

in the output layer of the network. The return value is meaningless and can be ignored.[40]

## Support Vector Machine

The support vector machine (SVM) is a classification algorithm that, in its basic form, is used to separate two classes based on a set of exemplars. Extensions of the SVM algorithm can be used to implement multiclass ($N_c > 2$) classification. The concept that underlies the SVM is the use of *kernels* by means of which the data points in some particular number of dimensions $N_d$ are mapped into a space of much higher dimension $N_{KS}$, called the *kernel space*.[41] In that space, a linear classifier can often be found that separates the two classes, even if no such linear separation was possible in the original, lower dimensional space. The SVM is called a *maximum margin classifier* because it selects a hyperplane in the kernel space that not only separates the two classes, but does so with the largest amount of distance (*margin*) to those exemplars of each class that are closest to the hyperplane. Those exemplars that are close to the hyperplane, and which define its location, are called the *support vectors*. The significance of the support vectors is that once they have been identified, only they need to be retained in order to make a decision about a future data point whose class identity is to be predicted.

More formally, this decision hyperplane can be described by the equation $\vec{w} \cdot \vec{x} + b = 0$ for a linear SVM. Note that although this decision plane is linear in the higher-dimensional space, it can be very nonlinear in the original, lower-dimensional space, as shown in Figure 21-12. In this case, the $\vec{x}$ are points in the kernel space and the vector $\vec{w}$ defines the normal to the hyperplane. By convention, $\vec{w}$ is normalized such that the distance between the decision hyperplane and the support vectors is $1 / \| \vec{w} \|$ in either direction. The value $b$ gives an offset of the hyperplane (relative to the parallel hyperplane that passes through the origin).[42] Given such a parameterization, the support vectors themselves will lie on the planes defined by $\vec{w}.\vec{x} + b = +1$ and $\vec{w}.\vec{x} + b = -1$.

40 It is only there to conform to the interface specified by `cv::ml::StatModel::predict()`.

41 It is also often called the *feature space*. This terminology is confusing, however, as other authors use it to refer to the lower-dimensional space of the original input features. To prevent this confusion, we will avoid this terminology.

42 More precisely, the offset is $b / \| \vec{w} \|$ .

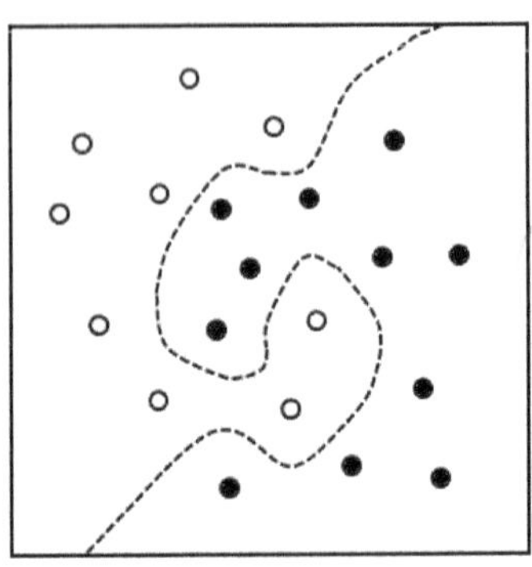

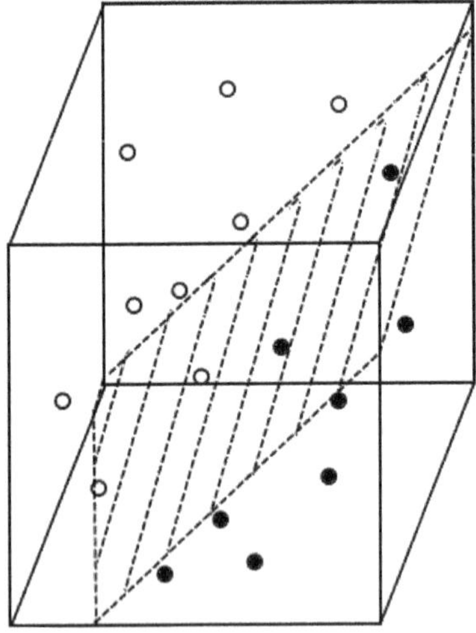

*Figure 21-12. The points in the original low-dimensional space (left) are separated by first being mapped into a higher-dimensional space (right). Though the decision surface is linear (a hyperplane) in the high-dimensional space, it can be very nonlinear in the low-dimensional space*

Though the full argument for this is beyond the scope of this book, looking at Figure 21-13 should make it at least seem reasonable that the vector $\vec{w}$ can be expressed in terms only of the support vectors themselves—the logic being that ultimately the hyperplane is defined only by these elements of the training set. This is analogous to the "front line" on a battlefield being defined only by the people near the opposition—it doesn't really matter where the people in the back arrange themselves. Somewhat less intuitively, it can be shown that the vector $\vec{w}$ can be expressed specifically by a linear combination of the support vectors. Similarly, the value of $b$ can be computed directly from the support vectors. Once we have these two parameters that define the decision hyperplane, any new point $\vec{x}$ can be passed to the classifier and can be determined to be in one of four regions: well inside of class 1 (i.e., $\vec{w}.\vec{x} + b \geq +1$), marginally inside of class 1 ($\vec{w}.\vec{x} + b > 0$), marginally inside of class 2 ($\vec{w}.\vec{x} + b < 0$), or strongly inside of class 2 ($\vec{w}.\vec{x} + b \geq +1$).

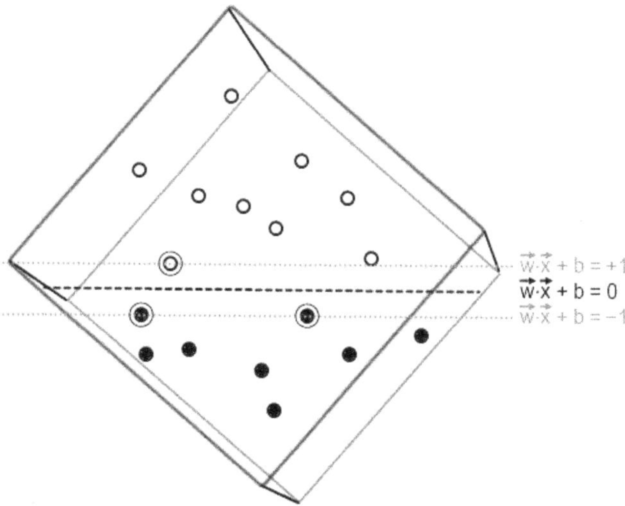


*Figure 21-13. The same kernel space from Figure 21-12 is shown here rotated so that we are looking at the edge of the decision hyperplane. The support vectors (circled) are equidistant on both sides of the hyperplane*

## About kernels

The kernel is the mapping that takes points from the native space of the input training vectors to the higher-dimensional feature space and back again. Remember that it is in this higher-dimensional space that we intend to find the separating hyperplane. The kernel is composed of two parts: the part that takes us from the space of the input features to the kernel space and the inverse transformation that takes us back. Customarily, the kernel is written as $K(\vec{x}, \vec{x}\,')$, while the mappings are written as $\vec{\varphi}(\vec{x})$ and $\vec{\varphi}^T(\vec{x}')$, respectively. The kernel, by definition, can then be expressed as:

$$K(\vec{x}, \vec{x}\,') = \vec{\varphi}^T(\vec{x})\vec{\varphi}(\vec{x}\,')$$

It is easiest to understand a kernel by way of example. One of the kernels available to OpenCV's SVM is called the *polynomial kernel*. This kernel has the form:

$$K(\vec{x}, \vec{x}\,') = (\vec{x} \cdot \vec{x}' + c)^q$$

for some value of $c > 0$ and integer $q > 0$. The mapping $\overrightarrow{\varphi_{q,c}}(\vec{x})$ that corresponds to this kernel can have a very large number of components. Even for the simplest case where $q = 2$ and $c = 0$, given an input vector of three dimensions ($N_d = 3$), with $\vec{x} = (x_1, x_2, x_3)$:[43]

$$\vec{\varphi}_{2,0}(x_1, x_2, x_3) = \left(x_3^2, x_2^2, x_1^2, \quad \sqrt{2}x_3x_2, \sqrt{2}x_3x_1, \sqrt{2}x_2x_1\right)$$

For comparison, for $c = 1$:

$$\vec{\varphi}_{2,1}(x_1, x_2, x_3) = \left(x_3^2, x_2^2, x_1^2, \quad \sqrt{2}x_3x_2, \sqrt{2}x_3x_1, \sqrt{2}x_2x_1, \sqrt{2}x_3, \sqrt{2}x_2, \sqrt{2}x_1, 1\right)$$

Notice that for the first choice of kernel, $N_d = 3$ corresponds to a kernel space of dimension of $N_{KS} = 6$; for the second kernel with the same input dimension, $N_{KS} = 10$.

What is particularly important about these kernels, however, is that the mappings $\vec{\varphi}(\vec{x})$ never need to be computed in the course of computing the separation hyperplane for the SVM. The reason for this critical fact is that all of the computations in the feature space that are needed only involve dot products between the points. This means that, wherever we find $\vec{\varphi}(\vec{x})$, it will be in the form $\vec{\varphi}^T(\vec{x})\vec{\varphi}(\vec{x}\,')$ for some pair of points $\vec{x}$ and $\vec{x}\,'$. Worded differently, we never need to evaluate $\vec{\varphi}(\vec{x}\,')$; we only need to evaluate $K(\vec{x}, \vec{x}\,')$. The implications of this are profound, since even though the dimension of the kernel space might be very high (even infinite!), we need only be able to evaluate $K(\vec{x}, \vec{x}\,')$ for vectors of the input feature space dimension to train and use the SVM. This property is called the *kernel trick* in SVM literature.

In practice then, one only needs to select a kernel and from there, the SVM can do the work. The actual kernel selected will have a heavy influence on how the constructed hyperplane is generated, how well it separates the training data, and how well that separation generalizes to future data. OpenCV currently provides six different kernels you can use in your support vector machines (see Table 21-5).

*Table 21-5. The available kernel functions for the OpenCV SVM implementation*

| Kernel | OpenCV name (kernel_type) | Kernel function | Parameters |
|---|---|---|---|
| Linear | `cv::ml::SVM::LINEAR` | $K(\vec{x}, \vec{x}') = \vec{x}^T\vec{x}'$ | |
| Polynomial | `cv::ml::SVM::POLY` | $K(\vec{x}, \vec{x}') = (\gamma\vec{x}^T\vec{x}' + c_0)^q$ | `degree` ($q$),<br>`gamma` ($\gamma > 0$),<br>`coef0` ($c_0$) |
| Radial basis functions | `cv::ml::SVM::RBF` | $K(\vec{x}, \vec{x}') = e^{-\gamma\|\vec{x}^T\vec{x}'\|^2}$ | `gamma` ($\gamma > 0$) |
| Sigmoid | `cv::ml::SVM::SIGMOID` | $K(\vec{x}, \vec{x}') = \tanh(\gamma\vec{x}^T\vec{x}' + c_0)$ | `gamma` ($\gamma > 0$),<br>`coef0` ($c_0$) |

43 It is not hard to verify that $\vec{\varphi}_{2,c}^T(\vec{x})\vec{\varphi}_{2,c}(\vec{x}') = \vec{\varphi}_{2,c}(\vec{x}) \cdot \vec{\varphi}_{2,c}(\vec{x}') = K(\vec{x}, \vec{x}\,') = (\vec{x} \cdot \vec{x}' + c)^2$; you can just multiply it out.

| Kernel | OpenCV name (kernel_type) | Kernel function | Parameters |
|---|---|---|---|
| Exponential chi-squared | `cv::ml::SVM::CHI2` | $K(\vec{x}, \vec{x}') = e^{-\gamma \frac{x-x'}{x-x'}}$ | gamma ($\gamma > 0$) |
| Histogram intersection | `cv::ml::SVM::INTER` | $K(\vec{x}, \vec{x}') = \min(\vec{x}, \vec{x}')$ | |

### Handling outliers

OpenCV supports two different variants of SVM called *C*-vector SVM and *ν*-vector SVM.[44] These extensions support the possibility of *outliers*. In this context, an outlier is a data point that cannot be assigned to the correct class (or equivalently a data set that cannot be separated by a hyperplane in the kernel space). In addition to outliers, the implementation of both of these methods in OpenCV allows for multiclass ($N_c > 2$) classification.

The *C*-vector SVM, also known as a *soft margin* SVM, allows for the possibility that any particular point is an outlier by penalizing the outlier by an amount proportional to the distance that point is found to be past the decision boundary. Such distances are customarily called *slack variables*, while this constant of proportionality is conventionally called *C* (the latter being the origin of the name for this method).

The second extension, which also allows for the possibility of outliers, is called the *ν*-vector SVM [Schölkopf00]. In this case, the proportionality constant analogous to *C* takes a constant (predetermined) value. Instead, however, there is the new parameter *ν*. This parameter sets an upper bound on the number of training errors (i.e., misclassifications), as well as a lower bound on the fraction of training points that are support vectors. The parameter *ν* must be between `0.0` and `1.0`.

Comparing the two, the *C*-vector SVM has a more straightforward implementation, and so is often faster to train. However, because there is no natural interpretation of *C*, it is difficult to find a good value other than through trial and error. On the other hand, the parameter of the *ν*-vector SVM has a natural interpretation, and so in practice is often easier to use.

### Multiclass extension of SVM

Both the *C*-vector SVM and the *ν*-vector SVM, as implemented in OpenCV, support the possibility of more than two classes. Though there are many possible ways to do this, OpenCV uses the method called *one-versus-one*.[45] In this method, if there are *k* classes, a total of $k(k - 1)/2$ classifiers are trained. When an object is to be classified,

44 That is a "nu" as in the Greek letter, but is pronounced like the English word "new." The pun is presumably intentional.

45 The "other" option is *one-versus-many*, which is not implemented in OpenCV.

every classifier is run and the class that has the majority of wins is taken to be the result.

### One-class SVM

Another interesting variant of the SVM is the *one-class* SVM. In this case, all of the training data is taken to be exemplary of a single class, and a decision boundary is sought that separates that one class from everything else. As with the multiclass extensions just described, the one-class SVM also supports outliers with a mechanism of slack variables similar to the *C*-vector SVM (and having the same parameter).

### Support vector regression

In addition to extensions for single class and multiclass, support vector machines have also been extended to allow for the possibility of regression (as opposed to classification). In the regression case, the input results are sequential values, rather than class labels, and the output is an interpolation (or extrapolation) based on the input values which were provided. OpenCV supports two different algorithms for support vector regression (SVR): $\varepsilon$-SVR [Drucker97] and $\nu$-SVR [Schölkopf00].

In its most abstract form, the SVR and the SVM are essentially the same, with the exception that the error function that is being minimized in SVM always has the output equal to one of two values (typically `+1` and `-1`), while the SVR has a different objective for each input example. The nuance, however, arises from how the slack variables are handled, and what points are considered "outliers." Recall that in the SVM, points were assigned to classes that were separated by the maximum margin hyperplane. Thus, being an outlier was a statement about the location of a point, given its label, relative to the hyperplane. In the SVR case, the hyperplane is the model for the function. (Recall that this plane in the kernel space can be very complicated in the input feature space.) Thus, it is the points that are more than some distance from the hyperplane that are outliers (in either direction). It is how this distance is handled that differentiates the two algorithms we have available to us for SVR in OpenCV.

The first form algorithm, $\varepsilon$-SVR, uses a parameter called $\varepsilon$ that defines a space around the hyperplane within which no cost is assigned to predictions being off by this much or less. Beyond this distance, a cost is assigned that grows linearly with distance. This is true for points both above and below the hyperplane.

The second algorithm, $\nu$-SVR, has a relationship with $\varepsilon$-SVR similar to that between $\nu$-SVM and *C*-SVM. Specifically, the $\nu$-SVR uses the parameter $\nu$ to set the minimum fraction of input vectors that will be support vectors. In this manner, the meaning of the parameter $\nu$ in the $\nu$-SVR, as with the $\nu$-SVM, is substantially more intuitive, and therefore easier to set to a sensible value.

### Using support vector machines and cv::ml::SVM()

The SVM classification interface is defined in OpenCV by the `cv::ml::SVM` class in the ML library. `cv::ml::SVM`, like the other objects we have looked at in this section, is derived from `cv::ml::StatModel` class, and itself serves as the interface to an implementation class, in this case called `cv::ml::SVMImpl`:

```
// Somewhere above..
// namespace cv {
//   namespace ml {
//
class SVMImpl: public SVM {    // cv::ml::SVM is derived
                               // from cv::ml::StatModel
public:

  // (Inherited from cv::ml::SVM)
  //
  //enum Types {
  //  C_SVC     = 100,
  //  NU_SVC    = 101,
  //  ONE_CLASS = 102,
  //  EPS_SVR   = 103,
  //  NU_SVR    = 104
  //};
  //
  //enum KernelTypes {
  //  CUSTOM    = -1,
  //  LINEAR    = 0,
  //  POLY      = 1,
  //  RBF       = 2,
  //  SIGMOID   = 3,
  //  CHI2      = 4,
  //  INTER     = 5
  //};
  //
  //enum ParamTypes {
  //  C         = 0,
  //  GAMMA     = 1,
  //  P         = 2,
  //  NU        = 3,
  //  COEF      = 4,
  //  DEGREE    = 5
  //};

  class Kernel : public Algorithm {

  public:

    int getType() const;
    void calc(
      int          vcount,
      int          n,
```

```
    const float* vecs,
    const float* another,
    float*       results
  );
};

int getType() const;                         // get the SVM type (C-SVM, etc.)
double getGamma() const;                     // get the gamma param of kernel
double getCoef0() const;                     // get the coeff0 param of kernel
double getDegree() const;                    // get degree param of kernel
double getC() const;                         // get C param of C-SVM or eps-SVR
double getNu() const;                        // get nu param of nu-SVM or nu-SVR
double getP() const;                         // get P param of eps-SVR
cv::Mat getClassWeights() const;             // get the class priors
cv::TermCriteria getTermCriteria() const;    // get training term criteria
int getKernelType() const;                   // get the kernel type

void setType( int val );                     // set the SVM type (C-SVM, etc.)
void setGamma( double val );                 // set the gamma param of kernel
void setCoef0( double val );                 // set the coeff0 param of kernel
void setDegree( double val );                // set degree param of kernel
void setC( double val );                     // set C param of C-SVM or eps-SVR
void setNu( double val );                    // set nu param of nu-SVM or nu-SVR
void setP( double val );                     // set P param of eps-SVR
void setClassWeights( const Mat &val );      // set the class priors
void setTermCriteria( const TermCriteria &val ); // set training term criteria
void setKernel( int kernelType );            // set the kernel type

...

// set the custom SVM kernel if needed
//
void setCustomKernel( const Ptr<Kernel> &_kernel );

Mat      getSupportVectors() const;
Mat      getUncompressedSupportVectors() const;
double   getDecisionFunction(
  int         i,
  OutputArray alpha,
  OutputArray svidx
);

ParamGrid getDefaultGrid( int param_id ); // return default grid for any param
                                          // Choose from SVM::ParamTypes above
bool trainAuto(
  const Ptr<TrainData>& data,
  int                   kFold      = 10,
  ParamGrid             Cgrid      = getDefaultGrid( C ),
  ParamGrid             gammaGrid  = getDefaultGrid( GAMMA ),
  ParamGrid             pGrid      = getDefaultGrid( P ),
  ParamGrid             nuGrid     = getDefaultGrid( NU ),
  ParamGrid             coeffGrid  = getDefaultGrid( COEF ),
```

```
    ParamGrid               degreeGrid = getDefaultGrid( DEGREE ),
    bool                    balanced   = false
  );

  ...

  // from class SVM
  //
  //static Ptr<SVM> create(); // The algorithm "constructor",
  //                          //  returns Ptr<SVMImpl>
};
```

As usual, you first create the instance of the class using the static `create()` method, then configure its parameters using the "setters," and then run the train method. The first method, `setType()`, determines the SVM or SVR algorithm that is to be used. It can be any of the five values shown in Table 21-6. The next important method is `setKernelType()`, which may be any of the values from Table 21-5.

*Table 21-6. The available types for the SVM and the corresponding values for the setType() method of cv::ml::SVM. The rightmost column lists the related properties of cv::ml::SVM that could be set for the SVM type along with the values to which they correspond (as described in the previous section)*

| SVM type | OpenCV name (svm_type) | Parameters | |
|---|---|---|---|
| $C$-SVM classifier | cv::ml::SVM::C_SVC | C ($C$) | |
| $\nu$-SVM classifier | cv::ml::SVM::NU_SVC | Nu ($\nu$) | |
| One-class SVM | cv::ml::SVM::ONE_CLASS | C ($C$), | Nu ($\nu$) |
| $\varepsilon$-SVR | cv::ml::SVM::EPS_SVR | P ($\varepsilon$), | C ($C$) |
| $\nu$-SVR | cv::ml::SVM::NU_SVR | Nu ($\nu$), | C ($C$) |

Once you have selected the type of SVM and the kernel type you would like to use, you should configure the kernel parameters with their associated `set*()` methods. The Parameters columns of Tables 21-5 and 21-6 indicate which parameters are needed in which cases, and which parameters in the equations of the previous sections are associated with each parameter. The default values of `degree`, `gamma`, `coef0`, `C`, `Nu`, and `P` are `0.0`, `1.0`, `0.0`, `1.0`, `0.0`, and `0.0`, respectively.

The `setClassWeights()` method allows you to supply a single-column array that provides an additional weighting factor for the slack variables. This is used only by the *C*-SVM classifier. The values you supply will multiply *C* for each individual training vector. In this way, you can assign a greater importance to some subset of the training examples, which will in turn help to guarantee that if any training vectors cannot be classified correctly, it will not be those. By default, class weights are not used.

The final method is `setTermCriteria()`, which can almost always be safely left at its default of `1000` iterations and `FLT_EPSILON`.

At this point, you might be looking at all of the parameters to the SVM and the kernels and wondering how you could possibly choose the right values for all of these things. If so, you would not be alone. In fact, it is common practice to simply step through ranges of all of these parameters and find the one that works the very best on the available data. This process is automated for you by the `cv::ml::SVM::trainAuto()` method.

When you are using `cv::ml::SVM::trainAutio()`, the train data you supply is the same as you would supply to `cv::ml::StatModel::train()`. The `kFold` argument controls how the validation is done for each parameter set. The method used is *k-fold cross-validation* (so the parameter should probably just be called *k*), by which the data set is automatically divided into *k*-subsets and then run *k* times; each time it is run, a different subset is held back for validating after training on the other ($k - 1$) subsets.

The next six parameters control the *grids*, which are the sets of values that will be tested for each of the parameters. A grid is a simple object with three data members: `minVal`, `maxVal`, and `logStep`. If you set the step to something less than `1`, then the grid will not be used and instead the relevant value from `params` will be used for all runs. Typically, you will not grid-search over every parameter. You can construct a grid with the constructors:

```
cv::ml::ParamGrid::ParamGrid() {
  minVal  = maxVal = 0;
  logStep = 1;
}
cv:: ml::ParamGrid::ParamGrid(
  double minVal,
  double maxVal,
  double logStep
);
```

or you can use the `cv::ml::SVM` method `cv::ml::SVM::getDefaultGrid(int)`. In this latter case, you need to tell `getDefaultGrid()` what parameter you would like a grid for. This is because the default grid is different for each argument. The valid values you can pass to `getDefaultGrid` are `cv::ml::SVM::C`, `cv::ml::SVM::GAMMA`, `cv::ml::SVM::P`, `cv::ml::SVM::NU`, `cv::ml::SVM::COEF`, or `cv::ml::SVM::DEGREE`. If you just want to create the grids yourself, you can do that as well, of course. Note that the argument `logStep` is interpreted as a multiplicative scale, so if you set it (for example) to `2.0`, each value tried will be double the previous value until `maxVal` is reached (or exceeded).

Finally, now that you have trained your classifier, you can make predictions using the `cv::ml::StatModel::predict()` interface as usual:

```
float cv::ml::SVM::predict(
  InputArray  samples,
  OutputArray results = noArray(),
  int         flags   = 0
) const;
```

As usual, the `predict()` method expects one or more samples, one per row, and returns the resulting predictions. In the case of one-class or two-class classification it's possible to retrieve the computed value instead of the chosen class label. In order to do that, pass the `cv::ml::StatModel::RAW_OUTPUT` flag. The returned signed value will correlate with the distance between `sample` and the decision surface; this is what you want in two class classification problems. Otherwise, a class label will be returned (for multiclass). In the case of regression, the return value will be the estimated value of the function, regardless of the flag.

### Additional members of cv::ml::SVM

The `cv::ml::SVM` object also provides a few utility functions, which allow you to get at the data in the object. This includes data you put in at training time, but also includes useful things like the computed support vectors as well as the default grids. The available functions are:

```
// get all the support vectors
//
cv::Mat cv::ml::SVM::getSupportVectors()  const;

// get uncompressed support vectors as found by the training procedure
//
cv::Mat cv::ml::SVM::getUncompressedSupportVectors() const;

// get the i-th decision function (out of n*(n-1)/2 in the case of n-class problem
//
double cv::ml::SVM::getDecisionFunction(
  int             i,
  cv::OutputArray alpha,
  cv::OutputArray svidx
) const;
```

The method `cv::ml::SVM::getSupportVectors()` gives you all the support vectors used to compute the decision hyperplane or hypersurface. In the case of linear SVM, all the support vectors for each decision plane can be compressed into a single vector that will basically describe the separating hyperplane. That's why linear SVM is super-fast at the prediction. However, users may be interested in looking at the original support vectors, which can be accessed via `cv::ml::SVM::getUncompressedSupportVectors()`. In the case of nonlinear SVM, uncompressed and compressed support vectors are the same thing. Then it's possible to get access to each decision hyperplane or hypersurface. We do so using the method `cv::ml::SVM::getDecisionFunction`. In the case of regression, or one-class or two-class classification, there will be just one decision function, and so `i=0`. In the case of *N*-class classification, there will be `N*(N-1)/2` decision functions, and so `0<=i<N*(N-1)/2`. The method returns the coefficients for the support vectors used in the particular function, the indices of the support vectors (within the matrix returned by `cv::ml::SVM::getSupportVectors()`), and the value `b` (see the preceding formulas), which is added to the weighted sum before the decision is made.

In addition to the training and prediction methods, `cv::ml::SVM` supports the usual `save()`, `load()`, and `clear()` methods.

# Summary

We began this chapter by learning about `cv::ml::StatModel`, which is the standard object-based interface used by OpenCV to encapsulate all of the modern learning methods. We saw how training data was handled and how predictions could be made from the model once the training data was learned. Thereafter, we looked at various learning techniques that have been implemented using that interface.

Along the way, we learned that both `cv::ml::TrainingData` and the individual classifiers derived from `cv::ml::StatModel` use a construction by which an interface is specified at one class level, and a derived *implementation class* is provided that does all of the work. This implementation class was something largely invisible to the users, because the `create()` member of classifier `X` always returned an instance of the implementation class `XImpl`. For this reason, when we looked at class definitions for important classifier objects, it was the `*Impl` objects that we actually looked at.

We learned that the naïve Bayes classifier assumes that all the features are independent from one another and that it was surprisingly useful for multiclass discriminative learning. We saw that binary decision trees use what is called an *impurity metric*, which it tries to minimize as it builds the tree. Binary decision trees were also useful for multiclass learning and have the useful feature that different weights could be assigned to misclassifications for different classes. Boosting and random trees use binary decision trees internally, but build larger structures containing many such trees. Both are capable of multiclass learning, but random trees also introduce a notion of similarity between input data points relative to the classes.

Expectation maximization is an unsupervised technique that was similar to K-means, but different in that a responsibility assigned to a data point is not necessarily to the closest cluster center. The K-nearest neighbors method is a classifier that can be very effective, but requires storing the entire training set; as a result, it can use a lot of memory and be quite slow. Multilayer perceptrons are a biologically inspired technique that can be used for classification or regression. They are notably very slow to train, but can be very fast to operate, and achieve state-of-the-art performance on many important tasks. Finally, we introduced support vector machines (SVM), a robust technique used typically for two-class classification, but which has variants for multiclass learning. SVMs operate by evaluating data in a kernel space where separation between classes can be easily expressed. Because they require only a small subset of the training data when used for classification, they can be very fast and have a small memory requirement. Finally, OpenCV is being extended to support deep neural networks; see the Appendix B repositories *cnn_3dobj* and *dnn*.

# Exercises

1. Figure 21-14 depicts a distribution of "false" and "true" classes. The figure also shows several potential places (a, b, c, d, e, f, g) where a threshold could be set.

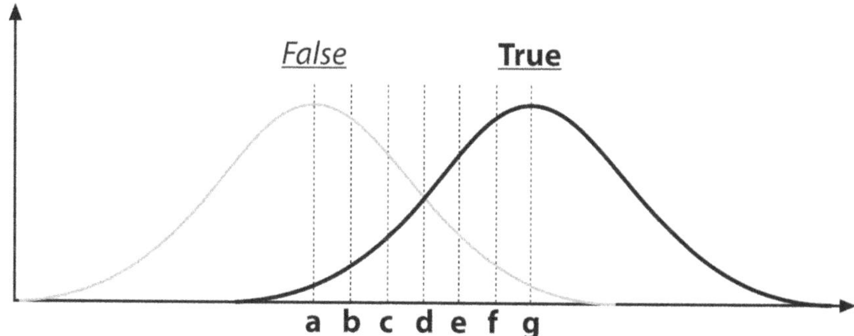


*Figure 21-14. A Gaussian distribution of two classes, "false" and "true"*

   a. Draw the points a–g on an ROC curve.

   b. If the "true" class is poisonous mushrooms, at which letter would you set the threshold?

   c. How would a decision tree split this data?

2. Refer back to Figure 20-2.

   a. Draw how a decision tree would approximate the true curve (the dashed line) with three splits (here we seek a regression, not a classification model).

The "best" split for a regression takes the average value of the data values contained in the leaves that result from the split. The output values of a regression-tree fit thus look like a staircase.

   b. Draw how a decision tree would fit the true data in seven splits.

   c. Draw how a decision tree would fit the noisy data in seven splits.

d. Discuss the difference between (b) and (c) in terms of overfitting.

3. Why do the splitting measures (e.g., Gini) still work when we want to learn multiple classes in a single decision tree?

4. Review Figure 20-6, which depicts a two-dimensional space with unequal variance at left and equalized variance at right. Let's say that these are feature values related to a classification problem. That is, data near one "blob" belongs to one of two classes, while data near another blob belongs to the same or another of two classes. Would the variable importance be different between the left or the right space for:

   a. Decision trees?

   b. K-nearest neighbors?

   c. Naïve Bayes?

5. Modify the sample code for data generation in Example 20-1, near the top of the outer `for` loop in the K-means section, to produce a randomly generated labeled data set. We'll use a single normal distribution of 10,000 points centered at pixel (63, 63) in a 128 × 128 image with standard deviation (`img.cols/6, img.rows/6`). To label these data, we divide the space into four quadrants centered at pixel (63, 63). To derive the labeling probabilities, we use the following scheme. If $x < 64$ we use a 20% probability for *class A*; else if $x \geq 64$ we use a 90% factor for *class A*. If $y < 64$ we use a 40% probability for *class A*; else if $y \geq 64$ we use a 60% factor for *class A*. Multiplying the $x$ and $y$ probabilities together yields the total probability for *class A* by quadrant with values listed in the 2 × 2 matrix shown. If a point isn't labeled *A*, then it is labeled *B* by default. For example, if $x < 64$ and $y < 64$, we would have an 8% chance of a point being labeled *class A* and a 92% chance of that point being labeled *class B*. The four-quadrant matrix for the probability of a point being labeled *class A* (and if not, it's *class B*) is:

   | | |
   |---|---|
   | $0.2 \times 0.6 = 0.12$ | $0.9 \times 0.6 = 0.54$ |
   | $0.2 \times 0.4 = 0.08$ | $0.9 \times 0.4 = 0.36$ |

   Use these quadrant odds to label the data points. For each data point, determine its quadrant. Then generate a random number from 0 to 1. If this is less than or equal to the quadrant odds, label that data point as *class A*; else, label it *class B*. We will then have a list of labeled data points together with $x$ and $y$ as the features. The reader will note that the x-axis is more informative than the y-axis as to which class the data might be. Train random forests on this data and calculate the variable importance to show $x$ is indeed more important than $y$.

6. Using the same data set as in Exercise 5, use discrete AdaBoost to learn two models: one with `weak_count` set to 20 trees and one set to 500 trees. Randomly select a training and a test set from the 10,000 data points. Train the algorithm and report test results when the training set contains:

   a. 150 data points;

   b. 500 data points;

   c. 1,200 data points;

   d. 5,000 data points.

   Explain your results. What is happening?

7. Repeat Exercise 5, but use the random trees classifier with 50 and 500 trees.

8. Repeat Exercise 5, but this time use 60 trees and compare random trees versus SVM.

9. In what ways is the random tree algorithm more robust against overfitting than decision trees?

10. The "no free lunch" theorem states that no classifier is optimal over all distributions of labeled data.

    a. Describe a labeled data distribution over which no classifier described in this chapter would work well.

    b. What distribution would be hard for naïve Bayes to learn?

    c. What distribution would be hard for decision trees to learn?

    d. How would you preprocess the distributions in Parts b and c so that the classifiers could learn from the data more easily?

11. Use the mushroom data set *agaricus-lepiota.data*, but take the data and remove all the labels. Then duplicate this data, but randomly shuffle each column with replacement. Now label the original data *class A* and the shuffled data *class B*. Split the data into a large training set and a smaller test set. Train a random tree classifier on the training data.

    a. How well can the random trees classifier tell *class A* and *class B* apart on the test set?

    b. Run variable importance and note what features are important.

    c. Train a network on the unaltered mushroom data set. Run variable importance. Are the choices of the important variables similar? This is an example of how one might handle unsupervised data for learning invented by Leo Brieman.

12. Back propagation uses the calculus chain rule in order to compute how to change each weight in a neural network so as to reduce the loss function error at the output. Using Figure 21-11, the back-propagation discussion, and the chain rule, derive the weight update equations given in the back propagation section.

13. The nonlinear function (the `f` or `g` function in Figure 21-11) is often a "sigmoid" or S-shaped function that rises from near zero for large negative values to one for large positive values. One typical form of this function is $\sigma(x) = \frac{1}{1 - e^{-x}}$. Prove that the derivative $\sigma'(x) = \frac{d\sigma(x)}{dx}$ takes the form $\sigma'(x) = \sigma(x)(1 - \sigma(x))$.

CHAPTER 22

# Object Detection

In the previous two chapters, we covered the basics of machine learning and then moved on to investigate, in some depth, a large number of techniques that the OpenCV library provides for discriminative and generative learning. Now it's time to put it all together, to combine the computer vision techniques we have been learning throughout the book with the machine learning techniques, and to actually apply learning to practical problems in computer vision. One of the most important such problems is *object detection*—the process of determining whether an image contains some particular object and, where possible, the localization of that object in pixel space. In this chapter, we will look at several methods that achieve these goals, in every case by making use of the lower-level machine learning techniques from the previous chapter.

## Tree-Based Object Detection Techniques

Having looked at many of the lower-level methods for machine learning in the library, we now turn to some higher-level functions that make use of those various learning methods in order to detect objects of interest in images. There are currently two such detectors ented on OpenCV. The first is the *cascade classifier*, which generalizes the very successful algorithm of Viola and Jones [Viola01] for face detection, and the second is the *soft cascade*, a further evolution of that algorithm that uses a new approach to give what is, in most cases, a more robust classification than the cascade classifier. Both algorithms have been very successfully used for detection of many object classes other than faces. In general, objects with rigid structure and rich texture tend to respond well to these methods.

These methods not only encapsulate the machine learning components on which they are based, but they also involve other stages that actually condition the input for learning or post-process the output of the learning algorithm. Not surprisingly, these object detection algorithms do not have such a uniform interface as the core machine learning algorithms. This is the case both because of the greater natural variation in the needs and results of such higher-level methods, but also because—as a matter of practice—these algorithms were often contributed to the library by their creators, and have interfaces more like their original implementations.

## Cascade Classifiers

The first such classifier we will look at is a tree-based technique called the *cascade classifier*, which is built on the important concept of the *boosted rejection cascade*. It has a different format from the bulk of the ML library in OpenCV because it was originally developed as a full-fledged face-detection application, and later modernized (somewhat) to be a little more general than the original implementation. In this section we will cover it in detail and show how it can be trained to recognize faces and other rigid objects.

Computer vision is a broad and fast-changing field, so the parts of OpenCV that implement a specific technique—rather than a component algorithmic piece—are more at risk of becoming out of date. The original face detector (back then called the Haar classifier) that was part of OpenCV for many years was in this "risk" category. However, face detection is such a common need that it was worth having a baseline technique that worked well. As the technique was built on the well-known and often-used field of statistical boosting, it actually had more general utility. Since that time, several companies have engineered the "face" detector in OpenCV to detect "mostly rigid" objects (faces, cars, bikes, human bodies) by training new detectors on many thousands of selected training images for each view of the object. This technique has been used to create state-of-the-art detectors, although with a different detector trained for each view or pose of the object. Thus, this classifier is a valuable tool to keep in mind for such recognition tasks. In its current form in the library, some of this generality is more manifest, and effort has been made to make the implementation more extensible as future advancements are made.

The cascade classifier in OpenCV implements a version of the technique for face-detection first developed by Paul Viola and Michael Jones, commonly known as the *Viola-Jones detector* [Viola01]. Originally, this technique, and its OpenCV implementation, supported only one particular set of features, the Haar wavelets.[1] The technique was later extended by Rainer Lienhart and Jochen Maydt [Lienhart02] to use

1 The Haar wavelet is the first known wavelet basis, and was originally proposed by Alfred Haar in 1909.

what are known as *diagonal features* (more on this distinction to follow), which were then incorporated into the OpenCV implementation. This extended set of features is commonly referred to as the "Haar-like" features. In OpenCV 3.x, cascades are further extended to work with *local binary patterns*, or LBP [Abonen04].

As implemented, the OpenCV version of the Viola-Jones detector operates in two layers. The first layer is the feature detector, which encapsulates and modularizes the feature computation. The second layer is the actual boosted cascade, which uses sums and differences over rectangular regions of the computed features; it is agnostic about how the features were computed.

### Haar-like features

The Haar-like features used in the classifier by default are shown in Figure 22-1. At all scales, these features form the "raw material" that will be used by the boosted classifiers. They all have the feature that they can be rapidly computed from an integral image (see Chapter 12) taken from the original grayscale image.

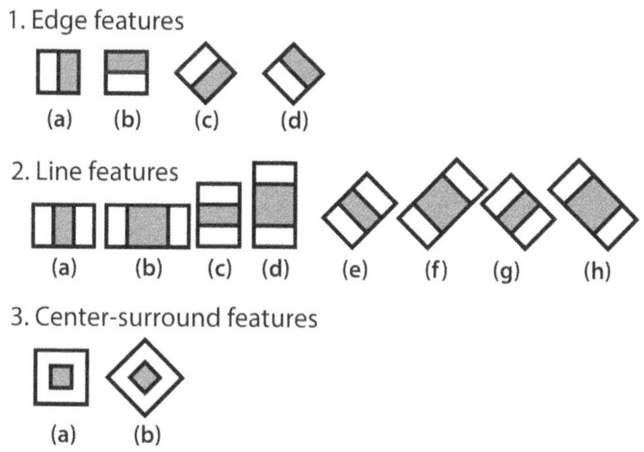


*Figure 22-1. Haar-like features from the OpenCV source distribution (the rectangular and rotated regions are easily calculated from the integral image). In this diagrammatic representation of the wavelets, the light region is interpreted as "add that area" and the dark region as "subtract that area"*

Currently, there are two distinct feature sets supported. These include both the "original" Haar wavelet features (including the "diagonal" features) and the alternate feature, LBP. In the future, other feature types may also be supported, or you may write or use other features (since the cascade feature interface is now fully extensible) by following how LBP was added.

### Local binary pattern features

The LBP feature was originally proposed by [Ojala94] and intended as a kind of texture descriptor. It was only later adapted for use in the boosted cascade environment of the Viola-Jones object detection algorithm. Recall that the Haar wavelet feature is associated with a small patch (e.g., 11 × 11 pixels) and assigns the "feature vector" of that patch to be the wavelet transform (projection onto the Haar basis) of the pixels in that patch. As seen in Figure 22-2, the LBP feature has a very different way of constructing that feature vector. It takes a rectangle whose width and height are both divisible by 3, which it then splits into a 3 × 3 array of nonoverlapping tiles. For each tile, it computes the sum of pixels (using the integral image). Finally, it compares sums of pixels in each of the eight noncentral tiles with the sum of pixels in the central tile and thus constructs an 8-bit pattern. This 8-bit pattern is used as descriptor of the rectangle. The 8-bit value is then used as a categorical value passed to the classifier (as long as the corresponding rectangle is chosen during the training as the discriminative-enough feature).

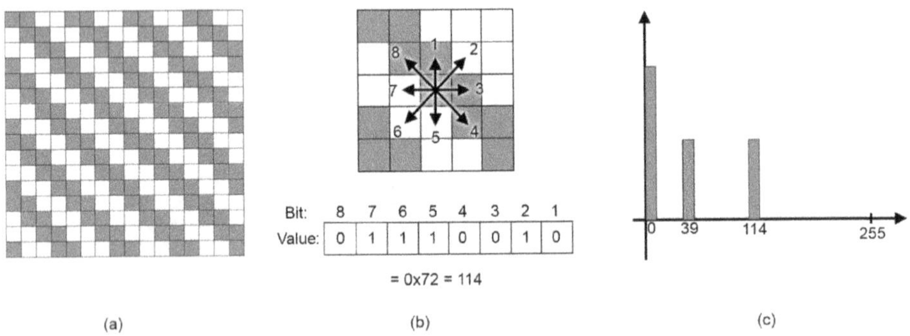


*Figure 22-2. The LBP feature is computed for an example region of texture (a). For each pixel in the region, we compute a binary representation by comparing that pixel to its neighbors (b). The LBP feature is the histogram of these computed values (c). In this example, there are only two nonzero bins in the histogram (a result of the simple and repetitive structure of the texture)*

### Training and pretrained detectors

OpenCV ships with a set of pretrained object-recognition files, but there is also code that allows you to train and store new object models for the detector. If these are not enough for your uses, there is also an application in the bundle (`opencv_traincascade` in *.../opencv/apps*) that you can use to train detectors for just about any rigid object, though its suitability will vary substantially from object to object.

The pretrained objects that come with OpenCV for this detector are in the *.../opencv/data/haarcascades* and *.../opencv/data/lbpcascades* directories. Currently, the model that works best for frontal face detection is *haarcascade_frontalface_alt2.xml*. Side face views are harder to detect accurately with this technique (as we will describe shortly), and those shipped models work less well but have been much improved due to work during Google Summer of Code 2012. If you end up training good object models, perhaps you will consider contributing them as open source back to the community.

## Supervised Learning and Boosting Theory

The cascade classifier that is included in OpenCV is a supervised classifier (these were discussed in Chapter 21). In this case we typically present histogram- and size-equalized image patches to the classifier, which are then labeled as containing (or not containing) the object of interest, which for this classifier is most commonly a face.

The Viola-Jones detector uses AdaBoost, but inside of a larger context called a *rejection cascade*. This "cascade" is a series of nodes, where each node is itself a distinct multitree AdaBoosted classifier. The basic operation of the cascade is that subwindows from an image are sequentially tested against all of the nodes, in a particular order, and those windows that "pass" every classifier are deemed to be members of the class being sought.

To make this possible, each node is designed to have a high (say, 99.9%) detection rate (low false negatives, or missed faces) at the cost of a low (near 50%) rejection rate (high false positives, or "nonfaces" wrongly classified). For each node, a "not in class" result at any stage of the cascade thus safely terminates the computation, and the algorithm then declares that no face exists at that location.

A true class detection is declared only if the area under consideration makes it through the entire cascade. When the true class is rare (e.g., a face in a picture), rejection cascades can greatly reduce total computation because most of the regions being searched for a face terminate quickly in a nonclass decision. This is further enhanced by the placement of the simplest nodes (fastest to compute) at the beginning of the cascade.

### Boosting in the Haar cascade

For the Viola-Jones rejection cascade, each node is itself a collection of weak classifiers that are combined, through boosting, to form one strong classifier node. These individual weak classifiers are themselves decision trees that often are only one level deep (i.e., *decision stumps*). A decision stump is allowed just one decision of the following form: "Is the value $v$ of a particular feature $h$ above or below some threshold $t$"; then, for example, a "yes" indicates face and a "no" indicates no face:

$$h_w = \begin{cases} +1 & v_w \geq t_w \\ -1 & v_w < t_w \end{cases}$$

The number of Haar-like or LBP features that the Viola-Jones classifier uses in each weak classifier can be set in training, but one mostly sticks with the single feature stump; at most about three features may be used in some contexts.[2] Boosting then iteratively builds up the strong classifier node as a weighted sum of these kinds of weak classifiers. The Viola-Jones classifier uses the classification function:

$$H = sign(\alpha_1 h_1 + \alpha_2 h_2 + \ \dots \ + \alpha_{N_w} h_{N_w})$$

Here, the sign function returns `-1` if the number is less than zero, `0` if the number equals zero, and `+1` if the number is positive. On the first pass through the data set, we learn the threshold $t_w$ for each $h_w$ that best classifies the input. Boosting then uses the resulting errors to calculate the weighted vote, $\alpha_w$. As in traditional AdaBoost, each feature vector (data point) is also reweighted low or high according to whether it was classified correctly[3] in that iteration of the classifier. Once a node is learned this way, the surviving data from higher up in the cascade is used to train the next node, and so on.

**Rejection cascades.** Figure 22-3 visualizes the Viola-Jones rejection cascade, composed of many boosted classifier groups. In the figure, each of the nodes $F_j$ contains an entire boosted cascade of groups of decision stumps (or trees) trained on the features from faces and nonfaces (or other objects the user has chosen to train on). Recall that one minimizes computational cost by ordering the nodes from least to most complex. Typically, the boosting in each node is tuned to have a very high detection rate (at the usual cost of many false positives). When training on faces, for example, almost all (99.9%) of the faces are found, but many (about 50%) of the nonfaces are erroneously "detected" at each node. This is OK however, because using, say, 20 nodes will still yield a face detection rate (through the whole cascade) of $0.999^{20} \approx 98\%$ with a false positive rate of only $0.5^{20} \approx 0.0001\%$!

2 This should all look familiar from the discussion of AdaBoost in Chapter 21.

3 There is sometimes confusion about boosting *lowering* the classification weight on points it classifies correctly in training and *raising* the weight on points it classified wrongly. The reason is that boosting attempts to focus on correcting the points that it has "trouble" on and to ignore points that it already "knows" how to classify. One of the technical terms for this is that boosting is a *margin maximizer*.

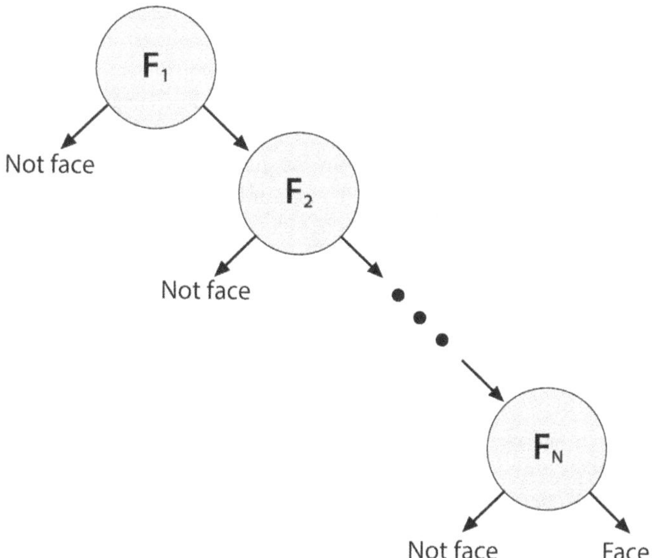


*Figure 22-3. Rejection cascade used in the Viola-Jones classifier: each node represents a multitree boosted classifier ensemble tuned to rarely miss a true face while rejecting a possibly small fraction of nonfaces; however, almost all nonfaces have been rejected by the last node, leaving only true faces*

In the run mode, a search region of different sizes is swept over the original image. In practice, 70–80% of nonfaces are rejected in the first two nodes of the rejection cascade, where each node uses about 10 decision stumps. This quick and early "attentional reject" vastly speeds up face detection and is the basis of the practicality of the Viola-Jones algorithm.

As mentioned earlier, this technique implements face detection but is not limited to faces; it also works fairly well on some other (mostly rigid) objects that have distinctive views. Front views of faces work well; backs, sides, or fronts of cars work well; but side views of faces or "corner" views of cars work less well—mainly because these views introduce variations in the template that the "blocky" features used in this detector don't handle well. For example, a side view of a face must catch part of the changing background in its learned model in order to include the profile curve. To detect side views of faces, you may try *haarcascade_profileface.xml*, but to do a better job you should really collect much more data than this model was trained with and perhaps expand the data with different backgrounds behind the face profiles. In Google Summer of Code 2012, *lbpcascade_profileface.xml* was introduced with improved performance.

Profile views are hard for the Haar cascade classifier because it uses block features and so is forced to attempt to learn the background variability that "peeks through" the informative profile edge of the side view of faces. This problem is a somewhat general one that affects both the Haar and, to a lesser extent, LBP features. In most cases, you are best off training the detector to find a square region that is a subset of the object you are looking for. This is precisely how the pretrained cascades in the package deal with the fact that heads are round: they look for the square region that is circumscribed by the face.

In training, it's more efficient to learn only one profile view (e.g., just the right side). Then the test procedure would be to (1) run the right-side-profile detector, and then (2) flip the image on its vertical axis and run the right-side-profile detector again to detect left-facing profiles.[4]

To recap, detectors based on these Haar-like features as well as the LBP features work well with "blocky" features—such as eyes, mouth, and hairline—but less well with tree branches (for example) or when the object's outline shape is its most distinguishing characteristic (as with a coffee mug). LBP has some advantage here because the background regions may just add some gradient noise into the LBP histogram, whereas Haar-like features are influenced by summations (offset level) over chunks of background.

All that being said, if you are willing to gather lots of good, well-segmented data on fairly rigid objects, then this classifier can still compete with the best, and its construction as a rejection cascade makes it very fast to run (though not to train). Here, "lots of data" means thousands of object examples and tens of thousands of nonobject examples. By "good" data we mean that one shouldn't mix, for instance, tilted faces with upright faces; instead, keep the data divided and use two classifiers, one for tilted and one for upright. "Well-segmented" data means data that is consistently boxed. Sloppiness in box boundaries of the training data will often lead the classifier to try to correct for fictitious variability in the data. For example, different placement of the eye locations in the face data location boxes can lead the classifier to assume that eye locations are not a geometrically fixed feature of the face and so can move around. Performance is almost always worse when a classifier attempts to adjust to things that aren't actually in the real data.

4 This relates to a more general principle in machine learning. It is never profitable to force an algorithm to learn a symmetry that you already know to exist. It is always better to train the algorithm on a canonically broken instance of that symmetry and then to map the input according to the symmetry before giving it to the algorithm.

### Viola-Jones classifier summary

The Viola-Jones classifier employs AdaBoost *at each node* in the cascade to learn a multitree (mostly multistump) classifier. In addition to this, the algorithm incorporates several other innovative features:[5]

- It uses features that can be computed very quickly (i.e., the Haar feature is a threshold applied to sums and differences of rectangular image regions).
- Its *integral image* technique enables rapid computation of the value of rectangular regions or such regions rotated 45 degrees (see Chapter 6). This data structure is used to accelerate computation of the input features.
- It uses statistical boosting to create binary (face/not face) classification nodes characterized by high detection and weak rejection.

It organizes these classifier nodes into a rejection cascade. In other words, the first group of classifiers is selected that best detects image regions containing an object while allowing many mistaken detections; the next classifier group is the second-best at detection with weak rejection; and so forth. In test mode, an object is detected only if it makes it through the entire cascade.

### The cv::CascadeClassifer object

As with many of the routines in the Machine Learning library from Chapter 21, the cascade classifier is implemented in OpenCV as an object. This object is called `cv::CascadeClassifier`, and stores the loaded (or trained) cascade, as well as providing the interface for running a detection pass on an image.

The constructor for the `cv::CascadeClassifier` object is:

```
cv::CascadeClassifier::CascadeClassifier(
const String& filename
);
```

This constructor takes just one argument: the name of the file in which your cascade is stored. There is also a default constructor you can use, if you would like to load the cascade later with the `load()` member.

5 Well, most of them are now standard tricks in the toolkit of a modern computer vision researcher or practitioner.

### Searching an image with detectMultiScale()

The function that actually implements the cascade classification is the `detectMultiScale()` method of the `cv::CascadeClassifier` object:

```
cv::CascadeClassifier::detectMultiScale(
const cv::Mat&    image,                           // Input (grayscale) image
vector<cv::Rect>& objects,                         // Output boxes (boxen?)
double            scaleFactor  = 1.1,              // Factor between scales
int               minNeighbors = 3,                // Required neighbors to count
int               flags        = 0,                // Flags (old style cascades)
cv::Size          minSize      = cv::Size(),       // Smallest we will consider
cv::Size          maxSize      = cv::Size()        // Largest we will consider
);
```

The first input, `image`, is a grayscale image of type `CV_8U`. The `cv::CascadeClassifier:: detectMultiScale()` function scans the input image for faces at all scales. Objects successfully located will be returned in the vector `objects`, in the form of their bounding rectangles. Setting the `scaleFactor` parameter determines how big of a jump there is between each scale; setting this to a higher value means faster computation time at the cost of possible missed detections if the scaling misses faces of certain sizes. The `minNeighbors` parameter is a control for preventing false detection. Actual face locations in an image tend to get multiple "hits" in the same area because the surrounding pixels and scales often indicate a face. Setting this to the default (3) in the face-detection code indicates that we will decide a face is present in a location only if there are at least three overlapping detections.

The `flags` parameter is ignored at this time, unless you are using a cascade that was created with the older OpenCV 1.x cascade tools. In that case, it may be set to the (also OpenCV 1.x vintage) value: `CV_HAAR_DO_CANNY_PRUNING`. In that case, the Canny edge detector will be used to reject some regions.

The final parameters, `minSize` and `maxSize`, are the smallest and largest region sizes in which to search for a face. Setting these values will reduce computation at the cost of missing faces that are either unusually small or unusually large. (This is desirable in many practical cases, as you often have an expectation for how much frame faces will occupy in your images. Anything else you find would probably just be noise anyway.) Figure 22-4 shows results for using the face-detection code on a scene with faces.

*Figure 22-4. Face detection in a park scene: even tilted faces are detected; for the 1,111 × 827 image shown, more than a million sites and scales were searched to achieve this result in about 0.25 seconds on a 3 GHz machine*

There are two other similar detection methods:

```
void detectMultiScale(
  cv::InputArray    image,
  vector<cv::Rect>& objects,
  vector<int>&      numDetections,
  double            scaleFactor        = 1.1,
  int               minNeighbors       = 3,
  int               flags              = 0,
  cv::Size          minSize            = cv::Size(),
  cv::Size          maxSize            = cv::Size()
);

void detectMultiScale(
  cv::InputArray    image,
  vector<cv::Rect>& objects,
  vector<int>&      rejectLevels,
  vector<double>&   levelWeights,
  double            scaleFactor        = 1.1,
  int               minNeighbors       = 3,
  int               flags              = 0,
  cv::Size          minSize            = cv::Size(),
  cv::Size          maxSize            = cv::Size(),
  bool              outputRejectLevels = false
);
```

The first of these is essentially identical to the constructor we saw before, but adds the new argument `numDetections`. This is an output that contains the same number of entries as `objects`. For each entry in `numDetections`, the value indicates the number of object detections that contributed to the corresponding entry in `objects`.

The second alternative form of `detectMultiScale()` has three additional parameters: `rejectLevels`, `levelWeights`, and `outputRejectLevels`. The former two will only be returned, however, if the latter is set to `true`. Both `rejectLevels` and `levelWeights` are vectors with one entry for each entry in `objects`. In the case of `rejectLevels`, this entry contains the level at which the subimage was rejected from the cascade. The `levelWeights` array contains the weighted sum of the weak classifiers for the last level, whether accepted or rejected. This allows the caller to handle marginal cases in whatever manner they choose, by considering this additional information.

#### Face detection example

The `detectAndDraw()` code shown in Example 22-1 will detect faces and draw their found locations in different-colored rectangles on the image. As shown in the comment lines, this code presumes that a previously trained classifier cascade has been loaded and that memory for detected faces has been created.

*Example 22-1. Detecting and drawing faces*

```
// Detect and draw detected object boxes on image
//
// Presumes 2 Globals:
//
//  cascade is loaded by something like:
//     cv::Ptr<CascadeClassifier> cascade( new CascadeClassifier( cascade_name ) );
//
void detectAndDraw(
cv::Mat&                         img,            // Input image
cv::Ptr<cv::CascadeClassifier> classifer,      // Preloaded classifier
  double                           scale    = 1.3 // resize image by...
){

  // Just some pretty colors to draw with
  //
  enum { BLUE, AQUA, CYAN, GREEN };
  static cv::Scalar colors[] = {
    cv::Scalar(  0,   0, 255 ),
    cv::Scalar(  0, 128, 255 ),
    cv::Scalar(  0, 255, 255 ),
    cv::Scalar(  0, 255,   0 )
  };

  // IMAGE PREPARATION:
  //
```

```
  cv::Mat gray( img.size(), CV_8UC1 );
  cv::May small_img(
    cvSize( cvRound(img.cols/scale), cvRound(img.rows/scale)),
    CV_8UC1
  );
  cv::cvtColor( img, gray, cv::BGR2GRAY );
  cv::resize( gray, small_img, cv::INTER_LINEAR );
  cv::equalizeHist( small_img, small_img );

  // DETECT OBJECTS IF ANY
  //
  vector<cv::Rect> objects;
  classifier->detectMultiScale(
    small_img,                  // The input image
    objects,                    // A place for the results
    1.1,                        // Scale Factor
    2,                          // Minimum number of neighbors
    cv::HAAR_DO_CANNY_PRUNING,  // (old format cascades only)
    cv::Size(30, 30)            // Throw away detections smaller than this
  );

  // LOOP THROUGH FOUND OBJECTS AND DRAW BOXES AROUND THEM
  //
  for( vector<cv::rect>::iterator r=objects.begin(); r!=objects.end; ++r ) {
    Rect r_ = (*r)*scale;
    cv::rectangle( img, r_, colors[i%4] );
}

}
```

For convenience, in this code the `detectAndDraw()` function has a static vector of colors `colors[]` that can be indexed to draw found faces in different colors. The classifier works on grayscale images, so the color BGR image `img` passed into the function is converted to grayscale via `cv::cvtColor()` and then optionally resized in `cv::resize()`.

This is followed by histogram equalization via `cv::equalizeHist()`, which spreads out the brightness values. This is a very important step. It is necessary because the integral image features are based on differences of rectangle regions and, if the histogram is not balanced, these differences might be skewed by overall lighting or exposure of the test images. (This is also important to do to the input data when training a cascade.)

The actual detection takes place just above the `for` loop, and then the loop steps through the found-face rectangle regions and draws them in different colors using `cv::rectangle()`.

## Learning New Objects

We've seen how to load and run a previously trained classifier cascade stored in an XML file; we loaded it either during our initial call to the `cv::CascadeClassifier` constructor, or afterward with the `cv::CascadeClassifier::load()` method. Once we had the classifier loaded, we were then able to actually detect objects with the `cv::CascadeClassifier::detectMultiScale()` function. We now turn to the question of how to train our own classifiers to detect other objects such as eyes, walking people, cars, and so on. We do this with the OpenCV *traincascade* application,[6] which creates a classifier given a training set of positive and negative samples. The four steps of training a classifier are:

1. Gather a data set consisting of examples of the object you want to learn (e.g., front views of faces, side views of cars). These may be stored in one or more directories indexed by a text file in the following *collection description file* format:

   ```
   <path>/<img_name_1> <count_1> <x11> <y11> <w11> <h11> <x12> <y12> ...
   <path>/<img_name_2> <count_2> <x21> <y21> <w21> <h21> <x22> <y22> ...
   ...
   ```

   Each of these lines contains the path (if any) and filename of the image containing the object(s). This is followed by the count of how many objects are in that image, and then a list of rectangles containing the objects. The format of the rectangles is the x- and y-coordinates of the upper-left corner followed by the width and height in pixels.

   To be more specific, if we had a data set of faces located in the directory *data/faces/*, then the collection description file *faces.dat* might look like this:

   ```
   data/faces/face_000.jpg 2 73 100 25 37 133 123 30 45
   data/faces/face_001.jpg 1 155 200 55 78
    . . .
   ```

   If you want your classifier to work well, you will need to gather high-quality data, and a lot of it (1,000–10,000 positive examples). "High quality" means that you've removed all unnecessary variance from the data. For example, if you are learning faces, you should align the eyes (and preferably the nose and mouth) as much as possible. The intuition here is that otherwise you are teaching the classifier that eyes need not appear at fixed locations in the face but instead could be anywhere within some region. Since this is not true of real data, your classifier will not perform as well. One strategy is to first train a cascade on a subpart, say

6 You can still train cascades with the older `haartraining` application, but the resulting cascades will be of the legacy type, so this is not recommended. Among other things, only `traincascade` supports the LBP features (in addition to the Haar features). In addition, only `traincascade` supports TBB for multithreading, and so is much faster (assuming you built the library with TBB support—i.e., with `-D WITH_TBB=ON`).

"eyes," which are easier to align. Then use eye detection to find the eyes and rotate/resize the face until the eyes are aligned. For asymmetric data, the "trick" of flipping an image on its vertical axis was described previously in the subsection "Rejection cascades" on page 884.

2. Once you have your data set, use the utility application `createsamples` to build a "vector" output file of the positive samples. Using this file, you can repeat the upcoming training procedure on many runs, trying different parameters while using the same computed vector file. Here is an example of how to use `createsamples`:

   ```
   createsamples -info faces.dat -vec faces.vec -w 30 -h 40
   ```

   This command reads in the *faces.dat* file described in Step 1 and outputs a formatted *vector file*, in this case called *faces.vec*. Internally, `createsamples` extracts the positive samples from the images, normalizes them, and resizes them to the specified width and height (in this example, 30 × 40 pixels). Note that you can also use `createsamples` to synthesize data by applying geometric transformations, adding noise, altering colors, and so on. This procedure is particularly useful when you have only a single archetype, like a corporate logo, and you want to take just this one image and put it through various distortions that might appear in real imagery. (More details on these options will be covered shortly.)

3. Generate your set of counterexamples. The training process will use these "no" samples to learn what does *not* look like our object. For training purposes, any image that does not contain the object of interest can be turned into a negative sample. It is best to take the "no" images from the same type of data we will test on; that is, if we want to learn faces in online videos, for best results we should take our negative samples from comparable frames (other frames from the same video). However, we can still achieve respectable results using negative samples taken from just about anywhere (e.g., Internet image collections). Again, we put the images into one or more directories and then make a *collection file* consisting of a list of these image filenames, with their paths, one per line.[7] For example, we might create an image collection file and call it *backgrounds.dat* and its contents might include the following paths and filenames of images:

   ```
   data/vacations/beach.jpg
   data/nonfaces/img_043.bmp
   data/nonfaces/257-5799_IMG.JPG
   ...
   ```

7 A *collection file* (used for counterexamples) is a file containing just a list of filenames, while a *collection description file* (used for positive examples) is a file containing a list of filenames and information about where to find objects of interest in each file.

4. Train the cascade. Here is an example of what you might type on a command line in order to create a trained cascade called *face_classifier_take_3.xml*:

```
traincascade                                              /
  -data face_classifier_take_3                            /
  -vec opencv/data/vec_files/trainingfaces_24-24.vec      /
  -w 24 -h 24                                             /
  -bg backgrounds.dat                                     /
  -nstages 20                                             /
  -nsplits 1                                              /
  [-nonsym]                                               /
  -minhitrate 0.998                                       /
  -maxfalsealarm 0.5
```

The *.xml* file extension will automatically be added to the `-data` argument, in this case to create the output file *face_classifier_take_3.xml*. Here, *trainingfaces_24-24.vec* is the set of positive samples (sized to width-by-height of 24 × 24), while random images extracted from *backgrounds.dat* will be used as negative samples. The cascade is set to have 20 (`-nstages`) stages, where every stage is trained to have a detection rate (`-minhitrate`) of 0.998 or higher. The false hit rate (`-maxfalsealarm`) has been set at 50% (or lower) for each stage to allow for the overall hit rate of 0.998. The weak classifiers are specified in this case as "stumps," which means they can have only one split (`-nsplits`); we could ask for more, and this might improve the results in some cases. For more complicated objects one might use as many as six splits, but mostly you want to keep this number smaller, using no more than three splits.

Even on a fast machine, training may take several hours to days depending on the size of the data set. The training procedure must test approximately 100,000 features within the training window over all positive and negative samples. This search is parallelizable and can take advantage of multicore machines (using TBB). This parallel version is the one shipped with OpenCV (assuming you built the library with TBB support—i.e., with `-D WITH_TBB=ON`).

#### Detailed arguments to createsamples

As we saw, in order to arrange your positive samples such that they can be ingested by `traincascade`, you will need to use the `createsamples` program.[8] The `createsamples` program not only crops out the individual samples from the images they are found in, it can also generate automatically modified representations of those samples that have slightly different orientation, lighting, and other characteristics. Here

8 You do not need to do this for the negative samples because `traincascade` is going to dynamically select random examples from the list of negative files you provide. Thus, since there is no detailed cropping or preparation to be done, you don't need to go through this extra step for the negatives.

we will look in detail at the options available to you when you call `createsamples` and what those options do.

When you call `traincascade`, you will run in one of four modes. The mode is determined by which options you select.[9] In particular, the options `-img`, `-info`, `-vec`, and `-bg` collectively determine the run mode. The four run modes are:

*Mode 1: Create training samples from a single image (by applying distortions)*

Starting from a single image (specified by `-img`), generate some number of new test images (specified by `-num`) by applying distortions to the single input image and then pasting it into images from the background set (specified by the `-bg` argument). Because training images are being generated, the output will be a vector file (specified by the `-vec` argument).

```
createsamples -img <image_file> -vec <vector_file> \
-bg <collection_file> -num <n_samples> ...
```

*Mode 2: Create test samples from a single image (by applying distortions)*

This form is very similar to Mode 1, differing primarily in the output file format. It is used to create new images you can use to test your detector on. It applies rotations and distortions to the input image (`-img` argument) from the objects in your collection description file and then generates new images by combining the distorted originals with background images (the collection file given by the `-bg` argument).[10] The point of this is to have an image where the object of interest (albeit a distorted one) is placed at a known location so that you can test your detector. The generated files will have filenames like *<number>_<x>_<y>_<width>_<height>.jpg*, where the values of *<x>*, *<y>*, and so on, specify the location and size of the object that was injected into the (otherwise background) image. Because test samples are being generated, the results will be in a collection description file (your `-info` argument), which will contain the generated filenames and the locations of the inserted objects in those files.

```
createsamples -img <image_file> -bg <collection_file> \
-info <collection_description_file> ...
```

*Mode 3: Create training samples from an image collection (no distortion)*

This is the usage we described in the previous section; it can be thought of as a file format conversion. It simply collects all of the images specified by the `-info` file, crops them as specified, and builds the vector file specified by `-vec`.

9 Which is to say, the mode is not selected by some argument with a name like "`-mode`," but rather it is inferred from the other arguments you supply.

10 The results of this process would look pretty strange to your eye—for example, disembodied slightly rotated heads floating in the middle of whatever background scenes you provided—but this is necessary to ensure that the rotation and distortions do not introduce black or empty pixels into the final training images.

```
createsamples -info <collection_description_file> \
    -vec <vector_file>                            \
-w <width> -h <height> ...
```

*Mode 4: View samples from the .vec file*

In this mode, all of the samples in the *.vec* file will be displayed to the screen one-by-one. This is mainly for debugging and as an aid to understanding the process.

```
createsamples -vec <vector_file>
```

Of these four modes, the one you probably want to use most is not on the list. In practice, you will typically find yourself with some number of exemplar images (maybe 1,000) and wanting to generate some much larger number of exemplars through distortion and transformation (maybe 7,000). In this case, what you want to do is actually use Mode 1 1,000 times, creating, for example, 7 new images for each original. In this case you will need to do a little of your own minor automation of the process. You will find yourself in the same situation if you want to generate a large collection of test images.

Here are the detailed descriptions of each option:

`-vec <file_name>`
: This is the name of the file that will be created by `createsamples`. It should have a *.vec* file extension.

`-info <file_name>`
: This is the name of the file that specifies the input collection of examples, including both the filenames as well as the location of the example objects in those images (i.e., the *faces.dat* file described previously).

`-img <file_name>`
: This is the alternative to `-info` (you must supply one or the other). Using `-img`, you can supply a single cropped positive exemplar. In the modes that use `-img`, multiple outputs will be created, all from this one input.

`-bg <file_name>`
: The `-bg` extension allows you to specify a file (again one with a *.dat* extension) that contains the names and ancillary information for the list of provided background images.

`-num <n_samples>`
: `-num` sets the number of positive samples that should be generated (i.e., by transformations on the input samples specified by `-vec`).

`-bgcolor <color>`
: This intensity value is interpreted as "transparent" in the input images. Note that grayscale images are assumed. This is used when you are overlaying positive exemplars onto alternate backgrounds.

`-bgthresh <delta>`
: In many practical cases, the input images will contain compression artifacts (e.g., *.jpg* files). Because of these artifacts, the background may not always be the same fixed color. Used in conjunction with `-bgthresh <color>`, `-bgthresh <delta>` will cause all pixels within the range [*color* – *delta*, *color* + *delta*] to be interpreted as transparent.

`-inv`
: If specified, all images will be inverted before the sample is extracted.

`-randinv`
: If specified, each image will be either inverted or not inverted (randomly) before the sample is extracted.

`-maxidev <deviation>`
: If specified, each image will randomly be (uniformly) lightened or darkened by up to this amount before extraction.

`-maxxangle <angle>`, `-maxyangle <angle>`, `-maxzangle <angle>`
: If specified, each image will be distorted by a random rotation by up to the given amount in each direction. This provides an approximation of possible viewpoint perspective shifts of the object (though, to define these transformations, it is assumed that the objects are effectively flat cards). The units of these rotations are radians.

`-show`
: If specified, each sample will be shown. Pressing the Esc key will continue the samples-creation process without showing further samples.

`-w <width>`
: The width (in pixels) of generated samples.

`-h <height>`
: The height (in pixels) of generated samples.

Much experimentation has been done to determine the best sizes to use for samples for face detection. In general, 18 × 18 or 20 × 20 seems to perform very well. For objects other than faces, you will likely need to experiment to find out what works best for your particular case.

As you can see, there are a lot of options for `createsamples`. The important thing to keep in mind is that most of these options are used to automatically create variants of the images you have provided in your available examples. Using these options you can (and typically will) turn hundreds or thousands of sample images into thousands or tens of thousands of images on which to actually train the classifier.

### Detailed arguments to traincascade

As with `createsamples`, there are myriad options that can be passed to `traincascade` in order to fine-tune its behavior. These include parameters to tune the cascade itself, the boosting method, the types of features used, and more. These parameters will heavily affect the training time for the cascade, but also the quality of the final result.

`-data` *`<classifier_file>`*
: The `-data` parameter specifies the name of the output-trained classifier file to be created. You do not need to provide the *.xml* file extension; it will be added for you.

`-vec` *`<vector_file>`*
: The `-vec` parameter specifies the filename for the input vector file of positive exemplars (i.e., created using `createsamples`).

`-bg` *`<collection_file>`*
: This parameter specifies the name of the background images collection file.

`-numPos` *`<n_samples>`*
: This is the number of positive examples that will be used in training each classifier stage (typically less than the number of examples that are provided).

`-numNeg <n_samples>`
: This is the number of negative examples that will be used in training each classifier stage (typically less than the number of examples that are provided).

`-numStages` *`<stages>`*
: The `-numStages` parameter specifies the number of cascade stages that will be trained for the classifier as a whole.

`-precalcValBufSize` *`<size-megabytes>`*
: This is the size of the buffer allocated for storage of precalculated feature values. The buffer size is specified in megabytes. This buffer is used by the Boost implementation to store results of feature evaluations so that they don't have to be recomputed every time they are needed. The net result is that making this cache larger will substantially improve the runtime of training. The current default value is 256 megabytes.

`-precalcIdxBufSize <size-megabytes>`
: This is similar to `-precalcValBufSize`, and also used by the Boost implementation. `-precalcIdxBufSize` sets the size of the cache used for the "buffer index values." What these things are exactly is not important to us; what is important about these objects is that the ability to cache them improves performance, similar to `-precalcValBufSize`. The current default value is 256 megabytes, and it is best to keep these two values the same if you choose to change either of them.

`-baseFormatSave <{true,false}>`
: This can be set to `true` if you are using the Haar-type features and want to save your cascade out in the "old style" format. The default value of this argument is `false`.

`-stageType <{BOOST}>`
: This argument sets the type of stages that will be used in the classifier training. At the moment, it has only one option, `BOOST` (meaning "boosted classifier cascade"), which is currently the default, so you can safely ignore it for now. This argument is here for future development.

`-featureType <{HAAR, LBP}>`
: Currently the cascade classifier supports two different feature types: the Haar(-like) features, and the local binary pattern features. You can select which one you would like to use by setting `-featureType` to `HAAR` or `LBP` (respectively).

`-w <sample_width-pixels>`
: The `-w` parameter tells `traincascade` the width of the samples that you provided to `createsamples`. The value passed to the `-w` parameter must be equal to the value used by `createsamples`. The units of `-w` are pixels.

`-h <sample_width-pixels>`
: The `-h` parameter tells `traincascade` the height of the samples that you provided to `createsamples`. The value passed to the `-h` parameter must be equal to the value used by `createsamples`. The units of `-h` are pixels.

`-bt <{DAB, RAB, LB, GAB}>`
: `traincascade` can train the cascade using any of four available variants of boosting (see our earlier discussion on boosting). The available options are Discrete AdaBoost (`DAB`), Real AdaBoost (`RAB`), LogitBoost (`LB`), and Gentle AdaBoost (`GAB`). The default value is `GAB`.

`-minHitRate <rate>`
: The minimum hit rate, `-minHitRate`, sets the target percentage of real occurrences in a window that should be flagged as hits. Of course, ideally this would be 100%. However, the training algorithm will never be able to achieve this. The

value of this parameter is unit-normalized, so the default value of `0.995` corresponds to 99.5%. This is the target hit rate per stage, so the final hit rate will be (approximately) this target raised to the power of the number of stages.

`-maxFalseAlarmRate` *<rate>*
: The maximum false alarm rate, `-maxFalseAlarmRate`, sets the target percentage of false occurrences in a window that can be expected to be (erroneously) flagged as hits. Ideally this would be 0%, but in practice it is quite large and we rely on the cascade to reject false alarms incrementally. The value of this parameter is unit-normalized, so the default value of `0.50` corresponds to 50%. This is the target false positive rate per stage, so the final hit rate will be (approximately) this target raised to the power of the number of stages.

`-weightTrimRate` *<rate>*
: We encountered this argument earlier as a parameter to the boosting algorithms. It is used to select which training samples to use in a particular boosting iteration. Only those samples whose weight is more than `1.0` minus the weight trim rate participate in the training on any given iteration. The default value for this parameter is `0.95`.

`-maxDepth` *<depth>*
: This parameter sets the maximum depth of the individual weak classifiers. Note that this is not the depth of the cascade, it is the depth of the individual trees that themselves comprise the elements of the cascade. The default value for this parameter is `1`, which corresponds to simple decision stumps.

`-maxWeakCount` *<count>*
: Like `-maxDepth`, the `-maxWeakCount` parameter is passed directly to the boosting component of the cascade classifier and sets the maximum number of weak classifiers that can be used to form each strong classifier (i.e., each stage in the cascade). The default value for this parameter is `100`, but remember that this doesn't mean that this number of weak classifiers will be used.

`-mode <BASIC | CORE | ALL>`
: The `-mode` parameter is used with Haar-like features and determines whether just the original Haar features are to be used (`BASIC` or `CORE`) or whether extended features are to be used (`ALL`). The features used are shown in Figure 22-5.

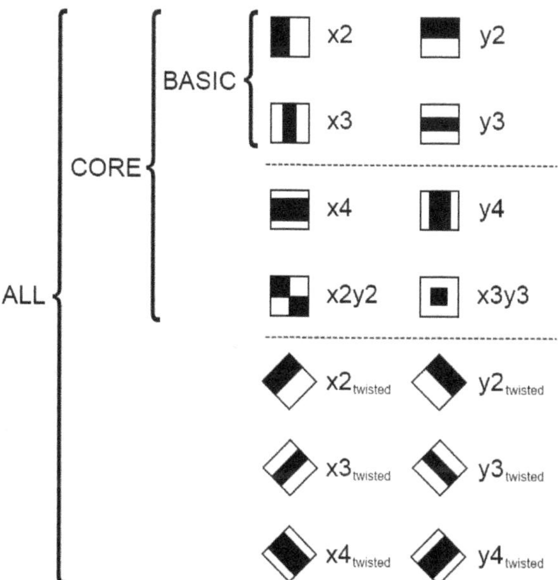


*Figure 22-5. The options for the -mode parameter. BASIC includes just the simplest possible set of Haar wavelets (one even and one odd wavelet in each direction). CORE includes four additional higher-order Haar wavelets, while ALL also includes diagonal elements that are rotated versions of (some of) the other wavelets*

Despite the potentially intimidating number of options available to `traincascade`, you will probably find that the default values will serve you well for many real-world situations. If you are training cascades for specific object types (not faces), it is worth a web or literature search to see if anyone else has tried to train a cascade for your particular object. Often, it will turn out that others have already done much of the hard work in optimizing the selection of training parameters for your object or, at the very least, a similar one.

# Object Detection Using Support Vector Machines

Similar to the object-detection techniques based on tree-based methods, there is another class of algorithms in OpenCV that use support vector machines as the basis of their learning strategies. But as we saw with the tree-based functions, there are a large number of additional components that are used to condition, organize, and handle both the data and the reusable representation of the training process.

In this section, we will look at two methods: the Latent SVM and the Bag of Words methods. These two methods are very different, despite their common reliance on the support vector machine. The Latent SVM is well suited to recognition of deformable objects (such as pedestrians), as it explicitly conceptualizes the idea of multiple subcomponents that are linked together by a deformable structure. The Bag of Words takes a different approach and ignores large-scale structure entirely, taking inspiration from techniques in document recognition in which only the list of components of the target is considered. In this way, Bag of Words generalizes beyond object detection alone, and can be used for entire scenes and for context analysis.

## Latent SVM for Object Detection

The Latent SVM algorithm, created by Pedro Felzenszwalb [Felzenszwalb10], is an algorithm for detecting (originally) pedestrians in images, but it generalizes well to many kinds of objects such as bicycles and automobiles. It builds on a well-known prior technique, HOG-SVM,[11] which was first proposed by Navneet Dalal and Bill Triggs [Dalal05]. HOG-SVM used a sliding window, analogous to the face detection cascade classifier that we saw earlier in this chapter. The algorithm identified pedestrians by subsectioning the window into smaller tiles and by computing a histogram of the orientations of image gradients in each tile. These histograms, often called "HOG" (histogram of oriented gradients) for short, could then be concatenated in order to form a feature vector that could then be passed to an SVM classifier.[12]

Felzenszwalb's technique, called either *part-based object detection* or *Latent SVM*, starts with a HOG feature similar to the one used by HOG-SVM. However, in addition to the detection of the whole object (as in HOG-SVM), it represents distinct parts of the object separately: those that might be expected to move relative to one another, or to the object as a whole (e.g., the arms, legs, and head of a pedestrian). In practice, the locations of the parts relative to the center of the image (called the *root node* by Felzenszwalb) are unknown; they are the *latent variables* in the model from which the algorithm derives its name. Once the root node has been located (by a detector very similar to that in HOG-SVM) and the parts have been located, an object hypothesis can be formed, taking into account the likelihoods that the found parts will be in the locations detected relative to the root node.

As implemented in OpenCV, the Latent SVM method can make use of already trained detectors that ship with the library.

11 The name *HOG-SVM* is actually a coinage; this algorithm was never given a name by its creators, but "the method of Dalal and Triggs" is a bit of a mouthful. The acronym HOG was coined by those authors and is now a standard term of art to refer to the features used by HOG-SVM, and since the algorithm is essentially applying an SVM to these HOG features, HOG-SVM seems like a reasonable enough name.

12 The actual HOG-SVM algorithm is implemented in OpenCV, but only as part of the GPU library. For more information, see `cv::gpu::HOGDescriptor` in the online documentation at OpenCV.org. There is no current CPU library implementation of HOG-SVM.

### Object detection with cv::dpm::DPMDetector

To use the Latent SVM classifier in OpenCV, you will first need to instantiate a `cv::dpm::DPMDetector` object (which resides in the *opencv_contrib/dpm* module, so you'd need to build OpenCV together with *opencv_contrib*).[13] This is done through the type of `create()` function we saw a lot of in Chapter 21 (complete with a derived implementation class that we will never really see: `DPMDetectorImpl`). Typically, when you instantiate this object, you will supply the cascades in the form of (fully qualified) filenames for the detectors themselves. Optionally, you can also supply class names for these detectors; if you don't, class names will be automatically induced from the detector filenames.

```
static cv::Ptr<cv::dpm::DPMDetector> create(
    std::vector<std::string> const &filenames,
    std::vector<std::string> const &classNames = std::vector<std::string>()
);
```

The filenames should be an STL-style vector of strings containing fully qualified filenames. Your class names, if present, should just be an STL-style vector of strings to be used as class names.

Once you have a model loaded, you can proceed to apply the model to find objects in your own images:

```
void cv::dpm::DPMDetector::detect(
  cv::Mat                        &image,
  std::vector<ObjectDetection> &objects
);
```

The `image` argument is your input image. The `objects` argument is an STL-style vector you provide that `detect()` will fill with `cv::dpm::DPMDetector::ObjectDetection` instances, which will contain all of the information about individual detections found in `image`.

Object detections returned to you will be in the form of `cv::dpm::DPMDetector::ObjectDetection` objects. These objects have the following definition:

```
class cv::dpm::DPMDetector {

public:

  ...

  struct ObjectDetection {
```

13 In general, we have not included work in *opencv_contrib* in this book. The reason for this exception is that this particular implementation, in addition to simply being very useful, has been available for quite a long time and is considered very stable.

```
        ObjectDetection();
        ObjectDetection( const cv::Rect& rect, float score, int classID = -1 );

        cv::Rect rect;
        float    score;
        int      classID;
    };

    ...

};
```

As you can see, this structure is very simple. The three elements it contains—`rect`, `score`, and `classID`—indicate the size and location of the window in which the object was found, the confidence assigned to that detection, and the integer class identifier for the particular class that was found, respectively. These IDs will correspond to the order in which you loaded the detectors when you first called `cv::dpm::DPMDetector::load()`.

#### Other methods of cv::dpm::DPMDetector

The `cv::dpm::DPMDetector` object also provides a few useful utility and accessor methods:

```
virtual bool isEmpty() const;

size_t getClassCount() const;

const std::vector<String>& getClassNames() const;
```

The `isEmpty()` method returns `true` if there are no detectors loaded.[14]

The `getClassCount()` method just returns the number of classes you loaded when you called `load()`, while the `getClassNames()` method returns an STL-style vector containing all of the names of the detectors. This is particularly useful when you did not supply these names yourself, but need to know what names were assigned by `cv::dpm::DPMDetector::load()` (based on the filenames).

#### Where to get models for cv::dpm::DPMDetector

At this time, OpenCV does not provide code that will allow you to train your own models for Latent SVM. Currently, *opencv_contrib/modules/dpm/samples/data/* contains a few pretrained models. If these are not enough for you, or you need something particularly unique, and you do have to train your own detector, the original creators

14 Internally, this is just turned into a call to the `std::vector<>::empty()` method of the vector member of `cv::dpm::DPMDetector` that contains the detectors.

of the Latent SVM maintain a website that contains their MATLAB implementation. This MATLAB implementation contains the necessary component (called "pascal") to train your own detectors. Once trained, the resulting *.xml* file can be loaded by the OpenCV detector.

## The Bag of Words Algorithm and Semantic Categorization

Also called *Bag of Keypoints*, the *Bag or Words* (BOW) algorithm is a method for *visual categorization*, or identifying the object content of a scene. The algorithm takes its inspiration from methods used in document categorization that attempt to organize documents into semantic categories by the presence of certain keywords that are identified as having a strong discriminatory capability between the classes. For example, in attempting to categorize medical documents, it might be found that the presence of the word *tumor* is an effective indicator that the document belongs to the category *cancer documents*. In this case, each document is not really read in any meaningful way, it is just treated as a collection ("bag") of words, and the relative frequencies of important words are the only thing used to discriminate the categories.

In the case of computer vision, it is possible to define features that have strong discriminating power to determine that an image is, for example, in the category *automobile* or *bicycle*. The BOW algorithm addresses both the problem of identifying which features are most salient, and of attempting to identify the features in a novel image and compare them to that database to categorize the image.

The first phase of the BOW algorithm is where it learns the features that it will subsequently use for categorization. During this phase, you provide the part of the algorithm called the *trainer* with images from every semantic category of interest to you (e.g., *cars*, *people*, *chickens*, or whatever you like). In the OpenCV implementation, you will actually first extract from each image your favorite flavor of keypoints (see Chapter 16) and give the resulting lists of descriptors to the BOW trainer.

At this point, it is not necessary to segregate the images in any way; the purpose of the trainer is to figure out which keypoints seem to form meaningful clusters. These clusters are groups of similar keypoint descriptors that came from the images you supplied and are close together in the vector space in which the keypoints are represented. These clusters are then abstracted into *keypoint centers*, which are essentially new keypoints, constructed by the BOW trainer, that live at the centers of the identified clusters. These keypoint centers play the role of words in the document-categorization analogy, and for this reason are often referred to as *visual words*.[15] Collectively, the set of visual words that have been identified is called the *vocabulary*.

Once the BOW trainer has generated the vocabulary, you can give it any image and it will convert that image into a *presence vector*. A presence vector is a vector of Boolean entries that represent the presence (or absence) of each word in the vocabulary. Note that this is a very high-dimensional vector; in practice, it is hundreds or even thousands of dimensions.

At this point, the normal process is to train a classification algorithm. Given all of the images in your data set, the BOW algorithm can convert them into presence vectors, which are then used to train the algorithm to produce the correct class label. You can use any classification algorithm you like that is capable of multiclass classification, but the canonical choice is either the naïve Bayes classifier or the support vector machine (SVM), both of which we have already come across in Chapter 21. The most important point, however, is that the BOW algorithm produces presence vectors from your known images that you can use to train a classifier, and it produces presence vectors from your novel images that you can give to that classifier so that it can associate a category with that image.

### Training with cv::BOWTrainer

The essential task of converting a large number of input feature descriptors into a manageable number of visual words can be done using any number of clustering techniques. The abstract base class `cv::BOWTrainer` defines the interface for any object that can perform this function for the BOW algorithm:

```
class cv::BOWTrainer {

public:

  BOWTrainer(){}
  virtual ~BOWTrainer(){}

  void                 add( const Mat& descriptors );
  const vector<Mat>& getDescriptors()                    const;
  int                  descriptorsCount()                const;

  virtual void         clear();
  virtual Mat          cluster()                         const = 0;
  virtual Mat          cluster( const Mat& descriptors ) const = 0;
```

15 Note that it is impractical to use every descriptor found in every image in the training phase. The purpose of the clustering is to aggregate similar descriptors in the inputs into a smaller, more manageable number. In principle, those that are clustered together have similar meaning. You could think of this in the analogy of document categorization as clustering words *walk*, *walked*, *walking*, and the like into a single category—rather than counting the occurrence of each separately.

```
    ...
};
```

The `cv::BOWTrainer::add()` method is used to add keypoint descriptors to the trainer. It expects an array that it will interpret such that each row is a separate descriptor. You can call `cv::BOWTrainer::add()` any number of times to accumulate your descriptors. At any point, you can find out how many descriptors have been added with `cv::BOWTrainer::descriptorsCount()`, or return all of them in one big array with `cv::BOWTrainer::getDescriptors()`.

Once you have loaded all of your descriptors in from all of your images, you can call `cv::BOWTrainer::cluster()`, which will actually compute the visual word vocabulary. It then returns the vocabulary to you as an array in which each row is interpreted as a separate visual word. There is also a one-argument version of `cv::BOWTrainer::cluster()` that expects a single array containing descriptors and will immediately compute the visual words for the set of descriptors in that array; it ignores any descriptors you might have already stored with `cv::BOWTrainer::add()`.

Finally, there is the method `cv::BOWTrainer::clear()`, which empties all of the loaded descriptors.

### K-means and cv::BOWKMeansTrainer

At this point, there is just one implementation of the `cv::BOWTrainer` interface—`cv::BOWKMeansTrainer`. The only member of `cv::BOWKMeansTrainer()` that is really new is the constructor, which has the following prototype:

```
cv::BOWKMeansTrainer::BOWKMeansTrainer(
  int                     clusterCount,
  const cv::TermCriteria& termcrit    = cv::TermCriteria(),
  int                     attempts    = 3,
  int                     flags       = cv::KMEANS_PP_CENTERS
);
```

The heart of the `cv::BOWKMeansTrainer()` implementation is the use of the K-means clustering algorithm. Recall that the K-means algorithm takes this large number of points and attempts to find a specific number of clusters that adequately explain the data. In the `cv::BOWKMeansTrainer()` constructor, this number of clusters is called `clusterCount`, which in this case is going to be the number of visual words generated, or the size of the vocabulary. Making this number too small will result in extremely poor classification results. Making it too large will make subsequent stages of the process operate very slowly and may render classification impossible.[16]

The remaining three arguments can be left at their default values if you are not an expert in the K-means algorithm. Should you choose to modify them, they have the same meaning to the constructor as they do to the K-means implementation we studied earlier in Chapter 20.

### Categorization with cv::BOWImgDescriptorExtractor

Once you have computed the cluster centers with the trainer, you can ask the BOW algorithm to try to convert the descriptors for an image into a presence vector that can be used for classification. Here is (the important part of) the class declaration for `cv::BOWImgDescriptorExtractor`, the routine responsible for this phase of the BOW algorithm:

```
class cv::BOWImgDescriptorExtractor {

public:

  BOWImgDescriptorExtractor(
    const cv::Ptr< cv::DescriptorExtractor >& dextractor,
    const cv::Ptr< cv::DescriptorMatcher >&   dmatcher
  );
  virtual ~BOWImgDescriptorExtractor() {;}

  void setVocabulary( const cv::Mat& vocabulary );
  const cv::Mat& getVocabulary() const;
  void compute(
    const cv::Mat&            image,
    vector< cv::KeyPoint >& keypoints,
    cv::Mat&                  imgDescriptor,
    vector< vector<int> >*  pointIdxsOfClusters = 0,
    cv::Mat*                  descriptors = 0
  );
  int descriptorSize() const;
  int descriptorType() const;

  ...

};
```

16 In one paper [Csurka04], a canonical reference for this algorithm, the authors have 1,776 images representing 7 classes. From each class they extract 5,000 keypoints from the images in that class, giving them a total of 35,000 keypoints. They find that setting the number of clusters (per class) to anywhere between 1,000 and 2,500 gives similar results, and so they opt to use 1,000 for their published results. One might fairly conclude from this that a number of clusters equal to a few percent of the number of points in the class is a reasonable number.

The first thing you will notice is that when we construct a `cv::BOWImgDescriptorExtractor` object, we need to provide it a descriptor extractor and a descriptor matcher. The extractor you provide should be the same as the one you used to extract the descriptors when you computed the cluster centers with the trainer. The matcher can be any one you like. For example:

```
cv::Ptr< cv::DescriptorExtractor >       descExtractor;
cv::Ptr< cv::DescriptorMatcher >         descMatcher;
cv::Ptr< cv::BOWImgDescriptorExtractor > bowExtractor;

descExtractor = cv::DescriptorExtractor::create( "SURF" );
descMatcher   = cv::DescriptorMatcher::create( "BruteForce" );
bowExtractor  = new cv::BOWImgDescriptorExtractor( descExtractor, descMatcher );
```

Once you have constructed your BOW descriptor extractor, you will need to give it the vocabulary you built using the trainer. The `vocabulary` input to `cv::BOWImgDescriptorExtractor::setVocabulary()` is exactly the output you got from `cv::BOWTrainer::cluster()`.[17]

Finally, once you have given a vocabulary to the descriptor extractor, you can then supply `cv::BOWImgDescriptorExtractor::compute()` with an image and it will compute `imageDescriptor`, which is the presence vector. The presence vector will have one row, and as many columns as there are rows in the vocabulary. Each element will be the number of elements matched to a particular cluster center.

The optional arguments `pointsIdxsOfClusters` and `descriptors` tell you something about the actual matching that happened to generate the presence vector. `pointsIdxsOfClusters` is a vector of vectors, with the first index relating to the cluster; thus, `pointsIdxsOfClusters[i]` refers to the *i*th cluster center (the *i*th entry in the vocabulary), and is itself a vector. The entries `pointsIdxsOfClusters[i]` list the indices for the descriptors in `image` that were matched to that cluster center. Those indexes indicate row numbers for `descriptors`. The `descriptors` array, if computed, is the list of the original descriptors (before association with cluster centers) extracted from the image by the feature extractor you gave the BOW extractor when you created it. To recap: if `pointsIdxsOfClusters[i][j] = q`, this means that `descriptors.row(i)` is the *j*th one of the descriptors, and was matched to `vocabulary.row(q)`.

#### Putting it together using a support vector machine

To actually implement the entire BOW algorithm, you will need one more step: to take the presence vectors and train a multiclass classifier with them. Of course, there

17 Under the hood, what is happening here is just that all of the descriptors in the vocabulary are being added to the train set of the matcher you gave the extractor when you created it.

are a lot of ways to implement multiclass classifiers; the one we will look at here is the support vector machine. Because SVMs don't natively support multiclass classification, we use the *one-against-many* approach.[18] In this approach, if one has $N_c$ classes, one trains $N_c$ different classifiers, each of which addresses the question of whether a query vector is in class $i$ or not in class $i$ (i.e., in any of the other classes $j \neq i$).

We train each SVM using the same set of presence vectors, but with different labelings for the responses. The following code fragment is from the samples included with the OpenCV release and is from the *bagofwords_classification.cpp* example code; it constructs the `trainData` and `responses` arrays that will be passed to the SVM training routine. This will be done once for each class:

```
cv::Mat trainData( (int)images.size(), bowExtractor->getVocabulary().rows, CV_32FC1 );
cv::Mat responses( (int)images.size(), 1, CV_32SC1 );

// Transfer bag of words vectors and responses across to the training data matrices
//
for( size_t imageIdx = 0; imageIdx < images.size(); imageIdx++ ) {

  // Transfer image descriptor (bag of words vector) to training data matrix
  //
  cv::Mat submat = trainData.row( (int)imageIdx );
  if( bowImageDescriptors[imageIdx].cols != bowExtractor->descriptorSize() ) {
    cout << "Error: computed bow image descriptor size "
         << bowImageDescriptors[imageIdx].cols
         << " differs from vocabulary size"
         << bowExtractor->getVocabulary().cols
         << endl;
    exit(-1);
  }
  bowImageDescriptors[imageIdx].copyTo( submat );

  // Set response value
  //
  responses.at<int>((int)imageIdx) = objectPresent[imageIdx] ? 1 : -1;
};
```

This is not quite the entire story, though. The possibility exists in the one-against-many method that an image will be found to be "not in this class" for every class, or "in this class" for more than one class. Recall, however, that the SVM works by building a linear decision boundary in the high-dimensional kernel space. Because this is a linear boundary in that space, once a query point is mapped into the kernel space, we can also compute rather easily which side of that decision boundary the point lies on. We can also compute how far the point is from that boundary. It is common to inter-

18 You might recall that we mentioned one-against-many briefly when we looked at support vector machines in detail. In our current context, we could just use the one-against-one approach supplied by OpenCV, but that is relatively slower, so we use one-against-many for this example.

pret this distance as a kind of confidence, and this is the key to resolving the problems of nonassociation and multiple association.

Typically, in the case of multiple association, the association with the largest margin is taken to be the correct association. In the nonassociation case, if one has prior knowledge that every image is in fact part of one of the known classes, then the one with the minimum negative margin can be selected. If images that are of unknown classes are possible, then one might set a threshold negative margin (possibly, but not necessarily, `0`) past which an image will be assigned to the "unknown" category if all classifiers return worse than this value.

# Summary

In this chapter we studied several methods that the OpenCV library provides for determining if an object is present in an image. In some cases, these methods provide some notion of the localization of the object in terms of the pixels that compose the object in the image. The tree-based methods, as well as the Latent SVM, had this property, which arose from their use of sliding windows. On the other hand, the Bag of Words method did not have this property. One advantage of the Bag of Words, however, was that it could be used for more abstract queries, like scene categorization.

All of these methods made use of the computer vision techniques we learned in the earlier chapters of this book combined with the machine learning methods of the most recent chapters. If we add to these detectors the methods of tracking and motion modeling we learned along the way, we now have all the tools necessary to approach very practical and contemporary problems in computer vision, such as the location, localization, and tracking of objects in a video sequence, as well as many others.

# Exercises

1. If you had a limited amount of training data, which would be more likely to generalize the test data better, a decision tree or an SVM classifier? Why?
2. Set up and run the Haar classifier to detect your face in a web camera.
   a. How much scale change can it work with?
   b. How much blur?
   c. Through what angles of head tilt will it work?
   d. Through what angles of chin down and up will it work?
   e. Through what angles of head yaw (motion left and right) will it work?
   f. Explore how tolerant it is of 3D head poses. Report on your findings.

3. Repeat Exercise 2, but this time for an LBP cascade. Where is it better or worse? And why is it better or worse?

4. Use blue or green screening to collect a thumbs-up hand gesture (static pose). Collect examples of other hand poses and of random backgrounds. Collect several hundred images and then:

   a. Train the Haar classifier to detect the thumbs-up gesture. Test the classifier in real time and compute its confusion matrix.

   b. Train and test a LBP cascade classifier and compute its confusion matrix on the thumbs-up data.

   c. Train and test a soft cascade classifier and compute its confusion matrix on the thumbs-up data.

   d. Compare and contrast the train time, run time, and detection results of these three algorithms.

5. Using the data set of Exercise 4 and three features of your own invention:

   a. Use random trees to recognize the thumbs-up data.

   b. Add a feature or features to improve the results.

   c. Use data analysis, variable importance, normalization, and cross validation to improve your results.

   d. Use your knowledge of random trees to improve your results.

6. Use the data set of Exercise 4 to train a `cv::dpm::DPMDetector` to recognize the thumbs-up gesture.

   a. Create a confusion matrix of the results.

7. Collect some streetview imagery (one image each) from your local neighborhood —say, 10 different places.

   a. Train a BOW classifier to recognize these 10 different places. Test recognition from nearby streetviews.

8. One can imagine using a Latent SVM to also classify where you are in streetview imagery in addition to the BOW classifier approach. What are some relative advantages and disadvantages for these algorithms when used for map localization?

   Hint: there is a difference between topology and geometry. The first might be associated with imagery, the second with triangulation.

9. The cascade classifier produces a two-class classifier (face or not face, for example). If you had 10 classes, describe a method by which you could use a cascade classifier methodology to train and recognize all 10 classes. What are the advantages and disadvantages of such an approach?

CHAPTER 23

# Future of OpenCV

## Past and Present

OpenCV was launched in August 1999 at the Computer Vision and Pattern Recognition conference (and so turns 17 years old at the publication of this book). Gary Bradski founded OpenCV at Intel with the intention to accelerate both the research and use of real applications of computer vision in society. Few things in life go according to their original plan, but OpenCV is one of those few. As of this writing, OpenCV has nearly 3,000 functions, has had 14 million downloads, is trending well above 200,000 downloads per month, and is used daily in millions of cell phones, recognizing bar codes, stitching panoramas together, and improving images through computational photography. OpenCV is at work in robotics systems—picking lettuce, recognizing items on conveyor belts, helping self-driving cars see, flying quadrotors, doing tracking and mapping in virtual and augmented reality systems, helping unload trucks and pallets in distribution centers, and more—and is built into the Robotics Operating System (ROS). It is used in applications that promote mine safety, prevent swimming pool drownings, process Google Maps and streetview imagery, and implement Google X robotics, to name a few examples.

Since the previous edition of this book, OpenCV has been re-architected from C to modern, modular C++ compatible with STL and Boost. The library has been brought up to modern software development standards with distributed development on Git (*https://github.com/opencv/opencv*), continuous build bots (*http://bit.ly/cv-bots*), Google unit tests, comprehensive documentation (*http://docs.opencv.org*), and tutorials (*http://bit.ly/2gkuGN1*). OpenCV was intended to be cross-platform from the beginning when it spanned Windows, Linux, and Mac OS X. It continues active support for these desktop OSes, but now also covers mobile with Android and iOS versions. It has optimized versions for Intel architectures, ARM, GPU, NVidia GPUs, and Movidius chips but also works with Xylinx Zync FPGAs. In addition to efficient C++

source code, it has extensive interfaces in Python (compatible with NumPy), Java, and MATLAB.

OpenCV has also added a new, independent section maintained by users (*https://github.com/opencv/opencv_contrib*). In that repository, all routines are standalone and follow OpenCV style and documentation, as well as pass the Buildbot tests. With *opencv_contrib*, OpenCV keeps up with the latest algorithms and applications in computer vision; see Appendix B for a snapshot of the directory's contents.

## OpenCV 3.x

OpenCV started as a purely C library, and version 1.0 focused mostly on building useful algorithmic content. OpenCV 2.0's main focus was on bringing the library up to modern C++ development standards, including the move to Git, Google-style unit tests, compatibility with STD, and of course a C++ interface. All new development has been in C++, but the older C functions were just wrapped in C++. Along the way, complete interfaces in Python, Java, and MATLAB were added.

OpenCV 3.0 focuses on modularity; it is written entirely in native C++ so that only one code base needs to be maintained. Computer vision's increasing success has led to a problem that there are too many potentially useful algorithms to be maintained in one monolithic code base. OpenCV 3.x solved that problem by keeping a strong supported core and turning everything else into easy-to-create and easier-to-maintain small, independent modules that may be mixed and matched as desired. More and more computer vision students and research groups are releasing new algorithms built on OpenCV data structures. OpenCV 3.x makes it easy for them to produce a module complete with documentation, unit tests, and example code that can be easily linked into OpenCV (or not).

OpenCV 3.x's independent modules will also help cloud, embedded, and mobile applications by allowing for smaller, more focused computer vision memory footprints. One of the mission statements of OpenCV is to foster increasing use of computer vision in society; embedded vision devices will help spread the use of visual sensing in robotics, mobile, security, safety, inspection, entertainment, education, and automation. For such applications, memory use is a key consideration. On the other side, cloud computing also has memory constraints—as algorithms scale across large numbers of machines running a wide mix of jobs, memory use becomes a key bottleneck.

Our hope is that by making it easy to assemble a mix of independent modules, including perhaps one's own module, OpenCV 3.x will not only enable the aforementioned areas but also foster something that may look like a "vision app store" in *opencv_contrib*. Such a collection of well-defined modules that plug directly into OpenCV will allow much wider and more creative uses of vision-enabled applications. External modules might be open, closed, free, or commercial, all aimed at allowing developers who know very little about vision to infuse vision capability into their applications.

## How Well Did Our Predictions Go Last Time?

In the previous edition of this book, we made some predictions about OpenCV's future. How did we do? We said that OpenCV would support robotics and 3D; this clearly came true. One of the authors, Gary, launched a robotics company, Industrial Perception Inc., that used OpenCV and 3D vision routines to allow robots to handle boxes in distribution centers. Google bought that company in 2013. At the same time, the other author, Adrian, ran many industry, government, and military robotics contract projects incorporating OpenCV while he was working at Applied Minds.

Calibration was forecast to be expanded and to include passive and active sensing. True to form, OpenCV now includes ArUco augmented reality markers and the combination of checkerboard and ArUco patterns so you no longer need to see the whole board, and multiple cameras can see different pieces of the same calibration pattern (see Appendix C). All these routines now exist to solve more challenging calibration and multicamera pose problems.

We predicted new 3D object recognition and pose recognition capabilities, and these were also integrated—from human-defined features in `linemod` (*http://bit.ly/2fJ9WuS*) to deep network 3D object recognition and pose (*http://bit.ly/2gksqFs*). Indeed, *opencv_contrib* was itself predicted as a modular repository that would make user contribution much easier.

Most of the applications predicted in the previous book, from much better stereo vision algorithms to dense optical flow, have come true. Back then, we said that 2D features would be expanded and supported by an engine, all of which happened in `features2D()`, which covers a large percentage of the hand-crafted 2D point detectors and descriptors. Improved functionality with Google data structures is also under way. We also said that better support for approximate nearest neighbor techniques would be added, and it was with the incorporation of FLANN (Fast Library for Approximate Nearest Neighbor) into OpenCV. We have long since run developer workshops at computer vision conferences as outlined in the previous book. Finally, better documentation did finally show up (*http://docs.opencv.org*).

What were we wrong about? We did not yet get a more general camera interface for higher-bit or multispectral cameras. SLAM (Simultaneous Localization And Mapping) support is in, but not as a robust complete implementation. Bayesian networks were not pursued because deep networks outpaced them. We did not yet implement anything special for artists, but artists nevertheless have continued to expand the use of OpenCV.

## Future Functions

This book has mentioned OpenCV's past and detailed its present state. Here are some future directions:

*Deep learning*
: OpenCV can already read and run networks such as Caffe, Torch, and Theano. This code is at *https://github.com/opencv/opencv_contrib/tree/master/modules/cnn_3dobj*. You can expect to see OpenCV integrate a full deep-learning module focused on running and training in embedded systems and smart cameras built around and expanding on an external code base called *tiny_dnn*.

*Mobile*
: The growth in "computationally capable" cameras is still phenomenal. So, one obvious direction OpenCV will take is increasing support of mobile. This support includes algorithms as well as mobile hardware and mobile OS. OpenCV already has ports to iOS and Android, which we hope to support by allowing smaller static memory footprints.

*Glasses*
: Augmented reality glasses that overlay the incoming scene with data and objects will be an increasingly supported area. Tracking the user's head pose in 3D will also aid virtual reality localization within a room. Already, we've expanded ArUco AR tags to ChArUco (checkerboard with ArUco) that give a much more accurate pose. We have some contributors working on adding SLAM support for Google Cardboard.

*Embedded apps*
: Embedded applications are also growing in importance and will become a whole new device area. Seeing this trend, Xilinx already has a port of OpenCV to its Zync architecture. We can expect to see vision showing up in a range of items, from toys to security devices, automotive applications, manufacturing uses, and unmanned vehicles on land, underwater, and in the air. OpenCV wants to help enable these developments.

*3D*
: Depth sensors are under development by many companies and will increasingly show up in mobile. OpenCV has a growing number of dense-depth support routines, from computing fast normals, surface finding, and depth feature extraction to refinement.

*Light field cameras*
: This is an area dating back to 1910 but having intense activity in the 1990s. We predict it will become increasingly popular, with cheaper cameras and embedded processors allowing lens arrays to capture wide multipoint views, apertures, and fields of view, perhaps using different lens configuration. Expect to see support for such cameras as they come into existence and get less expensive.

*Robotics*
: All of the preceding features directly benefit robotics. New hardware, cheaper cameras, and radically more flexible robot arms, coupled with better planning and control algorithms, mark the start of a whole new industry in sensor-guided robotics. Several key contributors to OpenCV work in robotics, and you can expect to see continued growth in support of this area.

*Cloud*
: Over time, expect to see support to make it easier to work across arrays of embedded cameras interoperating with servers running the same processing stack and tightly integrated with OpenCV, deep neural networks, graphics, optimization, and parallel capable libraries. There will be some effort to have this working seamlessly on commercial providers such as Amazon and Google servers using C++ or Python.

*Online education*
: We would like to provide online courses that cover computer vision problem solving using OpenCV. We hope to expand our visibility at conferences and workshops and perhaps offer our own "things you need to know" conferences.

## Current GSoC Work

For the last several years, Google has been kind enough, through its Google Summer of Code (GSoC) program, to support interns working over the summer on OpenCV. You may view a wiki page (*https://github.com/opencv/opencv/wiki*) on these efforts. You can also view videos covering this new functionality at the following URLs:

- **2015:** *https://youtu.be/OUbUFn71S4s*
- **2014:** *https://youtu.be/3f76HCHJJRA*
- **2013:** *https://youtu.be/_TTtN4frMEA*

In 2015, 15 interns were supported. This support has been invaluable both to the interns (many of whom go on to prominent positions in the field) and to OpenCV. The topics covered in 2015, almost all with accepted pull requests into OpenCV trunk, were:

*Omnidirectional cameras calibration and stereo 3D reconstruction*
*opencv_contrib/ccalib* module (Baisheng Lai, Bo Li)

*Structure from motion*
*opencv_contrib/sfm* module (Edgar Riba, Vincent Rabaud)

*Improved deformable part-based models*
*opencv_contrib/dpm* module (Jiaolong Xu, Bence Magyar)

*Real-time multi-object tracking using kernelized correlation filter*
*opencv_contrib/tracking* module (Laksono Kurnianggoro, Fernando J. Iglesias Garcia)

*Improved and expanded scene text detection*
*opencv_contrib/text* module (Lluis Gomez, Vadim Pisarevsky)

*Stereo correspondence improvements*
*opencv_contrib/stereo* module (Mircea Paul Muresan, Sergei Nosov)

*Structured-light system calibration*
*opencv_contrib/structured_light* module (Roberta Ravanelli, Delia Passalacqua, Stefano Fabri, Claudia Rapuano)

*Chessboard + ArUco for camera calibration*
*opencv_contrib/aruco* module (Sergio Garrido, Prasanna Krishnasamy, Gary Bradski)

*Implementation of universal interface for deep neural network frameworks*
*opencv_contrib/dnn module* (Vitaliy Lyudvichenko, Anatoly Baksheev) [this may be replaced by *tiny-dnn* in the future]

*Recent advances in edge-aware filtering, improved SGBM stereo algorithm*
*opencv/calib3d* and *opencv_contrib/ximgproc* modules (Alexander Bokov, Maksim Shabunin)

*Improved ICF detector, Waldboost implementation*
*opencv_contrib/xobjdetect* module (Vlad Shakhuro, Alexander Bovyrin)

*Multitarget TLD tracking*
*opencv_contrib/tracking* module (Vladimir Tyan, Antonella Cascitelli)

*3D pose estimation using CNNs*
*opencv_contrib/cnn_3dobj* module (Yida Wang, Manuele Tamburrano, Stefano Fabri)

As of the final editing of this book, the following 13 new algorithms are being worked on for GSoC 2016:

- Adding *tiny-dnn* deep learning training and test functions into OpenCV (Edgar Riba, Yida Wang, Stefano Fabri, Manuele Tamburrano, Taiga Nomi, Gary Bradski)
- Enhancing the existing *dnn* module to read and run Caffe models (Vludv, Anatoly Baksheev)
- Better visual tracking, GOTURN tracker (Tyan Vladimir, Antonella Cascitelli)
- Accurate, dynamic structured light (Ambroise Moreau, Delia Passalacqua)
- Adding very fast, dense optical flow (Alexander Bokov, Maksim Shabunin)
- Extending the text module with deep word-spotting CNN (Anguelos, Lluis Gomez)
- Improvement of the dense optical flow algorithm (VladX, Ethan Rublee)
- Multilanguage support in OpenCV tutorials: Python, C++, and Java (Carucho, Vincent Rabaud)
- New image stitching pipeline (Jiri Horner, Bo Li)
- Adding better file storage for OpenCV (Myls, Vadim Piarevsky)

## Community Contributions

The OpenCV community has become much more active as well. During the time of GSoC 2015, the community contributed:

- A plotting module (Nuno Moutinho)
- Ni-black thresholding algorithm: *ximgproc* (Samyak Datta)
- Superpixel segmentation using linear spectral clustering, SLIC superpixels: *ximgproc* (Balint Cristian)
- HDF (HDF5) support module (Balint Cristian)
- Depth to external RGB camera registration: *rgbd* (Pat O'Keefe)
- Computing normals for a point cloud: *rgbd* (Félix Martel-Denis)
- Fuzzy image processing (Pavel Vlasanek)
- Rolling shutter guidance filter: *ximgproc* (Zhou Chao)
- 3× faster SimpleFlow: *optflow* (Francisco Facioni)
- Code and docs for CVPR 2015 paper "DNNs Are Easily Fooled" (Anh Nguyen)
- Efficient graph-based image segmentation algorithm: ximgproc (Maximilien Cuony)
- Sparse-to-dense optical flow: *optflow* (Sergey Bokov)

- Unscented Kalman filter (UKF) and augmented UKF tracking (Svetlana Filicheva)
- Fast Hough transform: *ximgproc*, *xolodilnik*
- Improved performance of `haartraining` (Teng Cao)
- Python samples made compatible with Python 3: *bastelflp*

We hope that Google and the community continue this great work!

## OpenCV.org

In the time between the publication of the book's previous edition and this one, OpenCV became a California nonprofit foundation aimed at advancing computer vision in general, promoting computer vision education, and providing OpenCV as a free and open infrastructure for furthering vision algorithms in particular. To date, the foundation has had support from Intel, Google, Willow Garage, and NVidia. In addition, DARPA (through Intel) provided funding for a "People's Choice Best Paper" award at CVPR (Computer Vision And Pattern Recognition) 2015, and Intel has sponsored this contest to run again in 2016. The results from 2015 are available online (*https://github.com/opencv/opencv/wiki/VisionChallenge*). The winning entries resulted in several new algorithms hosted in the *.../opencv_contrib* directory. Prebuilt code for OpenCV can be downloaded from the user site (*http://opencv.org/downloads.html*), while raw code can be obtained from the developer site; see *https://github.com/opencv/opencv* for the core library and *https://github.com/opencv/opencv_contrib* for the user-contributed modules. The wiki for OpenCV is at *https://github.com/opencv/opencv/wiki*. There is also a Facebook page (*https://www.facebook.com/opencvlibrary*).

As the writing of this book comes to a close, the founding author, Gary Bradski, is in the process of turning OpenCV.org into a federal nonprofit 501(c)(3) corporation. Previously, OpenCV had no paid staff (beyond summer mentor stipends provided by Google), no office, and no equipment, and had been trying to pay out within the same year everything that came in. Now, there is an effort under way to turn OpenCV.org into a robust, full-featured nonprofit. This will involve bringing on some dedicated board members (unpaid), raising funds to support some paid full-time staff, developing educational materials and contests, putting on annual conferences that would emphasize new, useful vision solutions, providing in-depth training tutorials, sponsoring or at least supporting greater sensing and autonomy in robotics leagues, providing support and education for learning and using computer vision at the high school level, offering support and training for using computer vision in the artist community, and more.

We also hope to add more cooperation with OpenCV in China (*http://www.opencv.org.cn*), founded by Prof. Riuzhen Liu. This is hosted at the Shanghai

Academy of Artificial Intelligence (also known as AIV: Artificial Intelligence Valley), which is sponsored by the Chinese Academy of Science and Fudan University. It is a subscriber organization aiming to be an independent research organization focused on artificial intelligence, automation, intelligent device control, and pattern recognition. Over time, we hope to increase similar links to other organizations around the world.

If OpenCV.org can generate enough funding, it is possible that OpenCV can offer full-time phone and web support, develop courseware in vision and machine learning (possibly including partnering with manufacturers to provide compatible development kits), and certify vision developers who can be trusted to build applications in computer vision and deep learning perception. We may also develop a certification program for other camera functionality offered by partners where "Certified by OpenCV" can become a trusted brand. In so doing, we look forward to vastly expanding the reach and scope of OpenCV!

# Some AI Speculation

We are clearly at a turning point in the development of artificial intelligence (AI). As of this writing, AlphaGo from Google's Deep Mind group has beat the world champion, Lee Sedol, at the very difficult "spatial strategy" game Go. Using AI, robots are learning to drive, fly, walk, and manipulate objects. Meanwhile, AI technology is making speech, sound, music, and image recognition natural in our devices and across the Web. Silicon Valley has seen many "gold rushes" since the original one for real gold in 1848. The winners and losers in this new AI gold rush remain to be seen, but it is clear that the world will never be the same. In its function of accelerating progress in perception, OpenCV plays a role in this historical movement toward sentient (self-aware) machines.

It's clear that deep neural networks have essentially solved the problem of feedforward recognition of patterns (one can say that they are superb function approximators), but such networks are nowhere near sentient or "alive." First, there is the problem of experience itself. We humans don't just see, say, a color; we experience it subjectively. How this subjective experience arises is called the problem of "qualia." Second, machines also don't seem to ever really be autonomously creative. They can generate new things within an explicit domain, but they don't invent new domains, nor actively drive experimentation and open-loop discovery.

What may be missing is "embodiment." Humans and many robots have a model of their own being, their "self," acting in the world. This self can be simulated in isolation for planning actions, but this simulated model of self is more often coupled to the world by sensors. Using this coupled model, the embodied mind gives causal meaning to the world (choosing where to walk, avoiding danger, observing consequences to its plan), and this gives the embodied mind a sense of meaning in relation to its model of itself.

We believe that such a world-coupled model of itself allows the AI to make metaphors [Lakoff08] that are used to generalize to later experience. When young, for example, humans experience putting things into and taking things out of containers. Later in life, this embodied experience informs what it means to be, say, "in" a garden; in other words, the early experience of playing with containers is used to generalize what it means to be in a garden. Causal experience of the model of self, coupled to the world, allows an entity to attach meaning to things. Such meaning stabilizes perception since categories don't just come and go—they have causal and time-stable consequences to our simulated model of self within the world. It is complete speculation, but qualia, or subjective experience, may arise from simulating how our model of the world affects our simulated model of the self; that is, we experience our model's simulated reaction to the world, not the world itself.

Stanley, the robot that won the $2 million prize for the 2005 DARPA Grand Challenge robot race across the desert, used many sensors such as GPS, accelerometers, gyros, laser range finders, and vision to sense the world and fused these sensed results into a computationally efficient "bird's-eye view" world model. The model consisted of a tilted plane reflecting the general angle of the terrain that was then marked with drivable, un-drivable, and unknown regions derived from the sensor readings. In this world model, Stanley ran physics simulations of itself driving in the general direction of the next few GPS waypoints. The resulting paths were rated to find the most efficient path that would not tip the robot over. Stanley's brain was sufficient to win the DARPA Grand Challenge, but consider what it wasn't sufficient to do: it could not represent love, politics, astrophysics, or Shakespeare very well. If Stanley could ask us what a Shakespeare play meant, at best we could say it was something like the boundaries between the drivable and unknown areas in a difficult map. Stanley's model of itself and its interaction with the world are too sparse for understanding most of the things in the world. It seems obvious that we ape-like beings that live mostly in low-lying temperate watersheds are similarly limited in our ultimate ability to even detect what we don't know about the universe. In this way, "we are all Stanley."

We humans find it pretty easy to understand something such as the need for food (a natural part of our model), but we find it extremely difficult to figure out how to create a more intelligent AI or to fathom what qualia is. As another example, if we raised a kitten and had it listen to Shakespeare all day until it was grown, we wouldn't expect our grown cat to understand a sonnet. If we want to explain to the cat what a sonnet means, the best we could do is use a metaphor from the cat's natural models, such as, "Shakespeare's sonnet is like a kitten that inevitably gets lost in bad places." The cat might think, "Now I understand," but it has no means by which to even understand what it does not understand! Again, we humans must also be similarly limited. Perhaps we can build more powerful machines to which the problem of qualia is simple. But when we ask the machine to explain it to us, it might get flustered and then finally say, "Qualia is like a kitten that inevitably gets lost in bad places."

In Stanley the robot, the nature of its perception is entirely in terms of its model. Stanley doesn't perceive the world; instead, its cameras and sensors transduce signals that populate a causal model of the robot in the world. But the model is only like the real world in terms of the navigational needs of the robot car. By way of another example: in a laptop computer, you might see a GUI and conclude, for example, that there's a trash can inside the computer since you see one on the screen. A more clever physicist might look closely at the screen and cry out, "Everything is made up of quanta" (pixels)! But, in fact, the GUI is only a causal model to a linear Von Neumann machine reality inside. What is real is that things dragged into the trash are erased—the causal consequences of the model are real. Again, we humans must be similarly limited in what we can know of our own universe, since we've inherited a causal model mainly directed at our direct physical and social experience. But our machines may see further.

Today, there is a lot of debate about the dangers of AI. People confuse their metaphors around this. They think the "AI," the intelligence, is what drives the behaviors of the larger system. But look to ourselves. Our "programming language" as humans isn't our intellect, but our moods and drives—our emotions! Stanley's goal was to safely traverse GPS points as fast as possible in the correct order. It found an orderly following of GPS waypoints to be attractive and so that's what it employed its intelligence to do. Our programming isn't "thought," it's emotion. The emotions guide "what" to do; the intellect guides "how." The same will be true of our future machines. Design the motives well, and the machines will pursue them.

Ah, but the reader may worry that perhaps those machines will alter the goals given to the next generation of machines? First of all, it will be no easier for a greater machine intelligence to understand and create the minds of a yet greater machine than it is for us to create the first generation. The problem just shifts upward. The intelligent machines also will face the same dangers to themselves from their next evolution of AI that we do from the first evolution, and they will tend to program goals and emotions accordingly. In the end, AIs will be consumed with their own goals, which have evolved from our goals as embodied beings. The danger from AIs is not from the possible malevolence of their goals, but from the nonhuman differences in their goals, which to us may seem like indifference or even hostility. This is what made H. P. Lovecraft's alien monsters so interesting and frightening in his fiction—they were not so much malevolent as driven by wholly different motives and so wholly indifferent to our fate. We, for example, give little thought to the lives of ants and thereby sometimes bring them to harm.

However, and for the record, the authors don't fear AI, but rather see it as absolutely essential to solve many of humanity's vexing problems, such as providing reliable health care for all people, ending hunger, curing diseases, providing and maintaining new energy generation and storage techniques while protecting the environment, and helping run our ever-more-complex world. Rather than a threat, in AI we see purpose. We would not want the spark of self-aware intelligent life to die out with our world, but instead would rather see intelligence grow outward in space and in time. This, in a way, may be (or could be chosen to be[1]) humanity's ultimate purpose, and so is also, hereby, an indirect purpose of OpenCV!

# Afterword

We've covered a lot of theory and practice in this book, and we've described some of the plans for what comes next. Of course, as we're developing the software, the hardware is also changing. Cameras are now increasingly cheaper and more capable and have proliferated from cell phones to traffic lights and into factory and home monitoring. A group of manufacturers are aiming to develop cell phone projectors—perfect for robots, because most cell phones are lightweight, low-energy devices whose circuits already include an embedded camera. This opens the way for close-range portable structured light and thereby accurate detailed depth maps, which are just what we need, together with the development of light field cameras for robot manipulation and 3D object scanning.

Both authors participated in creating the vision system for Stanley, Stanford's robot racer that won the 2005 DARPA Grand Challenge. In that effort, a vision system coupled with a laser range scanner worked flawlessly for the seven-hour desert road race [Dahlkamp06]. For us, this drove home the power of combining vision with other perception systems: we converted the previously unsolved problem of reliable road perception into a solvable engineering challenge by merging vision with other forms of perception. It is our hope that—by making vision easier to use and more accessible through this book—others can add vision to their own problem-solving tool kits and thus find new ways to solve important problems. That is, with commodity camera hardware, cheap embedded processors, and OpenCV, people can start solving real problems, such as using stereo vision as an automobile backup safety system (or to make automotive improvements in general), monitoring all rail lines for people and vehicles on the tracks, implementing swimming safety measures, building new game

1 There is a sort of spiritual or religious sense to such thoughts, but they have an interesting inversion to "old time" religion. In the past, it was felt that a God chose a people and a purpose. In this new sense, people choose a purpose, which may result in a God-like AI (Google's algorithms and servers have what is to us almost omniscient knowledge, for example). In this sense, we move from a *chosen people* who hark back to some glorious past from a diminished present, to a *choosing people* who look out from a diminished present to a glorious future.

controls, developing new security systems, and so on. Be sure to keep an eye on *tiny-dnn*, a fully featured deep net library with a focus on embedded computing in *opencv_contrib*. Finally: get hacking!

Computer vision has a rich future ahead, and it seems likely to be one of the key enabling technologies for the 21st century. OpenCV seems likely to be (at least in part) one of the key enabling technologies for computer vision. Endless opportunities for creativity and profound contribution lie ahead. We hope that this book encourages, excites, and enables all who are interested in joining the vibrant computer vision community!

# APPENDIX A
# Planar Subdivisions

## Delaunay Triangulation, Voronoi Tesselation

*Delaunay triangulation* is a technique invented in 1934 [Delaunay34] for connecting points in a space into triangular groups such that the minimum angle of all the angles in the triangulation is a maximum. This means that Delaunay triangulation tries to avoid long skinny triangles when triangulating points. See Figure A-1 to get the gist of triangulation, which is done in such a way that any circle that is fit to the points at the vertices of any given triangle contains no other vertices. This is called the *circum-circle property* (see panel c).

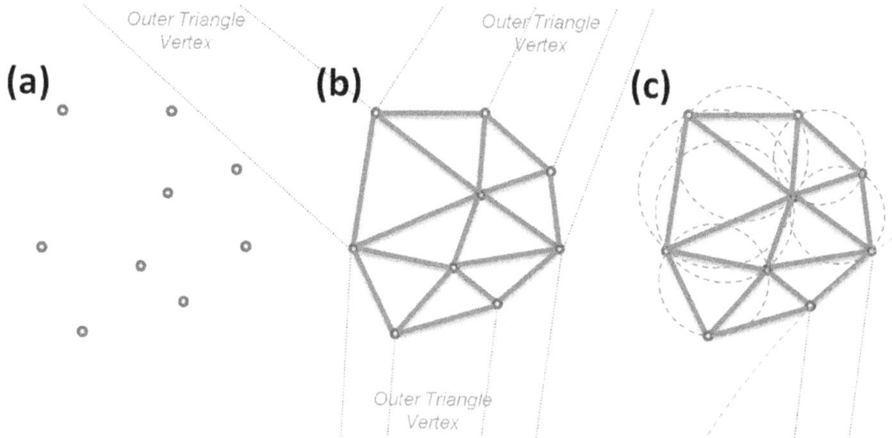


*Figure A-1. Delaunay triangulation: (a) set of points; (b) Delaunay triangulation of the point set with trailers to the outer bounding triangle; (c) example circles showing the circum-circle property*

For computational efficiency, the Delaunay algorithm invents a faraway outer bounding triangle from which the algorithm starts. Figure A-1(b) represents the fictitious outer triangle by dotted lines going out to its vertex. Figure A-1(c) shows some examples of the circum-circle property, including one of the circles (the one at the bottom-right corner) linking two outer points of the real data to one of the vertices of the fictitious external triangle.

There are now many algorithms to compute Delaunay triangulation; some are very efficient but with difficult internal details. The gist of one of the simpler algorithms is as follows:

1. Add the external triangle.
2. Add an internal point, and then search over all the triangles' circum-circles containing that point and remove those triangulations.
3. Retriangulate the graph, including the new point in the circum-circles of the just-removed triangulations.
4. Return to Step 2 until there are no more points to add.

The order of complexity of this algorithm is $O(n^2)$ in the number of data points. The best algorithms are (on average) as low as $O(n \log \log n)$.

Great—but what is it good for? For one thing, remember that this algorithm started with a fictitious outer triangle. Thus, all of the real outside points are actually connected to one (or two) of the fictitious triangle's vertices. Now recall the circum-circle property: circles that are fit through any two of the real outside points and to an external fictitious vertex contain no other inside points. This means that a computer may directly look up exactly which real points form the outside of a set of points by looking at which points are connected to the three outer fictitious vertices. In other words, we can find the convex hull of a set of points almost instantly after a Delaunay triangulation has been done.

We can also find who "owns" the space between points—that is, which coordinates are nearest neighbors to each of the Delaunay vertex points. Thus, using Delaunay triangulation of the original points, you can immediately find the nearest neighbor to a new point. Such a partition is called a *Voronoi tessellation* (see Figure A-2). This tessellation is the dual image of the Delaunay triangulation, because the Delaunay lines define the distance between existing points and so the Voronoi lines "know" where they must intersect the Delaunay lines in order to keep equal distance between points. These two methods, calculating the convex hull and nearest neighbor, are important basic operations for many operations that require clustering or classifying of points and point sets.

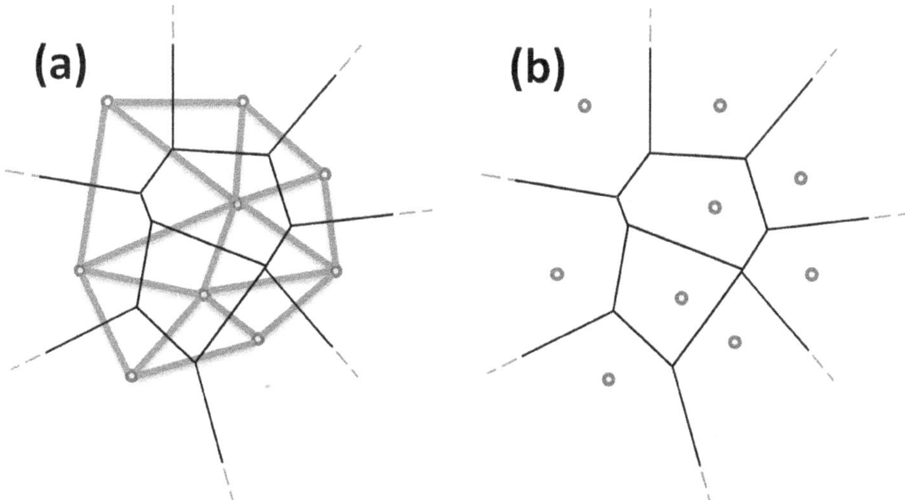


*Figure A-2. Voronoi tessellation, whereby all points within a given Voronoi cell are closer to their Delaunay point than to any other Delaunay point: (a) the Delaunay triangulation in bold with the corresponding Voronoi tessellation in fine lines; (b) the Voronoi cells around each Delaunay point*

If you're familiar with 3D computer graphics, you may recognize that Delaunay triangulation is often the basis for representing 3D shapes. If we render an object in three dimensions, we can create a 2D view of that object by its image projection and then use the 2D Delaunay triangulation to analyze and identify this object and/or compare it with a real object. Delaunay triangulation is thus a bridge between computer vision and computer graphics. However, one deficiency of OpenCV (soon to be rectified, we hope; see Chapter 23) is that OpenCV performs Delaunay triangulation only in two dimensions. If we could triangulate point clouds in three dimensions—say, from stereo vision (see Chapter 19)—then we could move seamlessly between 3D computer graphics and computer vision. Nevertheless, 2D Delaunay triangulation is often used in computer vision to register the spatial arrangement of features on an object or a scene for motion tracking, object recognition, or matching views between two different cameras (as in deriving depth from stereo images). Figure A-3 shows a tracking and recognition application of Delaunay triangulation [Göktürk01; Göktürk02] wherein key facial feature points are spatially arranged according to their triangulation.

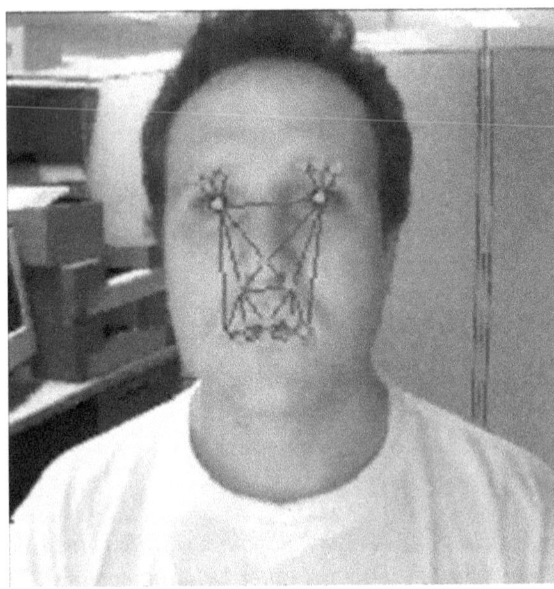

*Figure A-3. Delaunay points can be used in tracking objects; here, a face is tracked using points that are significant in expressions so that emotions may be detected*

Now that we've established the potential usefulness of Delaunay triangulation once given a set of points, how do we derive the triangulation? OpenCV ships with example code for this in the *.../opencv/samples/cpp/delaunay2.cpp* file. OpenCV refers to Delaunay triangulation as a Delaunay *subdivision*, whose critical and reusable pieces we discuss next.[1]

## Creating a Delaunay or Voronoi Subdivision

First we'll need someplace to store the Delaunay subdivision in memory. We'll also need an outer bounding box (remember, to speed computations, the algorithm works with a fictitious outer triangle positioned outside a rectangular bounding box). To set this up, suppose the points must be inside a 600 × 600 image:

```
// STRUCTURE FOR DELAUNAY SUBDIVISION
//
...
cv::Rect       rect(0, 0, 600, 600);         // Our outer bounding box
cv::Subdiv2D   subdiv(rect);                 // Create the initial subdivision
```

1 According to Merriam-Webster's dictionary (*http://www.merriam-webster.com/dictionary/subdivision*), the term *subdivision* can mean either "one of the parts into which something is divided" or "an area of land that has been divided into smaller areas on which houses are built." Though perhaps unintuitively, the use of this word in OpenCV always refers to the latter (i.e., it means a collection of parts, not the parts themselves).

This code creates an initial subdivision with a triangle containing the specified rectangle.

Next we'll need to know how to insert points. These points must be either of 32-bit floating-point type, or points with integer coordinates (i.e., `cv::Point`). In the latter case, they will be automatically converted to floating-point format. We add the points using the `cv::Subdiv3D::insert()` function:

```
cv::Point2f fp;    //This is our point holder

for( int i = 0; i < as_many_points_as_you_want; i++ ) {

     // However you want to set points
     //
     fp = your_32f_point_list[i];

     subdiv.insert(fp);
}
```

Now that we have entered the points, we can obtain a Delaunay triangulation. To compute triangles from the Delaunay triangulation, we use the `cv::Subdiv3D:: getTriangleList()` function:

```
vector<cv::Vec6f> triangles;
subdiv.getTriangleList(triangles);
```

After the call returns, each `Vec6f` in `triangles` will contain three triangle vertices: (*x1*, *y1*, *x2*, *y2*, *x3*, *y3*). We can compute and retrieve the associated Voronoi tessellation with the `cv::Subdiv2D::getVoronoiFacetList()` function:

```
vector<vector<cv::Point2f> > facets;
vector<cv::Point2f> centers;
subdiv.getVoronoiFacetList(vector<int>(), facets, centers);
```

The first output vector of vectors will contain Voronoi facets—polygons outlining the "proximity" regions for the previously inserted points. The second vector will contain the corresponding points—the region centers.

It's important to note that Delaunay triangulation is built iteratively; that is, each time you insert a new point the triangulation is updated, so it's always up-to-date. However, the Voronoi tessellation is built in batch mode when you call `cv::Subdiv2D::calcVoronoi()`. (This function does not take any parameters and does not return anything; it just updates the internal subdivision representation.) Alternatively, you can call the aforementioned `cv::Subdiv2D::getVoronoiFacetList()` (which calls `calcVoronoi()` internally). If you insert a new point after the Voronoi tessellation is computed, the tessellation is marked as invalid and it's recomputed from scratch the next time you need it (the compute cost for this is $O(N)$, where $N$ is the number of vertices).

Now that we can create Delaunay subdivisions of two-dimensional point sets and their corresponding Voronoi tessellations, the next step is to learn how to navigate across the subdivisions. Though simply having the triangulation or tessellation in enough in many cases, it is also often useful to be able to step from edge to point or from edge to edge in a subdivision. We describe how to do this in the next section.

## Navigating Delaunay Subdivisions

The basic data element of the planar subdivision is an *edge*. The edge is accessed via its index, and other neighboring edges can be accessed via this index and an additional parameter that indicates what new edge we would like relative to the one we started with. Every edge is associated with the points at either end (called the *origin* and the *destination*, respectively). It is also associated, through these points, with other edges that share those points. Finally, it is associated with its corresponding (i.e., "dual") edge; for every edge in the Delaunay triangulation there is an associated edge in the Voronoi diagram, and vice versa. The library supplies mechanisms to get from any edge to any of these related elements.

Note that, in the `cv::Subdiv2D` interface, edges are always treated as being directional. This is actually just done for convenience; there is no intrinsic sense of orientation for edges in either the Delaunay triangulation or the Voronoi diagram. However, this construction is very helpful when one is navigating the data structures that describe these entities. Notably, it is a way to distinguish things at one end of a particular edge (the origin) from the other (the destination).

### Points from edges

The simplest thing we can do with a Delaunay or Voronoi edge is to find the location of the points at either end. Each edge, regardless of whether it is a Delaunay or Voronoi edge, has two points associated with it: its origin point, and its destination point. You can obtain either of these points by using:

```
int cv::Subdiv2D::edgeOrg( int edge, cv::Point2f* orgpt = 0 ) const;
int cv::Subdiv2D::edgeDst( int edge, cv::Point2f* dstpt = 0 ) const;
```

These methods give you the indices of the respective vertices and optionally the points themselves. Given the vertex index, it's possible to retrieve the point and the associated edge:

```
cv::Point2f cv::Subdiv2D::getVertex( int vertex, int* firstEdge = 0 ) const;
```

Note that, just like with the edges, every point has an index. Points also have a location. It is important to keep track of which one is required in any particular context. The `cv::Subdiv2D` interface is designed with the idea that you will primarily use the indices for points and edges in most of the interface's functions.

### Locating a point within a subdivision

It does happen, however, that you may have the location of a particular point and want to look up the index of that point in the subdivision. Similarly, you may have a point that is not actually a vertex in the subdivision at all, but you would like to find the triangle or facet that contains this point. The method `cv::Subdiv2D::locate()` takes one point as input and returns either one of the edges on which this point lies, or one of the edges on the triangle or facet that contains the point (if the point is not a vertex). Note, however, that in this case it is not necessarily the closest `edge` that is returned; it is simply one of the edges in the containing triangle or facet. When the point is a vertex, `cv::Subdiv2D::locate()` will also return the vertex ID that has been assigned to it.

```
int cv::Subdiv2D::locate(
  cv::Point2f pt,
  int&        edge,
  int&        vertex
);
```

This function's return value tells us where the point landed, as follows:

`cv::Subdiv2D::PTLOC_INSIDE`
: The point falls into some facet; `*edge` will contain one of edges of the facet.

`cv::Subdiv2D::PTLOC_ON_EDGE`
: The point falls onto the edge; `*edge` will contain this edge.

`cv::Subdiv2D::PTLOC_VERTEX`
: The point coincides with one of subdivision vertices; `*vertex` will contain a pointer to the vertex.

`cv::Subdiv2D::PTLOC_OUTSIDE_RECT`
: The point is outside the subdivision reference rectangle; the function returns and no pointers are filled.

`cv::Subdiv2D::PTLOC_ERROR`
: One of input arguments is invalid.

### Orbiting around a vertex

Given an edge, you may want to get new edges that are associated with a particular point on that edge—either the beginning or the end of it. The way this works is that we specify the edge we are starting with and then we retrieve either the next edge clockwise or counterclockwise around the head (the *destination*) or the next edge going clockwise or counterclockwise around the tail (called the *origin*). This arrangement is illustrated in Figure A-4. We do this with the `cv:Subdiv2D::getEdge()` function:

```
int cv:Subdiv2D::getEdge(
   int edge,
   int nextEdgeType        // see text below
) const;
```

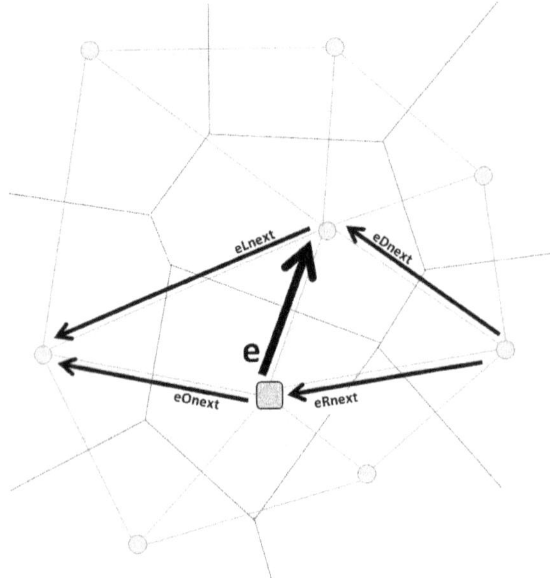


*Figure A-4. A cv::Subdiv2D::Vertex point and its associated edge e along with other associated edges that may be accessed via cvSubdiv2DGetEdge()*

When calling `cv:Subdiv2D::getEdge()` we provide the current `edge` and the argument `nextEdgeType`, which must take one of the following values:

- `cv::Subdiv2D::NEXT_AROUND_ORG`, next around the edge origin (`eOnext`)
- `cv::Subdiv2D::NEXT_AROUND_DST`, next around the edge destination vertex (`eDnext`)
- `cv::Subdiv2D::PREV_AROUND_ORG`, previous around the edge origin (reversed `eRnext`)
- `cv::Subdiv2D::PREV_AROUND_DST`, previous around the edge destination (reversed `eLnext`)

Depending on how you like to think about the way you are navigating, you can also specify the step motions using the following (ultimately equivalent) values:

- `cv::Subdiv2D::NEXT_AROUND_LEFT`, next around the left facet (`eLnext`)
- `cv::Subdiv2D::NEXT_AROUND_RIGHT`, next around the right facet (`eRnext`)

- cv::Subdiv2D::PREV_AROUND_LEFT, previous around the left facet (reversed eOnext)
- cv::Subdiv2D::PREV_AROUND_RIGHT, previous around the right facet (reversed eDnext)

We can use these to step around a Delaunay triangle if we're on a Delaunay edge or to step around a Voronoi cell if we're on a Voronoi edge.

Alternatively, where convenient and appropriate, you can also use the slightly simplified cv:Subdiv2D::nextEdge() function:

```
// equivalent to getEdge(edge, cv::Subdiv2D::NEXT_AROUND_ORG)
//
int cv:Subdiv2D::nextEdge(
   int edge
) const;
```

Calling cv:Subdiv2D::nextEdge() is exactly equivalent to calling cv:Subdiv2D::getEdge() with nextEdgeType set to cv::Subdiv2D::NEXT_AROUND_ORG. This construction is handy when, for example, we are given an edge associated with a vertex and we want to find all other edges from that vertex. This is helpful for finding things like the convex hull starting from the vertices of the (fictitious) outer bounding triangle.

### Rotating an edge

Assuming that you have a particular index in hand, either because you got it from some other function or because you are just starting arbitrarily at some particular index and wandering around the graphs, you can get from that edge on the Delaunay triangulation to an edge on the associated Voronoi diagram (or the reverse) with the following function:

```
int cv::Subdiv2D::rotateEdge(
  int edge,
  int rotate      // get other edges in the same quad-edge: modulo 4 operation
) const;
```

In this case edge is the index of your current edge, and the rotate parameter indicates what edge you would like. You specify that next edge by using one of the following arguments (see Figure A-5):

- 0, the input edge (e in Figure A-5 if e is the input edge)
- 1, the rotated edge (eRot)
- 2, the reversed edge (reversed e)
- 3, the reversed rotated edge (reversed eRot)

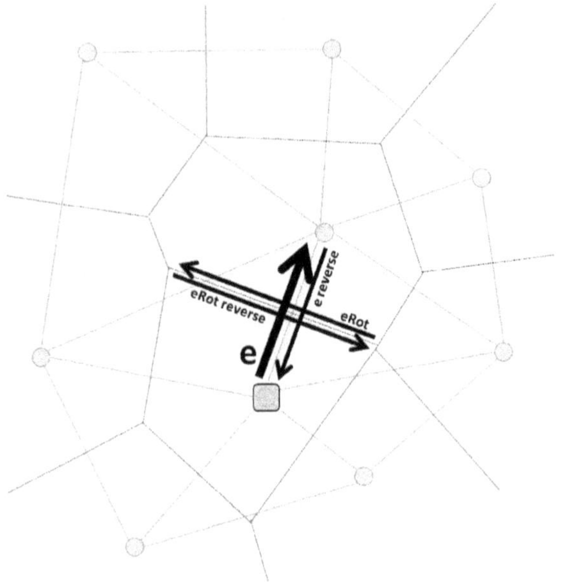


*Figure A-5. Quad edges that may be accessed by cv::Subdiv2DRotateEdge() include the Delaunay edge and its reverse (along with their associated vertex points) as well as the related Voronoi edges and points*

## More About Vertices and Edges

### Vertices and Their Numbering

Because of the way that the Delaunay triangulation is initialized, it will always be the case that:

1. The 0th vertex is a null vertex having no location. (It is just a bookkeeping abstraction.)
2. The next three vertices are the "virtual" vertices outside of the given bounding rectangle, each having made up locations far from the input points.
3. All subsequent vertices are part of the set provided to the `cv::Subdiv2D` object.

### Edges and Their Numbering

Every edge in the `cv::Subdiv2D` object is identified by an integer. These integers are constructed such that every sequential set of four correspond to:

`edge % 4 == 0`
: A Delaunay edge

`edge % 4 == 1`
: The Voronoi edge perpendicular to the original edge

`edge % 4 == 2`
: The original Delaunay edge, with reverse orientation

`edge % 4 == 3`
: The above Voronoi edge, with reverse orientation

**Virtual and Null Edges**

The 0th edge is a null edge pointing to nowhere (or, more accurately having the 0th—also null—vertex at either end). Edges 1, 2, and 3, will always be the virtual Delaunay edges connecting the virtual vertices; we will refer to these as *unanchored* virtual edges, as both vertices are virtual. The null edge will always report a location for their origin and destination of (0,0).

Any attempt to retrieve a rotated edge from the null edge will give another null edge. The "first edge" from the null vertex is also a null edge, as are any subsequent edges retrieved with `cv:Subdiv2D::nextEdge()`. The "first edge" retrieved from any of the virtual vertices will always connect to another virtual vertex.

### Identifying the bounding triangle

Given that, conveniently for us, when we create a Delaunay subdivision of a set of points, it is always the first three points that form the outer triangle (not including point 0; see the sidebar "More About Vertices and Edges" on page 936). We can then access these three vertices in the following way:

```
Point2f outer_vtx[3];
for( int i = 0; i < 3; i++ ) {
  outer_vtx[i] = subdiv.getVertex(i+1);
}
```

We can also obtain the three sides of the outer bounding triangle:

```
int outer_edges[3];
outer_edges[0] = 1*4;
outer_edges[1] = subdiv.getEdge(outer_edges[0], Subdiv2D::NEXT_AROUND_LEFT);
outer_edges[2] = subdiv.getEdge(outer_edges[1], Subdiv2D::NEXT_AROUND_LEFT);
```

Now that we know how to get on the graph and move around, we can investigate questions like when we're on the outer edge or boundary of the points.

### Identifying the bounding triangle or edges on the convex hull and walking the hull

Recall that we used a bounding rectangle `rect` to initialize the Delaunay triangulation with the constructor call `cv::Subdiv2D(rect)`. In this case, the following statements hold:

- If you are on an edge where both the origin and destination points are out of the `rect` bounds, then that edge is on the fictitious bounding triangle of the subdivision. These are what we call *unanchored* virtual edges.
- If you are on an edge with one point inside and one point outside the `rect` bounds, then the point in bounds is on the convex hull of the set; each point on the convex hull is connected to two vertices of the fictitious outer bounding triangle, and these two edges occur one after another. We will call those edges that connect a point inside the rectangular boundary to a virtual point outside *anchored* virtual edges.

If we wish to find the convex hull of the set of points, we can make use of these facts and quickly generate that hull.[2] For example, starting with the vertices 1, 2, and 3, which we know to be the virtual vertices, we can quickly generate, using `cv:Subdiv2D::nextEdge()`, the set of all anchored virtual edges (by simply rejecting the unanchored virtual edges). A quick call to `cv:Subdiv2D::rotateEdge(2)` flips that around, and a call or two to `cv:Subdiv2D::nextEdge` puts you on the convex hull of the point set.[3] There is precisely one such hull edge for each such anchored virtual edge, and the union of all of these edges is the convex hull of the point set.

We now know how to initialize Delaunay and Voronoi subdivisions, find the initial edges, and step through the edges and points of the graph. In the next section we present some practical applications.

## Usage Examples

We can use `cv::Subdiv2D::locate()` to step around the edges of a Delaunay triangle. In this example, we write a function that does "something" to every edge around the triangulation that contains some given point:

```
void locate_point(
  cv::Subdiv2D&      subdiv,
  const cv::Point2f& fp,
  ...
) {
  int e;
  int e0 = 0;
  int vertex = 0;
```

2 There are actually many ways to do this; the description given here is just one illustration of how this could be done.

3 Recall that because some outer vertices are connected to two vertices of a fictitious triangle, one call to `cv::Subdiv2D::nextEdge()` may not be enough. It would be best to check the resulting edge and verify that the destination (`cv::Subdiv2D::edgeDst()` on the new edge) is not one of the outer vertices. If it is, one more call to `cv::Subdiv2D::nextEdge()` will be required.

```
    subdiv.locate( fp, e0, vertex );
    if( e0 > 0 ) {
      e = e0;
      do // Always 3 edges -- this is a triangulation, after all.
      {
        // [Insert your code here]
        //
        // Do something with e ...
         e = subdiv.getEdge( e, cv::Subdiv2D::NEXT_AROUND_LEFT );
      }
      while( e != e0 );
    }
  }
```

We can also find the closest point to an input point by using:

```
int Subdiv2D::findNearest(
    cv::Point2f  pt,
    cv::Point2f* nearestPt
);
```

Unlike `cv::Subdiv2D::locate()`, `cv::Subdiv2D::findNearest()` will return the integer ID of the nearest vertex point in the subdivision. This point is not necessarily on the facet or triangle that the point lands on. Note that this is a nonconstant method, because it computes the Voronoi tessellation if it's missing or not up-to-date.

Similarly, we could step around a Voronoi facet (here we draw it) using:

```
void draw_subdiv_facet(
  cv::Mat&      img,
  cv::Subdiv2D& subdiv,
  int           edge
) {

  int t = edge;
  int i, count = 0;
  vector<cv::Point> buf;

  // Count number of edges in facet
  do{
      count++;
      t = subdiv.getEdge( t, cv::Subdiv2D::NEXT_AROUND_LEFT );
  } while (t != edge );

  // Gather points
  //
  buf.resize(count);
  t = edge;
  for( i = 0; i < count; i++ ) {
      cv::Point2f pt;
      if( subdiv.edgeOrg(t, &pt) <= 0 )
          break;
```

```
        buf[i] = cv::Point(cvRound(pt.x), cvRound(pt.y));
        t = subdiv.getEdge( t, cv::Subdiv2D::NEXT_AROUND_LEFT );
    }

    // Around we go
    //
    if( i == count ){
      cv::Point2f pt;
      subdiv.edgeDst(subdiv.rotateEdge(edge, 1), &pt);
      fillConvexPoly(
        img, buf,
        cv::Scalar(rand()&255,rand()&255,rand()&255),
        8, 0
      );
      vector< vector<cv::Point> > outline;
      outline.push_back(buf);
      polylines(img, outline, true, cv::Scalar(), 1, cv::LINE_AA, 0);
      draw_subdiv_point( img, pt, cv::Scalar(0,0,0) );
    }
  }
```

# Exercises

1. Modify the *.../opencv/samples/cpp/delaunay2.cpp* code to allow mouse-click point entry (instead of via the existing method where points are selected at a random). Experiment with triangulations on the results.
2. Modify the *delaunay2.cpp* code again so that you can use a keyboard to draw the convex hull of the point set.
3. Do three points in a line have a Delaunay triangulation?
4. Is the triangulation shown in Figure A-6(a) a Delaunay triangulation? If so, explain your answer. If not, how would you alter the figure so that it *is* a Delaunay triangulation?
5. Perform a Delaunay triangulation by hand on the points in Figure A-6(b). For this exercise, you need not add an outer fictitious bounding triangle.

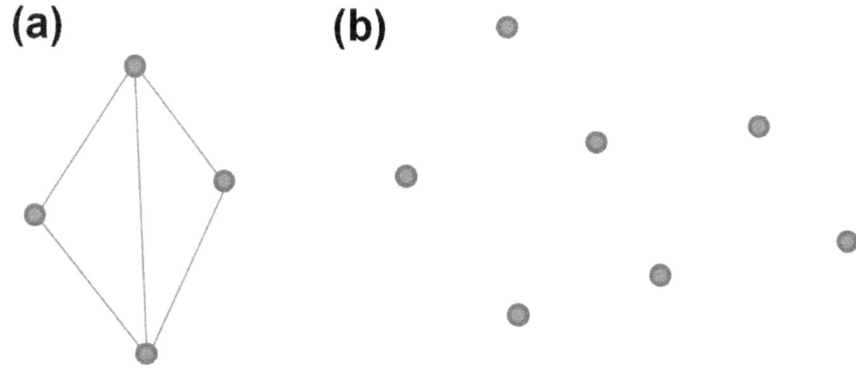


*Figure A-6. Exercises 4 and 5*

APPENDIX B

# opencv_contrib

## An Overview of the opencv_contrib Modules

The *opencv_contrib* repository is where most new user-generated content goes; it often contains more complete vision applications than you will find in OpenCV itself. It is composed of many modules that have no dependencies between them. Each module is required to have documentation, unit tests, and sample code, and many also have tutorials. Each module has to comply with all the other formatting, Buildbot tests, unit tests, and so on that OpenCV's core modules have to pass. Each one is documented in the same way as regular OpenCV functions and thus together they form a fairly self-maintaining superset of more advanced computer vision capability, ready to be used.

The *opencv_contrib* directory is located at *https://github.com/opencv/opencv_contrib* and needs to be built separately from the main OpenCV library. You can find documentation for these modules (and the regular OpenCV modules) at the nightly build site (*http://docs.opencv.org/master*). The content summary as of this writing appears next.

### Contents of opencv_contrib

This following list gives an overview of all modules available inside the *opencv_contrib* repository as of this writing. These modules must be downloaded and built separately. If you do decide to build this repository, but do not want to build all of the functions, you can turn off building any given function by replacing `<reponame>` in the build code that follows with the name of the function.

```
$ cmake -D OPENCV_EXTRA_MODULES_PATH=<opencv_contrib>/modules \
  -D BUILD_opencv_<reponame>=OFF                              \
  <opencv_source_directory>
```

The functions in *opencv_contrib* as of this writing are:

`aruco`
: ArUco and ChArUco markers. Includes augmented reality ArUco markers and ChARUco markers (ArUco markers embedded inside the white areas of the checkerboard).

`bgsegm`
: Background segmentation. Improved adaptive background mixture model and use for real-time human tracking under variable lighting conditions.

`bioinspired`
: Biological vision. A biologically inspired vision model providing methods to minimize noise and luminance variance, handle transient event segmentation, and perform high-dynamic-range (HDR) tone mapping.

`ccalib`
: Custom calibration. Patterns for 3D reconstruction, omnidirectional camera calibration, random pattern calibration, and multicamera calibration.

`cnn_3dobj`
: Deep object recognition and pose. Uses Caffe deep neural net library to build, train, and test a CNN model of visual object recognition and pose.

`contrib_world`
: *opencv_contrib* holder. `contrib_world` is the module that, when built, contains all other *opencv_contrib* modules. It may be used for the more convenient redistribution of OpenCV binaries.

`cvv`
: Computer vision debugger. Simple code that you can add to your program that pops up a GUI allowing you to interactively and visually debug computer vision programs.

`datasets`
: Data sets reader. Code for reading existing computer vision databases and samples of using the readers to train, test, and run using that data set's data.

`dnn`
: Deep neural networks (DNNs). This module can read in image-recognition networks trained in the Caffe neural network library and run them efficiently on CPU.

`dnns_easily_fooled`
: Subvert DNNs. This code can use the activations in a network to fool the networks into recognizing something else.

`dpm`
: Deformable part model. Felzenszwalb's cascade with deformable parts object recognition code.

`face`
: Face recognition. Face recognition techniques include Eigen, Fisher, and local binary pattern histograms (LBPH) methods.

`fuzzy`
: Fuzzy logic in vision. Fuzzy logic image transform and inverse; Fuzzy image processing.

`hdf`
: Hierarchical data storage. This module contains I/O routines for hierarchical data format (*https://en.m.wikipedia.org/wiki/Hierarchical_Data_Format*); meant to store large amounts of data.

`line_descriptor`
: Line segment extract and match. Methods of extracting, describing, and latching line segments using binary descriptors. One of the authors, Gary, built a robotic box handling company (Industrial Perception Inc.) out of a modification of this algorithm.

`matlab`
: MATLAB interface. OpenCV MATLAB Mex wrapper code generator for certain OpenCV core modules.

`optflow`
: Optical flow. Algorithms for running and evaluating deepflow, simpleflow, sparsetodenseflow, and motion templates (silhouette flow).

`plot`
: Plotting. The plot module allows you to easily plot data in 1D or 2D.

`reg`
: Image registration. Pixel-based image registration for precise alignment. Follows the paper by Richard Szeliski [Szeliski04].

`rgbd`
: RGB depth processing module. Linemod 3D object recognition; fast surface normal, and 3D plane finding. 3D visual odometry.

`saliency`
: Saliency API. Where humans would look in a scene. Has routines for static, motion, and "objectness" saliency.

`sfm`
: Structure from motion. This module contains algorithms to perform 3D reconstruction from 2D images. The core of the module is a light version of Libmv.

`stereo`
: Stereo correspondence. Stereo matching done with different descriptors: Census, CS-Census, MCT, BRIEF, and MV.

`structured_light`
: Structured light use. How to generate and project gray code patterns and use them to find dense depth in a scene.

`surface_matching`
: Point pair features. Implements 3D object detection and localization using multimodal point pair features.

`text`
: Visual text matching. In a visual scene, detect text, segment words, and recognize the text!

`tracking`
: Vision-based object tracking. Use and/or evaluate one of five different visual object tracking techniques.

`xfeatures2d`
: Features2D extra. Extra 2D features framework containing experimental and paid 2D feature detector/descriptor algorithms: SURF, SIFT, BRIEF, Censure, Freak, LUCID, Daisy, and Self-similar.

`ximgproc`
: Extended image processing. Includes structured forests, domain transform filter, guided filter, adaptive manifold filter, joint bilateral filter, and superpixels.

`xobjdetect`
: Boosted 2D object detection. Uses a Waldboost cascade and local binary patterns computed as integral features for 2D object detection.

`xphoto`
: Extra computational photography. Provides additional photo processing algorithms: color balance, denoising, and inpainting.

APPENDIX C

# Calibration Patterns

## Calibration Patterns Used by OpenCV

There are many different kinds of calibration patterns. Each pattern or marker could be used in a calibration procedure or just to find the 3D pose of that marker. Figures C-1 through C-7 show seven different patterns or markers.

*Figure C-1. Chessboard pattern (9 × 6 corners) for camera calibration or pose. This pattern can be used with the standard calibration technique described in Chapter 18, or in the camera calibration tutorial available online at opencv.org*

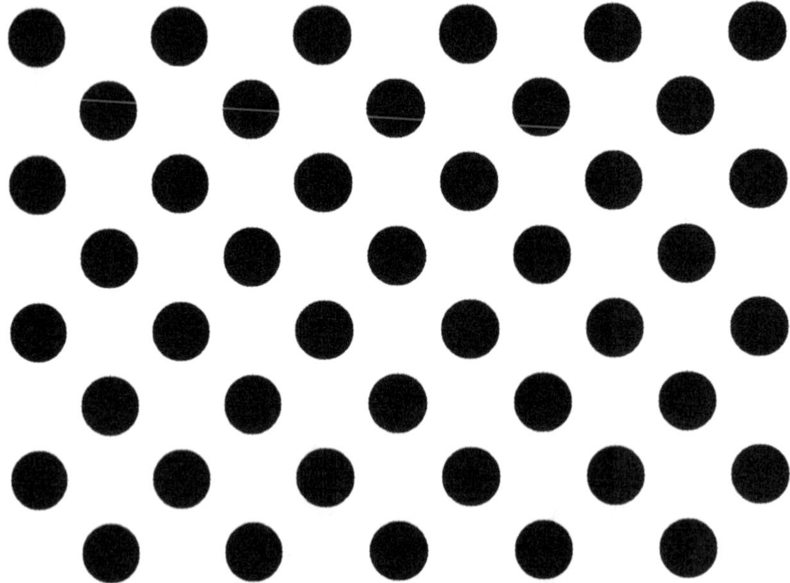

*Figure C-2. Circular features calibration or pose pattern. For this we can use the standard calibration technique described in this book or in tutorial_camera_calibration.html online (http://docs.opencv.org) using the findCirclesGrid() function*

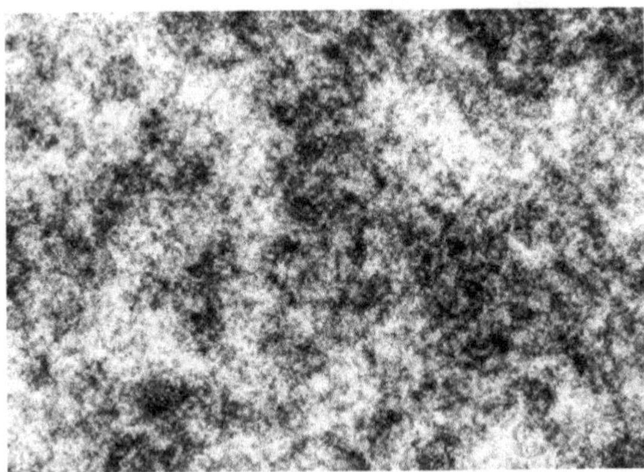

*Figure C-3. Random pattern for calibration or pose. You can find an example of how to use this pattern by following the code in .../opencv_contrib/modules/ccalib/samples/random_pattern_calibration.cpp*

*Figure C-4. ArUco board for calibration or pose. See the "Calibration with ArUco and ChArUco" tutorial online at opencv.org for how to detect and calibrate with this board*

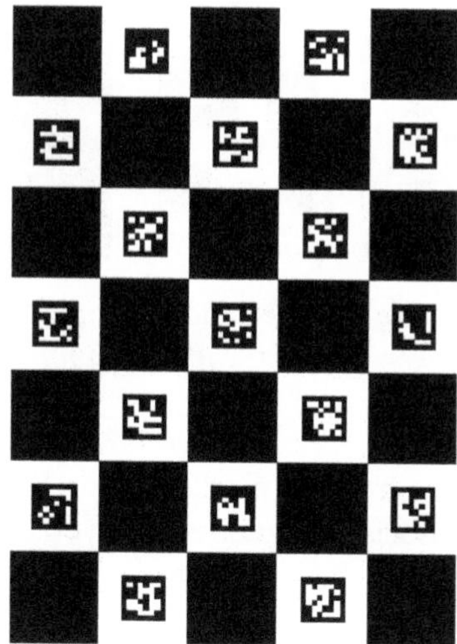

*Figure C-5. ChArUco board for calibration or pose; see the "Calibration with ArUco and ChArUco" tutorial online for how to detect and calibrate with this board*

*Figure C-6. ArUco marker; see the "Detection of ArUco Markers" tutorial online for how to detect an ArUco marker*

*Figure C-7. ChArUco marker. See the "Detection of Diamond Markers" tutorial for how to detect a ChArUco marker. This is the diamond marker that appears on the front cover of this book*

Figure C-7 shows a diamond ChArUco marker. If you build *opencv_contrib* modules with OpenCV, then in the associated *bin* directory, you can run the following to create this marker:

```
./example_aruco_create_diamond -bb=1 \
  -d=0 -sl=200 -ml=130 -ids=1,2,3,4  \
  -m=10 -si=true diamond.png
```

This code creates a diamond ChArUco marker where `-bb=1` creates a marker border as wide as the marker bits; `-d=0` creates a 4- × 4-bit marker (there are only four markers; we don't need markers to be able to represent large numbers); `-sl=200` draws a chessboard square of 200 × 200 pixels; `-ml=130` draws the ArUco marker to be a 130

× 130-pixel square; `-ids=1,2,3,4` just sets the ArUco markers to have coded values of 1, 2, 3, and 4, respectively; `-m=10` puts a 10-pixel-wide border around the diamond marker; and `-si=true` tells the program to show the generated image onscreen. Finally, *diamond.png* is the name of the output image.

To detect this image and also the one on the cover of this book, use the following code:

```
./example_aruco_detect_diamonds -as=1.0 -ci=0 -d=0 -ml=130 -sl=200
```

This code runs a camera that finds the marker shown both on the cover and in Figure C-7. Once found, the code decodes the ArUco markers and shows detected boxes and corners on the image. This code should work on the marker on the front cover of the book.[1] In this command line, `-as=1.0` tells the program to autoscale the detection, `-ci=0` tells the code to look for default system camera, `-d=0` tells the code to use the 4 × 4 ArUco detection/decode library, and `-ml=130` and `-sl=200` give the relative sizes of the ArUco and checkerboard squares (since we are autoscaling, it needs only relative sizes).

1 It is left as a final exercise to the reader to use the detected marker on the cover of this book to create some interesting augmented reality overlay!

# Bibliography

[Ahonen04] Ahonen, Timo, Abdenour Hadid, and Matti Pietikäinen. "Face recognition with local binary patterns." European conference on computer vision. Springer Berlin Heidelberg, 2004.

[Acharya05] Acharya, T., and A. Ray. *Image Processing: Principles and Applications.* New York: Wiley, 2005.

[Adelson84] Adelson, E. H., C. H. Anderson, J. R. Bergen, P. J. Burt, and J. M. Ogden. "Pyramid methods in image processing," *RCA Engineer* 29 (1984): 33–41.

[Agarwal08] Agrawal, M., K. Konolige, and M. R. Blas. "CenSurE: Center Surround Extremas for Realtime Feature Detection and Matching," *European Conference on Computer Vision*, 2008.

[Ahmed74] Ahmed, N., T. Natarajan, and K. R. Rao. "Discrete cosine transform," *IEEE Transactions on Computers* 23 (1974): 90–93.

[Alahi12] Alahi, Alexandre, Raphael Ortiz, and Pierre Vandergheynst. "Freak: Fast retina keypoint." *Computer vision and pattern recognition (CVPR), 2012 IEEE conference on.* IEEE, 2012.

[Arfken85] Arfken, G. "Convolution theorem," in *Mathematical Methods for Physicists,* 3rd ed. (pp. 810–814), Orlando, FL: Academic Press, 1985.

[Arraiy] Arraiy Corporation, *http://www.arraiy.ai.*

[Arthur07] Arthur, D., and S. Vassilvitskii., "*k*-means++: the advantages of careful seeding." *Proceedings of the eighteenth annual ACM-SIAM symposium on discrete algorithms.* (2007) pp. 1027–1035.

[Bajaj97] Bajaj, C. L., V. Pascucci, and D. R. Schikore. "The contour spectrum," *Proceedings of IEEE Visualization 1997* (pp. 167–173), 1997.

[Ballard81] Ballard, D. H. "Generalizing the Hough transform to detect arbitrary shapes," *Pattern Recognition* 13 (1981): 111–122.

[Ballard82] Ballard, D., and C. Brown. *Computer Vision*. Englewood Cliffs, NJ: Prentice-Hall, 1982.

[Bardyn84] Bardyn, J. J. et al. "Une architecture VLSI pour un operateur de filtrage median," *Congres reconnaissance des formes et intelligence artificielle* (vol. 1, pp. 557–566), Paris, 25–27, January 1984.

[Bay06] Bay, H., T. Tuytelaars, and L. V. Gool. "SURF: Speeded up robust features," *Proceedings of the Ninth European Conference on Computer Vision* (pp. 404–417), May 2006.

[Bay08] Bay, Herbert, Andreas Ess, Tinne Tuytelaars, and Luc J. Van Gool. "SURF: Speeded up robust features," *Computer Vision and Image Understanding (CVIU)* 110 (3) (2008) 346–359.

[Bayes1763] Bayes, T. "An essay towards solving a problem in the doctrine of chances. By the late Rev. Mr. Bayes, F.R.S. communicated by Mr. Price, in a letter to John Canton, A.M.F.R.S.," *Philosophical Transactions, Giving Some Account of the Present Undertakings, Studies and Labours of the Ingenious in Many Considerable Parts of the World* 53 (1763): 370–418.

[Bazargani15] Bazargani, Hamid, Olexa Bilaniuk, and Robert Laganiere. "A fast and robust homography scheme for real-time planar target detection." *Journal of Real-Time Image Processing* (2015): 1–20.

[Belongie02] Belongie, S., J. Malik, and J. Puzicha. "Shape matching and object recognition using shape contexts," in *IEEE Transactions on Pattern Analysis and Machine Intelligence*, vol.24, no. 4, pp. 509–522, Apr 2002.

[Bhattacharyya43] Bhattacharyya, A. "On a measure of divergence between two statistical populations defined by probability distributions," *Bulletin of the Calcutta Mathematical Society* 35 (1943): 99–109.

[BirchfieldTomasi99] Birchfield, Stan, and Carlo Tomasi. "Depth discontinuities by pixel-to-pixel stereo." *International Journal of Computer Vision* 35.3 (1999): 269–293.

[Bishop07] Bishop, C. M. *Pattern Recognition and Machine Learning*. New York: Springer-Verlag, 2007.

[Black92] Black, M. J. "Robust incremental optical flow" (YALEU-DCS-RR-923), Ph.D. thesis, Department of Computer Science, Yale University, New Haven, CT, 1992.

[Black93] Black, M. J., and P. Anandan. "A framework for the robust estimation of optical flow," *Fourth International Conference on Computer Vision* (pp. 231–236), May 1993.

[Black96] Black, M. J., and P. Anandan. "The robust estimation of multiple motions: Parametric and piecewise-smooth flow fields," *Computer Vision and Image Understanding* 63 (1996): 75–104.

[Bobick96] Bobick, A., and J. Davis. "Real-time recognition of activity using temporal templates," *IEEE Workshop on Applications of Computer Vision* (pp. 39–42), December 1996.

[Borgefors86] Borgefors, G., "Distance transformations in digital images," *Computer Vision, Graphics and Image Processing* 34 (1986): 344–371.

[Bouguet04] Bouguet, J.-Y. "Pyramidal implementation of the Lucas Kanade feature tracker description of the algorithm," *http://robots.stanford.edu/cs223b04/algo_tracking.pdf.*

[Boykov01] Boykov, Yuri, Olga Veksler, and Ramin Zabih. "Fast approximate energy minimization via graph cuts." *IEEE Transactions on pattern analysis and machine intelligence* 23.11 (2001): 1222-1239.

[Bracewell65] Bracewell, R. "Convolution" and "Two-dimensional convolution," in *The Fourier Transform and Its Applications* (pp. 25–50 and 243–244). New York: McGraw-Hill, 1965.

[Bradski] Gary Bradski, founder of OpenCV in 1999 and maintainer ever since. *https://en.wikipedia.org/wiki/Gary_Bradski.*

[Bradski00] Bradski, G., and J. Davis. "Motion segmentation and pose recognition with motion history gradients," *IEEE Workshop on Applications of Computer Vision*, 2000.

[Bradski98a] Bradski, G. R. "Real time face and object tracking as a component of a perceptual user interface," *Proceedings of the 4th IEEE Workshop on Applications of Computer Vision*, October 1998.

[Bradski98b] Bradski, G. R. "Computer video face tracking for use in a perceptual user interface," *Intel Technology Journal* Q2 (1998): 705–740.

[Breiman01] Breiman, L. "Random forests," *Machine Learning* 45 (2001): 5–32.

[Breiman02] Breiman, Leo. "Manual on setting up, using, and understanding random forests v3. 1." Statistics Department, University of California Berkeley, CA, USA 1 (2002).

[Breiman84] Breiman, L., J. H. Friedman, R. A. Olshen, and C. J. Stone. *Classification and Regression Trees.* Monterey, CA: Wadsworth, 1984.

[Brown71] Brown, D. C. "Close-range camera calibration," *Photogrammetric Engineering* 37 (1971): 855–866.

[Buades05] Mahmoudi, Mona, and Guillermo Sapiro. "Fast image and video denoising via nonlocal means of similar neighborhoods." *IEEE signal processing letters* 12.12 (2005): 839–842.

[Burt83] Burt, P. J., and E. H. Adelson. "The Laplacian pyramid as a compact image code," *IEEE Transactions on Communications* 31 (1983): 532–540.

[Calonder10] Calonder, M., V. Lepetit, C. Strecha, and P. Fua. Brief: Binary robust independent elementary features. In *European Conference on Computer Vision (ECCV)*, 2010.

[Canny86] Canny, J. "A computational approach to edge detection," *IEEE Transactions on Pattern Analysis and Machine Intelligence* 8 (1986): 679–714.

[Carpenter03] Carpenter, G. A., and S. Grossberg. "Adaptive resonance theory," in M. A. Arbib (Ed.), *The Handbook of Brain Theory and Neural Networks*, 2nd ed. (pp. 87–90), Cambridge, MA: MIT Press, 2003.

[Carr04] Carr, H., J. Snoeyink, and M. van de Panne. "Progressive topological simplification using contour trees and local spatial measures," *15th Western Computer Graphics Symposium*, Big White, British Columbia, March 2004.

[Chambolle04] Chambolle, Antonin. "An algorithm for total variation minimization and applications." *Journal of Mathematical imaging and vision* 20.1–2 (2004): 89–97.

[Chen05] Chen, D., and G. Zhang. "A new sub-pixel detector for x-corners in camera calibration targets," *WSCG Short Papers* (2005): 97–100.

[Chu07] Chu, C.-T., S. K. Kim, Y.-A. Lin, Y. Y. Yu, G. Bradski, A. Y. Ng, and K. Olukotun. "Map-reduce for machine learning on multicore," *Proceedings of the Neural Information Processing Systems Conference* (vol. 19, pp. 304–310), 2007.

[Ciresan11] Ciresan, Dan Claudiu, et al. "Convolutional neural network committees for handwritten character classification." 2011 International Conference on Document Analysis and Recognition. IEEE, 2011.

[Colombari07] Colombari, A., A. Fusiello, and V. Murino. "Video objects segmentation by robust background modeling," *International Conference on Image Analysis and Processing* (pp. 155–164), September 2007.

[Comaniciu99] Comaniciu, D., and P. Meer. "Mean shift analysis and applications," *IEEE International Conference on Computer Vision* (vol. 2, p. 1197), 1999.

[Comaniciu03] Comaniciu, D. "Nonparametric information fusion for motion estimation," *IEEE Conference on Computer Vision and Pattern Recognition* (vol. 1, pp. 59–66), 2003.

[Cooley65] Cooley, J. W., and O. W. Tukey. "An algorithm for the machine calculation of complex Fourier series," *Mathematics of Computation* 19 (1965): 297–301.

[Criminisi13] Criminisi, Antonio, and Jamie Shotton, eds. *Decision forests for computer vision and medical image analysis.* Springer Science & Business Media, 2013.

[Csurka04] Csurka, Gabriella, et al. "Visual categorization with bags of keypoints." *Workshop on statistical learning in computer vision, ECCV*. Vol. 1. No. 1–22. 2004.

[Dahlkamp06] Dahlkamp, H., A. Kaehler, D. Stavens, S. Thrun, and G. Bradski. "Self-supervised monocular road detection in desert terrain," *Robotics: Science and Systems*, Philadelphia, 2006.

[Dalal05] Dalal, N., and B. Triggs. "Histograms of oriented gradients for human detection," *Computer Vision and Pattern Recognition* (vol. 1, pp. 886–893), June 2005.

[Davis97] Davis, J., and A. Bobick. "The representation and recognition of action using temporal templates" (Technical Report 402), MIT Media Lab, Cambridge, MA, 1997.

[Davis99] Davis, J., and G. Bradski. "Real-time motion template gradients using Intel CVLib," *ICCV Workshop on Framerate Vision*, 1999.

[Delaunay34] Delaunay, B. "Sur la sphère vide," *Izvestia Akademii Nauk SSSR, Otdelenie Matematicheskikh i Estestvennykh Nauk* 7 (1934): 793–800.

[DeMenthon92] DeMenthon, D. F., and L. S. Davis. "Model-based object pose in 25 lines of code,"*Proceedings of the European Conference on Computer Vision* (pp. 335–343), 1992.

[Dempster77] Dempster, A., N. Laird, and D. Rubin. "Maximum likelihood from incomplete data via the EM algorithm," *Journal of the Royal Statistical Society, Series B* 39 (1977): 1–38.

[Douglas73] Douglas, D., and T. Peucker. "Algorithms for the reduction of the number of points required for represent a digitized line or its caricature," *Canadian Cartographer* 10(1973): 112–122.

[Drucker97] Drucker, Harris, Chris J.C. Burges, Linda Kaufman, Alex Smola, and Vladimir Vapnik. "Support vector regression machines." *Advances in neural information processing systems* 9 (1997): 155-161.

[Duda72] Duda, R. O., and P. E. Hart. "Use of the Hough transformation to detect lines and curves in pictures," *Communications of the Association for Computing Machinery* 15 (1972): 11–15.

[Duda73] Duda, R. O., and P. E. Hart. *Pattern Recognition and Scene Analysis.* New York: Wiley, 1973.

[Duda00] Duda, R. O., P. E. Hart, and D. G. Stork. *Pattern Classification.* New York: Wiley, 2001.

[Farin04] Farin, D., P. H. N. de With, and W. Effelsberg. "Video-object segmentation using multi-sprite background subtraction," *Proceedings of the IEEE International Conference on Multimedia and Expo*, 2004.

[Farnebäck03] Farnebäck, Gunnar. "Two-frame motion estimation based on polynomial expansion." *Scandinavian conference on image analysis.* Springer Berlin Heidelberg, 2003.

[Faugeras93] Faugeras, O. *Three-Dimensional Computer Vision: A Geometric Viewpoint.* Cambridge, MA: MIT Press, 1993.

[Felzenszwalb2004] Felzenszwalb, Pedro, and Daniel Huttenlocher. *Distance transforms of sampled functions.* Cornell University, 2004.

[Felzenszwalb2006] Felzenszwalb, Pedro F., and Daniel P. Huttenlocher. "Efficient belief propagation for early vision." *International Journal of Computer Vision*, 70(1), October 2006.

[Felzenszwalb2010] Felzenszwalb, Pedro F., et al. "Object detection with discriminatively trained part-based models." *IEEE transactions on pattern analysis and machine intelligence* 32.9 (2010): 1627–1645.

[Fischler81] Fischler, M. A., and R. C. Bolles. "Random sample consensus: A paradigm for model fitting with applications to image analysis and automated cartography," *Communications of the Association for Computing Machinery* 24 (1981): 381–395.

[Fitzgibbon95] Fitzgibbon, A. W., and R. B. Fisher. "A buyer's guide to conic fitting," *Proceedings of the 5th British Machine Vision Conference* (pp. 513–522), Birmingham, 1995.

[Fix51] Fix, E., and J. L. Hodges. "Discriminatory analysis, nonparametric discrimination: Consistency properties" (Technical Report 4), USAF School of Aviation Medicine, Randolph Field, Texas, 1951.

[Forsyth03] Forsyth, D., and J. Ponce. *Computer Vision: A Modern Approach.* Englewood Cliffs, NJ: Prentice-Hall, 2003.

[FourCC85] Morrison, J. "EA IFF 85 standard for interchange format files," *http://www.martinreddy.net/gfx/2d/IFF.txt.*

[Fourier] "Joseph Fourier," *http://en.wikipedia.org/wiki/Joseph_Fourier.*

[Freeman95] Freeman, W. T., and M. Roth. "Orientation histograms for hand gesture recognition," *International Workshop on Automatic Face and Gesture Recognition* (pp. 296–301), June 1995.

[Freund97] Freund, Y., and R. E. Schapire. "A decision-theoretic generalization of on-line learning and an application to boosting," *Journal of Computer and System Sciences* 55 (1997): 119–139.

[Fryer86] Fryer, J. G., and D. C. Brown. "Lens distortion for close-range photogrammetry," *Photogrammetric Engineering and Remote Sensing* 52 (1986): 51–58.

[Fukunaga90] Fukunaga, K. *Introduction to Statistical Pattern Recognition.* Boston: Academic Press, 1990.

[Fukushima80] Fukushima, Kunihiko. "Neocognitron: A self-organizing neural network model for a mechanism of pattern recognition unaffected by shift in position." *Biological cybernetics* 36.4 (1980): 193-202.

[Galton] "Francis Galton," *http://en.wikipedia.org/wiki/Francis_Galton.*

[Gao03] Gao, Xiao-Shan, Xiao-Rong Hou, Jianliang Tang, and Hang-Fei Cheng. "Complete solution classification for the perspective-three-point problem," *IEEE Transactions Pattern Analysis and Machine Intelligence* 25 (2003), 930–943.

[Garrido-Jurado] Garrido-Jurado, S., R. Munoz-Salinas, F. J. Madrid-Cuevas and M. J. Marin-Jimenez. "Automatic generation and detection of highly reliable fiducial markers under occlusion," *Pattern Recognition* 47, no. 6 (June 2014).

[Göktürk01] Göktürk, S. B., J.-Y. Bouguet, and R. Grzeszczuk. "A data-driven model for monocular face tracking," *Proceedings of the IEEE International Conference on Computer Vision* (vol. 2, pp. 701–708), 2001.

[Göktürk02] Göktürk, S. B., J.-Y. Bouguet, C. Tomasi, and B. Girod. "Model-based face tracking for view-independent facial expression recognition," *Proceedings of the Fifth IEEE International Conference on Automatic Face and Gesture Recognition* (pp. 287–293), May 2002.

[Goresky03] Goresky, Mark, and Andrew Klapper. "Efficient multiply-with-carry random number generators with maximal period." *ACM Transactions on Modeling and Computer Simulation* (TOMACS) 13.4 (2003): 310-321.

[Grossberg87] Grossberg, S., "Competitive learning: From interactive activation to adaptive resonance," *Cognitive Science* 11 (1987): 23–63.

[Harris88] Harris, C., and M. Stephens. "A combined corner and edge detector," *Proceedings of the 4th Alvey Vision Conference* (pp. 147–151), 1988.

[Hartley98] Hartley, R. I. "Theory and practice of projective rectification," *International Journal of Computer Vision* 35 (1998): 115–127.

[Hartley06] Hartley, R., and A. Zisserman. *Multiple View Geometry in Computer Vision.* Cambridge, UK: Cambridge University Press, 2006.

[Hastie01] Hastie, T., R. Tibshirani, and J. Friedman. *The Elements of Statistical Learning: Data Mining, Inference and Prediction*. New York: Springer-Verlag, 2001.

[Heckbert90] Heckbert, P. *A Seed Fill Algorithm* (Graphics Gems I). New York: Academic Press, 1990.

[Heikkila97] Heikkila, J., and O. Silven. "A four-step camera calibration procedure with implicit image correction," *Proceedings of the 1997 Conference on Computer Vision and Pattern Recognition* (p. 1106), 1997.

[Hinton06] Hinton, G. E., S. Osindero, and Y. Teh. "A fast learning algorithm for deep belief nets," *Neural Computation* 18 (2006): 1527–1554.

[Hirschmuller 08] Hirschmuller, H. "Stereo Processing by Semiglobal Matching and Mutual Information," *Pattern Analysis and Machine Intelligence PAMI* 30, No. 2, February 2008, pp. 328–341.

[Ho95] Ho, T. K. "Random decision forest," *Proceedings of the 3rd International Conference on Document Analysis and Recognition* (pp. 278–282), August 1995.

[Horn81] Horn, B. K. P., and B. G. Schunck. "Determining optical flow," *Artificial Intelligence* 17 (1981): 185–203.

[Hough59] Hough, P. V. C. "Machine analysis of bubble chamber pictures," *Proceedings of the International Conference on High Energy Accelerators and Instrumentation* (pp. 554–556), 1959.

Huttenlocher, Daniel P., Gregory A. Klanderman, and William J. Rucklidge. "Comparing images using the Hausdorff distance." *IEEE Transactions on pattern analysis and machine intelligence* 15.9 (1993): 850-863.

[Intel] Intel Corporation, *http://www.intel.com/*.

[Inui03] Inui, K., S. Kaneko, and S. Igarashi. "Robust line fitting using LmedS clustering," *Systems and Computers in Japan* 34 (2003): 92–100.

[IPP] Intel Integrated Performance Primitives, *https://software.intel.com/en-us/intel-ipp*.

[Itseez] A computer vision company that grew out of the original OpenCV project and one of the key maintainers of the free and open OpenCV.org. Now sold to Intel Corporation.

[Jaehne95] Jaehne, B. *Digital Image Processing*, 3rd ed. Berlin: Springer-Verlag, 1995.

[Jaehne97] Jaehne, B. *Practical Handbook on Image Processing for Scientific Applications*. Boca Raton, FL: CRC Press, 1997.

[Jain77] Jain, A. "A fast Karhunen-Loeve transform for digital restoration of images degraded by white and colored noise," *IEEE Transactions on Computers* 26 (1997): 560–571.

[Jain86] Jain, A. *Fundamentals of Digital Image Processing.* Englewood Cliffs, NJ: Prentice-Hall, 1986.

[Johnson84] Johnson, D. H. "Gauss and the history of the fast Fourier transform," *IEEE Acoustics, Speech, and Signal Processing Magazine* 1 (1984): 14–21.

[KaewTraKulPong2001] KaewTraKulPong, P., and R. Bowden. "An Improved Adaptive Background Mixture Model for Realtime Tracking with Shadow Detection," *Proc. 2nd European Workshop on Advanced Video Based Surveillance Systems,* AVBS01. Sept 2001.

[Kalman60] Kalman, R. E. "A new approach to linear filtering and prediction problems," *Journal of Basic Engineering* 82 (1960): 35–45.

[Kim05] Kim, K., T. H. Chalidabhongse, D. Harwood, and L. Davis. "Real-time foreground-background segmentation using codebook model," *Real-Time Imaging* 11 (2005): 167–256.

[Kimme75] Kimme, C., D. H. Ballard, and J. Sklansky. "Finding circles by an array of accumulators," *Communications of the Association for Computing Machinery* 18 (1975): 120–122.

[Kiryati91] Kiryati, N., Y. Eldar, and A. M. Bruckshtein. "A probablistic Hough transform," *Pattern Recognition* 24 (1991): 303–316.

[Koller09] Koller, Daphne, and Nir Friedman. *Probabilistic graphical models: principles and techniques.* Cambridge, MA: MIT Press, 2009.

[Konolige97] Konolige, K., "Small vision system: Hardware and implementation," *Proceedings of the International Symposium on Robotics Research* (pp. 111–116), Hayama, Japan, 1997.

[Kopf07] Kopf, Johannes, et al. "Joint bilateral upsampling." *ACM Transactions on Graphics (TOG).* Vol. 26. No. 3. ACM, 2007.

[Kreveld97] van Kreveld, M., R. van Oostrum, C. L. Bajaj, V. Pascucci, and D. R. Schikore. "Contour trees and small seed sets for isosurface traversal," *Proceedings of the 13th ACM Symposium on Computational Geometry* (pp. 212–220), 1997.

[Lakoff08] Lakoff, G., and M. Johnson. "Metaphors we live by," University of Chicago Press, 2008.

[Laughlin81] Laughlin, S. B. "A simple coding procedure enhances a neuron's information capacity," *Zeitschrift für Naturforschung* 9/10 (1981): 910–912.

[LeCun98a] LeCun, Y., L. Bottou, Y. Bengio, and P. Haffner. "Gradient-based learning applied to document recognition," *Proceedings of the IEEE* 86 (1998): 2278–2324.

[LeCun98b] LeCun, Y., L. Bottou, G. Orr, and K. Muller. "Efficient BackProp," in G. Orr and K. Muller (Eds.), *Neural Networks: Tricks of the Trade*. New York: Springer-Verlag, 1998.

[Leutenegger11] Leutenegger, Stefan, Margarita Chli, and Roland Y. Siegwart. "BRISK: Binary robust invariant scalable keypoints." *2011 International conference on computer vision*. IEEE, 2011.

[Lienhart02] Lienhart, Rainer, and Jochen Maydt. "An extended set of haar-like features for rapid object detection." *Proceedings 2002 International Conference on Image Processing*. Vol. 1. IEEE, 2002.

[Liu07] Liu, Y. Z., H. X. Yao, W. Gao, X. L. Chen, and D. Zhao. "Nonparametric background generation," *Journal of Visual Communication and Image Representation* 18 (2007): 253–263.

[Lloyd57] Lloyd, S. "Least square quantization in PCM's" (Bell Telephone Laboratories Paper), 1957. ["Lloyd's algorithm" was later published in *IEEE Transactions on Information Theory* 28 (1982): 129–137.]

[Lloyd82] Lloyd, Stuart. "Least squares quantization in PCM." *IEEE transactions on information theory* 28.2 (1982): 129-137.

[Lowe04] Lowe, D. G. "Distinctive image features from scale-invariant keypoints," *International Journal of Computer Vision* 60, no. 2 (2004): 91–110.

[LTI] LTI-Lib, Vision Library, *http://ltilib.sourceforge.net/doc/homepage/index.shtml.*

[Lucas81] Lucas, B. D., and T. Kanade. "An iterative image registration technique with an application to stereo vision," *Proceedings of the 1981 DARPA Imaging Understanding Workshop* (pp. 121–130), 1981.

[Lucchese02] Lucchese, L., and S. K. Mitra. "Using saddle points for subpixel feature detection in camera calibration targets," *Proceedings of the 2002 Asia Pacific Conference on Circuits and Systems* (pp. 191–195), December 2002.

[Lv07] Lv, Q., W. Josephson, Z. Wang, M. Charikar, and K. Li. "Multiprobe LSH: efficient indexing for high-dimensional similarity search." In *VLDB*, pages 950–961, 2007.

[Mahalanobis36] Mahalanobis, P. "On the generalized distance in statistics," *Proceedings of the National Institute of Science* 12 (1936): 49–55.

[Mair10] Mair, Elmar, et al. "Adaptive and generic corner detection based on the accelerated segment test." European conference on Computer vision. Springer Berlin Heidelberg, 2010.

[Maron61] Maron, M. E. "Automatic indexing: An experimental inquiry," *Journal of the Association for Computing Machinery* 8 (1961): 404–417.

[Marr82] Marr, D. *Vision.* San Francisco: Freeman, 1982.

[Marsaglia00] Marsaglia, George, and Wai Wan Tsang. "The ziggurat method for generating random variables." *Journal of statistical software* 5.8 (2000): 1-7.

[Martins99] Martins, F. C. M., B. R. Nickerson, V. Bostrom, and R. Hazra. "Implementation of a real-time foreground/background segmentation system on the Intel architecture," *IEEE International Conference on Computer Vision Frame Rate Workshop*, 1999.

[Matas00] Matas, J., C. Galambos, and J. Kittler. "Robust detection of lines using the progressive probabilistic Hough transform," *Computer Vision Image Understanding* 78 (2000): 119–137.

[Meer91] Meer, P., D. Mintz, and A. Rosenfeld. "Robust regression methods for computer vision: A review," *International Journal of Computer Vision* 6 (1991): 59–70.

[Merwe00] van der Merwe, R., A. Doucet, N. de Freitas, and E. Wan. "The unscented particle filter," *Advances in Neural Information Processing Systems*, December 2000.

[Meyer78] Meyer, F. "Contrast feature extraction," in J.-L. Chermant (Ed.), *Quantitative Analysis of Microstructures in Material Sciences, Biology and Medicine* [Special issue of *Practical Metallography*], Stuttgart: Riederer, 1978.

[Meyer92] Meyer, F. "Color image segmentation," *Proceedings of the International Conference on Image Processing and Its Applications* (pp. 303–306), 1992.

[Minsky61] Minsky, M. "Steps toward artificial intelligence," *Proceedings of the Institute of Radio Engineers* 49 (1961): 8–30.

[Moreno-Noguer07] Moreno-Noguer, F., Lepetit, V., Fua, P. "Accurate Non-Iterative O(n) Solution to the PnP Problem," *ICCV 2007. IEEE 11th International Conference on Computer Vision*, pp. 1–8, 2007.

[Morse53] Morse, P. M., and H. Feshbach. "Fourier transforms," in *Methods of Theoretical Physics* (Part I, pp. 453–471), New York: McGraw-Hill, 1953.

[Muja09] Muja, Marius, and David G. Lowe. "Fast Approximate Nearest Neighbors with Automatic Algorithm Configuration." *VISAPP (1)* 2.331-340 (2009): 2.

[Neapolitan04] Neapolitan, Richard E. *Learning Bayesian Networks.* Upper Saddle River, New Jersey: Pearson, 2004.

[O'Connor02] O'Connor, J. J., and E. F. Robertson. "Light through the ages: Ancient Greece to Maxwell," *http://www-groups.dcs.st-and.ac.uk/~history/HistTopics/Light_1.html.*

[Ojala94] Ojala, T., M. Pietikäinen, and D. Harwood. "Performance evaluation of texture measures with classification based on Kullback discrimination of distributions," Pattern Recognition, 1994. Vol. 1 [Oliva06] A. Oliva and A. Torralba, "Building the gist of a scene: The role of global image features in recognition visual perception," *Progress in Brain Research* 155 (2006): 23–36.

[OpenCV] Open Source Computer Vision Library (OpenCV) (free, BSD license), *http://opencv.org.*

[opencv_contrib] Newer content and higher functionality is separated into this repository. *https://github.com/opencv/opencv_contrib.*

[Papoulis62] Papoulis, A. *The Fourier Integral and Its Applications.* New York: McGraw-Hill, 1962.

[Pascucci02] Pascucci, V., and K. Cole-McLaughlin. "Efficient computation of the topology of level sets," *Proceedings of IEEE Visualization 2002* (pp. 187–194), 2002.

[Pearson] "Karl Pearson," *http://en.wikipedia.org/wiki/Karl_Pearson.*

[Pollefeys99a] Pollefeys, M. "Self-calibration and metric 3D reconstruction from uncalibrated image sequences," Ph.D. thesis, Katholieke Universiteit, Leuven, 1999.

[Pollefeys99b] Pollefeys, M., R. Koch, and L. V. Gool. "A simple and efficient rectification method for general motion," *Proceedings of the 7th IEEE Conference on Computer Vision,* 1999.

[Porter84] Porter, T., and T. Duff. "Compositing digital images," *Computer Graphics* 18 (1984): 253–259.

[Ranger07] Ranger, C., R. Raghuraman, A. Penmetsa, G. Bradski, and C. Kozyrakis. "Evaluating mapreduce for multi-core and multiprocessor systems," *Proceedings of the 13th International Symposium on High-Performance Computer Architecture* (pp. 13–24), 2007.

[Reeb46] Reeb, G. "Sur les points singuliers d'une forme de Pfaff completement integrable ou d'une fonction numerique," *Comptes Rendus de l'Academie des Sciences de Paris* 222 (1946): 847–849.

[Riedmiller93] Riedmiller, Martin, and Heinrich Braun. "A direct adaptive method for faster backpropagation learning: The RPROP algorithm." *1993 IEEE International Conference on Neural Networks..* IEEE, 1993.

[Rodgers88] Rodgers, J. L., and W. A. Nicewander. "Thirteen ways to look at the correlation coefficient," *American Statistician* 42 (1988): 59–66.

[Rosenfeld73] Rosenfeld, A., and E. Johnston. "Angle detection on digital curves," *IEEE Transactions on Computers* 22 (1973): 875–878.

[Rosenfeld80] Rosenfeld, A. "Some Uses of Pyramids in Image Processing and Segmentation," *Proceedings of the DARPA Imaging Understanding Workshop* (pp. 112–120), 1980.

[Rosten06] Rosten, Edward. "FAST corner detection." *Engineering Department, Machine Intelligence Laboratory, University of Cambridge*, 2006.

[Rother04] Rother, Carsten, Vladimir Kolmogorov, and Andrew Blake. "Grabcut: Interactive foreground extraction using iterated graph cuts." *ACM transactions on graphics (TOG)*. Vol. 23. No. 3. ACM, 2004.

[Rousseeuw84] Rousseeuw, P. J. "Least median of squares regression," *Journal of the American Statistical Association*, 79 (1984): 871–880.

[Rousseeuw87] Rousseeuw, P. J., and A. M. Leroy. *Robust Regression and Outlier Detection*. New York: Wiley, 1987.

[Rublee11] Rublee, E., V. Rabaud, K. Konolige, and G. Bradski. "ORB an efficient alternative to SIFT or SURF." *Proceedings of the 2011 IEEE International Conference on Computer Vision*, 2011.

[Rubner00] Rubner, Y., C. Tomasi, and L. J. Guibas. "The earth mover's distance as a metric for image retrieval," *International Journal of Computer Vision* 40 (2000): 99–121.

[Rumelhart88] Rumelhart, D. E., G. E. Hinton, and R. J. Williams. "Learning internal representations by error propagation," in D. E. Rumelhart, J. L. McClelland, and PDP Research Group (Eds.), *Parallel Distributed Processing. Explorations in the Microstructures of Cognition* (vol. 1, pp. 318–362), Cambridge, MA: MIT Press, 1988.

[Russ02] Russ, J. C. *The Image Processing Handbook*, 4th ed. Boca Raton, FL: CRC Press, 2002.

[Sánchez13] Sánchez, Javier, Enric Meinhardt-Llopis, and Gabriele Facciolo. "TV-L1 optical flow estimation." *Image Processing On Line 2013* (2013): 137-150.

[Scharr00] Scharr, Hanno. "Optimal operators in digital image processing." Diss. 2000.

[Schiele96] Schiele, B., and J. L. Crowley. "Object recognition using multidimensional receptive field histograms," *European Conference on Computer Vision* (vol. I, pp. 610–619), April 1996.

[Schmidt66] Schmidt, S. "Applications of state-space methods to navigation problems," in C. Leondes (Ed.), *Advances in Control Systems* (vol. 3, pp. 293–340), New York: Academic Press, 1966.

[Schölkopf00] Schölkopf, Bernhard, et al. "New support vector algorithms." *Neural computation* 12.5 (2000): 1207-1245.

[Schwartz80] Schwartz, E. L. "Computational anatomy and functional architecture of the striate cortex: A spatial mapping approach to perceptual coding," *Vision Research* 20 (1980): 645–669.

[Schwarz78] Schwarz, A. A., and J. M. Soha. "Multidimensional histogram normalization contrast enhancement," *Proceedings of the Canadian Symposium on Remote Sensing* (pp. 86–93), 1978.

[Semple79] Semple, J., and G. Kneebone. *Algebraic Projective Geometry*. Oxford, UK: Oxford University Press, 1979.

[Serra83] Serra, J. *Image Analysis and Mathematical Morphology*. New York: Academic Press, 1983.

[Sezgin04] Sezgin, M., and B. Sankur. "Survey over image thresholding techniques and quantitative performance evaluation," *Journal of Electronic Imaging* 13 (2004): 146–165.

[Shannon49] Shannon, C. E. (*http://en.wikipedia.org/wiki/Claude_E._Shannon*). "Communication in the presence of noise," *Proc. Institute of Radio Engineers*, vol. 37, no. 1, pp. 10–21, Jan. 1949. Reprint as classic paper in: Proc. IEEE, vol. 86, no. 2, (Feb. 1998), *http://www.stanford.edu/class/ee104/shannonpaper.pdf.*

[Shapiro02] Shapiro, L. G., and G. C. Stockman. *Computer Vision*. Englewood Cliffs, NJ: Prentice-Hall, 2002.

[Shaw04] Shaw, J. R. "QuickFill: An efficient flood fill algorithm," *http://www.codeproject.com/gdi/QuickFill.asp.*

[Shi94] Shi, J., and C. Tomasi. "Good features to track," *9th IEEE Conference on Computer Vision and Pattern Recognition*, June 1994.

[Smith79] Smith, A. R. "Painting tutorial notes," Computer Graphics Laboratory, New York Institute of Technology, Islip, NY, 1979.

[Sobel73] Sobel, I., and G. Feldman. "A 3 x 3 Isotropic Gradient Operator for Image Processing," in R. Duda and P. Hart (Eds.), *Pattern Classification and Scene Analysis* (pp. 271–272), New York: Wiley, 1973.

[Sochman05] Sochman, J., and J. Matas. "WaldBoost - learning for time constrained sequential detection," in *Computer Vision and Pattern Recognition*, 2005. CVPR 2005.

[Steinhaus56] Steinhaus, H. "Sur la division des corp materiels en parties," *Bulletin of the Polish Academy of Sciences and Mathematics* 4 (1956): 801–804.

[Sturm99] Sturm, P. F., and S. J. Maybank. "On plane-based camera calibration: A general algorithm, singularities, applications," *IEEE Conference on Computer Vision and Pattern Recognition*, 1999.

[Suzuki85] Suzuki, S., and K. Abe, "Topological structural analysis of digital binary images by border following," *Computer Vision, Graphics and Image Processing* 30 (1985): 32–46.

[Swain91] Swain, M. J., and D. H. Ballard. "Color indexing," *International Journal of Computer Vision* 7 (1991): 11–32.

[Szeliski04] Szeliski, R. "Image Alignment and Stitching: A Tutorial," *http://research.microsoft.com/apps/pubs/default.aspx?id=70092*. October 2004.

[Tao12] Tao, Michael, et al. "SimpleFlow: A Non-iterative, Sublinear Optical Flow Algorithm." *Computer Graphics Forum*. Vol. 31. No. 2pt1. Blackwell Publishing Ltd, 2012.

[Teh89] Teh, C. H., and R. T. Chin. "On the detection of dominant points on digital curves," *IEEE Transactions on Pattern Analysis and Machine Intelligence* 11 (1989): 859–872.

[Telea04] Telea, A. "An image inpainting technique based on the fast marching method," *Journal of Graphics Tools* 9 (2004): 25–36.

[Thrun05] Thrun, S., W. Burgard, and D. Fox. *Probabilistic Robotics: Intelligent Robotics and Autonomous Agents*, Cambridge, MA: MIT Press, 2005.

[Thrun06] Thrun, S., M. Montemerlo, H. Dahlkamp, D. Stavens, A. Aron, J. Diebel, P. Fong, J. Gale, M. Halpenny, G. Hoffmann, K. Lau, C. Oakley, M. Palatucci, V. Pratt, P. Stang, S. Strohband, C. Dupont, L.-E. Jendrossek, C. Koelen, C. Markey, C. Rummel, J. van Niekerk, E. Jensen, P. Alessandrini, G. Bradski, B. Davies, S. Ettinger, A. Kaehler, A. Nefian, and P. Mahoney. "Stanley, the robot that won the DARPA Grand Challenge," *Journal of Robotic Systems* 23 (2006): 661–692.

[Titchmarsh26] Titchmarsh, E. C. "The zeros of certain integral functions," *Proceedings of the London Mathematical Society* 25 (1926): 283–302.

[Tomasi98] Tomasi, C., and R. Manduchi. "Bilateral filtering for gray and color images," *Sixth International Conference on Computer Vision* (pp. 839–846), New Delhi, 1998.

[Tou77] Tou, J., and R. Gonzales. *Pattern Recognition Principles*. Addison Wesley Publishing (1977), p. 377.

[Toyama99] Toyama, K., J. Krumm, B. Brumitt, and B. Meyers. "Wallflower: Principles and practice of background maintenance," *Proceedings of the 7th IEEE International Conference on Computer Vision* (pp. 255–261), 1999.

[Trucco98] Trucco, E., and A. Verri. *Introductory Techniques for 3-D Computer Vision*. Englewood Cliffs, NJ: Prentice-Hall, 1998.

[Tsai87] Tsai, R. Y. "A versatile camera calibration technique for high accuracy 3D machine vision metrology using off-the-shelf TV cameras and lenses," *IEEE Journal of Robotics and Automation* 3 (1987): 323–344.

[Vandevenne04] Vandevenne, Lode. "Lode's computer graphics tutorial, flood fill." 2004.

[Vapnik95] Vapnik, V. *The Nature of Statistical Learning Theory*. New York: *Springer-Verlag*, 1995.

[Viola01] Viola, Paul, and Michael Jones. "Rapid object detection using a boosted cascade of simple features." *Computer Vision and Pattern Recognition, 2001. CVPR 2001. Proceedings of the 2001 IEEE Computer Society Conference on*. Vol. 1. IEEE, 2001.

[Viola04] Viola, P., and M. J. Jones. "Robust real-time face detection," *International Journal of Computer Vision* 57 (2004): 137–154.

[Welsh95] Welsh, G., and G. Bishop. "An introduction to the Kalman filter" (Technical Report TR95-041), University of North Carolina, Chapel Hill, NC, 1995.

[Wharton71] Wharton, W., and D. Howorth. *Principles of Television Reception*. London: Pitman, 1971.

[Wu08] Wu, K., O. Ekow and K. Suzuki. Two Strategies to Speed up Connected Component Labeling Algorithms, *http://escholarship.org/uc/item/5pc9s496*, 06-02-2008.

[Xu96] Xu, G., and Z. Zhang. *Epipolar Geometry in Stereo, Motion and Object Recognition*. Dordrecht: Kluwer, 1996.

[Zach07] Zach, Christopher, Thomas Pock, and Horst Bischof. "A duality based approach for realtime TV-L 1 optical flow." *Joint Pattern Recognition Symposium*. Springer Berlin Heidelberg, 2007.

[Zhang96] Zhang, Z. "Parameter estimation techniques: A tutorial with application to conic fitting," *Image and Vision Computing* 15 (1996): 59–76.

[Zhang99] Zhang, R., P.-S. Tsi, J. E. Cryer, and M. Shah. "Shape form shading: A survey," *IEEE Transactions on Pattern Analysis and Machine Intelligence* 21 (1999): 690–706.

[Zhang99] Zhang, Z. "Flexible camera calibration by viewing a plane from unknown orientations," *Proceedings of the 7th International Conference on Computer Vision* (pp. 666–673), Corfu, September 1999.

[Zhang00] Zhang, Z. "A flexible new technique for camera calibration," *IEEE Transactions on Pattern Analysis and Machine Intelligence* 22 (2000): 1330–1334.

[Zhang04] Zhang, H. "The optimality of naive Bayes," *Proceedings of the 17th International FLAIRS Conference*, 2004.

[Zivkovic04] Zivkovic, Z. "Improved adaptive Gaussian mixture model for background subtraction," *International Conference Pattern Recognition*, UK, August 2004.

[Zivkovic06] Zivkovic, Z., and F. van der Heijden. "Efficient Adaptive Density Estimation per Image Pixel for the Task of Background Subtraction," *Pattern Recognition Letters*, vol. 27, no. 7, pages 773–780, 2006.

# Index

## A

## B

## C

## D

## G

## H

## I



## J



## K

## L

## M

## N



## O

## P

## T

## U



## V

## W



## X



## Y



## Z

## About the Authors

**Gary Bradski** founded and still directs the Open Source Computer Vision library, OpenCV (now a nonprofit organization) starting from his work at Intel Labs as a principal engineer. While there he abstracted some of the machine learning algorithms from the manufacturing effort he was leading into the machine learning module (ml) in OpenCV. From there, he helped start VideoSurf, one of the first video search engines, which sold to Microsoft in 2011. He organized the vision team for Stanley, the robot that won the $2 million DARPA Grand Challenge robot race across the desert. This car, which launched the ongoing wave of self-driving automotive research, now sits in the Smithsonian. As a consulting professor at Stanford University's Computer Science Department, Gary cofounded the Stanford AI Robotics program (STAIR). Out of this grew the PR1 robot and indirectly Willow Garage (which Gary later joined). Willow produced the PR2 robot and the ROS robot operating system. From there, Gary organized and cofounded Industrial Perception Inc., a sensor-guided robotics company focused on distribution centers that sold to Google in 2013. Along the way, Gary made sure OpenCV stayed current with the mobile and now deep neural network revolutions. Gary has been involved in founding, running, and advising startups ever since.

Born in 1973, **Adrian Kaehler** is a scientist, inventor, and engineer, whose work spans a wide variety of disciplines. His fields of expertise include robotics, physics, electrical engineering, computer algorithms, machine vision, biometrics, machine learning, computer games, system engineering, human machine interface, numerical programming, and design. At the age of 14, he enrolled in UC Santa Cruz, studying mathematics, computer science, and physics, graduating at 18 with a BA in physics. He went on to Columbia University, where he received his PhD in 1998 under professor Norman Christ for his work in lattice gauge theory and on the QCDSP supercomputer project. From 1994 through 1998, Adrian worked on the QCDSP supercomputer project. The QCDSP supercomputer was one of the first Teraflop-scale supercomputers every built. For this, Adrian and his team were awarded the Gordon Bell Prize in 1998. In the 2005 DARPA Grand Challenge, Adrian was on Stanford's winning team, where he designed the computer vision system that played a central role in the team's victory. Adrian went on to found and run the many robotics and machine learning efforts at Applied Minds, a high-end research consulting firm, and is now a Fellow of Applied Invention, a spinout of Applied Minds. Adrian now focuses on advising and creating startups in Silicon Valley. He is also a founder of The Silicon Valley Deep Learning Group, an educational nonprofit focused on expanding, empowering, and connecting the growing community of deep learning practitioners and entrepreneurs.

## Colophon

The animal on the cover of *Learning OpenCV 3* is a giant, or great, peacock moth (*Saturnia pyri*). Native to Europe, the moth's range includes southern France and Italy, the Iberian Peninsula, and parts of Siberia and northern Africa. It inhabits open landscapes with scattered trees and shrubs and can often be found in parklands, orchards, and vineyards, where it rests under shade trees during the day.

The largest of the European moths, giant peacock moths have a wingspan of up to six inches; their size and nocturnal nature can lead some observers to mistake them for bats. Their wings are gray and grayish-brown with accents of white and yellow. In the center of each wing, giant peacock moths have a large eyespot, a distinctive pattern most commonly associated with the birds they are named for.

Many of the animals on O'Reilly covers are endangered; all of them are important to the world. To learn more about how you can help, go to *animals.oreilly.com*.

The cover image is from Cassell's *Natural History*, Volume 5. The cover fonts are URW Typewriter and Guardian Sans. The text font is Adobe Minion Pro; the heading font is Adobe Myriad Condensed; and the code font is Dalton Maag's Ubuntu Mono.